LIFE: THE SCIENCE OF BIOLOGY

FOURTH EDITION

LIFE

The Science of Biology

William K. Purves
Harvey Mudd College
Claremont, California

Gordon H. Orians
The University of Washington
Seattle, Washington

H. Craig Heller
Stanford University
Stanford, California

SINAUER ASSOCIATES, INC.

W. H. FREEMAN AND COMPANY

THE COVER

Elephants at a water hole in northern Botswana, Africa.
Photograph by Frans Lanting/Minden Pictures.

THE FRONTISPIECE

Scarlet ibis and cattle egrets at Hato el Frio, Venezuela.
Photograph by Art Wolfe.

LIFE: THE SCIENCE OF BIOLOGY, Fourth Edition

Address editorial correspondence to Sinauer Associates, Inc.,
Sunderland, Massachusetts 01375 U.S.A.

Address orders to W. H. Freeman and Co. Distribution Center,
4419 West 1980 South, Salt Lake City, Utah 84104 U.S.A.

Library of Congress Cataloging-in-Publication Data

Purves, William. K. (William Kirkwood), 1934-
Life, the science of biology / William K. Purves,
Gordon H. Orians, H. Craig Heller. -- 4th ed.
p. cm.
Includes bibliographical references and index.
ISBN 0-7167-2629-7
1. Biology. I. Orians, Gordon H. II. Heller, H. Craig.
III. Title.
QH305.2.P87 1995
574--dc20 94-24802
CIP

ABOUT THE BOOK

Editor: Andrew D. Sinauer
Project Editor: Carol J. Wigg
Developmental Editor: Elmarie Hutchinson
Copy Editor: Stephanie Hiebert
Production Manager: Christopher Small
Book Layout and Production: Janice Holabird
Art Editing and Illustration Program: J/B Woolsey Associates
Photo Research: Jane Potter
Book and Cover Design: Rodelinde Graphic Design
Composition: DEKR Corporation
Color Separations: Vision Graphics, Inc.
Prepress: Lanman Lithotech
Cover Manufacture: John P. Pow Company
Book Manufacture: R. R. Donnelley & Sons Company, Willard, OH

Printed in U.S.A.

4 3 2 1

To Jean, Betty, and Renu

ABOUT THE AUTHORS

William K. Purves

Bill Purves is Stuart Mudd Professor of Biology as well as founder and chair of the Department of Biology at Harvey Mudd College in Claremont, California. He received his Ph.D. from Yale University in 1959 under Arthur Galston. A Fellow of the American Association for the Advancement of Science, Professor Purves has served as head of the Life Sciences Group at the University of Connecticut, Storrs, and as chair of the Department of Biological Sciences, University of California, Santa Barbara, where he won the Harold J. Plous Award for teaching excellence. His research interests focus on the chemical and physical regulation of plant growth and flowering.

Professor Purves has taught introductory biology each year for over thirty years and considers teaching the course the most interesting and important of his professional activities. "I can't imagine a year without teaching it," he says. In describing his teaching philosophy, Purves states, "Students learn biological concepts much more rapidly and effectively if they understand where the concepts come from—what the experimental and conceptual background is. 'Facts' by themselves can be boring or incomprehensible, but they become exciting if given a context."

Gordon H. Orians

Gordon Orians is Professor of Zoology at the University of Washington. He received his Ph.D. from the University of California, Berkeley, in 1960 under Frank Pitelka. Professor Orians has been elected to the National Academy of Sciences and the American Academy of Arts and Sciences. He was President of the Organization for Tropical Studies from 1982 to 1994, and is currently President-elect of the Ecological Society of America. He is a recipient of the Brewster Medal from the American Ornithologists' Union, and in 1994 he received the Distinguished Service Award of the American Institute of Biological Sciences.

Professor Orians is a leading authority in ecology and evolution, with research interests in behavioral ecology, plant–herbivore interactions, community structure, and the biology of rare species. Like the other authors, he draws from his research to bring an added dimension to his teaching and writing. "Teachers who understand research because they are engaged in it can more easily communicate the excitement researchers feel as they discover new things," Orians says. "All three authors of *Life* have spent considerable time doing research and we have tried throughout the book to show the sources of our current understanding of biology."

H. Craig Heller

Craig Heller is Lorry Lokey/ Business Wire Professor of Biological Sciences and Human Biology and Associate Dean of Research at Stanford University, and is a popular lecturer on animal and human physiology. He received his Ph.D. from Yale University in 1970 and did postdoctoral research at the Scripps Institution of Oceanography on brain regulation of body temperature in mammals. He has continued this research since coming to Stanford in 1972, studying a variety of phenomena ranging from hibernating squirrels to sleeping college students to diving seals to meditating yogis. Professor Heller is a Fellow of the American Association for the Advancement of Science and a recipient of the Walter J. Gores Award for Excellence in Teaching.

"A first course in biology requires the student to learn more new words than a first course in a foreign language," Heller says. "The secret to teaching and learning biology is to focus on central and overarching concepts. Once you grasp the concept of how something works, you have a framework on which the facts and vocabulary fall into place. Conceptual understanding also helps you relate what you learn to the real world."

PREFACE

In revising this book, we have once again examined our goals and our hopes for it. Above all, we want to help students understand biological concepts and see where the concepts originate. For this reason we display biology as an experimental and observational science. Frequently we offer the student a chance to think—to figure out the next step rather than wait passively to learn it from us. For example, we ask the student to interpret experimental data used in elucidating the genetic code, in understanding the role of homeotic genes in flower development, and in working out the Calvin–Benson cycle. In Chapter 21, students are presented with real species with identified character traits and work through how to construct a phylogeny using cladistic methods. We use the cladistic material again in Chapter 27, where we explore human relationships with chimpanzees and gorillas. As yet another example, we lead the student to discover the physiological differences between reptiles and mammals by taking the reader through a series of experiments on the thermal biology of a mouse and a lizard.

Even when we present topics directly, we prefer to explain new material rather than serving it up as a cut-and-dried collection of "facts." Still, the study of biology requires exposure to a stunning number of new facts, and students are easily deterred by a bewildering excess of information. How can we deal with this problem? Our approach is to emphasize fascinating examples wherever possible, using these to engage the student's interest so that she or he wants to learn other related material.

UP-TO-DATE COVERAGE REFLECTS NEW DEVELOPMENTS IN BIOLOGY

The creation of a new edition provides opportunities to rethink how best to present existing material, as well as what from the exploding array of new information to include. Current advances in biomedical sciences present special challenges to the organizers of a course or a textbook. Diseases such as cancer and AIDS are of deep interest in relation to a wide variety of biological topics, including immunology, genetics, evolution, membrane biology, and virology; instructors often want to use the diseases as examples in discussing these topics. Yet students and instructors also want to see a disease and its biology considered in a single place in the book. For this edition, we have adopted an approach that we hope is effective. Recent advances in gene therapy and in the cloning of genes for particular diseases lend themselves to consideration in an all-new Chapter 15, "Genetic Disease and Modern Medicine." We consider some general principles and some widely applicable techniques in this chapter, which also includes much of our coverage of cancer. We build from the close of this chapter to begin the next, "Defenses against Disease," with an overview of AIDS as a worldwide problem. Before the end of Chapter 16, the student knows enough about the immune system to understand AIDS in greater biological detail.

Developmental biology continues to be one of the fastest-moving areas of biology; the newer work is reflected in Chapter 17. We pay particular attention to recent developments in the *Drosophila* larva and *Caenorhabditis elegans* systems. After giving genetic "instructions" for building a fly, we conclude the molecular section of the book with a transition from *Drosophila* larval genetics to evolutionary biology.

In the section on evolution we have expanded the coverage of methods of reconstructing phylogenies. We provide new treatments of cladistic methods and show how phylogenies are used to shed light on a wide variety of evolutionary questions. We use phylogenetic trees a number of times in subsequent chapters in the book to show how traits of organisms evolve under the influence of evolutionary agents.

Our restructuring of the chapters on plant anatomy and physiology resulted in the creation of a new Chapter 31, "Environmental Challenges to Plants," which deals with some of the ways plants cope with harsh environments, predators, and pathogens. We have updated the plant chapters to include material on patch clamping, homeotic mutations, and other important phenomena and techniques.

Three aspects of the animal biology chapters that were significant in the third edition have been strengthened. First, we frequently use a comparative approach to help students understand basic principles and mechanisms as well as their evolutionary variations. Second, we emphasize experimental approaches so that the student learns *how* we know as well as *what* we know. Third, a capstone to the treatment of each physiological system is a discussion of how its contributions to homeostasis are controlled and regulated. Chapter 45, "Animal Behavior," has been revised to focus more strongly on the physiological mechanisms underlying behavior.

Because ecology is an increasingly experimental science, we have added descriptions of well-designed experiments that have been performed to demonstrate the causes of the evolution of traits used in courtship among animals and to assess the importance of predation and competition in structuring ecological communities. To keep pace with the increasing importance of environmental problems, we have expanded our treatment of lake eutrophication and of overexploitation of commercially important species.

PEDAGOGICAL INNOVATIONS ENRICH THE LEARNING PROCESS

Both as textbook authors and as teachers we want to help students in every way possible. In the next section ("To the Student") we offer some helpful advice we'd like students to consider as they begin their study of biology. This advice has helped many of our own students. Following "To the Student" is "*Life* at a Glance," in which we illustrate some of the pedagogic improvements that we discuss in the next few paragraphs.

A major source of the success of the third edition of this textbook was the exciting new art program developed by J/B Woolsey Associates. We have upgraded that already fine art for this edition. We have added many entirely new drawings and graphs, and virtually all the drawings in the book have been improved in one way or another by the development of a new artistic vocabulary for this edition.

Because learning is a visual as well as a verbal process, we have given much attention to creating illustrations that explain biological concepts clearly. To facilitate learning we use color consistently from illustration to illustration; for example, the outside of the cell is always represented by light red and the cytoplasm by pale blue. We have developed a set of icons (see "*Life* at a Glance") to represent biologically important molecules (such as water or ATP), active forms of enzymes, or activation and inhibition of pathways. We have used blocks of color to distinguish major pieces of information from details or to separate an illustration into parts representing discrete ideas; we think of these as "visual paragraphs." Finally, we have flagged via marginal arrowheads (also shown in "*Life* at a Glance") illustrations that are particularly significant. These figures illustrate and synthesize important concepts—protein synthesis, for example, or the life cycle of a flowering plant.

We have developed a new type of chapter-end summary, one that should help students in at least three ways. First, the student can skim the summary for orientation before reading the chapter. Second, after reading the chapter, the student can use the summary to review the material. Finally, each summary identifies the illustrations that give the best overview of the chapter and its most important concepts.

Each chapter starts with a brief introduction intended to catch the reader's interest by discussing a fascinating bit of biology. Each of these new introductions is supported by a striking photograph; an example is shown in "*Life* at a Glance."

Before beginning this edition, we asked 36 biologists, most of whom were using the third edition, to maintain "diaries" and record their ideas for improving the book. These diarists were enormously helpful in getting us started on the right foot for the new edition. Their guidance influenced the decisions to create the two new chapters mentioned earlier and to give priority to simplifying our prose. We are indebted to them.

CAREFULLY REVIEWED WITH STUDENTS' NEEDS AND TEACHERS' CONCERNS IN MIND

As with the first three editions, many of our colleagues reviewed chapters or entire sections of this edition in manuscript. They and the diarists are listed below. The reviewers were helpful, thoughtful, and clearly dedicated to the success of this book, and we thank them all. We particularly thank Bob Cleland, Richard Cyr, Pat DeCoursey, Rob Dorit, Art Dunham, Margaret Fusari, Harry Green, Ray Huey, Bob Jeffries, Jim Manser, William Milsom, Ron O'Dor, Dianna Padilla, Ronald Patterson, Zoe Roizen, Seri Rudolph, Michael Ryan, Iain Taylor, and David Woodruff. They gave us explicit recommendations for extensive improvements, helped simplify our writing style, and did so in ways that encouraged us to do our very best to live up to their expectations.

THE EFFORTS OF MANY PEOPLE HELPED THE REVISION

The third edition profited greatly from the prodigious efforts of its outstanding developmental editor, Elmarie Hutchinson. We were delighted when Elmarie agreed to take an even stronger role in the development of the fourth edition. Among her innovations are the new type of chapter summary and its suggested use as a chapter preview. Elmarie also helped us respond to diarists' requests for simpler language, better topic sentences, and more restrained use of boldface terms. Her suggestions and guidelines were implemented by our copyeditor, Stephanie Hiebert, whose sharp and prescriptive line editing has helped streamline the book's prose.

In a book like this the illustrations are as important as the prose, and here again Elmarie gave us outstanding input, scrutinizing every figure with an eye for its internal consistency and its agreement with the text. Her suggestions were incorporated by artists John Woolsey and Patrick Lane as they met with the authors to reconceptualize artwork. The task of coordinating and checking the changes made by editors, artists, and authors fell to Carol Wigg, who got the job once again of putting all the pieces together. In addition, previous users of the book will see a significant improvement in the photography program. Jane Potter has tapped important new sources, and we have gone to great lengths to seek out new photographs to illustrate important concepts and enliven the book's appearance.

We wish to thank W. H. Freeman's entire marketing and sales group. Their enthusiasm for *Life* helped bring the book to a wider audience and the efforts of several of the sales representatives put us in touch with a number of our colleagues who had specific questions or criticisms of the book. This contact has been fruitful, and we look forward to more of this "firing line" interaction with the fourth edition.

Finally, the opportunity to work with a publishing company whose president provides frequent personal contacts and feedback that is scientifically useful as well as production-wise, is a great privilege for us. Andy Sinauer is the ideal person for authors to deal with—firm but kind, involved but not overbearing, and friendly—hence, motivating. He also has a superb eye for good associates—the Sinauer team is first-rate!

William K. Purves **Gordon H. Orians** **H. Craig Heller**

September 1994

REVIEWERS FOR THE FOURTH EDITION

Andrew R. Blaustein, Oregon State University
Daniel R. Brooks, University of Toronto
Andrew G. Clark, Pennsylvania State University
Rolf E. Christoffersen, University of California / Santa Barbara
Esther Chu, Occidental College
Patricia J. DeCoursey, University of South Carolina
T. A. Dick, University of Manitoba
Robert L. Dorit, Yale University
Arthur E. Dunham, University of Pennsylvania
Gisela Erf, Smith College
Russell G. Foster, University of Virginia
Michael Feldgarden, New Haven, Connecticut
William D. Fixsen, Harvard University
Gwen Freyd, Harvard University
William Friedman, University of Georgia
Margaret H. Fusari, University of California / Santa Cruz
Stephen A. George, Amherst College
Elizabeth A. Godrick, Boston University
Linda J. Goff, University of California / Santa Cruz
Harry W. Greene, University of California / Berkeley
Richard K. Grosberg, University of California / Davis
Marty Hanczyc, New Haven, Connecticut
Albert A. Herrera, University of Southern California
Raymond B. Huey, University of Washington
Robert L. Jeffries, University of Toronto
Cynthia Jones, University of Connecticut
David Kirk, Washington University / St. Louis
Will Kopachik, Michigan State University
Arthur P. Mange, University of Massachusetts / Amherst
Jim Manser, Harvey Mudd College
Charles W. Mims, University of Georgia
Ron O'Dor, Dalhousie University
Judith A. Owen, Haverford College
Dianna K. Padilla, University of Wisconsin / Madison
Daniel Papaj, University of Arizona
Murali Pillai, Bodega Marine Laboratory
Martin Poenie, University of Texas
Roberta Pollack, Occidental College
Seri Rudolph, Bates College
Michael J. Ryan, University of Texas
Joan Sharp, Simon Fraser University
Dwayne D. Simmons, University of California / Los Angeles
Iain E. P. Taylor, University of British Columbia
Mark Wheelis, University of California / Davis
Gene R. Williams, Indiana University
David S. Woodruff, University of California / San Diego
Ron Ydenberg, Simon Fraser University
Charles Yocum, University of Michigan / Ann Arbor

DIARISTS FOR THE FOURTH EDITION

Alice Alldredge, University of California / Santa Barbara
Daniel R. Brooks, University of Toronto
Margaret Burton, Memorial University of Newfoundland
Iain Campbell, University of Pittsburgh
Mark A. Chappell, University of California / Riverside
Robert E. Cleland, University of Washington
Richard J. Cyr, Pennsylvania State University
Wayne Daugherty, San Diego State University
William Davis, Rutgers University
Emma Ehrdahl, Northern Virginia Community College
Stuart C. Feinstein, University of California / Santa Barbara
Rachel Fink, Mt. Holyoke College
William Fixsen, Harvard University
R. M. Grainger, University of Virginia
Daniel Klionsky, University of California / Davis
James L. Koevenig, University of Central Florida
William Z. Lidicker, Jr., University of California / Berkeley
J. Richard McIntosh, University of Colorado / Boulder
William K. Milsom, University of British Columbia
Deborah Mowshowitz, Columbia University
Todd Newberry, University of California / Santa Cruz
Peter H. Niewiarowski, University of Pennsylvania
Larry D. Noodén, University of Michigan / Ann Arbor
Scott Orcutt, University of Akron
Ronald J. Patterson, Michigan State University
Zoe Roizen, Oakland, California
Steve Rothstein, University of California / Santa Barbara
Mark Shannon, University of Tennessee
Joan Sharp, Simon Fraser University
Steve Strand, University of California / Los Angeles
Brian Taylor, Texas A & M University
Peter Webster, University of Massachusetts / Amherst
Nathaniel T. Wheelwright, Bowdoin College
Brian White, Massachusetts Institute of Technology
Thomas Wilson, University of Vermont
Ronald C. Ydenberg, Simon Fraser University

TO THE STUDENT

Welcome to the study of life! In our student days—and ever since—we have enjoyed studying the fascinating and fast-changing field of biology, and we hope that you will, too.

There are a few things you can do to help you get the most from this book and from your course. For openers, read the book actively—don't just read passively, but do things that force you to think as you read. If we pose questions, stop and think about them. If a passage reminds you of something that has gone before, think about that, or even check back to refresh your memory. Ask questions of the text as you go. Do you understand what is being said? Does it relate to something you already know? Is it supported by experimental or other evidence? Does that evidence convince you? How does this passage fit into the chapter as a whole? Annotate the book—write down comments in the margins about things you don't understand, or about how one part relates to another, or even when you find an idea particularly interesting. The point of doing these things is that they will help you learn. People remember things they think about much better than they remember things they have read passively. Highlighting is passive; copying is drudge work; questioning and commenting are active and well worthwhile.

For this edition we have developed new ways to help you read the book actively. The chapter-end summaries have been redesigned so that they may be used as both summaries and previews. To find out what a chapter covers, try reading the summary at the end of the chapter before you begin reading the chapter itself. Don't worry about unfamiliar terms in the summary, but notice them as terms you will need to learn. Just read all the statements as an overview and preview without studying the cited illustrations. Then, after reading the chapter, use the summary as a framework for your review. It is essential that you do study the cited illustrations and their captions as you review because important information that is covered in illustrations has been left out of the summary statements. Add concepts and details to the framework by reviewing the text.

Take advantage of our use of color and symbols in the illustrations. We generally use colors to mean the same thing from illustration to illustration, and we have developed a set of symbols (see the section "*Life* at a Glance") to represent biologically important molecules and phenomena. In many of the illustrations you will see blocks of color used to help you separate the illustration into parts representing discrete ideas. Also, some figures are identified by an arrow in the margin. These figures are particular significant; they illustrate and synthesize important concepts and retell visually the story you read in the text. Studying them will help you learn important biological concepts and systems. Going back to review them will help you to remember these concepts.

The chapter summaries will help you quickly review the high points of what you have read. A summary identifies particular illustrations that you should study to help organize the material in your mind. A way to review the material in slightly more detail after reading the chapter is to go back and look at the boldfaced terms. You can use the boldfaced terms to pose questions—and see if you can answer those questions. The boldfacing will probably be more useful on a second reading than on the first.

Use the self-quizzes and study questions at the end of each chapter. The self-quizzes are meant to help you remember some of the more detailed material and to help you sort out the information we have laid before you. Answers to all self-quizzes are in the Appendix. The study questions, on the other hand, are often fairly open-ended and are intended to cause you to reflect on the material.

Two parts of a textbook that are, unfortunately, often underused or even ignored are the glossary and the index. Both can help you a great deal. When you are uncertain of the meaning of a term, check the glossary first—there are more than 1,500 definitions in it. If you don't find a term in the glossary, or if you want a more thorough discussion of the term, use the index to find where it's discussed.

What if you'd like to pursue some of the topics in greater detail? At the end of each chapter there is a short, annotated list of supplemental readings. We have tried to choose readings from books and magazines, especially *Scientific American*, that should be available in your college library.

Most students occasionally have difficulty in courses, including biology courses. If you find that you are slipping behind in the course, or if a particular topic is giving you an unreasonable amount of trouble, here are some useful steps you might take. First, the basics: attend class, take careful lecture notes, and read the textbook assignments. Second, note that one of the most important roles of studying is to discover what you don't know, so that you can do something about it. Use the index, the glossary, the chapter summaries, and the text itself to try to answer any questions you have and to help you organize the material. Make a habit of looking over your lecture notes within 24 hours of when you take them—find out right away what points are unclear, and get them straightened out in your mind. We also call your attention to the Study Guide that accompanies *Life*. It is by Jon Glase at Cornell and Jerry Waldvogel at Clemson. It parallels this textbook and each chapter contains learning objectives, key concepts, activities, and questions with full answers and explanations.

If none of these self-help remedies does the trick, get help! Other students are often a good source of help, because they are dealing with the material at the same level as you are. Study groups can be very useful, as long as the participants are all committed to learning the material. Tutors are almost always helpful and useful, as are faculty members. The main thing is to get help when you need it. It is not a good idea to be strong and silent and drift into a low grade.

But don't make the grade the point of this or any other course. You are in college to learn, to pursue interesting subjects, and to enjoy the subjects you are pursuing. We hope you'll enjoy the pursuit of biology.

Bill Purves **Gordon Orians** **Craig Heller**

SUPPLEMENTS

Life, Fourth Edition, is accompanied by a comprehensive set of supplements:

Study Guide . . . reviewed by students and extensively revised by Jon Glase of Cornell University and Jerry Waldvogel of Clemson University to help students master the textbook material.

Instructor's Manual . . . by Roberta Meehan of the University of Northern Colorado, featuring chapter objectives, chapter outlines, teaching hints and strategies, references, resources, and key terms.

Overhead transparencies and **slides** . . . a package of 300 full-color images from the book.

Transparency masters . . . of all text art figures not included in the overhead transparency / slide set.

Test bank . . . revised and updated, with at least 10 new questions per chapter and over 4,000 questions in total. Available in printed form, and in IBM and Macintosh formats.

Videodisc . . . for the first time the magnificent art program in *Life* comes to your classroom via laserdisc technology. The disc includes all the line art from the new edition, over 1,500 carefully selected still images, and more than 25 outstanding motion and animation sequences ranging from traffic through the membrane and electrophoresis to ecological succession.

Laboratory options . . . chosen from the following:

- The complete Abramoff and Thomson's *Laboratory Outlines in Biology VI*. The popular, critically acclaimed lab manual from W. H. Freeman, now in its new (1995) sixth edition.
- *Laboratory separates*. Select only those experiments you need from Abramoff and Thomson.
- *Customized laboratory package*. The separates of your choice, combined with your own laboratory exercises, notes, and other materials.

For information regarding policy on the educational use of these supplements, please contact your local W. H. Freeman representative.

Videodisc Focus Group Participants
Tad Day, University of West Virginia
Guy Cameron, University of Houston
Valerie Flechtner, John Carroll University
Arnold Karpoff, University of Louisville
William Eickmeir, Vanderbilt University
Paul Ramp, University of Tennessee

Videodisc Reviewers
Stephen C. Adolph, Harvey Mudd College
Sally S. De Groot, St. Petersburg Junior College (retired)
Rachel Fink, Mt. Holyoke College
Nancy V. Hamlett, Harvey Mudd College
Brian A. Hazlett, University of Michigan
Martinez J. Hewlett, University of Arizona
Dan Lajoie, University of Western Ontario
Alfred R. Loeblich III, University of Houston
James R. Manser, Harvey Mudd College
Catherine S. McFadden, Harvey Mudd College
T. J. Mueller, Harvey Mudd College

Study Guide Reviewer
Wayne Hughes, University of Georgia

Instructor's Manual Reviewers
Erica Bergquist, Holyoke Community College
Nels H. Granholm, South Dakota State University

Transparency Reviewers
William S. Cohen, University of Kentucky
Anne M. Cusic, University of Alabama
Bruce Felgenhauer, University of Southwestern Louisiana
Alice Jacklett, State University of New York at Albany
Susan Koptur, Florida International University
Charles H. Mallery, University of Miami
Stephen P. Vives, Georgia Southern University

In addition, Tad Day, William Eickmeier, and Paul Ramp conducted student reviews of the videodisc, and Jon Glase had his students review the Study Guide. We greatly appreciate their efforts.

LIFE

They Are Not All the Same
When observed closely, the individuals in a population of red and green macaws vary a great deal.

19

The Mechanisms of Evolution

426

We are aware that no two people (unless they are identifical twins) look exactly alike. We also recognize our pets as distinct individuals. But we have great difficulty in seeing differences among individuals of most other species of organisms. The brilliant red and green macaws feeding on a clay cliff in the Peruvian jungle may all appear identical to the untrained eye; scientists who study them closely, however, realize that each is unique. The colored feathers display many slight variations in pattern, and the black-and-white feathers that surround the birds' eyes form patterns that, like human fingerprints, are unique to the individual. Members of many groups, particularly among behaviorally sophisticated animals such as vertebrates, readily recognize one another and adjust their behavior accordingly.

Differences among individuals in local populations, even if they are subtle, are the raw material upon which evolutionary mechanisms act to produce the striking variability revealed by the multitude of organisms living on Earth today. A good fossil record can reveal much about when and how the forms of organisms changed. Fossils may also provide clues about the reasons for those changes, but they provide only indirect evidence of the causes of evolutionary change. To obtain direct evidence we must study evolutionary changes happening today. The study of variability is at the heart of investigations into the mechanisms of evolution.

In this chapter we discuss the agents of evolution and the short-term studies designed to investigate them. By testing hypotheses observationally and experimentally we can answer key questions about the processes guiding evolutionary changes. In later chapters we consider how we use this information to explain longer-term features of the evolutionary record.

Although ideas about evolution have been put forth for centuries, until the last one hundred years none of the h
tionary chang
his hypotheses
1), but he did
basis of evolut
vided by Greg
and Darwin's
twentieth cent
evolutionary h
test them.

WHAT IS EVOL

The fossil recor
over time. The
anisms shared

Intriguing chapter openers combine arresting photos and text to spark interest at the outset, setting the stage for the material that follows.

Consistent icons and color-coding clarify illustrations. Icons are used to represent biologically important molecules, active forms of enzymes, and the activation and inhibition of pathways. Color is used consistently throughout the text.

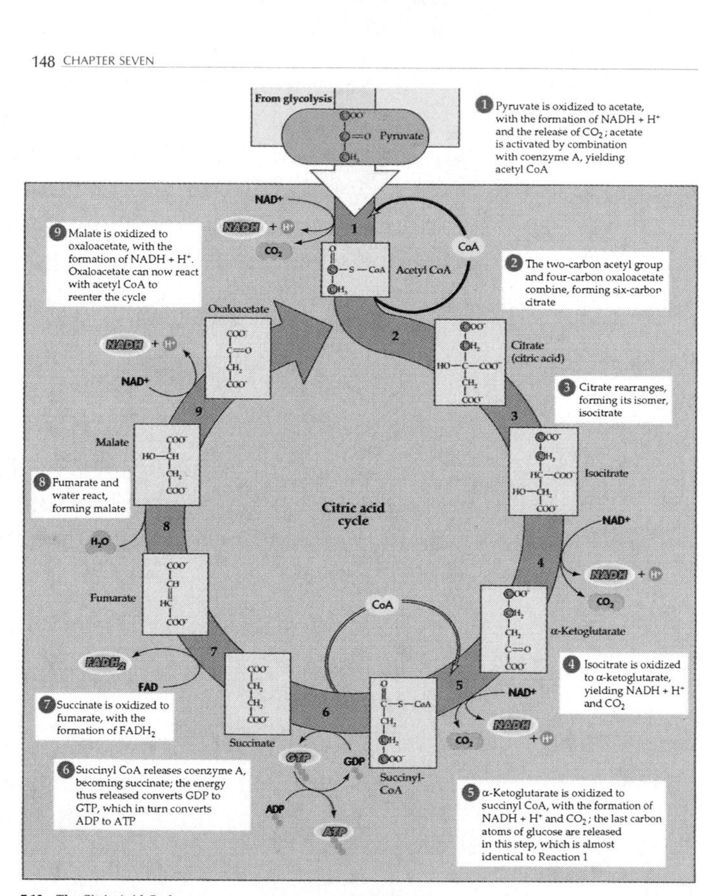

7.13 The Citric Acid Cycle
The first reaction produces the two-carbon acetyl CoA. Notice that the two carbons from acetyl CoA are traced with color through reaction 5, after which they may be at either end of the molecule (note the symmetry of succinate and fumarate). Reactions 1, 4, 5, 7, and 9 accomplish the major overall effect of the cycle—the storing of energy—by passing electrons to the carrier molecule NAD. Reaction 6 also stores energy.

AT A GLANCE

Dynamic visual paragraphs dramatize important concepts in a way that the text alone cannot. These are flagged with marginal arrows so the student is able to refer to them readily.

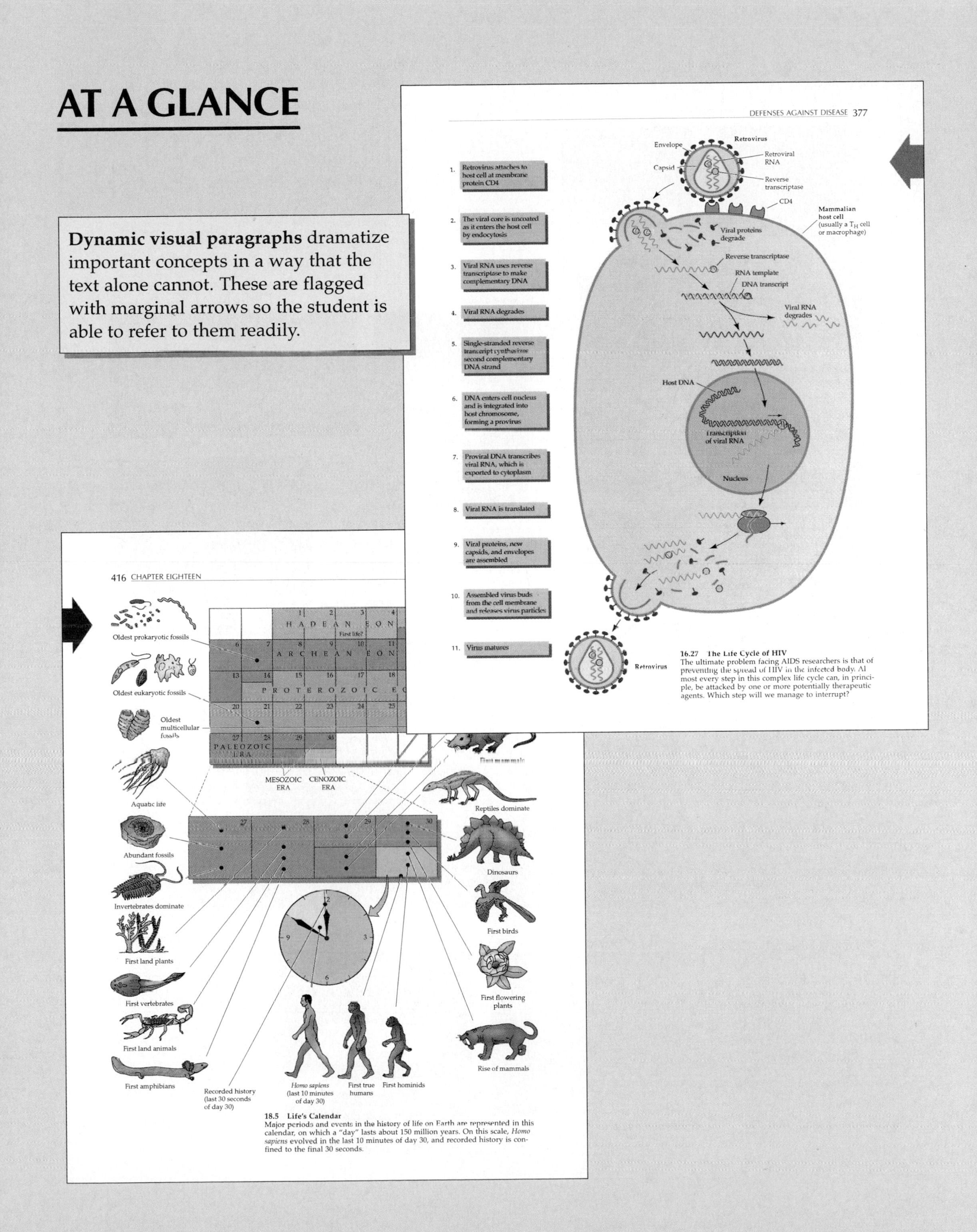

16.27 The Life Cycle of HIV
The ultimate problem facing AIDS researchers is that of preventing the spread of HIV in the infected body. Almost every step in this complex life cycle can, in principle, be attacked by one or more potentially therapeutic agents. Which step will we manage to interrupt?

18.5 Life's Calendar
Major periods and events in the history of life on Earth are represented in this calendar, on which a "day" lasts about 150 million years. On this scale, *Homo sapiens* evolved in the last 10 minutes of day 30, and recorded history is confined to the final 30 seconds.

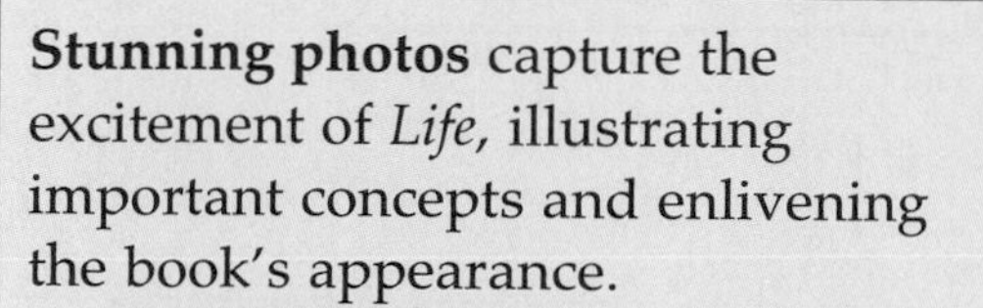

Stunning photos capture the excitement of *Life,* illustrating important concepts and enlivening the book's appearance.

(*a*) (*b*)

(*c*)

25.31 Dicots
(*a*) The cactus family is a large grou
about 1,500 species in the Americas.
takes its name from its scarlet flowe
Anterrhinum majus. (*c*) These wood
stone National Park are members of
as are the familiar roses from your
25.25 and 25.26*a* show other dicots.

pected, primarily
phyta have vess
angiosperms. In
by the light-micr
tilization in *Ephe*
The cycadeoids,
same time as di
portant characte
angiosperms. Th
cycadeoids, altho
with naked seed
flower of *Magnol*
The next grea
the question, Wh

Origin and Evolution of the Angi

How did the angiosperms aris
analyses (see Chapter 21) have s
ing question. It is widely agreed
and two groups of gymnosper
and the long-extinct cycadeoids
cycads, arose from a single ances
rise to no other groups. A close
the angiosperms and the Gneto

(*a*)

Hydrogen bond between water molecules

(*c*)

Hydrog
in a pro

(*b*)

(*c*)

lineage gave rise to the prosimians—lemurs, tarsiers, pottos, and lorises (Figure 27.29). Prosimians were formerly found on all continents, but today they are restricted to Africa, tropical Asia, and Madagascar. All mainland species are arboreal and nocturnal, but on Madagascar, where there has been a remarkable prosimian radiation, there are also diurnal and terrestrial species. Until the recent arrival of humans, there were no other primates on Madagascar.

The anthropoids—monkeys, apes, and humans—evolved from an early primate stock about 55 million years ago in Africa or Asia. New World monkeys have been evolving separately from Old World monkeys long enough that they could have reached South America from Africa when those two continents were still connected. Perhaps because tropical America has

27.29 Prosimians
(*a*) The sifaka lemur, *Propithecus verreauxi,* is one of many lemur species of Madagascar, where they are part of a unique assemblage of plants and animals. (*b*) *Loris tardigradis,* the slender loris, of southern India. (*c*) In the rainforests of Borneo, this tarsier (*Tarsius bancanus*) seems otherworldly to our eyes.

(*a*)

(*b*)

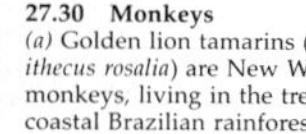

27.30 Monkeys
(*a*) Golden lion tamarins (*Leontopithecus rosalia*) are New World monkeys, living in the trees of the coastal Brazilian rainforest. (*b*) Some Old World species, such as these Japanese macaques (*Macaca fuscata*) live and travel in groups.

(*a*)

2.18 Surface Tension
(*a*) A water strider "skates" along, supported by the surface tension of the water that is its home. (*b*) Surface tension demonstrated by a soap bubble on a teacup.

(*b*)

Boxes describe fascinating biological phenomena, highlighting special interest topics and expanding the discussion of many issues.

BOX 27.A

The Four-Minute Mile

Many mammals can run much faster, yet we humans are proud to have achieved a four-minute mile. Terrestrial vertebrates did not achieve such speeds easily. Amphibians and reptiles fill and empty their lungs using some of the same muscles they use for walking. In addition, because the limbs protrude laterally, their movement generates a strong lateral force that bends the body from side to side. Recent studies have shown that these animals cannot breathe while they walk or run. Therefore, they can operate aerobically only briefly. Because they depend upon anaerobic glycolysis while running, they tire rapidly.

In the lineage leading to dinosaurs and birds and in the lineage leading to mammals, the legs assumed more vertical positions, which reduced the lateral forces on the body during locomotion. Special ventilatory muscles that can operate independently of locomotory muscles also evolved. These muscles are visible in living birds and mammals. We can infer their existence in dinosaurs from the structure of the vertebral column and the capability of many dinosaurs for bounding, bipedal (using two legs) locomotion. The ability to breathe and run simultaneously, a capability we take for granted, was a major innovation in the evolution of terrestrial vertebrates.

A future four-minute miler?

Figure 28.12). The ability to move actively on land was not achieved easily. The first terrestrial vertebrates probably moved only very slowly, much more slowly than their aquatic relatives. The reason is that they apparently could not walk and breathe at the same time. Not until evolution of the lineages leading to the mammals, dinosaurs, and birds did special muscles evolve enabling the lungs to be filled and emptied while the limbs moved (Box 27.A). This ability enabled its bearers to maintain steady, high levels of activity, which generated enough heat to result in

subclass Aves)
e embodies an
scendants of a
evolved in the
(Figure 27.22),
s and modern
pteryx was cov-
eloped wings,
much reduced
may have been

vs features re-
he modern

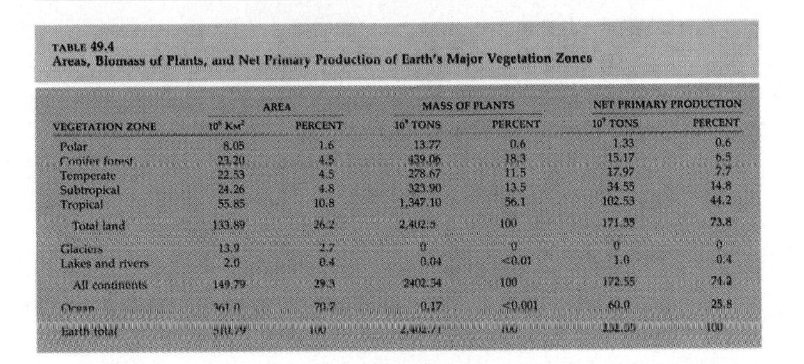

TABLE 49.4
Areas, Biomass of Plants, and Net Primary Production of Earth's Major Vegetation Zones

	AREA		MASS OF PLANTS		NET PRIMARY PRODUCTION	
VEGETATION ZONE	10^6 KM^2	PERCENT	10^9 TONS	PERCENT	10^9 TONS	PERCENT
Polar	8.05	1.6	13.77	0.6	1.33	0.6
Conifer forest	23.20	4.5	439.06	18.3	15.17	6.5
Temperate	22.53	4.5	278.67	11.5	17.97	7.7
Subtropical	24.26	4.8	323.90	13.5	34.55	14.8
Tropical	55.85	10.8	1,347.10	56.1	102.53	44.2
Total land	133.89	26.2	2,402.5	100	171.35	73.8
Glaciers	13.9	2.7	0	0	0	0
Lakes and rivers	2.0	0.4	0.04	<0.01	1.0	0.4
All continents	149.79	29.3	2402.54	100	172.55	74.2
Ocean	361.0	70.7	0.17	<0.001	60.0	25.8
Earth total	510.79	100	2,402.71	100	232.55	100

from the vents. Most of the other organisms of these ecosystems live directly or indirectly on the sulfur-oxidizing bacteria (see Figure 22.2).

This overview of the global pattern of biological production on Earth is sufficient to identify which processes limit primary production and nutrient cycling in different climatic zones and how they operate, but it does not give you a picture of what these ecosystems look like and how they function. Describing ecosystems is one of the goals of the next chapter.

SUMMARY of Main Ideas about Ecosystems

Ecosystems are powered by solar energy that first enters living organisms via photosynthesis at rates controlled by temperature and precipitation.
Review Figure 49.1

Food webs summarize who eats whom in ecological communities.
Review Figures 49.2 and 49.3

Because much of the energy taken in by an organism is used for maintenance and is eventually dissipated as heat, the efficiency of energy transfer to higher trophic levels is usually very low.
Review Figures 49.4 and 49.5

The main elements of living organisms—carbon, nitrogen, phosphorus, sulfur, hydrogen, and oxygen—cycle between organisms and other compartments of the global ecosystem.
Review Figures 49.10, 49.11, 49.12, 49.13, and Table 49.3

Human activity greatly modifies cycles of basic minerals on local, regional, and global scales.
Review Figures 49.14, 49.15, and 49.16

Earth's climate is determined primarily by the pattern of solar energy input at different latitudes and by Earth's rotation on its axis.

The directions of prevailing winds differ over the surface of Earth.
Review Figure 49.19

Surface winds drive global oceanic circulation.
Review Figure 49.20

The distribution of primary production on Earth is determined primarily by Earth's climate.
Review Figure 49.21

New chapter summaries synthesize information in the text and the art. Important concepts are encapsulated in clearly written summary sentences, with references back to key illustrations.

CONTENTS

IN BRIEF

V. THE BIOLOGY OF VASCULAR PLANTS

VI. THE BIOLOGY OF ANIMALS

VII. ECOLOGY AND BIOGEOGRAPHY

CONTENTS

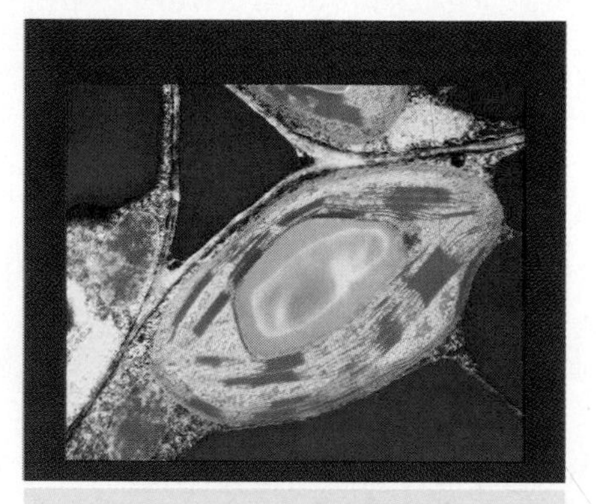

PART ONE
The Cell

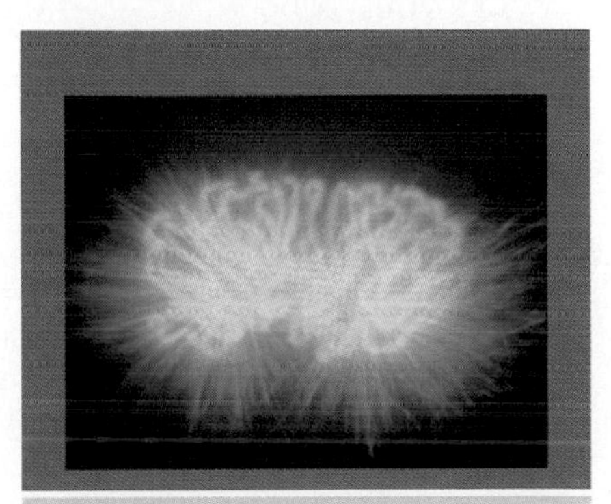

PART TWO
Information and Heredity

PART THREE Evolutionary Processes

PART FOUR The Evolution of Diversity

PART FIVE The Biology of Vascular Plants

PART SIX
The Biology of Animals

37. ANIMAL REPRODUCTION 824

38. NEURONS AND THE NERVOUS SYSTEM 849

39. SENSORY SYSTEMS 881

40. EFFECTORS 909

PART SEVEN Ecology and Biogeography

46. BEHAVIORAL ECOLOGY 1057

47. STRUCTURE AND DYNAMICS OF POPULATIONS 1080

48. INTERACTIONS WITHIN ECOLOGICAL COMMUNITIES 1100

49. ECOSYSTEMS 1127

PART FIVE

The Biology of Vascular Plants

Leaves are conspicuous parts of most flowering plants. As major sites of photosynthesis, leaves are the food sources for the rest of the plant and, in fact, for much of the living world. We all recognize most leaves as such when we see them. Their basic form—flattened, and with veins—helps them perform their nutritive function. The reactions of photosynthesis proceed in chloroplasts within certain leaf cells, and the photosynthetic products are distributed to other parts by a plumbing system that includes the leaf's veins. Some raw materials travel to the photosynthesizing cells through the veins; other materials travel from pores in the leaf surface through air spaces within the leaf. The structure of the leaf lets the photosynthesizing cells get the light they need; the flattened shape, coupled with the orientation of the leaf, minimizes internal shading. The structure of the leaf thus supports its photosynthetic function. What about the rest of the plant? The stems and roots of flowering plants also have interesting and complex structures, with structure and function well matched.

Part Five deals with plant structure and function. There are many aspects of plant function to consider. Plants—even the tallest trees—transport water from the soil to their tops, and they transport the products of photosynthesis from the leaves to the roots and other parts. Plants interact with their environment, both living and nonliving. They defend themselves against bacteria, fungi, animals, and other plants. Some plants can cope with hostile environments such as deserts, salt marshes, or sites polluted by mining and other human activities. Plants must obtain nutrients—not only the raw materials of photosynthesis, but also mineral elements such as potassium and calcium. Plants respond to environmental cues as they develop. They produce chemical signals that cause structural and functional changes appropriate to the environmental cues. Among the most important changes are those that lead to reproduction. Because we can understand plant function only in terms of the underlying structure, this chapter focuses on the structure of the plant body.

FLOWERING PLANTS

Recall that flowering plants are vascular plants characterized by double fertilization, by endosperm, and by seeds enclosed in modified leaves; their xylem contains vessel elements and fibers. If you have not been thinking about plants for a while, you might want to review the sections entitled "Two Groups within the Plant Kingdom" and "The Angiosperms: Flowering Plants" in Chapter 25 before reading the rest of Chapter 29.

Frosted Sites of Photosynthesis

29

The Flowering Plant Body

Flowering plants consist of a few important organs whose life-supporting functions can be understood in terms of their large-scale structure, as well as the microscopic structure of their component cells. The cells are grouped into tissues, and the tissues are grouped into organs. In this chapter we will present some anatomical features common to many flowering plants. As always in biology, it is important to remember that there are differences between organisms of the same species as well as between species. Let us begin, then, by looking at four important or familiar species: coconut palm, red maple, rice, and soybean.

FOUR EXAMPLES

Coconut Palm

In some cultures the coconut palm (*Cocos nucifera;* Figure 29.1) is called the Tree of Life because every aboveground part of the plant has value to humans. People use the stem (the trunk) of this tropical coastal lowland tree as lumber. They dry the sap from its trunk for use as a sugar, or they ferment it to drink. They use the leaves to thatch their homes and to make hats and baskets. They eat the apical bud at the top of the trunk in salads. The coconut fruit serves many purposes. The hard shell can be used as a container or burnt as fuel; the fibrous middle layer, or coir, of the fruit wall can be made into mats and rope. The seed of the coconut palm has both a liquid endosperm (coconut milk) and a solid endosperm (coconut meat). Because the refreshing and delicious milk contains no bacteria or other pathogens, it is a particularly important drink wherever the water is not fit for drinking. Millions of people get most of their protein from coconut meat. Much coconut meat is dried and marketed as copra, from which coconut oil is pressed. Coconut oil is the most widely used vegetable oil in the world; it is used in the manufacture of a range of products from hydraulic brake fluid to synthetic rubber and, although nutritionally poor, as food. Ground copra serves as fertilizer and as food for livestock.

The trunk of a coconut palm differs in three basic ways from the trunks of many other familiar trees. The most striking difference is that it bears no branches, and all the leaves are borne in a cluster at the top of the trunk. Second, the coconut trunk tapers little from the base of the tree to the top—even the youngest part of the trunk is essentially as thick as the base. We will discuss this phenomenon later in the chapter. Third, a cross section of the trunk reveals no annual rings.

Each coconut palm tree has separate male and female flowers; both are small and inconspicuous. The male flowers have six stamens. The leaves of the coconut palm are large and made up of numerous long, narrow leaflets, each having veins running parallel to one another.

Red Maple

One of the most familiar native trees in the eastern United States is the red, or scarlet, maple (*Acer rub-*

(*a*)

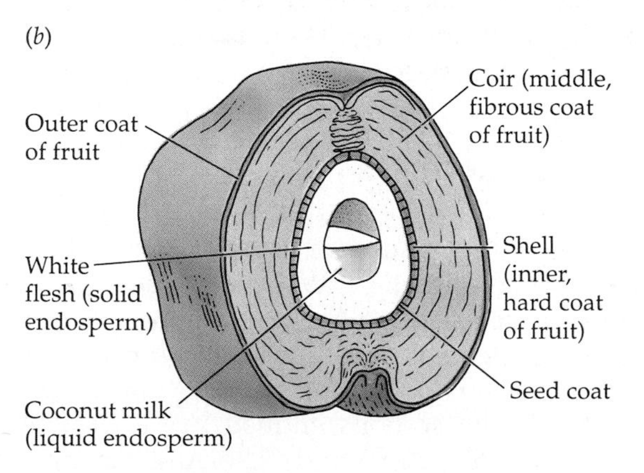

29.1 Coconut Palm
(*a*) A coconut plantation in the South Pacific. Palms are among the only monocot trees. (*b*) A cross section of the coconut's fruit.

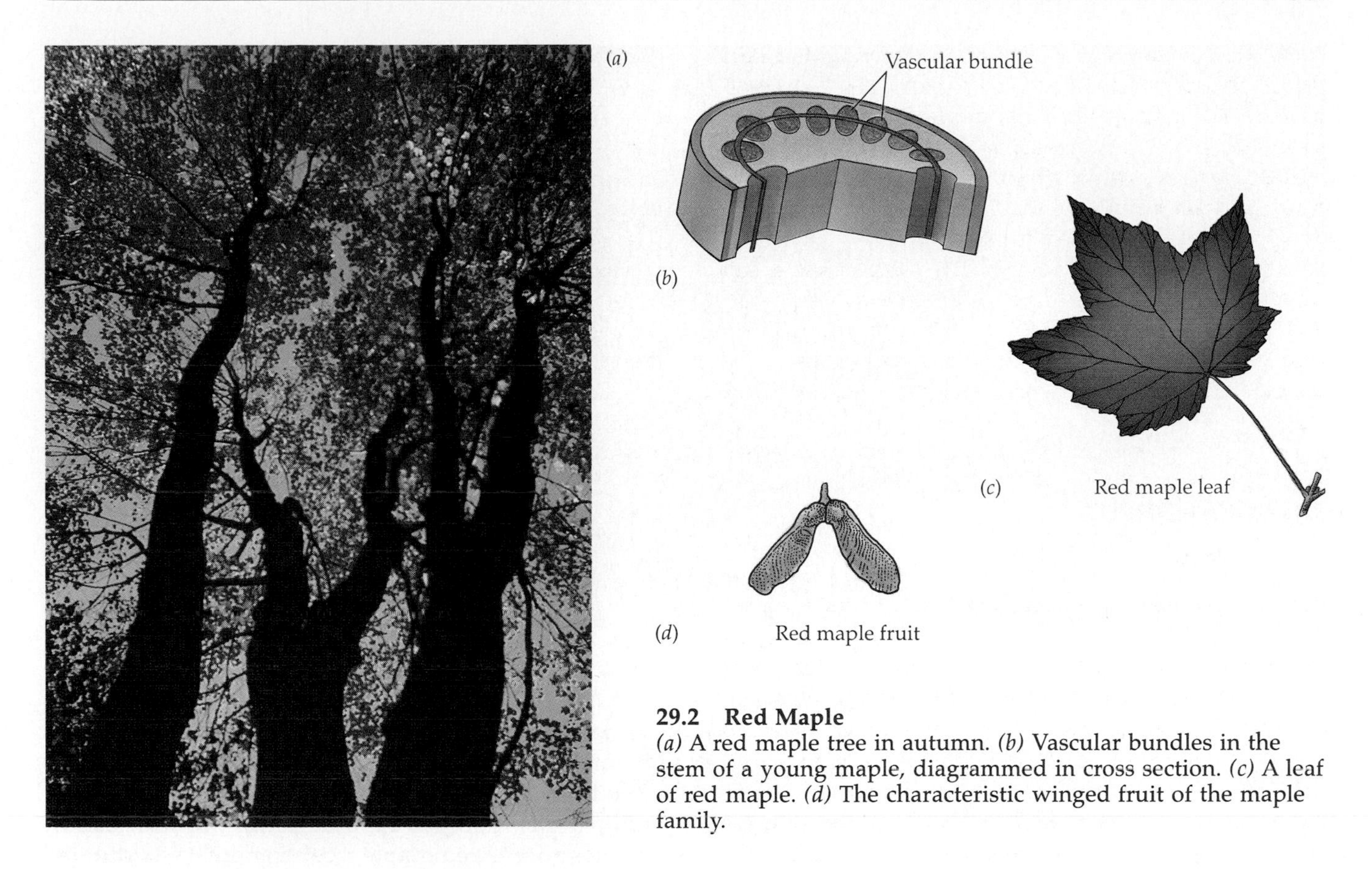

29.2 Red Maple
(a) A red maple tree in autumn. *(b)* Vascular bundles in the stem of a young maple, diagrammed in cross section. *(c)* A leaf of red maple. *(d)* The characteristic winged fruit of the maple family.

rum; Figure 29.2). Unlike the coconut palm, the red maple does not provide us with a great variety of useful products, but it enriches us by its beauty. Not only is it abundant in forests, but we admire it in parks and as a street tree growing to 10 to 30 meters tall. We use its wood as lumber, although the sugar maple is a more important commercial source of maple wood.

Microscopic examination of a very young maple stem reveals vascular bundles of water- and food-conducting cells, arranged in a cylindrical pattern. Like the palm, the mature maple tree has a thick, massive trunk, but a cross section of the trunk shows that the wood is made up of many annual rings. The roots, too, are woody. The red maple leaf—the symbol of Canada—consists of a single blade with three to five lobes, with veins that radiate from a single focal point. These leaves are among the brilliant contributors to the fall colors of eastern forests. The scarlet flowers have four sepals, four petals, eight stamens, and one pistil. The distinctive, winged fruit of the maple family contains two seeds.

Rice

More than half of the world's human population derives the bulk of its food energy from the seeds of a single plant: rice (*Oryza sativa;* Figure 29.3). Rice is particularly important in the diets of people in the Far East, where it has been cultivated for nearly 5,000 years. People use rice straw in many ways, such as thatching for roofs, food and bedding for livestock, and clothing. Rice hulls also have many uses, ranging from fertilizer to fuel.

Rice is a fast-growing plant, yielding more than one crop per year. Some rice is fed to livestock, but most is eaten by humans. When milled for human consumption, rice is an incomplete food because milling removes the bran that contains B vitamins. Even unmilled rice is a poor source of protein; thus, it should be eaten with other, supplementary foods such as soybeans or fish. Most rice varieties are grown submerged in water for the bulk of the growing season. Fish are often raised in rice paddies, where they serve as supplemental food for humans, add fertilizer to the paddy, and control the mosquito population.

The rice plant looks much like other cereal grain plants. The leaves are long, narrow, flat, and more than half a meter in length, with veins running parallel to one another along the length of the leaf. Rice stems do not thicken and become woody as do the stems of trees and shrubs. Rice flowers have six stamens and one ovary. The vascular bundles in the rice stem are scattered, rather than lying in a ring as in the red maple stem.

29.3 Rice
(a) Terraced rice paddies in Bali, Indonesia. *(b)* A rice plant.

Soybean

Soybeans (*Glycine max*; Figure 29.4) were first grown in China thousands of years ago, but today the United States is the largest single producer. Soybeans are featured in many foods and sauces. They also yield a commercially important oil, used in the manufacture of adhesives, paints, inks, and plastics. After oil has been squeezed from the seeds, the residue may be fed to livestock or made into soy flour. Soybean stems may be used for straw.

29.4 Soybean
Their leaves dominate this farmer's field of soybeans in summer.

The soybean plant stands from less than a meter to more than 2 meters in height. Soybean leaves have three lobes, with veins radiating in a netlike pattern. The vascular bundles of the young soybean stem, like those of the red maple, are arranged in a cylindrical pattern. The flowers are small, either white or blue, and consist of five sepals, five petals, ten stamens, and one pistil. Soybean plants tend to be drought-resistant because they have richly branching root systems, which often extend more than 1.5 meters below the soil surface.

CLASSES OF FLOWERING PLANTS

Comparison of the features of these four plants—coconut palm, red maple, rice, and soybean—suggests at least two ways in which they may be classified. We may divide them into trees (coconut palm, red maple) and herbs (rice, soybean) based on their growth plan, or we may divide them into monocots (coconut palm, rice) and dicots (red maple, soybean) based on one clearly distinguishing character (possession of one or two seed leaves—cotyledons—in their embryo) as well as on several important anatomical characters (Figure 29.5). Monocots are generally narrow-leaved flowering plants such as grasses (including rice), lilies, orchids, and palms. Dicots are broad-leaved flowering plants such as soybeans, roses, sunflowers, and maples. As we learned in Chapter 25, the monocots and dicots are the two classes, Monocotyledones and Dicotyledones, that make up the phylum Anthophyta—flowering plants (angiosperms). We'll consider other parts of plants in addition to cotyledons, but first let's consider some overall organizing concepts.

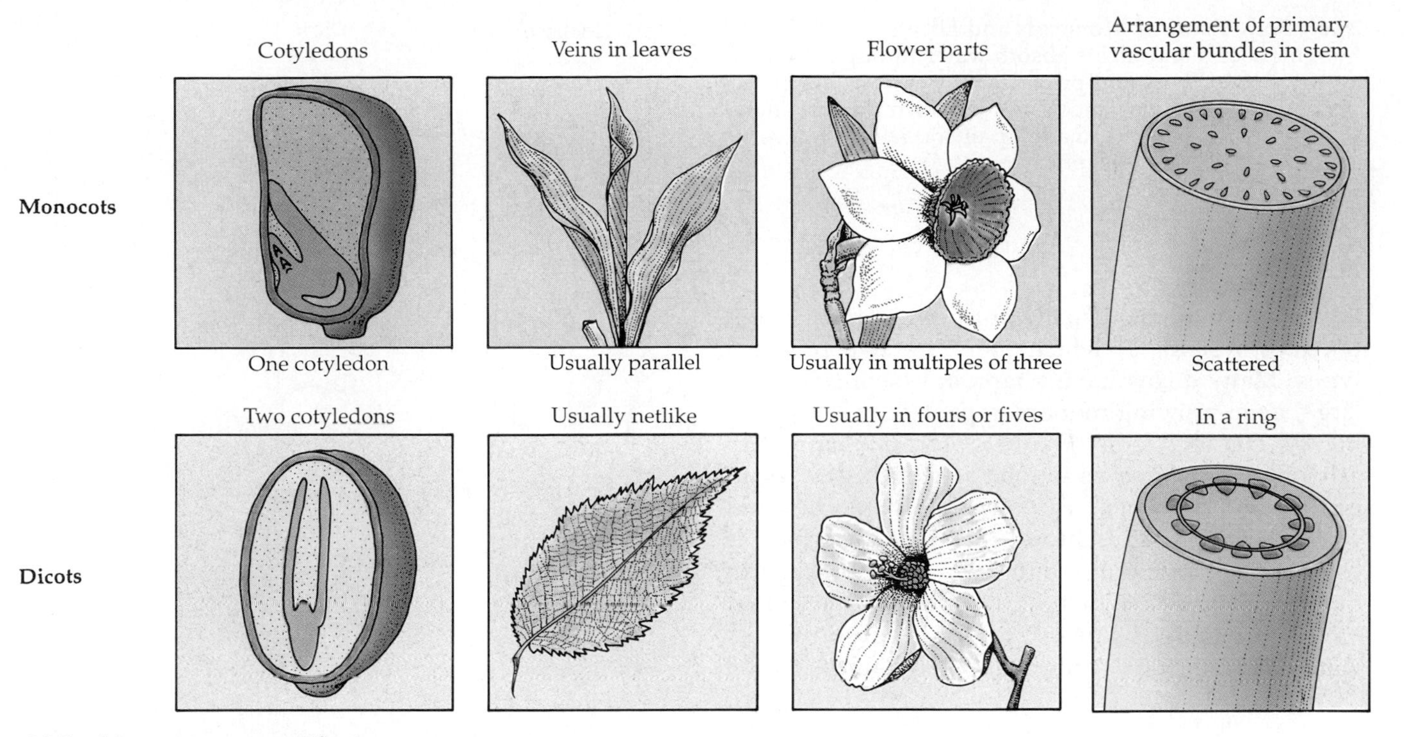

29.5 Monocots versus Dicots

AN OVERVIEW OF THE PLANT BODY

As the plant body grows, it may lose parts, and it forms new parts that may grow at different rates. Each branch of a plant may be thought of as a unit in many ways independent of the other branches. A branch of a plant does not bear the same relationship to the remainder of the body as does an arm to the remainder of the human body. Each branch lives out its own history, and branches grow independently, exploring different parts of the surrounding environment. Branches may respond differently to gravity, some growing more or less vertically and others horizontally. Leaves are units of another sort, produced in fresh batches to take over the constant function of feeding the plant. Often the shapes of different leaves of the same plant differ depending on their differing local environments. Leaves are much shorter-lived than are branches. Branch roots are semi-independent units.

The partial independence of plant parts results in a decentralization of control systems. Branches experiencing different local environments send differing reports to the rest of the plant. If a plant has more than one stem, each one may receive a different report as to the availability of water and minerals because each is served by different roots. In spite of this decentralization, the plant functions as a coherent unit.

Animals grow all over; that is, all parts of the body grow as the individual develops from embryo to adult. Most plant growth, by contrast, is in specific regions of active cell division and cell expansion. The regions of cell division are called **meristems**. Meristems at the tips of the root and stem produce the plant body by dividing to make the cells that compose the parts of the plant. All plant organs arise from cell divisions in the meristems, followed by cell expansion. As you read this chapter, notice the emphasis on the activities of meristems.

ORGANS OF THE PLANT BODY

The bodies of most vascular plants are divided into three principal organs: the **leaves**, the **stem**, and the **root system** (Figure 29.6). A stem and its leaves, taken together, are called a shoot. The **shoot system** of a plant consists of all stems and all leaves. Broadly speaking, the leaves are the chief organs of photosynthesis. The stem holds and displays the leaves to the sun, maximizing the photosynthetic yield, and provides transport connections between the roots and leaves. The points where leaves attach to the stem are called **nodes**, and the stem regions between nodes are **internodes** (see Figure 29.6). Roots anchor the plant in place, and their extreme branching and fine form adapt them to absorb water and mineral nutrients from the soil.

Each of the principal organs can best be understood in terms of its function and its structure. By structure we mean both gross form and microscopic anatomy—the component tissues as well as their arrangement.

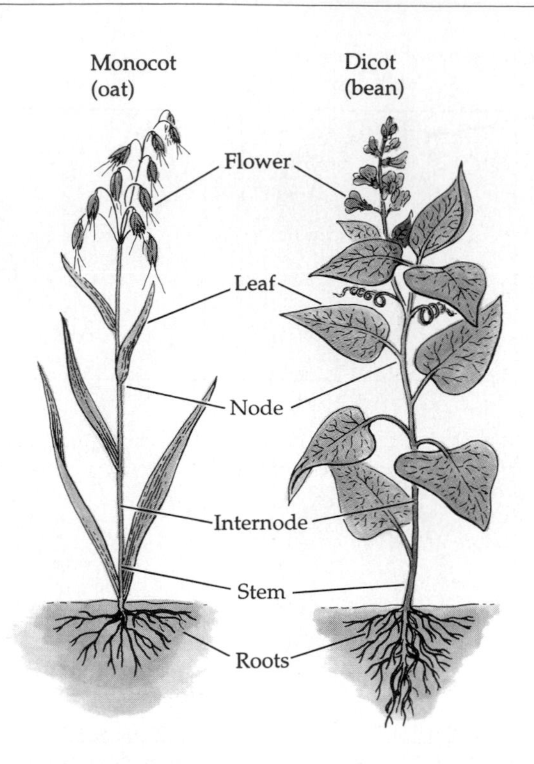

29.6 Body Plans of Monocots and Dicots
Both monocots and dicots absorb water through a root system that anchors and provides nutrients for a shoot system made of stems and leaves in which photosynthesis takes place. Flowers, made up of specialized leaves, are adapted for sexual reproduction.

Roots

Water and minerals usually enter the plant through the root system, of which there are two principal types. Many dicots have a **taproot system:** a single, large, deep-growing root accompanied by less prominent secondary roots (Figure 29.7*a*). The taproot itself often functions as a food-storing organ, as in carrots and radishes. By contrast, monocots and some dicots have a **fibrous root system**, which is composed of numerous thin roots roughly equal in diameter (Figure 29.7*b–e*). Fibrous root systems often have a tremendous surface area for the absorption of water and minerals. A fibrous root system holds soil

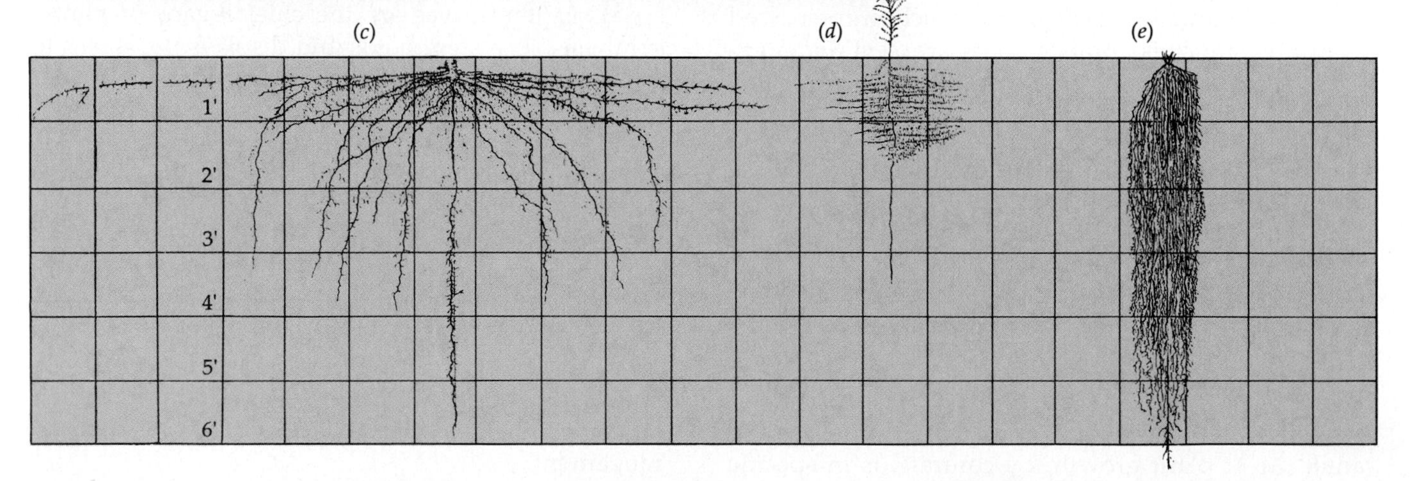

29.7 Root Systems
The taproot system of a dandelion *(a)* contrasts with the fibrous root system of grasses *(b)*. Fibrous root systems are diverse *(c–e)*, with forms adapted to the different environments in which they grow.

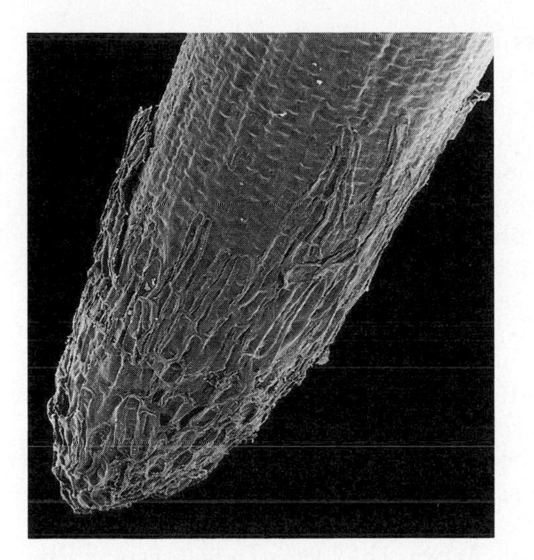

29.8 The Root Cap Protects the Root Tip
As the root grows through the soil, the root cap wears off and is replaced by cells from the root apical meristem.

very well, giving grasses with such systems a protective role on steep hillsides where runoff from rain could cause erosion.

A tissue composed of rapidly dividing cells is located at the tip of the root proper, just behind the root cap. This tissue is the **root apical meristem**, which produces all the cells that contribute to growth in the length of the root. Some of the daughter cells from the root apical meristem are contributed to the **root cap** that protects the delicate growing region of the root as it pushes through the soil (Figure 29.8). Cells of the root cap are often damaged or scraped away and must therefore be replaced constantly. The root cap is also the structure that detects the pull of gravity and thus causes the root to grow downward. Most root cells—those that are produced at the other end of the meristem—elongate. Following elongation, these cells differentiate, giving rise to the various tissues of the mature root.

Some plants have roots that arise from points along the stem, or from the leaves. Known as **adventitious roots**, they also form (in many species) when a piece of shoot is cut from the plant. Adventitious rooting enables the cutting to establish itself in the soil. Some plants—corn, for example—use adventitious roots as props to help support the young shoot.

Stems

Unlike most roots, a stem may be green and capable of photosynthesis. A stem bears leaves at its nodes, and where each leaf meets the stem there is a **lateral bud**, which develops into a branch if it becomes active (Figure 29.9*a*). A branch is also a stem. The branching patterns of plants are highly variable, depending upon the species, environmental conditions, and a gardener's pruning activities.

(*a*)

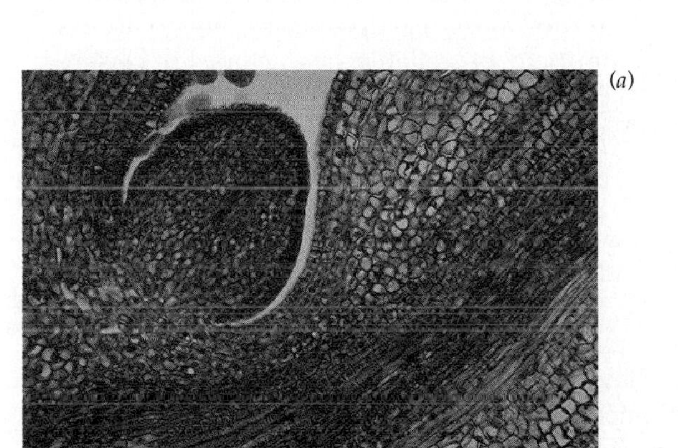

(*b*)

(*c*)

29.9 A Selection of Stems
(a) Microscopic view of a lateral bud developing at the junction between leaf and stem of a lilac; the bud contains vascular tissue and may develop into a branch. *(b)* A potato is a modified stem called a tuber; the sprouts that grow from its eyes are branches. *(c)* Runners (red) of beach strawberry are horizontal stems; such a stem produces roots at intervals, providing a local water supply and allowing rooted portions to live independently if the runner is cut.

Some stems are highly modified. The potato **tuber**, for example, is actually a portion of the stem. Its eyes (Figure 29.9*b*) contain lateral buds, and a sprouting potato is just a branching stem. The runners of strawberry plants and Bermuda grass are horizontal stems from which roots grow at frequent intervals (Figure 29.9*c*). If the links between the rooted portions are broken, independent plants can develop from each side of the break. This is a form of asexual reproduction, which we will discuss in Chapter 33.

Stems bear buds—embryonic shoots—of various types. We have already mentioned lateral buds, which give rise to branches. At the tip of each stem or branch is an **apical bud** containing a **shoot apical meristem**, which produces the cells for the growth and development of the stem. The shoot apical meristem also produces **leaf primordia**, which expand to become leaves in the apical bud (see the upper micrograph in Figure 29.16). At times that vary from species to species, buds are formed that develop into flowers.

Shoot systems have various forms, in which branches take on different relationships to the plant as a whole (Figure 29.10). Some shoots branch underground, and their branches emerge from the soil looking like separate plants.

Leaves

In most plants the leaves are responsible for most of the photosynthesis, producing food for the plant and releasing oxygen gas. Leaves also carry out metabolic reactions that make nitrogen available to the plant for the synthesis of proteins and nucleic acids (see Chapter 32). Leaves are important food-storage organs in some species; in others—the succulents—the leaves store water. The thorns of cacti are modified leaves. Certain leaves of poinsettias, dogwood, and some other plants are brightly colored and help attract pollinating animals to the often less-striking flowers. Many plants, such as peas and squash, have tendrils—modified leaves that support the plant by wrapping around other plants. A less obvious but often crucial function of leaves is to shade neighboring plants. Like all other organisms, plants compete; if a plant can reduce the photosynthetic capability of its neighbors by intercepting sunlight, it can obtain a greater share of the available water and mineral nutrients. Finally, as we will see in Chapter 33, the "timer" by which some plants measure the length of the night is located in the leaves.

Leaves are marvelously adapted to serve as light-gathering, photosynthetic organs. Typically, the **blade** of a leaf is flat, and during the daytime it is held by its stalk, or **petiole**, at an angle almost perpendicular to the rays of the sun. Some leaves track the sun, moving so that they constantly face it. If leaves were thicker, the outer layers of cells would absorb so much of the light that the interior layers of cells would be too dark and thus would be unable to photosynthesize.

The different leaves of a single plant may have quite different shapes. The form of a leaf results from a combination of genetic, environmental, and developmental influences. Most species, however, bear leaves of some broadly defined type (Figure 29.11). A leaf may be **simple**, consisting of a single blade, or **compound**, in which blades, or leaflets, are arranged along an axis or radiate from a central point. In a simple leaf, or in a leaflet of a compound leaf, the veins may be parallel to one another or in a netlike arrangement. The general development of a specific leaf pattern is programmed in the individual's genes and is expressed by differential growth of the leaf veins and of the tissue between the veins. As a result plant taxonomists have often found leaf forms (outline, margins, tips, bases, and patterns of arrangement) to be reliable characters for classification and identification. At least some of the forms in Figure 29.11 probably look familiar to you.

LEVELS OF ORGANIZATION IN THE PLANT BODY

Newly formed cells expand to their final size, and then they differentiate, that is, become structurally or chemically specialized for particular functions. A tissue is an organized group of cells, working together as a functional unit. Simple tissues are composed of a single type of cell; compound tissues are composed of several cell types. Plant tissues are or-

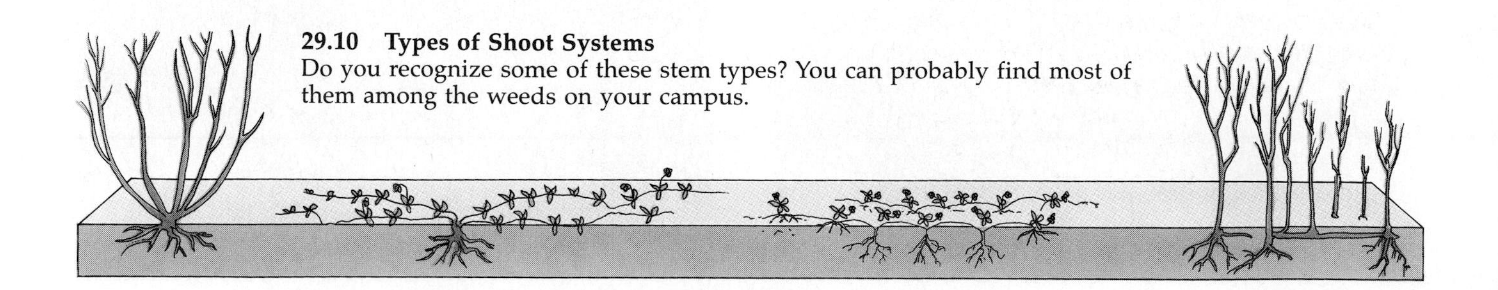

29.10 Types of Shoot Systems
Do you recognize some of these stem types? You can probably find most of them among the weeds on your campus.

Leaf shapes

Margins

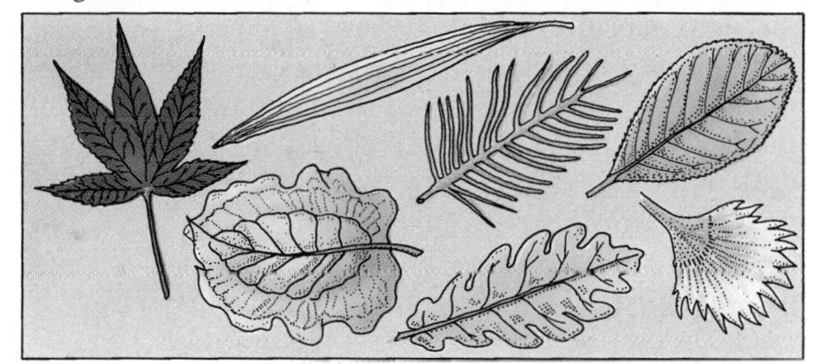

Apices and bases

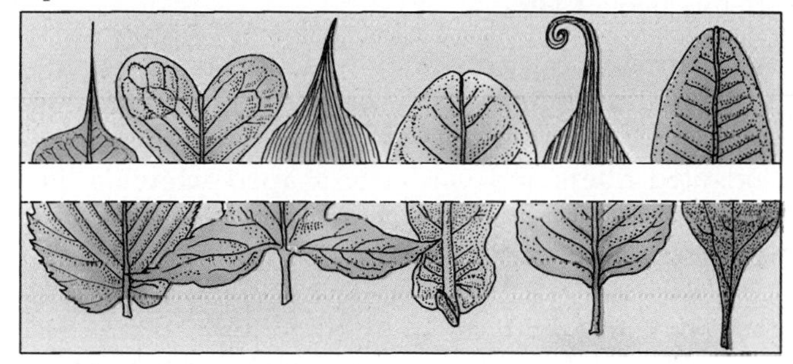

Arrangements on stem

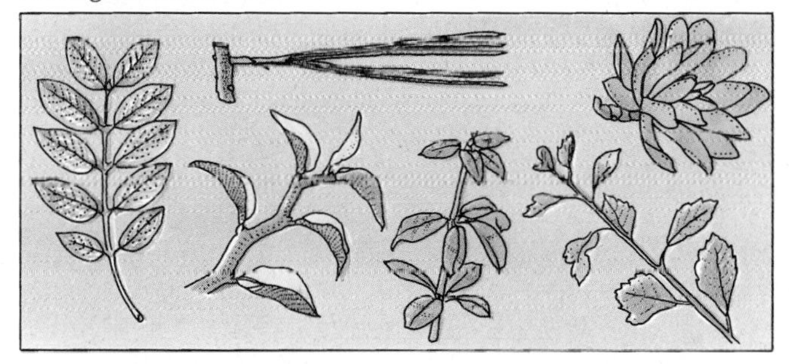

Parts and types

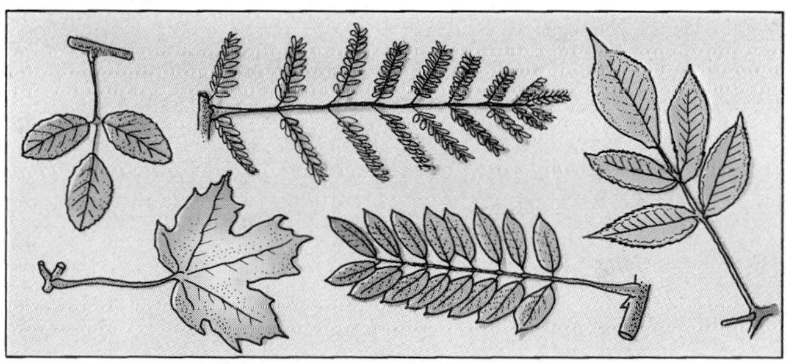

29.11 The Diversity of Leaves

ganized into three tissue systems that extend throughout the body of the plant, from organ to organ. These three systems are the vascular tissue system (xylem and phloem), which conducts materials from one part of the body to another; the dermal tissue system, which protects the body surface; and the ground tissue system, which plays many roles, including producing and storing food materials.

To understand the structures and functions of the tissue systems, we must know the nature of their building blocks. Some cells are alive when functional; others function only after their living parts have died and disintegrated. Some cells develop chemical capabilities not demonstrated by other cells. Several cell types differ dramatically in the structure of their cell walls. We will first consider the types of cells that make up the plant body and then see how aggregations of cells form functioning tissues and tissue systems.

PLANT CELLS

Living plant cells have all the essential organelles common to eukaryotes (see Chapter 4). In addition, they have some structures and organelles not shared by cells of the other kingdoms. Some plant cells contain chloroplasts and microbodies. Many plant cells contain large central vacuoles. Every plant cell is surrounded by a cellulose-containing cell wall.

Cell Walls

The division of a plant cell is completed when cell walls form, separating the daughter cells. The first barrier to form is the **middle lamella** (Figure 29.12*a*). The formation of this layer is followed by the secretion of structural materials, including cellulose, by the newly separated cells. Each daughter cell, as it expands to its final size, secretes more cellulose and other polysaccharides to complete formation of the **primary wall** (Figure 29.12*b*).

Once cell expansion stops, a plant cell may deposit more polysaccharides and other materials—such as lignin, characteristic of wood, or suberin, characteristic of cork—in one or more layers internal to the primary wall. These layers collectively form the **secondary wall** (Figure 29.12*c*), which often serves supporting or waterproofing roles.

Although the cell wall lies outside the plasma membrane of the cell, it is not a chemically inactive region. Chemical reactions in the wall play an important role in cell expansion. Cell walls may thicken or be sculpted or perforated as part of the differentiation into various cell types.

Except where the secondary wall is waterproofed, the structure is porous to water and to most small

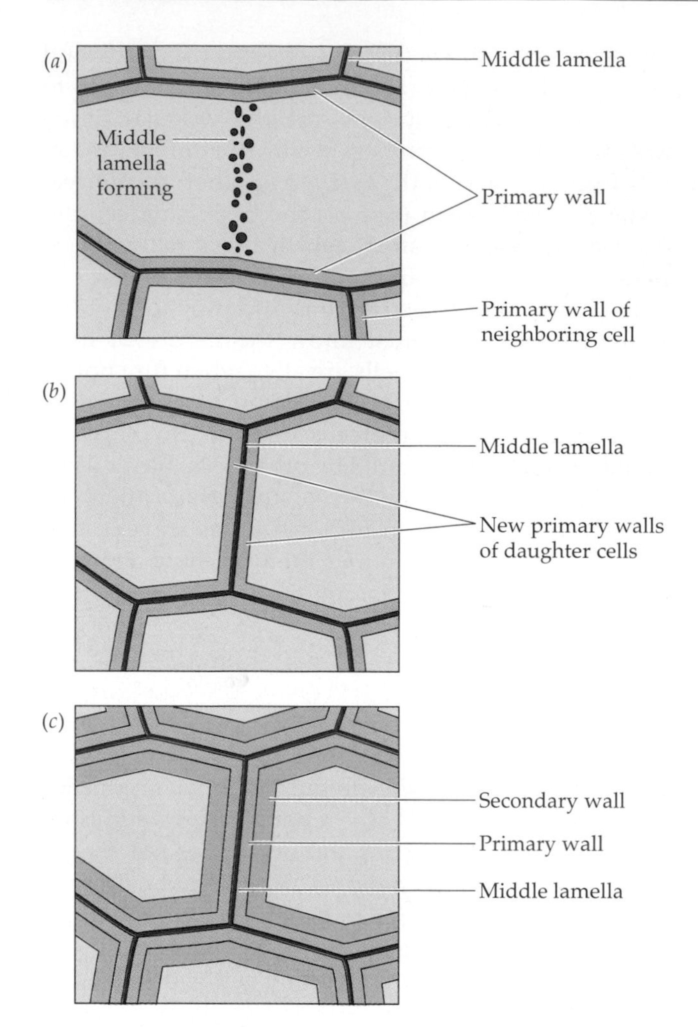

29.12 Cell Wall Formation
The middle lamella is the first wall layer to form. Each daughter cell secretes a primary wall. Once cell expansion stops, the cell may secrete more layers, forming secondary walls.

molecules. Water and dissolved materials can move directly from cell to cell without passing into the cell wall space because plant cells have structures called **pit pairs** connected by strands of cytoplasm called plasmodesmata (see Chapter 4). A pit is a thinning in the primary wall of a cell at a place where the secondary wall either is absent or is separated from the primary wall by a space. Where there is a pit in the wall of one cell, there is usually a corresponding pit in the adjacent cell's wall; together they are a pit pair. Plasmodesmata pass through pit pairs and the middle lamella between them, allowing molecules with molecular weights of about 850 daltons or less to pass freely from one cell to the other.

Parenchyma Cells

The most numerous cells in the young plant body are the **parenchyma** cells (Figure 29.13*a*). Parenchyma cells are alive when they perform their functions in the plant. They usually have thin walls, consisting only of a primary wall and the shared middle lamella. Many parenchyma cells have shapes similar to those of soap bubbles crowded into a limited space—figures with 14 faces. They are not elongated or otherwise asymmetrical. Most have large central vacuoles.

Many parenchyma cells store various substances, such as starch or lipids. In the cytoplasm of these cells, starch is often stored in specialized plastids called leucoplasts (see Chapter 4). Lipids may be stored as oil droplets, also in the cytoplasm. Other parenchyma cells appear to serve as packing material and play a vital role in supporting the stem. Leaves have a particularly important type of parenchyma cell that is specialized for photosynthesis and is equipped with abundant chloroplasts. Some other parenchyma cells—but not these photosynthetic cells—retain the capacity to divide and hence may give rise to new meristems, as when a branch root forms within a region of parenchyma cells inside a taproot.

Sclerenchyma Cells

Sclerenchyma cells function when dead. A heavily thickened secondary wall performs their function: support. There are two types of sclerenchyma cells: elongated **fibers** and variously shaped **sclereids**. Fibers, often organized into bundles, provide a relatively rigid support both in wood and in other parts of the plant (Figure 29.13*b*). The bark of trees owes much of its mechanical strength to long fibers. Sclereids may pack together very densely, as in a nut's shell or other types of seed coats (Figure 29.13*c*). Isolated clumps of sclereids, called stone cells, in pears and some other fruits give them their characteristic gritty texture.

Collenchyma Cells

Another type of supporting cell, the **collenchyma** cell, remains alive even after laying down thick cell walls (Figure 29.13*d*). Collenchyma cells are generally elongated. In these cells the primary wall thickens and no secondary wall forms. Collenchyma provides support to petioles, nonwoody shoots, and growing organs. Tissue made of collenchyma cells, although resistant to bending, is more flexible than is sclerenchyma; stems and leaf petioles strengthened by collenchyma can sway in the wind without snapping as they might if they were strengthened by sclerenchyma.

Water-Conducting Cells of the Xylem

The xylem of vascular plants contains cells called tracheary elements, which die before they assume

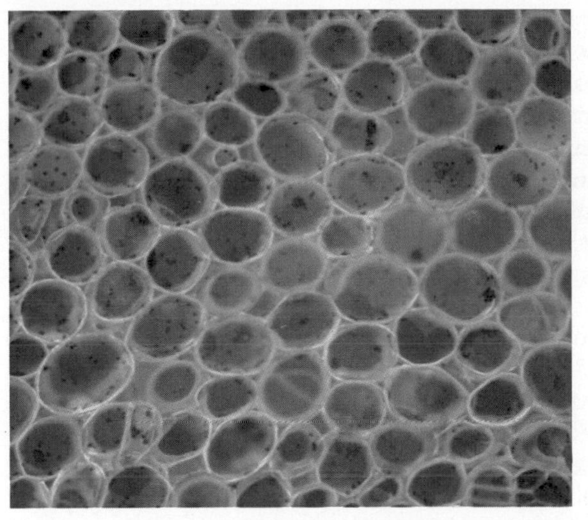

(a)

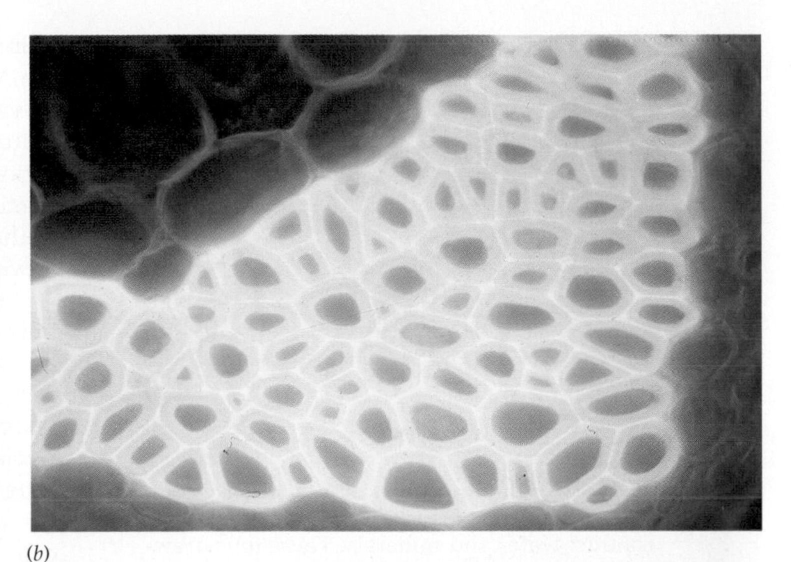

(b)

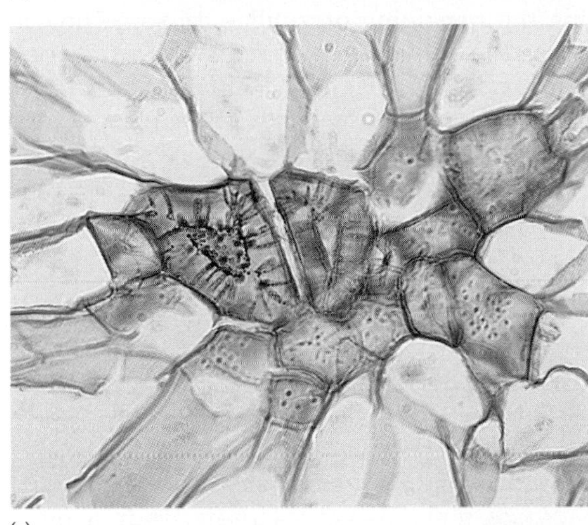

(c)

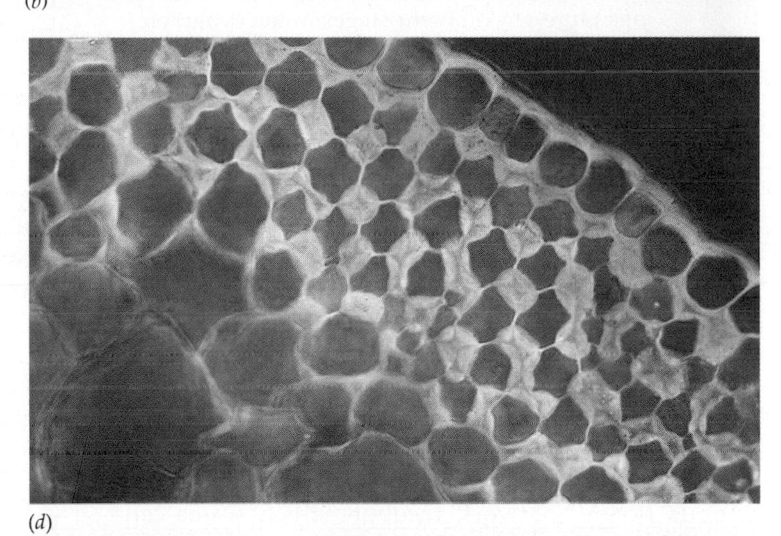

(d)

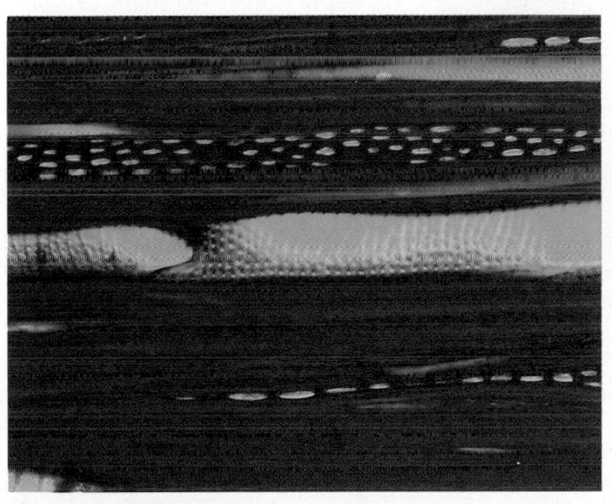

(e)

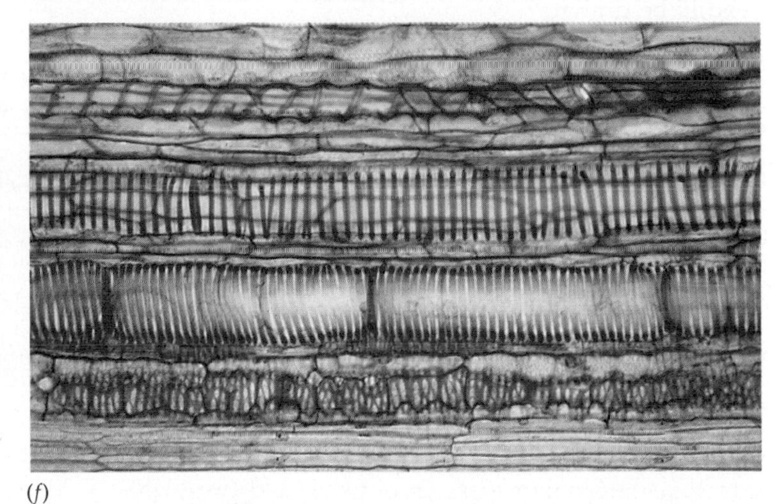

(f)

29.13 Plant Cells
(a) Parenchyma cells in the leaf of a primrose plant; note the uniform cell walls. *(b)* Sclerenchyma: Fibers in a broad bean pod. A stain causes the heavily thickened walls to fluoresce a brilliant yellow. *(c)* Sclerenchyma: Thick-walled sclereids; these extremely thick secondary cell walls are laid down in layers. They provide support and a hard texture to structures such as nuts and seeds. *(d)* Collenchyma cells make up the five outer cell layers of this spinach leaf vein. They are recognizable because their cell walls are very thick at the corners of the cells and thin elsewhere. *(e)* Tracheids appear deep red in this micrograph of bassswood; note the complexity of the cell walls. *(f)* Vessel elements in the stem of a squash. The secondary walls are stained red; note the different patterns of thickening, including rings and spirals. Which cells in this figure function when they are alive and which only when they are dead?

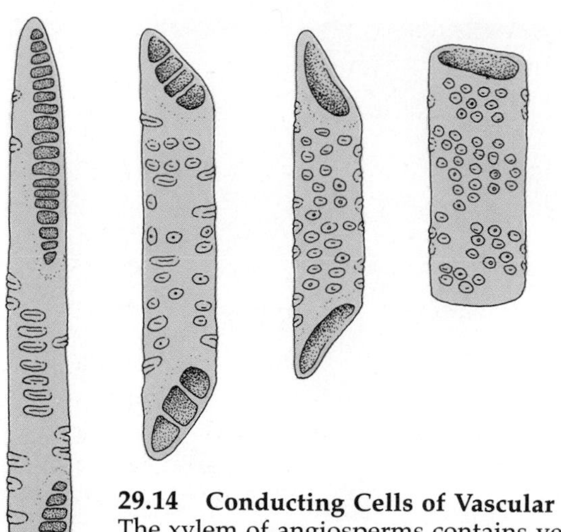

29.14 **Conducting Cells of Vascular Systems**
The xylem of angiosperms contains vessels that conduct water and minerals. These four drawings represent different stages in the evolution of the vessel element; the one on the left is the most ancient, and the one at the right is the most recently evolved.

their ultimate function of transporting water and dissolved minerals. The tracheary elements of gymnosperms and angiosperms differ significantly. The tracheary elements of gymnosperms are **tracheids**—spindle-shaped cells interconnected by numerous pits in their cell walls (Figure 29.13*e*). Because the cell contents—the nucleus and cytoplasm—disintegrate upon death, a group of dead tracheids forms a continuous hollow network through which water can readily be drawn.

Flowering plants evolved a water-conducting system made up of vessels. The individual cells, called **vessel elements**, also die before they become functional. Vessel elements are generally larger in diameter than tracheids; they are laid down end-to-end and lose all or part of their end walls, so that each vessel is a continuous hollow tube consisting of many vessel elements and providing a clear pipeline for water conduction (Figure 29.13*f*). In the course of angiosperm evolution, vessel elements have become shorter, and their end walls have become less and less obliquely oriented and less obstructed (Figure 29.14). The xylem of many angiosperms also includes tracheids.

Sieve Tube Elements

The phloem, in contrast to xylem, consists primarily of living cells. In flowering plants the characteristic cell of the phloem is the **sieve tube element** (Figure 29.15*a*). Like vessel elements, these cells meet end-to-end and form long sieve tubes, which transport foods from their sources to tissues that consume or store them. In plants with mature leaves, for example, excess products of photosynthesis move from leaves to tissues in roots. As sieve tube elements mature during their development, a chemical drilling action expands small holes in the end walls, connecting the contents of neighboring cells. The result is that the end walls look like sieves and are called **sieve plates** (Figure 29.15*b*).

As the holes in the sieve plates expand, the membrane around the central vacuole disappears, allowing some of the cytosol and the vacuole's contents to mingle and form a single fluid; this mixture can be forced from cell to cell along the sieve tube. The nucleus and some of the other organelles in the sieve tube element also break down and thus do not clog the holes of the sieve. A "fixed," stationary layer of cytoplasm remains, however, lining the cell wall and confining the remaining organelles. In some flowering plants, the sieve tube elements have adjacent **companion cells** (see Figure 29.15*a*). A parent cell divides, thereby producing a sieve tube element and its companion cell. Companion cells retain all their organelles and may, through the activities of their nuclei, regulate the performance of the sieve tube elements.

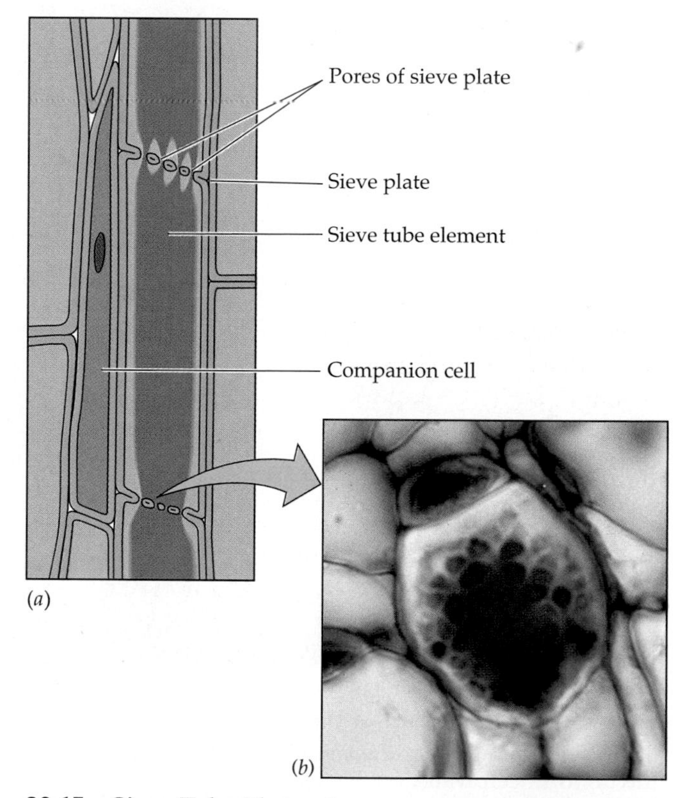

29.15 **Sieve Tube Elements**
(a) Sieve tube elements are usually accompanied by companion cells. *(b)* Micrograph of a sieve plate at the end of a sieve tube element. Phloem sap passes through the holes in sieve plates from one sieve tube element to the next.

PLANT TISSUES AND TISSUE SYSTEMS

Parenchyma cells make up parenchyma tissue, a simple tissue—that is, one composed of only one type of cell. Sclerenchyma and collenchyma are other simple tissues. Cells of various types also combine to form complex tissues. Xylem and phloem are complex tissues, composed of more than one type of cell. All xylem contains parenchyma cells, which store food. The xylem of angiosperms contains vessel elements, as well as thick-walled sclerenchyma fibers that provide considerable mechanical strength to the xylem. In most gymnosperms, tracheids serve both in water conduction and in support because vessels and fibers are absent. In addition, old xylem that is no longer active in transport becomes compacted at the center of the tree trunk and continues to contribute support for the tree. As a result of its cellular complexity, xylem can perform a variety of functions, including transport, support, and storage. The phloem of angiosperms includes sieve tube elements, companion cells, fibers, sclereids, and parenchyma cells.

The **vascular tissue system**, which includes the xylem and phloem, is the conductive, or "plumbing," system of the plant. All living cells require a source of energy and chemical building blocks. As already mentioned, the phloem transports food from the sites of production (called sources; commonly the leaves) to sites of utilization or storage (called sinks) elsewhere in the plant. The xylem distributes water and mineral ions taken up by the roots to the stem and leaves.

The **dermal tissue system** is the outer covering of the plant. All parts of the young plant body are covered by an **epidermis**, either a single layer of cells or several layers. The shoot epidermis secretes a layer of wax, the **cuticle**, that helps retard water loss. The protective covering of the stems and roots of older woody plants is the **periderm**, which is composed of cork and other tissues that will be discussed later in this chapter.

The **ground tissue system** makes up the rest of a plant and consists primarily of parenchyma tissue, often supplemented by collenchyma or sclerenchyma tissue. The ground tissues function primarily in storage, support, and photosynthesis. Let's look at how the tissue systems are organized in the different organs of a flowering plant, as well as how this organization develops as the plant grows.

GROWTH AND MERISTEMS

At the tip of each shoot or branch is a shoot apical meristem, and at the tip of each root is a root apical meristem. Growth from the apical meristems is called **primary growth**. These meristems give rise to the entire body of many plants (Figure 29.16).

Other plants develop what we commonly refer to as wood and bark. These complex tissues are derived from other meristems. One, called **vascular cambium**, is a cylindrical tissue consisting primarily of vertically elongated cells that divide frequently, producing derivative cells both to the inside of the vascular cambium layer, forming new xylem, and to the outside, forming new phloem. As trees grow in diameter, the outermost layers of the stem are sloughed off. Without the activity of **cork cambium**, which in a tree forms continuously in the bark, this sloughing off would expose the tree to potential damage, including excessive water loss or invasion by microorganisms. Cork cambium produces new cells, primarily in the outward direction. The walls of these cells become impregnated with the waxy substance suberin, thus augmenting the dermal tissue system. Growth in the diameter of stems and roots, produced by the vascular and cork cambia, is called **secondary growth**. It is the source of wood and bark.

In some plants, meristems may remain active for years—even centuries. Such plants grow in size, or at least in diameter, throughout their lifetimes. This phenomenon is known as **indeterminate growth. Determinate growth**, which stops at some point, is characteristic of most animals, as well as some plant parts, such as leaves, flowers, and fruits. The life cycles of plants fall into three categories: annual, biennial, and perennial. **Annuals**, such as many food crops, live less than a year. **Biennials**—carrots and cabbage, for example—grow for all or part of one year and live on into a second year, during which they flower, set seed, and die. **Perennials**, such as oak trees, live for a few to many years.

THE MERISTEMS AND THEIR PRODUCTS

The Young Root

Cell divisions in the root apical meristem produce both the protective root cap and the other primary tissues of the growing root. When a meristematic cell divides, the products initially take up no more volume between them than did the dividing cell. One of the products of each cell division develops into another meristematic cell the size of its parent, while the other product develops differently. The products above the apical meristem—away from the root cap—constitute three cylindrical primary meristems that give rise to the three tissue systems of the root. The innermost primary meristem, the **procambium**, gives rise to the vascular tissue system; the **ground meristem** gives rise to the ground tissue system; and the outermost, the **protoderm**, gives rise to the dermal tissue system (Figure 29.17). The apical and primary

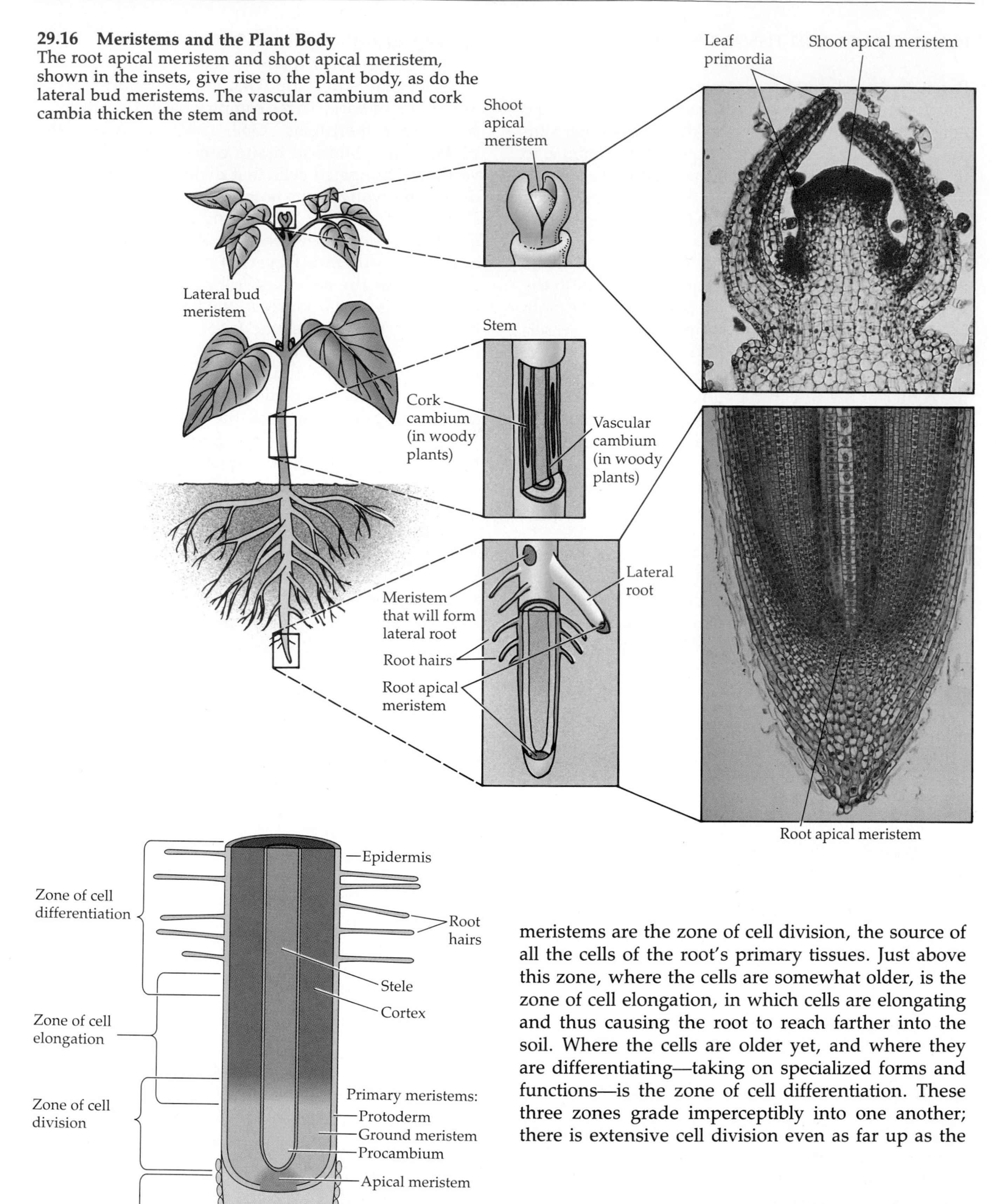

29.16 Meristems and the Plant Body
The root apical meristem and shoot apical meristem, shown in the insets, give rise to the plant body, as do the lateral bud meristems. The vascular cambium and cork cambia thicken the stem and root.

meristems are the zone of cell division, the source of all the cells of the root's primary tissues. Just above this zone, where the cells are somewhat older, is the zone of cell elongation, in which cells are elongating and thus causing the root to reach farther into the soil. Where the cells are older yet, and where they are differentiating—taking on specialized forms and functions—is the zone of cell differentiation. These three zones grade imperceptibly into one another; there is extensive cell division even as far up as the

29.17 Tissues and Regions of the Root Tip
Cells divide in the root apical meristem. Some of the daughter cells become part of the root cap, which is constantly being eroded away, but most daughter cells develop on the side away from the tip and differentiate into the primary tissues of the root.

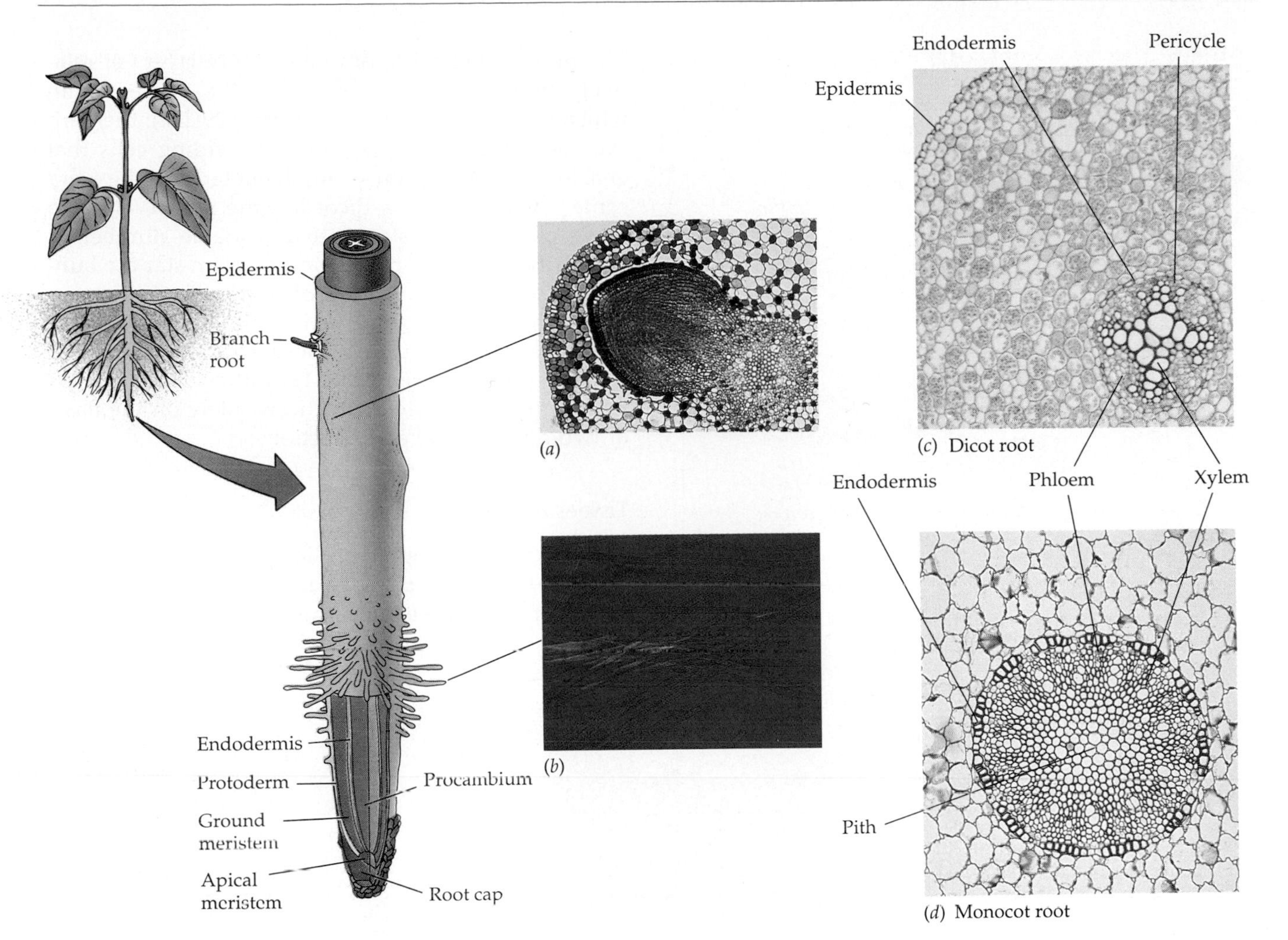

29.18 Root Anatomy
The drawing at the left shows a generalized root structure. *(a)* A branching root tip. Cells in the pericycle divide and the products differentiate, forming the tissues of a branch root. *(b)* Root hairs, viewed under polarized light. *(c, d)* The primary root tissues of a dicot and a monocot. The monocot (an orchid) has a central pith region; the dicot (ranunculus) does not.

zone of cell differentiation, and some cells differentiate even in the zone of cell division.

The protoderm gives rise to the outer layer of root cells, the epidermis, which is adapted for protection and for the absorption of mineral ions and water (Figure 29.18). Epidermal cells are flattened, and many of them produce amazingly long, delicate **root hairs** that vastly increase the surface area of the root (Figure 29.18*b*). It has been estimated that a mature rye plant has a total root surface of more than 1,500 square kilometers (600 square miles), all contained within about 6 liters of soil. Root hairs grow out among the soil particles, probing nooks and crannies and taking up water and minerals.

Internal to the root's epidermis is a region of ground tissue many cells in thickness, called the **cortex** (see Figure 29.17). The cells of the cortex are relatively unspecialized and often function in food storage. In many plants, but especially in trees, epidermal and sometimes cortical cells form an association with a fungus. This association, called a **mycorrhiza**, increases the absorption of minerals and water by the plant (see Box 24.A). Some plant species have poorly developed root hairs or no root hairs. These plants cannot survive unless they develop mycorrhizae that help in mineral absorption.

Proceeding inward, we come to the **endodermis** of the root, a single layer of cells that is the innermost cell layer of the cortex. Endodermal cells differ markedly in structure from the rest of the cortical cells; parts of their walls contain suberin, a waxy substance that forms a waterproof seal. The endodermal cells control the access of water and dissolved substances to the inner, vascular tissues (see Figure 30.6). Elsewhere in the root, water can pass freely through cell walls and between cells.

Once past the endodermis, we enter the domain produced by the procambium. This domain, the vascular cylinder or **stele**, consists of three tissues: the pericycle, the xylem, and the phloem (Figure 29.19).

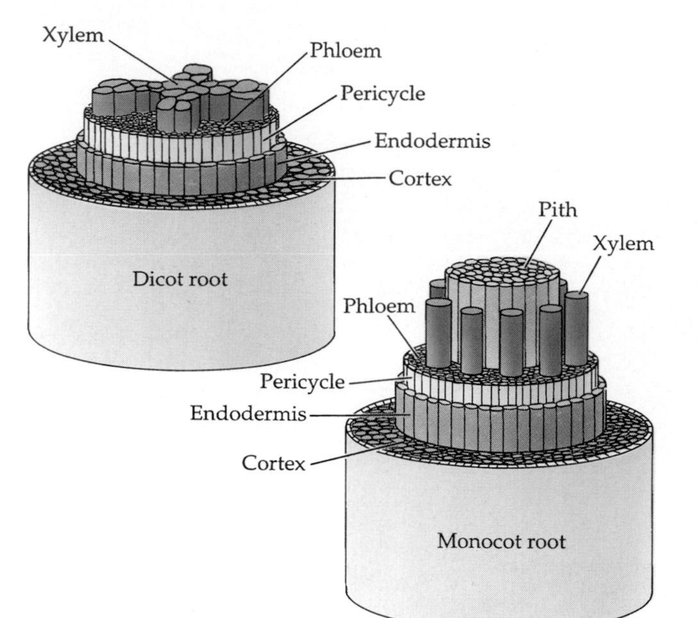

29.19 The Stele
The distribution of tissues in the stele—the region internal to the endodermis—differs in the roots of dicots and monocots.

The **pericycle** consists of one or more layers of relatively undifferentiated cells. It is the tissue within which branch roots arise (see Figure 29.18*a*); the pericycle also provides a few of the dividing cells that enable the root to grow in diameter. At the very center of the root of a dicot lies the xylem—seen in cross section as a star with a variable number of points. Between the points of the xylem star are bundles of phloem. In a monocot root, a region of parenchyma cells, the **pith**, lies internal to the xylem. It is useful to try picturing these structures in three dimensions, as in Figure 29.19, rather than attempting to understand their functions solely on the basis of two-dimensional cross sections.

Tissues of the Stem

The shoot apical meristem, like the root apical meristem, forms three primary meristems: the procambium, ground meristem, and protoderm, which in

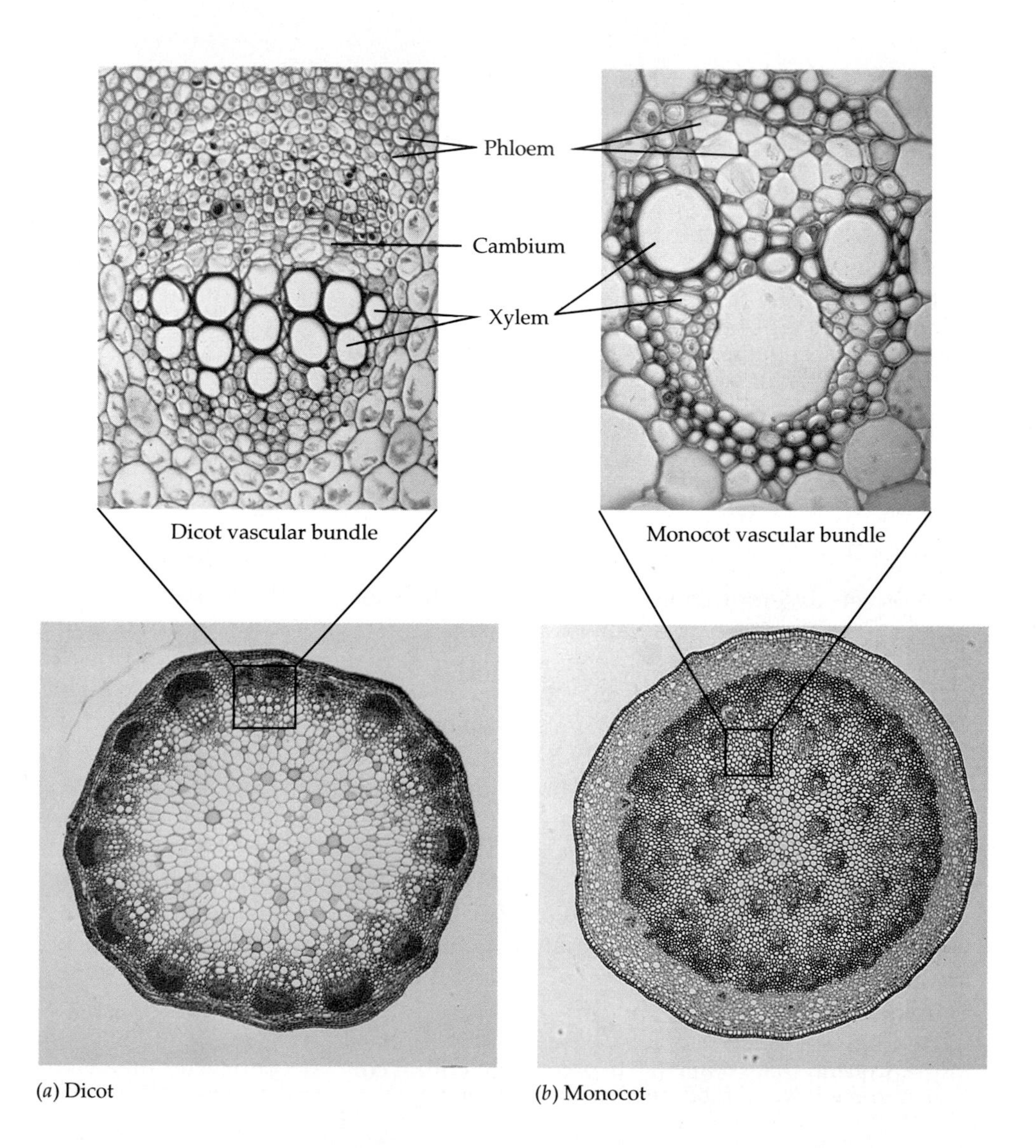

29.20 Vascular Bundles in Stems
The vascular tissues in stems are organized into bundles. *(a)* In dicots the vascular bundles are arranged in a circle with pith in the center and cortex outside the ring, as in this young sunflower stem. *(b)* This cross section shows the scattered arrangement of bundles typical of monocot stems. In both monocots and dicots, the bundles are oriented so that xylem is toward the center of the stem and phloem is to the outside.

turn give rise to the three tissue systems. Leaves arise from leaf primordia that form as cells divide on the sides of shoot apical meristems (see Figure 29.16). The growing stem has no cap analogous to the root cap, but the leaf primordia can act as a protective covering. Dicot stems extend in a region of elongation below the shoot apical meristem. Grasses and some other monocots, however, elongate at the bases of internodes and leaves, where there is some meristematic tissue. Lawn and range grasses can grow back after mowing or grazing because they grow from basal meristems close to the soil surface.

The plumbing of angiosperm stems differs from that of roots. In roots, the vascular tissue lies in the middle, with the xylem at or near the very center. The vascular tissue of a young stem, however, is divided into discrete **vascular bundles**. Vascular bundles generally form a cylinder in the dicots (Figure 29.20*a*) but are seemingly scattered throughout the cross section of the stem in the monocots (Figure 29.20*b*). Each vascular bundle contains both xylem and phloem.

The stem contains other important tissues in addition to the vascular tissues. Internal to the vascular bundles of dicots is a storage tissue, pith, and to the outside lies a similar storage tissue, the cortex. The cortex may contain strengthening collenchyma cells with thickened walls (Figure 29.21*a*). In many monocots the pith is hollowed out (Figure 29.21*b*). The pith, the cortex, and the regions between the vascular bundles in dicots—called pith rays—constitute the ground tissue system of the stem. The outermost cell layer of the young stem is the epidermis, which functions primarily to minimize the loss of water from the cells within.

Secondary Growth of Stems and Roots

Some stems and roots show little or no growth in diameter, remaining slender, but many others, all dicots, undergo considerable thickening. This thickening is of great importance and interest because it gives rise to wood and bark, as well as making the support of large trees possible. Secondary growth results from the activity of two meristematic tissues, vascular cambium and cork cambium (see Figure 29.16). Vascular cambia consist of cells that divide to produce new—secondary—xylem and phloem cells, while cork cambia produce mainly waxy-walled cork cells.

Initially, the vascular cambium is a single layer of cells between the primary xylem and the primary phloem. The root or stem increases in diameter when the cells of the vascular cambium divide, producing secondary xylem cells toward the inside of the root or stem and secondary phloem cells toward the outside (Figure 29.22*a*). In a stem, cells of the pith rays between the vascular bundles also divide, forming a continuous cylinder of vascular cambium running the length of the stem. This cylinder in turn gives rise to complete cylinders of secondary xylem—wood—and secondary phloem—bark (Figure 29.22*b*).

As the vascular cambium produces secondary xylem and phloem, its principal products are vessel elements and supportive fibers in the xylem, and sieve tube elements, companion cells, and fibers in the phloem. Not all xylem and phloem cells are adapted for transport or support; some store materials in the stem or root. Living cells such as these storage cells must be connected to the sieve tubes of the phloem, or they would starve to death. The con-

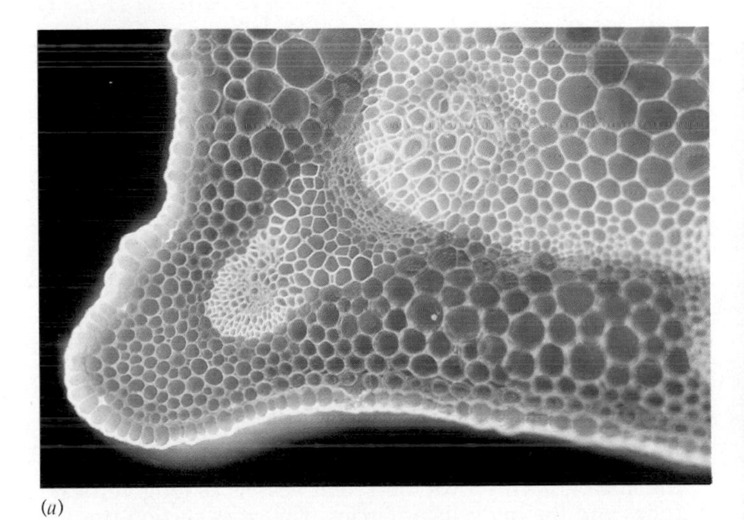

(*a*)

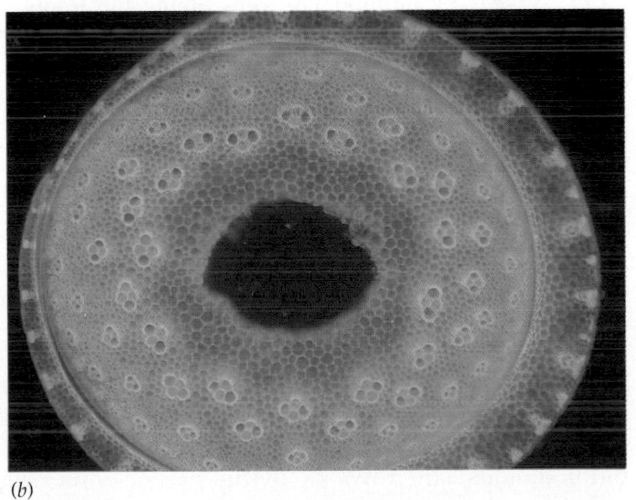

(*b*)

29.21 Other Stem Structures
(a) The stem of a broad bean resists bending but is not brittle. Collenchyma, the tissue in bright blue in the projection to the lower left of this stem cross section, provides flexible support. The next tissue toward the interior, shown in gold, consists of phloem fibers that are stiffer than the collenchyma. *(b)* This bamboo looks like a "typical" monocot, except for its hollowed-out pith.

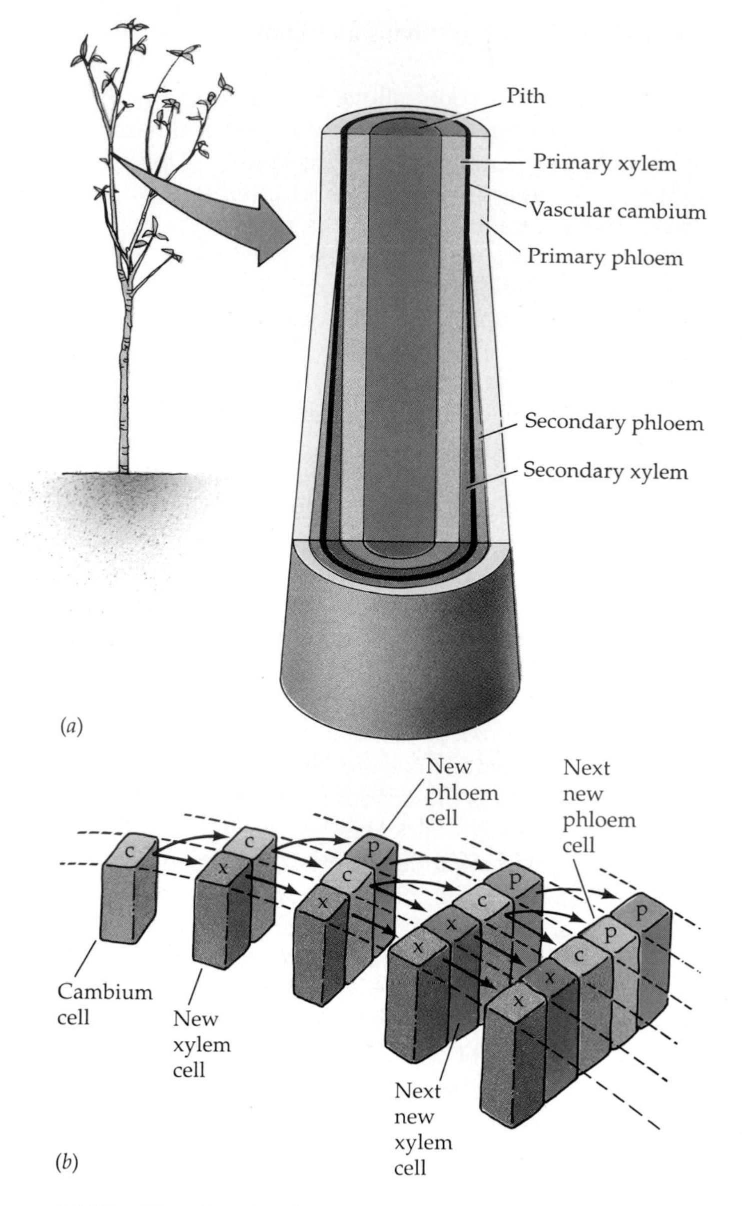

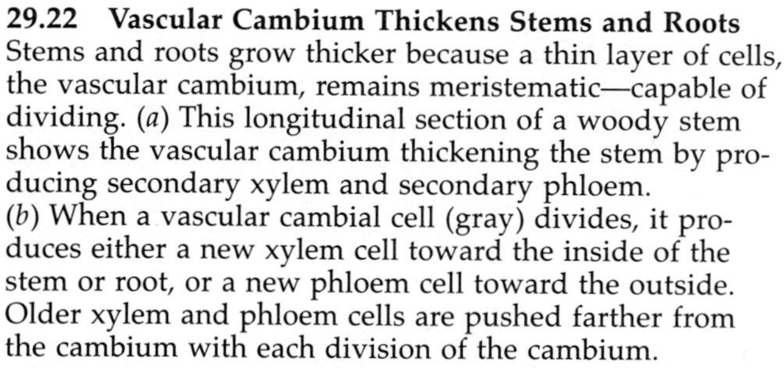

29.22 Vascular Cambium Thickens Stems and Roots Stems and roots grow thicker because a thin layer of cells, the vascular cambium, remains meristematic—capable of dividing. (*a*) This longitudinal section of a woody stem shows the vascular cambium thickening the stem by producing secondary xylem and secondary phloem. (*b*) When a vascular cambial cell (gray) divides, it produces either a new xylem cell toward the inside of the stem or root, or a new phloem cell toward the outside. Older xylem and phloem cells are pushed farther from the cambium with each division of the cambium.

nections are provided by **vascular rays**, which are composed of cells derived from the vascular cambium. The rays, laid down progressively as the cambium divides, are rows of living parenchyma cells running perpendicular to the xylem vessels and phloem sieve tubes (Figure 29.23). As the root or stem continues to increase in diameter, new vascular rays are initiated so that this storage and transport tissue continues to meet the needs of the bark and of the living cells in the xylem. The cambium itself increases in circumference with the growth of the root or stem, for if it did not, it would split. The vascular cambium grows by the division of some of its cells in a plane at right angles to the plane that gives rise to secondary xylem and phloem. The products of each of these divisions lie within the vascular cambium itself.

Many dicots have vascular cambia and cork cambia and thus undergo secondary growth. In the rare cases in which monocots form thickened stems—palm trees, for example—a greater girth is achieved by quite a different mechanism.

WOOD. Most trees in temperate-zone forests have **annual rings** (Figure 29.24), which result from changing environmental conditions during the growing season. In the springtime, when water is relatively plentiful, the tracheids or vessel elements produced by the vascular cambium tend to be large in diameter and thin-walled. As water becomes less available during the summer, narrower cells with thicker walls

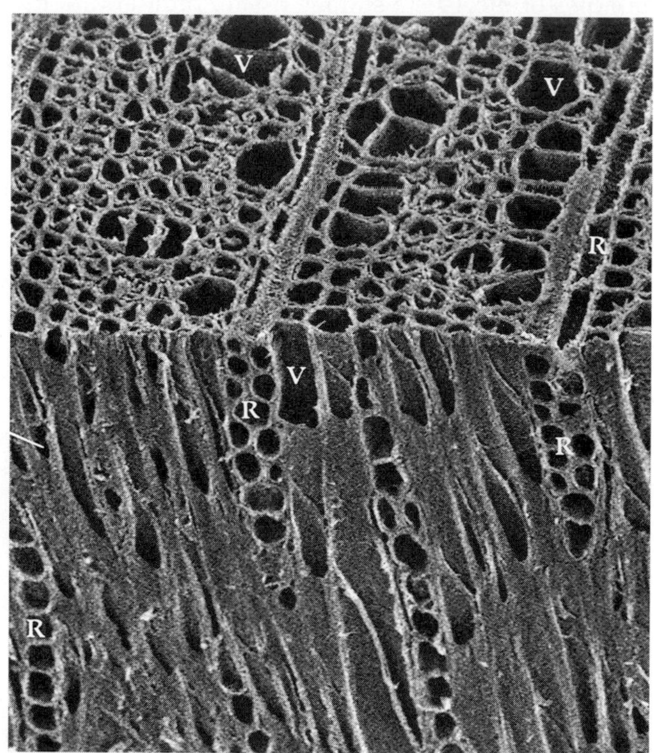

29.23 Vascular Rays
Wood of tulip poplar, showing that the orientation of xylem vessels (V) is perpendicular to that of vascular rays (R). The longitudinal section of the stem (lower half of the micrograph) shows the xylem vessels as long vertical tubes and the ray cells as circles. The cross section (top), which is perpendicular to the longitudinal section, shows the xylem vessels as circles and the rays as tubes. Rays transport food horizontally from the phloem to storage cells; xylem vessels conduct water vertically.

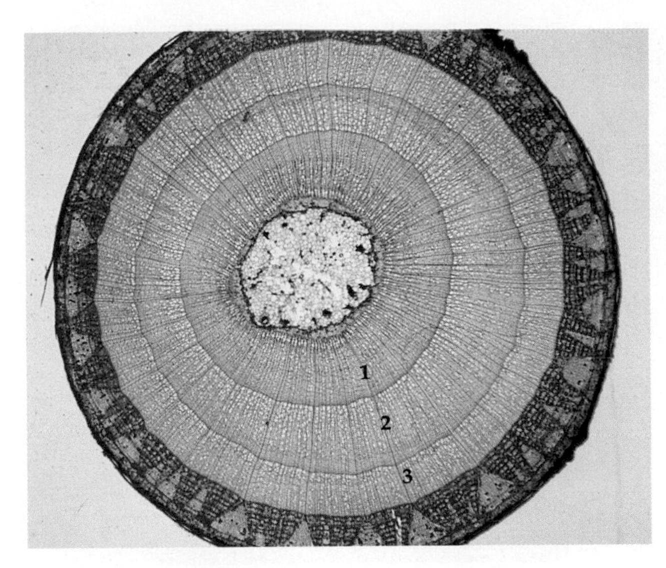

29.24 Annual Rings
Three rings of xylem vessels are the major feature of this cross section from a three-year-old basswood stem.

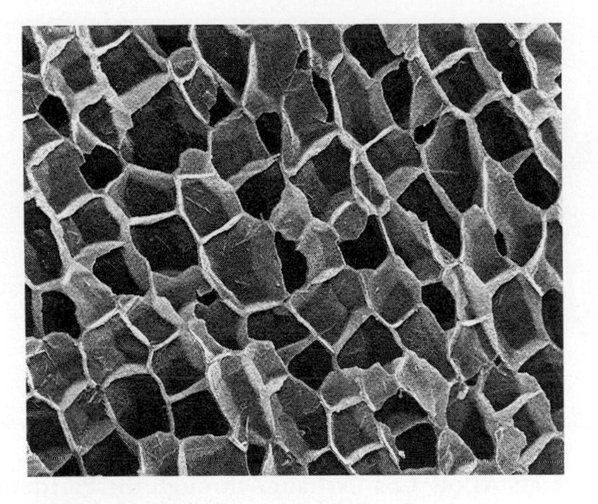

29.25 Cork
Commercial cork from the cork oak is seen in this scanning electron micrograph.

are produced; this summer wood is darker and perhaps more dense. Thus each year is usually recorded in a tree trunk by a clearly visible annual ring consisting of one light and one dark layer. Trees in the wet tropics do not lay down such obvious, regular rings.

The difference between old and new regions also contributes to the appearance of wood. As a tree grows in diameter, the xylem toward the center becomes clogged with resins and ceases to conduct water and minerals. This heartwood is darker; the sapwood—that portion that is actively conducting all water and minerals in the tree—is lighter and more porous. Knots—which we find attractive in knotty pine but regard as a defect in structural timbers—are branches: As a trunk grows, a branch extending out of it becomes buried in new wood and appears as a knot when the trunk is cut lengthwise.

PERIDERM. Obviously, as secondary growth continues, something has to give. Expansion of the vascular tissues stretches and breaks the epidermis and cortex, which are ultimately lost. Derivatives of the phloem then become the outermost tissue of the stem. Woody roots behave similarly. Because the epidermis is specialized in part for the retention of water, how does the plant cope if this tissue is shed? Before layers of epidermal cells are broken away, cells lying near the surface begin to divide and produce layers of cork, a tissue composed of cells with thickened, waterproof walls (Figure 29.25). The dividing cells, derived from the phloem, form a cork cambium. Sometimes cells are also produced to the inside by the cork cambium; these cells constitute what is known as the phelloderm. Cork is waterproofed by suberin. The cork soon becomes the outermost tissue of the stem or root. Cork, cork cambium, and phelloderm—if present—make up the periderm of the secondary body. As the vascular cambium continues to produce secondary vascular tissue, the corky layers are in turn lost, but a similar process of cell division in the underlying phloem gives rise to new corky layers.

As periderm forms, there is still a need for gas exchange with the environment. Carbon dioxide must be released and oxygen must be taken up for cellular respiration. **Lenticels** are spongy regions that allow such gas exchange (Figure 29.26).

Leaf Anatomy

Figure 29.27*a* shows a typical dicot leaf in cross section. Generally a leaf has two zones of photosynthesizing tissues referred to as **mesophyll**, meaning "middle of the leaf." The upper layer or layers of mesophyll consist of roughly cylindrical cells. This zone is referred to as palisade mesophyll. The lower layer or layers consist of irregularly shaped cells called spongy mesophyll. Within the mesophyll is a great deal of air space through which carbon dioxide can diffuse to surround all photosynthesizing cells. Vascular tissue branches extensively in the leaf, forming a network of **veins**. These veins extend to within a few cell diameters of all the cells of the leaf, ensuring that the mesophyll cells are well supplied with water. The products of photosynthesis are loaded into the phloem of the veins for export to the rest of the plant (Figure 29.27*b*).

Covering the entire leaf is a layer of nonphotosynthetic cells constituting the epidermis. To retard water loss, the epidermal cells and their overlying waxy cuticle must be highly impermeable, but this

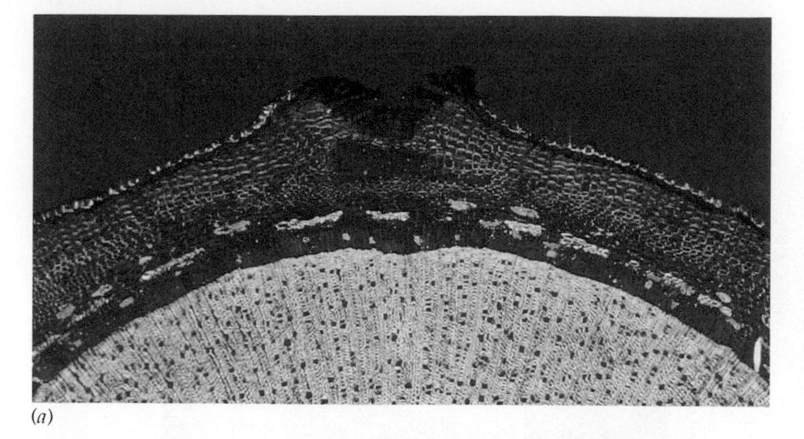

(*a*)

(*b*)

29.26 Lenticels Allow Gas Exchange through the Periderm
(*a*) The region at the top that appears broken open is a lenticel in a year-old elder twig; note the spongy tissue that constitutes the lenticel. (*b*) The rough areas on the trunk of this Chinese plum tree are lenticels. Most tree species have lenticels much smaller than these.

impermeability poses a problem: While keeping water within the leaf, the epidermis keeps carbon dioxide, the raw material of photosynthesis, out. The problem of balancing water retention and carbon dioxide availability is solved by an elegant regulatory system that will be discussed in more detail in Chapter 30. This system is based on pairs of **guard cells**, modified epidermal cells that change shape, thereby opening or closing pores called **stomata** (singular: stoma) between the guard cells (Figure 29.27*c*; see also Figure 30.10). When the stomata are open, carbon dioxide can enter, but water can be lost.

In Chapter 8 we described C_4 plants, which can fix carbon dioxide efficiently even when the carbon dioxide supply falls to a level at which the photosynthesis of C_3 plants is inefficient. One adaptation that helps C_4 plants do this is a modified leaf anatomy, as shown in Figure 8.28. Notice that the photosynthetic cells in the C_4 leaf are grouped around the veins in concentric layers: an outer mesophyll layer and an inner bundle sheath. These layers each contain different types of chloroplasts, leading to the biochemical division of labor described in Chapter 8.

SUPPORT IN A TERRESTRIAL ENVIRONMENT

Water buoys up aquatic plants, but terrestrial plants must either sprawl on the ground or somehow be supported against gravity. There are two principal types of support for terrestrial plants, one based on the osmotic properties of cells and the other based on tissues stiffened by specialized cell walls. One type of support is the pressure potential (sometimes called turgor) of the cells in the body. A small plant can maintain an erect posture if its cells are turgid, but it collapses—wilts—if the pressure potential falls too low. Think about the difference between a wilted plant and a turgid one; the distinction dramatically illustrates the role of pressure potential in supporting the body. Support by the pressure potential is often augmented by the second type of support, the presence of strengthening tissues such as collenchyma and sclerenchyma. Collenchyma is more flexible than sclerenchyma, which provides a more rigid, stronger support.

The most important support found in many plants is wood—a mass of secondary xylem. Wood is such a strong yet lightweight material that we have used it in buildings, furniture, and other structures for millennia. All dicot trees are supported by their woody stems. Not all wood is the same, however. Let's consider some of the special adaptations of secondary xylem.

Reaction Wood

As the branches of growing trees grow longer and heavier, why don't they simply sag to the ground? This problem is averted by means of a gravity-induced asymmetry in wood structure: Specialized **reaction wood**, differing from normal wood, keeps the limb straight. Angiosperms and gymnosperms have different kinds of reaction wood, and in different places. In gymnosperms, **compression wood** forms on the lower side of a branch. It is prestressed—that is, it is laid down under compressive stress—and expands, thus tending to push the branch upward (Figure 29.28*a*). Compression wood contains thicker and shorter tracheids, with more lignin and less cellulose in their walls than normal wood has. By contrast, the reaction wood of angiosperms, called **ten-**

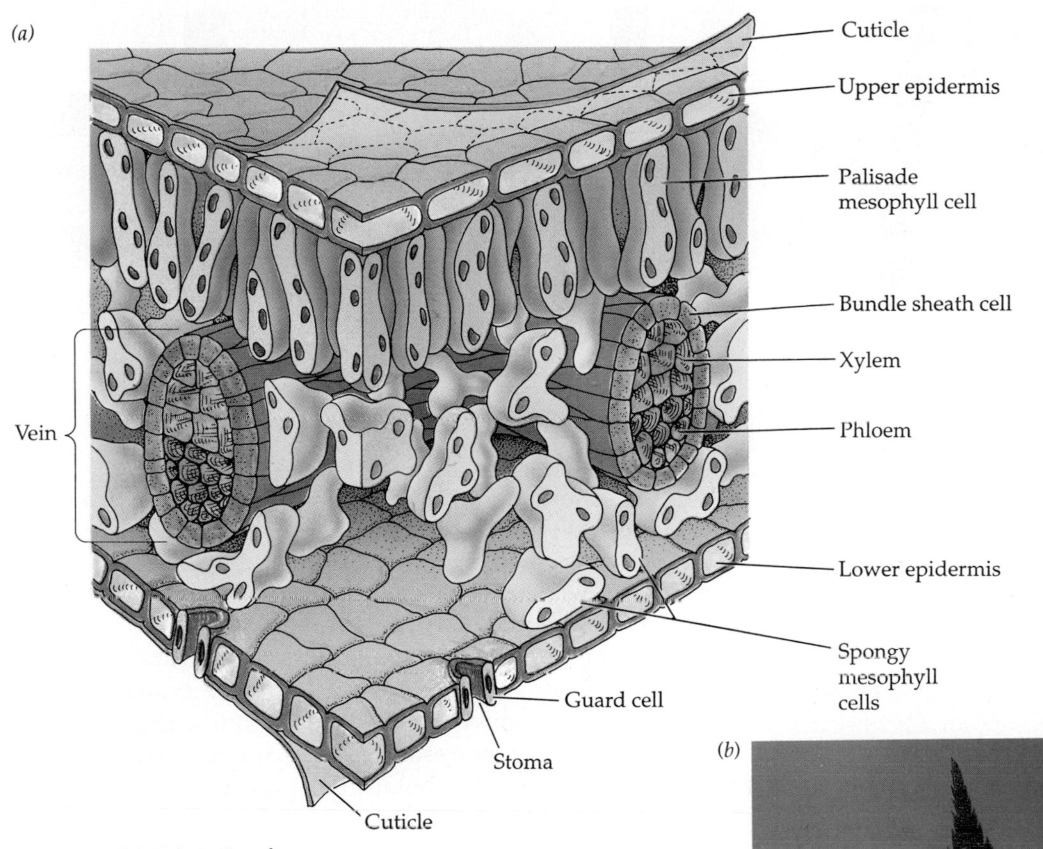

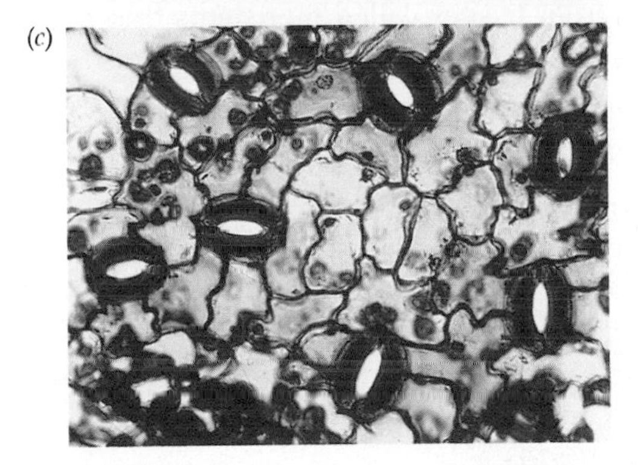

29.27 The Dicot Leaf
(a) Cross section of a dicot leaf. *(b)* The network of fine veins in this Japanese maple leaf (*Acer palmatum*) carries water to the mesophyll cells and carries photosynthetic products away from them. *(c)* The lower epidermis of a dicot leaf, stained. The small, heavily stained, paired cells are guard cells; the gaps between them are stomata, through which carbon dioxide enters the leaf.

sion wood, is formed on the upper side of the branch. It is laid down under tension and shrinks, thus tending to pull the branch upward, or at least to resist downward bending (Figure 29.28*b*). In tension wood the fibers have more heavily thickened walls, containing less lignin and more cellulose, and there are fewer and smaller vessels than in normal wood.

That a gravitational stimulus, not the sagging of the branch, determines which type of reaction wood forms is illustrated by an experiment performed by the Australian plant physiologist A. B. Wardrop (Figure 29.28*c*). Wardrop bent the trunk of a young angiosperm sapling into a circle and allowed reaction wood to develop. Reaction wood formed on the top side of the top of the loop, where it was under tension, and also on the top side of the bottom of the loop, where it was under compression.

Trees grown indoors tend to be much more spindly than their outdoor counterparts, apparently because indoor trees are not subjected to buffeting by wind. They develop a firmer trunk if they are simply

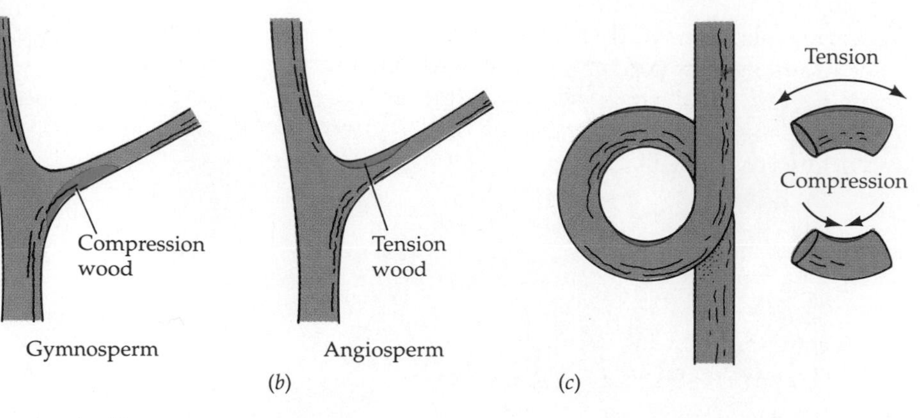

29.28 Reaction Wood Reaction wood reduces the tendency of branches to sag. *(a)* Compression wood, the reaction wood of gymnosperms, is heavily lignified wood that forms on the lower sides of branches. *(b)* Tension wood, the reaction wood of angiosperms, forms on the upper sides of branches. *(c)* An angiosperm sapling was bent into a loop and tied in place. Tension wood formed on the tops of both horizontal regions; as shown in the cutouts, one region of tension wood was under tension, but the other was actually under compressive stress. This result indicates that the stimulus to tension wood formation is gravitational (tension wood forms on the upper side) rather than a response to the stress itself.

shaken or pounded with a padded mallet from time to time. The change in wood deposition caused by such treatments may be akin to reaction wood formation.

The variety in the structure and development of woods has resulted in different economic uses for different woods, uses that parallel the functions of the woods.

Quonset Huts and Marble Palaces

People put up Quonset huts for cheap, relatively short-term shelter. They build palaces of marble if they want to create a monument for the ages. We find similar contrasts if we compare the wood of trees growing under very different conditions. Consider balsa and mahogany: one is extremely light and soft, the other dark and hard. Like most species with very soft woods, balsa is a fast-growing tree, frequently found in areas recently burnt or cut to the ground. Balsa wood has cells with relatively large diameters, and the wood fibers have very thin walls. For a given volume of wood, the amount of structural material laid down is slight. Thus the rapidly growing balsa plant can display its foliage to the sun without a great commitment of resources to structural support. In short, balsa is the botanical equivalent of a Quonset hut.

Mahogany grows extremely slowly. It is very sturdy. Its fine-textured wood has tiny cells, and the wood fibers have thick walls. In contrast to balsa, mahogany wood is "expensive" to form—it has much more dry weight per volume. As a result of its hardness, mahogany wood can support the plant as a long-lived tree in rainforests. Mahogany contains impregnating materials that darken the wood and help render it resistant to fungal attack. It is the botanical equivalent of a palace.

SUMMARY of Main Ideas about the Flowering Plant Body

Plants grow from tissues called meristems that retain the capacity of cell division.

Stems and roots grow from apical meristems.
Review Figure 29.16

The meristems that thicken stems and roots are the vascular cambium and the cork cambia.
Review Figures 29.16 and 29.22

The shoot and root systems make up the plant body.
Review Figures 29.7, 29.9, 29.10, 29.17, 29.18, and 29.19

Secondary growth thickens the stem, producing wood and periderm.
Review Figures 29.22, 29.23, 29.24, 29.25, and 29.26

Leaves exchange materials with the rest of the plant by means of vascular tissues in the veins, and they exchange gases with the environment by means of stomata.
Review Figure 29.27

The two classes of flowering plants, monocots and dicots, differ in plant body characteristics.
Review Figures 29.5 and 29.6

Flowering plants have cells specialized for support, conduction of water and minerals, conduction of food materials, photosynthesis, storage, and cell division.
Review Figures 29.13, 29.14, and 29.15

Flowering plants have three tissue systems: the vascular tissue system (xylem and phloem), the dermal tissue system (epidermis and periderm), and the ground tissue system (parenchyma, collenchyma, and sclerenchyma cells).

Wood provides support for many plants; reaction wood provides support for branches, counteracting the tendency of the branches to sag.
Review Figure 29.28

SELF-QUIZ

1. Which of the following is *not* a difference between monocots and dicots?
   a. Dicots more frequently have broad leaves.
   b. Monocots commonly have flower parts in multiples of three.
   c. Monocot stems do not generally undergo secondary thickening.
   d. The vascular bundles of monocots are commonly arranged as a cylinder.
   e. Dicot embryos commonly have two cotyledons.
2. Roots
   a. always form a fibrous root system that holds the soil.
   b. possess a root cap at their tip.
   c. form branches from lateral buds.
   d. are commonly photosynthetic.
   e. do not show secondary growth.
3. The plant cell wall
   a. lies immediately inside the plasma membrane.
   b. is an impenetrable barrier between cells.
   c. is always waterproofed with either lignin or suberin.
   d. consists of a primary wall and secondary wall, separated by a middle lamella.
   e. contains cellulose and other polysaccharides.
4. Which statement about parenchyma cells is *not* true?
   a. They are alive when they perform their functions.
   b. They typically lack a secondary wall.
   c. They often function as storage depots.
   d. They are the most numerous cells in the primary plant body.
   e. They are found only in stems and roots.
5. Tracheids and vessel elements
   a. die before they become functional.
   b. are important constituents of all bryophytes and tracheophytes.
   c. have walls consisting of middle lamella and primary wall.
   d. are always accompanied by companion cells.
   e. are found only in the secondary plant body.
6. Which statement is *not* true of sieve tube elements?
   a. Their end walls are called sieve plates.
   b. They die before they become functional.
   c. They link end-to-end, forming sieve tubes.
   d. They form the system for translocation of foods.
   e. They lose the membrane that surrounds their central vacuole.
7. The pericycle
   a. separates the stele from the cortex.
   b. is the tissue within which branch roots arise.
   c. consists of highly differentiated cells.
   d. forms a star-shaped structure at the very center of the root.
   e. is waterproofed by Casparian strips.
8. Secondary growth of stems and roots
   a. is brought about by the apical meristems.
   b. is common in both monocots and dicots.
   c. is brought about by vascular cambia and cork cambia.
   d. produces only xylem and phloem.
   e. is brought about by vascular rays.
9. Periderm
   a. contains lenticels that allow for gas exchange.
   b. is produced during primary growth.
   c. is permanent; once formed it lasts as long as the plant does.
   d. is the innermost part of the plant.
   e. contains vascular bundles.
10. Which statement about leaf anatomy is *not* true?
   a. Stomata are controlled by paired guard cells.
   b. The cuticle is secreted by the epidermis.
   c. The veins contain xylem and phloem.
   d. The cells of the mesophyll are packed together, minimizing air space.
   e. C_3 and C_4 plants differ in leaf anatomy.

FOR STUDY

1. When a young oak was 5 meters tall, a thoughtless person carved his initials in its trunk at a height of 1.5 meters above the ground. Today that tree is 10 meters tall. How high above the ground are those initials? Explain your answer in terms of the manner of plant growth.
2. Consider a newly formed sieve tube element in the secondary phloem of an oak tree. What kind of cell divided to produce the sieve tube element? What kind of cell divided to produce that parent cell? Keep tracing back in this manner until you arrive at a cell in the apical meristem.

3. Distinguish between sclerenchyma cells and collenchyma cells in terms of structure and function.

4. Distinguish between primary and secondary growth. Do all angiosperms undergo secondary growth? Explain.

5. What anatomical features make it possible for a plant to retain water as it grows? Describe the tissues and how and when they form.

READINGS

Esau, K. 1977. *Anatomy of Seed Plants,* 2nd Edition. John Wiley, New York. A comprehensive treatment; particularly good on secondary growth.

Feldman, L. J. 1988. "The Habits of Roots." *BioScience,* vol. 38, pages 612–618. Considers many aspects of the biology of roots, including structure, competition, associations with soil microorganisms, and others.

Mangelsdorf, P. C. 1986. "The Origin of Corn." *Scientific American,* August. What was the ancestry of this popular vegetable? The gross anatomy of some possible ancestors is compared.

Niklas, K. J. 1989. "The Cellular Mechanics of Plants." *American Scientist,* vol. 77, pages 344–349. This fine article, subtitled "How Plants Stand Up," details how cell walls and other aspects of stem architecture enable terrestrial plants to stand erect.

Raven, P. H., R. F. Evert and S. Eichhorn. 1991. *Biology of Plants,* 5th Edition. Worth, New York. An excellent general botany textbook.

Swaminathan, M. S. 1984. "Rice." *Scientific American,* January. This article deals primarily with ways to increase the yield of rice, but it also describes the structure of the plant.

Wilson, B. F. and R. R. Archer. 1979. "Tree Design: Some Biological Solutions to Mechanical Problems." *BioScience,* vol. 29, pages 293–298. As trees grow, they constantly redesign themselves. This article examines aspects of this process, such as reaction wood formation, from an engineering viewpoint.

Life first arose and flourished in the oceans, but the vascular plants arose on land. Plants were the first eukaryotes to face the challenges of life out of water. The mechanisms that terrestrial organisms require for taking in and conserving water differ from those of aquatic organisms. The cells and tissues of aquatic organisms are bathed in the water and minerals that they require, but a terrestrial plant must have a transport system to distribute water and minerals throughout its body. And because leaves are the sites of food production, all plants except the smallest ones need a system to transport food from the leaves to other parts of the body.

Terrestrial organisms also need ways to support their bodies. In a watery environment a giant kelp—a marine algal protist—can spread out like an enormous tree because of the buoying action of the surrounding water. On land, large organisms can resist the pull of gravity only with the help of rigid materials such as wood or bone. As you learned in the previous chapter, the pressure of water in the tissues provides much of the support for nonwoody terrestrial plants, which wilt in the absence of water. And, of course, wood is the most important source of support for many large terrestrial plants. The cellular anatomy that gave us an understanding of the support systems of terrestrial plants is also central to an understanding of how water, minerals, and food are transported within the bodies of plants.

How Do Water and Minerals Get Up There? Wood not only supports these California coast redwoods but is also the tissue through which water and minerals are transported more than 100 meters into the air.

30 Transport in Plants

UPTAKE AND TRANSPORT OF WATER AND MINERALS

Terrestrial plants obtain both water and mineral nutrients from the soil, usually by way of their roots. You know that leaves are loaded with chloroplasts and that water is one of the ingredients for food production by photosynthesis. How do leaves, high in a tree, obtain water from the soil? What are the mechanisms by which water and minerals enter the plant body through the dermal tissue of the root, pass through the ground tissue, enter the stele, and ascend as sap in the xylem? Because neither water nor minerals can move through the plant into the xylem without crossing at least one plasma membrane, we will first consider two membrane phenomena, osmosis and the uptake of minerals.

Osmosis

Osmosis, the movement of a solvent such as water through a membrane in accordance with the laws of diffusion, was discussed in Chapter 5. Recall that the **osmotic potential** of a solution results from the presence of dissolved solutes. The greater the solute concentration, the more negative the osmotic potential and hence the greater the tendency of water to move into that solution from another solution of lower solute concentration. The two solutions must be separated by a differentially permeable membrane (permeable to water but impermeable to the solute). Solutions—or cells—with identical osmotic potentials are isotonic; if two solutions differ in osmotic potential, the one with the less negative osmotic potential is hypotonic to the other. Recall, too, that osmosis is a passive process—ATP is not required.

Unlike animal cells, plant cells are surrounded by a relatively rigid cell wall. After a certain amount of water enters a plant cell, the entry of more water is resisted by an opposed **pressure potential**, sometimes called turgor pressure, owing to the rigidity of the wall. As more and more water enters, the pressure potential becomes greater and greater. The pressure potential is analogous to the air pressure in an automobile tire; it is a real pressure that can be measured with a pressure gauge. Cells with walls do not burst when placed in distilled water because of the rigidity of the walls. Water enters by osmosis until the pressure potential exactly balances the osmotic potential. At this point, the cell is quite turgid—that is, it has a high pressure potential.

The overall tendency of a solution to take up water from pure water is called the **water potential**. The water potential is the sum of the (negative) osmotic potential and the (positive) pressure potential. For distilled water under no applied pressure, all three of these potentials are defined as equal to zero. In all cases in which water moves between two cells, or between a cell and its environment, or between two solutions separated by a membrane, the following rule of osmosis applies: *Water always moves toward the region of more negative water potential* (Figure 30.1).

Osmotic phenomena are of great importance. The turgor of most plants is maintained by the pressure potentials of their cells; if the pressure potential is lost, a plant wilts (Figure 30.2). The movement of water within a plant follows a gradient of water potential, and as we will see, the flow of phloem sap through the sieve tubes is driven by a gradient in pressure potential.

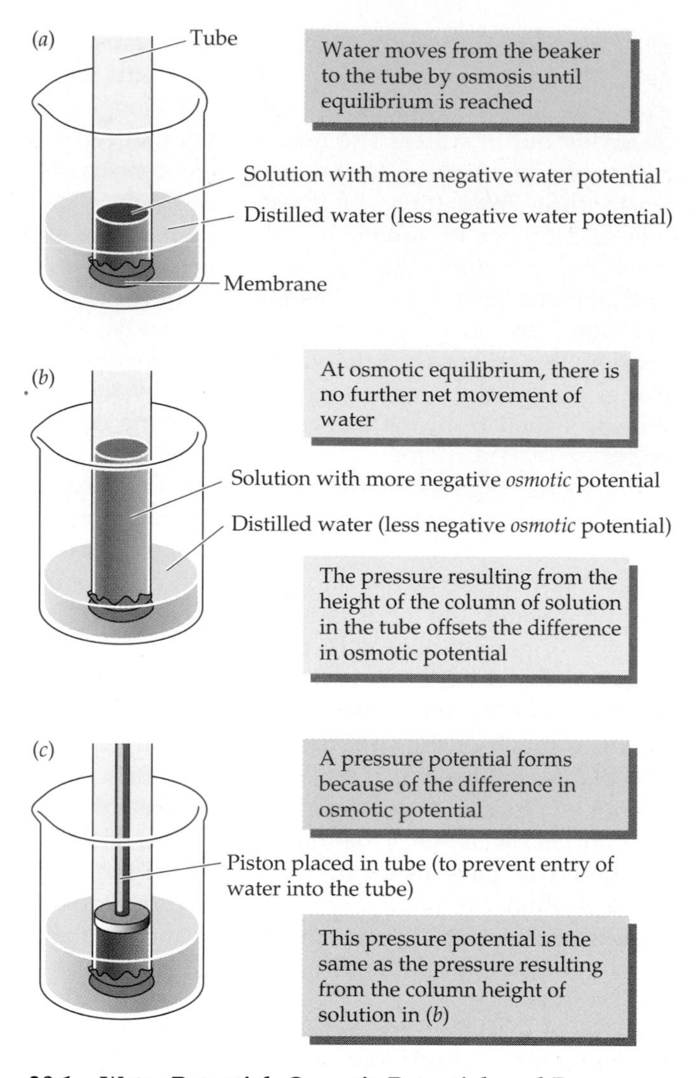

30.1 Water Potential, Osmotic Potential, and Pressure Potential
(a) The solution in the tube has a negative osmotic potential owing to the presence of dissolved solutes; the pressure potential = 0; thus the water potential is negative. The beaker contains distilled water (water potential = 0). *(b)* Water moves through the membrane into the tube because of the difference in water potentials. *(c)* If the water potentials are made equal by applying pressure with a piston, water does not enter the tube.

Uptake of Minerals

Mineral nutrient ions are taken up across plasma membranes with the help of proteins. (You may wish to review the section "Crossing the Membrane Barrier" in Chapter 5.) Some of these proteins are carriers for the facilitated diffusion of particular ions. Facilitated diffusion does not require ATP. The concentrations of some ions in the soil solution are lower than those required inside the plant, however. Thus the plant must take up these ions against a concentration gradient. Such active transport is an energy-requiring process, and it depends upon cellular respiration as a source of ATP. Active transport, too, requires specific carrier proteins.

Plants do not have a sodium–potassium pump for active transport. Rather, plants have a **proton pump** that uses energy obtained from ATP to push protons out of the cell against a proton concentration gradient (Figure 30.3*a*). Because protons (H^+) are positively charged, their accumulation on one side of a membrane has two results. First, the region outside the membrane becomes positively charged with respect to the region inside. Second, there is a proton con-

(a) (b)

30.2 Turgor in Plants
(a) This coleus plant remains turgid as long as the pressure potential of its cells is high. *(b)* When cells lose too much water, their pressure potential drops and the plant wilts.

centration difference. Each of these results has consequences for the movement of other ions. Because of the charge difference across the membrane, the movement of positively charged ions such as K^+ into the cell through their membrane channels is enhanced because these positive ions are moving into a region of negative charge (Figure 30.3*b*). The proton concentration difference can drive a form of secondary active transport in which negatively charged ions such as Cl^- are moved into the cell against a concentration gradient by a symport that couples their movement with that of H^+ (Figure 30.3*c*). In all, there is a vigorous traffic of ions across plant membranes. How do biologists measure these ion movements?

A technique called **patch clamping** allows us to monitor the flow of ions through just a few carrier proteins—or even just one—at a time. First we remove the cell wall, by digesting it with enzymes, to expose the plasma membrane. Then we immobilize the naked cell by pulling it part way into a very fine glass micropipette (Figure 30.4*a*). Next, we press a still finer glass micropipette against the exposed plasma membrane and apply a slight suction so that a tiny patch of the membrane is effectively isolated from the rest of the surface (Figure 30.4*b*). Once a tight seal has been made, we can proceed in one of various ways. In one approach, by pulling very carefully, we can tear the patch away from the rest of the

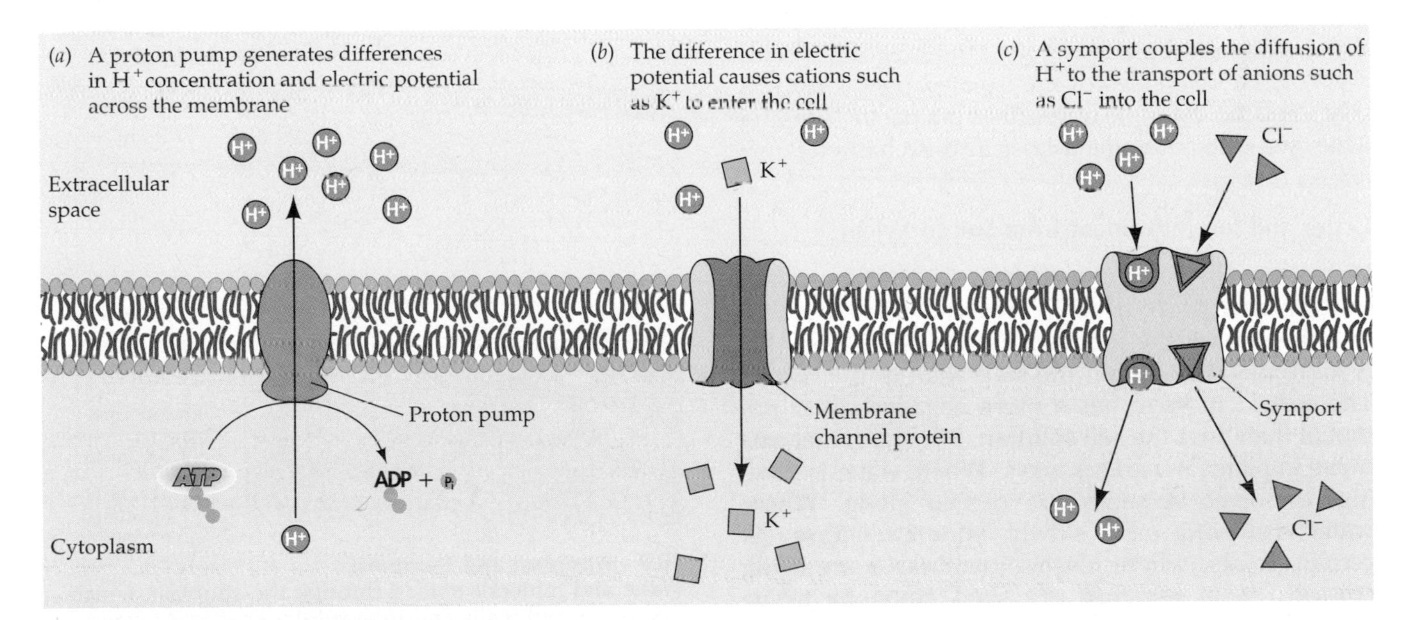

30.3 The Proton Pump and Its Effects
(a) The buildup of hydrogen ions transported across the cell membrane by the proton pump triggers the movement of both cations *(b)* and anions *(c)*.

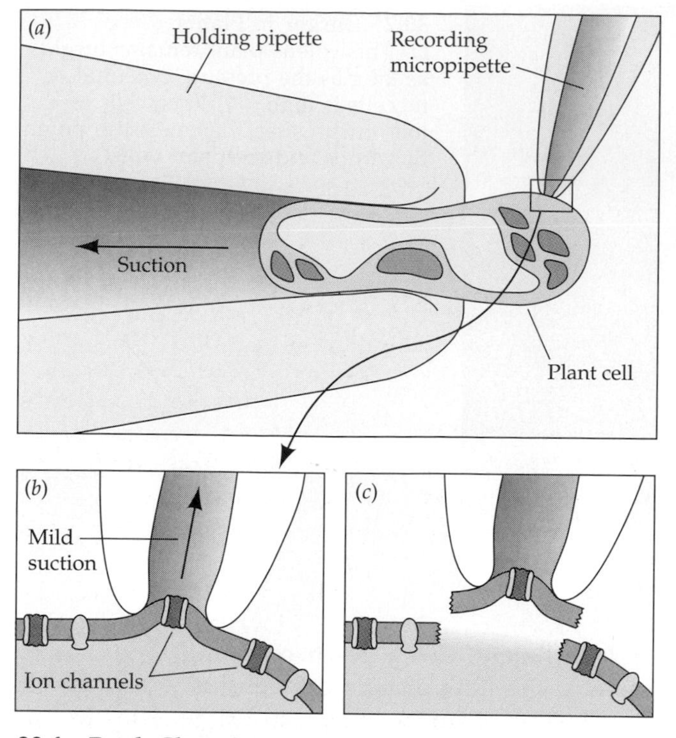

30.4 Patch Clamping
(a) The naked cell, lacking a wall, is trapped in the tip of a holding micropipette. *(b)* A part of the membrane is sucked into a smaller, recording micropipette. *(c)* The part of the plasma membrane has been torn away. It contains only a single ion channel, which can be studied by measuring electric currents across it as ions flow.

membrane and study the flow of ions through the carrier or carriers contained in just that patch (Figure 30.4*c*). Because ions are electrically charged, we can measure their movement through the patch by recording the tiny electric current that flows. We can also experiment by altering the contents of the solutions on the two sides of the isolated patch. We will give a specific example of results from patch clamping when we discuss stomata later in this chapter.

Water and Ion Movement from Soil to Xylem

Water moves along a gradient of water potential, toward ever more-negative regions. Water moves into the stele of the root because the water potential is more negative within the stele than in the cortex. The cortex, in turn, has a more negative water potential than does the soil solution. Minerals enter and move in plants in various ways. Where water is flowing, dissolved minerals are carried along. Where water is moving more slowly, minerals diffuse. At certain points, where plasma membranes are being crossed, some minerals are sped along by active transport.

Water and minerals from the soil may pass through the dermal and ground tissues to the stele via two plant parts: the apoplast and the symplast. Plant cells are surrounded by cell walls that lie outside the plasma membrane, and intercellular spaces are common in many tissues. The walls and intercellular spaces together constitute the **apoplast** (from the Greek for "away from living material"). The apoplast is a continuous meshwork through which water and dissolved substances can flow or diffuse without ever having to cross a membrane. Movement of materials through the apoplast is thus unregulated. The remainder of the plant body is the **symplast** (from the Greek for "together with living material")—that is, the plant body enclosed by membranes, the continuous meshwork consisting of the living cells, connected by plasmodesmata (Figure 30.5). The selectively permeable plasma membranes of the cells control access to the symplast, so movement of water and dissolved substances into the symplast is tightly regulated.

Water and minerals can pass from the soil solution through the apoplast to the inner border of the cortex. As you may recall from Chapter 29, to enter the stele, water and minerals must pass through the endodermis (the inner cell layer of the cortex; see Figure 29.18). What distinguishes the endodermis from the rest of the ground tissues is the presence of **Casparian strips**. These waxy, suberin-containing structures line the endodermal cells at their tops, bottoms, and

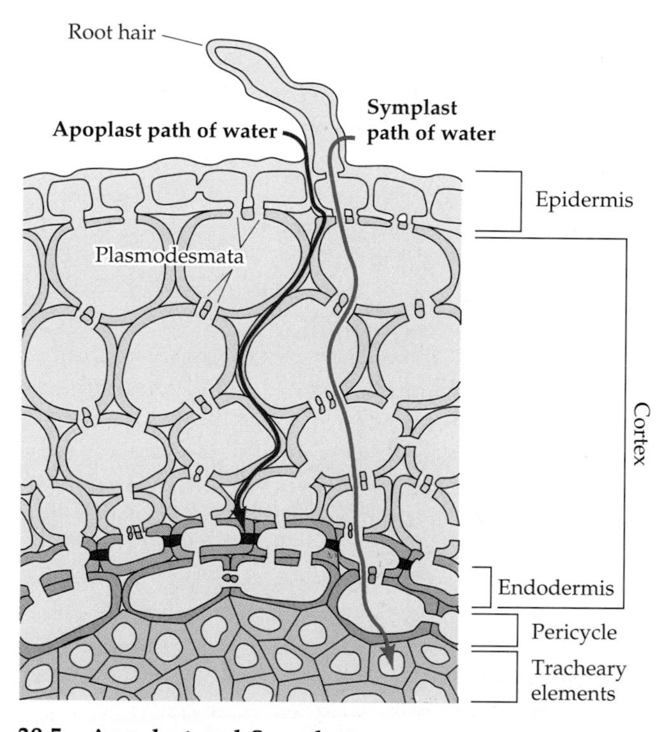

30.5 Apoplast and Symplast
Water and minerals spread through the apoplast, which consists of cell walls and intercellular spaces. At some point, water and minerals must enter the symplast—the remainder of the plant body—or they will be unable to pass into the stele.

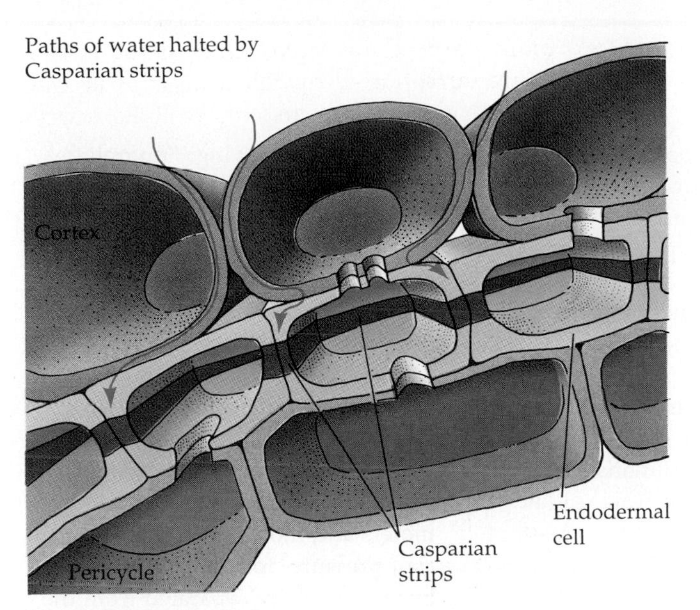

30.6 Casparian Strips
Casparian strips prevent water in the apoplast from passing between the endodermal cells and into the stele. Water must first enter the living endodermal cells; by entering the symplast, it can evade the Casparian strips.

sides, acting as a gasket that prevents water and ions from moving between them (Figure 30.6). The endodermis thus completely separates the apoplast of the cortex from the apoplast of the stele. The Casparian strips do not obstruct the outer or inner faces of the endodermal cells. Accordingly, water and ions can enter the stele only by way of the symplast—that is, by entering and passing through the cytoplasm of the endodermal cells. Thus transport proteins in membranes between the apoplast and symplast determine which minerals pass, and at what rates.

Once they have passed the endodermal barrier, water and minerals leave the symplast. Parenchyma cells in the pericycle or xylem help minerals move back into the apoplast. Some of these parenchyma cells, called **transfer cells**, are structurally modified for transporting mineral ions from their cytoplasm (part of the symplast) into their cell walls (part of the apoplast). The wall that receives the transported ions has many knobby growths extending into the transfer cell, increasing the surface area of the plasma membrane, the number of transport proteins, and thus the rate of transport. Transfer cells also have many mitochondria that produce the ATP needed to power the active transport of mineral ions. As mineral ions move into the solution in the walls, the water potential of the wall solution becomes more negative; thus water moves out of the cells into the wall solution by osmosis. Active transport of ions moves the ions directly, and water follows passively. The end result is that water and minerals end up in the xylem, where they constitute the sap.

Ascent of Sap in Xylem: The Magnitude of the Problem

The water and minerals in the xylem must be transported to the entire shoot system, all the way to the highest leaves and apical buds. Before we consider the mechanisms underlying this transport, we should know what needs to be explained: How much sap is transported, and how high must it go?

In answer to the first question, consider the following example: A single maple tree 15 meters tall was estimated to have some 177,000 leaves, with a total leaf surface area of 675 square meters. On a summer day, that tree lost 220 liters of water *per hour* to the atmosphere by evaporation from the leaves. To prevent wilting, xylem transport in that tree needed to provide 220 liters of water to the leaves every hour.

How high must the xylem sap be transported? The question may be rephrased: How tall are the tallest trees? The tallest gymnosperms, the coast redwoods—*Sequoia sempervirens*—exceed 110 meters in height, as do the tallest angiosperms, the Australian *Eucalyptus regnans*. Any successful explanation of transport in the xylem must account for transport over these great distances.

Early Models of Transport in the Xylem

Some of the earliest models to explain the rise of sap in the xylem were based on a hypothetical pumping action by living cells in the stem. Experiments published in 1893 by the German botanist Eduard Strasburger definitively ruled out such models (Box 30.A).

Another early suggestion was a model based on capillary action, the rising of watery solutions in very thin tubes or in woven materials like paper. At first glance this theory seems reasonable. The diameters of vessel elements and tracheids are tiny, and the narrower the tube, the higher water will rise by capillary action. However, the diameters of tracheids are only small enough to support a capillary column of about 40 centimeters—shorter than many shrubs, let alone a giant eucalyptus towering over us to a height greater than the length of a football field.

Root Pressure

After the capillary model was questioned, some plant physiologists turned to a model based on root pressure—a pressure exerted by the root tissues that would force liquid up the xylem. The basis for root pressure is a higher solute concentration, and accordingly a more negative water potential, in the xylem sap than in the soil solution. This negative potential draws water into the stele; once there, the water has nowhere to go but up.

BOX 30.A

There Are No Pressure Pumps in the Xylem

Eduard Strasburger was the leading plant cytologist of the late nineteenth century. He was one of those who established that the nucleus is the carrier of hereditary information, and he is best remembered today for pioneering work that led to the discovery of meiosis. He also performed some of the first important experiments that led to our current understanding of water movement in the xylem. His contemporaries generally believed that living cells in the xylem played a key role, probably by acting as pressure pumps, pushing the sap upwards.

Strasburger worked with trees about 20 meters in height. He sawed them through at their bases and plunged the cut ends into buckets containing solutions of poisons such as picric acid. The solutions rose through the trunks, as was readily evident from the progressive death of the bark higher and higher up. When the solutions reached the leaves, the leaves died, too, and the solutions were no longer transported; the liquid levels in the buckets stopped dropping at that point. This simple experiment established three important points: (1) Living, "pumping" cells are not responsible for the upward movement of the solutions, for the solutions themselves killed all living cells with which they came in contact. (2) The leaves play a crucial role in causing the transport. As long as they were alive, the solutions continued to be transported upward; when the leaves died, transport ceased. (3) Transport in these experiments, which covered distances of 20 meters and more, was not caused by root pressure, for the trunks had been completely separated from the roots. It is interesting, if disappointing, that theories of xylem transport based on root pressure were entertained for some years following Strasburger's definitive experiments.

There is good evidence for root pressure—for example, the phenomenon of **guttation**, in which liquid water is forced out through openings in leaves (Figure 30.7). Guttation occurs only under conditions of high atmospheric humidity and plentiful water in the soil. Root pressure is also the source of the sap that oozes from the cut stumps of some plants, such as *Coleus,* when their tops are removed. However, root pressure cannot account for the ascent of sap in trees. Root pressures seldom exceed one or two times atmospheric pressure, and they are actually less at times when transport in the xylem is most rapid. If root pressure were driving sap up the xylem, we should observe a positive pressure potential in the xylem at all times. In fact, as we are about to see, the xylem sap is under a tension—a negative pressure potential—when it is ascending. Furthermore, as Strasburger had already shown, materials can be transported upward in the xylem even when the roots have been removed.

30.7 Guttation
Root pressure forces water through openings in the tips of this strawberry leaf.

The Evaporation–Cohesion–Tension Mechanism

To understand how sap rises in the xylem, even to the tops of the tallest trees, we must begin by looking at the final step in the process of water movement from soil to root to leaf and out to the atmosphere. At the end of the line, water evaporates from the moist walls of mesophyll cells, diffuses through the air spaces of the leaf, and finally leaves as water vapor through the open stomata.

The evaporation of water from mesophyll cells makes their water potential more negative—effectively, it increases the solute concentration of the cells—so more water enters osmotically from the nearest tiny vein. The removal of water from the xylem of the veins establishes a tension, or pull, on the entire column of water contained within the xylem, so the column is drawn upward all the way from the roots. The ability of water to be pulled

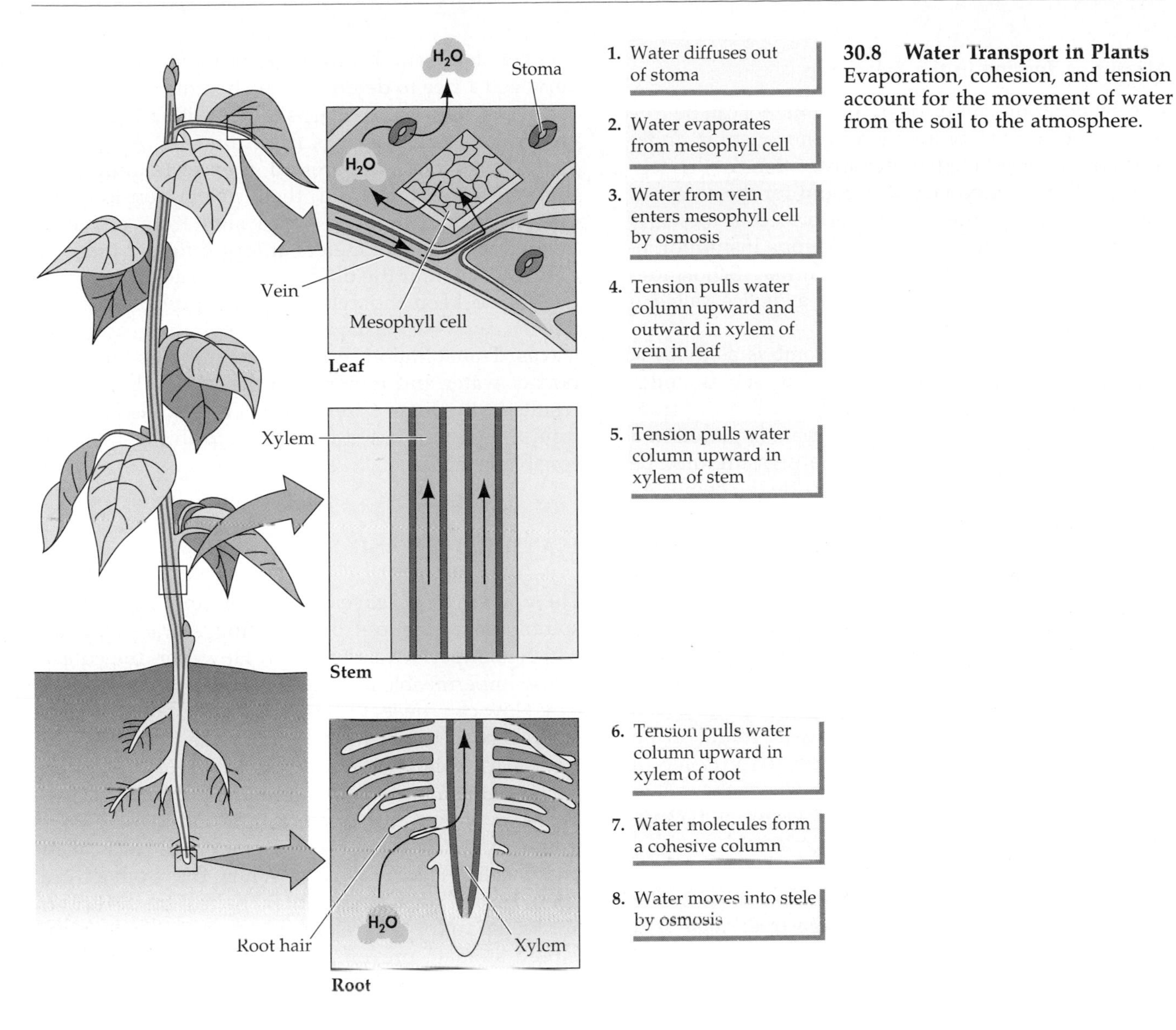

30.8 Water Transport in Plants Evaporation, cohesion, and tension account for the movement of water from the soil to the atmosphere.

upward through a tiny tube results from the remarkable cohesiveness of water—the tendency of water molecules to adhere to one another through hydrogen bonding (see Chapter 2). The narrower the tube, the greater the tension the water column can withstand without breaking. As the water column in the xylem is pulled upward, more water enters the xylem in the root by osmosis from surrounding cells.

In summary, the key elements of water transport in the xylem are *evaporation* from the moist cells in the leaves and a resulting *tension* in the remainder of the xylem's water owing to its *cohesion*, which pulls up more water to replace that which has been lost (Figure 30.8). All this requires no work on the part of the plant. At each step water moves passively to a region with a more strongly negative water potential. Dry air has the most negative water potential; the soil solution has the least negative water potential; xylem sap has a water potential more negative than that of cells in the root but less negative than that of mesophyll cells in the leaf.

Mineral ions contained in the xylem sap rise passively as the solution ascends from root to leaf. In this way the nutritional needs of the shoot are met. Some of the mineral elements brought to the leaves are subsequently redistributed to other parts of the plant by way of the phloem, but the initial delivery from the roots is through the xylem.

The evaporative loss of water from the shoot is called **transpiration**. In addition to promoting the transport of minerals, transpiration contributes to temperature regulation. As water evaporates from mesophyll cells, heat is taken up from the cells, and the leaf temperature drops. This cooling effect may be important in enabling plants to live in certain environments.

Measuring Tension in the Xylem Sap

The evaporation–cohesion–tension model can be true only if the column of solution in the xylem is under tension. The most elegant demonstrations of this tension, and of its adequacy to account for the ascent of sap in the tallest trees, were performed in the early 1960s by Per Scholander of the Scripps Institution of Oceanography in La Jolla, California. Scholander measured tension in stems with a device called a pressure bomb.

The principle of the pressure bomb is as follows: Consider a stem in which the xylem sap is under tension. If the stem is cut, the sap pulls away from the cut, into the stem. Now the tissue is placed in a cylinder—the bomb—in which the pressure may be raised. The cut surface remains outside the bomb. As pressure is applied, the xylem sap is forced back to the cut surface. When the sap first becomes visible again at the cut surface, the pressure in the bomb is recorded. This pressure is the same as the tension that was originally present in the xylem (Figure 30.9).

Scholander used the pressure bomb to study dozens of plant species, from diverse habitats, growing under a variety of conditions. In all cases in which the xylem sap was ascending, it was found to be under tension. The tension disappeared in some of the plants at night. In developing vines, the xylem sap was not under tension until leaves formed. Once leaves appeared, transport in the xylem began, and tensions were recorded.

Suppose you wanted to measure tensions in the xylem at various heights in a large tree, such as a Douglas fir more than 80 meters tall, to confirm that the tensions were sufficient to account for the rate at which sap was moving up the trunk. How would you get stem samples for measurement? Scholander surveyed a tree to determine the heights of particular twigs and then had a sharpshooter with a high-powered rifle shoot the twigs from the tree. As quickly as the twigs fell to the ground, they were inserted in the pressure bomb and their xylem tensions recorded. Scholander had twigs shot from a tree at heights of 27 and 79 meters at four different times of day. At each hour the difference in tensions was great enough to keep the xylem sap ascending, and that tension was established by transpiration from the leaves. Transpiration provides the impetus for transport of water and minerals in the xylem, but it also results in the loss of tremendous quantities of water from the plant. How do plants keep this loss to reasonable levels?

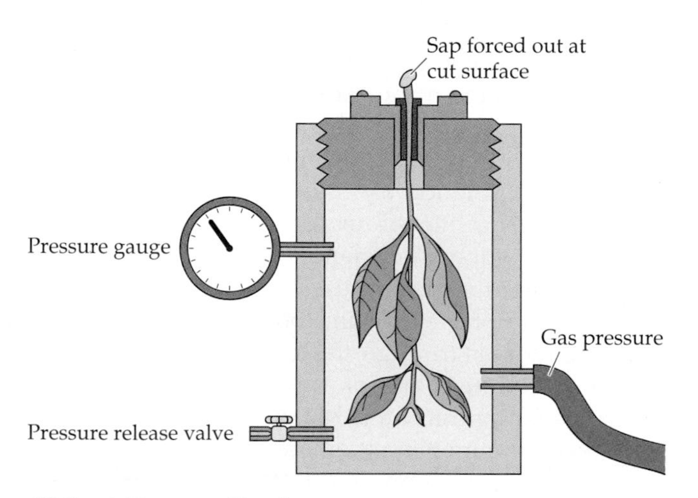

30.9 A Pressure Bomb
By applying just enough pressure so that xylem sap is pushed back to the cut surface of a plant sample, a scientist can determine the tension on the sap in the living plant.

TRANSPIRATION AND THE STOMATA

The epidermis of leaves and stems minimizes transpirational water loss by secreting a waxy **cuticle**, which is impermeable to water. However, the cuticle is also impermeable to carbon dioxide, posing a problem: How can the leaf balance its need to retain water with its need to obtain carbon dioxide for photosynthesis? An elegant compromise has evolved, in the form of stomata (singular: stoma). A **stoma** is a gap in the epidermis; its opening and closing are controlled by a pair of specialized epidermal cells called **guard cells** (Figure 30.10*a*). When the stomata are open, carbon dioxide can enter the leaf by diffusion, but water vapor may also be lost in the same way. Closed stomata prevent water loss but also exclude carbon dioxide from the leaf. Most plants compromise by opening the stomata only when the light intensity is sufficient to maintain a good rate of photosynthesis. At night, when darkness precludes photosynthesis, the stomata remain closed; no carbon dioxide is needed at this time, and water is conserved. Even during the daytime, the stomata close if water is being lost at too great a rate.

The stoma and guard cells in Figure 30.10*a* are typical of dicots. Monocots typically have specialized epidermal cells associated with their guard cells. The principle of operation, however, is the same for both monocot and dicot stomata.

The mechanism by which stomata open and close is now at least partially understood. The guard cells control the size of the stomatal opening. When the stomata are about to open, potassium ions (K^+) are actively transported into the guard cells from the surrounding epidermis. (The redistribution of potassium can be visualized by means of an instrument called an electron microprobe, and it can be measured by patch clamping.) The accumulation of potassium ions makes the water potential of the guard cells

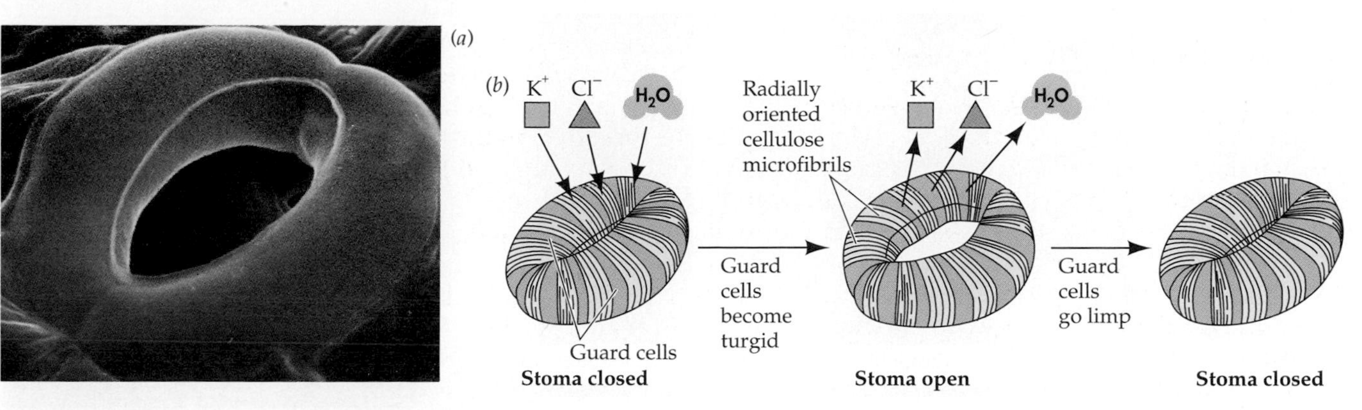

30.10 Stomata
(a) Mesophyll cells and air space inside this dicot leaf are visible through the gaping stoma between the two sausage-shaped guard cells. (b) Potassium ions are actively transported into the guard cells; their resulting negative water potential causes them to take up water and stretch so that a gap—the stomatal opening—appears between them. As K^+ then diffuses passively out of the guard cells, water follows by osmosis, the guard cells go limp, and the stoma closes. Negatively charged ions traveling with K^+ maintain electrical balance and contribute to the changes in osmotic potential that affect the guard cells.

more negative. Water enters the guard cells by osmosis, making them more turgid and stretching them in such a way that a gap, the stoma, appears between them. The pattern of stretching is controlled by the orientation of cellulose microfibrils in the walls of the guard cells. The stoma closes by the reverse process: Potassium ions diffuse passively out of the guard cells, water follows by osmosis, turgidity is lost, and the guard cells collapse together and seal off the stoma (Figure 30.10*b*). Negatively charged chloride and organic ions also move along with the potassium ions, maintaining electrical balance and contributing to the change in osmotic potential of the guard cells.

What controls the movement of potassium into and out of guard cells? The control system is complex, with more than one type of sensor system. For one thing, the level of carbon dioxide in the spaces inside the leaf is monitored; a low level favors opening of the stomata, thus allowing an increased carbon dioxide level and an enhanced rate of photosynthesis. On the other hand, certain cells monitor their own water potentials. If they are too dry—that is, if their water potential is too negative—they release a substance called abscisic acid. According to one hypothesis, abscisic acid then acts on the guard cells, causing them to release potassium ions, thus closing the stomata and preventing further drying of the leaf. (Some scientists think that abscisic acid serves only to keep the stomata closed, rather than causing the closure; further experiments on the timing of abscisic acid production are needed to resolve this point.)

Light also controls the opening of the stomata, which makes sense in view of the fact that most plants conduct photosynthesis only in the light; we would expect stomata to be closed in the dark, thus preventing unnecessary water loss. It was recently discovered that, under certain conditions, brief exposures to blue light cause guard cells to acidify their environment—that is, to pump out protons. Patch clamping experiments revealed further details about the relationship between blue light and proton pumping (Figure 30.11). The proton pump enables the guard cell to take up K^+ and Cl^- ions as described earlier in this chapter (see Figure 30.3).

CRASSULACEAN ACID METABOLISM AND THE STOMATAL CYCLE

Most plants change their stomatal openings on a schedule shown by the blue curve of Figure 30.12: The stomata are typically open for much of the day and closed at night (they may, however, close during very hot days). But not all plants follow this pattern. Many **succulent plants**—fleshy plants that live in dry areas or near the ocean—are members of the flow-

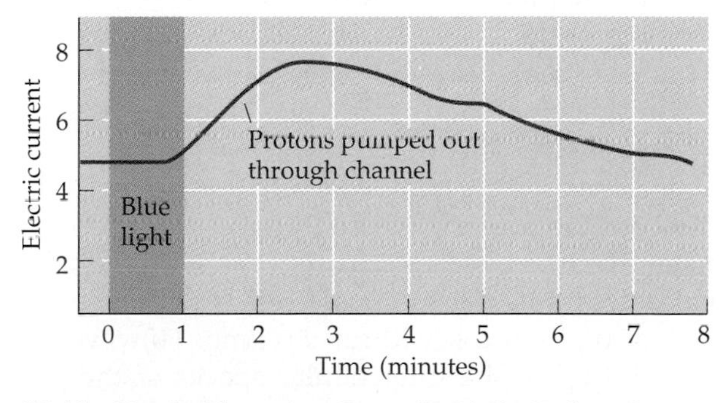

30.11 Patch Clamping Reveals Light-Induced Proton Pumping
A tracing of the tiny electric current that results from the flow of protons through a single channel in the plasma membrane of a guard cell. A brief exposure to blue light against a background of constant, dim red light causes protons to flow out of the cell for a few minutes.

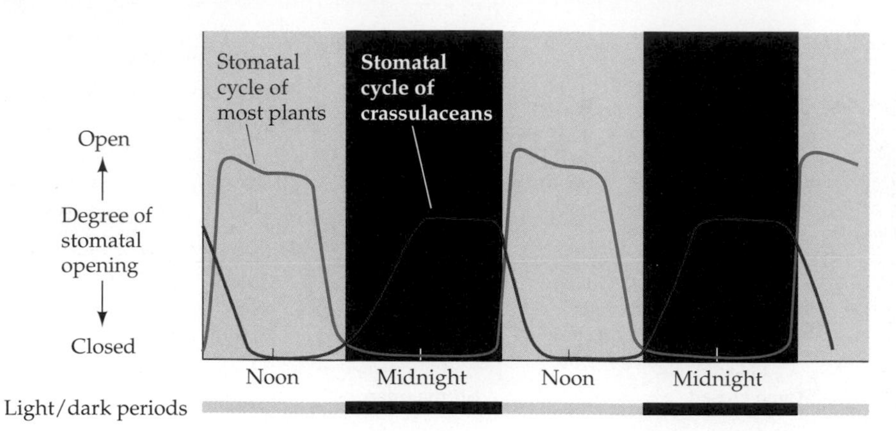

30.12 Stomatal Cycles
Most plants open their stomata during the daytime. Plants of the family Crassulaceae, such as the rock dudleya in the photograph, have evolved means to reverse this stomatal cycle. Crassulacean stomata open during the night.

ering plant family Crassulaceae. The crassulaceans have some unusual biochemical and behavioral features. One that was particularly surprising to its discoverers is their "backward" stomatal cycle: Their stomata are open at night and closed by day (red curve in Figure 30.12). It was then discovered that crassulacean leaf tissues become acidic at night and more neutral in the daytime.

The mystery was resolved with the following discoveries. At night, while the stomata are open, carbon dioxide diffuses freely into the leaf and reacts in the mesophyll cells with phosphoenolpyruvic acid to produce organic acids such as malic acid and aspartic acid (see Chapter 8). These acids accumulate to high concentrations in the vacuoles of the mesophyll cells. At daybreak the stomata close, thus preventing water loss. Throughout the day, the organic acids are broken down to release the carbon dioxide once again—behind closed stomata. Because the carbon dioxide cannot diffuse out of the plant, it is available for photosynthesis. This set of chemical reactions, discussed in Chapter 8, is referred to as crassulacean acid metabolism, or **CAM**. CAM and the accompanying stomatal behavior were subsequently observed in species of many other plant families besides the Crassulaceae.

Notice that the formation of organic acids is absolutely essential to the functioning of the reversed stomatal pattern of CAM plants. Without the acid formation, carbon dioxide could still be admitted to the leaf at night and saved for daytime. However, it could build up in the intercellular spaces of the leaf only to the same level—0.03 percent of the atmosphere—as in the surrounding air. This amount would be used up by the Calvin–Benson cycle of photosynthesis in a matter of minutes in the daytime. Instead, during the night a CAM plant makes the carbon dioxide into organic acids as fast as it comes in, thus allowing more carbon dioxide to enter. Acid formation in effect fills the leaf with carbon dioxide. CAM is well adapted to environments where water is scarce: A leaf with its stomata open at night loses much less water (because the environment is cooler then) than does a leaf with its stomata open by day.

In both CAM and non-CAM plants, the carbon dioxide is converted to the products of photosynthesis. How does the plant deliver these products to the parts of the plant that do not perform photosynthesis?

TRANSLOCATION OF SUBSTANCES IN THE PHLOEM

How substances in the phloem move from sources, such as leaves, to sinks, such as the root system, remains a topic of interest in plant physiology. Sugars, amino acids, some minerals, and a variety of other substances are translocated in the phloem. Any model to explain this translocation must account for a few important facts: (1) Translocation stops if the phloem tissue is killed by heating or other methods; thus the mechanism must be different from that of transport in the xylem. (2) Translocation often proceeds in both directions—up and down the stem or petiole—simultaneously. This bidirectional transport may be explained in terms of neighboring sieve tubes conducting in opposite directions, with each sieve tube transporting all its contents in a single direction. (3) Translocation is inhibited by compounds that inhibit cellular respiration and thus limit the ATP supply.

To answer some of the most pressing questions about translocation, plant physiologists needed to obtain samples of pure phloem sap from individual sieve tube elements. This task was simplified when

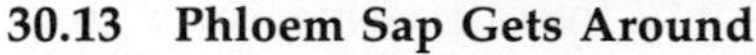

30.13 Phloem Sap Gets Around
Aphids—the white organisms with "sculpted" abdomens—are drilling into a plant to obtain phloem sap. A drop of the sap has formed at the anus of one of the aphids, and an ant is about to collect it.

scientists recognized that a common garden pest, the aphid, feeds by drilling into a sieve tube. An aphid inserts its stylet, or feeding organ, into a stem until the stylet enters a sieve tube. Within the sieve tubes, the pressure is much greater than in the surrounding plant tissues, so phloem sap is forced up the stylet and into the aphid's digestive tract. So great is the pressure that sugary liquid is forced out the insect's anus (Figure 30.13). At times, ants collect this sugary discharge as food, and some species of ants actually "farm" colonies of aphids, moving them from place to place and protecting them from enemies.

Plant physiologists use the aphid somewhat differently. When liquid appears on the aphid's abdomen, indicating a connection with a sieve tube, the physiologist freezes the aphid and cuts its body away from the stylet. Phloem sap continues to exude from the cut stylet, where it may be collected for analysis. Study of the sap gives accurate information about the chemical composition of the contents of a single sieve tube element over time. From that information one can infer such things as translocation rates. Data obtained by this and other means led to the general adoption of the pressure flow model as an explanation for transport in the phloem.

The Pressure Flow Model

Phloem sap flows, under pressure, through the sieve tubes. Two important steps in the flow are the active, ATP-requiring transport of sugars and other solutes *into* the sieve tubes in source areas and the *removal* of the solutes by active transport where the sieve tubes enter sinks. According to the **pressure flow model**, the solute concentration at the source end of a sieve tube is higher than at the sink end, so water has a greater tendency to enter the sieve tube by osmosis at the source end. In turn, this entry of water causes a greater pressure potential at the source end, so the entire fluid content of the sieve tube is, in effect, squeezed toward the sink end of the tube (Figure 30.14). This mechanism was first proposed more than half a century ago, but some of its features are still debated.

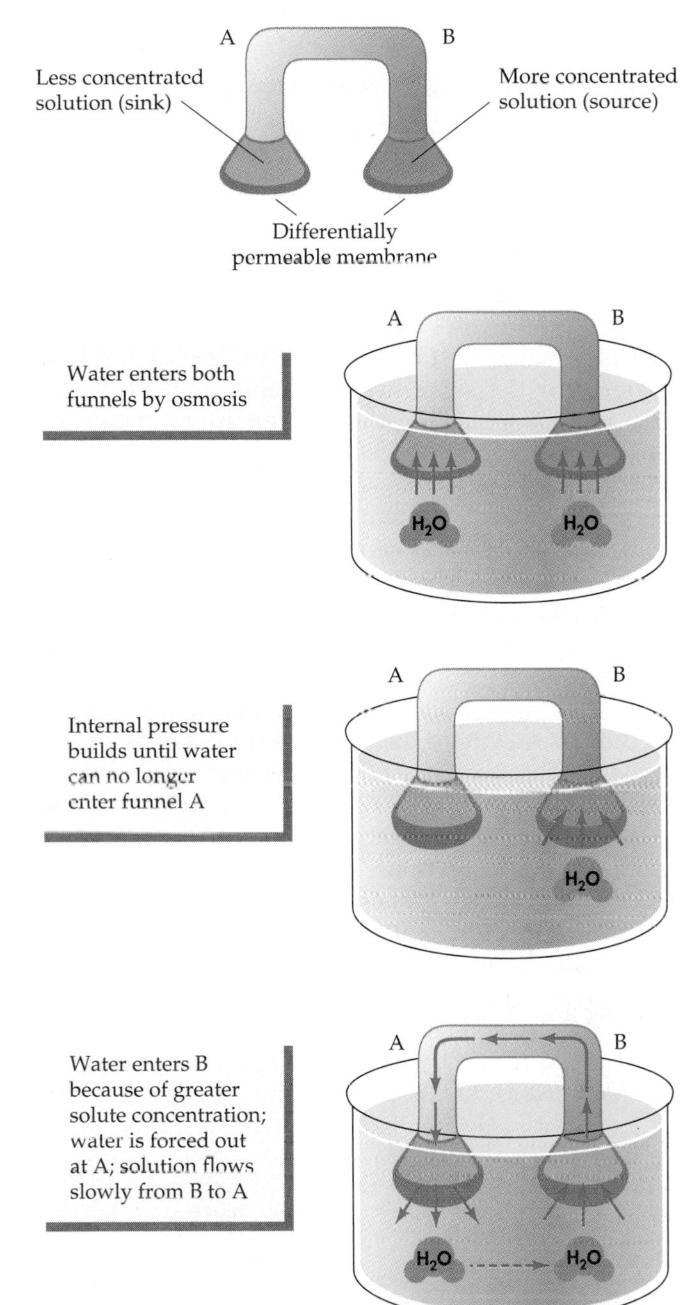

30.14 The Pressure Flow Model
This experimental demonstration of the pressure flow model describes how pressure potential and water potential combine to drive sugars and other solutes from the source (B) to the sink (A). Phloem sap may flow through sieve tubes in this manner.

Other mechanisms have been proposed to account for translocation in sieve tubes. Some have been disproved, and none of the rest have been supported by a weight of evidence comparable to that for the pressure flow model. The pressure flow model depends on two things: The sieve plates must be unclogged, so that bulk flow from one sieve tube element to the next is possible, and there must be an effective method for loading sucrose and other solutes into the phloem in source tissues and removing them in sink tissues. Are these conditions met?

Are the Sieve Plates Clogged or Free?

Early electron microscopic studies of phloem samples cut from plants produced results that seemed to contradict the pressure flow model. The pores in the sieve plates always appeared to be plugged with masses of a fibrous protein, suggesting that phloem sap could not flow freely. But what is the function of the fibrous protein? One possibility is that this protein is usually distributed more or less at random throughout the sieve tube elements until the sieve tube is damaged; then the sudden surge of sap toward the cut surface carries the protein into the pores, blocking the pores and effectively caulking the leak. That is, the protein does *not* block the pores unless the phloem is damaged. How might this be tested? How could we cut samples of phloem for microscopic observation without causing the sap to surge to the cut surface?

One way to prevent the surge of the sap is to freeze the tissue before cutting it. Another way is to let the tissue wilt so that there is no pressure in the phloem. These and related ideas were tested, and sure enough, when the tissue is cut in this way, the sieve plates are unclogged by protein! Thus, the first condition of the pressure flow model is met. Now, what about the need for an effective method for loading and unloading solutes?

Loading and Unloading of the Phloem

If the pressure flow model is correct, there must be mechanisms for loading sugars and other solutes into the phloem from source regions and for unloading them into sink regions. One mechanism of phloem loading has been demonstrated in a number of plants. Sugars and other solutes to be transported are passed from cell to cell through the symplast in the mesophyll. After these substances reach cells adjacent to the ends of leaf veins, they leave the mesophyll cells and enter the apoplast, sometimes with the help of transfer cells. Then specific sugars, amino acids, some mineral elements, and a few other compounds are actively transported into cells of the phloem, thus reentering the symplast (Figure 30.15).

Passage through the apoplast selects substances to be accumulated for translocation because substances can enter the phloem only upon passing through a differentially permeable membrane. In many plants, the cells thus loaded are companion cells (see Chapter 29), which then transfer the solutes to the adjacent sieve tube elements. Loading of the phloem with solutes results in a very negative water potential in the sieve tubes; thus water enters by osmosis from the surrounding tissue and maintains a high pressure potential within the sieve tubes. As Figure 30.15 shows, sucrose movement from the mesophyll to the sieve tube elements takes place entirely within the symplast in some species; that is, transfer of solutes from symplast to apoplast and back again is not a universal feature of phloem loading.

A form of secondary active transport (see Chapter 5) loads sucrose into the companion cells and sieve tube elements. Sucrose is carried through the plasma membrane from apoplast to symplast by a sucrose–proton symport—the entry of sucrose and of protons is strictly coupled. For this symport to work, there must be a high concentration of protons in the apoplast; the protons are supplied by a primary active transport system, the proton pump. The protons then "relax" back into the cell through the symport, bringing sucrose with them (Figure 30.16).

In sink regions, the transported solutes are actively transported out of the sieve tubes and into the surrounding tissues. This unloading serves two purposes: It helps to maintain the gradient of osmotic potential and hence of pressure potential in the sieve tubes, and it promotes the buildup of sugars and

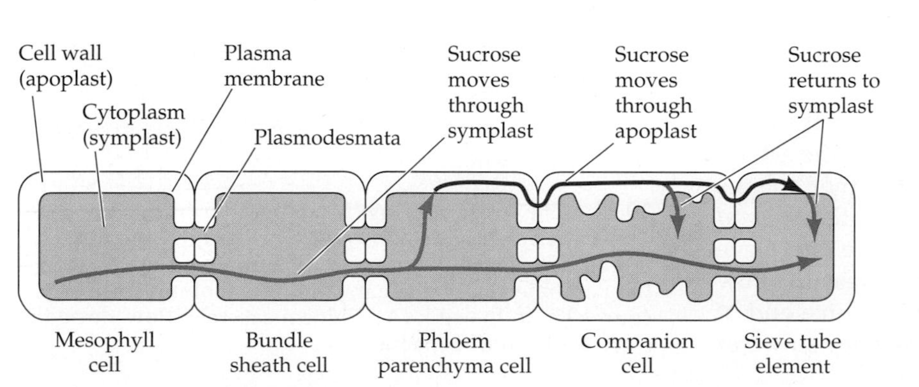

30.15 Solutes May Enter Sieve Tubes Via the Apoplast
Chloroplasts in the mesophyll produce sucrose, which moves primarily through the symplast (blue arrow) as it passes from cell to cell on its way to the sieve tube elements. In many species, however, sucrose exits into the apoplast (red arrow), from which it is loaded into the companion cells or sieve tube elements.

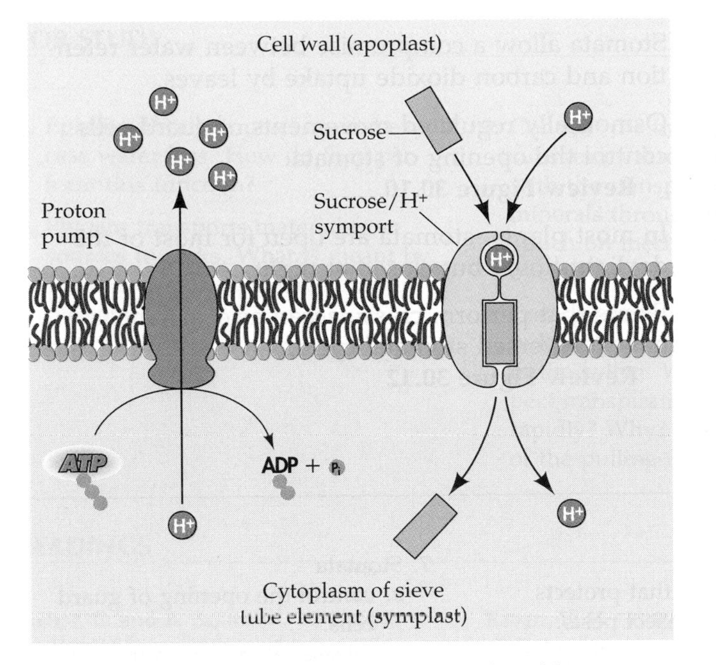

30.16 Sucrose–Proton Symport Loads the Phloem
The proton pump builds a gradient of proton concentration, and the relaxation of the gradient through the sucrose–proton symport carries sucrose, against its own concentration gradient, into the companion cells and sieve tube elements.

30.17 Collecting Maple Sap
Sap from the blue tap flows through tubes to a central location for collection.

starch to high concentrations in storage regions, such as developing fruits and seeds.

Sucrose in the Xylem

The xylem sap, as well as the phloem sap, occasionally may contain sugars. In sugar maples and many other deciduous trees and shrubs of the temperate zones, excess photosynthate produced in late summer and early fall is stored as starch in living xylem cells of the trunk and twigs. Later, in early spring, the starch is digested into sugars that appear initially in the xylem sap, which may be collected and concentrated into syrup (Figure 30.17). The activities of plants may vary with the time of year, and patterns of transport and storage may change accordingly.

SUMMARY of Main Ideas about Transport in Plants

Water and solutes move through tissues in the apoplast and symplast.

In the root, water and solutes may pass between the cortex and stele only in the symplast.

Casparian strips in the endodermis block water and solute movement in the apoplast.
Review Figures 30.5 and 30.6

Water moves between cells by osmosis.

The water potential of a cell or solution is the sum of the osmotic potential and the pressure potential.

Water always moves toward the cell or solution with a more negative water potential.
Review Figure 30.1

A proton pump mediates the active transport of many solutes across membranes in plants.
Review Figures 30.3 and 30.16

Patch clamping is a powerful method for studying the movement of ions through membranes.
Review Figures 30.4 and 30.11

Water and minerals are transported through tracheids or vessel elements in the xylem.

Root pressure plays a minor role in the transport of water and minerals in the xylem.

Evaporation from moist-walled cells in the leaf lowers the water potential of those cells and thus pulls water—held together by its cohesiveness—up through the xylem from the root.
Review Figure 30.8

A Beetle Disarms Laticifers
This beetle is inactivating a milkweed's defense system by cutting supply lines.

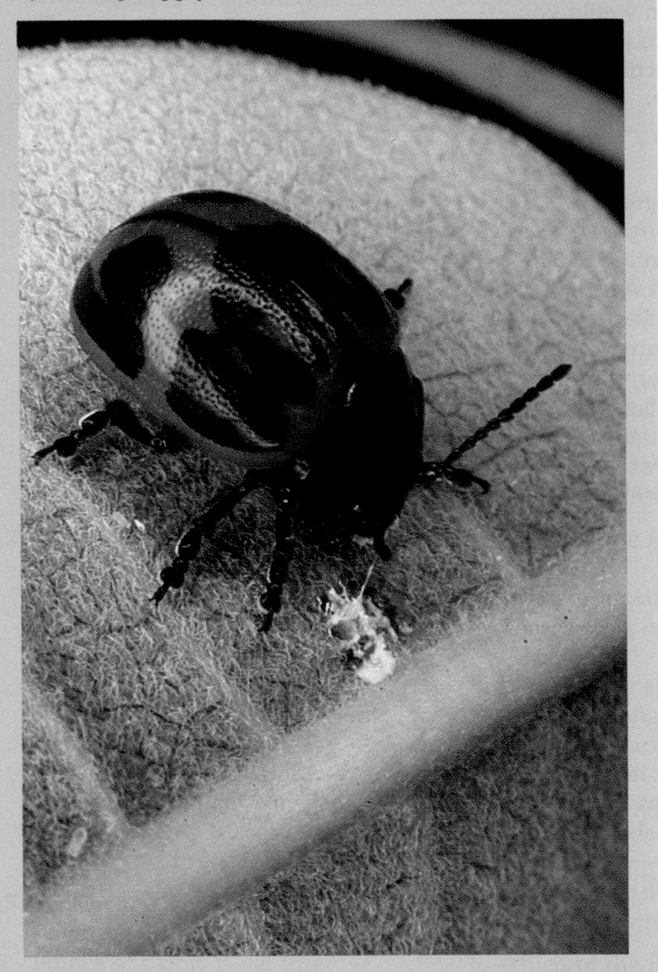

31 Environmental Challenges to Plants

What we see today is only one frame of a long movie of interactions among organisms on Earth. Some of these interactions represent adaptations, visible in this single frame, that are being challenged and perhaps further refined. Let's begin our discussion of the challenges that the environment poses to plants by looking at the current status of a deadly "game" between a plant and one of its insect enemies.

Milkweeds are latex-producing, or laticiferous, plants. When damaged, a milkweed releases copious amounts of a white, rubbery, toxic liquid—latex—from tubes called laticifers. Latex has long been suspected to deter insects from eating the plant because laticiferous plants are not attacked by insects that feed on neighboring plants of other species. This observed behavior is consistent with, but does not prove, the hypothesis that the latex keeps the insects at bay. Stronger support for the hypothesis was afforded in 1987 by David Dussourd, now of the University of Maryland, and Thomas Eisner of Cornell University, who studied field populations of *Labidomera clivicollis,* a beetle that is one of the few insects that feed on *Asclepias syriaca,* the field milkweed.

The two zoologists observed a remarkable prefeeding behavior: The beetles cut a few veins in the leaves before settling down to dine. In the undamaged plant the latex is under pressure, so cutting the veins, with their adjacent laticifers, causes massive leakage and depressurizes the system. By cutting a few veins, the beetles interrupt the latex supply to a downstream portion of the leaf. The beetles then move to the relatively latex-free portion and eat their fill. Some other insects that do not feed on undamaged milkweeds will eat parts of leaves that have had the latex supply cut off. When presented simultaneously with leaves on undamaged plants and leaf parts that have had their laticifers cut, *L. clivicollis* and other insects that share this vein-cutting behavior select the relatively latex-free leaf parts.

Does this behavior of the beetles negate the adaptational value of latex protection? Not at all. There are still great numbers of potential insect pests that are effectively deterred by the latex. And this is just one frame of the movie. It may be that, over time, milkweed plants producing higher concentrations of toxins will be selected by virtue of their ability to kill beetles that cut the laticifers.

THREATS TO PLANT LIFE

All plants, especially terrestrial ones, face environmental challenges, as we indicated in Chapters 25 and 30. Some environments, however, pose exceptional problems and thus drastically limit the kinds

31.1 Desert Annuals Evade Drought
Seeds of desert plants often lie dormant for long periods awaiting conditions appropriate for germination. When they do germinate, they grow and reproduce rapidly before the short wet season passes. They cover the desert landscape with color for only a few weeks, since water is inadequate at other times.

of plants that can live in them. The most challenging physical environments include ones that are very dry (deserts), ones at the other extreme that are waterlogged and thus limit the availability of oxygen, ones that are dangerously salty, and ones that contain high concentrations of toxic substances such as heavy metals. This chapter focuses in part on how some plants manage to thrive in such environments.

The biological environment is also a threat to all plants. It includes herbivores like *Labidomera clivicollis* that consume plants, as well as pathogenic fungi, bacteria, and viruses. The defenses of plants against these biological threats are the other major focus of the chapter.

DRY ENVIRONMENTS

Water for plants and other organisms is often in short supply in the terrestrial environment. Some terrestrial habitats, such as deserts, intensify this challenge, and many plants that inhabit particularly dry areas have one or more structural adaptations that allow them to conserve water. Plants adapted to dry environments are called **xerophytes**.

Some desert plants have no special *structural* adaptations for water conservation other than those found in almost all flowering plants. Instead they have an alternative *strategy.* Simply put, these plants carry out their entire life cycle—from seed to seed—during a brief period in which the surrounding desert soil is sufficiently moist (Figure 31.1). Through the long dry periods that intervene, only the seeds remain alive, until enough moisture is present to trigger the next life cycle. These desert annuals simply evade the periods of drought. Plants that remain active during the dry periods must have special adaptations that enable them to survive.

Special Adaptations of Leaves to Dry Environments

The secretion of a heavier layer of cuticle over the leaf epidermis to retard water loss is a common adaptation to dry environments. An even more common adaptation is a dense covering of epidermal hairs. Some species have stomata only in sunken cavities below the leaf surface, which reduces the drying effects of air currents; often these stomatal cavities contain hairs as well (Figure 31.2). Ice plants and their relatives have fleshy leaves in which water may be stored. Others, such as ocotillo, produce leaves only when water is abundant, shedding them as the soil dries out (Figure 31.3). Cacti and similar

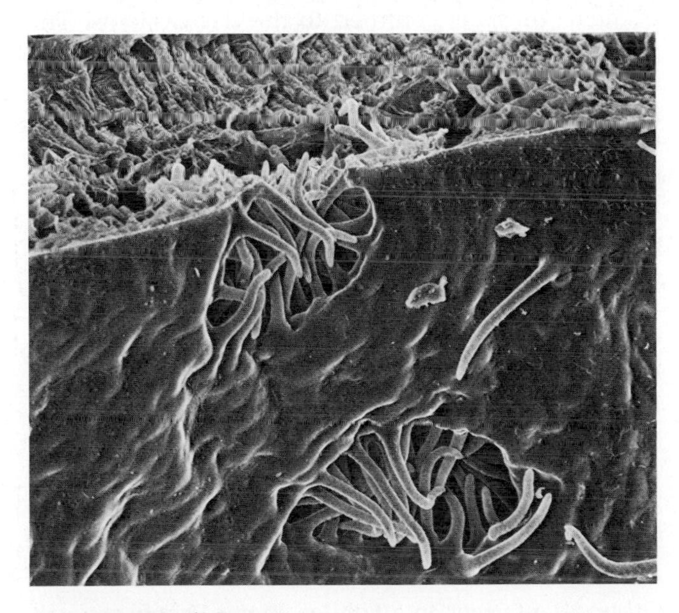

31.2 Stomatal Crypts
Stomata in some water-conserving leaves are in sunken pits called stomatal crypts. The hairs covering these two crypts presumably trap moist air. (Above the cut edge near the top of the photo, a section of the leaf's interior can be seen.)

31.3 Opportune Leaf Production
Plants in hot, dry environments lose great amounts of water through their leaves. The ocotillo, which lives in the lower deserts of the southwestern United States and northern Mexico, has leaves only when water is available in the soil. During dry periods, the thorny, leafless stems of an ocotillo appear almost dead *(a)*, but when water is on hand, leaves develop rapidly *(b)* and provide the plant with photosynthetic products.

plants have spines rather than typical leaves, and photosynthesis is confined to the fleshy stems. The spines may reflect incident radiation or perhaps dissipate heat. Corn and some related grasses have leaves that roll up during dry periods, thus reducing the leaf surface area through which water is lost. Some trees that grow in arid regions have leaves that hang vertically at all times, thus evading the midday sun. Characteristic examples are some eucalyptuses (Figure 31.4).

Xerophytic adaptations of leaves minimize water loss by the plant. However, such adaptations simultaneously minimize the uptake of carbon dioxide and thus limit photosynthesis. In consequence, most xerophytes grow slowly, but they utilize water more efficiently than do other plants; that is, they fix more grams of carbon by photosynthesis per gram of water lost to transpiration than other plants do.

Other Adaptations to a Limited Water Supply

Roots may also be adapted to environments low in water. The Atacama Desert in northern Chile often goes several years without receiving any measurable rainfall. The landscape there is almost barren save for a substantial number of surprisingly large mesquite trees of the genus *Prosopis*. How do these trees obtain water? They have taproots that grow to very great depths, sufficient to reach underground water supplies (Figure 31.5). These trees also obtain water from condensation on their leaves.

A more common adaptation of desert plants is to have root systems that grow each rainy season but die back during dry periods. Cacti, on the other hand, have shallow but extensive fibrous root systems that effectively trap water at the surface of the soil following even light rains.

Xerophytes and other plants receiving inadequate water may accumulate the amino acid proline to substantial concentrations in their vacuoles. As a consequence, the osmotic potential and water potential of the cells become more negative, thus tending to extract more water from the soil.

As we have seen, there are many ways in which some plants eke out an existence in environments with very little water. What happens if there is too much water?

WHERE WATER IS PLENTIFUL AND OXYGEN IS SCARCE

When soils become waterlogged, the availability of oxygen from the soil declines. Most plants cannot tolerate this situation for long. Some species, however, are adapted to life in a swampy habitat. Their

31.4 Shade at Midday
Because eucalyptus leaves hang vertically, their flat surfaces are not presented directly to the midday sun. This adaptation minimizes heating as well as water loss.

31.5 Mining Water with Deep Taproots
This California desert is not as arid as the Chilean Atacama, but *Prosopis juliflora*, a mesquite, still reaches far down in the sand dunes for its water supply.

roots grow slowly and hence do not penetrate deeply. With an oxygen level too low to support aerobic respiration, the roots carry on alcoholic fermentation (see Chapter 7), which provides a supply of ATP for the activities of the root system.

The root systems of some plants adapted to swampy environments have **pneumatophores**, extensions that grow out of the water and up into the air (Figure 31.6). Oxygen diffusing into pneumatophores aerates the submerged parts of the root system.

Submerged or partially submerged aquatic plants often have large air spaces in the leaf parenchyma and in the petioles. Tissue with such air spaces is called **aerenchyma** (Figure 31.7). Aerenchyma stores and permits the diffusion of oxygen, and it provides buoyancy.

Thus far we have considered water supply—either too little or too much—as a limiting factor in plant growth. Are there other substances that make an environment hospitable or inhospitable to plant growth?

SALINE ENVIRONMENTS

On a world scale, no toxic substance restricts plant growth more than does salt (sodium chloride). Sa-

31.6 Coming Up for Air
The roots of the mangroves in this tidal swamp obtain oxygen through their pneumatophores—extensions growing out of the water, under which the remainder of the root system is buried.

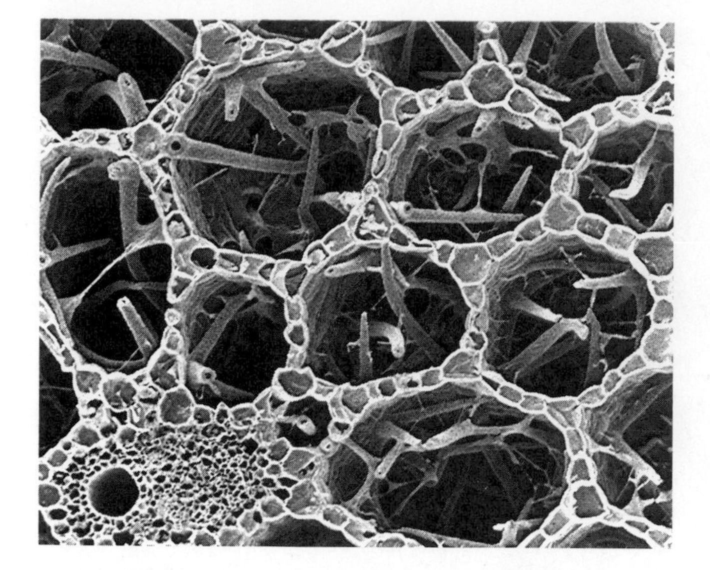

31.7 Aerenchyma Lets Oxygen Get to Submerged Tissues
This scanning electron micrograph, a cross section of a petiole of the yellow water lily, shows a vascular bundle and a number of open channels of the aerenchyma. The channels are lined by cells, including branched ones that send projections into the channels. Because aerenchyma has far fewer cells than does a comparable volume of most petiole tissue, less respiratory metabolism is carried on and the need for oxygen is much reduced.

line—salty—habitats support, at best, sparse vegetation. The **halophytes**, plants adapted to such habitats, belong to a wide variety of flowering plant groups. Saline environments themselves are diverse, ranging from hot, dry, salty deserts to moist, cool, salty marshes. Along the seashore are salty environments created by ocean spray. The ocean itself is a saline environment, as are river estuaries, where fresh and salt water meet and mingle. The salinization of agricultural land is an increasing world problem; where crops are irrigated, sodium ions in the water accumulate in the soil to ever greater concentrations. Biologists in Israel and elsewhere have had some success in breeding crops that can be watered with seawater or diluted seawater.

Saline environments pose an osmotic problem. Because of a high salt concentration, the environment has an unusually large negative water potential. To obtain water from such an environment, resident plants must have an even more negative water potential than that of plants in nonsaline environments; otherwise, the plants lose water and wilt. A second problem related to the saline environment is the potential toxicity of high concentrations of certain ions, notably sodium. Chloride ions may also be toxic at high concentrations. How can halophytes cope with a highly saline environment, while nonhalophytes cannot?

Salt Accumulation and Salt Glands

Most halophytes share one adaptation: They accumulate sodium and, usually, chloride ions and they transport these ions to the leaves. Nonhalophytes accumulate relatively little sodium, even when placed in a saline environment; of the sodium that is absorbed by their roots, very little is transported to the shoot. The increased salt concentration in halophytes makes their water potential more negative, so they can take up water from the saline environment. We still do not know how halophytes are able to tolerate such high internal sodium and chloride concentrations without being poisoned.

Some halophytes have other adaptations to life in saline environments. For example, some have salt glands in their leaves. These glands excrete salt, which collects on the leaf surface until it is removed by rain or wind (Figure 31.8). This adaptation, which reduces the danger of poisoning by accumulated salt, is found both in some desert plants, such as tamarisk, and in some mangroves growing in seawater in the tropics.

Salt glands can play multiple roles, as in the arid-zone shrub *Atriplex halimus*. This shrub has glands that secrete salt into small bladders on the leaves where, by increasing the gradient in water potential, it helps the leaves obtain water from the roots. At the same time, by making the water potential of the leaves more negative, the salt reduces the transpirational loss of water to the atmosphere.

These adaptations are specific to halophytes. Several other adaptations are shared by halophytes and xerophytes.

31.8 Secreting Salt
This salty mangrove has used special glands to secrete salt, which now appears as crystals on the leaves.

Adaptations Common to Halophytes and Xerophytes

Many halophytes accumulate the amino acid proline in their cell vacuoles. Unlike sodium, proline is relatively nontoxic. As in xerophytes, the accumulated proline makes the water potential more negative.

Succulence—the possession of fleshy, water-storing leaves—is an adaptation to dry environments. The same adaptation is common among halophytes, as might be expected, since saline environments also make water uptake difficult for plants. Succulence characterizes many halophytes occupying salt marshes. There the salt concentration in the soil solution may change throughout the day; while the tide is out, for instance, evaporation increases the salt concentration. Succulence may offer a reserve of water for the plant during the period of maximum salinity; when the salinity drops as the tide comes in, the leaf's store of water is replenished.

Other general adaptations to a saline environment are of the same sorts observed in xerophytes. These include high root-to-shoot ratios, sunken stomata, reduced leaf areas, and thick cuticles.

A Versatile Halophyte

The halophyte *Triglochin maritima* is unusual because it can adjust to a wide range of environmental salinities. Recent work in Toronto, Canada, has established that *T. maritima* plants change in many ways when they are shifted to environments with lower or higher salt concentrations. Researchers watered some plants with seawater and some with diluted seawater. The plants watered with pure seawater produced much smaller leaves, with smaller cells, than those watered with diluted seawater, but they retained their leaves longer. The rate of leaf production is the same regardless of salinity, but when the environment is changed, the pattern of leaf production changes accordingly.

These morphological differences in *T. maritima* leaves are accompanied by physiological changes as well. The leaf cells of the plants grown on undiluted seawater contained much higher concentrations of sodium and chloride ions as well as of proline. The rates of photosynthesis in the leaves of the plants watered with undiluted seawater also were higher.

Salt is not the only toxic solute in soils. Some other solutes are, in fact, more toxic than salt when presented at the same concentration.

HABITATS LADEN WITH HEAVY METALS

High concentrations of heavy metals, such as copper, lead, nickel, and zinc, poison most plants, even though plants require some heavy metals at low concentrations. Some sites are naturally rich in heavy metals as a result of normal geological processes. Acid rain leads to the release of toxic aluminum ions in the soil. Other human activities, notably the mining of metallic ores, leave localized areas—known as tailings—with substantial concentrations of heavy metals and low concentrations of nutrients. Such sites are hostile to most plants, and seeds falling on them generally do not produce adult plants.

31.9 Plant Life on a Mine Tailing
Although high concentrations of copper kill most plants, Bermuda grass is colonizing this copper mine tailing in Copperhill, Tennessee.

Mine tailings rich in heavy metals, however, generally are not completely barren (Figure 31.9). They may support healthy plant populations that differ genetically from populations of the same species on the surrounding normal soils. How can these plants survive? Within some species, a few individuals may have genotypes that allow them to survive in soils rich in heavy metals. Those individuals may grow poorly on such soils compared with their potential for growth on more normal soils but may nevertheless survive. Or, because most plants cannot survive in such habitats and hence the competition may be sharply reduced, the few plants growing in them may even thrive.

Initially, some plants were thought to tolerate heavy metals by excluding them: By not taking up the metal ions, the plant could avoid being poisoned. Measurements have shown, however, that tolerant plants growing on mine tailings do take up the heavy metals, accumulating them to concentrations that would kill most plants. Thus the tolerant plants must have a mechanism for dealing with the heavy metals they take up.

The British biologist D. Jowett made an interesting discovery about plants that tolerate heavy metals. In

31.10 Serpentine Barrens
The rocky serpentine barrens in the foreground support only a sparse vegetation of goldenrod and asters, contrasted with the green, forested hills in the background.

Wales and Scotland, bent grass (*Agrostis*) grows near many mines (see Figure 20.4). From mine to mine, the heavy metals in the soil differ. Jowett obtained samples of bent grass from several such sites and tested their ability to grow in various solutions, each containing only one of the heavy metals. In general, the plants tolerated a particular heavy metal—the one most abundant in their habitat—but were sensitive to other heavy metals. That is, their tolerance was for one or two heavy metals only, rather than for the heavy metals as a group.

Tolerant populations can evolve and colonize an area surprisingly rapidly. The bent grass population around a particular copper mine in Wales is resistant to copper and relatively abundant, even though the copper-rich soil dates from mining done late in the nineteenth century, only a century ago. If populations can evolve and cope with toxic soils, can they deal as well with soils in which nutrients are in short supply or in improper balance?

SERPENTINE SOILS

One unproductive soil type that is found in many parts of the world is derived from rock called **serpentine**. Calcium is in short supply in serpentine soils, as are some other essential plant nutrients. Magnesium is present in greater concentration than calcium, reducing plant growth. Chromium, nickel, and certain other heavy metals may be abundant. These factors make serpentine soils inhospitable to many plants. The vegetation on most serpentine soils differs dramatically from that on immediately adjacent nonserpentine soils, and the serpentine vegetation generally is more sparse and less diverse (Figure 31.10).

The shortage of calcium and the higher magnesium concentration probably are the principal challenges facing potential colonizers of a serpentine soil. A number of species have physiological adaptations to meet the challenges successfully. Different species exhibit striking differences in their response to calcium supply in the soil. Biologists divided some serpentine soil into several samples, adjusted the cal cium level of each, and grew jewel flower (*Streptanthus glandulosis*), which grows on serpentine, and tomato—a crop plant intolerant of serpentine—on each sample. The growth of the tomato plants was sharply dependent on calcium concentration, while that of the jewel flower was remarkably insensitive to it (Figure 31.11). Serpentine plants such as jewel flower can absorb calcium efficiently even from soils highly deficient in that element. These serpentine plants may also be able either to exclude excess magnesium or to tolerate high internal magnesium concentrations.

Having considered plant adaptations to their physical environment—water, oxygen, salt, and toxic substances—we will now turn to the biological environment. Specifically, we will consider how plants interact with the animals that eat them, and with fungi, bacteria, and viruses that produce plant diseases.

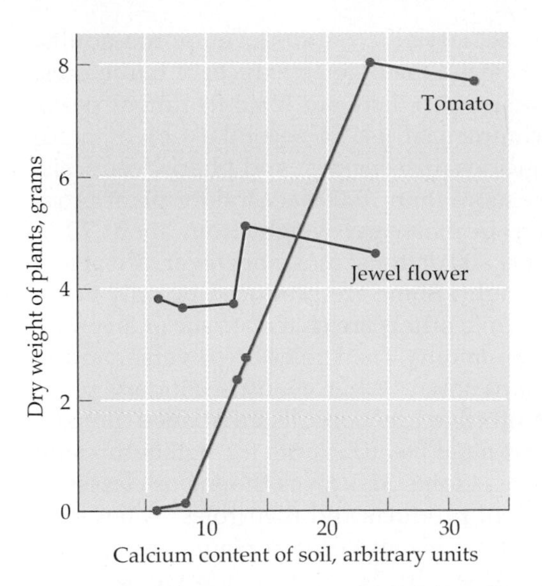

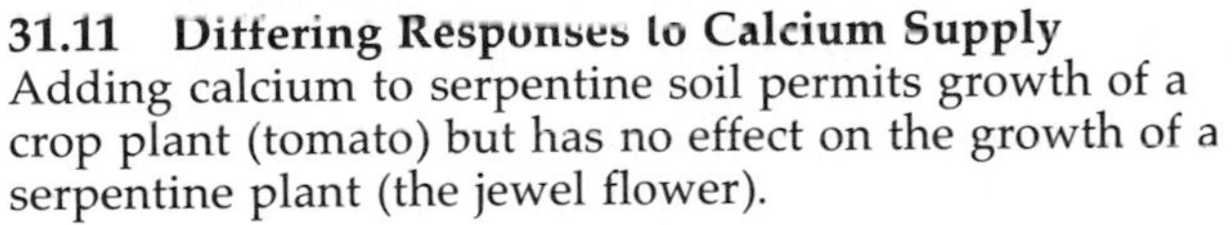
31.11 Differing Responses to Calcium Supply
Adding calcium to serpentine soil permits growth of a crop plant (tomato) but has no effect on the growth of a serpentine plant (the jewel flower).

PLANTS AND HERBIVORES

Herbivores—animals that eat plants—depend on plants for energy and nutrients. Plants have defense mechanisms to protect them against herbivores, as we will see, but first let's consider some examples in which herbivores have a positive effect on the plants that they eat.

Grazing and Plant Productivity

Consider the phenomenon of grazing, in which an animal predator eats part of its prey, such as the leaves of plants, without killing the prey organism, which has the potential to grow back. What are the consequences of grazing? Is it detrimental to the plants, or are they somehow adapted to their place in the food chain of nature? In fact, certain plants and their predators evolved together, each acting as the agent of natural selection on the other. Because of this coevolution, grazing increases photosynthetic production in certain plant species.

The removal of some leaves from a plant typically increases the rate of photosynthesis of the remaining leaves. This phenomenon probably is the result of several factors. First, nitrogen obtained from the soil by the roots no longer needs to be divided among so many leaves. Second, the transport of sugars and other photosynthetic products from the leaves may be enhanced because the demand for those products in the sinks—such as roots—is undiminished, while the sources—leaves—have been decreased. A third and particularly significant factor, especially in grasses, is an increase in the availability of light to the younger, more active leaves or leaf parts. The removal of older, dying—or even dead—leaves and leaf parts by a grazer decreases the shading of younger leaves, and unlike most other plants, which grow from their shoot and leaf tips, grasses grow from the base of the shoot and leaf.

Some grazed plants continue to grow until much later in the season than ungrazed but otherwise similar plants do. This longer growing season results in part because the removal of apical buds by the grazers stimulates lateral buds to become active, thus producing a more heavily branched plant. In addition, leaves on ungrazed plants may die earlier in the growing season than leaves on grazed plants.

A clear case of increased productivity due to grazing was reported in 1987 by Ken Paige of the University of Utah and Thomas Whitham of Northern Arizona University. Mule deer and elk graze many plants, including one called scarlet gilia. The grazing removes about 95 percent of the aboveground part of each scarlet gilia. However, each plant quickly regrows not one but four replacement stems. The cropped (grazed) plants produce three times as many fruits by the end of the growing season as do ungrazed plants (Figure 31.12). Paige and Whitham

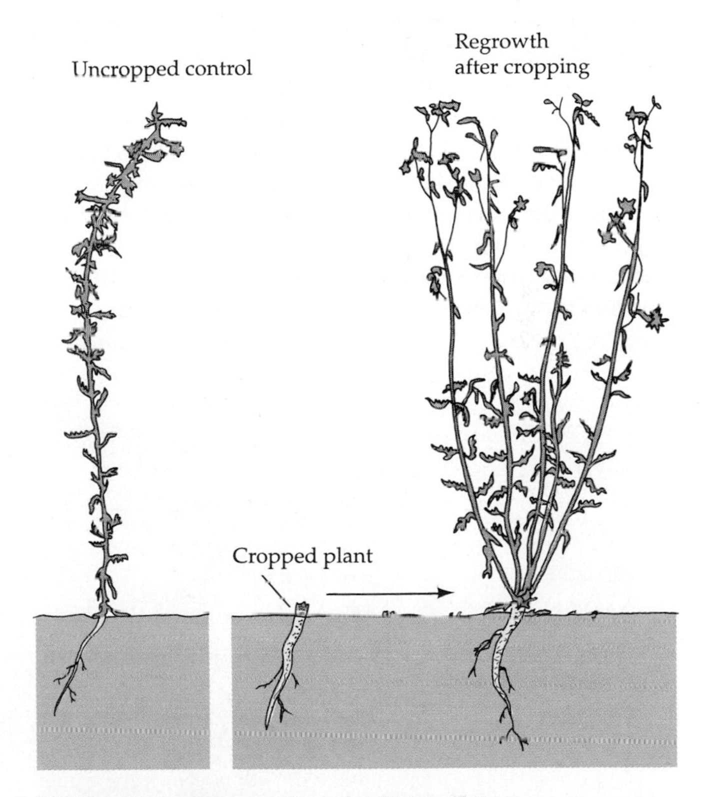

31.12 Overcompensation for Being Eaten
A scarlet gilia was cropped to the point indicated. It then grew four new stems and produced almost three times as many offspring as did uncropped plants like the one on the right.

cropped some scarlet gilia in the laboratory; these plants, too, produced more fruits than did uncropped plants. Not only does the productivity of cropped plants increase, but we may also conclude that grazing by herbivores increases the fitness of scarlet gilia, as the cropped plants pass their genes on to more surviving offspring than do uncropped plants.

Some other plants also profit from moderate herbivory. In addition to the increase in its productivity, a plant benefits by attracting animals that spread its pollen or that eat its fruit and thus distribute its seeds through their feces. However, resisting attack by fungi and herbivorous animals and inhibiting the growth of neighboring plants are also to the advantage of a plant.

Chemical Defenses against Herbivores

Plants attract, resist, and inhibit other organisms often by producing special chemicals known as **secondary products**. (Primary products are substances such as proteins, nucleic acids, carbohydrates, and lipids, that are produced and used by all living things.) Although different kinds of organisms share a biochemical heritage of primary products, they may also differ as radically in chemical content as in external appearance. Animals and fungi, for example, need various enzymes to digest their food, a need not shared by most plants. The plant kingdom is noteworthy for its profusion of secondary products that serve special functions. These compounds help plants compensate for being unable to move. Although a plant cannot flee its herbivorous enemies, it may be able to defend itself chemically.

The effects of defensive secondary products on animals are diverse. Some act upon the nervous systems of herbivorous insects, mollusks, or mammals. Others mimic the natural hormones of animals, causing some insect larvae to fail to develop into adults. Still others damage the digestive tracts of herbivores. Some secondary products are toxic to fungal pests. We make commercial use of secondary plant products as fungicides, insecticides, and pharmaceuticals.

There are more than 10,000 secondary plant products, ranging in molecular weight from about 70 to more than 400,000 daltons; most, however, are of low molecular weight. Some are produced by only a single species, while others are characteristic of an entire genus or even family. Their roles are diverse, and in most cases unknown. While many secondary products have protective functions, as mentioned already, others are essential as attractants for pollinators and seed dispersers. Table 31.1 gives the major classes of secondary plant products and their roles. A few proteins and amino acids also protect plants against herbivores (Box 31.A). In the next section we will look at a specific example of an insecticidal secondary product.

A Versatile Secondary Product

Some plants produce canavanine, an amino acid that is not found in proteins but that is closely similar to the amino acid arginine, which is found in almost all proteins (Figure 31.13). Canavanine has recently been found to have two important roles in plants that produce it in significant quantity. The first role is as a nitrogen-storing compound in seeds. The second role is a defensive one and is based on the similarity of canavanine to arginine. Many insect larvae that consume canavanine-containing plant tissue are poisoned: The canavanine is mistakenly incorporated into the insect's proteins in some of the places where the DNA has coded for arginine. Canavanine is different enough in structure from arginine that some of the proteins end up with modified tertiary struc-

TABLE 31.1
Secondary Plant Products

CLASS	SOME ROLES
Alkaloids	Affect herbivore nervous systems
Other nitrogen and sulfur compounds	Cause cancers, nerve damage, and pain in herbivores
Phenolics	Taste obnoxious to herbivores; affect herbivore nervous systems; act as fungicides
Quinones	Inhibit growth of competing plants
Terpenes	Act as fungicides and insecticides; attract pollinators
Steroids	Mimic animal hormones; prevent normal development of insect herbivores
Flavonoids	Attract pollinators and animals that disperse seeds

BOX 31.A

A Protein for Defense against Insects

A group of scientists at the ARCO Plant Cell Research Institute and the University of Wisconsin recently studied the abilities of seeds of wild and domesticated common beans (*Phaseolus vulgaris*) to resist attack by two species of bean weevils. The investigation began with the observation that some wild bean seeds show high resistance to the weevils, whereas no cultivated bean seeds show such resistance. The scientists discovered that all weevil-resistant bean seeds contain a specific seed protein, arcelin. This protein has never been found in cultivated bean seeds. Therefore, the scientists hypothesized that arcelin is responsible for the resistance of some seeds to predation by the weevils.

Because other differences between wild and cultivated beans might have been responsible for the resistance, two series of experiments were performed to test the relationship between resistance and the possession of arcelin. In one series, cultivated and wild bean plants were crossed. The progeny seeds of such crosses showed an absolute correlation between the presence of arcelin and resistance to weevils. In the other series of experiments, the scientists worked with "artificial" bean seeds made by removing seed coats of cultivated beans and grinding the remainder of the seeds into flour. Different concentrations of arcelin were added to different batches, and the flour was molded into artificial seeds. Bean weevils were then allowed to attack the artificial seeds. The more arcelin the artificial seeds contained, the more resistant they were to weevils.

The scientists then proceeded to prepare, clone, and sequence a cDNA (see complementary DNA, Chapter 14) for arcelin so that they could compare the structure of arcelin with that of other, possibly related proteins that may confer insect resistance on seeds of beans and other legumes. The goal of this ongoing work is to introduce genes for arcelin or other resistance-conferring proteins into agriculturally important crops such as beans. In preliminary tests, this group also showed that arcelin in cooked beans is not harmful to rats—a first step toward demonstrating whether arcelin is safe in food for humans.

tures and hence reduced biological activities. The defects in protein structure and function lead in turn to developmental abnormalities in the insect.

A few insect larvae are able to eat canavanine-containing plant tissue and still develop normally. How can this be? In these larvae the enzyme that charges the tRNA specific for arginine discriminates accurately between arginine and canavanine. The canavanine they ingest is thus not incorporated into the proteins they form. The corresponding enzyme in the susceptible larvae discriminates much less effectively between those two amino acids, so canavanine is frequently substituted for arginine.

As we have seen, some plants have potent weapons for their struggle with herbivores. Can plants also deal with smaller invaders—the pathogenic fungi, bacteria, and viruses?

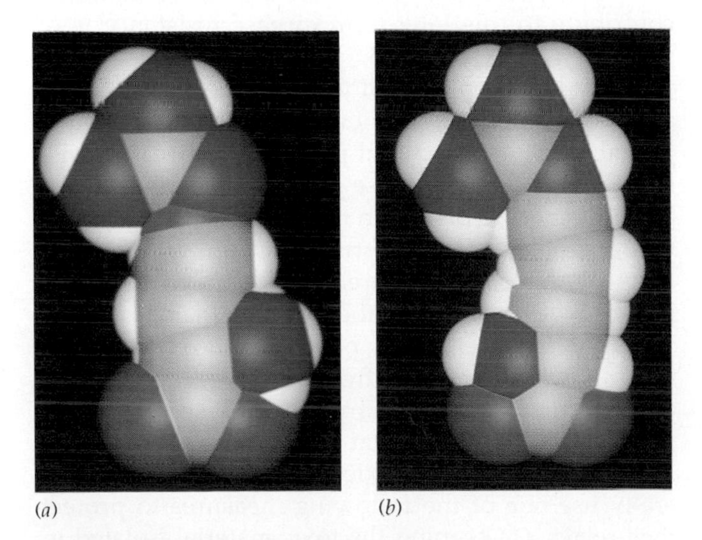

31.13 A Toxic Secondary Product and Its Analog
(a) Computer model of canavanine, a nitrogen-storing compound produced by some plants. *(b)* The amino acid arginine.

PROTECTION AGAINST FUNGI, BACTERIA, AND VIRUSES

Plants resist infection by pathogens by a variety of mechanical and chemical means. The outer surfaces of plants are protected by tissues such as the epidermis or cork, and these tissues are generally covered by cutin, suberin, or waxes. If pathogens pass these barriers, differences between the defense systems of plants and animals (see Chapter 16) become apparent. Animals generally repair tissues that have been infected; they heal, through appropriate developmental pathways. Plants, on the other hand, do not make repairs. Instead, they develop in ways that seal off the damaged tissue so that the rest of the plant

BOX 31.B

Take Two Aspirin and Call Me in the Morning

Aspirin—acetylsalicylic acid—is one of the best-selling drugs in the world and has been for many years. People commonly take aspirin to reduce fever and pain. In our bodies aspirin is hydrolyzed to produce the substance that actually causes the effects, salicylic acid. Since ancient times, people in Asia, Europe, and the Americas have used willow leaves and bark to relieve pain and fever. The active ingredient contained in willow (*Salix*) is salicylic acid, and it now appears that all plants contain at least some salicylic acid. This compound appears to play a number of roles in the plants themselves—notably a role in disease resistance.

The acquired resistance that sometimes follows the hypersensitive reaction is accompanied by the synthesis of pathogenesis-related proteins (PR proteins). Treatment of plants with salicylic acid or with aspirin leads to the production of PR proteins and to a resistance to pathogens. Salicylic acid treatment provides substantial protection against tobacco mosaic virus (a well-studied plant pathogen) and some other viruses.

Scientists strongly suspect that salicylic acid serves as a signal for disease resistance—that microbial infection in one part of a plant leads to the export of salicylic acid to other parts of the plant, where it causes the production of PR proteins before the infection can spread. The PR proteins would, according to this hypothesis, limit the extent of the infection.

Although, of course, the "medicinal" effect of salicylic acid in plants is not the same as that of aspirin in humans, temperature changes are part of the response in both. If we have a fever, aspirin lowers our body temperature. In some plant species salicylic acid or aspirin *causes* a "fever." In skunk cabbage, jack-in-the-pulpit, and other members of the family Araceae, for example, production of salicylic acid leads to a dramatic increase in the temperature of part of the plant, and the higher temperature leads to the release of an odorous compound that attracts insect pollinators. In a few plants salicylic acid has yet another effect: It induces flowering. It is not yet known whether these various effects—protection against disease, induction of elevated temperatures, and induction of flowering—are related or entirely unrelated. In any case, salicylic acid plays key roles in the lives of plants.

---

does not become infected. Trees seal off damaged tissue by producing new wood different in orientation and chemical composition from the previously deposited wood. Some of the new cells also contain substances that resist the growth of microorganisms and hence tend to protect the rest of the plant.

Sealing off damaged tissue is primarily a mechanical mechanism for healing. Many plants have chemical defenses as well. Certain fungi and bacteria, when they infect one of these plants, stimulate the host to produce substances called **phytoalexins**. These substances are not present until the plant is infected, but within hours of the onset of infection they are produced in the infected area and in immediately neighboring cells. Phytoalexins are toxic to many fungi and bacteria. Their antimicrobial activity is nonspecific: Phytoalexins can destroy many species of fungi and bacteria in addition to the one that originally triggered their production. Physical injuries, viral infections, and even certain chemicals can also induce the production of phytoalexins.

Some plants that are resistant to fungal, bacterial, or viral diseases owe this resistance to the **hypersensitive reaction**. In this reaction, cells around the site of microbial infection produce phytoalexins and other chemicals and then die, leaving a necrotic lesion—a "dead spot"—that contains what is left of the microbial invasion. The remainder of the plant remains free of the infecting microbe. The hypersensitive reaction can impart long-lasting resistance to subsequent attacks by pathogens. One of the chemicals that may contribute to this long-term disease resistance may surprise you; see Box 31.B.

There has been a great deal of recent excitement about the discovery of a new tool in the fight against fungal, viral, and bacterial pests: plant **disease resistance genes**. Some of these genes, which afford resistance to specific pathogen strains, have been identified and sequenced. The two most important questions about these genes are, how do they work? and how can we best transfer them to crop plants?

As we have seen, many plants produce toxic chemicals that protect them from herbivores and from pathogenic microbes. Why don't these secondary products kill the plants that produce them?

Plants that produce toxic secondary products generally use one of the following measures to protect themselves: (1) keeping the toxic material isolated in a special compartment, (2) producing the toxic substance only after the plant's cells have been damaged, or (3) using modified enzymes or modified receptors that do not recognize the toxic substance. The first method is the most common.

Plants using the first method store their poison in vacuoles if it is water-soluble. If hydrophobic, the poison is stored in laticifers (see the beginning of this chapter) or in waxes on the epidermal surface. The storage keeps the toxic substance away from mitochondria, chloroplasts, and other parts of the plant's own metabolic machinery.

Some plants store the precursors of toxic substances in one compartment, such as the epidermis, and store the enzymes that convert the precursors to the active poison in another compartment, such as the mesophyll. These plants use the second protective measure of producing the toxic substance only after being damaged. When an herbivore chews part of the plant, cells rupture and the enzymes come in contact with the precursors, releasing the toxic product that repels the herbivore. The only part of the plant that is damaged by the toxic material is that which was already damaged by the herbivore. Plants that respond to attack by producing cyanide—a potent inhibitor of cellular respiration in all organisms that respire—are among those using this protective measure.

The third protective measure is used by the canavanine-producing plants described earlier. These plants produce a tRNA-charging enzyme for arginine that does not bind canavanine. Some herbivores can evade being poisoned by canavanine because their enzymes, like that of the canavanine-producing plants, do not use canavanine by mistake.

Not all plants use protective chemicals to defend themselves against herbivores or pathogens. Should we encourage such defenses in the plants that we cultivate? That is, should we be breeding crop plants that make their own pesticides?

A TRADEOFF: PROTECTION OR TASTE?

A plant with sturdy chemical defenses may taste bad, make us sick, or even kill us. Not surprisingly, we have bred our food plants to minimize toxicity and obnoxious tastes—that is, to make the plants contain very little in the way of chemical defenses. As a result, we must take steps to save our relatively defenseless crops. A current goal of agricultural biotechnology is to develop crop plants that produce their own useful pesticides that are not harmful or offensive to us.

SUMMARY of Main Ideas about Environmental Challenges to Plants

Xerophytes are plants adapted to environments where water is scarce.

Review Figures 31.2, 31.3, 31.4, and 31.5

Some plants are adapted to environments that are waterlogged and provide little oxygen.

Review Figures 31.6 and 31.7

Halophytes are plants adapted to saline environments.

Review Figure 31.8

Some plants are resistant to the effects of normally toxic materials such as heavy metals.

Plants produce thousands of secondary products, some of which defend the plants against herbivores.

Review Table 31.1

SELF-QUIZ

1. Which statement about latex is *not* true?
   a. It is sometimes contained in laticifers.
   b. It is typically white in color.
   c. It is often toxic to insects.
   d. It is a rubbery solid.
   e. Milkweeds produce it.

2. Which of the following is *not* an adaptation to dry environments?
   a. A less negative osmotic potential in the vacuoles
   b. Hairy leaves
   c. A heavier cuticle over the leaf epidermis
   d. Sunken stomata
   e. Root systems that grow each rainy season and die back when it is dry

3. Some plants adapted to swampy environments meet their need for oxygen for their roots by means of a specialized tissue called
   a. parenchyma.
   b. aerenchyma.
   c. collenchyma.
   d. sclerenchyma.
   e. chlorenchyma.

4. Halophytes
   a. all accumulate proline in their vacuoles.
   b. have less negative osmotic potentials than other plants.
   c. are often succulent.
   d. have low root-to-shoot ratios.
   e. rarely accumulate sodium.

5. Which of the following is *not* a commonly toxic heavy metal?
   a. Copper
   b. Lead
   c. Nickel
   d. Potassium
   e. Zinc
6. Plants tolerant to heavy metals commonly
   a. grow poorly where the soil contains heavy metals.
   b. do not take up the heavy metal ions.
   c. are tolerant to all heavy metals.
   d. are slow to colonize an area rich in heavy metals.
   e. weigh more than plants that are sensitive to heavy metals.
7. Herbivory
   a. is predation by plants on animals.
   b. always reduces plant growth.
   c. usually increases the rate of photosynthesis in the remaining leaves.
   d. reduces the rate of transport of photosynthetic products from the remaining leaves.
   e. always is lethal to the grazed plant.
8. Which statement is *not* true of secondary plant products?
   a. Some attract pollinators.
   b. Some are poisonous to herbivores.
   c. Most are proteins or nucleic acids.
   d. Most are stored in vacuoles.
   e. Some mimic the hormones of animals.
9. Which of the following is *not* a common defense against bacteria, fungi, and viruses?
   a. New wood formation
   b. Phytoalexins
   c. A waxy covering
   d. The hypersensitive reaction
   e. Mycorrhizae
10. Plants sometimes protect themselves from their own toxic secondary products by
    a. producing special enzymes that destroy the toxic substances.
    b. storing precursors of the toxic substances in one compartment and the enzymes that convert precursors to toxic products in another compartment.
    c. storing the toxic substances in mitochondria or chloroplasts.
    d. distributing the toxic substances to all cells of the plant.
    e. performing crassulacean acid metabolism.

FOR STUDY

1. How might plant adaptations affect the evolution of herbivores? How might adaptations of herbivores affect plant evolution?
2. The stomata of the common oleander (*Nerium oleander*) are sunk in crypts in its leaves. Whether or not you know what an oleander is, you should be able to describe an important feature of its natural habitat.
3. Explain thoroughly why halophytes often use the same mechanisms for coping with their challenging environment as xerophytes do for theirs.
4. In ancient times, people used less sophisticated methods for mining than we use today. Thus ancient mines often yield substantial profits to modern-day miners who find and work them. Based on material in this chapter, how might you try to find an ancient mine site?
5. We mentioned the possibility of designing crop plants that produce their own pesticides. In Chapter 33 you will read about designing crop plants capable of detoxifying weed-killers, so that crops grow after farmers have destroyed competing vegetation. Discuss the likely usefulness and possible drawbacks of such applications of recombinant DNA technology.

READINGS

Barrett, S. C. H. 1987. "Mimicry in Plants." *Scientific American*, September. A discussion of how some plants use camouflage to avoid predation, and some weeds survive by mimicking crops so that humans will select them.

Dussourd, D. E. and T. Eisner. 1987. "Vein-Cutting Behavior: Insect Counterploy to the Latex Defense of Plants." *Science*, vol. 237, pages 898–901. Describes the phenomenon discussed at the beginning of this chapter. Clear, readable account of field observations and sound experimentation.

Goulding, M. 1993. "Flooded Forests of the Amazon." *Scientific American*, March. Describes the adaptations of plants and other organisms in a seriously threatened environment.

Lewin, R. 1987. "On the Benefits of Being Eaten." *Science*, vol. 236, pages 519–520. Describes the work on scarlet gilia cited (under "Grazing and Plant Productivity") in this chapter. Discusses other examples of cropping.

Rosenthal, G. A. 1986. "The Chemical Defenses of Higher Plants." *Scientific American*, January. Plants employ many chemicals that repel or poison herbivores or that retard the growth of herbivorous insects; some herbivores use these plant-derived compounds for their own defense.

Shigo, A. L. 1985. "Compartmentalization of Decay in Trees." *Scientific American*, April. A description of how trees defend themselves by sealing off the damage done to them.

Taiz, L. and E. Zeiger. 1991. *Plant Physiology*. Benjamin/Cummings, Redwood City, CA. A good general textbook. Chapters 13 and 14 focus on the topics of this chapter.

Pitcher plants live in wet, boggy regions with acidic soil. Their pitcher-shaped leaves collect small amounts of rainwater. Insects are attracted into these pitchers either by bright colors or by scent. Once an insect enters the pitcher, the stiff, downward-pointing hairs that line the plant prevent it from ever leaving. The insect eventually dies and is digested by a combination of enzymes and bacteria in the water; the plant uses the nutrients—especially nitrogen—from the insect protein to thrive in an environment where it is very difficult to obtain enough nitrogen from the soil.

Why do plants need nitrogen? The answer is simple if we recall the chemical structures of amino acids—and hence proteins—and nucleic acids. These vital components of all living things contain nitrogen, as do chlorophyll and many other important biochemical compounds. If a plant cannot get enough nitrogen, it cannot synthesize these compounds at a rate adequate to keep itself healthy.

Nitrogen deficiency is the most common mineral deficiency of plants; the visible symptoms of nitrogen deficiency include uniform yellowing, or chlorosis, of leaves because chlorophyll, which is responsible for the green color of leaves, contains nitrogen. Thus without nitrogen there is no chlorophyll, and without chlorophyll the leaves turn yellow.

ACQUIRING NUTRIENTS

Every living thing needs raw materials from its environment. These **nutrients** include the ingredients of macromolecules: carbon, hydrogen, oxygen, and nitrogen. Carbon and oxygen enter the living world through photosynthesis carried out by plants and by some bacteria and protists; these organisms obtain carbon and oxygen from atmospheric carbon dioxide. The principal source of hydrogen is water, usually taken up from the soil solution by plants. For hydrogen, too, photosynthesis is the gateway to the living world. Nitrogen, which constitutes about four-fifths of the atmosphere, exists as the virtually inert gas N_2 (dinitrogen). A large amount of energy is required to break the triple covalent bond linking the two nitrogen atoms in a molecule of nitrogen gas and to obtain a reasonably reactive form from which amino acids and other nitrogen-containing organic compounds may be synthesized. Movement of dinitrogen into organisms begins with processing by some highly specialized bacteria in the soil. The bacteria fix (convert into a form usable by other organisms) and oxidize dinitrogen, yielding materials that can be taken up by plants. The plants, in turn, provide organic nitrogen and carbon to animals, fungi, and many microorganisms.

A Meat-Eating Plant

32 Plant Nutrition

The proteins of organisms contain sulfur, and their nucleic acids contain phosphorus. There is magnesium in chlorophyll, and iron in many important compounds, such as the cytochromes. Within the soil, minerals dissolve in water, forming a solution that contacts the roots of plants. Plants take up most of these **mineral nutrients** from the soil solution in ionic form.

Autotrophs and Heterotrophs

The plant kingdom provides carbon, oxygen, hydrogen, and nitrogen to the rest of the living world. Plants and some protists and bacteria are autotrophs; that is, they make their own organic food from simple *inorganic* nutrients—carbon dioxide, water, nitrate or ammonium ions containing nitrogen, and a few soluble minerals (Figure 32.1). Organisms that require at least one of their raw materials in the form of *organic* compounds are called heterotrophs; herbivores depend directly and carnivores depend indirectly on autotrophs as their source of nutrition.

Most autotrophs are **photosynthetic;** that is, they use light as the source of energy for synthesizing organic compounds from inorganic raw materials. Some autotrophs, however, are **chemosynthetic,** deriving their energy not from light but from reduced inorganic substances such as hydrogen sulfide (H_2S) in their environment. All chemosynthesizers are bacteria. The activities of chemosynthetic bacteria that fix nitrogen are vital to the nutrition of plants. But how does a plant get to the bacterial products—or to its nutrients in general?

32.1 What Do Plants Need?
Plants require only light plus carbon dioxide, water, nitrate or ammonium ions, and several essential minerals. These parsley plants are growing on nothing more than a solution containing these ingredients. This is hydroponic culture.

How Does a Sessile Organism Find Nutrients?

An organism that is sessile (stationary) must exploit energy that is somehow brought to it. Most sessile animals depend primarily upon the movement of water to bring energy in the form of food to them, but a plant's supply of energy arrives at the speed of light! A plant's supply of essential materials, however, is strictly local, and the plant may deplete its local environment of water and minerals as it develops. How does a plant cope with such a problem? One answer is to extend itself by growing into new resources—growth is a plant's version of locomotion. Roots mine the soil. They grow to reach new sources of minerals and become more elaborate to obtain more water. Growth of leaves helps a plant secure light and carbon dioxide. A plant may compete with other plants for light by outgrowing them, both capturing more light for itself and also preventing the growth of its neighbors by shading them.

As it grows, the plant, or even a single root, must deal with environmental diversity. Animal droppings give high local concentrations of nitrogen. A particle of calcium carbonate in the soil may make a tiny area alkaline, while dead organic matter may make a nearby area acidic. Does the root take up whatever materials it encounters?

Bulk Ingestion versus Selective Uptake of Nutrients

Animals ingest their meals, taking in unneeded and sometimes toxic materials along with needed nutrients; what an animal ingests is not determined by its needs. Part of what animals ingest must eventually be disposed of as waste products, such as urea. Plants, by contrast, do not urinate or produce wastes in other obvious ways. Instead, they control their uptake of most substances, matching the uptake rates to their biochemical needs. The major waste products released to the environment by plants are carbon dioxide or, during active photosynthesis, oxygen gas.

TABLE 32.1
Elements Required by Plants

ELEMENT	SOURCE	ABSORBED FORM	MAJOR FUNCTIONS
Nonmineral elements			
Carbon (C)	Atmosphere	CO_2	In all organic molecules
Oxygen (O)	Atmosphere	CO_2	In most organic molecules
Hydrogen (H)	Soil	H_2O	In most organic molecules
Nitrogen (N)	Soil	NH_4^+ and NO_3^-	In proteins, nucleic acids, etc.
Mineral nutrients			
Macronutrients			
Phosphorus (P)	Soil	$H_2PO_4^-$	In nucleic acids, ATP, phospholipids, etc.
Potassium (K)	Soil	K^+	Enzyme activation; water balance; ion balance
Sulfur (S)	Soil	SO_4^{2-}	In proteins, coenzymes
Calcium (Ca)	Soil	Ca^{2+}	Affects the cytoskeleton, membranes, and many enzymes; second messenger
Magnesium (Mg)	Soil	Mg^{2+}	In chlorophyll; required by many enzymes; stabilizes ribosomes
Micronutrients			
Iron (Fe)	Soil	Fe^{3+}	In active site of many redox enzymes and electron carriers; needed for chlorophyll synthesis
Chlorine (Cl)	Soil	Cl^-	Photosynthesis; ion balance
Manganese (Mn)	Soil	Mn^{2+}	Activates many enzymes
Boron (B)	Soil	$H_2BO_3^-$, HBO_3^{2-}	May be needed for carbohydrate transport (poorly understood)
Zinc (Zn)	Soil	Zn^{2+}	Enzyme activation; auxin synthesis
Copper (Cu)	Soil	Cu^{2+}	In active site of many redox enzymes and electron carriers
Molybdenum (Mo)	Soil	MoO_4^{3-}	Nitrogen fixation; nitrate reduction

This is not to say that plants do not take up toxic substances from the soil. They do, at times, as discussed in Chapter 31. However, plants exert a more systematic control over what can enter their bodies than do animals. What important minerals do get admitted, and what are their roles?

WHICH NUTRIENTS ARE ESSENTIAL?

Table 32.1 lists the mineral elements essential for plants. They all come from the soil solution and derive ultimately from rock. The criteria for calling something an **essential element** are the following: (1) The element must be necessary for normal growth and reproduction. (2) The element cannot be replaceable by another element. (3) The requirement must be direct—that is, not the result of an indirect effect, such as the need to relieve toxicity caused by another substance.

Before a plant that is deficient in an essential element dies, it usually displays characteristic **deficiency symptoms.** Table 32.2 describes the symptoms of some common mineral deficiencies, and the opening figure in this chapter shows an example. Such symptoms help horticulturists diagnose nutrient deficiencies in plants.

Several essential elements fulfill multiple roles—some structural, others catalytic. Magnesium, as we have mentioned, is a constituent of the chlorophyll molecule and hence is essential to photosynthesis. It is also required as a cofactor by numerous enzymes in cellular respiration and other metabolic pathways. Iron is a constituent of many molecules, including some proteins, that participate in oxidation–reduction reactions. Phosphorus, usually in phosphate groups, is found in many compounds, particularly in pathways of energy metabolism such as photosynthesis and glycolysis. The transfer of phosphate groups is important in many energy-storing and energy-releasing reactions, notably those that use or produce ATP. Other roles of phosphate groups include the activation and inactivation of enzymes.

Plant tissues contain high concentrations of potassium, which plays a major role in maintaining electric neutrality of cells. Potassium ions (K^+) balance the negative charges of ionized carboxyl groups ($—COO^-$) of organic acids. Potassium also helps move water from cell to cell. There are no "pumps" for the active transport of water, yet water must be moved from place to place—for example, into and out of the guard cells surrounding stomata (see Chapter 30). Plants and animals achieve water movement by actively transporting K^+ from one cell to

TABLE 32.2
Some Mineral Deficiencies

DEFICIENCY	SYMPTOMS
Calcium	Growing points die back; young leaves are yellow and crinkly
Iron	Young leaves are white or yellow with green veins
Magnesium	Older leaves have yellow in stripes between veins
Manganese	Younger leaves are pale with stripes of dead patches
Nitrogen	Oldest leaves turn yellow and die prematurely; plant is stunted
Phosphorus	Plant is dark green to purple and stunted
Potassium	Older leaves have dead edges
Sulfur	Young leaves are yellow to white with yellow veins
Zinc	Older leaves have many dead spots

another. Chloride ions (Cl^-) follow the K^+ passively, maintaining electrical balance. Movement of these ions changes the water potential of the cells, and water then moves passively to maintain osmotic balance.

Calcium plays many roles in plants. Its function in the processing of hormonal and environmental cues is the subject of great current interest (the analogous function of Ca^{2+} in animal cells is discussed in Chapter 36). Calcium also affects membranes and cytoskeleton activity, participates in spindle formation for mitosis and meiosis, and is a constituent of the middle lamella of cell walls.

The essential minerals in Table 32.1 are divided into two categories: the macronutrients and the micronutrients. Plant tissues need **macronutrients** in concentrations of at least 1 milligram per gram of their dry matter, and they need **micronutrients** in concentrations of less than 100 micrograms per gram of their dry matter. (Dry matter, or dry weight, is what remains after all the water has been removed from a tissue sample.) The essential elements differ in that some, such as nitrogen, may move around within the plant, while others, such as iron, are not redistributed. *All* these elements are essential to the life of all plants. Other elements may be essential to some plants—and perhaps to all.

Plant physiologists identified most of the essential elements by the technique outlined in Figure 32.2. An element is considered essential if a plant does not grow, flower, or produce viable seed when deprived of that element.The technique is limited by the possibility that some elements thought to be absent from the solutions are present. Some of the chemicals used in early experiments were so impure that they provided micronutrients that the first investigators thought they had excluded. Some minerals are required in such tiny amounts that there may be enough in a seed to feed the embryo and the resultant plant throughout its entire lifetime and leave enough in the next seed to get the next generation well started. There was enough chloride on dust particles and water droplets in the air in Berkeley, California (where some of the work on essential elements was performed), for example, to provide the infinitesimal amounts needed to keep experimental plants growing. The essentiality of chlorine thus was not established until 1954, after special air filters had been installed in the laboratory. Simply touching a plant may give it a significant dose of chlorine in the form of chloride ions from sweat.

Only rarely are new essential elements reported now; either the list is virtually complete, or more

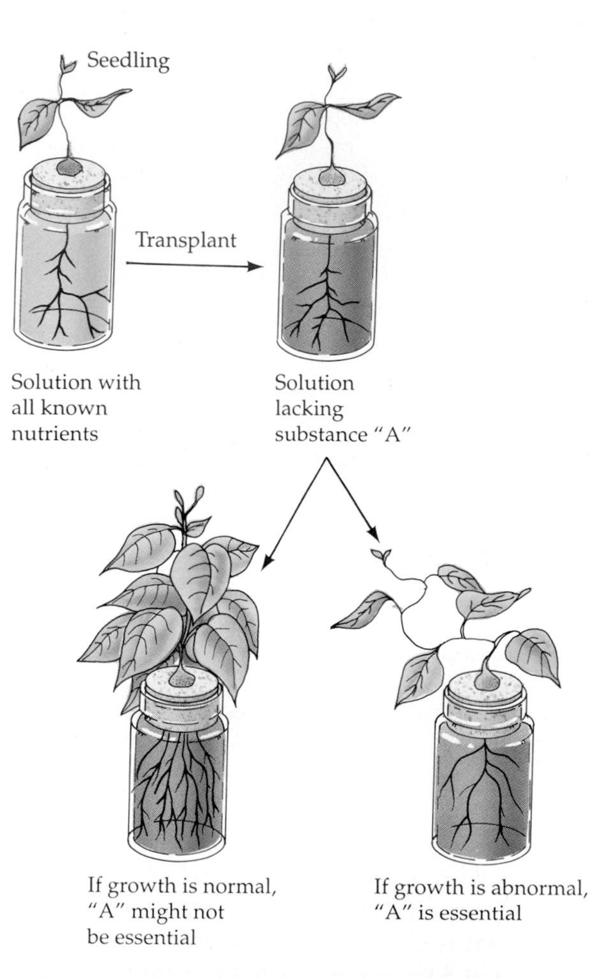

32.2 Identifying Essential Plant Nutrients
The plant physiologist transplants a seedling to a solution lacking only one of the ingredients thought to be essential for growth (substance "A" in this example). If the plant grows and reproduces normally after being transplanted, the missing ingredient is assumed to be nonessential. The experimental environment must be rigorously controlled because some essential nutrients are needed in only tiny amounts that may be present as contaminants of other materials or on objects.

likely, we will need more sophisticated techniques to find others. A new essential element likely to be announced soon—the first since chlorine—is nickel, which was shown in 1984 to be essential for legumes. Some minerals are essential for certain plants but apparently not for others. Continued research may show that only legumes need nickel or that all plants do.

Where does the plant find these essential minerals? How does the plant get the environment to yield the minerals to it?

TABLE 32.3
Soil Particles

SOIL TYPE	PARTICLE SIZE (mm)
Coarse sand	0.2–2.0
Fine sand	0.02–0.2
Silt	0.002–0.02
Clay	<0.002

SOILS

Soils are of great importance to plants, and plant interactions with the soil are complex. Plants obtain their mineral nutrients from the soil or the water in which they grow. Water for terrestrial plants also comes from the soil, as does the supply of oxygen for the roots; soil also provides mechanical support for plants on land. Soil harbors bacteria that perform chemical reactions leading to products required for plant growth; on the other hand, soil may also contain organisms harmful to plants.

Soils have living and nonliving components. The living components include plant roots, as well as populations of bacteria and fungi (Figure 32.3). The nonliving portion of the soil includes rock fragments ranging in size from large boulders to particles 2 μm and less in diameter (Table 32.3). As we will see in the next section, the **clay particles** play a special role in plant nutrition. Soils also contain minerals, water, gases, and organic matter from animals, plants, fungi, and bacteria. Soils change constantly because of both nonhuman natural causes—such as rain, high and low temperatures, and the activities of plants and animals—and human activities, farming in particular. Soils from different parts of the world differ dramatically in their chemical composition and physical structure because the temperature, water supply, and other factors during their formation differ from place to place.

The structure of any soil changes with depth, revealing a soil profile. Although soils differ greatly, virtually all soils consist of two or more **horizons**—recognizable horizontal layers—lying on top of one another (Figure 32.4). Mineral nutrients tend to be leached—that is, dissolved in rain or irrigation water and carried to deeper horizons. Other processes also move materials down—or up—in the soil. Soil sci-

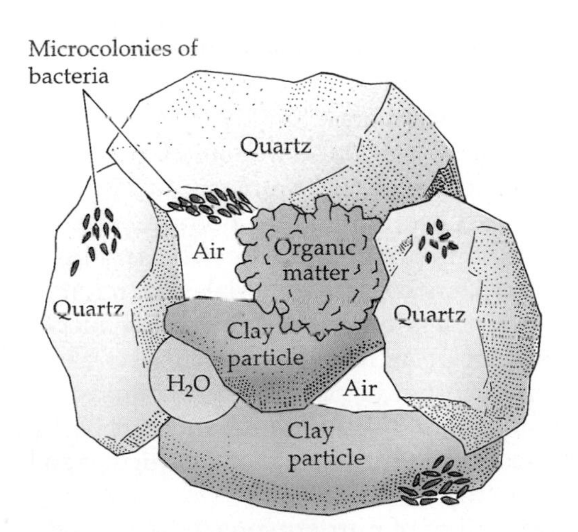

32.3 The Complexity of Soil
Soil consists of more than inorganic particles such as clay and quartz. It contains living organisms such as the bacteria shown here. Air and water are present in pores in soil crumbs like this one.

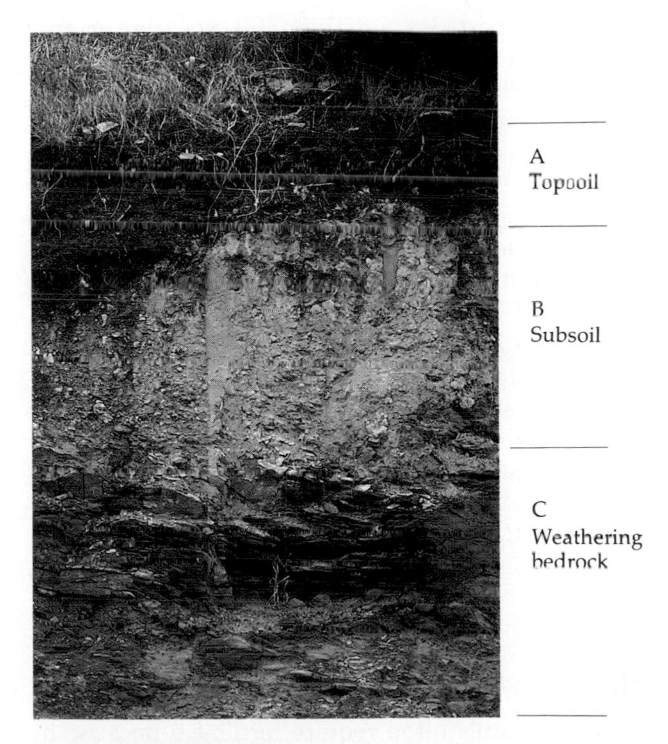

32.4 A Soil's Profile
The A, B, and C horizons can sometimes be seen in road cuts such as this one in Australia. The upper layers developed from the bedrock. The dark upper layer is home to most of the living organisms in the soil.

entists recognize three major zones in the profile of a typical soil. The A horizon is the zone from which minerals have been depleted by leaching. Most of the organic matter in the soil is in the A horizon, as are most roots, earthworms, soil insects, nematodes, and soil protists. Successful agriculture depends on the presence of a suitable A horizon. The B horizon is the zone of infiltration and accumulation of materials leached from above, and the C horizon is the original parent material from which the soil is derived. Some deep-growing roots extend into the B horizon, but roots rarely enter the C horizon.

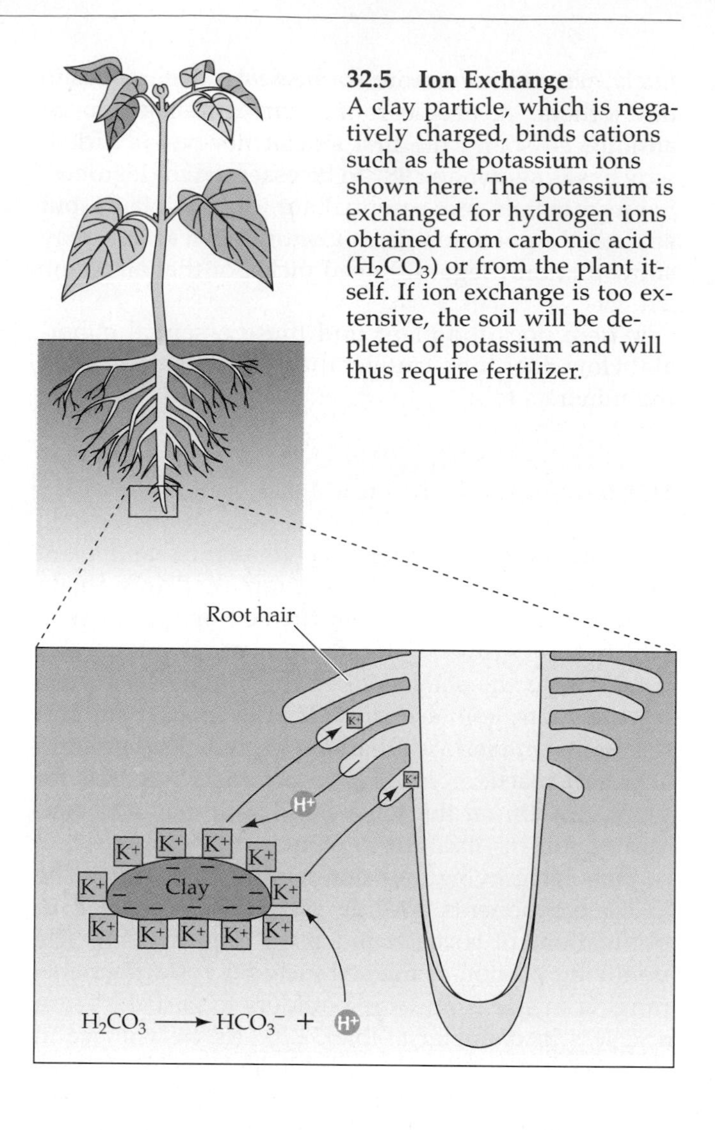

32.5 Ion Exchange A clay particle, which is negatively charged, binds cations such as the potassium ions shown here. The potassium is exchanged for hydrogen ions obtained from carbonic acid (H_2CO_3) or from the plant itself. If ion exchange is too extensive, the soil will be depleted of potassium and will thus require fertilizer.

Soils and Plant Nutrition

The supply of minerals to plants depends upon the presence of clay particles, which have a net negative charge. Many of the minerals that are important for plant nutrition, such as potassium, magnesium, and calcium, exist in soil as positive ions chemically attached to clay particles. To become available to plants, the positive ions must be detached from the clay particles, and this is accomplished by reactions with protons (hydrogen ions, H^+), which are released into the soil by roots or by the ionization of carbonic acid (H_2CO_3). (Carbonic acid is almost universally present in soils because it forms whenever CO_2 from respiring roots or from the atmosphere dissolves in water, according to the reaction $CO_2 + H_2O \rightarrow H_2CO_3$.)

The clay particles get their net negative charge from negatively charged ions that are permanently attached to them. Positively charged ions in solution associate reversibly with these attached negative ions. Protons then trade places with ions such as potassium (K^+) on the clay particles, thus putting the nutrients back into the soil solution. This trading of places is called **ion exchange** (Figure 32.5). The fertility of a soil is determined primarily by its ability to provide nutrients such as potassium, magnesium, and calcium in this manner.

Clay particles effectively hold and exchange positively charged ions, but there is no comparable exchanger of negatively charged ions. As a result, important negative ions such as phosphate, nitrate, and sulfate—the primary sources of phosphorus, nitrogen, and sulfur, respectively—leach rapidly from soil, whereas positive ions tend to be retained in the A horizon.

Fertilizers and Lime

Agricultural soils often require fertilizers because irrigation leaches mineral nutrients from the soil and the harvesting of crops removes the nutrients that the plants took up from the soil during growth. Crop yields fall if too much of any element is removed. Minerals may be replaced by organic fertilizers, such as rotted manure, or inorganic fertilizers of various types. The three elements most commonly added to agricultural soils are nitrogen (N), phosphorus (P), and potassium (K). The ratios of these elements vary among fertilizers, which are often characterized by their N–P–K percentages. A 5–10–5 fertilizer, for example, contains 5 percent nitrogen, 10 percent phosphate, and 5 percent potassium. Sulfur, in the form of sulfate, also is often added. Both organic and inorganic fertilizers can provide the necessary minerals. Organic fertilizers contain materials that improve the physical properties of the soil, providing air pockets for gases, root growth, and drainage. Inorganic fertilizers, on the other hand, provide an almost instantaneous supply of soil nutrients and can be formulated to meet the requirements of a particular soil and a particular crop.

The availability of nutrient ions, whether naturally present in the soil or added as fertilizer, is altered by changes in soil pH. Rainfall and the decomposition of organic substances in the soil lower the pH. Such acidification of the soil can be reversed by **liming**—

the application of calcium-containing material (usually calcium carbonate or calcium hydroxide). This practice is older than agricultural history. It is easy to guess how we learned the use of fertilizer; it didn't take much insight to notice improved plant growth around animal feces. Perhaps a similar observation about limestone, or chalk, or oyster shells—all sources of calcium—led to the practice of liming. Whatever its ancient source, the addition of Ca^{2+} allows H^+ to be released from soil particles by ion exchange and leached away, raising the soil pH (see Chapter 2 for a review of pH). Liming also increases the availability of Ca^{2+} to plants, which require it as a macronutrient. Sometimes a soil is not acidic enough; in this case, a farmer can add sulfate ions to decrease the pH. Iron and some other elements are more available at a slightly acid pH.

Spraying leaves with a nutrient solution is another effective way to deliver some essential elements to growing plants. Plants take up more copper, iron, and manganese when these elements are applied as foliar (leaf) sprays rather than as soil fertilizer. By adjusting the concentrations of nutrient ions and the pH to optimize uptake and to minimize toxicity and by controlling time of spraying to avoid "burning," one can achieve excellent results. Foliar application of mineral nutrients is increasingly used in wheat production, but fertilizer is still delivered most commonly by way of the soil.

Soil Formation

The type of soil in a given area depends on the rock from which it formed, the climate, the topography (features of the landscape), the organisms living there, and the length of time that soil-forming processes have been acting. Rocks are broken down in part by **mechanical weathering,** the physical breakdown—with no accompanying chemical changes—of materials by wetting, drying, and freezing. The most important parts of soil formation, however, include **chemical weathering,** the chemical alteration of at least some of the materials in the rocks. The key process is the formation of clay. Both the physical and the chemical properties of soils depend on the amount and kind of clay particles they contain. Just grinding up rocks does not produce a clay that swells and shrinks and is chemically active. The rock must be chemically changed as well. The initial step in the chemical weathering of most soil minerals is hydrolysis, as illustrated for feldspar, a common soil mineral, in the following formula:

$$\langle \text{Si, Al, O} \rangle K^+ + H^+OH^- \rightarrow \langle \text{Si, Al, O} \rangle H^+ + K^+OH^-$$

FELDSPAR — WATER — HYDROLYZED FELDSPAR — POTASSIUM HYDROXIDE

Two examples illustrate the diversity of soil-forming processes. In wet tropical regions, where rainfall and temperatures are high and the soils are usually moist, water moves rapidly downward through the soil. Silica and soluble nutrients are quickly leached, leaving insoluble iron and aluminum compounds in the A horizon. These are often oxidized, giving bright reddish colors to the soils. This type of soil formation is known as **laterization.** The resulting soil is very poor in nutrients. In semiarid regions, where water from rainfall evaporates rapidly, there is no net movement of water downward through the soil. Instead, water penetrates for a distance, stops, and then is drawn back up by the roots of plants and by evaporation from the soil surface. Under these conditions the soil remains rich in mineral nutrients, and a hard layer of calcium carbonate often forms in the B horizon. The B horizon may be so hard that it prevents deeper penetration of the soil by plant roots. These soils are very fertile, however, when supplied with additional water and nitrogen. Much of the success of irrigated agriculture depends on the high nutrient content of arid-zone soils.

Effects of Plants on Soils

How soil forms in a particular place also depends on the types of plants growing there. Plant litter is a major source of carbon-rich materials that break down to form **humus**—dark-colored organic material, each particle of which is too small to be recognizable with the naked eye.

Soils rich in exchangeable positive ions of mineral nutrients tend to support plants that extract large quantities of nutrients for incorporation into their tissues. When these tissues die and decompose, they produce a rich, alkaline humus called mull. Plants growing on nutrient-poor soils extract fewer nutrients and form tissues that yield a poor, acidic humus known as mor. The mor produced by conifers is particularly resistant to decay and may accumulate in thick layers on the surface of the soil.

Soils age. Young soils support rapidly increasing amounts of vegetation (biomass) that contribute materials for humus, which thus increases at the same rapid rate; the green and yellow lines in Figure 32.6 plot these increases for a hypothetical example. The increase in biomass comes at the expense of nutrients, which decline rapidly, as exemplified by the drop in calcium carbonate in Figure 32.6. The fraction of clay in the soil gradually increases, decreasing the availability of water for plant growth. The loss of mineral nutrients from the soil also contributes to the long-term decline in biomass.

Biologists were slow to recognize the importance of the long-term changes in soils because nearly all the soils of the northern temperate zone, where most

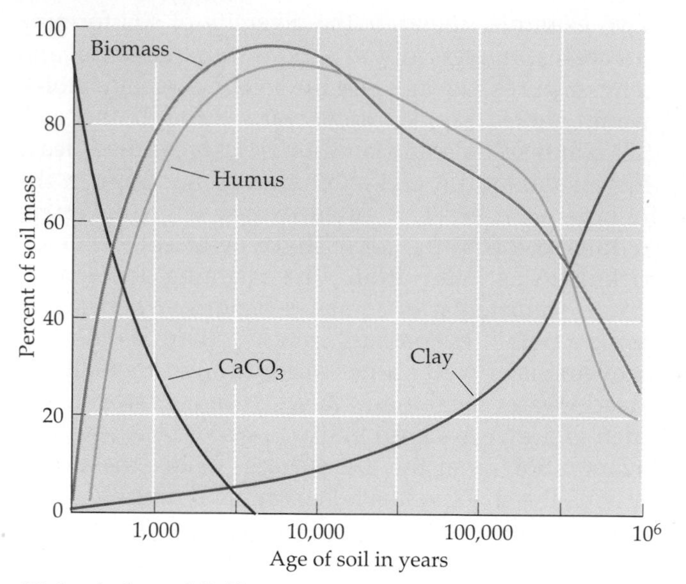

32.6 Aging of Soils
Weathering processes gradually remove most exchangeable nutrient ions from a soil, leaving an infertile residue. The pattern plotted for calcium carbonate ($CaCO_3$) is typical of many soil nutrients. In this hypothetical example, the rich humus and the biomass it supports decrease over time while the fraction of clay in the soil increases.

soil scientists live and work, are very young, dating from the last glacial period only a few thousand years ago. In large areas of the tropics and subtropics, however, especially away from areas of recent mountain formation, soils are ancient and have few remaining nutrients. The right-hand edge of Figure 32.6 suggests the makeup of such soils, which are unsuitable for agriculture unless heavily fertilized.

Nitrogen is the essential element most often required as fertilizer. In the next few sections we will look at how nitrogen is made available and at how certain soil bacteria participate in this process.

NITROGEN FIXATION

Plants cannot use nitrogen gas (dinitrogen: N_2) directly as a nutrient. Making nitrogen from N_2 available to plants takes a great deal of energy because N_2 is a highly unreactive substance. A few species of bacteria can convert it into a more useful form. These prokaryotic organisms, the nitrogen fixers, convert N_2 to ammonia (NH_3). Some of them must live in intimate association with specific eukaryotes before they develop functional nitrogen-fixing machinery. There are relatively few kinds of nitrogen fixers, and what few there are have a small biomass relative to the mass of other organisms on Earth. Without the nitrogen fixers, however, other organisms would not survive! This elite group of prokaryotes is just as essential in the biosphere as are the photosynthetic autotrophs.

Organisms fix approximately 90 million tons of atmospheric dinitrogen per year. Tens of millions of tons are fixed industrially, by a method called the Haber process. A smaller amount of nitrogen is fixed in the atmosphere by nonbiological means such as lightning, volcanic eruption, and forest fires; the products thus formed are brought to Earth by rainwater. By far the greatest share of total world nitrogen fixation, however, is that performed biologically by nitrogen-fixing organisms.

Nitrogen-fixing species are widely scattered in the kingdom Eubacteria. One group of microorganisms fixes nitrogen only in close association with the roots of certain seed plants; the best known of these bacteria belong to the genus *Rhizobium*. Some *Rhizobium* species live free in the soil, where they do not fix nitrogen. Others live in nodules on the roots of plants in the legume family, which includes peas, soybeans, alfalfa, and many tropical shrubs and trees (Figure 32.7). These nodule-inhabiting *Rhizobium* species do fix nitrogen. Some cyanobacteria fix nitrogen in association with fungi in lichens or with ferns, cycads, or bryophytes. Finally, the filamentous bacteria called actinomycetes fix nitrogen in association with root nodules on shrub species such as alder. How were the nitrogen-fixing roles of bacteria discovered?

The ancient Chinese, Greeks, and Romans, and probably members of other early civilizations, recognized that plants such as clover, alfalfa, and peas improve the soil in which they are grown. Two Ger-

32.7 Root Nodules
These large, round, tumorlike nodules are developing from a yellow wax bean root—the structure to the right. The nodules house nitrogen-fixing bacteria.

32.8 Equipped to Colonize
A retreating glacier in Alaska left this area of bare rock exposed. *Dryas drummondii*, the low-lying plant seen here in bloom, was one of the first plants to colonize the bare area. It and other rapid colonizers growing here share the characteristic of having nitrogen-fixing root nodules.

man chemists, Hellriegel and Wilfarth, first showed in 1888 that the root nodules on these plants are caused by bacteria and are sites of nitrogen fixation. These particular plant-infecting bacteria all belong to the genus *Rhizobium*, and the various species of *Rhizobium* show a fairly high specificity for the species of legume they nodulate.

In the oceans, various photosynthetic bacteria, including cyanobacteria, fix nitrogen; in fresh water, cyanobacteria are the principal nitrogen fixers. On land, free-living soil bacteria make some contribution to nitrogen fixation, but it is bacteria in the root nodules of plants that produce most of the fixed nitrogen. Unlike various free-living nitrogen fixers that fix what they need for their own uses and release the fixed nitrogen only upon their deaths, bacteria in root nodules release up to 90 percent of the nitrogen they fix to the plant and excrete some amino acids into the soil, making nitrogen immediately available to other organisms. Some farmers alternate their crops, planting clover or alfalfa occasionally to increase the useful nitrogen content of the soil.

Root nodules permit some *non*leguminous plants to be pioneers, to occupy environments having few or no other plants (Figure 32.8; see also Figure 48.25). Shrubs such as alder thrive in mountainous areas, with their roots grasping chunks of rock in the talus (debris) slopes below the cliffs. The western mountain lilac *Ceanothus* grows well in extremely gravelly soils that have little or no organic matter and hence no fixed nitrogen. Eastern sweet gale *Myrica* flourishes on almost pure sand; its growth is dependent on nitrogen from nodules. Pioneer plants, with their bacterial partners, make initially barren habitats available to other plants and to the animals that depend on them.

How does biological nitrogen fixation work?

Chemistry of Nitrogen Fixation

Nitrogen fixation progressively reduces the dinitrogen molecule by adding pairs of hydrogen atoms to cleave the three bonds between the nitrogen atoms:

$$\underset{\text{DINITROGEN}}{N{\equiv}N} \xrightarrow{2H} HN{=}NH \xrightarrow{2H} H_2N{-}NH_2 \xrightarrow{2H} \underset{\text{AMMONIA}}{2\,NH_3}$$

Throughout this series of reactions, the reactants are firmly bound to the surface of a single enzyme called **nitrogenase** (Figure 32.9). The reactions require a strong reducing agent (see Chapter 7) to transfer hydrogen atoms to dinitrogen and the intermediate products, as well as a great deal of energy, which is supplied by ATP. Depending on the species of nitrogen fixer, either respiration or photosynthesis may provide the necessary reducing agent and ATP.

Nitrogenase is extremely sensitive to oxygen—so

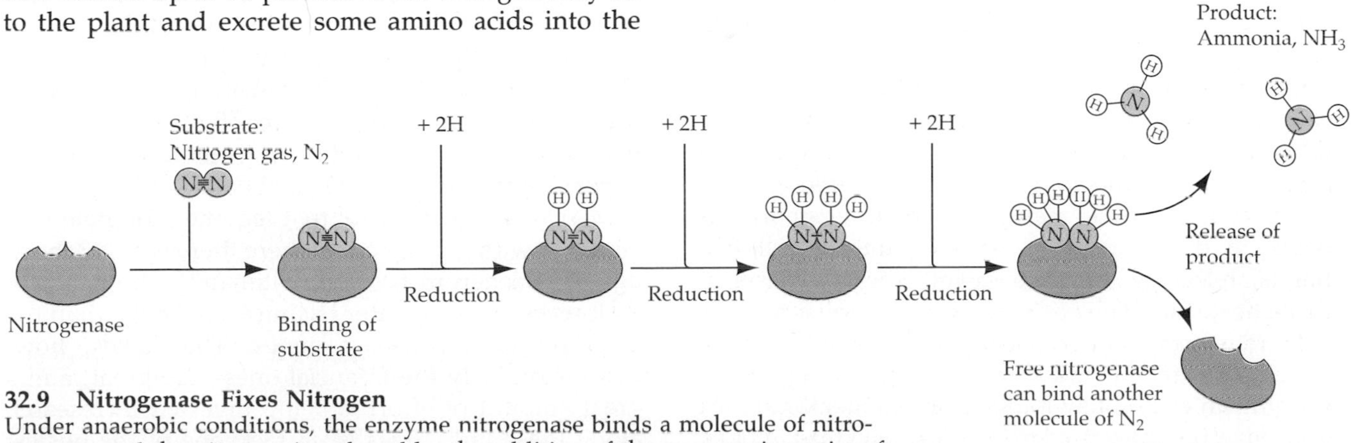

32.9 Nitrogenase Fixes Nitrogen
Under anaerobic conditions, the enzyme nitrogenase binds a molecule of nitrogen gas, and the nitrogen is reduced by the addition of three successive pairs of hydrogen atoms. The final products—two molecules of ammonia—are released, freeing the nitrogenase to bind another dinitrogen molecule.

much so that its discovery was delayed because investigators had not thought to seek it under anaerobic conditions, which are inconvenient to establish in the laboratory. Because nitrogenase cannot function in the presence of oxygen, it is not surprising that many nitrogen fixers are anaerobes. Free-living *Rhizobium* species respire aerobically, but they do not fix nitrogen under these conditions. Legumes respire aerobically, but their *Rhizobium*-containing, nitrogen-fixing root nodules maintain an anaerobic internal environment. Aerobic nitrogen fixers must decrease their internal oxygen levels drastically in order for the process to work. One means for doing this—the production of a special type of hemoglobin—will be described in the next section.

Symbiotic Nitrogen Fixation

The legume nodule provides an excellent example of symbiosis, in which two different organisms live in physical contact and, in association, do things that neither organism can do separately. In the form of symbiosis called mutualism, both organisms benefit from the relationship. Neither free-living *Rhizobium* species nor uninfected legumes can fix nitrogen. Only when the two are closely associated in root nodules does the reaction take place.

The establishment of this symbiosis between *Rhizobium* and a legume requires a complex series of steps with active contributions by both the bacteria and the root. First the root releases chemical signals that attract the *Rhizobium* to the vicinity of the root. The bacteria, in turn, produce substances that cause cell divisions in the root cortex, leading to the formation of a primary nodule meristem. Next comes the infection of the plant by the bacteria. The bacteria first attach to root hairs that project from the epidermal cells and then produce one or more growth substances that cause the cell walls of the root hairs to invaginate—fold inward. With help from the Golgi apparatus in the cells, the invagination proceeds inward through several cells as an infection thread; the bacteria in the thread continue to divide, although slowly (Figure 32.10).

At this stage the bacteria are still outside the plant in a sense, for the thread is lined with cellulose and other cell-wall materials. The thread grows into the cortex tissue of the root until it encounters cells in the primary nodule meristem. The meristem cells begin to divide rapidly, and the infection thread bursts, releasing the bacteria into the cytoplasm of these host cells. The bacteria now undergo a remarkable transformation, increasing about tenfold in size, developing an outside membranous envelope, and forming an elaborately folded internal membrane. At this stage the infecting bacteria are called bacteroids. The bacteroids are, in effect, organelles for nitrogen fixation; it has been suggested that, in the course of further evolution, such bacteroids may in the future give rise to permanent nitrogen-fixing organelles in some plant species.

The final step before the fixation of nitrogen can begin is that the plant produces leghemoglobin, which surrounds the bacteroids. Hemoglobin is an oxygen-carrying pigment that one seldom associates with plants, but some nodules contain enough of its close relative leghemoglobin to be bright pink when viewed in cross section. The leghemoglobin traps oxygen, keeping it away from the oxygen-sensitive bacteroids and nitrogenase.

The Need to Augment Biological Nitrogen Fixation

Bacterial nitrogen fixation does not suffice to support the needs of agriculture. Native Americans used to plant dead fish along with corn so that the decaying fish would release fixed nitrogen that the developing corn could use. Industrial nitrogen fixation is becoming ever more important to world agriculture because of the degradation of soils and the need to feed a rapidly expanding population. Research on biological nitrogen fixation is being vigorously pursued, with commercial applications very much in mind (for an example see Box 32.A).

An alternative to the current process for industrial nitrogen fixation is urgently needed because of the cost of energy and other economic factors associated with it. At present, manufacturing nitrogen-containing fertilizer takes more energy than any other aspect of crop production in the United States. One line of investigation centers on recombinant DNA technology as a means of "teaching" new plants to produce nitrogenase. Workers in many industrial and academic laboratories are working on the insertion of bacterial genes coding for nitrogenase into plasmids, and on the incorporation of such plasmids into the cells of angiosperms, particularly crop plants. However, developing crops that can fix their own nitrogen will take more than just the insertion of genes for nitrogenase; there must also be provisions for excluding free oxygen and for obtaining strong reducing agents and an energy source. Biological nitrogen fixation, like industrial nitrogen fixation, is extremely expensive in terms of energy. Looking to nature for evidence of this, we find that legumes compete successfully with grasses only where there is a real shortage of nitrogen in the soil. Ultimately, the need for ATP represents a greater technical challenge than the insertion of nitrogenase genes. The stakes, however—especially the financial ones—are great, and a great amount of effort is being invested in research along these lines. This is, in fact, one of the busiest areas in the burgeoning field of biotechnology.

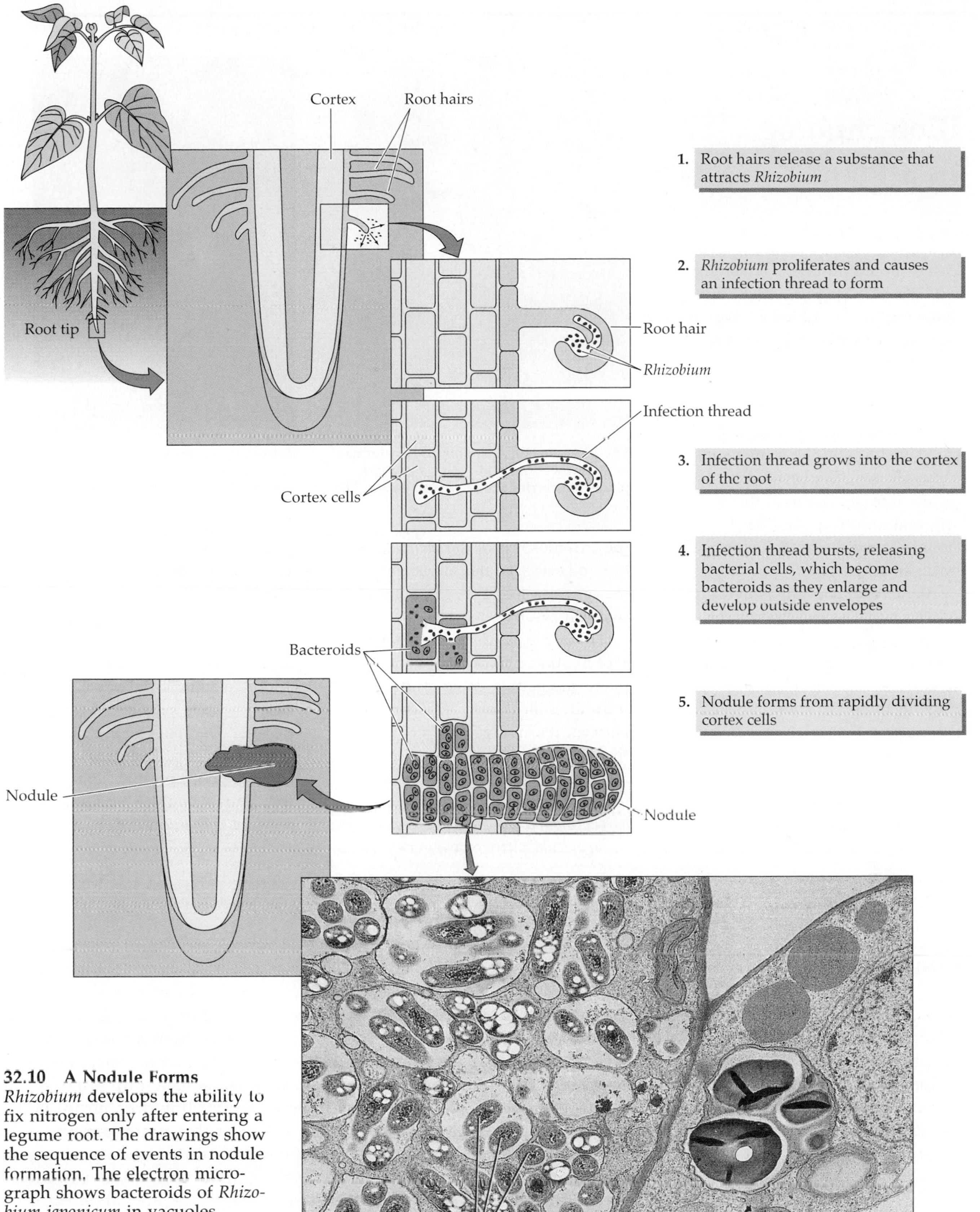

32.10 A Nodule Forms *Rhizobium* develops the ability to fix nitrogen only after entering a legume root. The drawings show the sequence of events in nodule formation. The electron micrograph shows bacteroids of *Rhizobium japonicum* in vacuoles within a soybean root cell. A portion of an uninfected root cell is seen to the right.

BOX 32.A

Biotechnology in a Plant

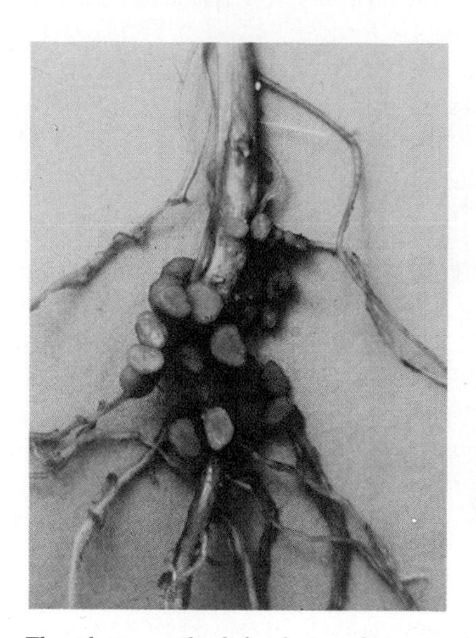

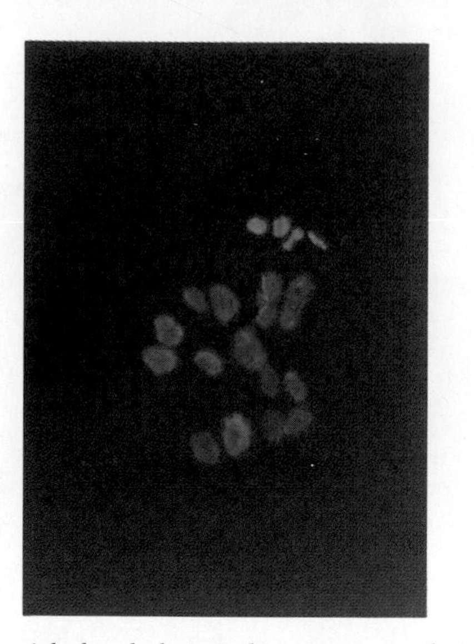

The photo on the left, shot under normal light, shows root nodules of a soybean plant that carries the luciferase genes. The right-hand photo is the same scene, shot in the dark in the presence of decanal fumes.

When should nitrogen-containing fertilizer be added to a crop? If too much fertilizer is added during the growing season, the farmer is wasting money, and leaching of nitrates from the overfertilized soil leads to serious pollution of the groundwater. If too little fertilizer is added, the crop's yield—and the farmer's income—is low. The best judge of a plant's nitrogen status is the plant itself, and one crop plant has been "taught" how to report its nitrogen status to scientists.

Aladar Szalay of the University of Alberta and Thomas Baldwin of Texas A & M University isolated genes from *Vibrio harveyi,* a luminescent marine bacterium. The genes code for the production of luciferase, the enzyme that catalyzes the light-producing reaction. These scientists then inserted the luciferase genes at a key spot—between the gene coding for nitrogenase production and the promoter of the nitrogenase gene—in the chromosome of *Rhizobium japonicum,* a bacterium that participates in the nitrogen-fixing root nodules of soybean plants.

The nitrogenase genes in soybean root nodules are inactive when the plants are getting enough nitrogen from the soil. If more nitrogen is needed, the nitrogenase promoter activates the nitrogenase gene, allowing the nodules to fix their own nitrogen directly. The insertion of *V. harveyi* luciferase genes between the *R. japonicum* nitrogenase gene and its promoter has the following result: When more nitrogen is needed, the nitrogenase promoter activates the neighboring genes, including those that code for luciferase. One can determine when luciferase is present by adding a suitable substrate, in this case a volatile substance known as decanal. Decanal vapor is absorbed directly by cells. If decanal and luciferase are both present, light is emitted. Thus when the modified *R. japonicum* is fixing nitrogen, it glows.

DENITRIFICATION

Because nitrogen fixation decreases the amount of nitrogen gas in the atmosphere, we should mention, if only briefly, the opposite process, called **denitrification.** Some normally aerobic bacteria, mostly species of the genera *Bacillus* and *Pseudomonas,* can use nitrate (NO_3^-) as a terminal electron acceptor in place of oxygen if they are kept under anaerobic conditions:

$$2\,NO_3^- + 10\,e^- + 12\,H^+ \rightarrow N_2 + 6\,H_2O$$

These bacteria are extremely common and, as the equation shows, they return dinitrogen to the atmosphere. Nature's nitrogen cycle is discussed in Chapter 49.

NITRIFICATION

Fixed nitrogen released into the soil by nitrogen fixers is primarily in the form of ammonia (NH_3) and ammonium ions (NH_4^+). Although ammonia is toxic to plants, ammonium ions can be taken up safely at low concentrations. Most plants, however, grow better with nitrate (NO_3^-) than with ammonium ions as a source of nitrogen because of the toxicity of ammonium ions. The form of nitrogen taken up is also affected by soil pH: Nitrate ions are taken up preferentially under more acidic conditions, ammonium ions under more basic ones.

Where do nitrate ions in the soil come from? Ammonia is oxidized to nitrate by the process of **nitri-**

fication. This process is carried out in the soil by bacteria. Bacteria of two genera, *Nitrosomonas* and *Nitrosococcus,* are capable of converting ammonia to nitrite ions (NO_2^-), and *Nitrobacter* bacteria oxidize nitrite to nitrate. These three genera of prokaryotes constitute a critical ecological link, taking the products of other crucial bacteria, the nitrogen fixers, and converting them into a form more available to plants, and hence to the rest of the biosphere.

What do these bacteria get in return for being so ecologically helpful? Actually, they carry on nitrification for their own selfish ends. These three genera are chemosynthetic autotrophs; that is, their chemosynthesis is powered by the energy released by oxidation of ammonia or nitrite. For example, by passing the electrons from nitrite through an electron transport chain, *Nitrobacter* can make ATP, and using some of this ATP, it can also make NADH. With the ATP and NADH, the bacterium can convert carbon dioxide and water to glucose and other foods. In short, the nitrifiers base their entire biochemistry—their entire lives—on the oxidation of ammonia or nitrite ions. *Nitrobacter* can convert 6 molecules of carbon dioxide to 1 molecule of glucose for every 78 nitrite ions that they oxidize—not terribly efficient, but efficient enough to keep *Nitrobacter* living, growing, and reproducing.

NITRATE REDUCTION

We have seen dinitrogen *reduced* to ammonia in nitrogen fixation and ammonia *oxidized* to nitrate in nitrification. We will now see that plants *reduce* the nitrate they take up all the way back to ammonia before using it further to manufacture amino acids (Figure 32.11). The reactions of **nitrate reduction** are carried on by the plant's own enzymes. The later steps, from nitrite to ammonia, take place in the chloroplasts; this conversion is not a part of photosynthesis. The final products of nitrate reduction are amino acids, from which the plant's proteins and all its other nitrogen-containing compounds are formed.

Nitrogen metabolism, in bacteria and in plants, is complex. It is also of great importance. Nitrogen atoms constitute approximately 1 to 5 percent of the dry weight of a leaf, and nitrogen-containing compounds constitute 5 to 30 percent of the plant's total dry weight. The nitrogen content of animals is even higher, and all the nitrogen in the animal world arrives by way of the plant kingdom. As we are about to see, plants also play an important part in delivering sulfur to animals.

SULFUR METABOLISM

All living things require sulfur, which is a constituent of two amino acids, cysteine and methionine, and hence of almost all proteins. Sulfur is also a component of other biologically crucial compounds, such as coenzyme A (see Chapter 7). Animals must obtain their cysteine and methionine from plants, but plants can start with sulfate ions (SO_4^{2-}) obtained from the soil or from a liquid environment. Interestingly, all of the most abundant elements in plants are taken up from the environment in their most oxidized forms—sulfur as sulfate, carbon as carbon dioxide, nitrogen as nitrate, phosphorus as phosphate, and hydrogen as water. In plants, sulfate is reduced and incorporated into cysteine; from this amino acid all the other sulfur-containing compounds in the plant are made. These important processes—sulfate reduction and the utilization of cysteine—are closely analogous to the reduction of nitrate to ammonia and the subsequent utilization of ammonia by plants.

Numerous bacteria base their metabolism on the modification of sulfur-containing ions and compounds in their environment. One group, for example, performs reactions analogous to nitrification. Just as the nitrifiers oxidize ammonia to nitrate, the chemosynthetic sulfur bacteria oxidize hydrogen sulfide

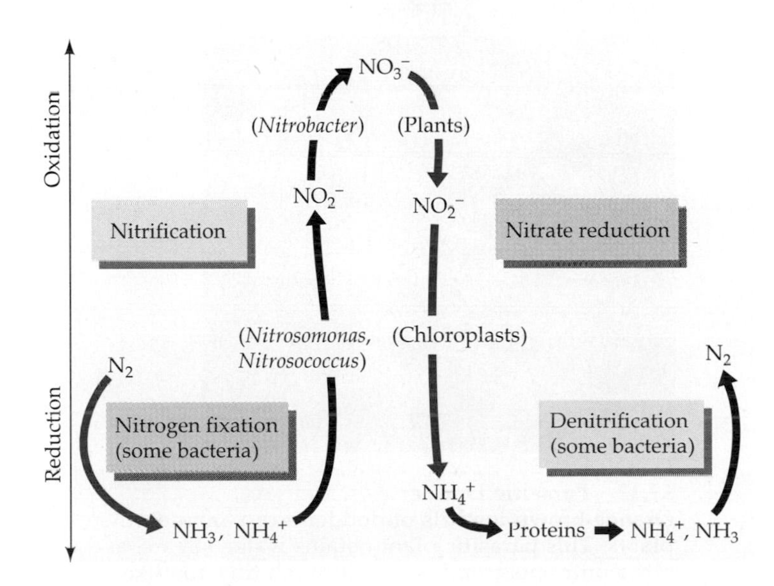

32.11 The Path of Nitrogen
Bacteria fix nitrogen from the atmosphere and conduct nitrification. Plants reduce nitrates back to ammonia, the form in which nitrogen is incorporated into proteins. Finally, some denitrifying bacteria can oxidize ammonia back to nitrogen gas, which returns to the atmosphere. Red arrows identify steps in which nitrogen compounds become reduced.

(H_2S) to sulfate, using the energy thus released to make ATP and fix carbon dioxide.

Thus far in this chapter we have considered the mineral nutrition of plants. As you already know, another crucial aspect of plant nutrition is photosynthesis—the principal source of energy and carbon for plants and for the biosphere as a whole. Not all plants, however, are photosynthetic autotrophs. A few, in the course of their evolution, have lost the ability to feed themselves by photosynthesis. How do these plants get their energy and carbon?

HETEROTROPHIC SEED PLANTS

A few plants are parasites that obtain their food directly from the living bodies of other plants (Figure 32.12). Perhaps the most familiar parasitic plants are the mistletoes and dodders. Mistletoes are green and carry on some photosynthesis, but they parasitize other plants for water and mineral nutrients and may derive photosynthetic products from them as well. Another parasitic plant, the Indian pipe (see Figure 25.1), once was thought to obtain its food from dead organic matter; it is now known to get its nutrients, with the help of fungi, from nearby actively photosynthesizing plants. Hence it too is a parasite.

32.13 A Carnivorous Sundew
A sundew has trapped an insect on its sticky hairs. Secreted enzymes will digest the carcass externally.

32.12 Parasitic Dodder
Orange-brown tendrils of dodder wrap around other plants. This parasitic plant obtains water, sugars, and other nutrients from its host through tiny, rootlike protuberances that penetrate the surface of the host.

Some other heterotrophic plants are the 450 or so carnivorous species—those that augment their nitrogen and phosphorus supply by capturing and digesting flies and other insects. The best-known plant carnivores are Venus's-flytrap (genus *Dionaea*), sundews (genus *Drosera*), and pitcher plants (genus *Sarracenia*). These plants are normally found in boggy regions where the soil is acidic. Most decay-causing organisms require a more neutral pH to break down the bodies of dead organisms, so relatively little available nitrogen is recycled into these acidic soils. Accordingly, the carnivorous plants have adaptations that allow them to augment their supply of nitrogen with animal proteins.

We discussed pitcher plants and how they capture insects at the opening of this chapter. Sundews have leaves covered with hairs that secrete a clear, sticky liquid high in sugar (Figure 32.13). An insect touching one of these hairs becomes stuck, and more hairs curve over the insect and stick to it as well. The plant secretes enzymes to digest the insect and later absorbs the carbon- and nitrogen-containing products of digestion. An insect entering the Venus's-flytrap springs a mechanical trap triggered by three hairs in the center of a partially closed leaf lobe. The two halves of the leaves close, and spiny outgrowths at the margins of the leaves interlock to imprison the insect. Enzymes secreted by the plant digest the trapped insect. None of the carnivorous plants must feed on insects; they grow adequately without insects, but in nature, they grow faster and are a darker green when insects are available to them. The extra supply of nitrogen is used to make more proteins and chlorophyll, as well as other nitrogen-containing compounds.

SUMMARY of Main Ideas about Plant Nutrition

Nearly all plant species are photosynthetic autotrophs.

A few plant species are parasitic heterotrophs that feed on other plants.

Carnivorous plant species are autotrophs that supplement their nitrogen supply by feeding on insects.

The only nutrients required by most plants are carbon dioxide, water, nitrate or ammonium ions, and several mineral salts.
Review Tables 32.1 and 32.2

Clay particles hold and exchange positively charged mineral nutrients, making them available to plants.
Review Figure 32.5

A few species of soil bacteria are responsible for all nitrogen fixation, the conversion of nitrogen gas to usable nitrogen compounds.
Review Figure 32.9

Some nitrogen-fixing bacteria live free in the soil; others, such as some species of *Rhizobium*, live symbiotically within the roots of plants.
Review Figure 32.10

Ammonium ions are the main product of nitrogen fixation in the soil, yet most plants preferentially take up nitrate ions under most conditions.

Nitrate ions are available because other species of soil bacteria carry on nitrification.

The nitrifying bacteria are chemosynthetic autotrophs.

Once in the plant, nitrate is reduced to ammonia, which is incorporated into organic compounds.
Review Figure 32.11

Sulfate ions, produced in the soil by bacteria that oxidize hydrogen sulfide, are reduced within the plant.

SELF-QUIZ

1. Which of the following is *not* an essential mineral element for plants?
a. Potassium
b. Magnesium
c. Calcium
d. Lead
e. Phosphorus

2. Fertilizers
a. are often characterized by their N–P–O percentages.
b. are not required if crops are removed frequently enough.
c. restore needed mineral nutrients to the soil.
d. are needed to provide carbon, hydrogen, and oxygen to plants.
e. are needed to destroy soil pests.

3. Which of the following is *not* an important step in soil formation?
a. The removal of bacteria
b. Mechanical weathering
c. Chemical weathering
d. Clay formation
e. Hydrolysis of soil minerals

4. Laterization
a. results in a very productive soil.
b. often takes place in mine tailings.
c. produces a soil rich in copper, lead, nickel, and zinc.
d. produces a soil rich in chromium and poor in calcium.
e. produces a soil rich in insoluble iron and aluminum compounds.

5. Nitrogen fixation
a. is performed only by plants.
b. is the oxidation of nitrogen gas.
c. is catalyzed by the enzyme nitrogenase.
d. is a single-step chemical reaction.
e. is possible because N_2 is a highly reactive substance.

6. Nitrification
a. is performed only by plants.
b. is the reduction of ammonium ions to nitrite and nitrate ions.
c. is the reduction of nitrate ions to nitrogen gas.
d. is catalyzed by the enzyme nitrogenase.
e. is performed by certain bacteria in the soil.

7. Nitrate reduction
a. is performed by plants.
b. takes place in mitochondria.
c. is catalyzed by the enzyme nitrogenase.
d. includes the reduction of nitrite ions to nitrate ions.
e. is known as the Haber process.

8. Which of the following statements about sulfur is *not* true?
   a. All living things require it.
   b. It is a component of DNA and RNA.
   c. It is a constituent of two amino acids.
   d. Its metabolism is similar to the metabolism of nitrogen.
   e. Many bacteria base their metabolism on reactions of sulfur-containing ions.

9. Which of the following is a parasite?
   a. Venus's-flytrap
   b. Pitcher plant
   c. Sundew
   d. Dodder
   e. Tobacco

10. All heterotrophic seed plants
    a. are parasites.
    b. are carnivores.
    c. are incapable of photosynthesis.
    d. derive their nutrition from animals.
    e. develop from multicellular embryos.

FOR STUDY

1. Methods for determining whether a particular element is essential have been known for over a century. Since the methods are so old and well established, why is it that the essentiality of some elements was discovered only recently?

2. If a Venus's-flytrap were to be deprived of soil sulfates and hence made unable to synthesize the amino acids cysteine and methionine, would it die from lack of proteins? Discuss.

3. Soils are dynamic systems. What changes might result when land is subjected to heavy irrigation for agriculture, after being relatively dry for many years? What changes in the soil might result when a virgin, deciduous forest is replaced by crops that are harvested each year?

4. What is the significance of nitrification to plants? to the bacteria that carry on nitrification?

READINGS

Brill, W. J. 1981. "Agricultural Microbiology." *Scientific American*, September. A look at how microbes can be used to support plant growth.

Epstein, E. 1984. "Rhizostats: Controlling the Ionic Environment of Roots." *BioScience*, November. Computers, roots, and plant nutrition.

Power, J. F. and R. F. Follett. 1987. "Monoculture." *Scientific American*, March. There is an increasing tendency to grow the same crop year after year. This article discusses how this practice affects soils, and whether it is good or bad in general.

Raven, P. H., R. F. Evert and S. Eichhorn. 1991. *Biology of Plants*, 5th Edition. Worth, New York. An excellent general botany textbook.

Taiz, L. and E. Zeiger. 1991. *Plant Physiology*. Benjamin/Cummings, Redwood City, CA. An authoritative textbook with good chapters on mineral nutrition and related topics.

What causes the different parts of a plant to take on distinctive forms? Why do roots and shoots differ in structure? What causes roots to form? Hormones, which are chemical signals, play important roles in determining patterns of plant development. Recall, for example, that plants need not start as seeds; a piece of shoot (a cutting) can form roots and give rise to an intact plant as, in the extreme case, can a small piece of tissue or even a single cell. Gardeners commonly induce cuttings to root. How do they do it? How do they change the course of development in this way?

The gardener, or a biologist, or you, can do what we did to produce this photograph—we treated the cut stems with a hormone called auxin. A tiny amount of this hormone causes some stem cells to develop not into stem structures but, instead, into root apical meristems and thus into new roots. By modifying the amount of auxin applied, we can control the number and lengths of new roots. This simple experiment shows the basis of our discussion in this chapter—that hormones profoundly affect plant development.

What Made the Difference?
The cutting with more roots was treated with auxin, a hormone discussed in this chapter. The other cutting is an untreated control.

33 Regulation of Plant Development

WHAT REGULATES PLANT DEVELOPMENT?

The development of a plant—the progressive changes that take place throughout its life—is regulated in complex ways. There are four major players in regulation: the *environment, hormones,* the pigment *phytochrome,* and the plant's *genome.*

Environmental cues, such as light or temperature changes, trigger some important developmental events, such as flowering and the onset and end of dormancy. Hormones and phytochrome mediate the effects of environmental cues.

Hormones—compounds produced in small quantity in one part of an organism and then moved to other parts where they produce effects—mediate many developmental phenomena, such as stem growth and autumn leaf fall. The plant hormones include **auxin, gibberellins, cytokinins, abscisic acid,** and **ethylene.** Each is produced in one or more specific parts of the plant's body. Each plays multiple regulatory roles, affecting several different aspects of development (Table 33.1).

Like the hormones, phytochrome regulates many processes. Unlike the hormones, it does not travel from one part of the plant to another but acts within the cells that produce it. Light (an environmental cue) acts directly on phytochrome, which in turn regulates developmental processes such as the many changes accompanying the growth of a young plant out of the soil and into the light.

TABLE 33.1
Plant Hormones

		ACTIVITY				
HORMONE	SITE OF PRODUCTION	SEED DORMANCY	SEED GERMINATION	SEEDLING GROWTH	APICAL DOMINANCE	LEAF ABSCISSION
Gibberellins	Embryo, young leaves, root and shoot apices	Breaks	Promotes	Promotes cell division and expansion	—	—
Auxin	Embryo, young leaves, shoot apical meristem	—	—	Promotes cell expansion	Inhibits lateral buds	Inhibits
Cytokinins	Roots	—	—	Promotes cell division	Promotes lateral buds	Inhibits
Ethylene	Ripening fruit, senescing tissue, stem nodes	—	—	—	—	Promotes
Abscisic acid	Root cap, older leaves, stem	Imposes	Inhibits	—	—	—

No matter what cues direct development, ultimately the plant's genome determines how the plant and its parts develop. The genome is the master plan of the plant, but it is interpreted differently depending on the status of the environment. The genome encodes phytochrome and the enzymes that catalyze the formation of the hormones and mediate some of their actions; it is also the target for some hormone actions. For several decades hormones and phytochrome were the focus of most work on plant development, but recent advances in molecular genetics now allow us to focus on underlying processes.

AN OVERVIEW OF DEVELOPMENT

Let's now review the life history of a flowering plant, from seed to death, focusing on how the developmental events are regulated. Keep in mind that as plants develop, the environment, hormones, and phytochrome affect three fundamental processes: cell division, cell expansion, and cell differentiation. Try to envision how the division, expansion, and differentiation of cells contribute to different developmental phenomena.

From Seed to Seedling

Consider a seed. All developmental activity may be suspended in this seed; in other words, it may be **dormant.** Cells in dormant seeds do not divide, expand, or differentiate. Seed dormancy may be broken by one of several physical mechanisms—mechanical abrasion, fire, leaching of inhibitors by water—described later in this chapter. As the seed **germinates** (begins to develop), it first imbibes (takes up) water. The growing embryo must then obtain building blocks—monomers—by digesting the polymeric reserve foods stored either in the cotyledons or in the endosperm. The embryos of some plant species secrete gibberellins that direct the mobilization of the reserves.

If the seed germinates underground, it must elongate rapidly and cope with life in the absence of light. Phytochrome controls this stage and ends it when the shoot reaches the light. The growth of the seedling, both in darkness and light, is also regulated by auxin and gibberellin, and auxin is known to regulate tissue and organ formation. Thus auxin affects cell differentiation. Other information regulating these phenomena comes from cytokinins (Figure 33.1).

Reproductive Development

Eventually the plant flowers. Flowering may be initiated when the plant reaches an appropriate age or size. Some plant species, however, flower at particular times of the year, meaning that the plant must sense the appropriate date. These plants are photoperiodic (see Chapter 34); they measure the length of the night (shorter in the summer, longer in the winter) with great precision. Although we don't know *how* it works, we do know a lot about photoperiodism. We know that the plant measures the length of

WINTER DORMANCY	FLOWERING	FRUIT DEVELOPMENT
Breaks	Stimulates in some plants	Promotes
—	—	Promotes
—	—	—
—	—	Promotes ripening
Imposes	—	—

darkness, that there is a biological clock—itself a mystery, and that light absorbed by phytochrome can affect the time-measuring process.

Once a leaf has determined, by measuring the night length, that it is time for the plant to flower, that information must be transported to the places where flowers will form. How this comes about remains a mystery, but it seems likely that a "flowering hormone"—named florigen, but not yet discovered—travels from the leaf to the point of flower formation. Perhaps information travels in some other form, but this form has not been discovered either.

After flowers have formed, hormones, including auxin and gibberellin, play further roles. Hormones and other substances control the growth of a pollen tube down the style of a pistil. Following fertilization, a fruit develops, controlled in several ways by gibberellin and auxin (Figure 33.2). The ripening of the fruit is also under chemical control, commonly by the gaseous hormone ethylene.

Dormancy, Senescence, and Death

Some perennials have buds that enter a state of winter dormancy during the cold season. (Perennial plants are those that grow for a number of years.) Abscisic acid helps maintain such dormancy. Finally, the plant **senescences** (deteriorates because of aging) and dies. Death, which may be under environmental control, follows senescent changes that are controlled by regulators such as ethylene (Figure 33.3).

The elements of this brief overview will be considered in detail in this chapter and the next. Let's begin with regulation at the start of the life cycle.

SEED DORMANCY AND GERMINATION

Some seeds are, in effect, instant plants, because all they need for germination is water. Many other species have seeds whose germination is regulated in more complex ways because they are initially dor-

33.1 From Seed to Seedling
Environmental factors, hormones, and phytochrome regulate the first stages of plant growth.

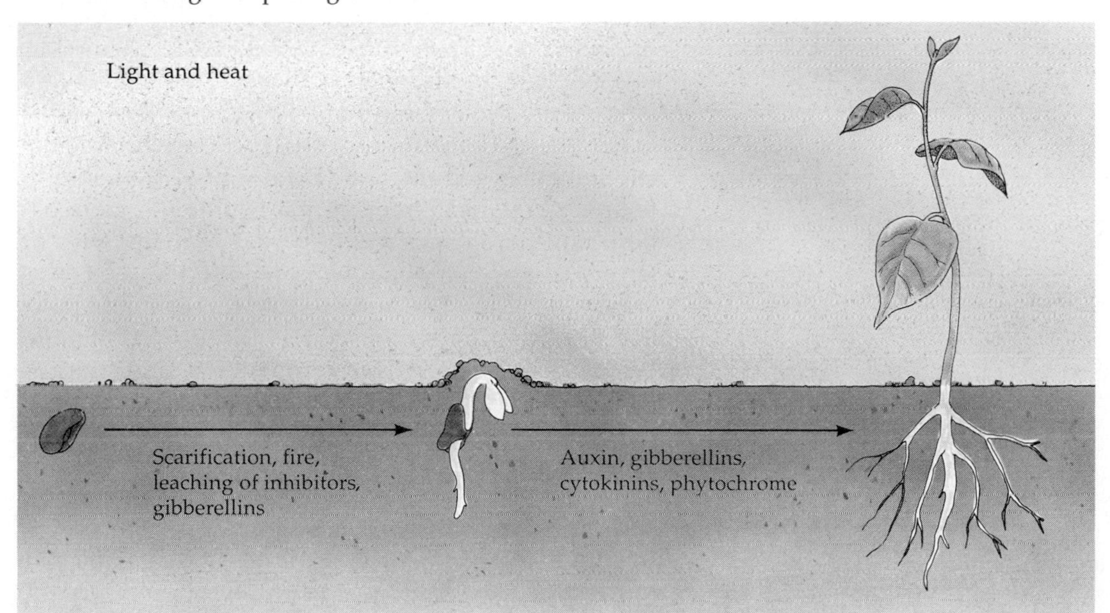

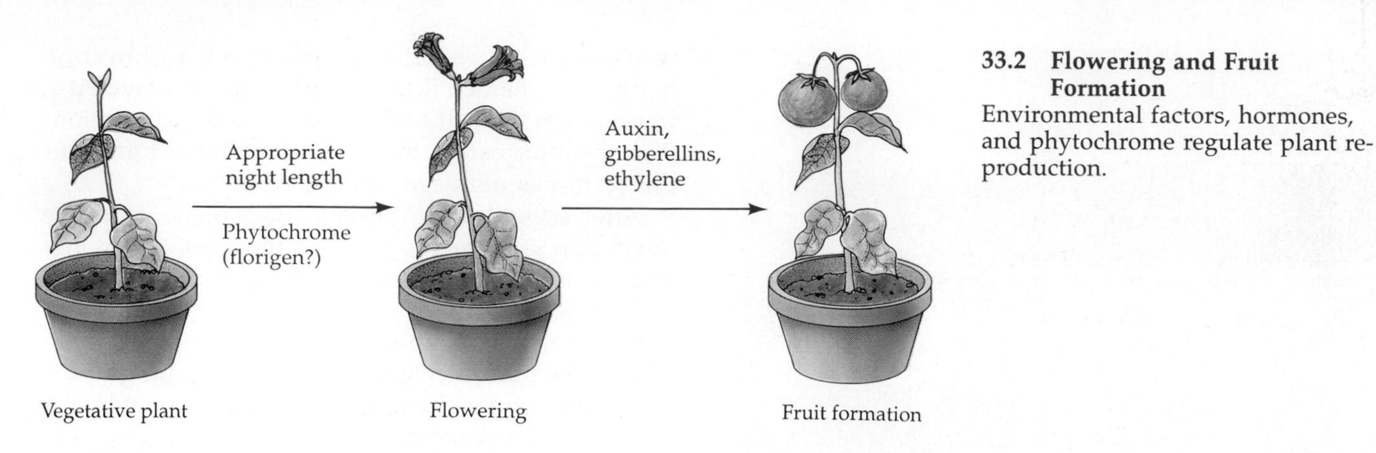

33.2 Flowering and Fruit Formation
Environmental factors, hormones, and phytochrome regulate plant reproduction.

mant. Seed dormancy may last for weeks, months, years, or even decades! The mechanisms of dormancy are numerous and diverse, but three principal strategies dominate: exclusion of water or oxygen from the embryo, mechanical restraint of the embryo, and chemical inhibition of embryo development.

Some seeds exclude water or, sometimes oxygen, by having an impermeable seed coat. The breaking of dormancy in such seeds depends on the abrasion of the seed coat, perhaps by the seed's tumbling across the ground or through creek beds, by its passing through the digestive tracts of birds, or by other means. Such modification of the seed coat is called **scarification.** The environment regulates germination of these seeds by providing or withholding the means of scarification.

The seed coat may also impose dormancy by the simple mechanical restraint of the embryo; if the embryo cannot expand, the seed cannot germinate. In the laboratory we can promote germination of such a seed by simply cutting away part of the coat or by partially dissolving it with strong acid. In nature, soil microorganisms probably play a major role in softening seed coats of this type, and the action of digestive enzymes in the guts of birds or other animals is also important. Another agent of scarification to release mechanical restraint is fire, which causes significantly increased germination in some natural habitats. Fire can also melt wax in seed coats, making water available to the embryo (Figure 33.4; see also Figure 47.16).

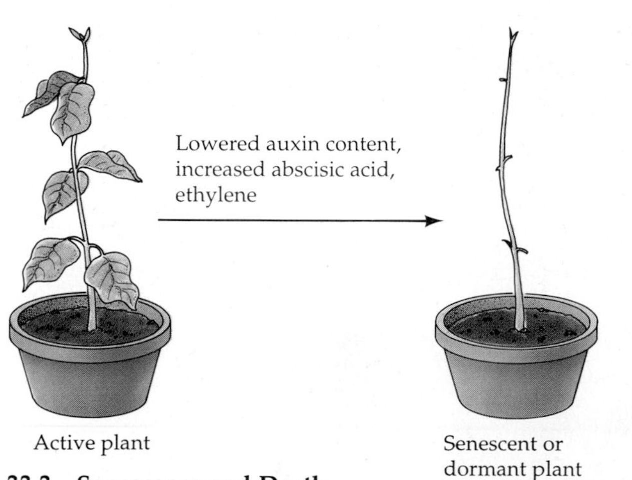

33.3 Senescence and Death
Environmental factors and hormones regulate the final stages of plant growth.

The action of chemical germination inhibitors is another mechanism of seed dormancy. As long as the concentration of inhibitor is high, the seed remains dormant. One means of reducing the level of inhibitor is by leaching, that is, prolonged exposure to water. Another is the scorching of the seeds by fire, which breaks down some inhibitors. Usually inhibitors of germination are already present in the dry seed, but in a few cases they are produced only after the seed has begun to take up water. The most common chemical inhibitor of seed germination is abscisic acid. In some seeds the level of abscisic acid or other inhibitors does not decline during germination; rather, the effect of the inhibitor is overcome by gradually increasing the concentrations of growth promoters.

There are still other mechanisms for breaking dormancy. Some seeds, such as those of the tomato and lima bean, remain dormant until they have dried extensively. Temperature can be an environmental cue initiating germination. Even a brief exposure to temperatures near freezing may stimulate germination, but more commonly a period of many days or weeks of low temperature is required to end dormancy. In agriculture and forestry, it is common practice to refrigerate seeds such as those of conifers to hasten germination. This refrigeration procedure is known as **stratification;** typically, it consists of a month or two at 5°C. The effects of cold treatment vary from species to species, but one result may be a gradual decrease in the content of germination inhibitors such as abscisic acid. Although the means vary, the environment is a potent regulator of germination.

33.4 Fire and Seed Germination This fireweed germinated and flourished after a great fire in Denali National Park, Alaska.

Adaptive Advantages of Seed Dormancy

Why might regulating the germination of seeds be a good thing? For many species, dormancy results in germination at a favorable time. Seeds that require a long cold period for germination commonly germinate in the spring, when water is usually abundant—germination in the dry days of late summer could be risky. Some other seeds will not germinate until a certain amount of time has passed, regardless of how they are treated. This period of **afterripening** prevents germination while the seed of a cereal grain, for example, is still attached to the parent plant, and it tends to favor dispersal of the seed.

Regulating germination may increase the likelihood of a seed's germinating in the right place. For example, some cypress trees grow in standing water, and their seeds germinate only if leached extensively by water (Figure 33.5). Many weeds must have their seed coats damaged before they will germinate, and other weed seeds will not germinate unless they have been exposed to light. Either type of weed germinates best in disturbed soils. You may have noticed how a freshly cultivated patch of soil quickly teems with weeds that are then free from competition with other plants. Seeds that must be scorched by fire in order to germinate also avoid competition; they germinate only when the area has been largely cleared by fire. Light-requiring seeds, which germinate only at or near the surface of the soil, are generally tiny seeds with few food reserves. The germination of some seeds is inhibited by light; these germinate only when well buried. Light-inhibited seeds are generally large and well stocked with nutrients.

Seed dormancy helps annual species counter the effects of year-to-year variations in the environment. Some seeds remain dormant throughout an unfavorable year, and other seeds germinate at different times during the year. Seed dormancy can also contribute to the dispersal of a plant species. Seeds that remain dormant until they have passed through the guts of birds or other animals will likely be carried some distance before they are deposited. Seeds carried by birds in their digestive tracts can give rise to the first plants on newly formed volcanic islands, for example.

33.5 Leaching of Germination Inhibitors The seeds of *Nyssa aquatica* germinate only after being leached by water, which increases the chances that they germinate in a situation suitable for their growth.

33.6 The Radicle Emerges
The tip of this lima bean's radicle has just broken through its protective sheath. The appearance of the radicle is one of the first externally visible events in seed germination.

Seed Germination

Imbibition—the uptake of water—is the first step in the germination of a seed. Typically, the seed is dry prior to the start of germination. Its water potential (see Chapter 30) is very negative, and water can be taken up readily if the seed coat allows it. The magnitude of the water potential is demonstrated by the force exerted by seeds expanding in water. Cocklebur seeds that are imbibing can exert a pressure of up to 1,000 atmospheres against a restraining force. As a seed takes up water, it undergoes metabolic changes: Certain preformed enzymes become activated, RNA and then new enzymes are synthesized, the rate of cellular respiration increases, and other metabolic pathways become activated. Interestingly, there is no DNA synthesis and no cell division during these early stages of seed germination. Emergent growth arises solely from expansion of small, preformed cells. DNA is synthesized only after the radicle—the embryonic root—begins to grow and poke out beyond the seed coat (Figure 33.6).

Mobilizing Food Reserves

Until the young plant (the **seedling**) becomes able to carry on photosynthesis, it depends on built-in reserves from the seed to meet its needs for energy and materials. The principal reserve of energy and carbon in the seeds of some species is the carbohydrate starch. More species, however, store lipids—fats or oils—as reserves in their seeds. Typically, the endosperm of the seed holds amino acid reserves in the form of proteins, rather than as free amino acids. Plant species differ in what their reserves are.

The giant molecules of starch, lipids, and proteins must be digested into monomers before they can enter the cells of the embryo to be used as building blocks and energy sources. Starch is a polymer of glucose, the starting point for glycolysis and cellular respiration. Starch is digested to glucose. The lipids can be digested to release glycerol and fatty acids, both of which yield energy through cellular respiration. Glycerol and fatty acids can also be converted to glucose, which permits fat-storing plants to make all the building blocks they need for growth. Lipid-storing seeds can pack more energy in a smaller space than starch-storing seeds can because lipids contain more calories per unit of weight than does starch. Some species that store lipids in their seeds store starch in their roots or tubers, where space is not a concern. Proteins are digested as well. The growing embryo can break down the proteins to obtain the amino acids it needs to assemble into its own myriad proteins. Plant species differ in their patterns of digestion of reserve polymers.

Germinating barley and other cereal seeds digest proteins and starch as follows: As the embryo becomes active, it secretes gibberellins. These diffuse through the endosperm to a surrounding tissue called the **aleurone layer,** which lies inside the seed coat. The gibberellins trigger a crucial series of events in the aleurone layer. First, protein-containing bodies called aleurone grains break down, releasing amino acids. The aleurone layer then uses the amino acids in the assembly of digestive enzymes, including amylases (starch-degrading enzymes), proteases (protein-degrading enzymes), and ribonucleases (RNA-degrading enzymes). These enzymes, along with certain others already present in the aleurone layer, are next secreted into the endosperm, where they catalyze the release of sugars and amino acid monomers from reserve polymers for use by the growing embryo (Figure 33.7).

We will now consider each of the major plant hormones, beginning with the gibberellins. Here is a study hint for the rest of this chapter. We have seen in a brief overview that information about hormones can be organized by considering the development of each part of the plant one after another. Now, to take a closer look at the hormones, we will organize the information by considering each hormone in turn. Try outlining the material in the first way, showing in detail how the development of roots, stems, buds, and leaves are regulated. In general, this is an excellent way to study: Look for a different way to organize the material presented in lecture or in the book.

GIBBERELLINS

We just encountered the gibberellins in the mechanism by which the germinating seeds of barley and

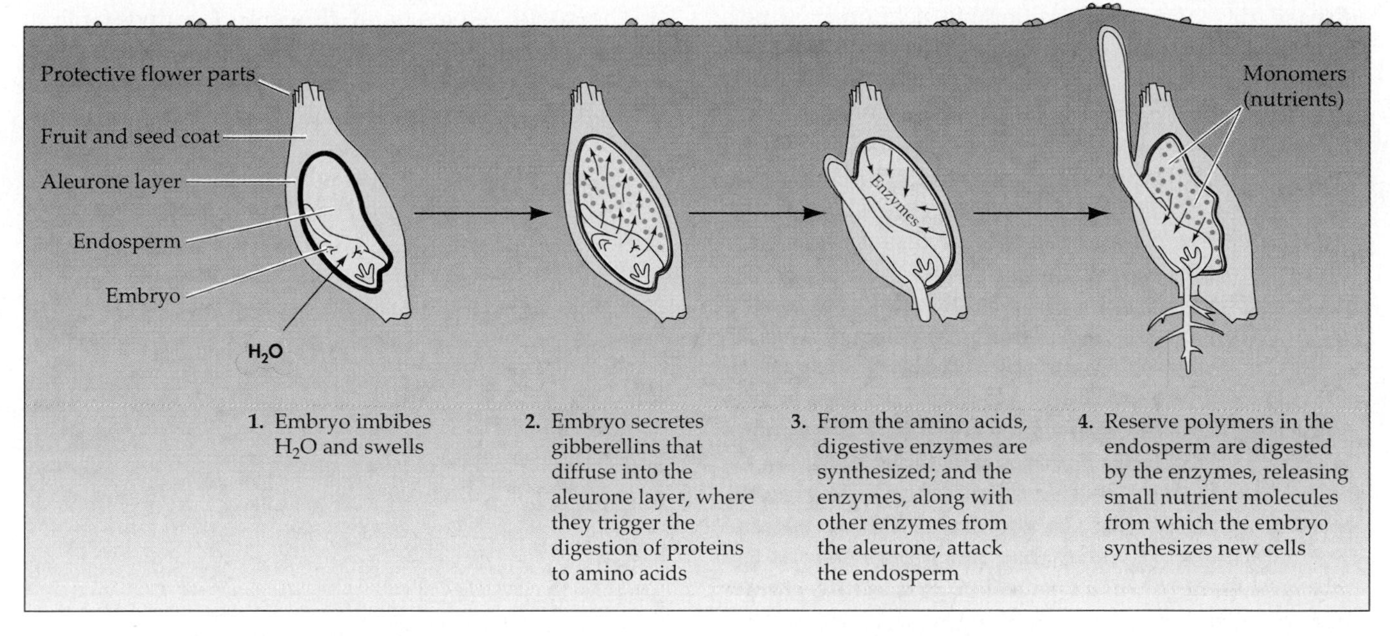

33.7 Embryos Mobilize Polymer Reserves
Seed germination in cereal grasses consists of a cascade of processes. Gibberellin signals the conversion of reserve polymers into monomers that can be used by the developing embryo.

other cereals convert their reserve proteins and starch into soluble monomers (see Figure 33.7). This is an elegant sequential mechanism, in which gibberellin functions as a *signal* to give the embryo access to the nutrients it needs. Gibberellins produce a wide variety of effects on plant development in addition to this triggering of digestive enzyme synthesis.

Discovery of the Gibberellins

The gibberellins are a large family of closely related compounds (Figure 33.8*a,b*), some found in plants and others in a pathogenic (disease-causing) fungus, where they were first discovered. The discovery of the gibberellins followed a crooked path, beginning with a book dictated in 1809 by a Japanese farmer named Konishi. In it, he described the symptoms of *ine bakanae-byo*, the "foolish seedling" disease of rice. Seedlings affected by the disease grow more rapidly than healthy plants, but the rapid growth gives rise to spindly plants that die before producing seed. The

33.8 Plant Hormones
Chemical structures of some representative compounds.

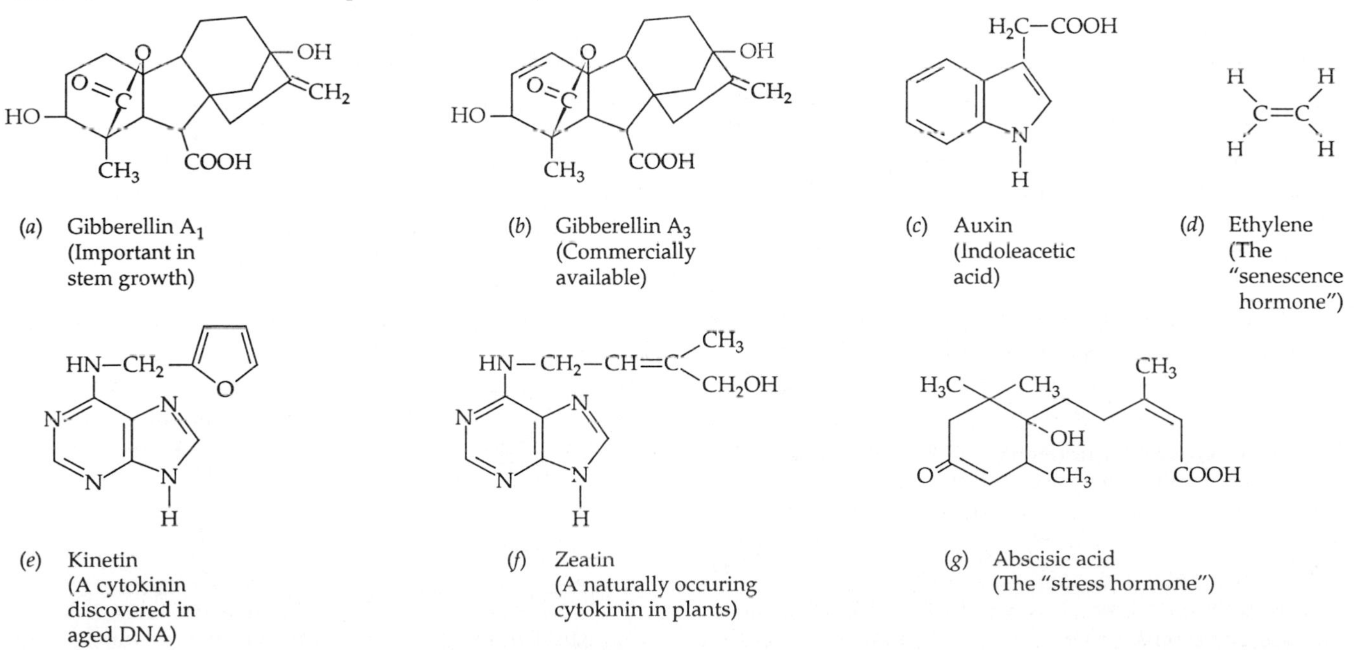

disease has been of considerable economic importance in several parts of the world. In 1898 it was learned that the *bakanae* disease, including both its growth-promoting and toxic effects, was caused by a fungus now known as *Gibberella fujikuroi*.

In 1925 the Japanese biologist Eiichi Kurosawa studied how *G. fujikuroi* caused the excessive spindly growth characteristic of the *bakanae* disease. He grew the fungus on a liquid medium and then separated the fungus from the medium by filtering. He heated the medium to kill any remaining fungus. The resulting heat-treated filtrate still stimulated the growth of uninfected rice seedlings. He found no such effects using medium that had never contained the fungus (Figure 33.9). Thus, Kurosawa established that *G. fujikuroi* produces a chemical substance with growth-promoting properties. In the late 1930s it became clear that there was more than one gibberellin, as the new growth substance was called. It was also shown that the toxic effects of the fungus were caused by another, inhibitory substance.

Were the gibberellins simply exotic products of an obscure fungus, or did they play a more general role in the growth of plants? Bernard O. Phinney of the University of California, Los Angeles, partially answered this question in 1956, when he reported the spectacular growth-promoting effect of gibberellins on certain dwarf strains of corn. These plants were known to be genetic dwarfs; each phenotype was produced when a particular recessive allele was present in the homozygous condition (see Chapter 10 if you need to review these genetic terms). Gibberellin applied to nondwarf—tall—corn seedlings had virtually no effect, but gibberellin applied to the dwarfs caused them to grow as tall as their normal relatives. (A comparable effect of gibberellin on dwarf mustard plants is shown in Figure 33.10.) This result suggested to Phinney that (1) gibberellins are normal constituents of corn and perhaps of all plants, and (2) the dwarfs are short because they cannot produce their own gibberellin. According to these hypotheses, nondwarf plants manufacture enough gibberellins to promote their full growth. Phinney and other scientists tested extracts from numerous plant species to see if they promoted growth in dwarf corn and they found that many such extracts did. These findings provided direct evidence that plants that are not genetic dwarfs contain gibberellinlike substances.

The roots, leaves, and flowers of a dwarf plant appear normal, but the stems are much shorter than their counterparts on other plants. What does this tell us? We know that all the cells of the dwarf plants

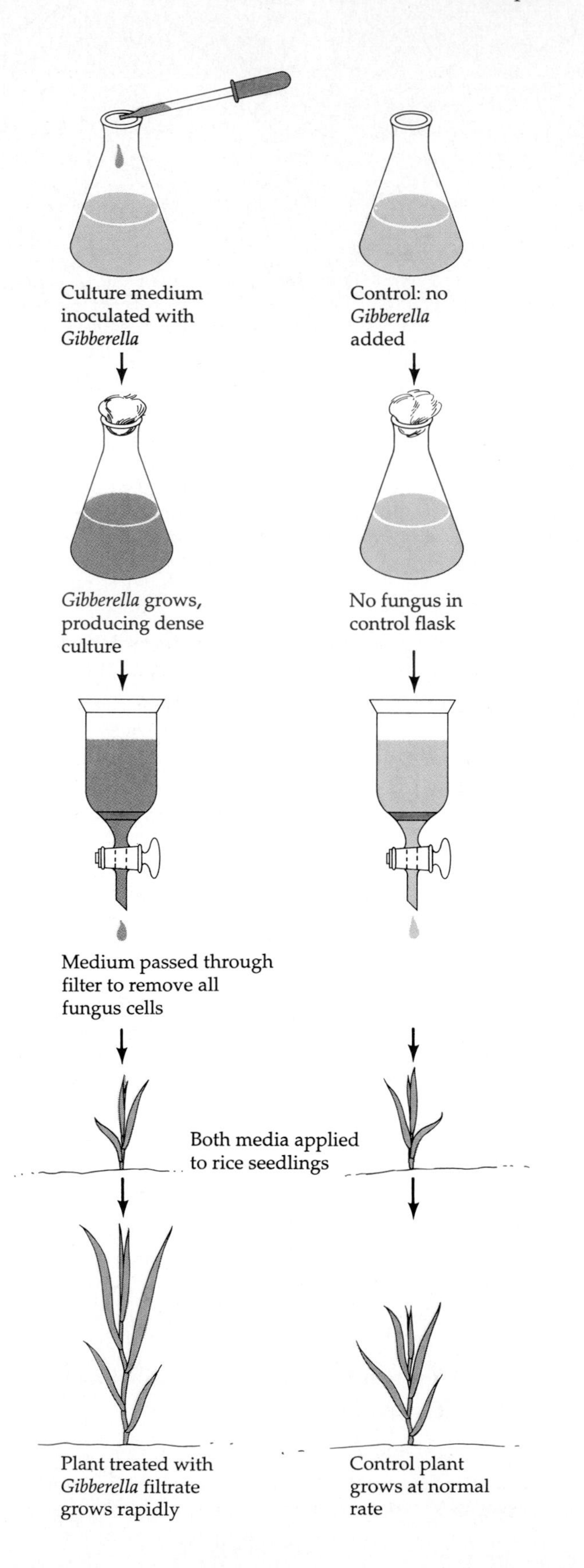

33.9 Kurosawa's Experiment
Kurosawa demonstrated that *bakanae* disease is caused by substances produced by *Gibberella fujikuroi*. Fungus-free medium previously used to culture the fungus made rice seedlings grow rapidly, as if they had the *bakanae* disease. As a control, Kurosawa determined that medium in which the fungus had never been cultured did not cause disease symptoms.

33.10 Gibberellin's Effect on a Dwarf Plant
In this experiment, a tiny amount of gibberellin was added to the dwarf mustard plant (*Brassica rapa*) on the right; on the left is an untreated control plant. After 22 days, the plant on the right reached a size more typical of a nondwarf plant, while the control remained dwarf.

have the same genome and all parts of the dwarf plants contain much less gibberellin than do the organs of other plants. From this we may infer that stem elongation *requires* gibberellin or the products of gibberellin action; on the other hand, gibberellin plays no comparable role in the development of roots, leaves, and flowers.

Why So Many Gibberellins?

We do not yet know how many different gibberellins exist. Each year brings reports of more, with several dozen now having been characterized. Some gibberellins are produced in the root system, others in young leaves. For many years plant physiologists were puzzled by the existence of such a great number of different gibberellins. Recent work, however, has led to the conclusion that only one gibberellin, gibberellin A_1, actually controls stem elongation; the other gibberellins found in stems are simply intermediates in the production of gibberellin A_1. As we will see in the next section, gibberellins affect processes other than stem elongation, but we do not yet know which gibberellin has any other particular effect.

Other Activities of Gibberellins

Gibberellins and other hormones regulate the growth of fruits. It has long been known that seedless varieties of grapes form smaller fruit than their seeded relatives. In one experiment, removal of seeds from very young seeded grapes prevented their normal growth, suggesting that the seeds are sources of a fruit growth regulator. It was then shown that spraying young seedless grapes with a gibberellin solution causes them to grow as large as seeded varieties. Subsequent biochemical studies showed that the developing seeds produce gibberellin, which diffuses out into the immature fruit tissue.

Some biennial plants respond dramatically to an increased level of gibberellin. Biennial plants grow vegetatively in their first year and flower and die in their second year. In their second year, in response either to the increasing length of days or to the winter cold period, the apical meristems of biennials produce elongated shoots that eventually bear flowers. This elongation is called **bolting.** When the plant senses the appropriate environmental cue—longer days or a sufficient winter chilling—it produces more gibberellins, raising the gibberellin concentration to a level that causes the shoot to bolt. Plants of some biennial species will bolt if sprayed with a gibberellin solution even though they have not experienced the environmental cue (Figure 33.11).

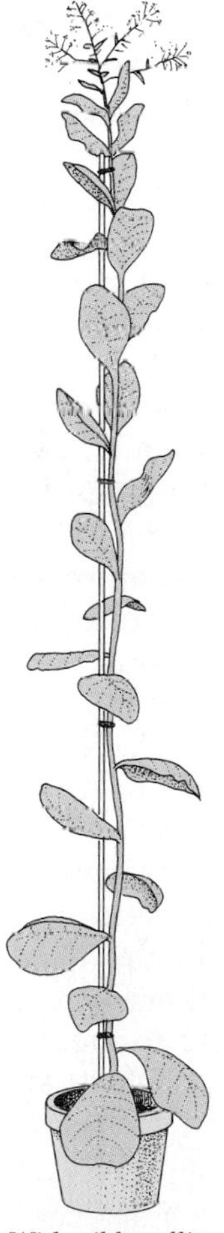

33.11 Bolting
Spraying with gibberellin causes cabbage and some other plants to bolt. While untreated control plants retain their compact leafy heads, the internodes of treated plants elongate dramatically, resulting in towering shoots.

Gibberellins also cause fruit to grow from unfertilized flowers, promote seed germination in lettuce and some other species, and in the spring help bring buds out of winter dormancy. Hormones usually have multiple effects within the plant and often interact with one another in regulating developmental processes. In controlling stem elongation, for example, gibberellins interact with another hormone, auxin.

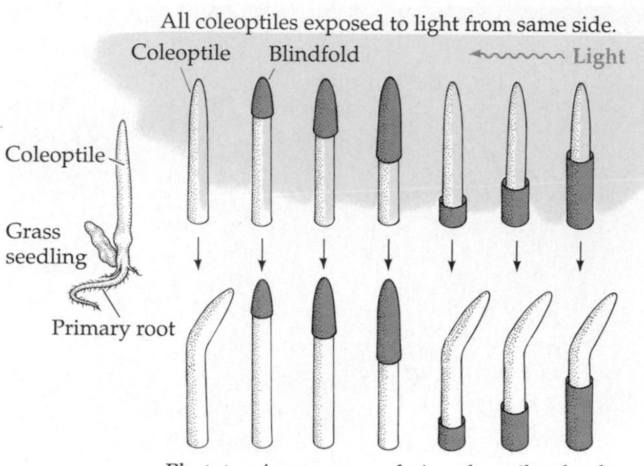

33.12 The Darwins' Experiment
The top drawings show some of the ways in which Charles and Francis Darwin "blindfolded" dark-grown grass seedlings; the lower drawings show what they observed in each case. Coleoptiles responded to light only when the top millimeter or so was exposed; they responded as they grew by bending toward the light a few millimeters below the tip. These observations suggested that the plant's "eye" (that which senses light) is in the tip, and that it sends a message from the tip to the region of bending.

AUXIN

If you pinch off the apical bud at the top of a bean plant, lateral buds that were once inactive become active. Similarly, pruning a shrub causes an increase in branching. If you cut off the blade of a leaf but leave its petiole attached to the plant, the petiole drops off sooner than it would if the leaf were intact. If a plant is kept indoors, its shoot system grows toward a bright window. What these diverse responses of shoot systems have in common is that they are mediated by a plant hormone called auxin—or **indoleacetic acid** in chemical terms (see Figure 33.8*c*).

Discovery of Auxin

The discovery of auxin and its numerous physiological activities traces back to work done in the 1880s by Charles Darwin and his son Francis, who were interested in plant movements. One type of movement they studied was **phototropism,** the growth of plant structures toward light (as in shoots) or away from it (as in roots).

An obvious question that they asked was, What part of the plant senses the light? To answer this question, the Darwins worked with canary grass seedlings grown in the dark. The dark-grown grass seedling has a coleoptile, or leaf sheath, covering the immature shoot. To find the light-receptive region of the coleoptile, the Darwins tried "blindfolding" it in various places and then illuminating it from one side (Figure 33.12). The coleoptile grew toward the light whenever its tip was exposed. If the top millimeter or more of the coleoptile was covered, however, there was no phototropic response. Thus the tip contains the photoreceptor. The bending, however, takes place in a growing region a few millimeters below the tip. Therefore, the Darwins reasoned, some type of message must travel within the coleoptile from the tip to the growing region.

Others later demonstrated that the message is a chemical substance; it cannot pass through a barrier impermeable to chemicals but does pass through certain nonliving materials, such as gelatin. The tip of the coleoptile produces a hormone that moves down the coleoptile to the growing region. If the tip is removed, the growth of the coleoptile is sharply inhibited; if the tip is then carefully replaced, growth resumes, even if the tip and base are separated by a thin layer of gelatin. The hormone moves down from the tip but does not move from one side of the coleoptile to the other. If the tip of an oat coleoptile is cut off and replaced so that it covers only one side of the cut end of the shoot, the coleoptile curves as the cells on the side below the replaced tip grow more rapidly than those on the other side.

With this information as a beginning, the Dutch botanist Frits W. Went succeeded, where many had failed, in isolating the hormone from oat coleoptiles. Went removed coleoptile tips and placed their cut surfaces on a block of gelatin, hoping the hormone would diffuse into the gelatin. Then he placed pieces of the gelatin block on decapitated coleoptiles—positioned to cover only one side, just as coleoptile tips had been placed in some of the earlier experiments. As they grew, the coleoptiles curved toward the side away from the gelatin. This curvature demonstrated that the hormone had indeed diffused into the gelatin block from the isolated coleoptile tips (Figure 33.13). The hormone had at last been isolated from the plant. It was later named auxin; still later, it was shown to be indoleacetic acid. Went's historic experiment was performed in 1926—the very year Kurosawa published his classic account of the isolation of a growth

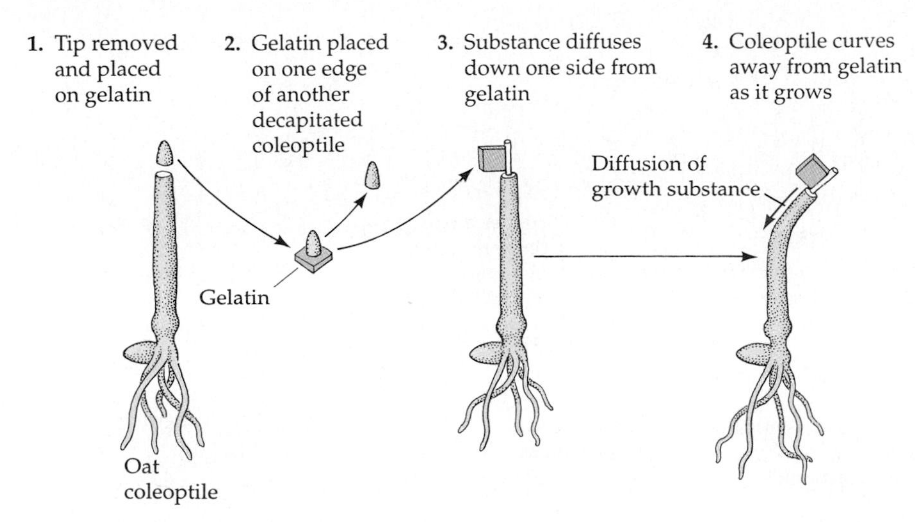

33.13 Went's Experiment
Went isolated the growth substance from coleoptile tips in a small block of gelatin. The gelatin block caused another coleoptile to bend, confirming that the block contained the growth substance. The substance is now called auxin.

substance from the fungus *Gibberella fujikuroi,* an accomplishment closely analogous to Went's.

Auxin Transport

Auxin was studied in a number of ways once it had been isolated from the plant. Early experiments showed that its movement through certain tissues is strictly polar, that is, unidirectional along a line from apex to base. By inverting the setups in half of the experiments, scientists determined that the apex-to-base direction of auxin movement has nothing to do with gravity; the polarity of this movement is a totally biological matter. Many plant parts show at least partial polarity of auxin transport. For example, auxin moves in leaf petioles from the blade end toward the stem end.

Phototropism and Gravitropism

The *lateral* movement of auxin in the apex affects the direction of plant growth. When light strikes a coleoptile from one side, auxin at the tip moves toward the shaded side. The imbalance thus established is maintained down the coleoptile, so that in the growing region below there is more auxin on the shaded side, causing the unequal growth that results in a coleoptile bent toward the light. This is phototropism (Figure 33.14*a*).

Similarly, but even in the dark, auxin moves to the lower side of a shoot that has been tipped over, causing more rapid growth in the lower side and, hence, an upward bend of the shoot. This phenomenon is **gravitropism,** the growth of a plant part in a direction determined by gravity (Figure 33.14*b*). The upward gravitropic response of shoots is defined as negative; the gravitropism of roots, which bend downward, is positive.

Auxin and Vegetative Development

Like the gibberellins, auxin has many roles in the plant. Cuttings from the shoots of some plants can produce roots and thus grow into entire new plants. For this to happen, certain undifferentiated cells in the *interior* of the shoot, originally destined to function only in food storage, must set off on an entirely new mission: to differentiate and become organized into the meristem of a root. These changes are very similar to those in the pericycle of a root when a branch root forms (see Chapter 29). Shoot cuttings of many species can be stimulated to grow roots profusely by dipping the cut surfaces into an auxin solution.

The effect of auxin on leaf **abscission,** the separation of old leaves from stems, is quite different. If the blade of the leaf is excised, the petiole abscises more rapidly than if the leaf had remained intact (Figure 33.15*a*). If the cut surface is treated with an auxin solution, however, the petiole remains attached to the plant, often longer than an intact leaf would have (Figure 33.15*b*). It appears that the time of abscission of leaves in nature is determined in part by a decrease in the movement of auxin, produced in the blade, through the petiole.

Auxin maintains **apical dominance,** the tendency of some plants to grow a single main stem with minimal branching. This phenomenon can be shown by an experiment with dark-grown pea seedlings. If the plant remains intact, the stem elongates and the lat-

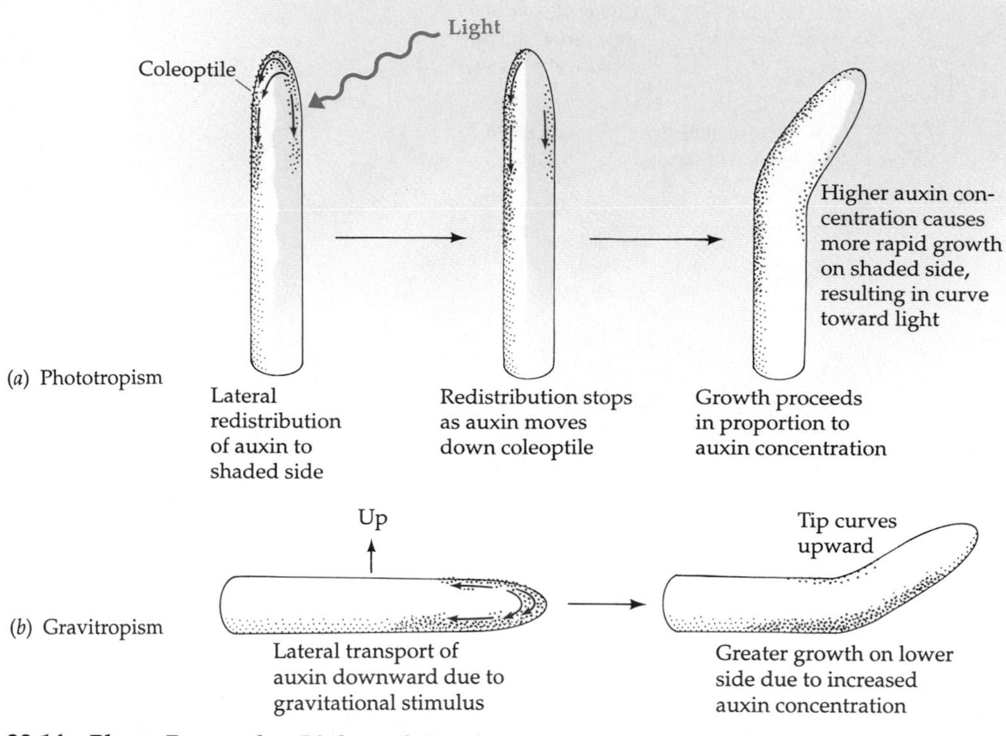

33.14 Plants Respond to Light and Gravity
Auxin in the coleoptile tip moves toward the shaded side, beginning the phototropic response. Auxin accumulates on the lower side of a horizontal coleoptile, beginning the gravitropic response.

eral buds remain inactive. Removal of the apical bud—the major site of auxin production—causes the lateral buds to grow out vigorously, but this growth is prevented if the cut surface of the stem is treated with an auxin solution (Figure 33.16). Apical buds of branches exert apical dominance, with their own lateral buds being inactive unless the apex is removed. Note that in the experiments on leaves and stems that we have discussed, removal of a particular part of the plant produces an effect—abscission or loss of apical dominance—and that the effect is prevented by treatment with auxin. These results are consistent with other data showing that the excised part of the leaf or stem is an auxin source and that auxin in the intact plant helps maintain apical dominance and delays the abscission of leaves.

Many synthetic auxins—chemical analogs of indoleacetic acid—have been produced and studied. One of them, 2,4-dichlorophenoxyacetic acid (2,4-D), has the striking property of being lethal to dicots at concentrations that are harmless to monocots. This property made 2,4-D a widely used **selective herbicide** that could be sprayed on a lawn or a cereal crop to kill the dicots, thus eliminating most of the weeds. Because 2,4-D takes a long time to break down, however, it pollutes the environment, so scientists are seeking new approaches to selective weed killing (Box 33.A).

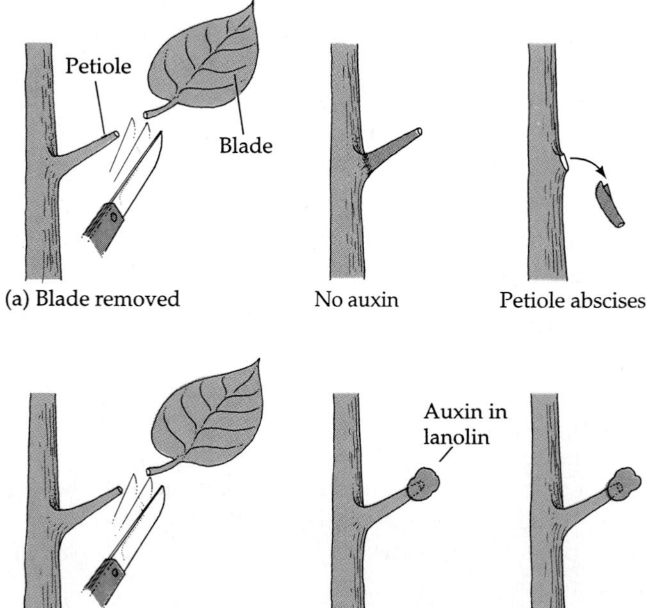

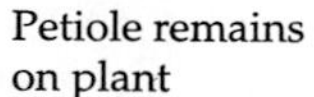

33.15 Auxin Delays Leaf Abscission
The leaf blade is a source of auxin throughout the growing season.

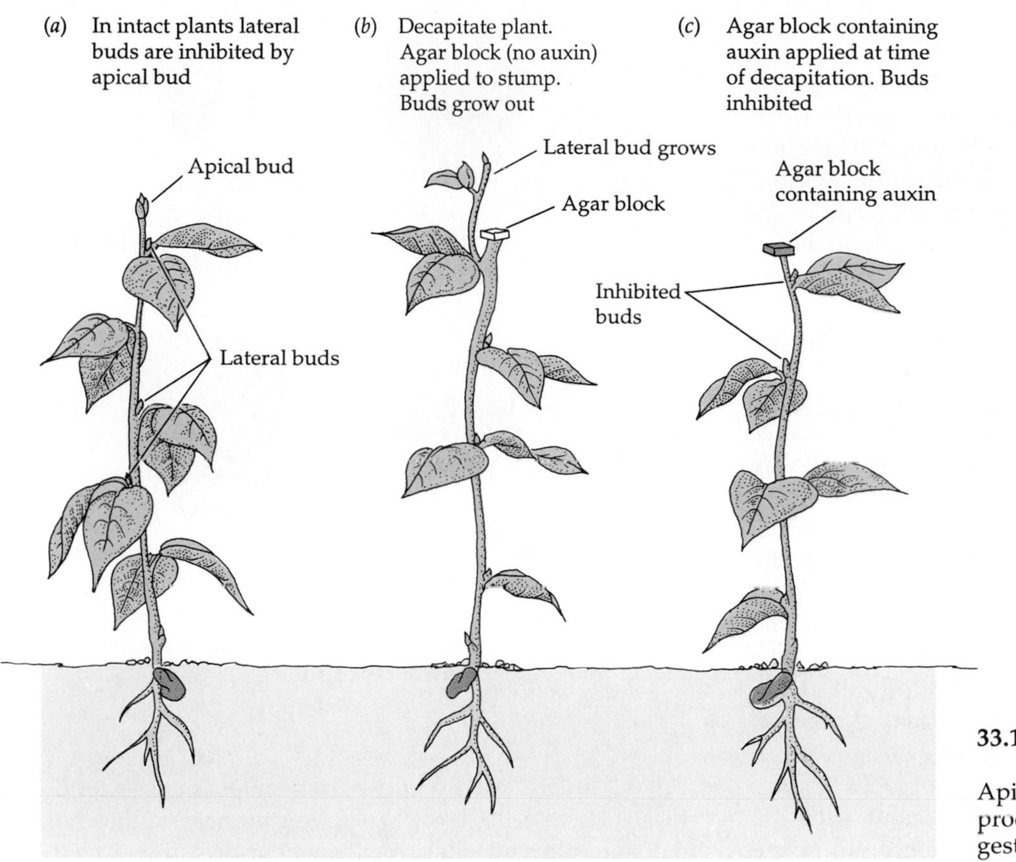

33.16 Auxin and Apical Dominance
Apical dominance results from auxin produced by the apical bud, as suggested by these experimental results.

BOX 33.A

Strategies for Killing Weeds

The first commercial herbicides—weed killers—were substances that killed any plant they touched. Today we use such general herbicides when it is not necessary to save other vegetation; for example, products containing 3-aminotriazole (3AT) are used to kill stands of poison oak or poison ivy. We cannot use such an all-out, nonspecific approach if we wish to kill certain plants but retain others.

The next approach was the use of selective herbicides such as 2,4-D. Although 2,4-D is an effective herbicide, its use is controversial because it pollutes the environment.

The advent of recombinant DNA technology opened entirely new avenues of herbicide research. One approach is to transfer genes that confer resistance to herbicides into plants we want to be unaffected by the herbicides. The concept is simple and attractive: First, develop a highly effective herbicide that will kill all the plants in a field; second, genetically engineer the chosen crop plants so that they are resistant to that herbicide; finally, spray the field and watch the weeds die while the crop, protected by its inserted gene, prospers. Biotechnology companies view this as a promising method for weed control, although some scientists fear its environmental and health consequences.

Scientists at Calgene, a biotechnology company in Davis, California, recently reported the results of such an experiment. Their goal was to render selected plants insensitive to bromoxynil, a potent inhibitor of photosynthesis. The Calgene scientists discovered that a particular soil bacterium, *Klebsiella ozaenae*, converts bromoxynil to an inactive product. From this bacterium they isolated the gene that codes for the enzyme that inactivates bromoxynil. They then inserted the *Klebsiella* gene into tobacco plants. To make sure the gene would be expressed in the tobacco plants, they inserted it along with and under the control of a light-activated promoter. The promoter normally controls the gene that codes for rubisco, the enzyme that fixes carbon dioxide in photosynthesis. The experiment was a success: Transgenic plants containing the bacterial gene grew vigorously after being sprayed with bromoxynil solutions that killed control tobacco plants.

Auxin and Fruit Development

Although fruit development normally depends on prior fertilization of the egg, in many species treatment of an unfertilized ovary with auxin or gibberellin causes **parthenocarpy**—fruit formation without fertilization of the egg. Parthenocarpic fruit form spontaneously in some plants, including dandelions, seedless grapes, and the cultivated varieties of pineapple and banana.

The strawberry is an unusual "fruit." What we commonly call the fruit is actually a modified stem, or receptacle, with the tiny, dry "seeds" being the true fruits, called achenes. The achenes produce auxin, and the auxin induces the growth of the fleshy receptacle, as the French botanist Jean-Pierre Nitsch demonstrated (Figure 33.17). When he removed all the achenes within three weeks after pollination, the stem tissue did not develop into a "strawberry" (Figure 33.17*b*). If he pollinated only one to three of the many pistils and kept the others virgin, he observed localized receptacle growth in the area of pollination (Figure 33.17*c*). He could induce normal expansion if, after removing all of the achenes, he spread an auxin-containing paste over the receptacle (Figure 33.17*d*). These three results are consistent with the hypothesis that the achenes cause the growth of the receptacle by producing auxin.

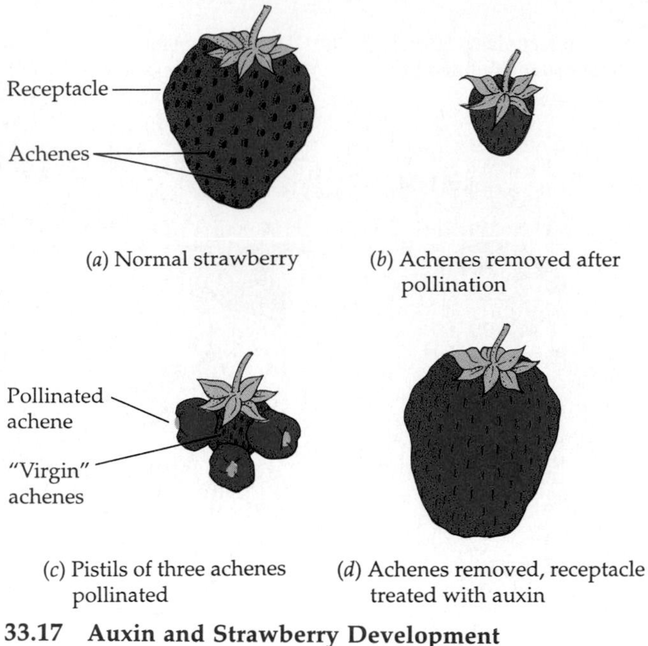

33.17 Auxin and Strawberry Development
Auxin produced by the achenes is responsible for the normal growth of the strawberry "fruit."

Plant hormones control other aspects of fruit physiology and of the development and senescence of flower parts as well. These activities illustrate again the great diversity of important roles that hormones play.

Are There Master Reactions?

Each of the known plant hormones causes a variety of responses that are often seemingly unrelated, thus raising one of the major questions of plant physiology: Do all the effects of a particular hormone arise from a single mechanism—a common master chemical reaction? We know so little about how the plant hormones act at the molecular level that this question cannot yet be answered. For one hormone—auxin—we are beginning to gain some insight into its central mechanisms, if indeed there is more than one molecular mechanism. To appreciate this mechanism, we must first briefly consider the architecture of the plant cell wall.

Cell Walls and Growth

The principal strengthening component of the plant cell wall is cellulose, a large polymer of glucose. In the wall, cellulose molecules tend to associate with one another, forming crystalline regions called micelles. Individual cellulose molecules may extend from one micelle across relatively noncrystalline regions to other micelles. Bundles of approximately 250 cellulose molecules, including many micelles, constitute microfibrils visible with an electron microscope. What makes the cell wall rigid is a network of cellulose microfibrils connected by bridges of other, smaller polysaccharides (Figure 33.18*a*). Peter Albersheim and his colleagues at the University of Colorado proposed a model for the molecular architecture of the cell wall, showing how the other polysaccharides may interconnect the cellulose microfibrils (Figure 33.18*b*).

The growth of a plant cell is driven primarily by the uptake of water, which enters the cytoplasm of the cell and its vacuole (see Figure 4.24). As the vacuole expands, the cell grows rapidly, with the vacuole often making up more than 90 percent of the volume of a mature cell. As the vacuole expands, it presses the cytoplasm against the cell wall, and the wall resists this force. For the cell to grow, its wall must loosen and be stretched. As the wall stretches, it should become thinner; however, because new polysaccharides are deposited throughout the wall and new cellulose microfibrils are deposited at the inner surface of the wall, the cell wall maintains its thickness. Thus the cellulose microfibrils in the outermost part of the wall are the oldest, and those in the innermost part the youngest.

The wall plays a key role in controlling the growth rate of a plant cell. How does the plant determine the behavior of its cell walls?

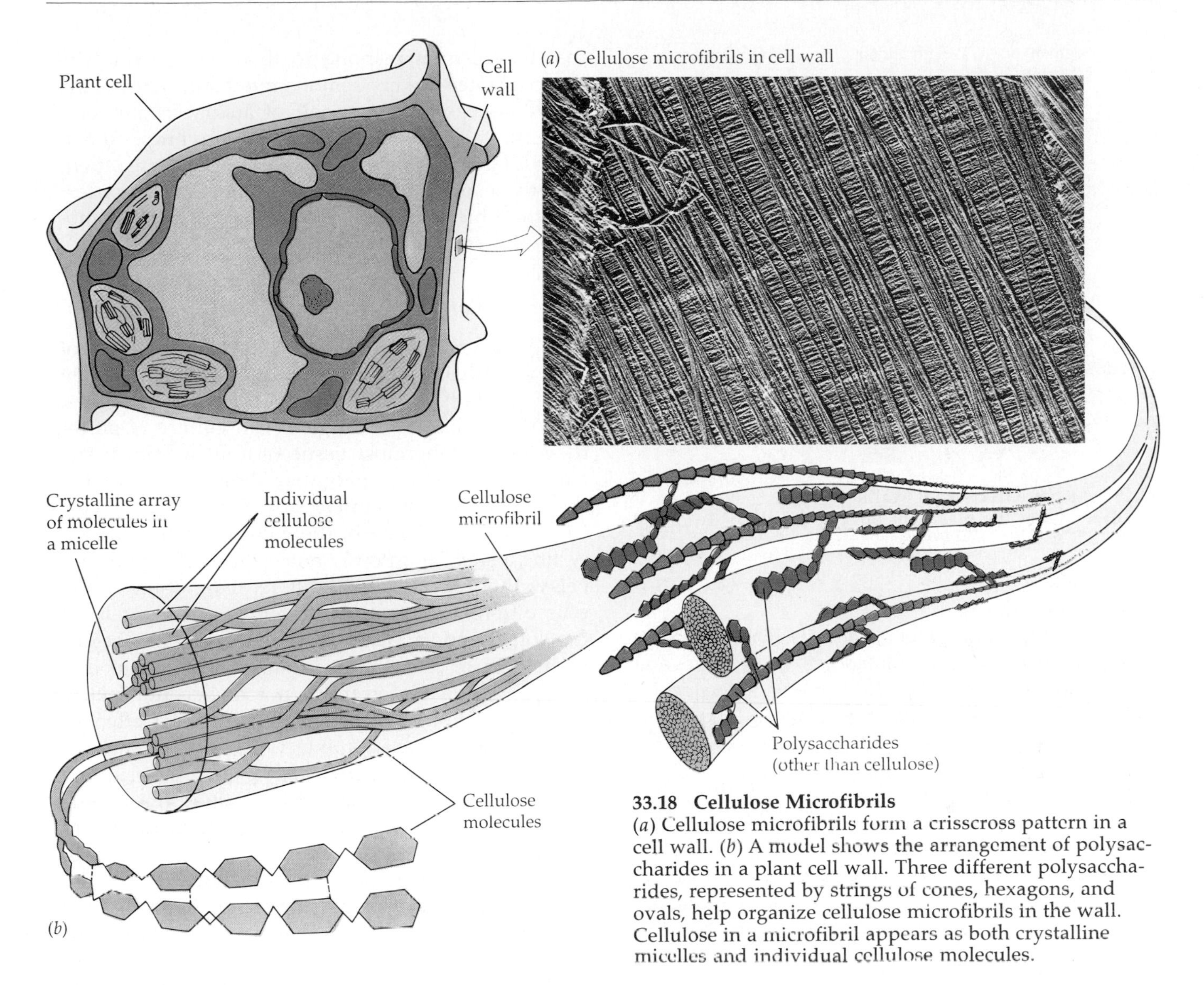

33.18 Cellulose Microfibrils
(*a*) Cellulose microfibrils form a crisscross pattern in a cell wall. (*b*) A model shows the arrangement of polysaccharides in a plant cell wall. Three different polysaccharides, represented by strings of cones, hexagons, and ovals, help organize cellulose microfibrils in the wall. Cellulose in a microfibril appears as both crystalline micelles and individual cellulose molecules.

Auxin and the Cell Wall

Auxin can loosen cell walls—make them more stretchable—as the Dutch physiologist A. J. N. Heyn first demonstrated half a century ago. Heyn hung segments of oat coleoptiles on pins and hung weights on the ends of the segments, causing them to bend. Then he removed the weights, allowing the segments to bend back. Recovery was incomplete; that is, some of the bending was not reversible. Heyn called the reversible bending **elasticity** and the irreversible bending **plasticity** (Figure 33.19). Pretreating the coleoptile segments with auxin significantly increased their plasticity; it loosened the wall. This result suggested that auxin-induced cell expansion might result from just such a loosening effect.

It was later shown that auxin itself does not loosen cell walls upon contact. Rather, there is an intervening step; auxin may cause the release of a "wall-loosening factor" from the cytoplasm. Work in the 1970s in the United States and in Europe indicated that the wall-loosening factor was simply hydrogen ions (protons, H^+). Acidifying the growth medium (that is, adding H^+) causes segments of stems or coleoptiles to grow as rapidly as segments treated with auxin, and treating coleoptile segments with auxin causes acidification of the medium. Treatments that block acidification by auxin also block auxin-induced growth.

It was proposed that hydrogen ions, secreted into the cell wall as a result of auxin action, activate some enzyme or enzymes in the wall. These enzymes might digest specific linkages connecting the cellulose microfibrils, or they might alter bonds in the matrix in which the microfibrils reside. The end result in either case would be a temporary loosening of the wall. In 1992, a group led by Daniel Cosgrove at Pennsylvania State University isolated two interest-

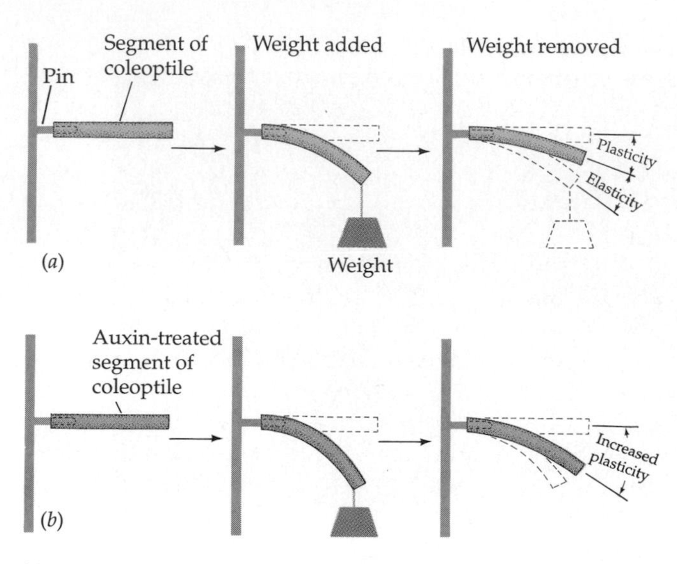

33.19 Auxin Affects Cell Walls
Auxin increases the plasticity, but not the elasticity, of cell walls.

ing proteins from cucumber cell walls. These proteins cause the extension of isolated cell walls of several species and appear to be the proteins activated by hydrogen ions. They seem not to digest linkages in polysaccharides such as cellulose, but rather to alter bonds in the matrix.

The cell wall is an important site for the major activities regulating plant development. Auxin is not the only agent that affects the properties of the cell wall. Gibberellins, too, can loosen the wall, which is not surprising in view of their growth-promoting activities. Other plant hormones also modify the cell wall.

Auxin Receptors

We will see in Chapter 36 that animal hormones begin to act only after binding with specific receptor proteins. Does the action of plant hormones also require their recognition by receptor proteins? Recently, it has been shown that several proteins can bind various plant hormones. It has not been shown, however, that these proteins function in the living plant as receptors that mediate the effects of the regulators.

In 1989 Glenn Hicks, David Rayle, and Terri Lomax, at Oregon State University, provided the first solid evidence for a connection between such apparent receptor proteins and an auxin-related plant response. They were working with the *diageotropica* (*dgt*) mutation of tomato. Plants homozygous for the *dgt* mutation fail to show normal gravitropism; instead of growing upright, they simply sprawl on the ground. They also show other symptoms indicating that they cannot respond to their own auxin. The Oregon State workers demonstrated that stems of the *dgt* homozygotes lack a pair of auxin-receptor proteins that are present in normally gravitropic tomatoes and in many other plant species. It seems, then, that these proteins participate in auxin responses, and that the absence of these proteins in *dgt* homozygotes accounts for their aberrant growth.

Differentiation and Organ Formation

What, within a plant, signals the different types of cells and organs to form? Much of the research on these questions has been done with cultured tissues. One tissue that is easily grown in culture is pith—the spongy, innermost tissue of a stem. Pith tissue cultures proliferate rapidly but show no differentiation; all the cells are similar, and similarly unspecialized. Cutting a notch in the cultured tissue and inserting a stem tip into the notch causes the pith cells below the inserted tip to differentiate. Some of the cells differentiate to form water-conducting cells of the sort found in xylem. Differentiation also begins if, instead of a stem tip, a mixture of auxin and coconut milk is placed in the notch. Coconut milk is a rich source of plant hormones. A similar effect of auxin can be observed in intact plants. If notches are cut in the stems of *Coleus blumei* plants, interrupting some of the strands of conducting tissues, the strands gradually regenerate from the top side of the cut to the lower (recall that auxin moves from the tip to the base of a stem). If the leaves above the cut are removed, regeneration is slowed; if, however, the missing leaves are replaced with an auxin solution, new conductive tissue regenerates. Auxin and other plant hormones signal the formation of specific cell types.

Other work with cultured tissues has helped clarify which hormones control organ formation. Undifferentiated cultures of tobacco pith form roots when treated with an appropriate concentration of auxin. Another group of hormones—the cytokinins—causes buds and then shoots to form in such cultures. The pattern of organ formation depends on the ratio of auxin to cytokinin in the medium. A high proportion of auxin favors roots and a high proportion of cytokinins favors buds, but both processes are most active when both hormones are present.

CYTOKININS

The cytokinins, which have a variety of effects besides stimulating bud formation, promote cell division in cultured tissues, an activity that led to their discovery.

Discovery of the Cytokinins

After studying cell division for many years, Folke Skoog, at the University of Wisconsin, and his associate Carlos Miller reasoned that, since a plant hormone (auxin) is what regulates cell *expansion*, a plant hormone must also regulate cell *division*. Because cell division requires DNA replication, Skoog and Miller suspected that the hypothetical hormone regulates the metabolism of nucleic acids and that it might even consist of DNA itself.

In 1955 they and other members of Skoog's group were studying the effects of various compounds on the rate of cell division in cultures of carrot root tissue. They took an old bottle of herring sperm DNA (the only DNA on hand) off the shelf and added a bit to the medium in which some of the cultures were growing. They were gratified to observe that these cultures began to proliferate much more rapidly than the controls. Determined to discover what component of the DNA preparation was the active material, they purchased some fresh herring sperm DNA, but they were disappointed to find that it was quite inactive. Deciding that the only difference between the two samples of DNA was age, they tried to "age" the new DNA by heating it in a sterilizer. This worked! The heated DNA preparation caused rapid cell division in the carrot cultures. Careful chemical work revealed that a single substance in the old or sterilized DNA preparations was the active material. Skoog's group named this substance kinetin (see Figure 33.8*e*) and suggested that it might be just one of a family of compounds, which are now called cytokinins.

For several years they and other investigators tried in vain to find kinetin in plant tissues. What they did find were two closely related compounds called zeatin (see Figure 33.8*f*) and isopentenyl adenine. These two are naturally occurring cytokinins; kinetin, on the other hand, may be considered a synthetic, since it has not been isolated from plant tissue.

Other Activities of the Cytokinins

Cytokinins are believed to form primarily in the roots and to move to other parts of the plant. Adding an appropriate combination of auxin and cytokinin to the medium yields rapid growth of plant tissues. Cytokinins can cause certain light-requiring seeds to germinate when the seeds are kept in constant darkness. Cytokinins usually inhibit the elongation of stems, but they cause lateral swelling of stems and roots; the fleshy roots of radishes are an extreme example. Cytokinins stimulate lateral buds to grow into branches; thus the balance between auxin and cytokinin levels controls the bushiness of a plant. Cytokinins increase the expansion of cut pieces of leaf tissue, so they may regulate normal leaf expansion. Cytokinins also delay the senescence of leaves. If leaf blades are detached from a plant and floated on water or a nutrient solution, they quickly turn yellow and show other signs of senescence. If instead they are floated on a solution containing a cytokinin, they remain green and senesce much more slowly. Cytokinins apparently regulate the redistribution of biologically active materials from one part of a plant to another. When one of a pair of leaves opposite each other on the stem of a bean plant is treated with a cytokinin, the treated leaf remains dark green and healthy. The untreated leaf opposite it, on the other hand, turns completely yellow and senesces rapidly as a result of its loss of nutrients to the treated leaf.

ETHYLENE

Whereas the cytokinins oppose or delay senescence, another plant hormone promotes it. This hormone is the gas ethylene, $H_2C{=}CH_2$, which is sometimes called the senescence hormone (see Figure 33.8*d*). Ethylene can be produced by all parts of the plant, and like all plant hormones, it exerts a number of effects. Back when streets were lit by gas rather than by electricity, leaves on trees near street lamps abscised earlier than those on trees farther from the lamps. We now know that it was ethylene, a combustion product of the illuminating gas, that caused the abscission. Auxin delays leaf abscission, but ethylene strongly promotes it; thus a balance of auxin and ethylene controls abscission (Figure 33.20).

Another effect of ethylene that is related to senescence is the ripening of fruit. The old saying, "One rotten apple spoils the barrel," is true. That rotten apple is a rich source of ethylene, which triggers the ripening and subsequent rotting of the others in the barrel. As the fruit ripens, it loses chlorophyll and its cell walls break down. Ethylene produced in the fruit tissue promotes both processes. Ethylene also causes an increase in its own production. Thus, once ripening begins, more and more ethylene is formed, and because it is a gas, ethylene diffuses readily throughout the fruit and even to neighboring fruit on the same or other plants.

Another gas, carbon dioxide, antagonizes the effects of ethylene. Commercial shippers and storers of fruit can thus precisely control the ripening of their wares. They hasten ripening by adding ethylene to the storage chambers; they slow ripening by adding carbon dioxide. This use of ethylene is the single most important use of a plant hormone in agriculture and commerce.

In some places autumn is a time of striking change

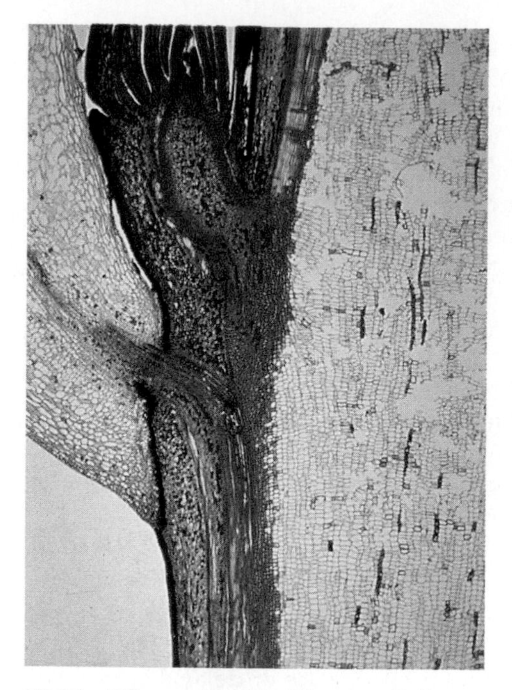

33.20 When a Leaf Is About to Fall
The petiole to the left is about to abscise from the stem by breaking away at the abscission layer (the black band). There is a bud in the axil of the leaf.

33.21 Leaf Senescence
Where leaves would be a liability under winter conditions, they senesce and die in autumn. Leaves are senescing in this forest in Vermont; only a skeleton of branches will remain to face the elements.

in the colors of the leaves of deciduous trees and shrubs. The display of colors is followed by the falling of the leaves, which have, in effect, passed through "old age" and died (Figure 33.21). Equally dramatic, in a different way, is the aging and death of entire plants, especially when they grow in great fields. Many crop plants grow vigorously throughout most of the year but then, after flowering and setting fruit, die together by the thousands. These examples show that senescence consists of irreversible, deteriorative changes controlled by internal factors.

Is senescence simply an undesirable but unavoidable fact of life, or does it play a useful role? Leaf senescence and the subsequent abscission are of real importance for the survival of the plant. Leaves senesce and abscise at the end of the growing season, shortly before the onset of the severe conditions of winter. Many species of plants have delicate leaves that could be a liability during a typical winter in the temperate zone: The temperature would be too low for efficient photosynthesis (and the ground might be frozen, making water unavailable), yet water could still be lost from the stomata in the leaves; damage from freezing would render the leaf unable to function normally during the next growing season in any case. Before the leaves die and are shed, their proteins are hydrolyzed to yield amino acids, which are then exported from the leaves to the stems—an important form of resource conservation. Controlled leaf abscission thus costs the plant little and benefits it greatly. In other parts of the world, plants shed their leaves during the harsh dry season and grow them during the wet periods, which are more favorable for growth.

What about the senescence and death of the entire plant that follows flowering and seed setting in some species? This process appears to be an adaptation for producing more offspring—by pumping so much energy (food) and so many nutrients into the seeds that the parent essentially starves itself to death.

Although primarily associated with senescence, ethylene is active at other stages of plant development as well. The stems of many dicot seedlings that have not yet seen light—as they grow upward in soil during germination—often form an apical hook (Figure 33.22). The apical hook is maintained through an asymmetric production of ethylene gas, which inhibits the elongation of cells on the inner surface of the hook. Ethylene inhibits stem elongation in general, promotes lateral swelling of stems (as do the cytokinins), and causes stems to lose their sensitivity to gravitropic stimulation.

ABSCISIC ACID

Abscisic acid is another hormone with multiple effects in the living plant (see Figure 33.8*g*). This com-

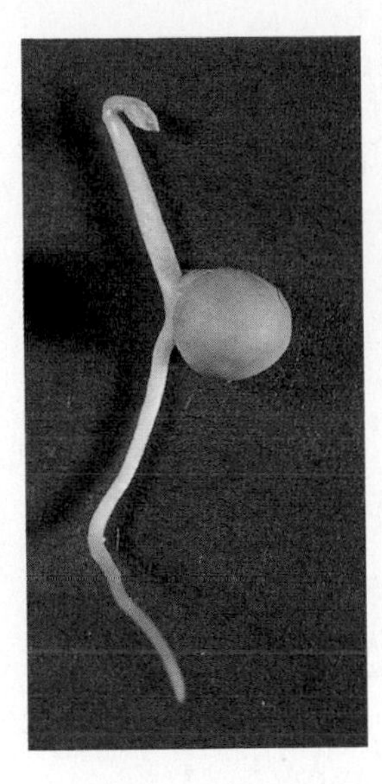

33.22 The Apical Hook of a Dicot
Asymmetric production of ethylene is responsible for the apical hook of this young pea seedling, which was grown in the dark.

pound inhibits stem elongation and is generally present in high concentrations in dormant buds and some dormant seeds. As we saw in Chapter 30, it also regulates gas and water vapor exchange between leaves and the atmosphere, through its effects on the stomata in the leaf surface. Abscisic acid is sometimes referred to as the stress hormone of plants because it accumulates when plants are deprived of water and because of its possible role in maintaining winter dormancy of buds.

In temperate zones the shoots of perennial plants do not grow constantly and in all seasons. At some time in the year the terminal buds of temperate-zone perennials become inactive, and growth ceases until the next spring. This dormancy minimizes damage to the plant during a harsh winter. Buds on a typical deciduous tree of the northern temperate zone undergo a number of changes well in advance of winter. These changes include the formation of thickened, overlapping **bud scales** that are covered with wax, which helps waterproof the bud contents—the leaf primordia and the growing point of the stem (Figure 33.23). The winter bud often contains an insulating material consisting of modified, cottony leaves.

Elsewhere in the plant are other changes. Leaves abscise, and the scars produced where leaves were formerly attached to the stems are sealed with a corky material. Lateral growth of the trunk ceases, and the solute concentrations in the transport systems increase, lowering the freezing point of the sap. These are several of the changes that constitute winter dormancy. In at least some species, some of these changes appear to be associated with an increased concentration of abscisic acid in the buds. Both the onset and termination of winter dormancy are precisely controlled.

ENVIRONMENTAL CUES

An environmental cue, the length of the night, determines the onset of winter dormancy. As summer wears on, the days get shorter—that is, the nights become longer. Leaves have a mechanism for measuring the length of the night, as we will see in the next section. This is a marvelous way to determine the season of the year. If a plant determined the season by the temperature, it could be fooled by a winter warm spell or by unseasonable cold weather in the summer. The length of the night, on the other hand, is determined by Earth's rotation around the sun and does not vary; plants use this accurate indicator to time several aspects of growth and development.

Length of night is one of several environmental cues detected by plants, or by individual parts such as leaves. Light—its presence or absence, its intensity, and its duration—provides various cues. Temperature, too, provides important environmental cues, both by its value at any particular time and by the distribution of warmer and colder stretches over a period of time. The plant "reads" an environmental cue and then "interprets" it, often by stepping up or decreasing its production of hormones.

LIGHT AND PHYTOCHROME

Light regulates many aspects of plant development. For example, some seeds will not germinate in darkness but do so readily after even a brief exposure to light. Studies have shown that blue and red light are highly effective in promoting germination, whereas green light is not. Of particular importance to plants is the fact that far-red light *reverses* the effect of a prior exposure to red light! Far-red is a very deep red, bordering on the limit of human vision and cen-

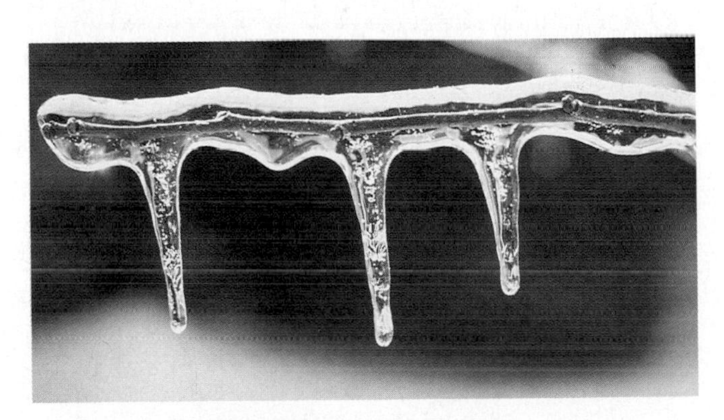

33.23 Winter Dormancy
As winter approaches, many deciduous plants cease growth, cover their buds with scales, and shed their leaves—all changes that aid survival in harsh conditions. This winter-dormant twig is coated with ice.

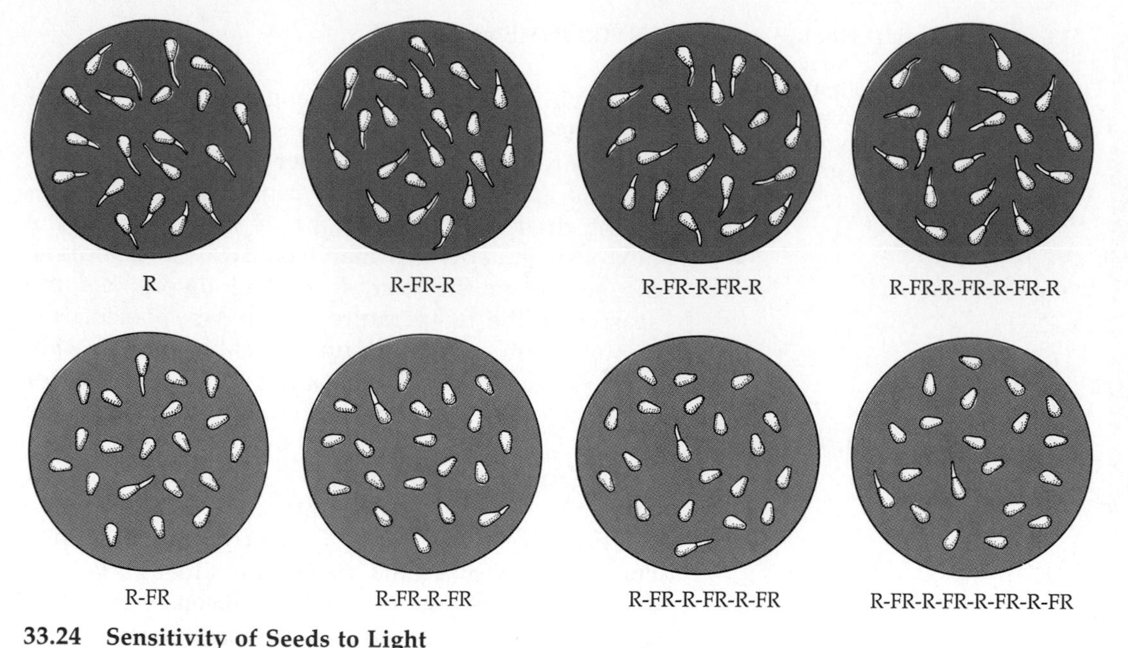

33.24 Sensitivity of Seeds to Light
Lettuce seeds were exposed to alternating periods of red light for 1 minute and far-red light for 4 minutes. Seeds germinated if the final exposure was to red (top), and remained dormant if the final exposure was to far-red (bottom). In each case the final exposure reversed the effect of the preceding exposure to the other wavelength of light.

tered upon a wavelength of 730 nm; red wavelengths are around 660 nm. If exposed to brief, alternating periods of red and far-red light in close succession, seeds respond only to the final exposure: If red, they germinate; if far-red, they remain dormant (Figure 33.24). This reversibility of the effects of red and far-red light regulates many other aspects of plant development.

The basis for the red and far-red effects resides in a bluish pigment—a protein called **phytochrome**—which exists in plants in two interconvertible forms. The pigment is blue because it absorbs red and far-red light (Figure 33.25), and reflects blue light. Red light converts phytochrome into one form; far-red converts the phytochrome back into the other form (Figure 33.26). Light drives the interconversion of the two forms of phytochrome, both in the test tube and in the living plant. The form that absorbs principally red light is called P_r. Upon absorption of a photon of red light, a molecule of P_r is converted into P_{fr}, the far-red absorbing form. P_{fr} has a number of important biological effects. As we have just seen, one of them is to initiate germination in certain seeds.

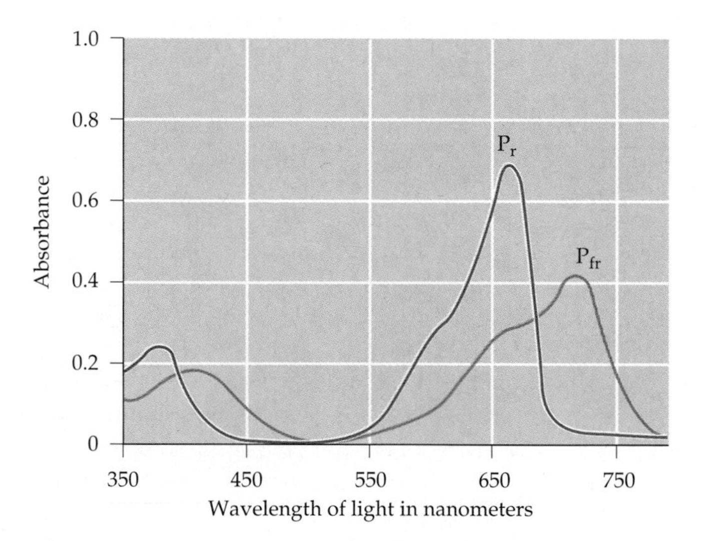

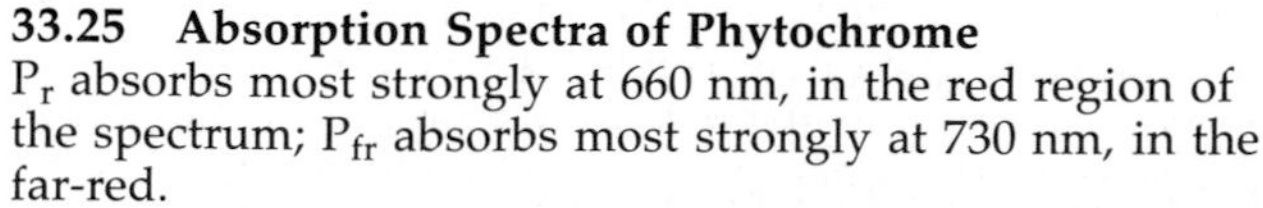

33.25 Absorption Spectra of Phytochrome
P_r absorbs most strongly at 660 nm, in the red region of the spectrum; P_{fr} absorbs most strongly at 730 nm, in the far-red.

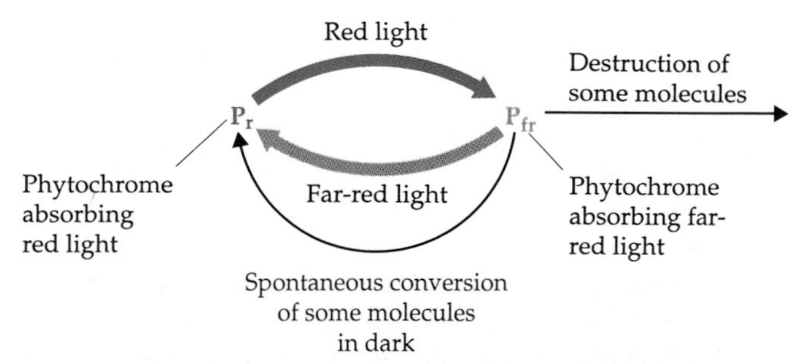

33.26 Behavior of Phytochrome
Phytochrome can exist either as P_r, which absorbs red light, or as P_{fr}, which absorbs far-red light. Each form is converted to the other by light of the appropriate color. In the living plant, some P_{fr} is spontaneously converted to P_r, and some P_{fr} is destroyed.

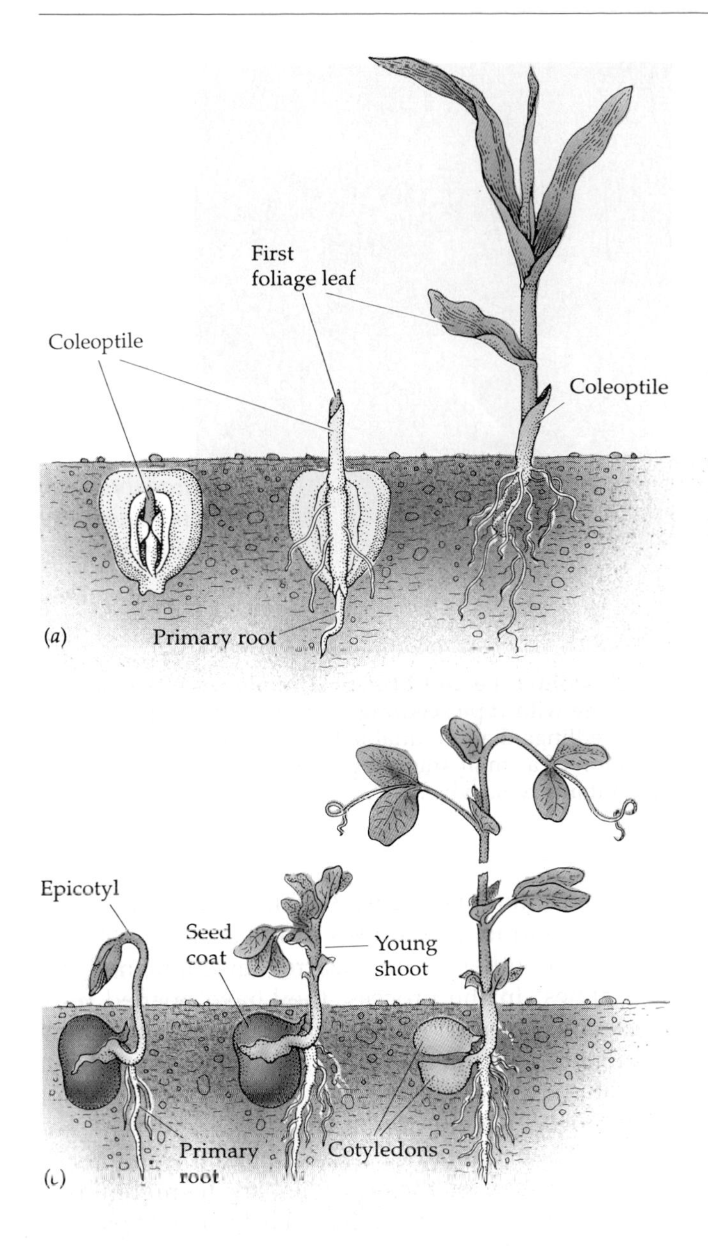

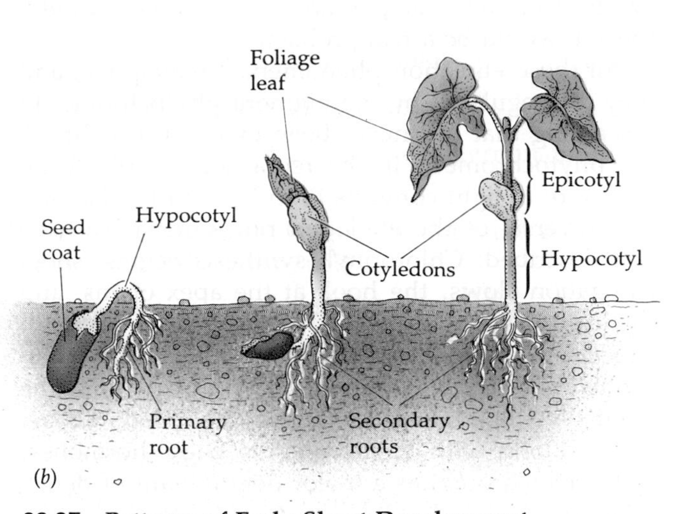

33.27 Patterns of Early Shoot Development
Shoots are protected as they penetrate the soil during germination. *(a)* A coleoptile covers the early shoot of corn and other monocots. After the shoot emerges from the soil, it pierces the coleoptile and grows out. *(b)* The shoot apex of most dicots, such as the bean shown here, is protected by the cotyledons as the upper part of the plant is pulled through the soil by the elongating hypocotyl. When the epicotyl elongates, the first foliage leaves emerge. *(c)* The cotyledons of other dicots, such as peas, remain in the soil. The shoot apex is pulled up as the bent epicotyl elongates.

We do not yet know how P_{fr} produces its many effects. Some of a plant's phytochrome may be included in membranes, where it may regulate how ions move into and out of cells and organelles. P_{fr} may function as a channel through which ions move until the conversion to P_r closes the channel. This and other ideas are currently undergoing active investigation.

Depending on whether any of a seedling's tissue is exposed to light, phytochrome helps regulate its early growth. The radicle, the embryonic root, is the first portion of the seedling to escape the seed coat. The shoot emerges later. Frequently seeds germinate below the soil surface, yielding a young plant that is **etiolated**—pale as a result of being kept in darkness. The seedling must reach the surface and begin photosynthesis before its food reserves are expended and it starves. Plants have evolved a variety of ways to cope with this problem. Etiolated flowering plants, for example, do not form chlorophyll. They turn green only upon exposure to light, thereby conserving precious resources, for chlorophyll would be of no use in the dark. Only when light is available does it "pay" to expend metabolic energy on the production of chlorophyll. An etiolated shoot elongates rapidly to hasten its arrival at the soil surface, where photosynthesis quickly begins.

Early shoot development varies among the flowering plants (Figure 33.27). In some monocots, such as grasses, the shoot is initially protected by a leaf sheath, the coleoptile. The developing shoot later grows out through the coleoptile. Dicots lack this protective structure. In most dicots, the hypocotyl elongates and the cotyledons are carried to the surface, where they become the first important photosynthetic structures. In other dicots, such as peas, the cotyledons remain below the soil surface, and tissue above them, the epicotyl, grows up through the soil. In all three cases, elongation—whether of the coleoptile, the hypocotyl, or the epicotyl—proceeds much more rapidly in the dark than in the light.

The shoot of etiolated dicot seedlings curves back into a hook at the apex during part of their development, in response to ethylene production (see Figure 33.22). The hook protects the tender apical bud

9. Ethylene
   a. is antagonized by carbon dioxide.
   b. is liquid at room temperature.
   c. delays the ripening of fruits.
   d. generally promotes stem elongation.
   e. inhibits the swelling of stems, in opposition to cytokinin effects.

10. Phytochrome
    a. is a nucleic acid.
    b. exists in two forms interconvertible by light.
    c. is a red or far-red colored pigment.
    d. is sometimes called the "stress hormone."
    e. is the photoreceptor for phototropism.

FOR STUDY

1. How may it be advantageous for some species to have seeds whose dormancy is broken by fire?
2. Cocklebur fruits contain two seeds each, and the two seeds are kept dormant by two different mechanisms. How may this be advantageous to cockleburs?
3. Corn stunt virus causes a great reduction in the growth rate of infected corn plants, so the diseased plants take on a dwarfed form. Since their appearance reminds you of the genetically dwarfed corn studied by Phinney, you suspect that the virus may inhibit the synthesis of gibberellins by the corn plants. Describe two experiments you might conduct to test this hypothesis, only one of which should require chemical measurement.
4. Whereas relatively low concentrations of auxin promote the elongation of segments cut from young plant stems, higher concentrations generally inhibit growth, as shown in the figure. In some plants, the inhibitory effects of high auxin concentrations appear to be secondary: High auxin concentrations cause the synthesis of ethylene, which is what causes the growth inhibition. Cobalt ions inhibit ethylene synthesis. How do you think the addition of cobalt ions to the solutions in which the stem segments grew would affect the appearance of the graph?
5. When carbon dioxide is applied to plants, it has effects opposite to those of ethylene. Carbon dioxide is sometimes added to the atmosphere around fruit being shipped to other parts of the country. What do you suppose is the purpose of this procedure?

READINGS

Bewley, J. D. and M. Black. 1994. *Seeds: Physiology of Development and Germination*, 2nd Edition. Plenum, New York. Lots of more advanced information; includes a chapter on agricultural and industrial aspects.

Evans, M. L., R. Moore and K.-H. Hasenstein. 1986. "How Roots Respond to Gravity." *Scientific American*, December. Classical and modern experimentation on the mechanisms of gravitropism.

Mandoli, D. F. and W. R. Briggs. 1984. "Fiber Optics in Plants." *Scientific American*, August. How plants guide light to regions of high phytochrome concentration. Includes an excellent description of the light environment in a wheat field.

Moses, P. B. and N.-H. Chua. 1988. "Light Switches for Plant Genes." *Scientific American*, April. How is light absorption by phytochrome transduced into developmental effects? Some stretches of DNA respond to phytochrome by turning on specific genes.

Raven, P. H., R. F. Evert and S. Eichhorn. 1991. *Biology of Plants*, 5th Edition. Worth, New York. A well-balanced general botany textbook.

Salisbury, F. B. and C. W. Ross. 1992. *Plant Physiology*, 4th Edition. Wadsworth, Belmont, CA. A sound textbook with excellent chapters on plant hormones and development.

Taiz, L. and E. Zeiger. 1991. *Plant Physiology*. Benjamin/Cummings, Redwood City, CA. Chapters 15 through 20 of this authoritative textbook deal with plant hormones, phytochrome, and other aspects of development.

34 Plant Reproduction

What Are All the Flowers For?

What are all these flowers for—decoration? A flower is a sexual reproductive structure, containing either female or male organs or both. The female and male organs produce, respectively, eggs and sperm. Petals produce neither eggs nor sperm, but the showy petals of many plants attract animals that carry pollen from plant to plant, assisting in cross-fertilization. Thus the petals are part of a plant's way of getting sperm and eggs together.

Biologists are still seeking answers to some important questions about flowering and reproduction. We want to know just how environmental clues lead to flower formation. The plants in this photo are of several species, each having flowers with characteristic structures. The kinds of flower parts are pretty much the same, but their numbers, shapes, colors, and arrangements are different—we want to know how flowers develop in general, and in the specific ways characteristic of different species. In this chapter we deal with several aspects of plant reproduction, including some for which we are still in search of answers.

MANY WAYS TO REPRODUCE

Plants have many ways of reproducing themselves—and with humans helping, there are even more ways. Flowers are sex organs. It is thus no surprise that almost all flowering plants reproduce sexually. But many reproduce asexually as well; some reproduce asexually most of the time. Which is the better way to reproduce? Most of the answers to this question relate to genetic diversity or genetic recombination because sexual reproduction produces new genetic combinations. The details of sexual reproduction differ among different species of flowering plants. In our discussion of Mendel's work (see Chapter 10), we saw that some plants can reproduce sexually either by cross-pollinating or self-pollinating. Self-pollination is possible because, as we explained in Chapter 25, in many species each individual has both male and female sex organs.

Both sexual and asexual reproduction are important in agriculture. Annual crops, including wheat, rice, millet, and corn—the great grain crops, all of which are grasses—as well as plants in other families, such as soybeans and safflower, are grown from seed—that is, sexually. Other crops begin asexually by slips, grafts, or other means. Orange trees, which have been under cultivation for centuries, are grown from seed except for one type, the navel orange. This plant has apparently arisen only once in history. Early in the nineteenth century, on a plantation on the Brazilian coast, one seed gave rise to one tree that had aberrant flowers. Parts of the flowers

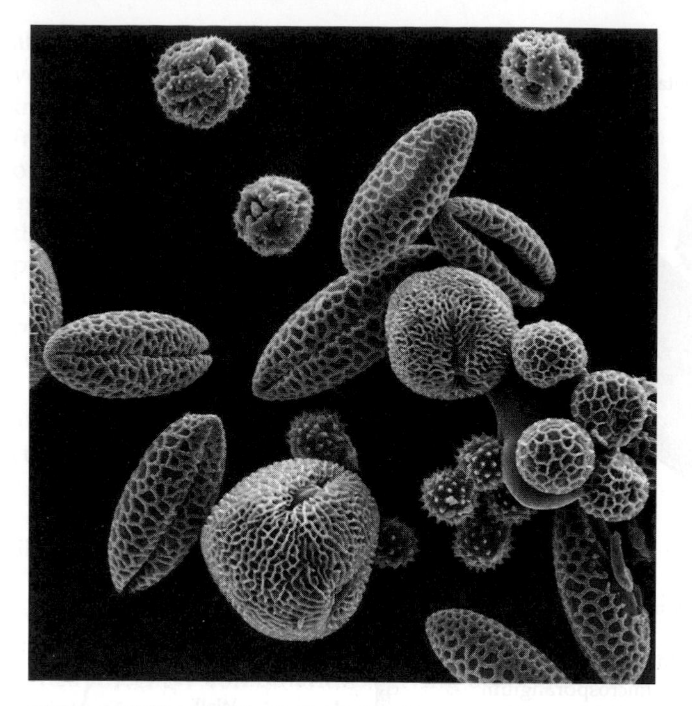

34.2 A Pollen Grain Sampler
Pollen grains of geranium, tiger lily, phlox, marigold, and dandelion. Each species has a characteristic size, shape, and pattern of wall sculpturing.

aquatic angiosperms are pollinated by water action, with water carrying pollen grains from plant to plant. Animals are important pollinators; the mutually beneficial aspects of such plant–animal associations are discussed in Chapter 48.

Double Fertilization

When a pollen grain lands on the stigma of a pistil, a pollen tube develops from the pollen grain and either grows downward on the inner surface of the style or digests its way down the spongy tissue of this female organ, growing millimeters or even centimeters in the process. The pollen tube follows a chemical gradient of calcium ions or other substances in the style until it reaches the micropyle. Of the two nuclei in the pollen grain, one is the **tube nucleus**, close to the tip of the pollen tube, and the other is the **generative nucleus** (Figure 34.4). The pollen tube eventually digests its way through megasporangial tissue and reaches the female gametophyte. The generative nucleus meanwhile has undergone one mitotic division to produce two **sperm nuclei**, *both* of which are released into the cytoplasm of one of the synergids.

From this synergid, which degenerates, each sperm nucleus enters a different cell. One sperm nucleus enters the egg cell and fuses with its nucleus, producing the diploid zygote. The other sperm nucleus enters the central cell of the embryo sac and fuses with the two polar nuclei to form a triploid ($3n$) nucleus. While the zygote nucleus begins division to form the new sporophyte embryo, the triploid nucleus undergoes rapid mitosis to form a specialized nutritive tissue, the **endosperm**. The female antipodal cells and synergids eventually degenerate. A recent research breakthrough, described in Box 34.A, will open the way for new studies of fertilization in plants.

Shortly after fertilization, highly coordinated growth and development of embryo, endosperm, integuments, and carpel ensues. As large amounts of nutrients are moved in from other parts of the plant, the endosperm begins accumulating starch, lipids, and proteins. The integuments develop into a double-layered seed coat, sometimes fleshy, and sometimes heavily lignified and hard. The carpel ultimately becomes the wall of the fruit that encloses the seed.

Of all the characteristic traits of the angiosperms, only one trait, the double fertilization mechanism just described, is found in *all* angiosperms and *only* in angiosperms—and in one tiny gymnosperm group. The origin of this process in geological time is wholly unknown.

Embryo Formation in Flowering Plants

The first step in the formation of the embryo is a mitotic division of the zygote, the fertilized egg in

34.3 Wind Pollination
A flowering inflorescence of a grass. The numerous anthers all point away from the stalk and stand free of the plant, promoting dispersal of the pollen by wind.

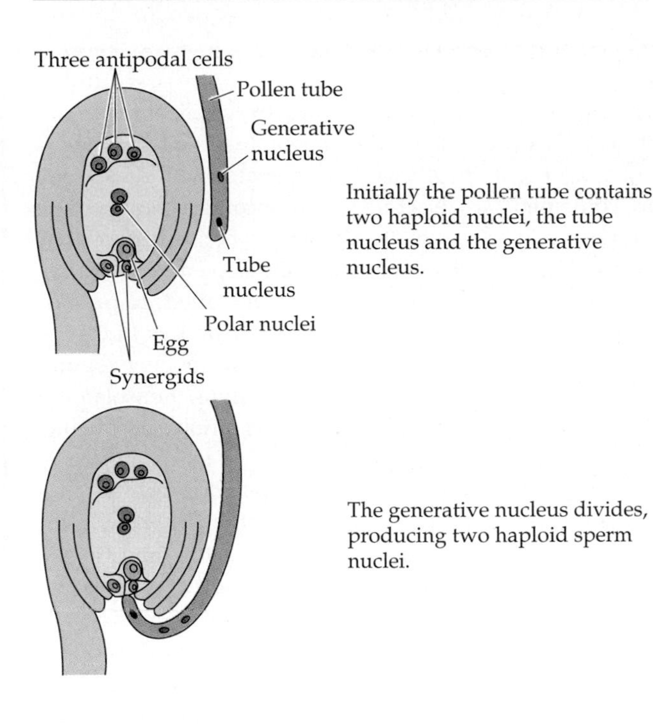

The sperm nuclei enter the cytoplasm of a synergid.

The synergid breaks down; one sperm nucleus fertilizes the egg, forming the zygote, the first cell of the $2n$ sporophyte generation. The other unites with the two polar nuclei, forming the first cell of the $3n$ endosperm.

34.4 Pollen Nuclei and Double Fertilization
The sperm nuclei contribute to the formation of the diploid zygote and the triploid endosperm. Double fertilization is a characteristic feature of angiosperm reproduction.

the embryo sac, giving rise to two daughter cells. Even at this stage the two cells face different fates. An asymmetric (uneven) distribution of contents within the zygote causes one end to produce the embryo proper and the other end to produce an early supporting structure, the **suspensor** (Figure 34.5). Polarity has been established, as has the longitudinal axis of the new plant. A filamentous suspensor and a globular embryo are distinguishable after just four mitotic divisions. The suspensor soon ceases to elongate, and, as development continues, the first organs take form within the embryo.

In dicots (monocots are somewhat different) the embryo soon takes on a characteristic heart-shaped form as the cotyledons start to grow. Further elongation of the cotyledons and of the main axis of the embryo gives rise to what is called the torpedo stage, during which some of the internal tissues begin to differentiate. The elongated region below the cotyledons is called the hypocotyl. At the top of the hypocotyl, between the cotyledons, is the shoot apex; at the other end is a root apex. Each of these apical regions contains an apical meristem whose dividing cells give rise to the organs of the mature plant. In many species, such as peas and peanuts, the cotyledons absorb the food reserves from the surrounding endosperm and grow very large in relation to the rest of the embryo (Figure 34.6*a*). In others, including the castor bean, the cotyledons remain thin (Figure 34.6*b*); they will draw on the reserves in the endo-

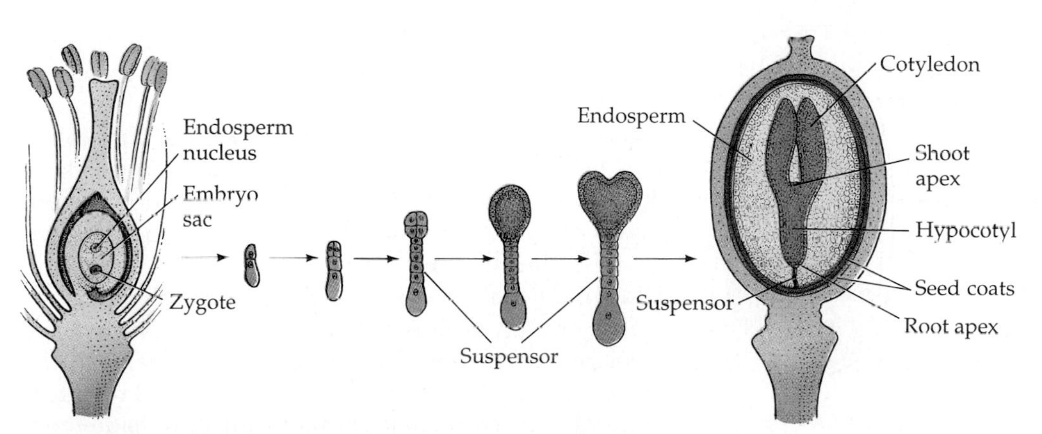

34.5 Early Development of a Dicot
The zygote nucleus divides mitotically, one daughter cell giving rise to the embryo proper and the other to the suspensor. The embryo develops through intermediate stages to form the torpedo stage (far right). The tissues surrounding the embryo sac develop into seed coats.

BOX 34.A

In Vitro Fertilization: Test-Tube Plants

Hundreds of children have been conceived by means of in vitro fertilization, the combination of egg and sperm in a petri plate rather than in the mother's body. Can we do the same thing with plant gametes—make test-tube plants? It sounds simple, but in vitro fertilization in plants did not succeed fully until 1993.

What made this achievement so difficult? In animals, sperm and eggs are set free and can unite readily. In plants the gametes are not free of other cells. The sperm are contained within the pollen tube, and osmotic treatments are needed to free them from the pollen. The egg presents an even greater problem because it is contained in the embryo sac. Treatment with enzymes that digest cell walls, followed by microdissection, can release the egg. However, the freed sperm and egg appear to be incapable of fusing spontaneously. What is wrong? One possibility is that fertilization depends on activities of other cells in the embryo sac; another possibility is that the egg nucleus is damaged by the enzymes used to remove the egg cell wall.

A combination of techniques has now made in vitro fertilization in plants possible. One key technique is to treat both gametes with bursts of electricity. The second technique is to use "feeder cells," derived from normal embryos and separated from the treated gametes by a filter, to provide as yet unknown substances that support development of the new zygote. Under these conditions the isolated gametes fuse successfully, producing zygotes that develop normally and give rise to normal adult plants.

In vitro fertilization will enable the study of some interesting questions about the reproductive development of plants. For example, analysis of the products of the cultured feeder cells will shed light on the role of the synergids in normal, in vivo fertilization in plants.

sperm as needed when the seeds germinate. In either case, the endosperm is the maternal plant's contribution to the nutrition of the next generation (Box 34.B).

Why does the embryo stop developing within the seed? In the late stages of embryonic development the seed loses water—sometimes as much as 95 percent of its original water content. In its dried state, the embryo is incapable of further development. It remains in this dormant state until the conditions are right for germination. (Recall from Chapter 33 that a necessary first step in seed germination is the massive imbibition of water.)

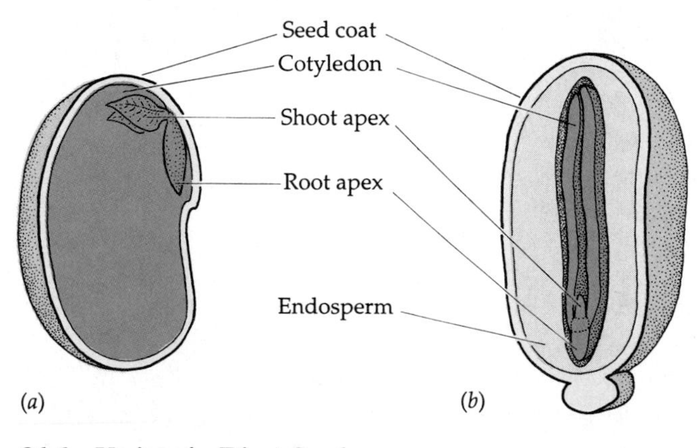

34.6 Variety in Dicot Seeds
In some dicots the cotyledons absorb much of the endosperm and fill most of the seed *(a)*. In others the endosperm remains separate and the cotyledons remain thin *(b)*.

Fruit

After fertilization, the ovary wall of a flowering plant—together with its seeds—develops into a fruit. A fruit may consist of only the mature ovary and its seeds, or it may include other parts of the flower or structures that are closely related to it. The major variations on this theme are illustrated in Figure 25.28.

Some fruits play a major role in reproduction because they help disperse seeds over substantial distances. A number of trees, including ash, elm, maple, and tree of heaven, produce a dry, winged fruit called a samara (see Figure 29.2*d*). A samara spins like a helicopter blade and, while whirling downward, holds the fruit aloft long enough for it to be blown some distance from the parent tree. The dandelion fruit is also marvelously adapted for dispersal by wind. Water disperses some fruits; coconuts have been spread in this way from island to island in the Pacific (Figure 34.7*a*). Still other fruits travel by hitching rides with animals—either inside or outside them (Figure 34.7*b*). Burdocks, for example, have hooks that adhere to animal fur, and many other plants have prickled, barbed, hairy, or sticky fruits. Fleshy fruits such as berries provide food for mammals or

(*a*)

(*b*)

34.7 Dispersal of Fruits
(*a*) These coconuts are germinating where they washed ashore on a Tahitian beach. (*b*) Many sticky fruits, such as the burrs of the common burdock, hitch rides on animals, traveling far from their parent plants.

BOX 34.B

Maternal "Care" in Plants

The mammalian mother has a special relationship with her embryonic and newborn offspring. The mammalian embryo receives its nourishment by way of the maternal bloodstream, and it is subjected to hormonal influences from the mother, but not from the father. Maternal *plants* also influence their offspring in important ways. Of course, the offspring receive half their genetic endowment by way of the egg, but the maternal parent has other, specific effects, especially relating to the size of seeds produced, because the seeds develop from and within tissues of the maternal flower.

Barbara A. Schaal, of Washington University, St. Louis, has studied maternal effects in the Texas bluebonnet, *Lupinus texensis*. She demonstrated that variation in seed mass in these plants could be attributed to responses of the maternal plant to environmental variation. Differences in seed mass in turn affected the subsequent performance of the offspring, as the figure shows. A higher percentage of larger seeds germinated, and they germinated earlier. Larger seeds gave rise to larger seedlings. The seedlings from larger seeds grew more leaves and survived better during the first month of life after germination. By six weeks of age, differences in size and survivorship between seedlings from larger and smaller seeds were no longer evident. However, between the ages of 80 and 130 days, the period of most intense reproduction in the bluebonnet, plants from larger seeds produced more ovules than did those from smaller seeds. Because the size of the seed is determined more by the maternal parent than by the genotype of the embryo, Schaal concluded that maternal effects were important influences on the development and reproduction of offspring.

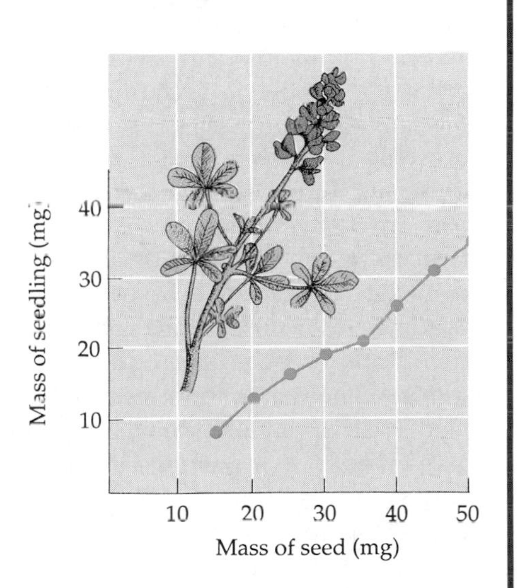

birds; their seeds travel safely through the animal's digestive tract or are regurgitated, in either case being deposited some distance from the parent plant.

Having discussed mechanisms for dispersing seeds and pollen, we should make it clear that most seeds, and most pollen grains, end up close to their sources. Long-range dispersal is the exception, not the rule. In cases of extensive dispersal, paternal genes usually travel farther than maternal genes because, although both maternal and paternal genes travel with the seed, only paternal genes travel with the pollen grain.

We have traced the sexual life cycle from the flower to the fruit to the dispersal of seeds. We discussed seed germination and vegetative development of the seedling in Chapter 33. Now let's complete the sexual life cycle by considering the transition from the vegetative to the flowering state, and how this transition is regulated.

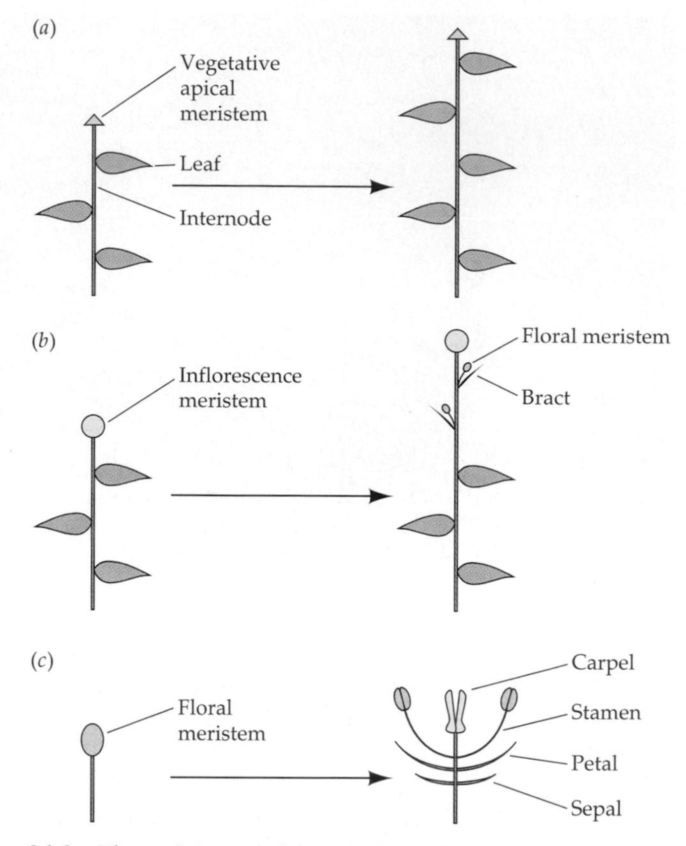

34.8 Flowering and the Apical Meristem
A vegetatively growing apical meristem continues to produce leaves and internodes. Inflorescence and floral meristems each give rise to a characteristic set of products.

TRANSITION TO THE FLOWERING STATE

Flowering may terminate, repeatedly interrupt, or accompany vegetative growth. The transition to the flowering state often marks the end of vegetative growth for a plant. If we view a plant as something produced by a seed for the purpose of bearing more seeds, then the act of flowering is one of the supreme events in a plant's life.

The first visible sign of the transition to the flowering state may be a change in one or more apical meristems in the shoot system. During vegetative growth an apical meristem continues to produce leaves, lateral buds, and internodes (Figure 34.8*a*); this growth is *indeterminate* (see Chapter 33), in contrast to the usually determinate growth of an animal to a standard size. However, if a vegetative meristem becomes an **inflorescence meristem** that will give rise to an inflorescence, it produces other structures. The inflorescence meristem generally produces smaller leafy structures called **bracts** separated by internodes, as well as new meristems in the angles between the bracts and the internodes (Figure 34.8*b*). These new meristems may be inflorescence meristems or **floral meristems**, which give rise to the flowers themselves (Figure 34.8*c*). Each floral meristem in turn typically produces four consecutive whorls of organs—the sepals, petals, stamens, and carpels—separated by very short internodes. In contrast to vegetative meristems and some inflorescence meristems, floral meristems are responsible for *determinate* growth—the limited growth of the flower.

How does the floral meristem give rise, in short order, to whorls of four different organs? This problem is not unlike that of accounting for the construction of a fruit fly larva (see Chapter 17). Recent work underscores some of the similarities. Most strikingly, a group of **homeotic genes**—organ-identity genes—work in concert to specify the successive whorls. We recognize the presence of homeotic genes because mutations in these genes, called homeotic mutations, lead to major alterations in flower structure. Table 34.1 gives a sense of some of these mutations. Do you see a pattern in the alterations? These mutations, along with analysis of the homeotic genes and their products, are leading us to a preliminary understanding of how normal flowers develop. Earlier work, which we will consider next, helped to explain how the transition to the flowering state is initiated.

Flowering is triggered in very different ways in different plant species. Environmental cues include seasonal changes in the lengths of days and nights and seasonal temperature changes.

PHOTOPERIODIC CONTROL OF FLOWERING

In 1920 W. W. Garner and H. A. Allard of the U.S. Department of Agriculture studied the behavior of a newly discovered mutant tobacco plant. The mutant

TABLE 34.1
Homeotic Mutations in Flower Development

	PHENOTYPE			
GENOTYPE	WHORL 1	WHORL 2	WHORL 3	WHORL 4
Wild type	Sepals	Petals	Stamens	Carpels
Mutant A	Carpels	Stamens	Stamens	Carpels
Mutant B	Sepals	Sepals	Carpels	Carpels
Mutant C	Sepals	Petals	Petals	Sepals

was named Maryland Mammoth because of its large leaves and exceptional height (and where it was found). The other plants in the field flowered, but the Maryland Mammoth continued to grow. Garner and Allard took cuttings of the Maryland Mammoth into their greenhouse, and the plants that grew from the cuttings finally flowered in December. Garner and Allard also noticed that some soybean plants all flowered at about the same time, in late summer, even though they had been planted at different times in the spring.

Garner and Allard guessed that both of these observations had something to do with the seasons. They tested a number of likely seasonal variables, such as temperature, but the key variable proved to be the length of day. They moved plants between light and dark rooms at different times to vary the length of day artificially and were able to establish a direct link between flowering and day length. The **critical day length** for Maryland Mammoth tobacco proved to be 14 hours (Figure 34.9). The plants did

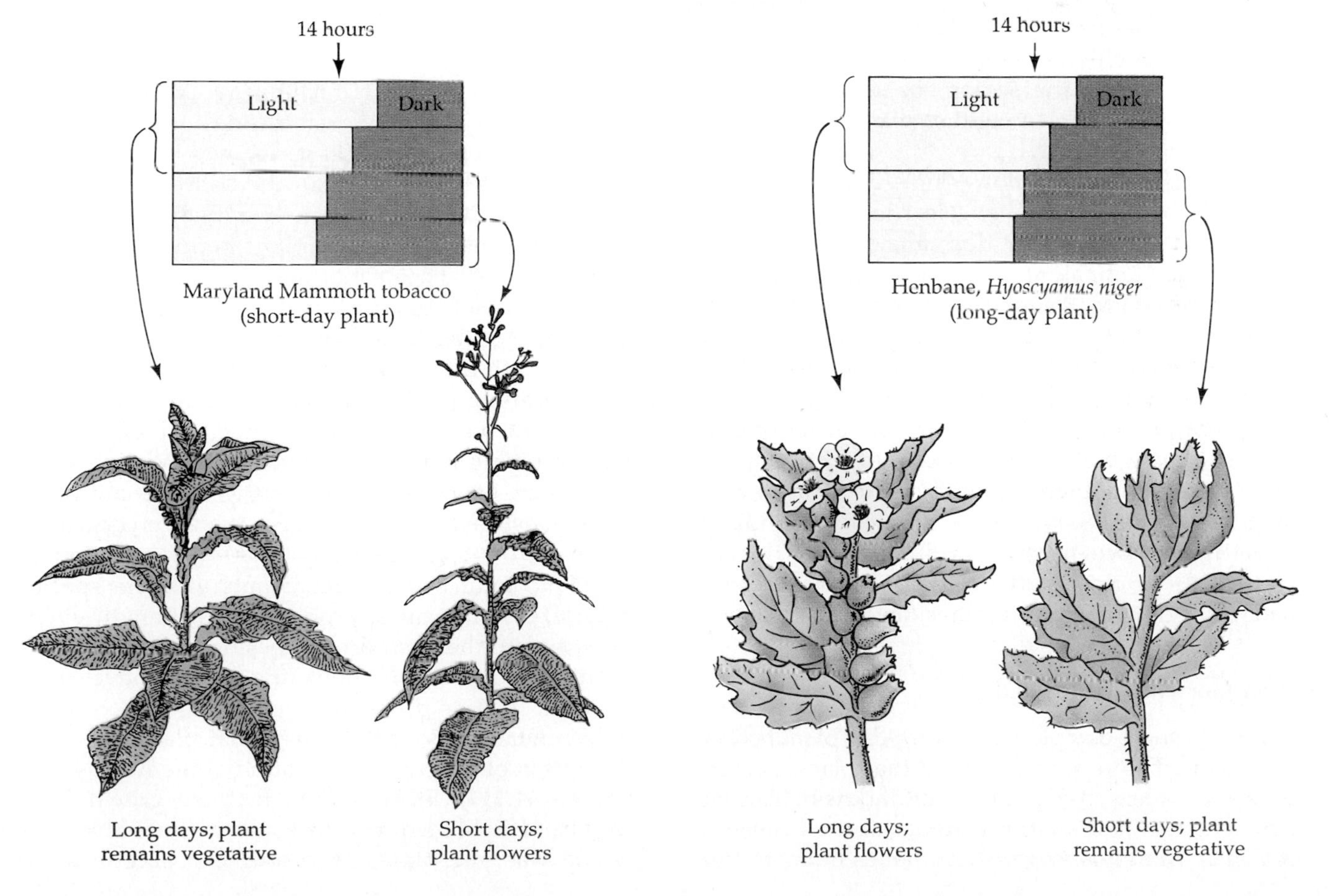

34.9 Day Length and Flowering
By varying the length of day, Garner and Allard showed that Maryland Mammoth tobacco flowers only when days are shorter than 14 hours; that is, the critical day length is 14 hours.

not flower if the light period was longer than 14 hours each day, but flowering commenced after the days became shorter than that. This phenomenon of control by lengths of days and nights is called **photoperiodism**. Soybeans and Maryland Mammoth tobacco are **short-day plants** (SDPs). Spinach and clover are examples of **long-day plants** (LDPs), which flower only when the day is longer than a critical minimum. Generally LDPs are triggered to flower in midsummer and SDPs in late summer, or sometimes in the spring.

It is a historical accident that we use the terms short-day plant and long-day plant. The SDPs could as well have been called long-night plants and the LDPs short-night plants, because the natural day has a fixed length of 24 hours. Do plants measure the length of day or the length of night? The answer will be given presently.

Other Patterns of Photoperiodism

Some plants require more-complex photoperiodic signals in order to flower. One group, the short-long-day plants, must first experience short days and then long ones. Accordingly, because they pass first through the short days of early spring and then through ever longer ones, they flower during the long days before midsummer. Another group, the long-short-day plants, cannot flower until the long days of summer have been followed by shorter ones, so they bloom only in the fall. Long-short-day plants will not bloom in the spring, in spite of its short days, nor will a short-long-day plant flower in late summer.

Other effects besides flowering are also under photoperiodic control. We have learned, for example, that the onset of winter dormancy is triggered by short days. (Animals, too, show a variety of photoperiodic behaviors; in aphids, for example, long days favor the development of sexually reproducing females, whereas females that reproduce asexually develop when days are short.)

It is important to note that the flowering of some angiosperms is not photoperiodic. In fact, there are more **day-neutral** plants than there are short-day and long-day plants. Some plants are photoperiodically sensitive only when young and become day-neutral as they grow older. Others require specific combinations of day length and other factors to flower.

Importance of Night Length

The terms short-day plant and long-day plant became entrenched before it was learned that plants actually measure the length of night, or of darkness. This fact was demonstrated by Karl Hamner of the University of California at Los Angeles and James Bonner of the California Institute of Technology. Working with cocklebur, an SDP, they ran a series of experiments in which either (1) the light period was kept constant, either shorter or longer than the critical day length, and the dark period was varied, or (2) the dark period was kept constant and the light period was varied (Figure 34.10). The plants flowered under all treatments in which the dark period exceeded 9 hours, regardless of the length of the light period. Thus it is the length of the *night* that matters; for cocklebur, the critical night length is about 9 hours.

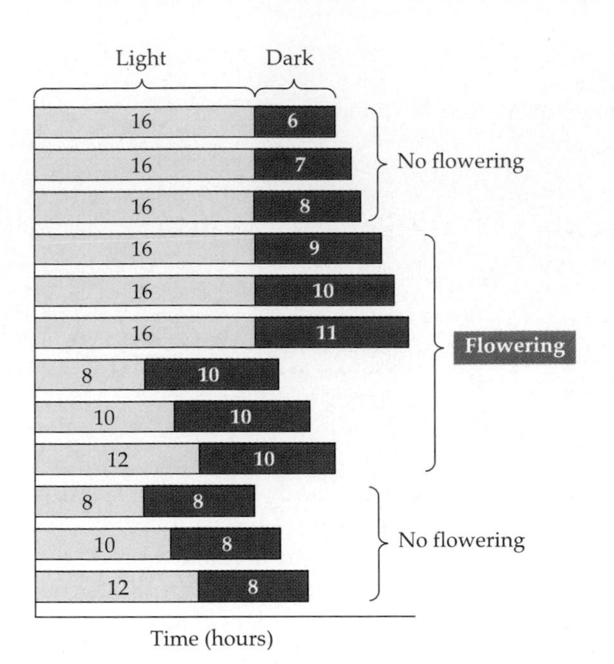

34.10 Night Length and Flowering
In the experiments symbolized by the six upper bars, plants were exposed to 16 hours of light, followed by dark periods of various duration. Only plants given 9 or more hours of dark flowered. In the 6 experiments indicated at the bottom, plants were exposed to various light periods followed by 8 or 10 hours of dark. Only plants given 10 hours of dark flowered. Experiments like these showed that short-day plants really should be called long-night plants.

In cocklebur, a single long night is enough of a photoperiodic stimulus to trigger full flowering some days later, even if the intervening nights are short ones. Most plants, less sensitive than the cocklebur, require from two to many nights of appropriate length to induce flowering. Plants of some species must experience an appropriate night length every night before they can flower. A single shorter night, even one day before flowering would have commenced, inhibits flowering.

Hamner and Bonner showed that plants measure the length of the night using another method as well (Figure 34.11*a*). SDPs and LDPs were grown on a variety of light regimes. In some regimes the dark period was interrupted by a brief exposure to light;

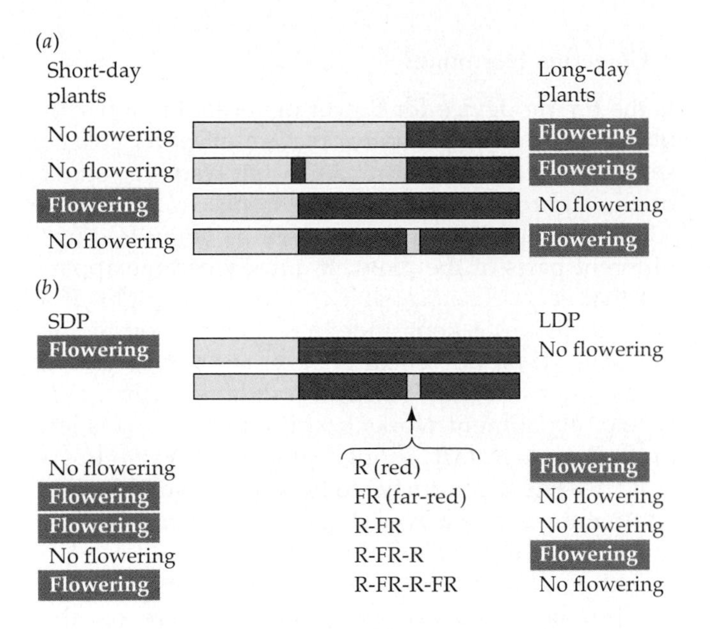

34.11 Interrupted Days and Nights
(a) Short-day plants require long, uninterrupted nights to flower. Long-day plants flower when the night is short; interrupting their long day has no effect but interrupting a long night with a brief period of light induces flowering. *(b)* When plants are exposed to red (R) and far-red (FR) light in alternation, the final treatment determines the effect of the light interruption, suggesting that phytochrome participates in photoperiodic responses.

in others, the light period was interrupted briefly by darkness. Interruptions of the light period by darkness had no effect on the flowering of either short-day or long-day plants. Even very brief interruptions of the dark period, however, completely nullified the effect of a long night. An SDP flowered only if the long nights were uninterrupted. An LDP experiencing long nights flowered if these were broken by exposure to light. Thus a plant must have a timing mechanism that measures the length of a continuous dark period and uses the result to trigger flowering or to remain vegetative. Despite much study, the nature of the timing mechanism is still unknown.

Phytochrome seems to participate in the photoperiodic timing mechanism. In the interrupted-night experiments, the most effective wavelengths of light were red (Figure 34.11*b*). The effect of a red-light interruption of the night was fully reversed by a subsequent exposure to far-red light. It was once thought that the timing mechanism might simply be the slow conversion of phytochrome during the night from the P_{fr} form—produced during the light hours—to the P_r form. But this suggestion is inconsistent with most of the experimental observations and must be wrong. Phytochrome must be only a photoreceptor. The time-keeping role is played by a biological clock.

Circadian Rhythms and the Biological Clock

It is abundantly clear that organisms have a way of measuring time and that they are well adapted to the 24-hour day–night cycle of our planet. Some sort of biological clock resides within the cells of all eukaryotes, and the major outward manifestations of this clock are known as **circadian rhythms** (Latin *circa*, = "about" and *dies*, = "day"). Plants provide innumerable examples of approximately 24-hour cycles. The leaflets of a plant such as clover or the tropical tree *Albizia* normally hang down and fold at night and rise and expand during the day. Flowers of many plants show similar "sleep" movements, closing at night and opening during the day. They continue to open and close on an approximately 24-hour cycle even when the light and dark periods are experimentally modified (Figure 34.12).

The circadian rhythms of protists, animals, fungi, and plants share some important characteristics. First, the **period** is remarkably insensitive to temperature, although the **amplitude** of the fluctuation may be drastically reduced by lowering the temperature (Figure 34.13 explains these terms). Second, circadian rhythms are highly persistent; they continue even in an environment in which there is no alternation of light and dark. Third, circadian rhythms can be **entrained**, within limits, by light–dark cycles that differ from 24 hours. That is, the period an organism expresses can be made to coincide with that of the light–dark regime. The period in nature is approximately 24 hours. If an *Albizia* tree, for example, were to be

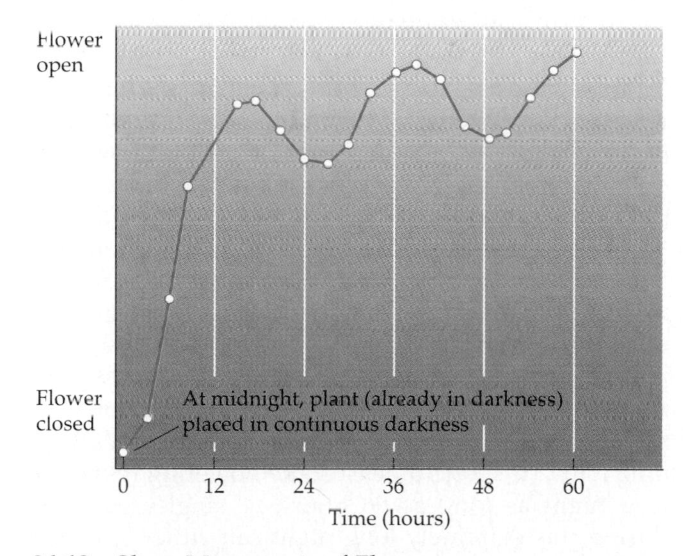

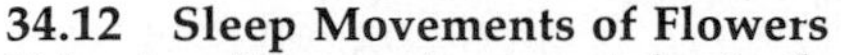

34.12 Sleep Movements of Flowers
Kalanchoe flowers close at night. In the middle of the night (time = 0), when the flowers were completely closed, biologists transferred a *Kalanchoe* plant to a dark box and kept it there for another 60 hours. During that time the flowers continued to open and partially close on a 24-hour cycle.

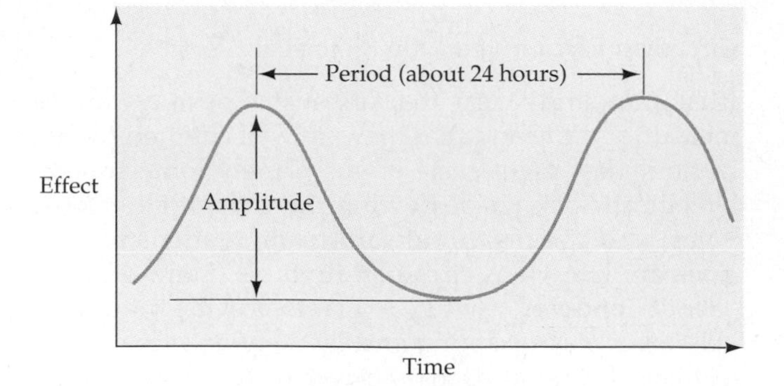

34.13 Features of Circadian Rhythms
Circadian rhythms are characterized on the basis of time, measured in periods of about 24 hours, and on the basis of the magnitude of the rhythmic effect, measured by the cycle's amplitude.

placed under artificial light on a day–night cycle totaling exactly 24 hours, the rhythm expressed would show a period of exactly 24 hours. If, however, an experimenter used a day–night cycle of, say, 22 hours, then the rhythm would be entrained to a 22-hour period.

If light–dark cycles can entrain circadian rhythms, it follows that light alone should also be able to shift the rhythm. If an organism is maintained under constant darkness, with its circadian rhythm expressed on the approximately 24-hour period, a brief exposure to light can make the next peak of activity appear either later or earlier than one would have predicted, depending on when the exposure is given. Moreover, the organism does not then return to its old schedule if kept in darkness. If the first peak is delayed by 6 hours, the subsequent peaks are all 6 hours late. Such phase shifts are permanent—until the organism receives more exposures to light.

Important questions about circadian rhythms remain to be answered. We do not know, for example, how light resets the biological clock. In fact, we still do not know what the clock's biochemical or biophysical basis is.

There is now ample evidence that the photoperiodic behavior of plants is based on the interaction of night length with the biological clock. How the clock is coupled with flowering, however, is unclear. Experiments with the small lawn weed *Chenopodium rubrum* provided one sort of evidence. As a short-day plant, *Chenopodium* will flower in response to a single long night. It shows at least some flowering response to a night as long as 96 hours. A single red flash during this extremely long night can either enhance or inhibit this response, depending on precisely when it is administered (Figure 34.14). The effect of the red flash oscillates on an approximately daily basis (like the clock), but just how the underlying daily rhythm relates to flowering remains to be learned.

A Flowering Hormone?

Is the timing device for flowering located in a particular part of an angiosperm, or are all parts able to sense the length of night? As in the Darwins' study of the light receptor for phototropism (see Figure 33.12), this question was resolved by "blindfolding" different parts of the plant. It quickly became apparent that each leaf is capable of timing the night. If a short-day plant is kept under a regime of short nights and long days, but a leaf is covered so as to give it long nights, the plant will flower (Figure 34.15*a*). This type of experiment works best if only one leaf is left on the plant. In fact, if one leaf is given a photoperiodic treatment conducive to flowering—an inductive treatment—other leaves kept under noninductive conditions will tend to inhibit flowering.

Although it is the leaves that sense an inductive night period, the flowers form elsewhere on the plant. Thus a message must be sent from the leaf to the site of flower formation. Three lines of evidence suggest that this message is a chemical substance—a flowering hormone. First, if a photoperiodically induced leaf is removed from the plant shortly after the inductive night period, the plant does not flower. If, however, the induced leaf remains attached for several hours, the plant flowers. This result suggests that something—the hypothetical hormone—must be synthesized in the leaf in response to the inductive night, then move out of the leaf to induce flowering.

The second line of evidence for the existence of a flowering hormone comes from grafting experiments. If two cocklebur plants are grafted together, and if one plant is given inductive long nights and its graft

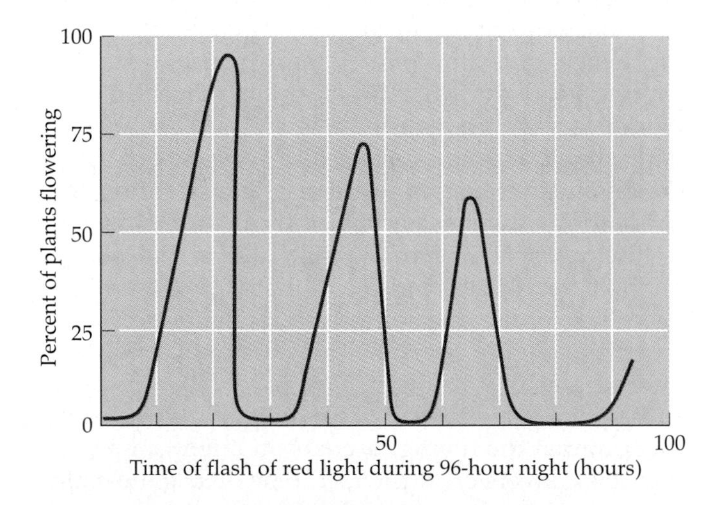

34.14 A Flowering Rhythm
The short-day plant *Chenopodium rubrum* flowers in response to a single 96-hour "night." If single flashes of red light are given during such a 96-hour dark period, they either enhance or inhibit the flowering response. The peaks of enhancement are on about a 24-hour cycle initiated by the light-to-dark transition that started the night.

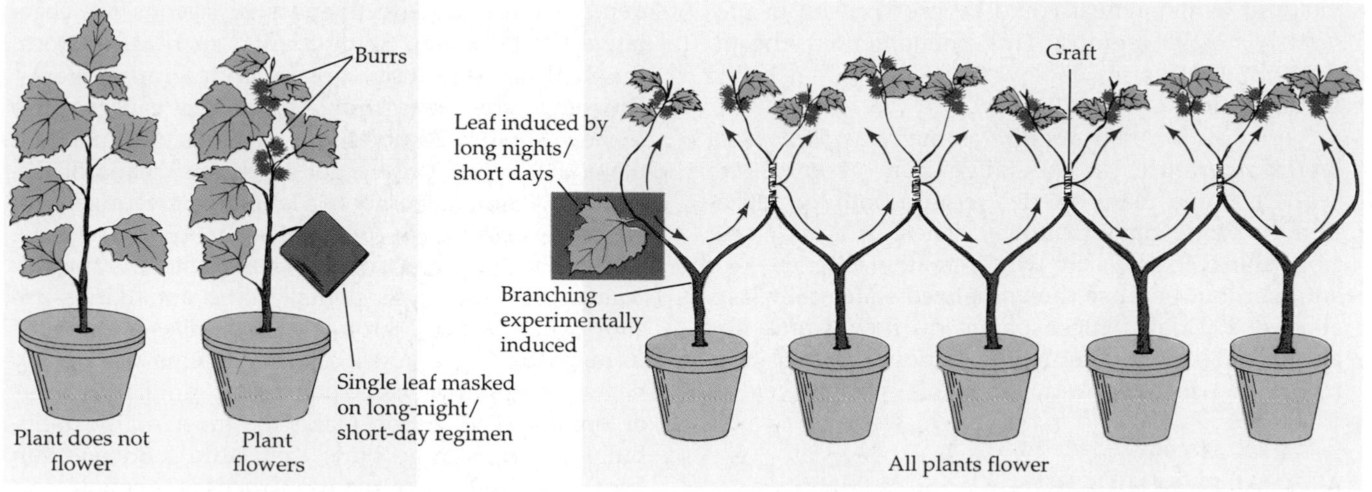

(*a*) Cockleburs on short nights/long days

(*b*) Grafted cockleburs on short nights/long days

34.15 Evidence for a Flowering Hormone
(*a*) Cocklebur, a short-day plant, will not flower if kept under long days and short nights. If even one leaf is masked for part of the day, however—thus shifting that leaf to short days and long nights—the plant will flower; note the burrs. Because the flowers are formed far from the induced leaf, some substance probably carries the flowering message from the leaf. (*b*) Five cocklebur plants grafted together and kept under long days and short nights, with most leaves removed. If a leaf on a plant at one end of the chain is subjected to long nights, all of the plants will flower. Arrows indicate the routes of the hypothetical flowering hormone from the induced leaf.

partner is given noninductive short nights, both plants flower (Figure 34.15*b*). Grafting experiments also provided the third line of evidence for a flowering hormone. Jan A. D. Zeevaart, a plant physiologist now at Michigan State University, exposed a single leaf of the SDP *Perilla* to a short-night/long-day regime, inducing the plant to flower. Then he detached this leaf and grafted it onto another, noninduced, *Perilla* plant—which responded by flowering. The same leaf grafted onto successive hosts caused each of them to flower in turn. As long as three months after the leaf was exposed to the short-night/long-day regime, it could still cause plants to flower.

The Search for Florigen

Experiments such as Zeevaart's suggest that the photoperiodic induction of a leaf causes a more or less permanent change in it, inducing it to start and to continue producing a flowering hormone that is transported to other parts of the plant, switching those target parts to the reproductive state. So reasonable is this idea that biologists have named the hormone, even though after decades of active searching the compound has yet to be isolated and characterized. The hormone is called **florigen**. The direct demonstration of florigen activity remains a cherished goal of plant physiologists. Gibberellins regulate the flowering of many species, especially of long-day plants, but these hormones do not have the properties of florigen.

As a final teaser, we will describe an experiment that suggests that the florigen of short-day plants is identical with that of long-day plants, even though SDPs produce it only under long nights and LDPs only under short nights. An SDP and an LDP were grafted together, and both flowered, as long as the photoperiodic conditions were inductive for one of the partners. Either the SDP or the LDP could be the one induced, but both would always flower. These results suggest the transfer of a flowering-inducing hormone, the elusive florigen, from one plant to the other.

VERNALIZATION AND FLOWERING

In both wheat and rye, we distinguish two categories of flowering behavior. Spring wheat, for example, is sown in the spring and flowers the same year. It is an annual plant. Winter wheat is biennial and must be sown in the fall; it flowers in its second summer. If winter wheat is not exposed to cold after its first year, it will not flower normally the next year. The implications of this finding were of great agricultural interest in Russia because winter wheat is a better producer than spring wheat, but it cannot be grown in parts of Russia because the winters there are too cold for its survival. A number of studies performed in Russia during the early 1900s demonstrated that if seeds of winter wheat were premoistened and prechilled, they would develop and flower normally the same year when sown in the spring. Thus, high-yielding winter wheat could be grown even in previously hostile regions. This phenomenon—the induction of flowering by low temperatures—is called **vernalization**.

Vernalization may require as many as 50 days of low temperature (in the range from about −2 to +12°C). Some plant species require both vernalization and long days to flower. There is a long wait from the cold days of winter to the long days of summer, but because the vernalized state easily lasts at least 200 days, these plants do flower once the appropriate night length is experienced. Thus vernalization, once induced, is a stable condition.

ASEXUAL REPRODUCTION

Although sexual reproduction takes up the bulk of the space in this chapter's discussion of how plants reproduce, asexual reproduction is responsible for an important fraction of the new plant individuals appearing on Earth. This suggests another answer to the question asked at the beginning of this chapter: In some circumstances, asexual reproduction is better. For example, think about genetic recombination. We have already noted that when a plant self-fertilizes there are fewer opportunities for genetic recombination than there are with cross-fertilization. With self-fertilization, the only genetic variation that can be arranged into new combinations is that possessed by the single parental plant. A plant that is heterozygous for a locus can produce among its progeny both kinds of homozygotes for that locus plus the heterozygote, but it cannot produce any progeny that carry alleles that it does not itself possess. Asexual reproduction goes a step further: It eliminates genetic recombination. When a plant reproduces asexually, it produces a clone of progeny with genotypes identical to its own. If a clone is highly adapted to its environment, the many copies of that genotype that may be formed by asexual reproduction may spread throughout that environment. This ability to exploit a particular environment is an advantage of asexual reproduction.

Asexual Reproduction in Nature

We call stems, leaves, and roots vegetative parts, distinguishing them from the reproductive parts. The modification of a vegetative part of a plant is what makes **vegetative reproduction** possible. The stem is the part modified in many cases. Strawberries and some grasses produce **stolons**, horizontal stems that form roots at intervals and establish potentially independent daughter plants (see Figure 29.9*c*). Other stolons are branches that sag to the ground and put out roots. The rapid multiplication of water hyacinths demonstrates the effectiveness of stolons for vegetative reproduction. Some plants, such as potatoes, form **tubers**, the fleshy tips of underground stems. **Rhizomes** are underground stems that can give rise to new shoots. Bamboo is a striking example of a plant that reproduces vegetatively by means of rhizomes. A single bamboo plant can give rise to a stand—even a forest—of plants constituting a single, physically connected entity. Whereas stolons and rhizomes are horizontal stems, bulbs and corms are short, vertical, underground stems. Lilies and onions form **bulbs** (Figure 34.16*a*), short stems with many fleshy, modified leaves, such as the familiar "scales" of onions. The leaves make up most of the bulb. Bulbs are thus large buds that store food and can later give rise to new plants. Crocuses, gladioli, and many other plants produce **corms**, underground stems that function very much as bulbs do. Corms are conical and consist primarily of stem tissue; they lack the fleshy scales characteristic of bulbs.

Not all vegetative reproduction arises from modified stems. Leaves may also be the source of new plantlets, as in the succulent plants of the genus *Kalanchoe* (Figure 34.16*b*). Many kinds of angiosperms, ranging from grasses to trees such as aspens and poplars, form interconnected, genetically homogeneous populations by means of **root suckers**—horizontal roots. What appears to be a whole stand of aspen trees, for example, is a clone derived from a single tree by root suckers (see Figure 47.9).

Plants that reproduce vegetatively often grow in physically unstable environments such as eroding hillsides. Plants with stolons or rhizomes, such as beach grasses, rushes, and sand verbena, are common pioneers on coastal sand dunes. Rapid vegetative reproduction enables the plants, once introduced, not only to multiply but also to survive burial by the shifting sand; in turn, the dunes are stabilized by the extensive network of rhizomes or stolons.

Dandelions and some other plants reproduce by **apomixis**, the asexual production of seeds. As you learned in Chapter 9, meiosis reduces the number of chromosomes in gametes, and fertilization restores the sporophytic number of chromosomes in the zygote. Some plants can skip over *both* meiosis and fertilization and still produce seeds. Apomixis produces seeds within the female gametophyte without the mingling and segregation of chromosomes and without the union of gametes. The ovule develops into a seed, and the ovary wall develops into a fruit. An apomictic embryo has the sporophytic number of chromosomes. The result of apomixis is a fruit with seeds genetically identical to the parent plant.

Interestingly, apomixis sometimes requires pollination. In some apomictic species a sperm nucleus must combine with the polar nuclei in order for the endosperm to form. In other apomictic species, the

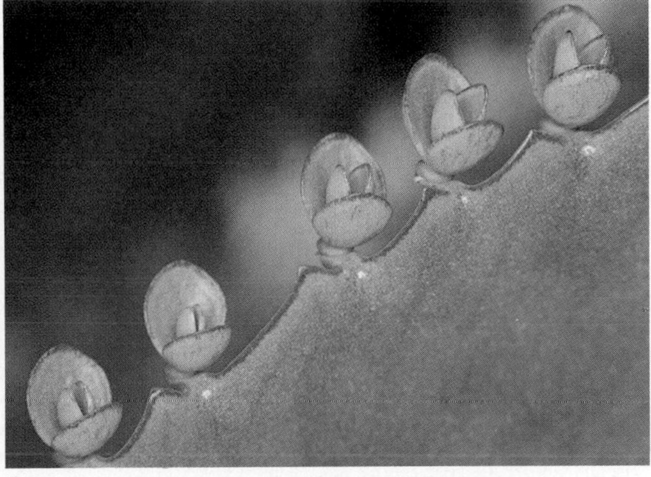

34.16 Vegetative Reproduction
(a) The short stem is visible at the bottom of this sectioned daffodil bulb. White storage leaves grow from the stem; the yellow parts contain flower buds. *(b)* The plantlets forming on the margin of this *Kalanchoe* leaf will fall to the ground and start independent lives.

pollen provides the signals for embryo and endosperm formation, although neither sperm nucleus participates in fertilization. Pollination and fertilization are not the same thing.

Asexual Reproduction in Agriculture

Farmers take advantage of some of these natural forms of vegetative reproduction. Farmers and scientists have also added new types of asexual reproduction by manipulating plants. One of the oldest methods of vegetative reproduction used in agriculture consists simply of making cuttings, or **slips**, of stems, inserting them in soil, and waiting for them to form roots and thus become autonomous plants. Rooting is sometimes hastened by treating the slips with a plant hormone, auxin, as described in Chapter 33.

Agriculturists reproduce many woody plants by **grafting**—attaching a piece of one plant to the stem or root of another plant. The part of the resulting plant that comes from the root-bearing "host" is called the **stock**; the part grafted on is called the **scion**. Figure 34.17 shows three types of grafts. In order for a graft to "take," the cambium of the scion must become associated with the cambium of the stock (see Chapter 29). The cambia of scion and stock both form masses of wound tissue. If the two masses meet and fuse, the resulting continuous cambium can produce xylem and phloem, allowing transport of water and minerals to the scion and of photosynthate to the stock. Grafts are most often successful when

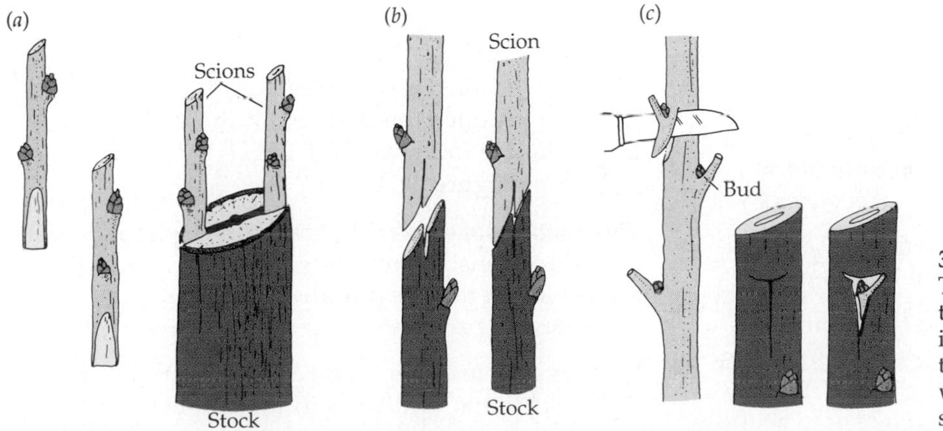

34.17 Grafting
The scions are shown in green and the stocks in brown. *(a)* Cleft grafting; the scions are placed so that their vascular cambia are aligned with the vascular cambium in the stock. *(b)* Whip grafting. *(c)* Budding.

the stock and scion belong to the same or closely related species. Grafting techniques are of great importance in agriculture; most fruit grown for the market in the United States is produced on trees grown from grafts.

There are many reasons for grafting plants for fruit production. The most common is the desire to combine a hardy root system with a shoot system that produces the best-tasting fruit. This motive is illustrated by the story of the wine grape *Vitis vinifera*. In 1863, plant lice of the genus *Phylloxera* inflicted great damage in French vineyards. The roots of vines on more than 2.5 million acres were destroyed. The problem was solved by importing great numbers of *V. vinifera* plants, which have *Phylloxera*-resistant root systems, from California. These plants were used as stocks to which French vines were grafted as scions. Thus the fine French grapes could be grown using roots resistant to the lice. The battle continues, however; in recent years, a new strain of *Phylloxera* has been damaging the grape vines in California.

Scientists in universities and industrial laboratories have been developing new ways to produce valuable plant materials. For example, gene splicing can provide plants with capabilities they previously lacked (see Chapter 14). By causing cells of different sorts to fuse, one can obtain plants with exciting new combinations of properties. By cloning—making genetically identical copies of—small bits of tissue, one can obtain large numbers of equally desirable plants. A problem remains: How can one efficiently take such small, delicate materials and get them to grow in the field? Plants in nature solved this problem long ago. The product of sexual and apomictic reproduction in flowering plants is a compact package, protectively wrapped, containing an embryonic member of the next generation along with a supply of the nutrients it needs to begin its independent existence. This package is the seed. What was needed as a tool of plant biotechnology, and what has actually been developed, is an artificial seed, containing the product of laboratory invention (Figure 34.18).

34.18 Artificial Seeds
These water-soluble capsules, developed by Plant Genetics, Inc., house somatic embryos along with nutrients and other chemicals.

Artificial seeds contain a multicellular "somatic embryo." This is not a sexually produced embryo, but an embryolike product of mitotic divisions in tissue culture. Individual cells or small clusters of cells isolated from the body of a suitable parent plant may develop in liquid culture into structures similar to normal embryos derived from zygotes. So that the somatic embryo does not dry out, and so that it may be stored and transported before planting, it is embedded in a water-soluble gel; then the combined embryo and gel are encapsulated in a protective plastic coat. The coat and gel dissolve away after the artificial seed is planted. Other materials may be added to the gel, among them suitable inorganic nutrients, fungicides, and pesticides. Such scientifically designed artificial seeds should be used more and more often as the remaining problems are solved and methods are perfected for the mass production of these tiny packages.

SUMMARY of Main Ideas about Plant Reproduction

In the sexual cycle of angiosperms, a gametophyte (gamete-producing) generation alternates with a sporophyte (spore-producing) generation.

The gametophytes develop in the flowers of the sporophytes.

The mature female gametophyte (embryo sac) typically contains eight nuclei in a total of seven cells.

The entire male gametophyte travels as a pollen grain.
Review Figure 34.1

In the embryo sac, one sperm nucleus unites with the egg, and the other unites with the two polar nuclei to produce the first cell of the endosperm; this is double fertilization.
Review Figure 34.4

The zygote develops into an embryo with its attached suspensor, then into a seedling, and ultimately into a mature plant.
Review Figure 34.5

Flowers develop into seed-containing fruits, which often play important roles in the dispersal of the species.

For a vegetatively growing plant to flower, an apical meristem in the shoot system must change.
Review Figure 34.8

Some angiosperms have photoperiodic flowering behavior.

Short-day plants flower only when the length of the nightly dark period exceeds a critical night length.
Review Figures 34.9, 34.10, and 34.11

Long-day plants flower only when the length of the nightly dark period is shorter than a critical night length.

Some angiosperms have more complex photoperiodic requirements than short-day or long-day plants have, but most are day-neutral.

The mechanism of photoperiodic control appears to include a biological clock and phytochrome.
Review Figures 34.12, 34.13, and 34.14

In some species, vernalization is required for the plants to flower.

There is evidence for a flowering hormone, florigen, but the substance has yet to be convincingly isolated from any plant.
Review Figure 34.15

Some angiosperms reproduce asexually.

Stolons, tubers, rhizomes, bulbs, corms, or root suckers are means by which plants may reproduce vegetatively.

Some species produce seeds by apomixis.

SELF-QUIZ

1. Which of the following does *not* participate in asexual reproduction?
a. Stolon
b. Rhizome
c. Fertilization
d. Tuber
e. Apomixis

2. Apomixis includes
a. sexual reproduction.
b. meiosis.
c. fertilization.
d. a diploid embryo.
e. no production of a seed.

3. Sexual reproduction in angiosperms
a. is by way of apomixis.
b. requires the presence of petals.
c. can be accomplished by grafting.
d. gives rise to genetically diverse offspring.
e. cannot result from self-pollination.

4. The typical angiosperm female gametophyte
a. is called a megaspore.
b. has 8 nuclei.
c. has 8 cells.
d. is called a pollen grain.
e. is carried to the male gametophyte by wind or animals.

5. Pollination in angiosperms
a. never requires water.
b. never occurs within a single flower.
c. always requires help by animal pollinators.
d. is also called fertilization.
e. makes most angiosperms independent of water for reproduction.

6. Which statement is *not* true of double fertilization?
a. It is found in all angiosperms.
b. It is found in no plants other than angiosperms.
c. One of its products is a triploid nucleus.
d. One sperm nucleus fuses with the egg nucleus.
e. One sperm nucleus fuses with two polar nuclei.

7. The suspensor
a. gives rise to the embryo.
b. is heart-shaped in dicots.
c. separates the two cotyledons of dicots.
d. ceases to elongate early in embryo development.
e. is larger than the embryo.

8. Which statement is *not* true of photoperiodism?
a. It is related to the biological clock.
b. Phytochrome plays a role in the timing process.
c. It is based on measurement of the length of the night.
d. Most plant species are day-neutral.
e. It is limited to the plant kingdom.

9. Although florigen has never been isolated, we think it exists because
a. night length is measured in the leaves, but flowering occurs elsewhere.
b. it is produced in the roots and transported to the shoot system.
c. it is produced in the coleoptile tip and transported to the base.
d. we think that gibberellin and florigen are the same compound.
e. it may be activated by prolonged (more than a month) chilling.

10. Which statement is *not* true of vernalization?
a. It may require more than a month of low temperature.
b. The vernalized state generally lasts for about a week.
c. Vernalization makes it possible to have two winter wheat crops each year.
d. It is accomplished by subjecting moistened seeds to chilling.
e. It was of interest to Russian scientists because of their native climate.

FOR STUDY

1. For a crop plant that reproduces both sexually and asexually, which method of reproduction might the farmer prefer?
2. Thompson seedless grapes are produced by vines that are triploid. Think about the consequences of this chromosomal condition for meiosis in the flowers. Why are these grapes seedless? Describe the role played by the flower in fruit formation when no seeds are being formed. How do you suppose Thompson seedless grape plants are propagated?
4. Poinsettias are popular ornamental plants that typically bloom just before Christmas. Their flowering is photoperiodically controlled. Are they long-day or short-day plants? Explain.
5. You plan to induce the flowering of a crop of long-day plants in the field by using artificial light. Is it necessary to keep the lights on continuously from sundown until the critical night length is reached? Explain.

READINGS

Barrett, S. C. H. 1987. "Mimicry in Plants." *Scientific American*, September. Some plants mimic insects, encouraging the mimicked insects to act as pollinators.

Cleland, C. E. 1978. "The Flowering Enigma." *BioScience*, April. Florigen eluded physiologists when this article was published, and it still does.

Cox, P. A. 1993. "Water-Pollinated Plants." *Scientific American*, October. An unusual environment calls for different pollination strategies, and it provides unusual opportunities for research.

Goodman, B. 1993. "A 'Shotgun Wedding' Finally Produces Test-Tube Plants." *Science*, vol. 261, page 430. A brief account of the discovery of in vitro fertilization in plants.

Handel, S. N. and A. J. Beattie. 1990. "Seed Dispersal by Ants." *Scientific American*, August. Some plants produce seeds bearing specialized fat bodies that are eaten by ants after the ants carry the seeds to their nests. This and other mechanisms are discussed.

Niklas, K. J. 1987. "Aerodynamics of Wind Pollination." *Scientific American*, July. Do the mechanics of pollination seem improbable to you? Wind-pollinated plants have many adaptations that favor successful pollination.

Raven, P. H., R. F. Evert and S. Eichhorn. 1991. *Biology of Plants*, 5th Edition. Worth, New York. A well-balanced general botany textbook.

Salisbury, F. B. and C. W. Ross. 1992. *Plant Physiology*, 4th Edition. Wadsworth, Belmont, CA. An authoritative textbook with good chapters relating to reproductive development.

Taiz, L. and E. Zeiger. 1991. *Plant Physiology*. Benjamin/Cummings, Redwood City, CA. Chapter 21 deals with the control of flowering.

PART SIX

The Biology of Animals

Animals live in amazing places, including the most extreme environments on our planet. The adaptations that make these lifestyles possible can be fascinating. Consider, for example, emperor penguins, the largest penguin species. They spend most of their lives feeding at sea, but they breed and raise their young far from the open sea during the bitter cold winter season of Antarctica. As the winter ice shelf forms off Antarctica, emperor penguins gather at its edge. They then walk over 100 kilometers inland across the ice, to breed in one of the coldest, most inhospitable places on Earth. After the female lays an egg, she walks back to the sea to feed; the male incubates the egg, then protects and feeds the chick until the female returns with her body fat replenished. The female then takes over the feeding of the chick, and the male walks back to the sea, having fasted for more than four months.

In Part Six you will learn about adaptations that allow birds to fly at altitudes at which humans cannot exist without technological assistance, seals to remain underwater for more than half an hour, mice to live without water in the hottest deserts, and bats to capture flying insects in total darkness. The explanations of how animals achieve such feats are found in physiology, the study of how animals work. Unusual adaptations are not just the spice of physiology; they are extensions of basic physiological mechanisms and therefore help us to understand the principles of normal physiological functions in humans and other animals.

Earlier in this book we learned how cells work. Because animals consist of cells, how cells function—including their energy metabolism, their needs for nutrients, their production of waste products such as carbon dioxide, and their osmotic balance—can be extended to whole animals. For the simplest animals, sponges and cnidarians, this extension is straightforward, since most of those species live in the sea and their bodies are only two cell layers thick. Seawater contains nutrients, it has a suitable composition of salts, and it provides a stable physical environment. Each cell of a sponge or a jellyfish is in direct contact with the environment and functions almost as an autonomous unit. It receives its nutrients directly from and releases its wastes directly into the seawater. With aquatic animals that are larger and more complex than sponges and cnidarians, not every cell of their bodies can be in direct contact with the external environment. With terrestrial animals, the cells of their bodies cannot be in direct contact with their external environment, air, because it would dry and kill them. For these reasons, most cells of most animals are served by an internal environment consisting of extracellular fluids. That internal environment provides appropriate physical conditions for the cells of the animal, supplies all the nutrients they need, and removes all their wastes.

A Long Walk to a Cold Breeding Ground
Emperor penguins migrate between the sea and their breeding grounds. The distance may be more than 100 kilometers.

35 Physiology, Homeostasis, and Temperature Regulation

HOMEOSTASIS

Most of animal physiology focuses on how the internal environment is maintained in a condition that meets the needs of the cells. The internal environment is obviously not as vast as the sea. Its nutrient content can be rapidly exhausted, and its physical conditions are altered by the metabolic activities of the cells it serves. Animal organs and organ systems keep various aspects of the internal environment at a steady state, a condition called **homeostasis**. Gas-exchange organs provide oxygen to and remove carbon dioxide from the extracellular fluids, digestive organs supply nutrients, and excretory systems eliminate wastes.

Homeostasis is an essential feature of complex animals, and it has enabled animals to adapt to nearly every environment on Earth. It has also enabled biochemical systems to become more efficient by being adapted to function over narrow ranges of physical parameters. If an organ fails to function properly, homeostasis of the internal environment is lost, and as a result cells become sick and die. The sick cells are not just those of the organ that functions improperly, but the cells of all other organs as well. Loss of homeostasis is therefore a serious problem that makes itself worse. To avoid loss of homeostasis, the activities of organs must be controlled and regulated in response to changes in the internal environment.

Control and regulation require information; hence the organ systems of information—the endocrine system and the nervous system—must be included in our discussions of every physiological function. For that reason, we treat the endocrine and nervous systems early in this part of the book. Subsequent chapters deal with the organ systems that are responsible for homeostasis of various aspects of the internal environment.

ORGANS AND ORGAN SYSTEMS

The diversity of adaptations that enable animals to live in just about any environment presents us with a bewildering number of details that can make the study of physiology seem daunting. Thus we begin our study with a road map of the organs, organ systems, and physiological functions of at least one species. That species might as well be *Homo sapiens*, but the road map also applies to most other vertebrates and to some invertebrates as well.

The Structure of Organs

Organs are made of tissues, and a tissue consists of cells with similar structure and function. Biologists who study cells and tissues recognize many types of cells but group them into only four general types of tissues: epithelial, connective, muscle, and nervous.

Most **epithelial tissues** are sheets of tightly connected cells such as those that cover the body surface and those that line various hollow organs of the body, including the digestive tract and the lungs. Some epithelial cells have secretory functions—for example, those that secrete mucus, digestive enzymes, or sweat. Other epithelial cells have cilia and help substances move over surfaces or through tubes. Since epithelial cells create boundaries between the inside and the outside of the body and between body compartments, they frequently have absorptive and transport functions. An epithelium can be stratified, as is the skin, which consists of many layers of cells, or it can be simple, as is the lining of the gut, which consists of a single layer of cells. We'll encounter epithelial tissues in our discussions of the linings and tubules of reproductive systems (see Chapter 37), the linings of gas-exchange systems (see Chapter 41), the linings of digestive tracts (see Chapter 43), and the tubules of excretory systems (see Chapter 44).

Connective tissues support and reinforce other tissues. Unlike epithelial tissues, which consist of densely packed, tightly connected populations of cells, most connective tissues consist of a dispersed population of cells embedded in an extracellular matrix. The properties of the matrix differ in different types of connective tissues. The connective tissue in skin contains many elastic fibers that can be stretched and then return to their original position. The connective tissues that connect muscles to bone and bones to one another have many collagen fibers with high tensile strength. Bone is a connective tissue in which the extracellular matrix has been hardened by mineral deposition (see Chapter 40). Two other major types of connective tissue are adipose tissue (fat cells) and the cellular components of the blood (see Chapter 42).

Muscle tissue consists of cells that can contract and therefore cause movements of organs, limbs, or most other parts of the body. Muscles are the most important **effectors** of the body; they enable it to do things (see Chapter 40). Since muscle tissues play important roles in most organs and organ systems, we'll encounter them in many places in Part Six.

Nervous tissue (see Chapters 38, 39, and 40) enables animals to deal with information. The cells of nervous tissue, **neurons**, are extremely diverse. Some respond to specific types of stimuli, such as light, sound, pressure, or certain molecules, by generating electric signals in their membranes. These electric signals can be conducted via long extensions of the cell to other parts of the body, where they are passed on to other neurons, muscle cells, or secretory cells. Because neurons are involved in controlling the activities of most organ systems, we will run across them frequently in Part Six.

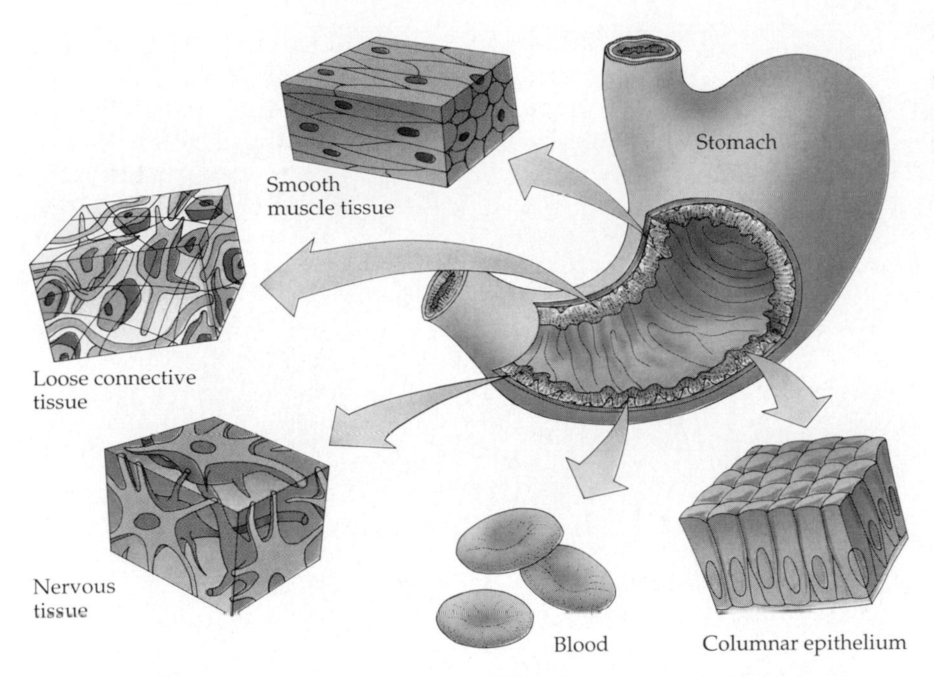

35.1 One Organ Contains Many Tissue Types
The human stomach—a section of the gut—contains many tissue types that enable it to perform its functions within the digestive system (see Figure 35.5*a*).

Organs are usually made up of more than one tissue type, and most organs include all tissue types. The gut—the organ that digests food and absorbs nutrients—is a good example (Figure 35.1). The gut is lined internally with a single layer of columnar epithelial cells. Some secrete mucus or enzymes and others mainly absorb nutrients. Beneath the gut lining is connective tissue, within which are glands and blood vessels. Concentric layers of muscle tissue move food through the gut and mix it with the secretions of the epithelial cells. Neurons extend between the layers of other tissues to control both the secretions and the movements of the gut.

An individual organ, such as the gut, and other organs with complementary functions may be the parts of an organ system. The major **organ systems** of the body are outlined in Figures 35.2 through 35.7. We'll discuss each of these systems in the chapters that follow.

Organs for Information and Control

The principal organ system that processes information and uses that information to control the physiology and behavior of the animal is the nervous system (Figure 35.2*a*). It consists of the brain and spinal cord (the central nervous system) along with peripheral nerves that conduct electric signals from sensors to the central nervous system and conduct signals from the central nervous system to effectors, which are either muscle tissue or secretory tissue. The sensors of the nervous system are diverse; they include eyes, ears, organs of taste and smell, and cells sensitive to temperature, touch, pressure, stretch, and pain.

The endocrine system (Figure 35.2*b*) also processes information and controls the functions of organs, but its messages are distributed mostly in the blood to the entire body as chemical signals called hormones. The principal organs of the endocrine system are ductless glands that secrete specific hormones into the blood. In addition, many other tissues contain individual cells that secrete hormones. There are strong interactions between the nervous system and

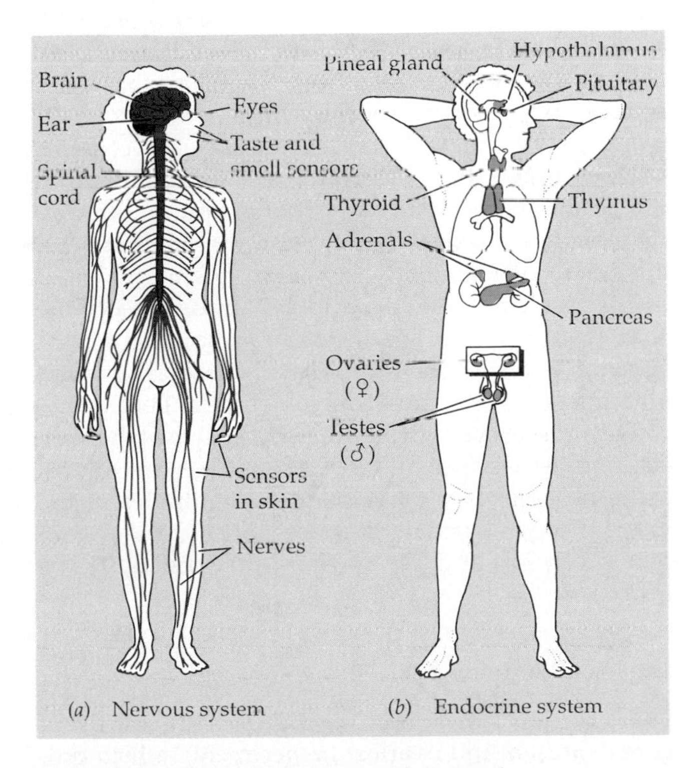

35.2 Organs of Information and Control

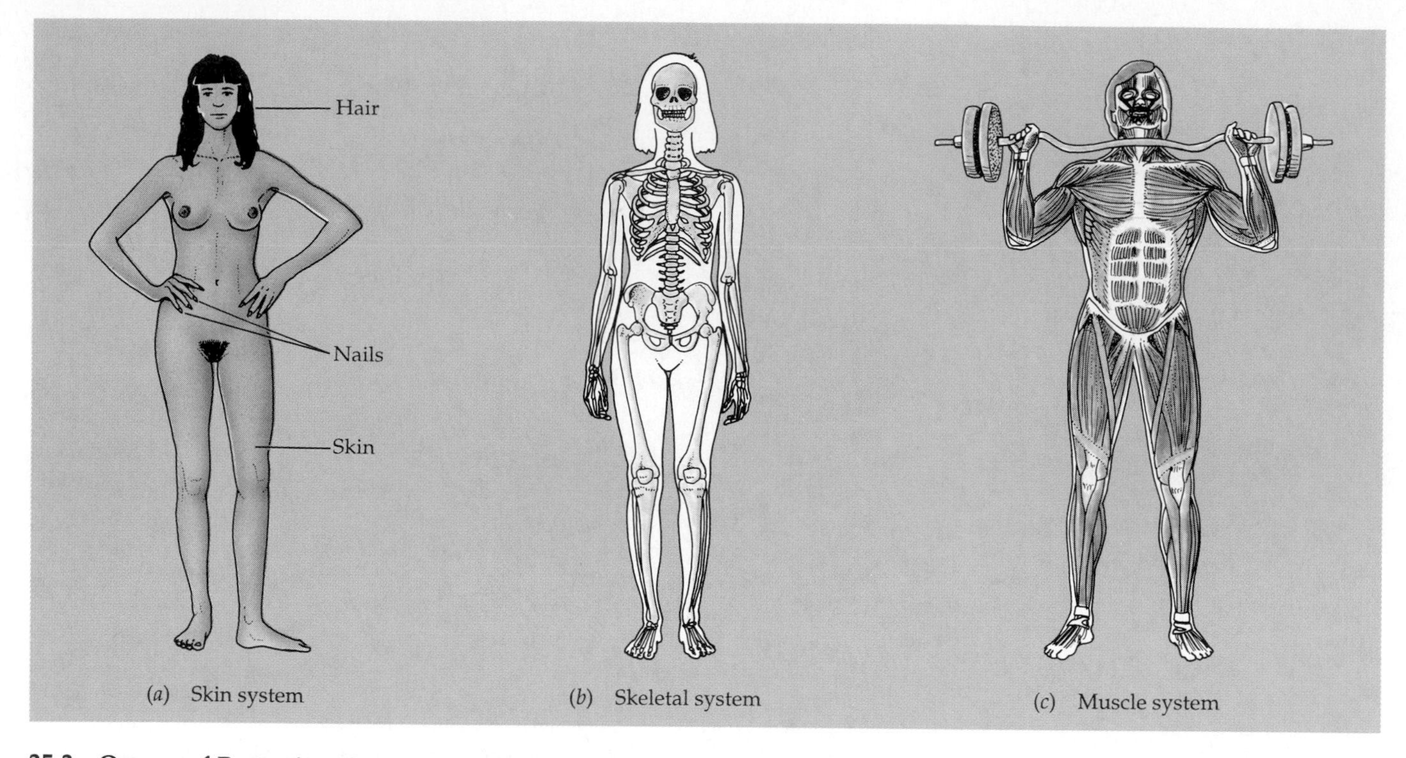

35.3 Organs of Protection, Support, and Movement

the endocrine system. Cells in the brain produce hormones that control parts of the endocrine system. In turn, there are cells in the brain that respond to the hormones produced by endocrine glands.

Organs for Protection, Support, and Movement

The largest organ of the body is the skin, along with its special elaborations, hair and nails (Figure 35.3*a*). The skin protects the body from organisms that cause disease, from the physical environment, and from excessive loss of water. Because the skin contains nervous tissue that is sensitive to various stimuli, it is a major sense organ. The skin is also an effector: As a route of heat exchange with the environment, it helps to regulate body temperature.

The skeletal system supports and protects the body (Figure 35.3*b*). The skeleton is also an important effector: It forms the supports and the levers that muscles pull on to cause movement and behavior.

The muscle system (Figure 35.3*c*) includes the skeletal muscles that are under our conscious control and cause all voluntary movements, the muscles of the internal organ systems that are not under our conscious control, and the muscles that constitute the heart.

Organs of Reproduction

The male and female reproductive systems consist of gonads (testes and ovaries, respectively), which produce sex cells (gametes). Additional organs deliver the sex cells to the site where fertilization takes place (Figure 35.4). The female reproductive system includes the uterus, the organ that supports the development of the embryo. The female's mammary glands provide nutrients for the infant. In the gonads of both sexes are tissues that secrete hormones that play roles in sexual development and reproduction.

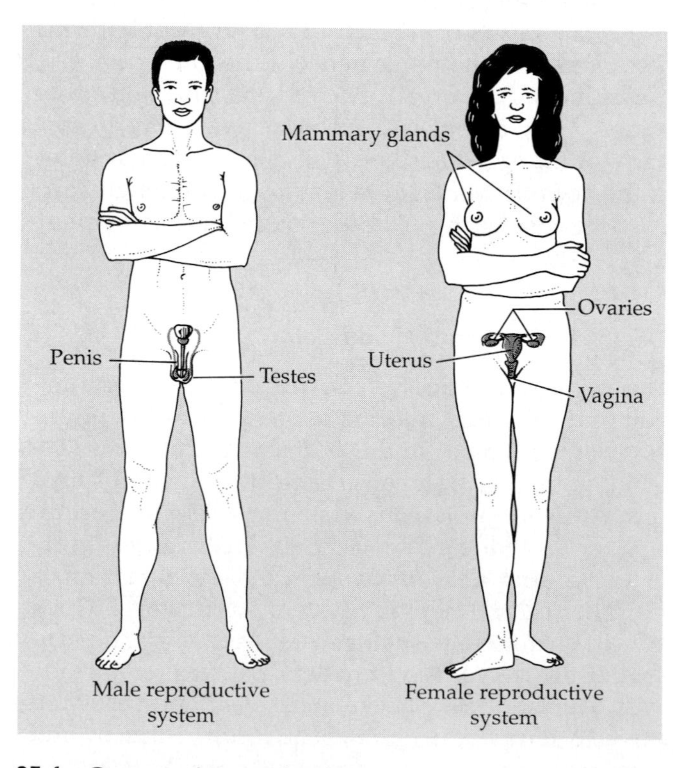

35.4 Organs of Reproduction

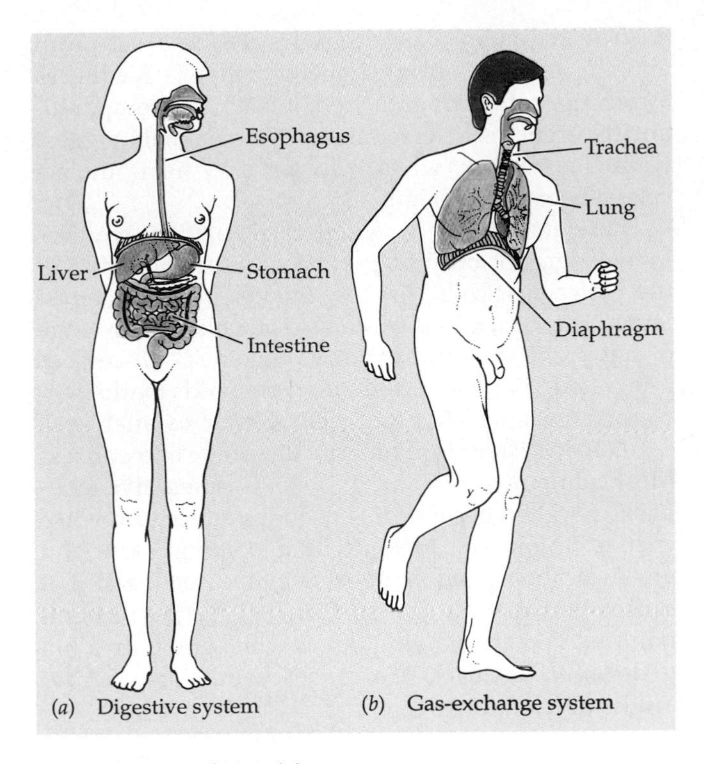

(*a*) Digestive system (*b*) Gas-exchange system

35.5 Organs of Nutrition

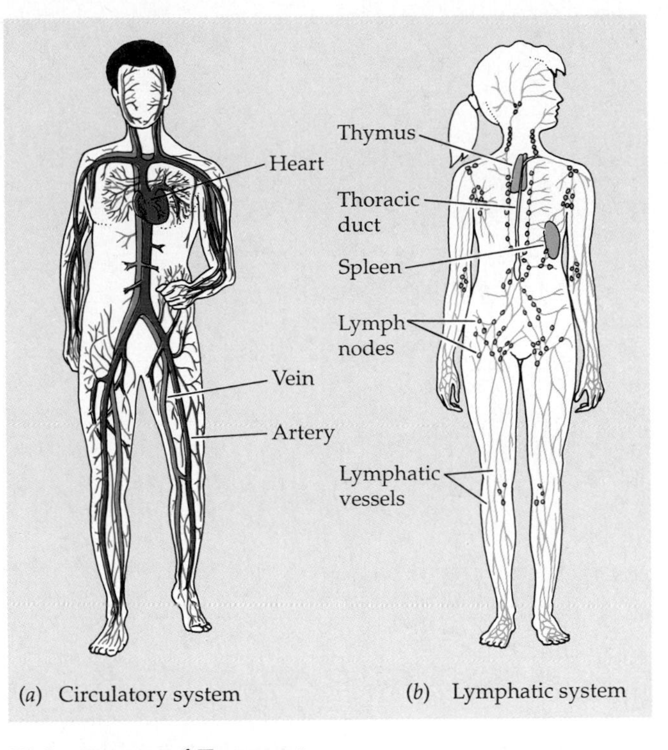

(*a*) Circulatory system (*b*) Lymphatic system

35.6 Organs of Transport

Organs of Nutrition

The digestive system is largely a continuous tubular structure that extends from mouth to anus (Figure 35.5*a*). This tube, also called the gut, is divided into different segments that serve different functions in the processing and digestion of food and the absorption of nutrients. Glands associated with the gut deliver into it digestive enzymes and other molecules that break down complex food molecules. The lower gut stores and periodically eliminates solid wastes and water.

The gas-exchange system, also called the respiratory system, provides oxygen, which is essential for cellular respiration (Figure 35.5*b*). Carbon dioxide, a waste product of cellular respiration, is eliminated by the gas-exchange system. The gas-exchange organs of humans are lungs, which are a system of progressively dividing airways leading to tiny but numerous sacs with membranous gas-exchange surfaces that have a very large combined surface area. The muscles that move air into and out of the lungs are another component of the gas-exchange system.

Organs of Transport

Oxygen must be transported from the lungs to the tissues of the body, and carbon dioxide must be transported from the tissues to the lungs. These gases are transported by the circulatory system, which includes a pump (the heart), a system of blood vessels (veins and arteries), and blood (Figure 35.6*a*). The circulatory system also transports nutrients from the gut, delivers nitrogenous wastes to the excretory system, transports hormones, transports heat, and generates mechanical forces. Blood is made up of cellular components in a liquid medium called plasma. The blood plasma is virtually continuous with the extracellular fluids that are the internal environment of the body.

The lymphatic system is another transport system consisting of a set of vessels that extend throughout the body, but, unlike the circulatory system, it does not include a pump and its vessels do not form a complete circuit (Figure 35.6*b*). The lymphatic system picks up extracellular fluid and delivers it to the blood circulatory system.

Organs of Excretion

Urine forms in the kidneys. Urine includes nitrogenous wastes from the metabolism of proteins and nucleic acids as well as excess salts and other substances that the body excretes. The kidneys play crucial roles in maintaining the correct water content of the body and the correct salt composition of the extracellular fluids. Urine passes to a bladder for storage until it is released to the exterior through the urethra (Figure 35.7).

In the chapters that follow we will explore physiological functions and the organ systems that accomplish them. In the remainder of this chapter we will discuss general principles of homeostasis, which have applications to all animal physiology. We'll illustrate these principles by considering in detail the

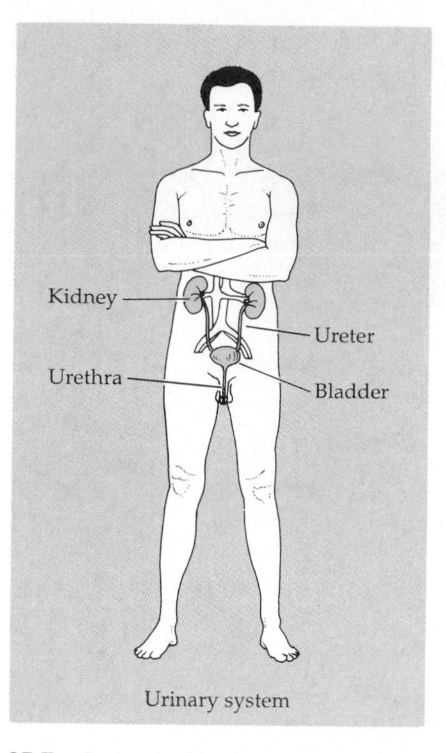

35.7 Organs of Excretion and Water/Salt Balance

regulation of body temperature. Temperature is an important physical parameter of the internal environment. It can be perturbed by the activities of cells and by changes in the outside environment. Animals have evolved a number of adaptations for dealing with changes in temperature.

GENERAL PRINCIPLES OF HOMEOSTASIS

Homeostasis refers to a constant state of internal conditions in the body. Homeostasis depends on the functions of the organs and organ systems of the body; these functions must be controlled and regulated to achieve a relatively constant internal environment. The terms *control* and *regulation* might seem interchangeable, but their meanings differ. Control implies the ability to *change* the rate of a reaction or process. Regulation—the more sophisticated and more specific physiological concept—refers to *maintaining* a variable within specific levels or limits. You control the speed of your car by using the accelerator and brake, but you regulate it by using the accelerator and brake to maintain a particular speed.

Set Points and Feedback

Regulation requires, in addition to control mechanisms, the ability to obtain and use information. You can regulate the speed of a car only if you know the speed at which you are traveling and the speed you wish to maintain. The desired speed is a **set point** and the reading on your speedometer is **feedback.** When the set point and the feedback are compared, any difference is an **error signal**. Error signals suggest corrective actions, which you make by using the accelerator or brake (Figure 35.8).

Understanding physiological regulation requires knowledge not only of the mechanisms of action of the molecules, cells, tissues, organs, and organ systems—the **controlled systems**—but also knowledge of how relevant information is obtained, processed, integrated, and converted into commands by the regulatory systems. A fundamental way to analyze a regulatory system is to identify its source of feedback. Most common in regulatory systems is **negative feedback**, so called because it is used to reduce or reverse change. In our car analogy, the recognition that you are over the speed limit is negative feedback if it causes you to slow down. Conversely, the recognition that you are slowing down while going up a hill is negative feedback if it causes you to step harder on the accelerator.

Regulatory Systems

To understand the features of a regulatory system, consider a thermostat—a relevant analogy, since we will be looking at the biological thermostats of vertebrate animals later in this chapter. The thermostat that is part of the heating–cooling system of a house is a regulatory system. It has upper and lower set points that you can adjust, and it receives feedback from a sensor. The circuitry of the thermostat converts any differences between the set points and the sensor into signals that activate the controlled systems—the furnace and the air conditioner. When room temperature rises above the upper set point, the thermostat activates the air conditioner, thus reducing room temperature below the set point. When room temperature falls below the lower set point, the thermostat activates the furnace, thus raising room temperature toward that set point. Hence the sensor of room temperature provides negative feedback (Figure 35.9).

Negative feedback makes good sense for physiological regulatory systems, so you may wonder if there is any such thing as positive feedback in physiology. Although not as common as negative feedback, it does exist. Rather than returning a system to a set point, **positive feedback** amplifies a response. One example is sexual behavior, in which a little stimulation can cause more behavior, which causes more stimulation, and so on. Positive feedback is not used by regulatory systems that maintain stability!

Feedforward information is another feature of regulatory systems. The function of feedforward is to change the set point. Seeing a deer ahead on the road

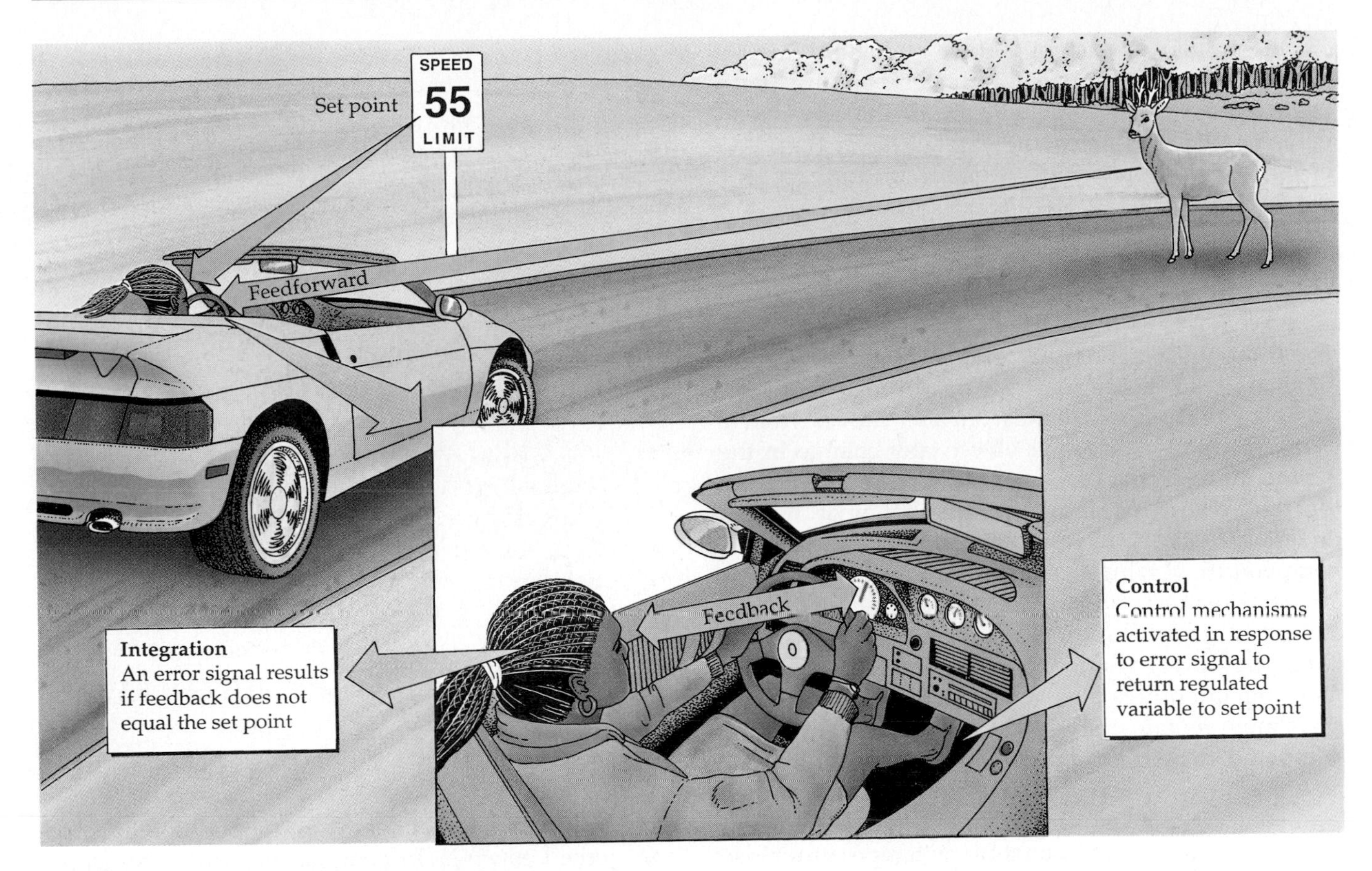

35.8 Control, Regulation, and Feedback
As you drive a car, the posted speed limit is your set point and the speedometer gives you feedback information. Comparing the speed limit to the speedometer reading gives you error signals that you convert into corrective actions by using the brakes and the accelerator to regulate the car's speed. The sight of a deer in the road ahead is feed*forward* information.

when you are driving is an example of feedforward; this information takes precedence over the posted speed limit, and you change your set point to a slower speed. If you want the temperature of your house to be lower at night than during the day, you can add a clock to the thermostat to provide feedforward information about time of day.

These general considerations about control, regulation, and regulatory systems help to organize our thinking about physiological systems, but the physiological systems can be far more complex than the thermostat and driving analogies. In some systems we do not even know the nature of the feedback. For most people, body weight is regulated, although it might not be at the level we prefer. Without consciously counting calories, the brain controls hunger so that food intake matches energy expenditure. We do not understand what information the brain uses to achieve this remarkable feat of regulation, but there are some interesting hypotheses and active research on the problem.

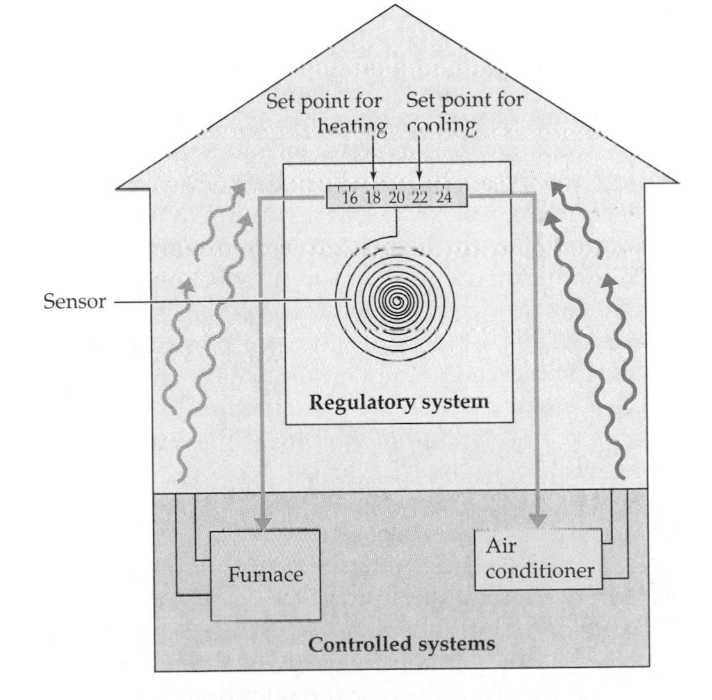

35.9 A Thermostat Regulates House Temperature
Changes in room temperature cause a sensor to move relative to set points on the thermostat and activate the furnace or the air conditioner. Room temperature, as detected by the sensor, is feedback to the regulatory system.

One regulatory system that we understand well is the system that regulates body temperature. Let's begin by asking two related questions: Why do organisms need to thermoregulate? What is the effect of temperature on living systems?

THE EFFECTS OF TEMPERATURE ON LIVING SYSTEMS

Over the face of Earth, temperatures vary enormously—from boiling hot springs to the frigid Antarctic plateau. Because heat always moves from a warmer object to a cooler object, any change in the temperature of the environment causes a change in the temperature of an organism in that environment, unless the organism does something to regulate its temperature. Living cells are restricted to a narrow range of temperatures. If cells cool to below 0°C, ice crystals can form within them; this can fatally damage their structures. Some cells are adaptated to prevent freezing and others to survive freezing, but generally cells must remain above 0°C to stay alive. The upper temperature limit is around 45°C for most cells. Some specialized algae can grow in hot springs at 70°C, and some bacteria can live at near 100°C, but in general, proteins begin to denature as temperatures rise above 45°C. As proteins denature, they lose their functional properties. Most cellular functions are limited to the range of temperatures between 0 and 45°C, which are considered the thermal limits for life.

The Q_{10} Concept

Even within the range of 0 to 45°C, temperature changes can create problems for animals. Like the biochemical reactions of which they are made up, most physiological processes are temperature-sensitive, going faster at higher temperatures (see Figure 6.27). The temperature sensitivity of a reaction or process can be described in terms of the **Q_{10}**, a quotient calculated by dividing the rate of a process or reaction at a certain temperature, R_T, by the rate of that process or reaction at a temperature 10°C lower, R_{T-10}:

$$Q_{10} = \frac{R_T}{R_{T-10}}$$

The Q_{10} can be measured for a simple enzymatic reaction or for a complex physiological phenomenon. If a reaction or process is not temperature-sensitive, it has a Q_{10} of 1. Most biological Q_{10}'s are between 2 and 3, which means the reaction rates double or triple as the temperature increases by 10°C (Figure 35.10).

Changes in temperature can be particularly disruptive to an animal's functioning because all the component reactions in the animal do not have the same Q_{10}. Individual reactions with different Q_{10}'s are linked together in complex networks that carry out physiological processes. Temperature changes shift the rates of some of the reactions more than those of others, thus disrupting the balance and integration that the processes require. To maintain homeostasis, organisms must be able to compensate for or prevent changes in temperature.

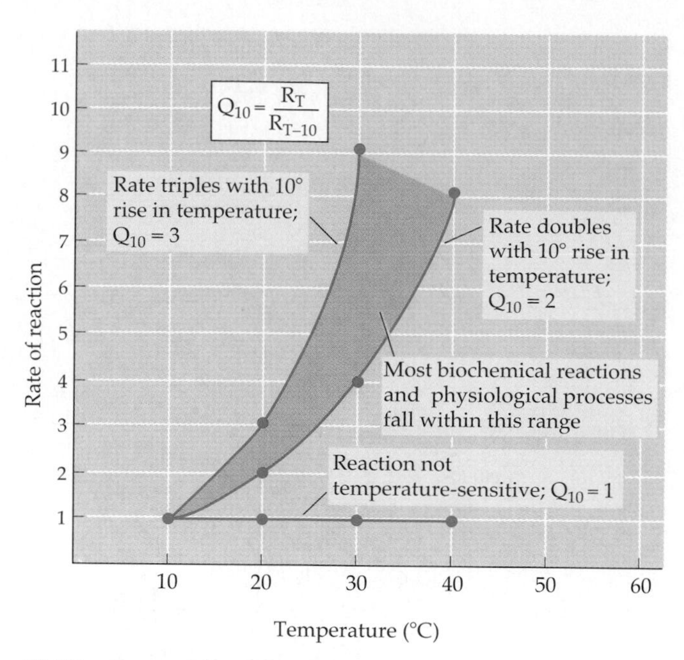

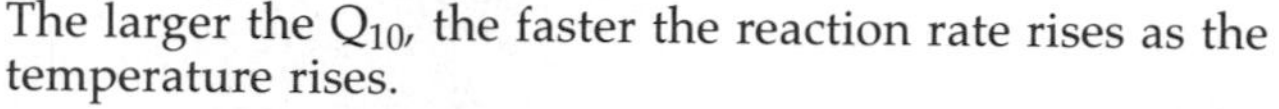

35.10 Q_{10} and Reaction Rate
The larger the Q_{10}, the faster the reaction rate rises as the temperature rises.

Metabolic Compensation

The body temperatures of some animals are tightly coupled to the temperature of the environment. Think of a fish in a pond in a highly seasonal environment. As the temperature of the pond water changes from 4°C in midwinter to 24°C in midsummer, the body temperature of the fish does the same. If we bring such a fish into the laboratory in the summer and measure its metabolic rate (the sum total of the energy turnover of its cells) at different water temperatures, we might calculate a Q_{10} of 2 and plot our data as shown by the red line in Figure 35.11. We predict from our graph that in winter, when the temperature is 4°C, the fish's metabolic rate will be only one-fourth of what it was in the summer. We then return the fish to its pond. When we bring the fish back to the laboratory in the winter and repeat the measurements, we find, as the blue line shows, that its metabolic rate at 4°C is not as low as we predicted, but is almost the same as it was at 24°C in the summer. If we repeat the measurement over a range of temperatures, we find that the fish's meta-

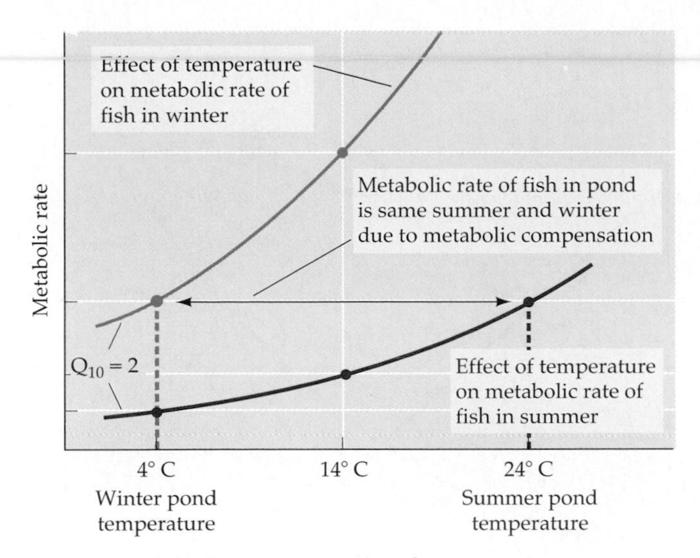

35.11 Metabolic Compensation for Seasonal Differences
When the metabolic rate of a fish is measured in summer and in winter, it is shown to be temperature-sensitive. At normal environmental temperatures in nature, however, the fish may have the same metabolic rate summer and winter. Metabolic compensation acclimatizes a fish to its changing environment.

bolic rate is always higher than the rate we measured at the same temperature in the summer-acclimatized fish. **Acclimatization** is the process of physiological and biochemical change that an animal undergoes in response to seasonal changes in climate.

The reason for the difference between our prediction from the summer data and our measurements on the winter fish is that seasonal acclimatization in the fish has produced **metabolic compensation**. Metabolic compensation readjusts the biochemical machinery to counter the effects of temperature. What might account for such a change? Look again at Figure 6.25, which suggests a hypothesis. If the fish we are studying have duplicate enzymes that operate at different optimal temperatures, they may compensate metabolically by catalyzing reactions with one set of enzymes in the summer and another set in the winter. The end result of such readjustment is that metabolic functions are much less sensitive to seasonal changes in temperature than they are to shorter-term thermal fluctuations.

THERMOREGULATORY ADAPTATIONS

Organisms have evolved numerous behavioral and physiological mechanisms for maintaining optimal body temperatures. The best way to discuss these thermoregulatory adaptations is to look at a thermoregulatory classification of the animals that possess them. Animals are commonly described as cold-blooded or warm-blooded. Taken at face value these terms could lead to some ridiculous errors. For example, desert reptiles could be classified as warm-blooded during the day and cold-blooded at night, many insects might be classified as warm-blooded during flight and cold-blooded at rest, and hibernating mammals would be very cold-blooded during most of the winter.

Biologists prefer a different set of terms. A **homeotherm** is an animal that regulates its body temperature at a constant level; a **poikilotherm** is an animal whose temperature changes. This system of classification says something about the biology of the animals, but it also presents problems. Should a fish in the deep ocean, where the temperature changes very little, be called a homeotherm? Should a hibernating mammal that allows its body temperature to drop to nearly the temperature of its environment be called a poikilotherm? The problem posed by the hibernator has been set aside by coining a third category, the **heterotherm**: an animal that regulates its body temperature at a constant level *some* of the time. Homeotherm, poikilotherm, and heterotherm are useful descriptive terms.

Another set of terms classifies animals on the basis of thermoregulatory mechanisms. **Ectotherms** depend largely on external sources of heat, such as solar radiation, to maintain their body temperatures above the environmental temperature. **Endotherms** can regulate their body temperature by producing heat metabolically or by mobilizing active mechanisms of heat loss. Mammals and birds behave as endotherms; animals of all other species behave as ectotherms most of the time.

Laboratory Studies of Ectotherms and Endotherms

Let's choose a small lizard to represent ectotherms and a mouse of the same body size as the lizard to represent endotherms. In the laboratory we put each animal in a small metabolism chamber that enables us to measure the body temperature of the animal and its metabolic rate as we change the temperature of the chamber from 0 to 35°C. The results obtained from the two species are very different (Figure 35.12). The body temperature of the lizard always equilibrates with that of the chamber, whereas the body temperature of the mouse remains at 37°C. The metabolic rate of the lizard increases with temperature. Below about 27°C the metabolic rate of the mouse increases as chamber temperature decreases (notice that you must read the graph right to left to see this). It seems the lizard cannot regulate its body temperature or metabolism independently of environmental temperature. The mouse regulates its body temperature by altering its rate of metabolic heat production.

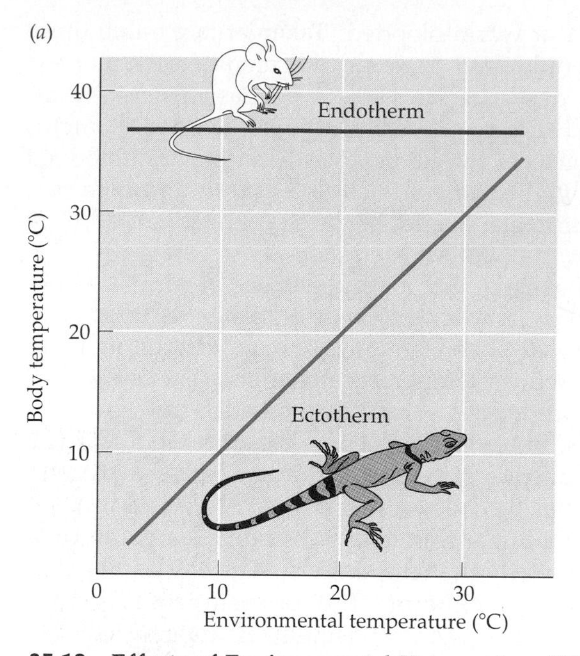

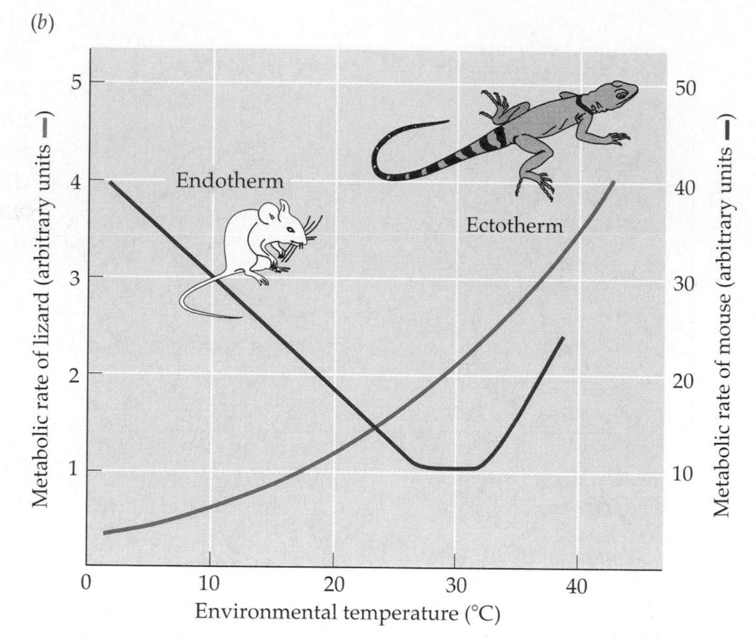

35.12 Effects of Environmental Temperature Differ
(a) The body temperatures of a lizard (blue line) and a mouse (red line) respond differently to changes in environmental temperature. *(b)* The reason for the difference is that, as the environmental temperature falls, the rate of metabolic heat production in the lizard decreases, whereas that of the mouse increases.

Field Study of an Ectotherm

A logical next step is to test in nature our laboratory conclusion that the lizard cannot regulate its body temperature. We can do this by implanting a capsule containing a radio telemeter in the lizard's body and then releasing the lizard in its desert environment. The radio telemeter measures the lizard's body temperature and then converts it to a radio signal that can be heard through a portable radio. We can thus observe the body temperature of the lizard as it goes about its normal behavior. Our prediction is that the temperature of the lizard will follow the temperature of the environment, which can change more than 40°C in a few hours.

The results of the experiment, however, differ strikingly from this prediction. At night the temperature in the desert may drop close to freezing, but the temperature of the lizard remains stable at 16°C. This is not difficult to explain; the lizard spends the night in a burrow where the soil temperature is a constant 16°C. Early in the morning, soon after sunrise, the lizard emerges from its burrow. The air temperature is still quite cool, but the body temperature of the lizard rises to 35°C in less than 30 minutes. The lizard achieves this rapid rise in temperature by basking on a rock with maximum exposure to the sun. As its dark skin absorbs solar radiation, its body temperature rises considerably above the surrounding air temperature. By altering its exposure to the sun, the lizard maintains its body temperature at around 35°C all morning as it seeks food, avoids predators, and interacts with potential mates or competitors. By noon the air temperature near the surface of the desert has risen to 50°C, but the lizard's body temperature remains around 35°C. It is now staying mostly in shade, frequently up in bushes where there is a cooling breeze. As afternoon progresses, air temperature declines, and the lizard again spends more of its time in the sun and on hot rocks to maintain its body temperature around 35°C. The lizard returns to its burrow just before sunset, and its body temperature rapidly drops to 16°C. Figure 35.13 reviews the patterns of the lizard's behavior and the temperature changes over the course of a day.

This field experiment shows that the lizard can regulate its body temperature quite well by behavioral mechanisms rather than by metabolic mechanisms. The deficiency in our laboratory experiment was that the lizard in the chamber could not use its thermoregulatory behavior. If we give a lizard access to a thermal gradient in the laboratory, it is capable of regulating its body temperature by selecting the right place on the gradient. If only a hot place and a cold place are available, it will shuttle back and forth. It will maintain a different body temperature during the night than during the day, and if it is infected with pathogenic bacteria it will give itself a fever by selecting higher temperatures on the gradient (see Box 35.A on page 795).

The lesson to be learned from the discrepancies between the results from the laboratory and field experiments on the lizard are encapsulated by a quotation from the German embryologist Hans Spe-

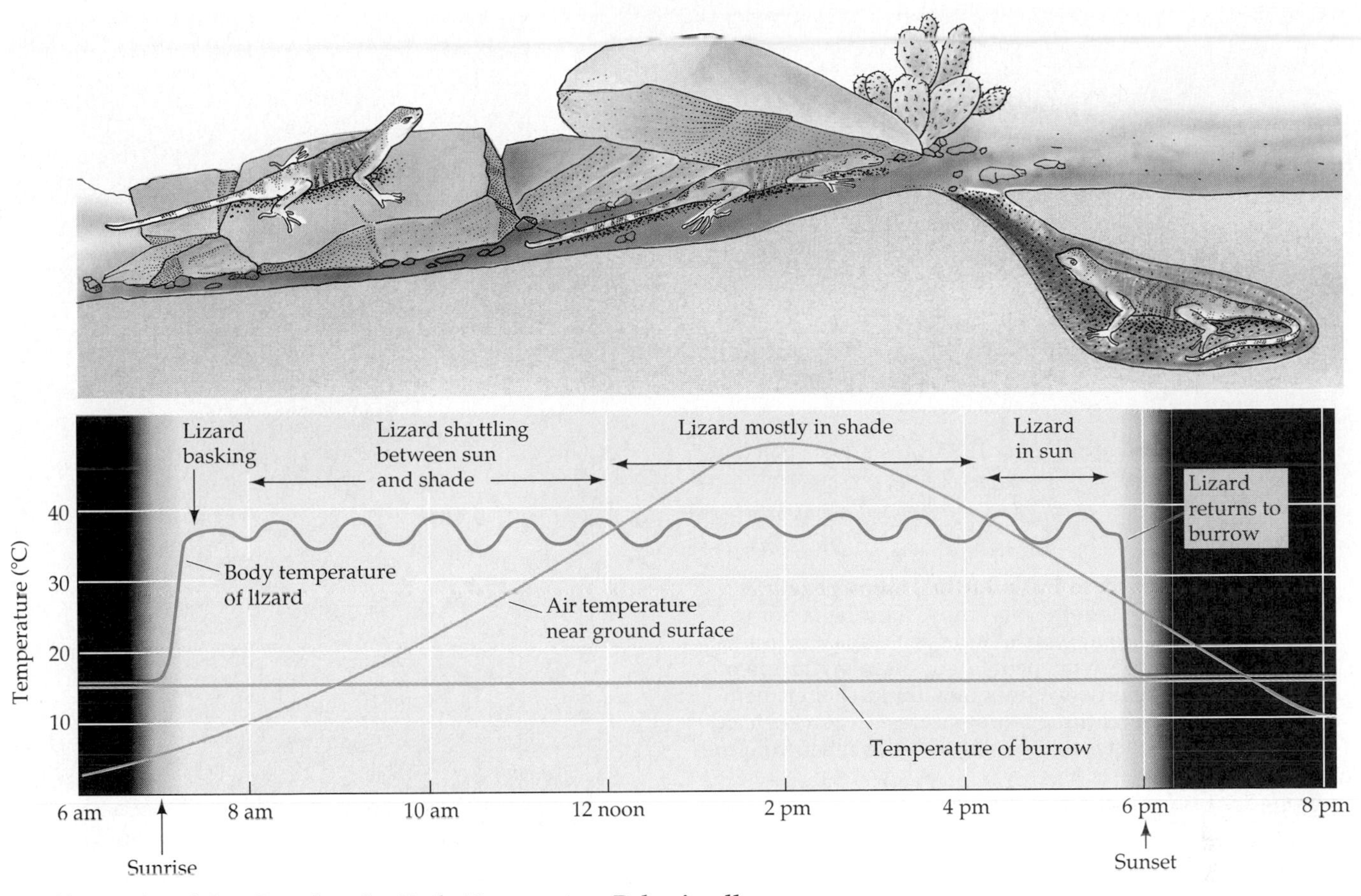

35.13 A Lizard Can Regulate Its Body Temperature Behaviorally

mann: "An experiment is like a conversation with an animal, but the animal must be permitted to answer in its own language."

Behavioral Thermoregulation

Behavioral thermoregulation is not the exclusive domain of ectotherms. It is also the first line of defense for endotherms. When the option is available, most animals select thermal microenvironments that are best for them. They may change their posture, orient to the sun, move between sun and shade, and move between still air and moving air, as demonstrated by the lizard in our field experiment. Examples of more complex thermoregulatory behavior are nest construction and social behavior such as huddling. In humans, the selection of clothing is quite important. Behavioral thermoregulation is widespread in the animal kingdom (Figure 35.14).

Control of Blood Flow to the Skin

Physiological thermoregulation is not the exclusive domain of endotherms. Ectotherms exhibit various physiological thermoregulatory adaptations. Both ectotherms and endotherms can alter the rate of heat exchange between their bodies and their environments by controlling the flow of blood to the skin. For example, when a person's body temperature rises as a result of exercise, blood flow to the skin increases, and the skin surface gets quite warm. The extra heat brought from the body core to the skin by the blood is lost to the environment, thus tending to bring body temperature back to normal. By contrast, when a person is exposed to cold, the blood vessels supplying the skin constrict, decreasing blood flow and heat transport to the skin, thus reducing heat loss to the environment.

The control of blood flow to the skin is an important adaptation for ectotherms like the marine iguana of the Galapagos Islands. The Galapagos are volcanic islands on the equator, bathed by cold oceanic currents. Marine iguanas are reptiles that bask on black lava rocks near the ocean and swim in the sea, where they feed on submarine algae. When the iguanas cool to the temperature of the sea, they are slower and more vulnerable to predators, and probably incapable of efficient digestion. They therefore alternate between feeding in the sea and basking on the rocks. It is advantageous for iguanas to retain body heat as long as possible while swimming and to warm up as fast as possible when basking. They adjust their cool-

35.14 Endotherms Use Behavior to Thermoregulate Humans and other endotherms adjust their behavior to the environmental temperature in many ways. *(a)* In the extreme cold of the Arctic, people put on many layers of insulating clothing. The ice huts they build shelter them from even colder nighttime temperatures. *(b)* An African elephant showers itself with water to bring relief from the heat.

ing and heating rates by changing the flow of blood to the skin.

Blood vessels to the skin constrict when an iguana is in the ocean and dilate when it is basking. In addition, an iguana's heart rate is slower when it is swimming than when it is basking. Slowed heart rate and constricted vessels when swimming mean that less blood is being pumped through the skin, and therefore less heat is being transported from deep in the iguana's body to its skin to be lost to the water. Faster heart rate and dilated vessels when the iguana is basking increase the transport of heat from the skin to the rest of its body. Of course, basking on black rocks under the equatorial sun can be too much of a good thing. When an iguana reaches an optimal body temperature, it lifts its body off the rocks and orients itself to minimize the surface area of its body that is directly exposed to solar radiation. Thus, the marine iguana uses both physiological and behavioral mechanisms to regulate its body temperature.

Metabolic Heat Production

The use of metabolic heat production to maintain a body temperature above that of the environment is surprisingly common among ectotherms. For example, the powerful flight muscles of many insects, such as dragonflies, moths, bees, and beetles, must reach a fairly high temperature (35 to 40°C) before the insects can fly, and they must maintain these high temperatures during flight, even at air temperatures around 0°C. Such insects use the flight muscles themselves to produce the required heat. These muscles are about 20 percent efficient; that is, about 20 percent of the energy they consume goes into useful work and 80 percent is lost as heat. Flight thus produces an enormous amount of heat, which keeps body temperature elevated. To reach flight temperature from resting temperature, the insects contract their flight muscles isometrically: The muscles that contract alternately during flight to produce the wingbeats contract simultaneously during warm-up. Even though the muscles are contracting and producing heat, the wings do not move (Figure 35.15). During warm-up some bees and moths appear to be shivering because the wings show movements of small amplitude.

The heat-producing ability of insects can be quite remarkable. It enables moths to fly at night when air temperatures are low and solar basking is not possible. The heat-producing ability of a species of scarab beetle that lives in the mountains north of Los Angeles, California, has made it possible for these beetles to have an unusual mating behavior. The beetles spend most of their life cycle in the soil, except for mating, at which time females and males emerge from the soil and the males fly in search of females. What is unusual is that they engage in this behavior in winter, at night, during snowstorms. The drop in

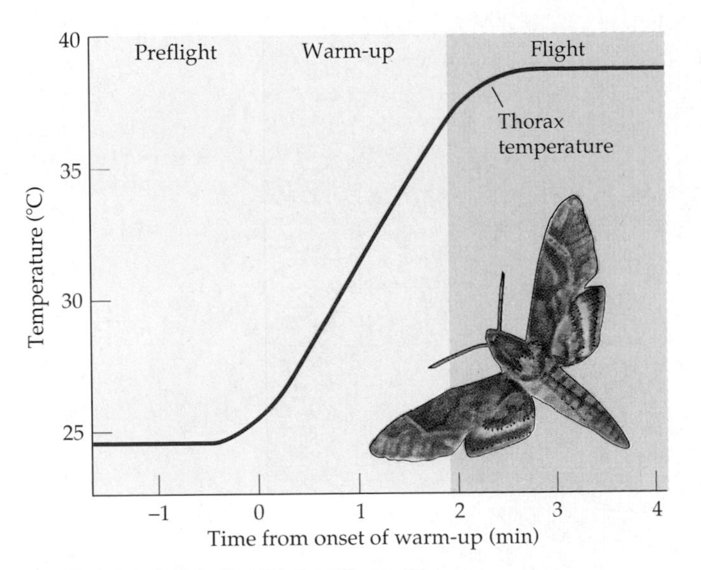

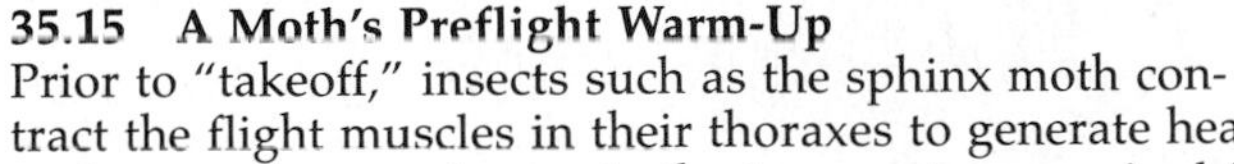

35.15 A Moth's Preflight Warm-Up
Prior to "takeoff," insects such as the sphinx moth contract the flight muscles in their thoraxes to generate heat and warm the muscles up to the temperature required for flight.

barometric pressure associated with a storm probably triggers the emergence from the soil. These beetles were long considered to be very rare because very few entomologists look for beetles in the mountains, in winter, at night, during snowstorms. Presumably the same is true for potential predators!

Honeybees regulate temperature as a group. They live in large colonies consisting mostly of female worker bees that maintain the hive and rear young that are hatched from eggs laid by the single queen bee in the colony. During winter, honeybee workers combine their individual heat-producing abilities to regulate the temperature of the brood. They cluster in the area of the hive where the brood is located and adjust their joint metabolic heat production and density of clustering so that the brood temperature remains remarkably constant, at about 34°C, even as outside air temperature drops below freezing.

Some reptiles use metabolic heat production to raise body temperature above air temperature. The female Indian python protects her eggs by coiling her body around them. If air temperature falls, she uses isometric contractions of her body wall muscles to generate heat. Like the use of flight muscles by insects, this adaptation of the python is analogous to shivering in mammals. The python is able to maintain the temperature of her body—and therefore that of her eggs—considerably above air temperature.

Biological Heat Exchangers

If heat that active muscles produce is not rapidly lost to the environment, it can be used to raise body temperature above the temperature of the surrounding air or water. It is particularly difficult for fish to slow the loss of body heat to the environment because blood pumped from the heart comes into close contact with water flowing over the thin gill membranes before it travels through the body. Therefore, any heat transferred to the blood from active muscles is lost rapidly to the environment. It is therefore surprising to find that some large, rapidly swimming fishes, such as bluefin tuna and great white and mako sharks, can maintain temperature differences as great as 10 to 15°C between their bodies and the surrounding water (Figure 35.16). The heat comes from their powerful swimming muscles, of course, but the ability to conserve that heat is due to remarkable arrangements of the blood vessels.

In the usual fish circulatory system, oxygenated blood from the gills collects in a large, dorsal vessel, the aorta, which travels through the center of the fish, distributing blood to all organs and muscles. "Hot" fish such as bluefin tuna have smaller central dorsal aortas. Most of their oxygenated blood is transported in large vessels just under the skin (Fig-

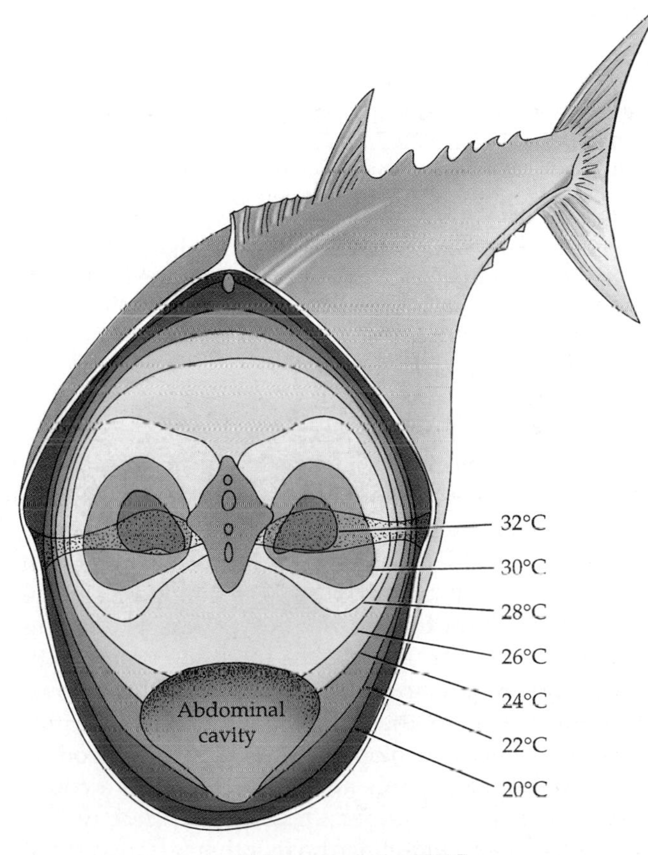

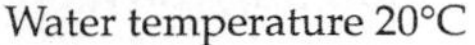

35.16 Cold Water, Warm Muscles
In an actively swimming fish such as the bluefin tuna, the muscles that power swimming generate heat that keeps the fish's internal body temperature much higher than that of the surrounding water.

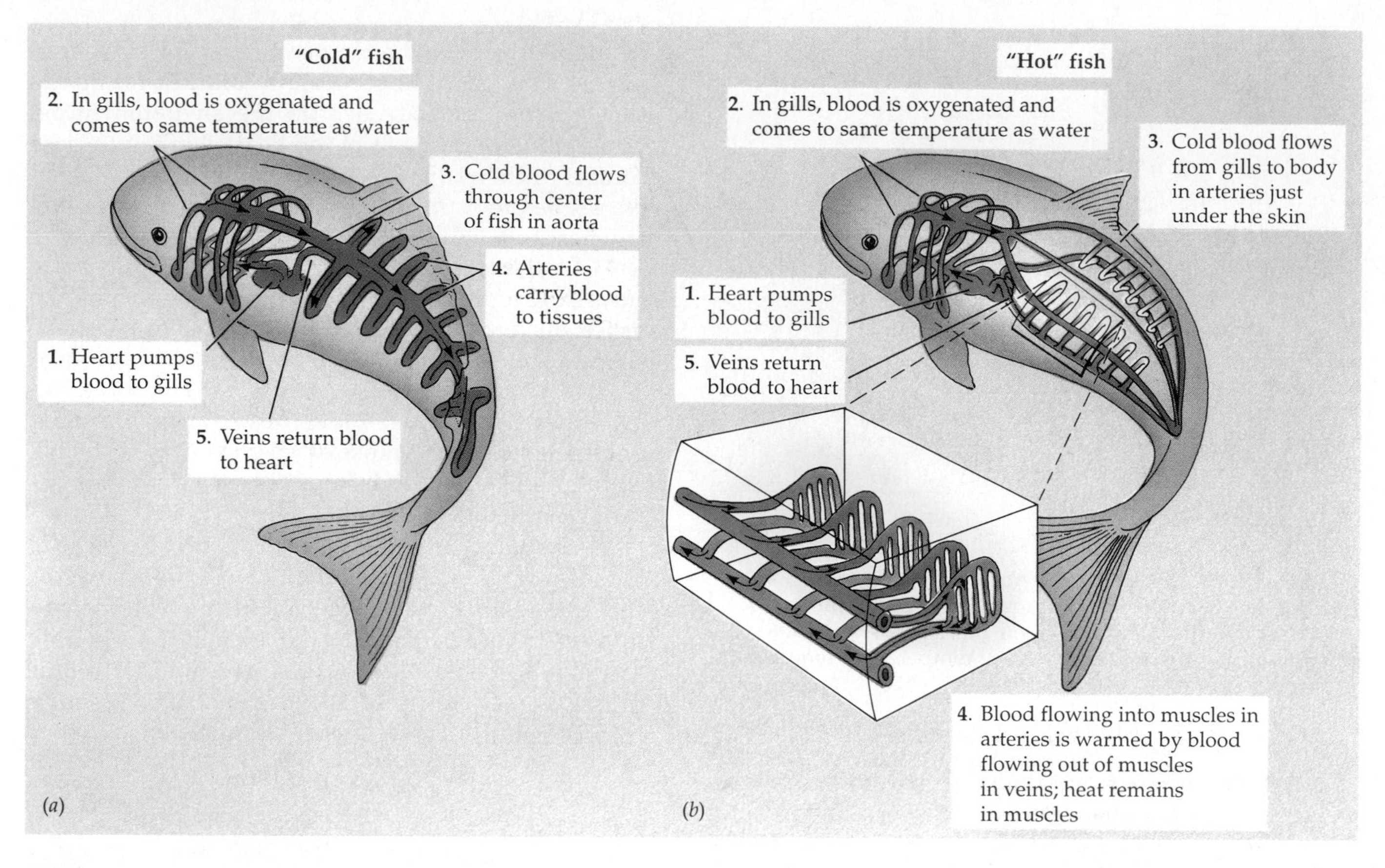

35.17 Hot and Cold Fish
(a) The circulatory systems of most fish conduct the oxygenated blood from the gills to the organs of the fish through a large dorsal aorta. Because the blood comes into equilibrium with water temperature in the gills, the blood the aorta carries through the interior of the fish's body is at water temperature. *(b)* "Hot" fish species such as the bluefin tuna (see Figure 35.16) retain the heat produced by their muscles because the anatomy of their blood vessels allows for heat exchange between the warm blood leaving the muscle and the cold blood entering the muscle.

ure 35.17). Hence the cold blood from the gills is kept close to the surface of the fish. Smaller vessels transporting this cold blood into the muscle mass run parallel to vessels transporting warm blood from the muscle mass back toward the heart. Vessels carrying cold blood into the muscle are in close contact with vessels carrying warm blood away, and heat flows from the warm to the cold blood. Because heat is exchanged between blood vessels carrying blood in opposite directions, this adaptation is called a countercurrent heat exchanger. It keeps the heat within the muscle mass, enabling the fish to have an internal body temperature considerably above the water temperature. Why is it advantageous for the fish to be warm? Each 10-degree rise in muscle temperature increases its sustainable power output almost threefold!

THERMOREGULATION IN ENDOTHERMS

An endotherm responds to changes in the temperature of its environment primarily by changing its metabolic rate. Within a narrow range of environmental temperatures called the **thermoneutral zone**, the metabolic rate of the endotherm is low and independent of temperature. The metabolic rate of a resting animal at a temperature within the thermoneutral zone is called the **basal metabolic rate**. It is usually measured on animals that are quiet but awake and that are not using energy in the digestive processes or for reproduction. A resting animal consumes energy at the basal metabolic rate just to carry out all of its metabolic functions other than thermoregulation.

The basal metabolic rate of an endotherm is about six times greater than the metabolic rate of a similarly sized ectotherm at rest and at the same body temperature (see Figure 35.12*b*). A gram of mouse tissue consumes energy at a much higher rate than does a gram of lizard tissue when both tissues are at 37°C. This difference is due to a basic change in cell metabolism that accompanied the evolution of endotherms from their ectothermic ancestors. The higher level of heat production by endotherms makes it easier for them to maintain a temperature difference between the body and the environment.

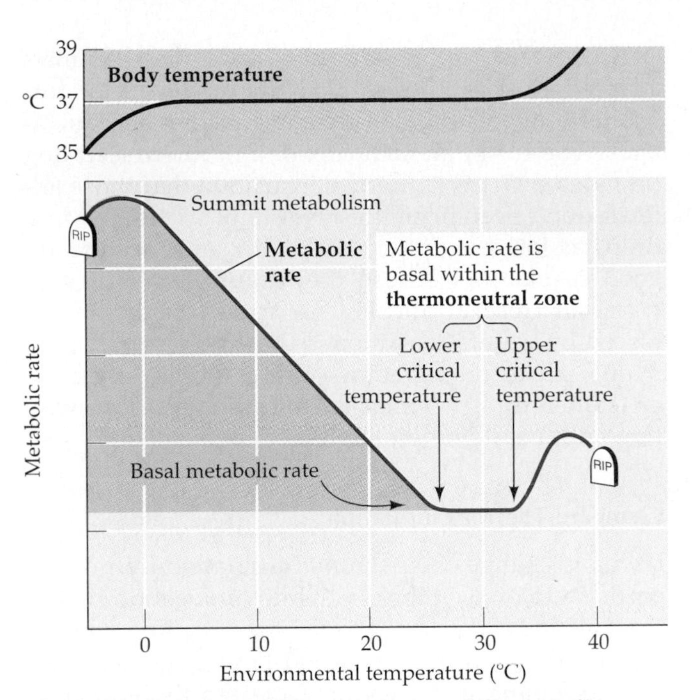

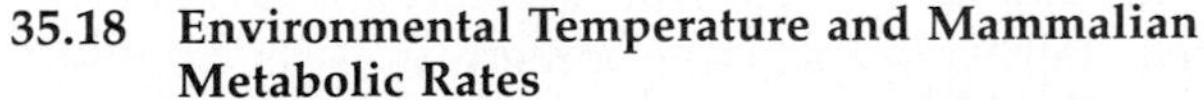

35.18 Environmental Temperature and Mammalian Metabolic Rates
Within a mammal's thermoneutral zone its metabolic rate is low and constant; the animal thermoregulates behaviorally and by changing its thermal insulation. Above the upper critical temperature the animal must expend energy to lose heat by panting or sweating, which makes its metabolic rate increase. Below the lower critical temperature, the animal produces metabolic heat to compensate for increased heat loss to the environment, as indicated by the dark-shaded area under the plotted line.

Active Heat Production and Heat Loss

The thermoneutral zone is bounded by a lower critical temperature and an upper critical temperature. Below the lower critical temperature an endotherm's metabolic rate increases as environmental temperature declines because the animal must produce more and more heat to maintain a constant body temperature as heat loss to the environment increases. As the environment gets colder, eventually the animal reaches its summit metabolism, or maximum possible thermoregulatory heat production. If the environmental temperature falls still lower, the animal's body temperature will begin to drop. On the other hand, when the environmental temperature goes above the upper critical temperature, the animal pants or sweats. Since these active heat loss responses require an increased expenditure of energy, the metabolic rate rises. A graph of metabolic rate as a function of environmental temperature illustrates the thermoregulatory responses of an endotherm (Figure 35.18).

Mammals use two mechanisms—shivering and nonshivering heat production—to create heat for thermoregulation. Birds use only shivering heat production. Shivering uses the contractile machinery of skeletal muscles to consume ATP without causing observable behavior. The muscles pull against each other so that little movement other than a tremor results. All of the energy from the conversion of ATP to ADP in this process is released as heat. Most nonshivering heat production occurs in specialized tissue called **brown fat**. It looks brown because of its abundant mitochondria and rich blood supply (Figure 35.19). In brown fat cells a protein called thermogenin

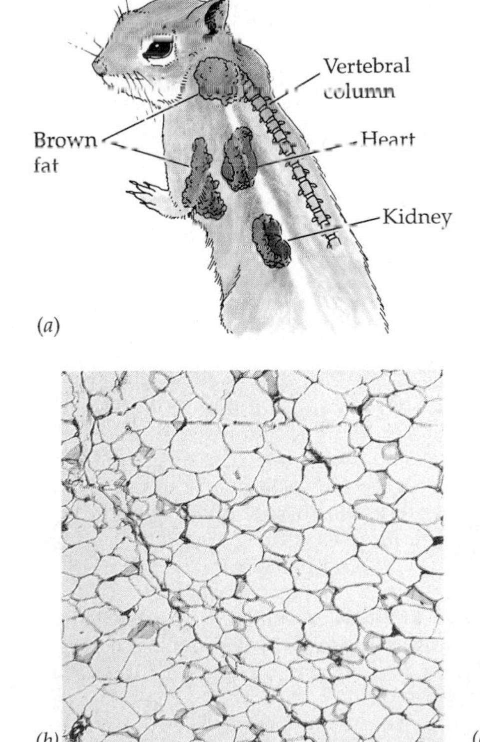

35.19 Brown Fat: A Heat-Producing Tissue
In many mammals, brown fat produces heat. *(a)* In a ground squirrel, brown fat occurs in specific anatomical locations. *(b)* White fat viewed through a light microscope. Each cell is filled with a globule of lipid and has few organelles. The tissue has few blood vessels. *(c)* Brown fat viewed through a light microscope at the same magnification reveals cells with many intracellular structures and multiple droplets of lipid. Numerous capillaries run through the tissue. *(d)* An electron micrograph of brown fat shows the tight packing of mitochondria in a brown fat cell. A portion of a lipid droplet is visible in the upper right; in the upper left is part of a blood capillary.

(b)

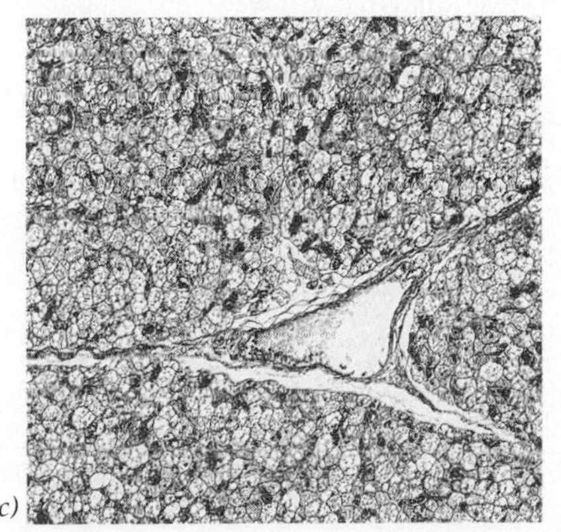

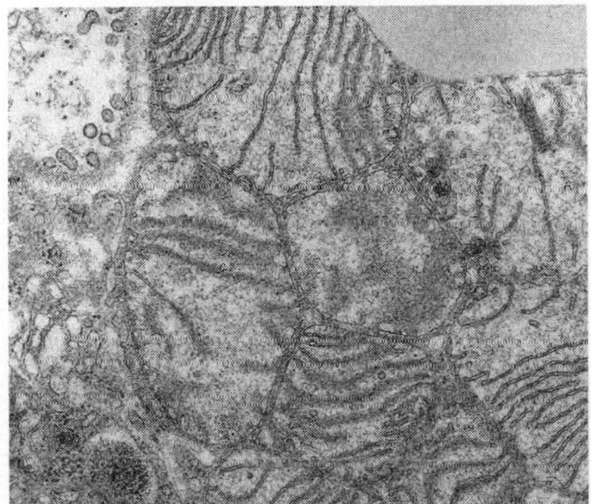

uncouples oxidative phosphorylation. Thus metabolic fuels are consumed to produce heat without the production of ATP. Brown fat is especially abundant in newborn infants of many mammalian species, in some adult mammals that are highly acclimatized to cold, and in mammals that hibernate.

Living in the Cold

The coldest habitats on Earth are in the Arctic, the Antarctic, and at the tops of high mountains. Many birds and mammals, but no reptiles or amphibians, live in the coldest habitats. The ability to produce a substantial amount of heat metabolically has enabled endotherms to exploit formidable, frigid environments. Most tropical species of birds and mammals, however, would not fare well in those environments. What adaptations besides endothermy characterize species that live in the cold?

The most important adaptations of endotherms to cold environments are those that reduce their heat loss to the environment. Since most heat loss is from the body surface, many cold-climate species have smaller surface areas than their warm-climate cousins, even when their body masses are the same. Rounder body shapes and shorter appendages reduce the surface area-to-volume ratios of some cold-climate species; compare, for example, the desert jackrabbit and the arctic hare (Figure 35.20). Another means of decreasing heat loss is to increase thermal insulation. Fur, feathers, and layers of fat decrease the loss of heat from an endotherm to the environment. You can experience the effectiveness of thermal insulation by comparing what it feels like to sit on a cold stone or metal bench while wearing only cotton shorts with what it feels like after you put a feather pillow or a wool blanket between you and the bench.

Arctic and alpine animals, and other animals adapted to cold, have much thicker layers of fur, feathers, or fat than do their warm-climate equivalents. The fur of an arctic fox or a northern sled dog provides such good thermal insulation that those animals don't even begin to shiver until air temperature drops as low as −20 to −30°C. Fur and feathers are good insulators because they trap a layer of still, warm air close to the skin surface. If that air is displaced by water, insulation is drastically reduced. In many species oil secretions spread through their fur or feathers by grooming is critical for resisting wetting and maintaining a high level of insulation.

Changing Thermal Insulation

Humans change their thermal insulation by putting on or taking off clothes. How do animals do it? We have already discussed one example, the marine iguana. By changing the blood flow to its skin, the marine iguana increases or decreases the exchange of heat between the environment and its body—it changes its thermal insulation. Increasing or decreasing blood flow to the skin is an important thermoregulatory adaptation for endotherms as well. In a hot environment, your skin feels hot because of the high rate of blood flow through it, but when you are sitting in an overly air-conditioned theater, your hands, feet, and other body surfaces feel cold as blood flow to those areas decreases. The wolf has an elegant mechanism for decreasing heat loss from its feet without the risk of freezing them. As long as the wolf's foot temperature is more than a few degrees above freezing, certain blood vessels in its foot are constricted and blood flow to the foot is minimal. As foot temperature approaches 0°C, these vessels open and allow more warm blood to flow through the foot, thus keeping it from freezing.

(a)

(b)

35.20 Adaptations to Hot and Cold Climates
(a) The desert jackrabbit has a large surface area for its body mass, largely due to its long extremities. The large ears serve as heat exchangers, passing heat from the rabbit's blood to the surrounding air. *(b)* The arctic hare has shorter extremities and therefore a smaller surface area for its body size. The fur of the arctic hare, longer and thicker than that of the desert hare, provides good insulation.

For highly insulated arctic animals and for many large mammals from all climates, getting rid of excess heat can be a serious problem, especially during exercise. Arctic species usually have a place on the body surface, such as the abdomen, that has only a thin layer of fur and can act as a window for heat loss. Large mammals, such as elephants, rhinoceroses, and water buffalo, have little or no fur and seek places where they can wallow in water when the air temperature is too high. Having water in contact with the skin greatly increases heat loss because water has a much greater capacity for absorbing heat than does air.

Evaporative Water Loss

The evaporation of water is a very effective means of dissipating heat. A gram of water absorbs about 580 calories when it evaporates. However, water is heavy, animals do not carry an excess supply of it, and hot environments tend to be arid places where water is a scarce resource. Therefore, evaporation of water by sweating or panting is usually a last resort for animals adapted to hot environments. Sweating and panting are active processes that require the expenditure of metabolic energy. That's why the metabolic rate increases when the upper critical temperature is exceeded (see Figure 35.18). A sweating or panting animal is producing heat in the process of dissipating heat. This can be a losing battle. Animals can survive in environments that are below their lower critical temperature much better than they can in those above their upper critical temperature.

THE VERTEBRATE THERMOSTAT

The thermoregulatory mechanisms and adaptations we have already discussed are the controlled systems for the regulation of body temperature. These controlled systems must receive commands from a regulatory system that integrates information relevant to the regulation of body temperature. A convenient name for the regulatory system in this case is thermostat. All animals that thermoregulate, both vertebrate and invertebrate, must have regulatory systems, but here we will focus on the vertebrate thermostat.

Where is the vertebrate thermostat? The major integrative center is at the bottom of the brain in a structure called the **hypothalamus**. If you slide your tongue back as far as possible along the roof of your mouth, it will be just a few centimeters below your hypothalamus. The hypothalamus is a part of many regulatory systems, so we will refer to it many times in the chapters to come. If the hypothalamus of a mammal's brain is damaged, the animal loses its ability to regulate its body temperature, which then rises in warm environments and falls in cold ones.

Set Points and Feedback

What information does the vertebrate thermostat use? In many species the temperature of the hypothalamus itself is a major source of feedback to the thermostat. Cooling the hypothalamus causes fishes and reptiles to seek a warmer environment, and heating the hypothalamus causes them to seek a cooler environment. In mammals, cooling the hypothalamus can stimulate constriction of blood vessels to the skin and increase metabolic heat production. Because of the activation of these thermoregulatory responses, body temperature rises when the hypothalamus is cooled. Conversely, warming of the hypothalamus stimulates dilation of blood vessels to the skin and sweating or panting, and the overall body temperature falls when the hypothalamus is warmed (Figure 35.21). The hypothalamus appears to generate a set point like a thermostat setting. When the temperature of the hypothalamus exceeds or drops below that set point, thermoregulatory responses (the controlled system) are activated to reverse the direction of temperature change. Hence hypothalamic temperature is a negative feedback signal.

An animal has separate set points for activating different thermoregulatory responses. If the hypothalamus of a mammal is heated and cooled, the vessels supplying blood to the skin constrict at a specific hypothalamic temperature. A slightly lower hypothalamic temperature initiates shivering, and a hypothalamic temperature two or three degrees higher initiates panting. We can describe the characteristics of hypothalamic control of each response. For example, if we measure metabolic heat production while heating and cooling the hypothalamus (see Figure 35.21), we can describe the results graphically (Figure 35.22). Within a certain range of hypothalamic temperatures, metabolic heat production remains low and constant, but cooling the hypothalamus below a certain level—the set point—stimulates increased metabolic heat production. The increase in heat production is proportional to how much the hypothalamus is cooled below the set point. This regulatory system is much more sophisticated than a simple on–off thermostat like the one in a house.

The vertebrate thermostat integrates other sources of information in addition to hypothalamic temperature. It integrates information about the temperature of the environment as registered by temperature sensors in the skin. Changes in skin temperature shift the hypothalamic set points for responses. As Figure 35.22 shows, in a warm environment you might have to cool the hypothalamus of a mammal to stimulate it to shiver, but in a cold environment you would

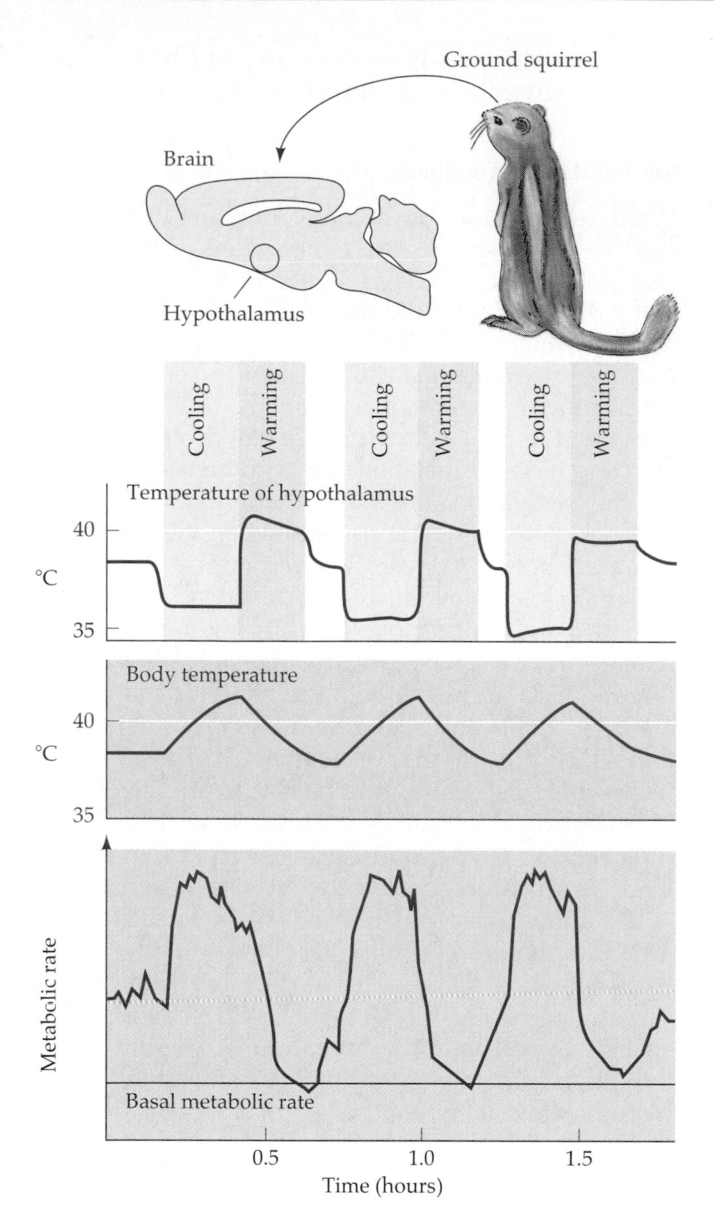

35.21 The Hypothalamus Acts as a Thermostat In an experiment a ground squirrel was maintained at low environmental temperatures so that its initial metabolic heat production was high. Cooling the hypothalamus increased metabolic heat production even further and the animal's body temperature rose. Heating the hypothalamus reduced metabolic rate, and the animal's body temperature fell.

have to warm the hypothalamus of the same animal to stop its shivering. The set point for the metabolic heat production response is higher when the skin is cold and lower when the skin is warm. Information from the skin can be considered feedforward that adjusts the hypothalamic set point. Many other factors also shift hypothalamic set points for responses. Set points are higher during wakefulness than during sleep, and they are higher during the active part of the daily cycle than during the inactive part, even if the animal is awake at both times.

Fever also causes shifts in set points. Fevers are rises in body temperature in response to substances called **pyrogens** derived from bacteria or viruses that invade the body. Injections of the killed bacteria or even the purified cell walls of the killed bacteria can also cause fever. The presence of the pyrogen in the body causes a rise in the hypothalamic set point for the heat production response. If you are the unlucky person developing a fever, you may feel unbearably cold (the chills) even though your body temperature is normal. You shiver, put on more clothes, and turn up your room temperature or electric blanket. As a result your body temperature rises until it matches the new set point. At the higher body temperature you no longer feel cold, and you may not feel as if you are hot, but someone touching your forehead will say that you are "burning up." If you take an aspirin, it lowers your set point to normal. Now you feel hot, take off clothes, and even sweat until your elevated body temperature returns to normal. Extreme fevers can be dangerous and must be reduced, but there is evidence that moderate fevers help the body fight an infection (Box 35.A). Perhaps we should not be too hasty in using medication to counter shifts in our hypothalamic set points.

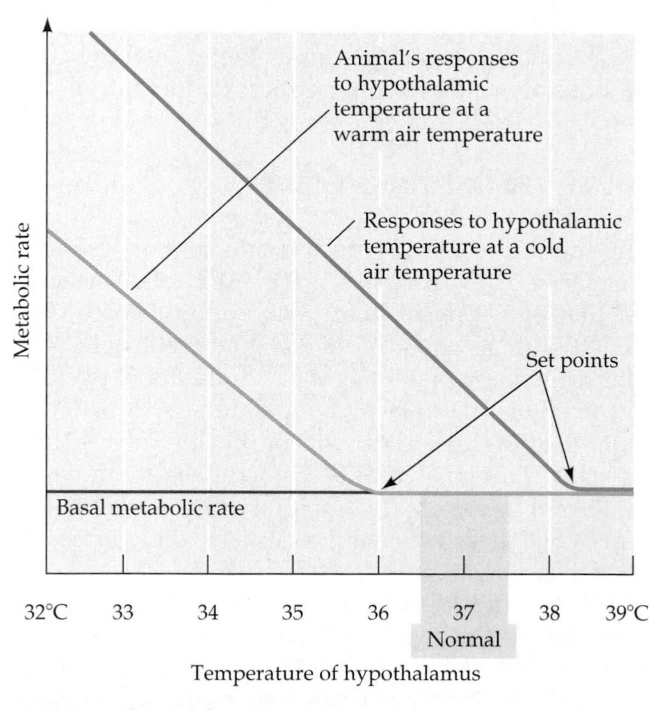

35.22 Adjustable Set Points Vertebrates have different a set point for the metabolic heat production response to hypothalamic temperature at different environmental temperatures. Other factors, such as being asleep or awake, the time of day, or the presence of a fever can also affect the set point.

BOX 35.A

Fevers and "Feeling Crummy"

You respond to many infectious illnesses by getting a fever and feeling crummy. You lose your appetite, you have no energy, your joints and muscles ache, you get the chills, and you feel like just putting on flannel pajamas and getting into bed. Are these well-known symptoms simply unfortunate side effects? To the contrary, scientists are beginning to think that getting a fever and feeling crummy are adaptive responses that help us fight diseases.

The immediate causes of these responses are chemical messages from the immune system. When an infectious virus or bacterium invades the body, scavenger cells called macrophages grab it. One of the things macrophages do is release chemicals called interleukins that sound the alarm to other cells of the immune system throughout the body. Interleukins cause many other responses as well. They make neurons transmitting pain more sensitive, and so we ache. They make us sleepy. They stimulate the hypothalamus to release corticotropin-releasing hormone, which initiates the stress responses of the body. Interleukins also cause a rise in the hypothalamic set point for thermoregulatory responses. Intracellular messengers activated by the interleukins include prostaglandins. A potent inhibitor of prostaglandin synthesis is aspirin, thus explaining how this drug can reduce fever and make us feel better. But is it a good idea to reduce fever and feel less crummy if these are adaptive responses to infection?

Some convincing evidence that fever is an adaptive response to infection came from experiments on lizards by Matthew Kluger at the University of Michigan. Lizards having access to a heat lamp maintain their body temperature at about 38°C by shuttling in and out of the vicinity of light. When injected with pathogenic bacteria, the lizards spend more time under the light and raise their body temperatures to between 40 and 42°C—they develop fevers. Does a fever help the lizard fight infection? To answer this question, groups of lizards receiving equal inoculations of bacteria were kept in incubators at 34, 36, 38, 40, and 42°C. All of the lizards at 34 and 36°C died, about 25 percent at 38°C survived, and about 75 percent at 40 and 42°C survived. Fevers do help.

Even though fever clearly seems to be an adaptive response to infection, high fevers can be dangerous and even lethal. Even modest fevers can be dangerous to people with weakened hearts and people who are chronically ill. A fetus can be endangered when a pregnant woman has a high fever. Drugs that reduce fever may be important in such cases, but perhaps they should not be taken by most people at the first sign of aches or chills.

TURNING DOWN THE THERMOSTAT

Having learned that fever results from turning up the hypothalamic thermostat, we can ask, Are there cases of turning down the thermostat so that body temperature is regulated at a lower level? The answer is yes, but not all decreases in body temperature are regulated. Hypothermia is the condition in which body temperature is below normal. It can result from a natural turning down of the thermostat, or from traumatic events such as starvation (lack of fuel), exposure, serious illness, or anesthesia. Because of Q_{10} effects, hypothermia slows metabolism, slows the heart, weakens muscle contractions (including those of the heart), decreases nerve conduction, and causes unconsciousness. This is not a happy state of affairs for most endotherms and can lead to death, but it can also be somewhat protective.

There are cases in which drowning victims have been under water for 10 or 15 minutes or more and have shown no pulse when pulled out of the water. Nevertheless, some—mostly small children drowned in cold water—were revived by paramedics and recovered to a remarkable extent. The rapid fall in their body temperature slowed metabolism and thereby slowed the progress of cell damage caused by lack of oxygen. In this way, the hypothermia prevented irreversible brain damage even though it was a pathological condition induced by drowning. We will see next that some animals can anticipate unfavorable circumstances and induce the hypothermia of torpor or hibernation as an adaptive, protective mechanism by turning down their thermostats.

Shallow Torpor

Hypothermia conserves metabolic energy. Many species of birds and mammals use regulated hypothermia as a means of surviving periods of cold and food scarcity. Because of their extremely high surface-to-volume ratios, very small endotherms such as hummingbirds and pocket mice may exhaust their metabolic reserves just to get through one day without food if they are at normal body temperature. Animals of such species can extend the period over which they can survive without food by dropping body temperature. This adaptive hypothermia is called **shallow torpor** or daily torpor because it usually occurs on a daily basis, with body temperature falling at the time of day the animal normally becomes inactive. Body temperature can drop 10 to 15°C during shallow torpor, resulting in an enormous saving of metabolic energy.

A small bird, the willow tit, studied by R. Reinertsen in Norway, provides an example of shallow torpor that shows how well regulated this process can be. Willow tits live through the winter above the Arctic Circle. In spite of their good thermal insulation, these tiny birds must become hypothermic to survive the long, cold Arctic nights. Each evening the bird lowers its metabolic rate to a level that it maintains all night, and its body temperature falls as a result of the decreased heat production. How low the bird's metabolic rate drops is different on different nights (Figure 35.23). The decrease in its metabolism depends on air temperature, on season (and hence on length of night), and on the bird's fat reserves at roosting time. Every morning the bird has depleted its fat reserves and must immediately feed on seeds it has stored nearby. This is living on the razor's edge! The brain of this small animal integrates all the relevant information, resulting in just the right resetting of its thermostat to get it through the night.

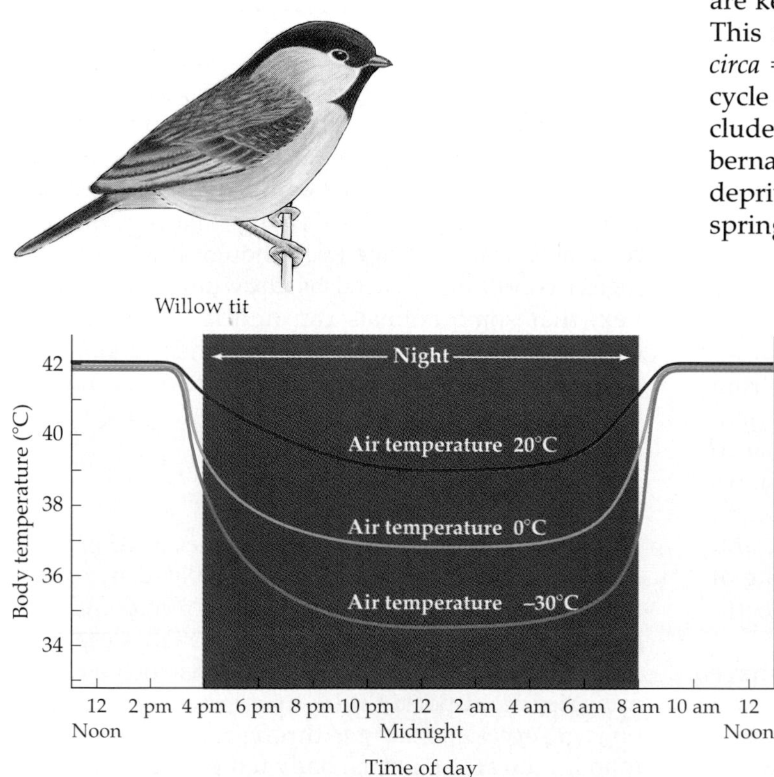

35.23 Hypothermia Deepens with Cold and Dark
The curves show how a willow tit's body temperature changes during long nights at different environmental temperatures. The colder the air, the deeper is the bird's hypothermia. Notice that the depth of hypothermia is set early in the night, and must therefore be a result of information available at that time—not simply a consequence of running out of fuel reserves faster during colder nights. If nights are made shorter (which is possible in the laboratory), the birds maintain higher body temperatures at these same air temperatures.

Hibernation

Regulated hypothermia can last for days or even weeks, with drops to very low temperatures; this phenomenon is called **hibernation**. Many diverse species of mammals hibernate, but only one species of bird, the poorwill, has been shown definitely to hibernate. For the deep sleep of hibernation, the body's thermostat is turned down to an extremely low level to maximize energy conservation. Body temperature falls during hibernation because the hypothalamic set point drops, and arousal from hibernation is due to a return of the hypothalamic set point to a normal mammalian level. Many hibernators maintain body temperatures around 2 to 4°C during hibernation. The metabolic rate needed to sustain an animal after this incredible drop in body temperature is only 1⁄30 to 1⁄50 of basal metabolic rate, an enormous saving of metabolic energy.

Animals hibernate when temperatures are low and food is scarce. Individual bouts of hibernation may last from less than a day to over a week (Figure 35.24). A bout terminates spontaneously when the hibernator's body temperature returns to normal. The animal may remain at its normal temperature for a few hours to a day, during which time it may eat. (Some hibernators store food in their well-insulated nests.) Then the animal enters another bout of hibernation.

The hibernation season is controlled by an internal biological clock (or calendar), which continues to run with a periodicity of about a year even when animals are kept under constant conditions in the laboratory. This is called a **circannual rhythm** (from the Latin *circa* = about and *annus* = year). A typical circannual cycle for a hibernator such as a ground squirrel includes an active season, during which it cannot hibernate even if exposed to cold temperatures and deprived of food. During the active season, usually spring through fall, the animals breed, raise their

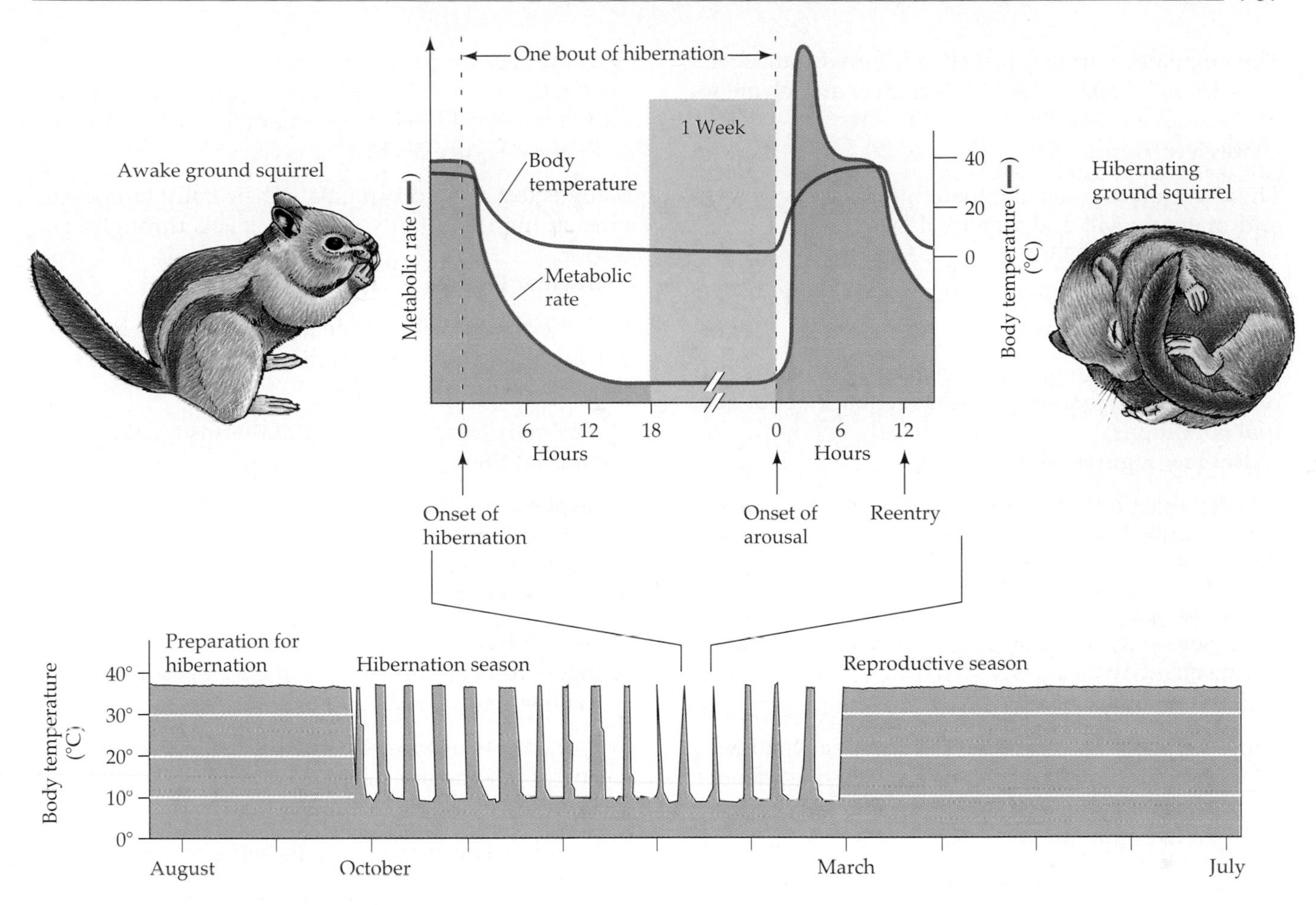

35.24 A Ground Squirrel Hibernates in Bouts
There are approximately three bouts per month during the hibernation season. A bout ends when the animal's body temperature returns to normal mammalian levels; it may remain there for a few hours or even a day or more before another bout begins.

young, prepare their nests for winter, fatten their bodies, and store food. During the hibernation season, as Figure 35.24 shows, animals hibernate in recurrent bouts. They progressively lose body weight even if food is available. Toward the end of the hibernation season the reproductive organs grow and become functional. The ability of hibernators to reduce the set point so dramatically probably evolved as an extension of the set point decrease that accompanies sleep even in nonhibernating species of mammals and birds.

SUMMARY of Main Ideas about Physiology, Homeostasis, and Temperature Regulation

The organs of an animal maintain a constancy, or homeostasis, of the animal's internal environment.

Organs are made of four types of tissues: epithelial, connective, muscle, and nervous.
Review Figure 35.1

Organs can be grouped into systems with common or complementary functions.

The nervous and endocrine systems process information and control physiological functions and behavior.
Review Figure 35.2

Organs that function in protection, support, and movement include the skin, the skeletal system, and the muscular system.
Review Figure 35.3

The reproductive system produces gametes, achieves fertilization, and supports and protects the developing fetus.
Review Figure 35.4

The organs of nutrition include the lungs and the digestive system.
Review Figure 35.5

The circulatory system provides internal transport of nutrients, wastes, heat, hormones, and elements of the immune system.
Review Figure 35.6

The excretory organs eliminate nitrogenous wastes and maintain salt and water balance.
Review Figure 35.7

Homeostasis is achieved by the control and regulation of organ systems.

A regulatory system, which includes set points that reflect optimal conditions, uses feedback about actual conditions.
Review Figures 35.8 and 35.9

Life can exist only within the range of temperatures between the freezing point of water and the temperature that denatures proteins.

Most biological processes and reactions are temperature-sensitive, and the Q_{10} is a measure of temperature sensitivity.
Review Figure 35.10

Some animals that cannot avoid seasonal changes in their body temperatures have biochemical adaptations that compensate for the changes.
Review Figure 35.11

Animals have evolved behavioral and physiological adaptations for controlling their body temperatures.

Homeotherms maintain a fairly constant body temperature most of the time; poikilotherms do not.

Endotherms can produce a significant amount of metabolic heat to elevate body temperature, but ectotherms depend on environmental sources of heat.
Review Figure 35.12

Many ectotherms can regulate their body temperatures at high and fairly constant levels through behavior.
Review Figures 35.13 and 35.15

Endotherms have basal metabolic rates that are much higher than the resting metabolic rates of similarly sized ectotherms.

The primary adaptation of endotherms to different climates is their level of insulation.

In response to cold, endotherms can elevate metabolic rate through shivering and nonshivering heat production.
Review Figure 35.19

Some fish have vascular countercurrent heat exchangers to conserve heat produced by muscle.
Review Figure 35.17

Evaporative water loss is an effective but costly means of dissipating heat.
Review Figure 35.18

The mammalian hypothalamus serve as a thermostat by sensing temperature, generating set points, and sending commands to effector organs.
Review Figures 35.21 and 35.22

The mammalian thermostat is turned down during shallow torpor and deep hibernation, thereby conserving energy.
Review Figure 33.24

SELF-QUIZ

1. If the Q_{10} of the metabolic rate of an animal is 2, then
   a. the animal is better acclimatized to a cold environment than if its Q_{10} were 3.
   b. the animal is an ectotherm.
   c. the animal consumes half as much oxygen per hour at 20°C as it does at 30°C.
   d. the animal's metabolic rate is not at basal levels.
   e. the animal produces twice as much heat at 20°C than at 30°C.

2. Which of the following statements is *true* of brown fat?
   a. It produces heat without producing ATP.
   b. It insulates animals acclimatized to cold.
   c. It is a major source of heat production for birds.
   d. It is found only in hibernators.
   e. It provides fuel for muscle cells responsible for shivering.

3. What is the most important and most general difference between mammals and birds adapted to cold climates in comparison to species adapted to warm climates?
   a. Higher basal metabolic rates
   b. Higher Q_{10}'s
   c. Brown fat
   d. Greater insulation
   e. Ability to hibernate

4. Which of the following would cause a *decrease* in the hypothalamic temperature set point for metabolic heat production?
   a. Entering a cold environment
   b. Taking an aspirin when you have a fever
   c. Arousing from hibernation
   d. Getting an infection that causes a fever
   e. Cooling the hypothalamus

5. Mammalian hibernation
   a. occurs when animals run out of metabolic fuel.
   b. is a regulated decrease in body temperature.
   c. is less common than hibernation in birds.
   d. can occur at any time of year.
   e. lasts for a period of several months, during which body temperature remains close to environmental temperature.

6. Which of the following is an important difference between an ectotherm and an endotherm of similar body size?
   a. Ectotherms have higher Q_{10}'s.
   b. Only ectotherms use behavioral thermoregulation.
   c. Only endotherms can constrict and dilate the blood vessels to the skin to alter heat flow.
   d. Only endotherms can get fevers.
   e. At body temperatures of 37°C, the ectotherm has a lower metabolic rate than the endotherm.

7. The function of the countercurrent heat exchanger in "hot" fish is:
   a. to trap heat in the muscles.
   b. to produce heat.
   c. to heat the blood returning to the heart.
   d. to dissipate excess heat generated by powerful swimming muscles.
   e. to cool the skin.

8. What is the difference between a winter- and a summer-acclimatized fish that is termed "metabolic compensation"?
   a. The winter-acclimatized fish has a higher Q_{10}.
   b. The winter-acclimatized fish develops greater insulation.
   c. The winter-acclimatized fish hibernates.
   d. The summer-acclimatized fish has a countercurrent heat exchanger.
   e. The summer-acclimatized fish has a lower metabolic rate at any given water temperature than does the winter fish.

9. Which of the following is an important characteristic of epithelial cells?
   a. They generate electric signals.
   b. They contract.
   c. They have an extensive extracellular matrix.
   d. They have secretory functions.
   e. They are found only on the surface of the body.

10. Negative feedback
    a. works in opposition to positive feedback to achieve homeostasis of a physiological variable.
    b. always turns off a process.
    c. reduces an error signal in a regulatory system.
    d. is responsible for metabolic compensation.
    e. is a feature of the thermoregulatory systems of endotherms but not of ectotherms.

FOR STUDY

1. Make a table that lists all of the properties of the internal environment that you think are critical to keep the cells of the body healthy. Next to each property list the organs or organ system responsible for maintaining it.

2. What are the major differences between ectotherms and endotherms? Compare and contrast their major thermoregulatory adaptations.

3. Why is an environment above the upper critical temperature of an endotherm more dangerous for that animal than is an environment below its lower critical temperature?

4. Why is it difficult for a fish to be endothermic? How do "hot" fish overcome these difficulties?

5. If the temperature of the hypothalamus of a mammal is the feedback information for its thermostat, why does the hypothalamic temperature scarcely change when that animal moves between environments hot enough and cold enough to stimulate the animal to pant and to shiver, respectively?

READINGS

Crawshaw, L. I., B. P. Moffitt, D. E. Lemons and J. A. Downey. 1981. "The Evolutionary Development of Vertebrate Thermoregulation." *American Scientist*, vol. 69, pages 543–550. All vertebrates thermoregulate, and the nervous system mechanisms involved appear to have a common origin even though the effector mechanisms may differ.

French, A. R. 1986. "The Patterns of Mammalian Hibernation." *American Scientist*. vol. 76, pages 569–575. Body size has important consequences for energy metabolism; therefore, a variety of patterns of hibernation have evolved, as this article discusses.

Heinrich, B. 1981. "The Regulation of Temperature in the Honeybee Swarm." *Scientific American*, June. When honeybees leave their hive in a swarm, they thermoregulate.

Heller, H. C., L. I. Crawshaw and H. T. Hammel. 1978. "The Thermostat of Vertebrate Animals." *Scientific American*, August. This article describes research on and properties of the brain mechanisms responsible for thermoregulation in vertebrates and the adaptations in those mechanisms that make hibernation possible.

Schmidt-Nielsen, K. 1981. "Countercurrent Systems in Animals." *Scientific American*, May. Countercurrent exchanges are the basis for a variety of physiological adaptations, some of which are presented in this article. Developed in special detail is the case of water conservation in the camel's nose.

Schmidt-Nielsen, K. 1990. *Animal Physiology: Adaptation and Environment*, 4th Edition. Cambridge University Press, New York. An excellent advanced textbook on comparative animal physiology. Chapter 8, "Temperature Regulation," expands on many of the topics presented in this chapter.

36

Animal Hormones

A Wimpy and a Macho Male Cichlid

A species of cichlid fish that lives in Lake Tanganyika in east central Africa gives biological meaning to "macho" and "wimpy." In shallow pools around the edge of the lake, big, brightly colored males stake out and vigorously defend territories against neighboring males. These "macho" males constantly patrol their territories and display their sexual adornments for the benefit of females who assemble in groups at the edge of the colony. The females are hard to see because they are inactive and protectively colored. When a female is ready to spawn and is impressed by a male's territory and display, she enters his territory and lays her eggs in a spawning pit the male has prepared. The male then fertilizes the eggs. At any one time, only about 10 percent of the males in the population are displaying and holding territories. All the other males are small and nondescript like the females, nonaggressive, and incapable of fertilizing eggs—that is, "wimpy." If, however, a macho male is removed by a predator, a group of wimpy males fight over the vacated territory. The winner rapidly turns into a macho male, brightly colored, big, aggressive, and able to attract females and fertilize eggs.

What accounts for this dramatic change in lifestyle of cichlid males? Soon after the wimpy male's victory, certain cells in its brain enlarge and secrete a chemical message that triggers a cascade of other cells to secrete chemical messages. These molecules circulating around the body help bring about a variety of changes in cells and tissues that convert the once wimp into a macho male. This is one example of how chemical messages, or hormones, released in this case by a behavioral stimulus, can produce and coordinate major developmental, physiological, and behavioral changes in an animal.

CHEMICAL MESSAGES

In this chapter we will examine the science of **endocrinology**—the study of hormones and their actions. A **hormone** is a substance that serves as a chemical message between cells of a multicellular organism. A chemical communication system that uses a hormone is made up of at least two cells: One cell produces and releases the hormone—the message—and a second cell with appropriate **receptors** receives the message. The receiving cell is called the **target cell**. The receipt of the message activates mechanisms within the target cell that interpret the message and respond to it. The response may be developmental, physiological, or behavioral.

A simple way to classify chemical communication systems is according to the distance over which the messages operate: Are their effects local or are they distributed throughout the body? (The effects of some chemical messages are even exerted on other organisms in the environment. We will learn about those substances—pheromones—in Chapter 45.)

Local Hormones

Some chemical messages, released from secreting cells into surrounding extracellular fluids, exert their effects locally (Figure 36.1*a*). These hormones are inactivated so rapidly by enzymes in the extracellular environment or taken up so completely by cells in the immediate vicinity that they usually do not exert effects on distant cells.

An example of a local hormone is **histamine**, one of the mediators of inflammation, a tissue response that can help protect the body from invasion by foreign organisms or materials. Histamine is released in damaged tissues by specialized cells called mast cells. When the skin is cut by a dirty object, the area around the cut becomes inflamed—red, hot, and swollen. Histamine causes this response by dilating the local blood vessels and making them more permeable, or leaky, allowing blood plasma, including protective blood proteins and white blood cells, to move into the damaged tissue.

Local responses to histamine are protective, but histamine responses spread over large areas of the body can cause problems, such as the symptoms of hay fever. When a person sensitized to a type of pollen inhales that pollen, it causes cells in the respiratory passages to release histamine. The histamine causes the tissues of the passages to swell and to increase their secretions of mucus, leading to congestion, coughing, sneezing, and a runny nose. Such allergic reactions are unpleasant but rarely dangerous. In a person who is extremely allergic, however, or in a person who has a blood-borne infection, histamine and other mediators of inflammation may be released in such large amounts that they enter the blood and circulate around the body. The resulting expansion of blood vessels and leakage of fluid from the circulatory system can cause blood pressure to drop severely. Fluid may leak into the lungs and the airways may become severely congested. The resulting failure of the circulatory and respiratory systems, termed **anaphylactic shock**, can be lethal. Some highly sensitive people can go into anaphylactic shock from a single bee sting, an injection of an antibiotic to which they are allergic, or ingestion of a food to which they are allergic.

Circulating Hormones

Traditionally we have thought of a hormone as a chemical message secreted by cells and distributed throughout the body by the circulatory system (Figure 36.1*b*). Wherever such a hormone encounters a cell with a receptor to which it can bind, it triggers a response. The nature of the response depends on the responding cell. The same hormone can cause different responses in different types of cells. For example, consider the hormone epinephrine (adrenaline). If a lion creeps up behind you and roars, your brain sends signals through your nervous system to epinephrine-containing cells, which immediately release epinephrine. The hormone diffuses into the blood and rapidly circulates around your body. What does it do for you?

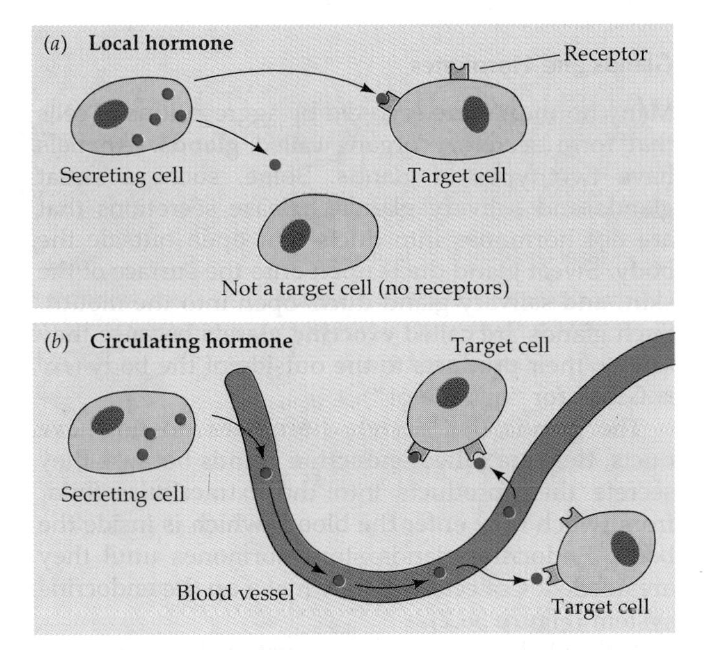

36.1 Chemical Signaling Systems
(*a*) Many cells in the body secrete hormones that influence only nearby target cells. (*b*) Some cells secrete hormones into the bloodstream, which carries them to target cells elsewhere in the body.

Epinephrine activates receptors in the heart to make the heart beat faster and pump more blood. Epinephrine activates receptors in the vessels supplying blood to your digestive tract, causing those vessels to constrict—you can digest lunch later. Your heart is pumping more blood, and a greater percentage of that blood is going to the muscles needed for your escape. In the liver, epinephrine stimulates the breakdown of glycogen into glucose for a quick energy supply. In fatty tissue, epinephrine stimulates the breakdown of fats as another source of energy. These are some of the many actions triggered by this one hormone. They all contribute to increasing your chances of escaping the lion. Whether a cell in your body responds to the surge of epinephrine depends on whether it has epinephrine receptors. The specific response of each cell with epinephrine receptors depends on the type of cell it is.

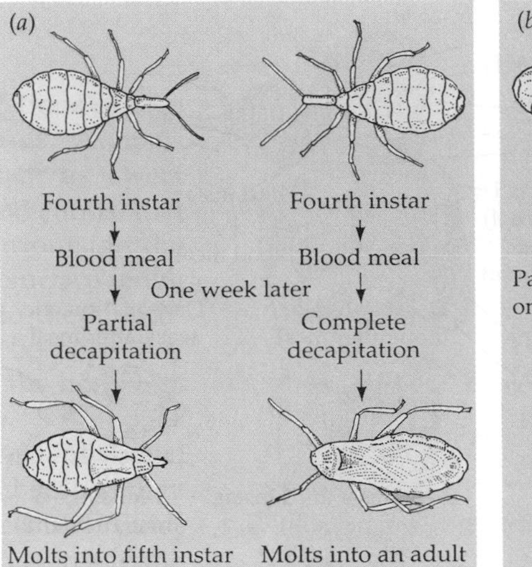

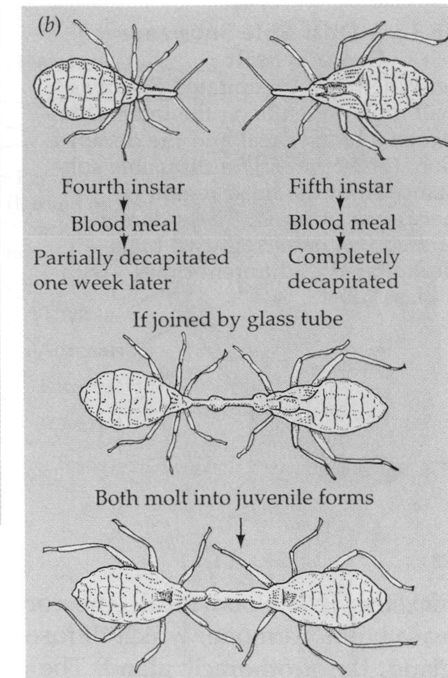

36.4 A Diffusible Substance Controls Development
(a) Decapitated *Rhodnius* molt into adult forms unless the posterior part of the brain is left intact. *(b)* The substance from the posterior part of the brain that maintains juvenile status can diffuse through a glass tube connecting two bugs.

36.5 Three Hormones Control Molt and Metamorphosis
Neurosecretory cells in the brain of the silkworm moth produce brain hormone that is stored in and released from the corpora cardiaca. Brain hormone stimulates the prothoracic gland to produce ecdysone. The corpora allata produce juvenile hormone. As long as juvenile hormone is abundant, the larva molts into a larger larva in response to ecdysone. As juvenile hormone wanes, the larva molts into a pupa. The pupa does not produce juvenile hormone, so it metamorphoses into an adult. The release of ecdysone is episodic, and each release stimulates a molt.

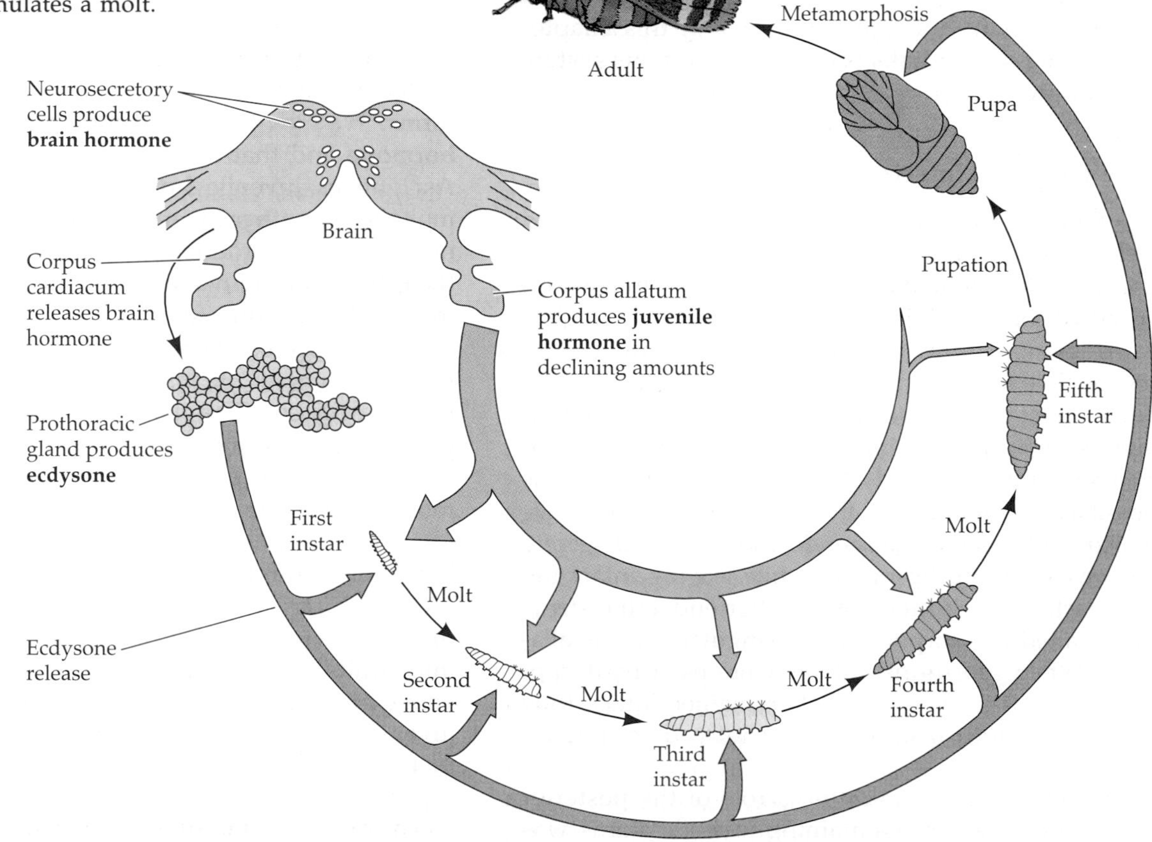

experimentally demonstrated many years before the hormones were identified chemically. That is not surprising when you consider the tiny amounts of certain hormones that exist in an organism. In one of the earliest studies of ecdysone, biochemists produced only 250 mg of pure ecdysone (about one-fourth the weight of an apple seed) from 4 tons of silkworms!

VERTEBRATE HORMONES

As endocrine systems have evolved, the same chemical messages have become coupled to new physiological responses. In many cases the same chemical substance is a hormone in widely divergent species but has completely different actions. Many vertebrate hormones have molecular structures similar or identical to those of invertebrate hormones, but their functions are different. The hormone thyroxine, for example, is found in animal species ranging from tunicates (sea squirts) to humans, but thyroxine's function differs greatly among these species. In mammals it elevates cellular metabolic rate; in frogs it is essential for metamorphosis from tadpole to adult. Another example is the hormone prolactin, which stimulates milk production in female mammals after they give birth. In pigeons and doves prolactin stimulates the production of crop milk for nourishment of the young. Crop milk is really not a milk secretion at all, but a sloughing off of cells lining the upper digestive tract. In amphibians prolactin causes the animals to prepare for reproduction by seeking water, and in fishes, such as salmon, that migrate between salt and fresh water, prolactin regulates the mechanisms that maintain osmotic balance with the changing environment.

The endocrine systems of vertebrates are varied and complex. We recognize at least nine endocrine glands (see Figure 36.2), most of which produce and release more than one hormone. In addition, cells in many other organs produce and release hormones. The list of chemical messages in the bodies of vertebrates is long and growing longer. To make the subject more manageable, we will focus mostly on the hormones of mammals—how they function and how they are controlled. Table 36.1 presents an overview of the hormones of humans. Notice that the column listing the target tissues of these hormones includes every organ system of the body.

We begin our survey by examining the "master gland," the **pituitary**, which produces many hormones, some of which target other hormone secreting cells elsewhere in the body. The pituitary gland sits in a depression at the bottom of the skull just over the back of the roof of the mouth (Figure 36.6). It is attached to the part of the brain called the hypothalamus, which we discovered in Chapter 35 is the location of the body's thermostat. The hypothalamus is involved in many homeostatic regulatory systems, including endocrine systems.

The pituitary has distinct anterior and posterior divisions that have separate origins during develop-

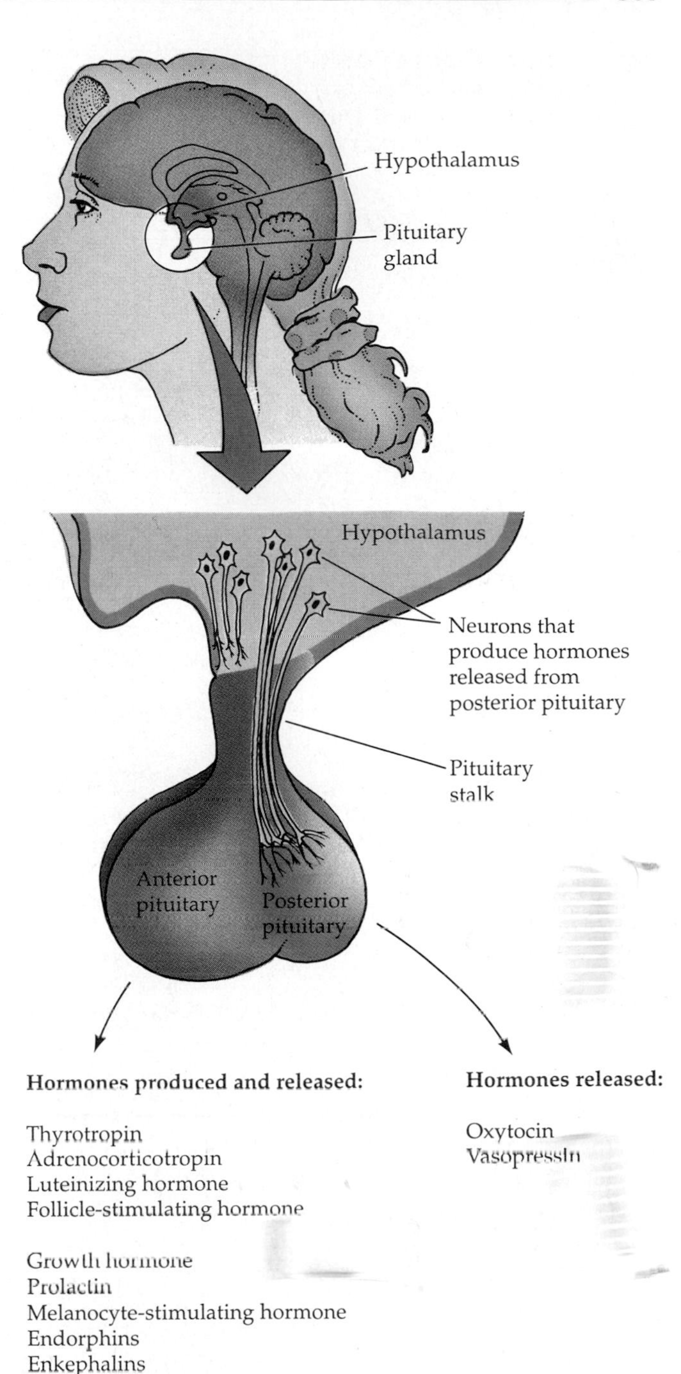

Hormones produced and released:

Thyrotropin
Adrenocorticotropin
Luteinizing hormone
Follicle-stimulating hormone

Growth hormone
Prolactin
Melanocyte-stimulating hormone
Endorphins
Enkephalins

Hormones released:

Oxytocin
Vasopressin

36.6 A Powerful Gland
The human pituitary gland is only the size of a blueberry, yet it secretes many hormones. The posterior pituitary secretes two hormones produced in the hypothalamus. Among the hormones produced and secreted by the anterior pituitary are four (listed first in the figure) that stimulate their target cells to secrete other hormones.

TABLE 36.1
Principal Hormones of Humans

SECRETING TISSUE OR GLAND	HORMONE	CHEMICAL NATURE	TARGETS	IMPORTANT PROPERTIES OR ACTIONS
Hypothalamus	Releasing and release-inhibiting hormones (see Table 36.2)	Peptides	Anterior pituitary	Control secretion of hormones of anterior pituitary
	Oxytocin, vasopressin	Peptides	(See Posterior pituitary)	Stored and released by posterior pituitary
Anterior pituitary: Tropic hormones	Thyrotropin	Glycoprotein	Thyroid gland	Stimulates synthesis and secretion of thyroxine
	Adrenocorticotropin	Polypeptide	Adrenal cortex	Stimulates release of hormones from adrenal cortex
	Luteinizing hormone	Glycoprotein	Gonads	Stimulates secretion of sex hormones from ovaries and testes
	Follicle-stimulating hormone	Glycoprotein	Gonads	Stimulates growth and maturation of eggs in females; stimulates sperm production in males
Anterior pituitary: Other hormones	Growth hormone	Protein	Bones, liver, muscles	Stimulates protein synthesis and growth
	Prolactin	Protein	Mammary glands	Stimulates milk production
	Melanocyte-stimulating hormone	Peptide	Melanocytes	Controls skin pigmentation
	Endorphins and enkephalins	Peptides	Spinal cord neurons	Decreases painful sensations
Posterior pituitary	Oxytocin	Peptide	Uterus, breasts	Induces birth by stimulating labor contractions; causes milk flow
	Vasopressin (antidiuretic hormone)	Peptide	Kidneys	Stimulates water reabsorption
Thyroid	Thyroxine	Iodinated amino acid derivative	Many tissues	Stimulates and maintains metabolism necessary for normal development and growth
	Calcitonin	Peptide	Bones	Stimulates bone formation; lowers blood calcium
Parathyroids	Parathormone	Protein	Bones	Absorbs bone; raises blood calcium
Thymus	Thymosins	Peptides	Immune system	Activate immune responses of T cells in the lymphatic system
Pancreas	Insulin	Protein	Muscles, liver, fat, other tissues	Stimulates uptake and metabolism of glucose; increases conversion of glucose to glycogen and fat
	Glucagon	Protein	Liver	Stimulates breakdown of glycogen and raises blood sugar

ment. The **anterior pituitary** originates as an outpocketing of the mouth region of the embryonic digestive tract, and the **posterior pituitary** originates as an outpocketing of the developing brain in the region that becomes the hypothalamus. Thus the posterior pituitary derives from nervous-system tissue, and the hormones it stores and releases are neurohormones produced in the hypothalamus. The anterior pituitary consists of epithelial cells that develop endocrine functions.

Posterior Pituitary Hormones

The posterior pituitary releases two neurohormones, **vasopressin** (also called antidiuretic hormone, or ADH) and **oxytocin**. These are small peptides synthesized in nerve cells in the hypothalamus. Vasopressin and oxytocin move down long extensions of these nerve cells that stretch down the pituitary stalk into the posterior pituitary, where the hormones are stored in the nerve endings (see Figure 36.6).

TABLE 36.1
Principal Hormones of Humans

SECRETING TISSUE OR GLAND	HORMONE	CHEMICAL NATURE	TARGETS	IMPORTANT PROPERTIES OR ACTIONS
Pancreas (continued)	Somatostatin	Peptide	Digestive tract; other cells of the pancreas	Inhibits insulin and glucagon release; decreases secretion, motility, and absorption in the digestive tract
Adrenal medulla	Adrenaline, noradrenaline	Modified amino acids	Heart, blood vessels, liver, fat cells	Stimulate "fight-or-flight" reactions: increase heart rate, redistribute blood to muscles, raise blood sugar
Adrenal cortex	Glucocorticoids (cortisol)	Steroids	Muscles immune system, other tissues	Mediate response to stress; reduce metabolism of glucose, increase metabolism of proteins and fats; reduce inflammation and immune responses
	Mineralocorticoids (aldosterone)	Steroids	Kidneys	Stimulates excretion of potassium ions and reabsorption of sodium ions
Stomach lining	Gastrin	Peptide	Stomach	Promotes digestion of food by stimulating release of digestive juices; stimulates stomach movements that mix food and digestive juices
Lining of small intestine	Secretin	Peptide	Pancreas	Stimulates secretion of bicarbonate solution by ducts of pancreas
	Cholecystokinin	Peptide	Pancreas, liver, gall bladder	Stimulates secretion of digestive enzymes by pancreas and other digestive juices from liver; stimulates contractions of gall bladder and ducts
	Enterogastrone	Polypeptide	Stomach	Inhibits digestive activities in the stomach
Pineal	Melatonin	Modified amino acid	Hypothalamus	Involved in biological rhythms
Ovaries	Estrogens	Steroids	Breasts, uterus, other tissues	Stimulate development and maintenance of female characteristics and sexual behavior
	Progesterone	Steroid	Uterus	Sustains pregnancy; helps to maintain secondary female sexual characteristics
Testes	Androgens	Steroids	Various tissues	Stimulate development and maintenance of male sexual behavior and secondary male sexual characteristics; stimulate sperm production
Most cells	Prostaglandins	Modified fatty acids	Various tissues	Have many diverse actions
Heart	Atrial natriuretic hormone	Peptide	Kidneys	Increases sodium ion excretion

The posterior pituitary increases its release of vasopressin whenever blood pressure falls or the blood becomes too salty. The main action of vasopressin is to increase the amount of water conserved by the kidneys. When vasopressin secretion is high, the kidneys reabsorb more water and produce only a small volume of highly concentrated urine. When vasopressin secretion is low, the kidneys produce a large volume of dilute urine. We will discuss the mechanism of vasopressin action in Chapter 44.

When a woman is about to give birth, her posterior pituitary releases oxytocin, which stimulates the contractions of the muscles that push the baby out of her body. Oxytocin also brings about the flow of milk from her breasts. The baby's suckling stimulates nerve cells in the mother, causing secretion of oxytocin. Even the sight and sounds of her baby can cause a nursing mother to secrete oxytocin and release milk from her breasts.

Anterior Pituitary Hormones

The anterior pituitary produces and secretes many peptide hormones, each of which is produced by a different type of pituitary cell. Four of these hormones that control the activities of other endocrine glands are **tropic hormones**. The four tropic hormones are thyrotropin, adrenocorticotropin, luteinizing hormone, and follicle-stimulating hormone. We will say more about these tropic hormones when we describe their target glands (thyroid, adrenal cortex, testes, and ovaries) later in this chapter and in the next. The other hormones produced by the anterior pituitary influence tissues that are not endocrine glands. These hormones are growth hormone, prolactin, melanocyte-stimulating hormone, and endorphins and enkephalins.

Growth hormone consists of about 200 amino acids and acts on a wide variety of tissues to promote growth directly and indirectly. One of its important direct effects is to stimulate cells to take up amino acids. Growth hormone promotes growth indirectly by stimulating the liver to produce growth-regulating chemical messages called **somatomedins**, which circulate in the blood and stimulate the growth of bone and cartilage. Overproduction of growth hormone in children causes gigantism, and underproduction causes dwarfism (Figure 36.7). High levels of growth hormone in adults cannot cause increased height because the shafts and the growth plates of the long bones have fused. Rather, abnormally high levels of growth hormone in adults cause thickening of the hands, feet, jaw, nose, and ears—a condition known as acromegaly.

36.7 Effects of Abnormal Amounts of Growth Hormone
(a) Gigantism results from the overproduction of growth hormone in childhood. In this news photo from 1939, a young man and his father visit New York on business; the son is over 8 feet tall, whereas his father is of average height: 5 feet, 11 inches. *(b)* When the anterior pituitary does not produce enough growth hormone during childhood, pituitary dwarfism results. The man on the left is P. T. Barnum, circus entrepreneur. With him is Charles Stratton, a dwarf who appeared in Barnum's circus under the name General Tom Thumb.

Beginning in the late 1950s, children diagnosed as having a serious deficiency of growth hormone, and therefore destined to become dwarfs, were treated with human growth hormone extracted from human pituitaries from cadavers. The treatment was successful in stimulating substantial growth, but it was extremely costly; a year's supply of human growth hormone for one individual required up to 50 pituitaries. In the mid-1980s scientists using genetic engineering technology isolated the gene for human growth hormone and introduced it into bacteria, which produced enough of the hormone to make it commercially available.

Preventing pituitary dwarfism is now feasible and affordable, but the availability of growth hormone raises new questions. Should every child at the lower end of the height charts be treated? Should a normal child whose parents want him or her to play basketball be given growth hormone? These types of questions are impossible to answer with scientific data alone. The controversy around growth hormone has become even more complex because of a recent study suggesting that when growth hormone is administered to older persons, it reverses some of the effects of aging. In comparison to a control group not receiving growth hormone, a group of elderly persons receiving growth hormone decreased their body fat, increased their muscle mass, and reported feeling more energetic. It is not known, however, if these changes will last beyond the period of treatment, or if the treatment has side effects.

Earlier in the chapter we described the evolutionary diversity of the functions of **prolactin**, another hormone produced by the anterior pituitary. In human females the major function of prolactin is to stimulate the production and secretion of milk. In some mammals prolactin also functions as an important hormone during pregnancy. In human males prolactin plays a role along with other pituitary hormones in controlling the endocrine function of the testes.

Melanocyte-stimulating hormone is produced in very low amounts by the human anterior pituitary, and its functions in humans are not well understood. Melanocytes are cells that contain melanin, a black pigment. In fishes, amphibians, and reptiles that can change their color, melanocyte-stimulating hormone changes the way melanin is distributed in the melanocytes, thereby darkening or lightening the tissue containing them.

Endorphins and **enkephalins**, the remaining hormones of the anterior pituitary, are referred to as the body's "natural opiates." These molecules help control pain. Interestingly, the production of endorphins and enkephalins in the pituitary is encoded by the same gene as are two other pituitary hormones. The gene codes for a large parent molecule called pro-opiomelanocortin. This large molecule is cleaved to produce several peptides, some of which have hormonal functions. Adrenocorticotropin, melanocyte-stimulating hormone, endorphins, and enkephalins all result from the cleavage of pro-opiomelanocortin.

Hypothalamic Neurohormones

The idea of the anterior pituitary as the "master gland" received quite a blow with the discovery that it is really a "middleman" controlled by the hypothalamus. The hypothalamus receives information about conditions in the body and in the external environment through the nervous system. If the connection between the hypothalamus and the pituitary is cut, pituitary hormones are no longer released in response to changes in the environment or in the body. If pituitary cells are maintained in culture, extracts of hypothalamic tissue stimulate some of those cells to release their hormones into the culture medium. Therefore, scientists hypothesized that secretions of the hypothalamic cells control the activities of the cells of the anterior pituitary. Although hypothalamic neurons do not extend into the anterior pituitary as they do into the posterior pituitary, a possible route by which such chemical messages could reach the anterior pituitary was known: a special set of blood vessels called **portal blood vessels** that run between the hypothalamus and the anterior pituitary (Figure 36.8). It was thus proposed that secretions from nerve endings in the hypothalamus are absorbed into the blood and are conducted down the portal vessels to the anterior pituitary, where they cause the release of anterior pituitary hormones.

In the 1960s two large teams of scientists, led by Roger Guillemin and Andrew Schally, initiated the search for the hypothalamic releasing neurohormones. Because the amounts of such hormones in any individual mammal would be tiny, "bucket biochemistry" was called for. The scientists set up teams in slaughterhouses to collect massive numbers of hypothalami from pigs and sheep. The resulting *tons* of tissue were shipped to laboratories in refrigerated trucks. One effort began with the hypothalami from 270,000 sheep and yielded only 1 mg of purified thyrotropin-releasing hormone. Biochemical analysis of this pure sample revealed that thyrotropin-releasing hormone contains only three amino acids; it is a tripeptide. Soon after discovering thyrotropin-releasing hormone, the scientists identified gonadotropin-releasing hormone, which controls the release of follicle-stimulating hormone and luteinizing hormone from the anterior pituitary. For these discoveries Guillemin and Schally received the 1972 Nobel prize in medicine. Because isolation techniques have been improved enormously, we now need only a few milligrams of tissue to isolate and characterize a peptide.

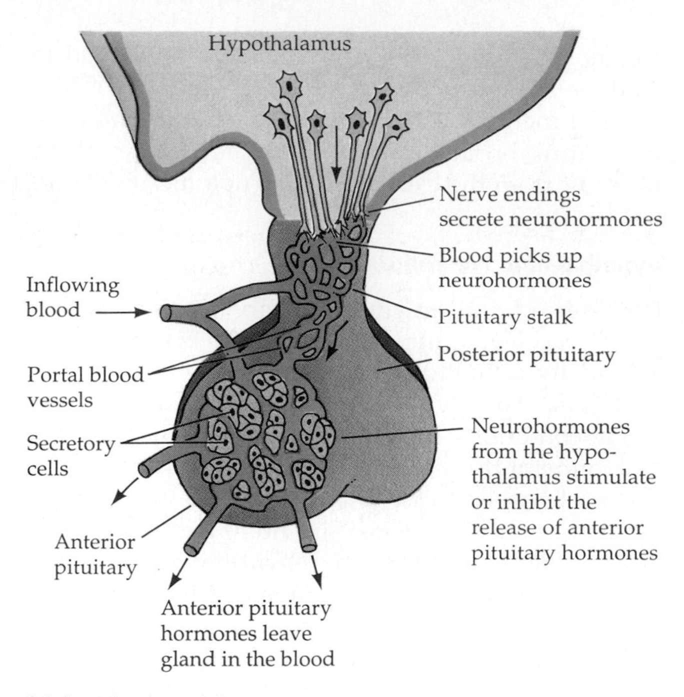

36.8 The Hypothalamus Controls the Anterior Pituitary
A system of blood vessels runs between the hypothalamus and the anterior pituitary. Tropic neurohormones produced by hypothalamic cells enter these blood vessels and are transported to the anterior pituitary, where they control the activity of the pituitary cells that synthesize and release hormones.

Many more hypothalamic neurohormones are now known, and they include **releasing hormones** and **release-inhibiting hormones** (Table 36.2).

Thyroid Hormones

The **thyroid gland** consists of two lobes, one on either side of the trachea (windpipe), connected by a strip of thyroid tissue that wraps around the front of the trachea. If you gently place your thumb and forefinger on either side of your trachea just below your Adam's apple and swallow, you will feel your thyroid gland move up and down under your fingertips.

The thyroid gland produces the hormones thyroxine and calcitonin. The **thyroxine** molecule consists of two molecules of the amino acid tyrosine and four atoms of iodine. (Another form of thyroid hormone has only three iodine atoms, but for convenience we will refer to both as thyroxine.) Thyroxine in mammals plays many roles in regulating cell metabolism. It elevates the metabolic rates of most cells and tissues and promotes the use of carbohydrates rather than fats for fuel. Exposure to cold for several days leads to an increased release of thyroxine and an increase in basal metabolic rate. Thyroxine is especially crucial during development and growth. It promotes amino acid uptake and protein synthesis by cells. Insufficient thyroxine in a human fetus or growing child greatly retards physical and mental growth, resulting in a condition known as cretinism.

Malfunction of the thyroid gland causes goiter, a condition in which the thyroid gland becomes very large (Figure 36.9). Goiter can be associated with either **hyperthyroidism** (high levels of thyroxine) or **hypothyroidism** (low levels of thyroxine). To understand how these opposite conditions can lead to the same symptom, we must consider the control of thyroid activity by the pituitary and the hypothalamus.

The tropic hormone **thyrotropin**, which is secreted into the blood by the anterior pituitary, determines the activity of the thyroid gland. Thyrotropin activates the thyroid gland cells that produce thyroxine. Thyrotropin-releasing hormone produced in the hypothalamus and transported to the pituitary through the portal blood vessels activates the thyrotropin-producing pituitary cells. The brain uses environmental information such as temperature or day length to determine whether to increase or decrease

TABLE 36.2
Releasing and Release-Inhibiting Neurohormones of the Hypothalamus

NEUROHORMONE	ACTION
Thyrotropin-releasing hormone (TRH)	Stimulates thyrotropin release
Gonadotropin-releasing hormone	Stimulates release of follicle-stimulating hormone and luteinizing hormone
Prolactin release-inhibiting hormone	Inhibits prolactin release
Prolactin-releasing hormone	Stimulates prolactin release
Somatostatin (growth hormone release-inhibiting hormone)	Inhibits growth hormone release; interferes with thyrotropin release
Growth hormone-releasing hormone	Stimulates growth hormone release
Adrenocorticotropin-releasing hormone	Stimulates adrenocorticotropin release
Melanocyte-stimulating hormone release–inhibiting hormone	Inhibits melanocyte-stimulating hormone release

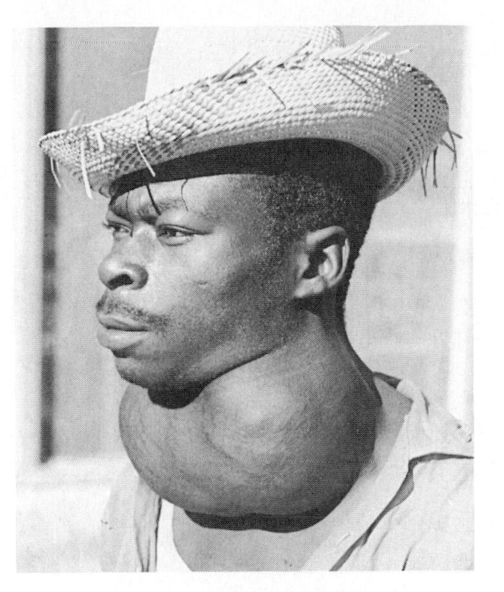

36.9 Goiter
A goiter is a greatly enlarged thyroid gland. Worldwide, goiter affects about 5 percent of the population. The addition of iodine to table salt has greatly reduced the incidence of the condition in industrialized nations, but goiter is still common in the less developed countries of the world.

the secretion of thyrotropin-releasing hormone. There is a very important negative feedback loop in this sequence of steps: Circulating thyroxine inhibits the response of the pituitary cells to thyrotropin-releasing hormone. Therefore, when thyroxine levels are high less thyrotropin is released, and when thyroxine levels are low more thyrotropin is released (Figure 36.10).

Hyperthyroid goiter results when the pituitary cells are not turned off by thyroxine. Thyrotropin levels remain high and the thyroid gland is activated so much that it grows bigger. Production of thyroxine by the thyroid stays abnormally high. Hyperthyroid patients have high metabolic rates, are jumpy and nervous, usually feel hot, and may have a buildup of fat behind the eyeballs, causing their eyes to bulge.

Hypothyroid goiter results when there is not enough circulating thyroxine to turn off thyrotropin production. Its most common cause is a deficiency of dietary iodide, without which the thyroid gland cannot make *functional* thyroxine. Without thyroxine, thyrotropin levels remain high, so the thyroid continues to produce large amounts of nonfunctional thyroxine and gets very large. The symptoms of hypothyroidism are low metabolism, intolerance of cold, and general physical and mental sluggishness. Hypothyroid goiter used to be extremely common in mountainous areas and regions far from the oceans, where there is little iodide in the soil or water. The addition of iodide to table salt has greatly reduced the incidence of the disease.

Another hormone of the mammalian thyroid gland is **calcitonin**. It is not produced by the same cells that produce thyroxine. Calcitonin helps regulate the levels of calcium circulating in the blood (Figure 36.11). Bone is a huge repository of calcium in the body and is continually being remodeled. Cells called **osteoclasts** break down bone and release calcium; **osteoblasts**, on the other hand, use circulating calcium to deposit new bone. Calcitonin decreases the activity of osteoclasts and stimulates the activity of osteoblasts, thus shifting the balance from adding calcium ions to the blood to removing calcium ions from the blood. The regulation of blood calcium levels is influenced more strongly by parathormone, which we consider in the next section, than it is by calcitonin, but calcitonin plays an important role in preventing bone loss in women during pregnancy.

Parathormone

The **parathyroid glands** are four tiny structures embedded on the surface of the thyroid gland. Their single hormone product is parathyroid hormone, or **parathormone**, a critical control element in the regulation of blood calcium levels. Growth and remodeling of bone require calcium; so do many cellular processes, such as nerve and muscle functions, which are sensitive to changes in calcium concentration. Muscle contraction and nerve function are severely impaired if the blood calcium level rises or falls by as little as 30 percent of normal values. A fall in blood calcium triggers the release of parathormone, which in turn stimulates actions that add calcium to the blood. Parathormone stimulates osteo-

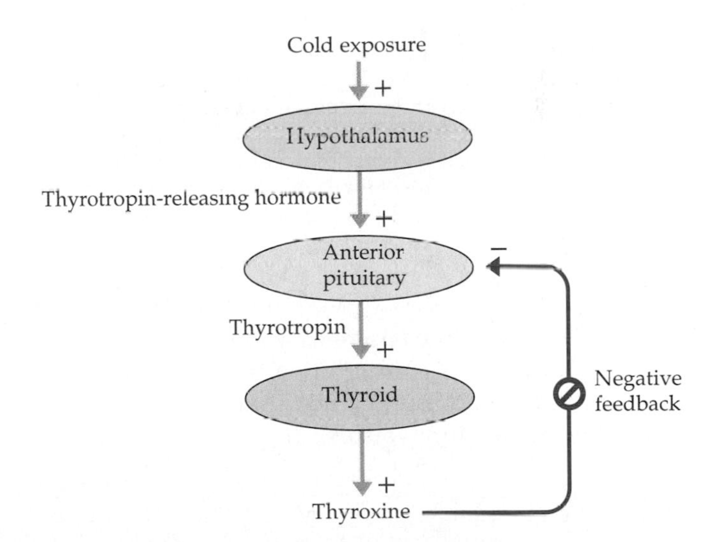

36.10 Regulation of Thyroid Function
Environmental cues such as exposure to cold stimulate the hypothalamus to produce thyrotropin-releasing hormone, which initiates a cascade of events that stimulate the thyroid to release thyroxine. High thyroxine levels act as negative feedback.

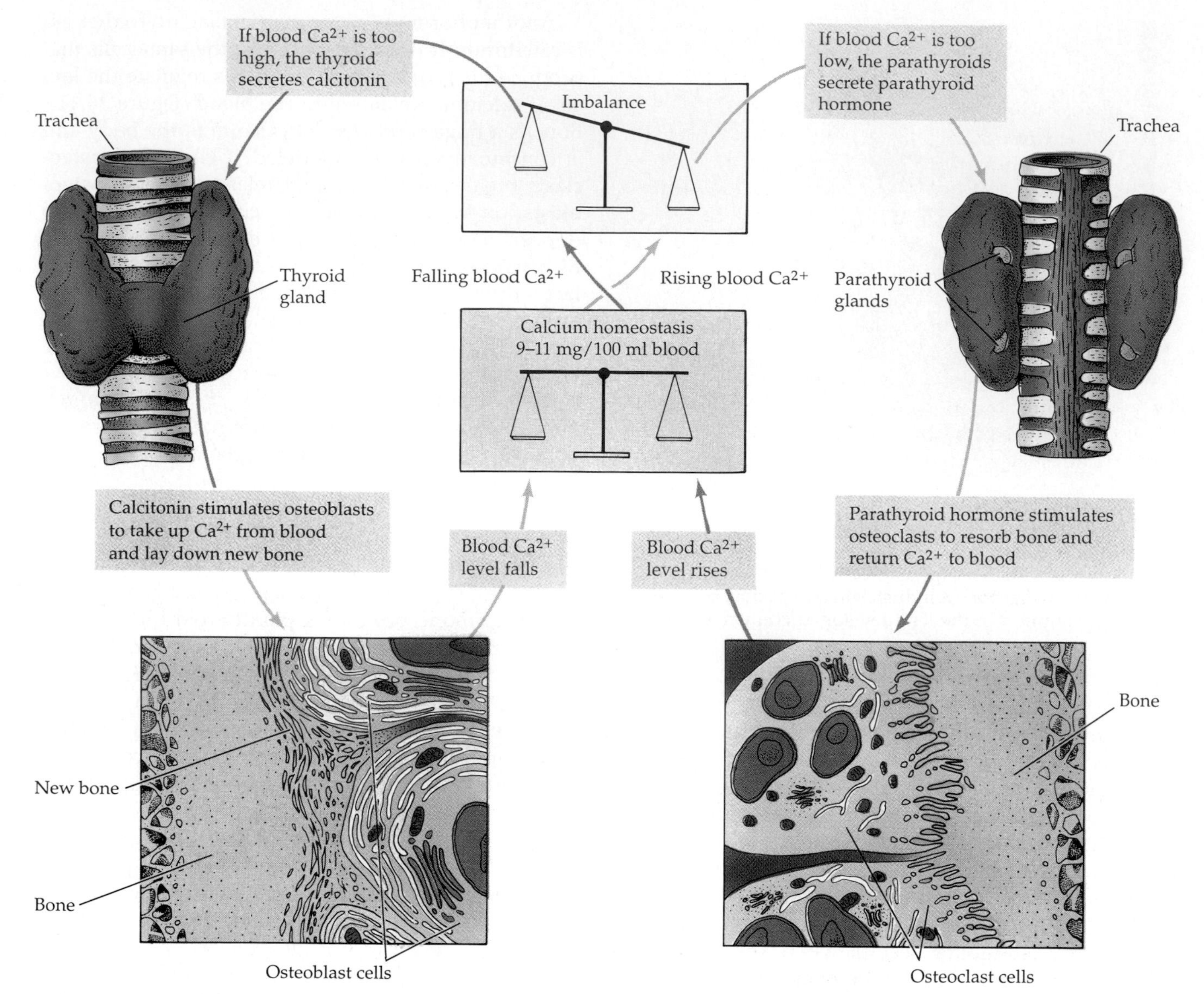

36.11 Calcium Balance
Calcitonin and parathyroid hormone help regulate blood calcium levels. Bone can be a source (site of production) of calcium or a sink (site of utilization or storage) for excess calcium. Osteoclasts break down bone and release calcium; osteoblasts build new bone using calcium from the blood.

clasts to dissolve bone and release calcium (see Figure 36.11). It also helps the digestive tract to absorb calcium from food and helps the kidneys reabsorb calcium before excreting wastes.

Pancreatic Hormones

Before the 1920s, diabetes mellitus was a fatal disease, characterized by weakness, lethargy, and body wasting. The disease was known to be connected with a gland located just below the stomach, the **pancreas**, and with abnormal glucose metabolism, but the link was not clear. Today we know that diabetes mellitus is caused by a lack of the hormone **insulin**. Insulin replacement therapy makes it possible for more than 1.5 million diabetics in the United States to lead almost normal lives.

Most cells in an untreated diabetic's body cannot use glucose in the blood for metabolic fuel. As a result, glucose accumulates in the blood until it is lost in the urine. High blood glucose causes water to move from cells into the blood by osmosis, and the kidneys increase urine output to excrete the excess fluid volume from the blood. The name *diabetes* refers to copious production of urine, and *mellitus* (from the Greek for "honey") reflects the fact that the urine of the untreated diabetic is sweet. Since the cells of the body cannot use blood glucose for fuel, they must burn fat and protein. As a result, the body of the untreated diabetic wastes away, and critical tissues and organs are damaged.

The change in the outlook for diabetics came almost overnight in 1921, when medical doctor Frederick Banting and medical student Charles Best of the University of Toronto discovered that they could reduce the symptoms of diabetes with an extract they prepared from pancreatic tissue. The work of Banting and Best led to enormous relief of human suffering. The active component of the extract Banting and Best prepared was a small protein hormone, insulin, consisting of 51 amino acids. Insulin is produced in clusters of cells in the pancreas, called **islets of Langerhans** after a German medical student who discovered them. Other cells in the islets produce two other hormones, glucagon and somatostatin. The rest of the pancreas produces enzymes and secretions that travel through ducts to the intestine, where they play roles in digestion. Thus the pancreas is both an endocrine and an exocrine gland.

Following a meal, the concentration of glucose in the blood rises as glucose is absorbed from the gut. This rise in glucose concentration stimulates the pancreas to release insulin. Insulin stimulates cells to use glucose as fuel and to convert it into storage products such as glycogen and fat. When there is no longer food in the gut, the blood's glucose concentration falls, and the pancreas stops releasing insulin. As a result, most of the cells of the body shift to using glycogen and fat rather than glucose for fuel. If the concentration of glucose in the blood falls below normal, cells in the islets release the hormone **glucagon**, which stimulates the liver to convert glycogen back to glucose to resupply the blood. These effects and conversions will be discussed in greater detail in Chapter 43.

Somatostatin, a hormone released from the pancreas in response to rapid rises of glucose and amino acids in the blood, inhibits the release of both insulin and glucagon and slows the digestive activities of the gut. Pancreatic somatostatin extends the period of time over which nutrients are absorbed from the gut and used by the cells of the body. Somatostatin is also secreted by cells in a different part of the body, and these secretions serve a different hormonal function. First discovered as a hypothalamic neurohormone that *inhibits* the release of growth hormone by the pituitary, it was called growth hormone release-inhibiting hormone, but somatostatin is a more convenient name.

Adrenal Hormones

An adrenal gland sits above each kidney. Functionally and anatomically an adrenal gland is a gland within a gland (Figure 36.12). The core, called the **adrenal medulla**, produces the hormone **epinephrine** (also known as adrenaline) and, to a lesser degree, **norepinephrine** (or noradrenaline). Surrounding the medulla (as an apricot surrounds its pit) is the **adrenal cortex**, which produces other hormones. The medulla develops from nervous system tissue and is under the control of the nervous system; the cortex is under hormonal control, largely by adrenocorticotropin from the anterior pituitary.

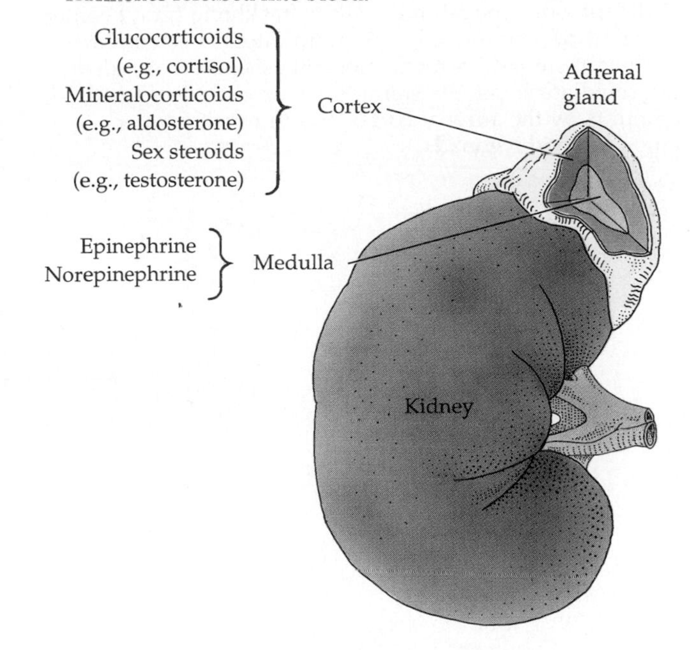

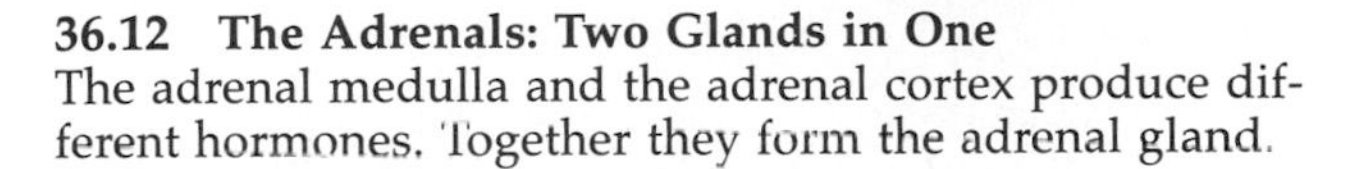

36.12 The Adrenals: Two Glands in One
The adrenal medulla and the adrenal cortex produce different hormones. Together they form the adrenal gland.

By producing epinephrine, which arouses the body to action, the adrenal medulla is involved in "fight-or-flight" reactions. As we saw earlier in the chapter, in stressful situations epinephrine increases heart rate, breathing rate, and blood pressure, and it diverts blood flow to active skeletal muscles and away from the gut.

All hormones produced by the adrenal cortex are steroids synthesized from cholesterol (Figure 36.13). In general they are called the **corticosteroids**, and they are divided into three functional classes. The **glucocorticoids** influence blood glucose concentrations as well as other aspects of fat, protein, and carbohydrate metabolism. The **mineralocorticoids** influence the ionic balance of the extracellular fluids. The **sex steroids** stimulate sexual development and reproductive activity. Sex steroids are secreted in only small amounts by the adrenal cortex and will be discussed further in the section on the gonads. Of the 30 or so different steroids produced by the adrenal cortex, the only two of great importance in human physiological functions under normal conditions are the mineralocorticoid aldosterone (which helps regulate salt concentration in the blood; see Chapter 44) and the glucocorticoid cortisol.

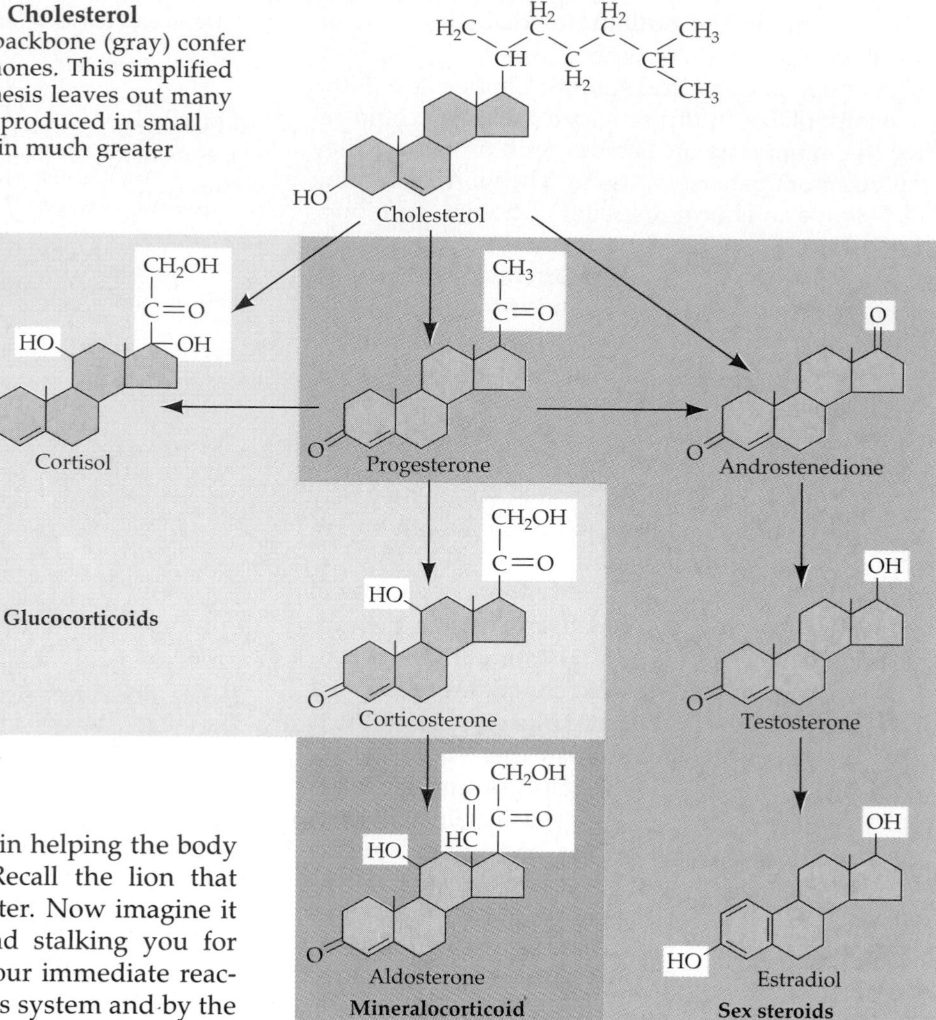

36.13 Steroid Hormones Begin as Cholesterol Different side groups on the sterol backbone (gray) confer different properties on steroid hormones. This simplified outline of steroid hormone biosynthesis leaves out many intermediate steps. Sex steroids are produced in small amounts by the adrenal cortex and in much greater amounts by the gonads.

Cortisol plays important roles in helping the body respond to short-term stress. Recall the lion that roared at you earlier in the chapter. Now imagine it chasing you into high grass and stalking you for hours as you try to get away. Your immediate reaction is stimulated by your nervous system and by the release of epinephrine. Your heart is beating faster, you are breathing faster, and your running muscles are getting maximal supplies of oxygen and glucose. This is not a sustainable situation, so within about 5 minutes cortisol levels rise and help you sustain your escape. Because you need a high level of blood glucose for your brain to function, cortisol stimulates the other cells of your body to decrease their use of glucose and start to metabolize fats and proteins for energy. This is not a time to feel sick, have allergic reactions, or heal wounds, so cortisol blocks immune system reactions. This is why cortisol is useful for reducing inflammations and allergies.

The effects of epinephrine and cortisol in reactions to short-term stress are beneficial, but the prolongation of these effects by the chronic, long-term stresses of modern life can be very damaging to the body. High blood pressure, poor gastrointestinal function, inhibition of protein synthesis, fat mobilization, and inhibition of the immune system are not healthy responses over long periods of time. Details on the interactions of stress, aging, and the cortisol response are discussed in Box 36.A.

Cortisol release is controlled by the pituitary hormone **adrenocorticotropin**, which in turn is controlled by the hypothalamic adrenocorticotropin-releasing hormone. Because the cortisol response to stress has this chain of steps, each involving secretion, diffusion, circulation, and cell activation, it is much slower than the epinephrine response to stress.

The Sex Hormones

The testes of the male and the ovaries of the female (that is, the gonads) produce hormones as well as gametes. Most of the gonadal hormones are steroids synthesized from cholesterol (see Figure 36.13). The male steroids are collectively called **androgens**, and the dominant one is **testosterone**. The female steroids are **estrogens** and **progesterone**. The dominant estrogen is **estradiol**. The sex steroids have important developmental effects: They determine whether a fetus develops into a female or a male. (A fetus is the latter stage of an embryo; a human embryo is called a fetus from the eighth week of pregnancy to the moment of birth.) After birth the sex steroids control the maturation of the reproductive organs and the development and maintenance of secondary sexual characteristics, such as breasts and facial hair.

The sex steroids begin to exert developmental ef-

BOX 36.A

The Rat Race

We usually assume that excessive stress leads to premature aging. Stress reactions *increase* blood pressure and fat metabolism while they *inhibit* digestion and immune system function. When prolonged, the effects of stress can contribute to cardiovascular disease, strokes, ulcers, and susceptibility to diseases. Therefore, assuming that stress accelerates aging is not unreasonable. The exact physiological interactions between stress and aging have only recently been elucidated, however, and they are quite interesting. Research by Robert Sapolsky at Stanford University has shown that old rats can initiate a stress response just as effectively as young rats, but they cannot turn it off as rapidly. All the harmful effects of stress persist longer in older rats than in younger ones. Why?

The answer involves the way negative feedback controls stress responses. A region of the brain called the hippocampus has cells with receptors for cortisol. When a rat experiences stress, cortisol levels rise in the blood. The cortisol activates these hippocampal cells, which inhibit the secretion of adrenocorticotropin-releasing hormone by the hypothalamus. The drop in adrenocorticotropin-releasing hormone causes a decline in adrenocorticotropin release from the pituitary, which induces the adrenal cortex to stop producing cortisol.

As rats age, they lose the hippocampal cells that function in this negative feedback loop. Apparently cortisol contributes to the demise of these cells. The more stress experienced, the more of these crucial hippocampal cells are lost. The result is the loss of the negative feedback mechanism that protects the body from the harmful effects of sustained stress responses. In rats the increased incidence of stress responses leads to many disorders usually associated with aging—strokes, cardiovascular disease, digestive system malfunction, and impaired immune system function that increases susceptibility to cancers and other diseases. These relationships between stress and aging are also believed to pertain to humans.

fects at about the seventh week of gestation in the human embryo. Until that time, the embryo can develop into either sex. The ultimate instructions for sex determination reside in the genes. Individuals receiving two X chromosomes normally become females, and individuals receiving an X and a Y chromosome normally become males. These genetic instructions, however, are carried out through the production and action of the sex steroids, and the potential for error exists.

The presence of the Y chromosome normally causes the embryonic, undifferentiated gonads to begin producing androgens in the seventh week. In response to the androgens, the reproductive system develops into that of a male (Figure 36.14). If the androgens are not produced at that time, or the androgen receptors do not function, the female reproductive structures develop even if the fetus is a male genetically. In humans, female development is the default, or neutral, course; a fetus develops female characteristics unless androgens are present to trigger male development. The opposite situation exists in some other vertebrates—male development is the default condition, which is switched to female development if estrogens are present.

Occasionally the hormonal control of sexual development does not work perfectly and intersex individuals are produced. The most extreme (but rare) case is a true **hermaphrodite**, who has both testes and ovaries. **Pseudohermaphrodites** have the gonads of one sex and the external sex organs of the other. For example, an XY fetus will develop testes, but if his tissues are insensitive to the androgens produced by those testes, the external sex organs and the secondary sexual characteristics of a female develop (Figure 36.15).

Sex steroids have dramatic effects at the time of puberty. Throughout childhood the production of sex steroids by the gonads is extremely low. As puberty approaches, the hypothalamus begins to produce and secrete gonadotropin-releasing hormone, which causes the pituitary to produce two gonadotropins—**luteinizing hormone** and **follicle-stimulating hormone**. In the preadolescent male, the increased level of luteinizing hormone stimulates groups of cells in the testes to synthesize androgens, which in turn initiate the profound physiological, anatomical, and psychological changes associated with adolescence. The voice deepens, hair begins to grow on the face and body, the testes and the penis grow, and skeletal muscles enlarge. Even an active program of weight lifting will not lead to massive muscle development in preadolescent boys or in women because such increase requires a level of androgens not normally found in individuals other than males past puberty. Taking synthetic androgens to enhance muscle development is a dangerous practice and a serious type of substance abuse (Box 36.B).

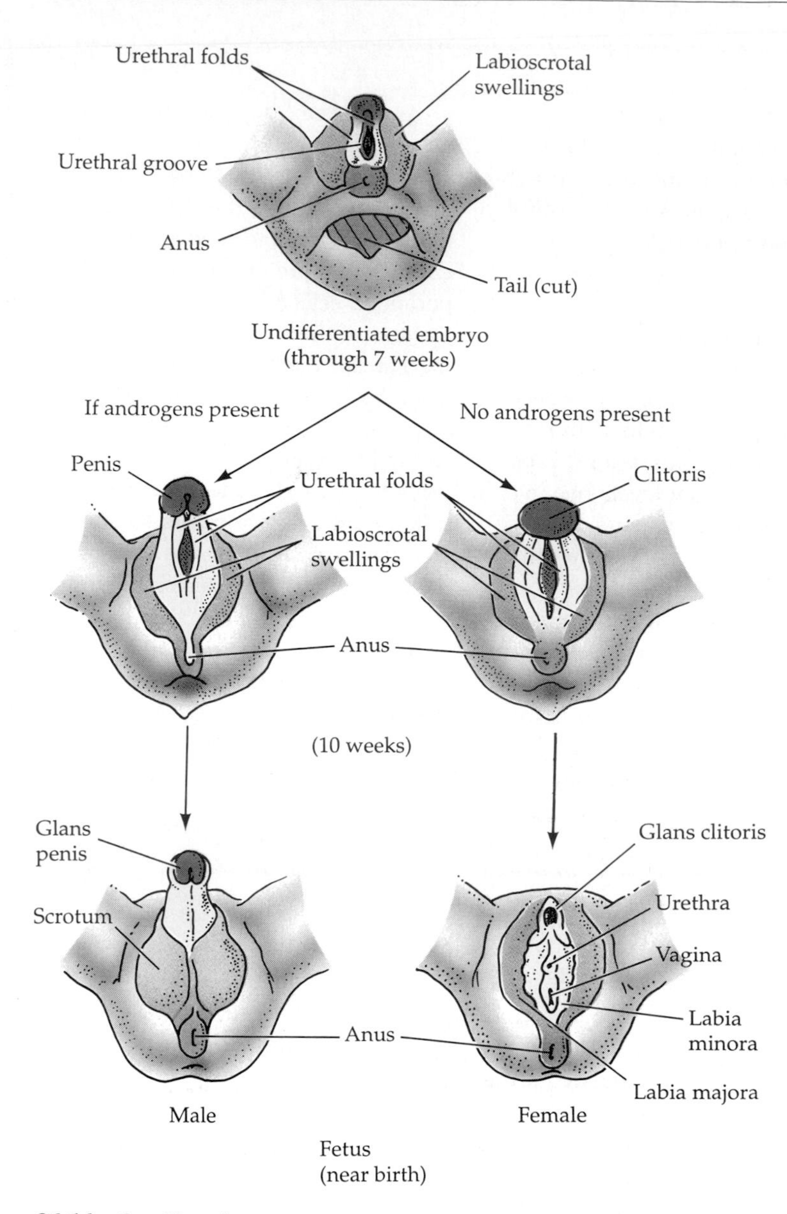

36.14 Sex Development
Hormones direct sex development along one of two pathways. The sex organs of early human embryos are similar (top). The testes of genetic males begin to secrete androgens about seven weeks after fertilization. Under the influence of androgens, a penis and scrotum form. Without the influence of androgens, female external organs develop.

36.15 A Pseudohermaphrodite
This person, who is genetically a male with an XY genotype, carries a mutation that leaves body cells unresponsive to male sex hormones. During early fetal life, the developing testes produced normal amounts of testosterone, but because of the mutation, the cells forming the sex organs could not respond. Development thus followed the female pattern. Fully functional testes developed within the body, but external female organs developed. The person does not have a uterus or ovaries, however, and the vagina ends blindly.

Puberty in the female is also ushered in by an increased release of gonadotropin-releasing hormone by the hypothalamus. The levels of luteinizing hormone and follicle-stimulating hormone rise, stimulating the ovaries to begin producing the female sex hormones. The increased circulating levels of these sex steroids initiate the development of the traits characteristic of a sexually mature woman: enlarged breasts, vagina, and uterus; a broad pelvis; increased subcutaneous fat; pubic hair; and the initiation of the menstrual cycle.

BOX 36.B

Muscle-Building Anabolic Steroids

The androgen testosterone helps skeletal muscles grow, especially when they are exercised regularly. The bulging biceps, triceps, pectorals, and deltoids of body builders are extreme examples of the skeletal muscle growth that occurs in every male past puberty. Natural muscular development can be exaggerated by both men and women who want to increase their maximum strength in athletic competition if they take synthetic androgens—**anabolic steroids.** However, anabolic steroids have serious negative side effects. In women, the use of artificial androgens causes the breasts and uterus to shrink, the clitoris to enlarge, menstruation to become irregular, facial and body hair to grow, and the voice to deepen. In men, the testes shrink, hair loss increases, the breasts enlarge, and sterility can result. Other effects are even more serious. For example, taking anabolic steroids greatly increases the risk of heart disease, certain cancers, kidney damage, and personality disorders such as depression, mania, psychoses, and extreme aggression. Most official athletic organizations, including the International Olympic Committee, ban anabolic steroid use. Competing athletes are tested frequently for the presence of steroids in their blood. In the 1988 Olympics, Canadian track competitor Ben Johnson was forced to forfeit his gold medal and the records he set because tests indicated that he had used anabolic steroids.

Other Hormones

We have discussed all of the major endocrine glands and "classical" hormones in this chapter, but there are many hormones we have not mentioned. Examples include the numerous hormones produced in the digestive tract that help organize the way the gut processes food. Even the heart has endocrine functions. When blood pressure rises and causes the walls of the heart to stretch, certain cells in the walls of the heart release **atrial natriuretic hormone**. This hormone increases the excretion of sodium ions and water by the kidneys, thereby lowering blood volume and blood pressure. As we discuss the physiology of the organ systems of the body in the chapters that follow, we will frequently mention hormones that those organs produce.

RECEIVING AND RESPONDING TO HORMONES

In discussing the functions of many hormones in this chapter, we have not yet addressed the question of how these chemical messages are read by the cells that receive them. How do hormones induce their actions at the cellular and molecular levels? We start to answer this question by dividing hormones into two groups according to where they bind their receptors. **Water-soluble hormones**, such as peptide and protein hormones, do not cross cell membranes readily, and their receptors are membrane proteins with binding regions that project from the cell surface. **Lipid-soluble hormones**, such as the steroids and thyroxine, can easily pass through cell membranes, and their receptors are in the cytoplasm or the nucleus of the cell.

Receptors for Water-Soluble Hormones and Second Messengers

Water-soluble hormones bind with receptors on the surface of the target cells. The receptors are glycoproteins that have a binding domain that projects beyond the outside of the cell membrane, a transmembrane domain, and a catalytic domain that extends into the cytoplasm of the cell. Directly or indirectly, the catalytic domain of most receptors of water-soluble hormones initiates cell responses by activating a protein kinase. There are many different kinds of protein kinases, but they all phosphorylate proteins; they catalyze the transfer of phosphate groups from ATP to specific proteins. In some cases the added phosphate activates the protein, and in other cases phosphorylation inactivates the protein. Thus the binding of the hormone to its receptor can result in activation or inhibition of a cellular process.

Some receptors for water-soluble hormones have their own protein kinase sites on their catalytic domains. When hormones bind to these receptors, the protein kinase sites are activated to catalyze the phosphorylation of cytoplasmic proteins. In this case the catalytic domain acts directly as a protein kinase.

Most receptors for water-soluble hormones act indirectly by way of **second messengers**. The hormone itself is the first messenger. When it binds to its receptor, the hormone stimulates a chain of reactions, producing small, diffusible molecules that serve as second messengers within the cell. The second messengers activate protein kinases. Cyclic AMP, or **cAMP** (cyclic adenosine monophosphate) is a well-studied second messenger that activates a wide range of protein kinases in many different kinds of cells.

The second messenger role of cAMP was discovered by E. W. Sutherland of Washington University, who began this work in the 1950s. He was investigating how epinephrine stimulates liver cells to break down carbohydrate storage molecules and liberate glucose. It became evident that there were a number of steps between the binding of the hormone by the receptor and the liberation of glucose. Sutherland was able to show that one of these steps was the control of enzyme activity through phosphorylation. This was the first demonstration of this common mechanism for regulating enzyme function. Sutherland then discovered that epinephrine could stimulate disrupted liver cells to release glucose as long as fragments of their plasma membranes were present. Showing hormone action in a cell-free system was also a major landmark in biochemistry. The third major discovery in this research program was that the hormone interacted with the membrane fragments to produce a small molecule that could stimulate the phosphorylation of enzymes in a liver cell mixture free of membranes. This molecule was identified as cAMP. For this work, E. W. Sutherland received the Nobel prize in 1971.

In the years after Sutherland's work, the list of systems known to be activated by cAMP grew rapidly. Many hormones in vertebrate tissues act via this second messenger. These systems include epinephrine's stimulation of the breakdown of stored carbohydrates and fats, adrenocorticotropin's stimulation of the production of glucocorticoids in the adrenals, luteinizing hormone's stimulation of androgen synthesis, and many more.

How can the same second messenger induce completely different responses in different cells? Different target cells have different protein kinases that are activated by the second messenger. The specificity of hormone action depends not only on the receptors that determine which cells respond to a given hormone, but also on what responding mechanisms a cell has.

The Role of G-Proteins in the Production of Second Messengers

A complex set of reactions takes place between binding of the hormone and production of the second messenger. When a receptor binds a hormone molecule, the receptor's shape (the tertiary structure; see Chapter 3) changes. In its new form, the receptor can interact with a second membrane protein, enabling that protein to bind a molecule of guanosine triphosphate (GTP). Proteins that bind GTP are called **G-proteins**.

Next, the subunit of the G-protein that is bound to the GTP separates and moves to another membrane protein, the enzyme **adenylate cyclase**. The complex consisting of G-protein and GTP activates the adenylate cyclase. The active adenylate cyclase catalyzes the conversion of ATP to cAMP within the cell. Thus, the G-protein is the link between the receptor and the production of the second messenger (Figure 36.16). As we will see in the next section, by activating a protein kinase that is otherwise inactive, the second messenger cAMP takes the next step leading to the target cell's response to the hormone.

There are many kinds of G-proteins. Some are even inhibitory, inactivating adenylate cyclase. The G-protein subunit that binds the GTP eventually hydrolyzes the GTP to GDP, thus inactivating itself and helping terminate hormone-induced responses.

G-proteins are important control elements in many cells and are the targets of some pathogenic organisms. Examples are the bacteria that cause the diseases cholera, whooping cough, and some forms of "traveler's diarrhea." Cholera bacteria produce a toxin that prevents the G-protein from hydrolyzing the GTP and inactivating itself. Thus this toxin causes cells to continue to produce high levels of cAMP. In the lining of the intestine where the cholera bacteria attack, cAMP stimulates the active transport of sodium ions into the gut. The toxin therefore causes massive sodium loss, with water following, resulting in severe, rapid dehydration and ionic imbalance that can lead to death in a day or two.

cAMP Targets and Response Cascades

The action of epinephrine on liver cells is an example of how cAMP-dependent protein kinases work (Figure 36.17). Epinephrine binds to its receptor on the plasma membrane, adenylate cyclase is activated, and cAMP is formed. cAMP activates a specific protein kinase that acts on two other enzymes, adding a phosphate group from ATP to each one. One of the newly phosphorylated proteins is the enzyme glycogen synthase. This enzyme catalyzes the joining of glucose molecules to synthesize the energy-storing molecule glycogen, but it is inactivated by the addi-

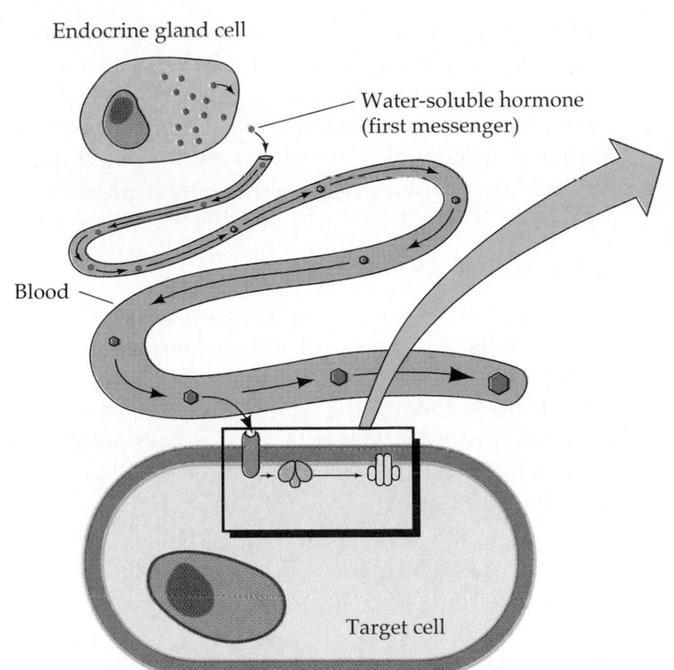

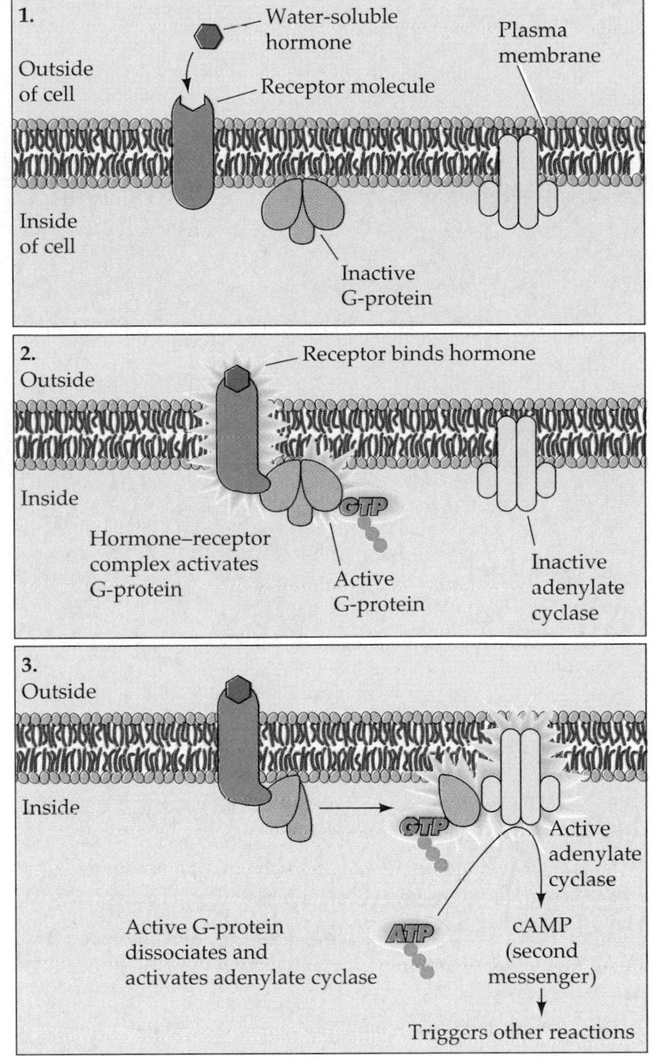

36.16 Second Messengers
Water-soluble hormones bind to receptors on the surface of their target cells. For receptors that act indirectly to phosphorylate protein kinases, this binding begins a chain of reactions that involves G-proteins and the production of a second messenger, such as cAMP.

tion of the phosphate group. The other enzyme phosphorylated by the cAMP-activated protein kinase is phosphorylase kinase, which is activated by the addition of the phosphate group. Phosphorylase kinase, itself a protein kinase, catalyzes the phosphorylation of the enzyme glycogen phosphorylase. Glycogen phosphorylase participates in the breakdown of glycogen to glucose. Thus cAMP, through its effects on two protein kinases, inhibits the storage of glucose as glycogen and promotes the release of glucose through glycogen breakdown. Both of these effects increase glucose levels in liver cells and hence in the blood as well.

This cascade of regulatory steps amplifies the effect of a single hormone molecule. A molecule of epinephrine binds to a single receptor molecule, but the activated receptor activates many molecules (let's say ten) of the G-protein. Each activated G-protein activates one molecule of adenylate cyclase, but adenylate cyclase is an enzyme and can catalyze the production of perhaps 100 molecules of cAMP. Each molecule of cAMP activates only one protein kinase molecule, but each protein kinase may activate 100 phosphorylase kinase molecules, each phosphorylase kinase may activate 100 glycogen phosphorylase molecules, and each glycogen phosphorylase may catalyze the production of 100 molecules of glucose from glycogen. Thus an amplification of $10 \times 100 \times 100 \times 100 \times 100$ is achieved; that is, each molecule of epinephrine can cause the production of about 1 billion molecules of glucose.

Unless there is a continuing supply of the hormone, it diffuses away from the receptor or is enzymatically degraded, allowing the receptor to revert to its inactive tertiary structure. In turn, the concentration of the complex of G-protein and GTP decreases as the enzyme inactivates itself by breaking down the GTP, and cAMP is no longer formed. The cAMP still present is quickly removed by the action of specific phosphodiesterases, enzymes that catalyze the conversion of cAMP to an inactive product.

In one other way the decrease in the second messenger, cAMP, causes the actions originally induced by the hormone to cease suddenly. Besides activating protein kinases, cAMP inhibits phosphoprotein phosphatase, whose role is to remove phosphate groups from proteins phosphorylated by protein kinases. Thus when cAMP levels fall, phosphoprotein phosphatase is no longer inhibited and is free to

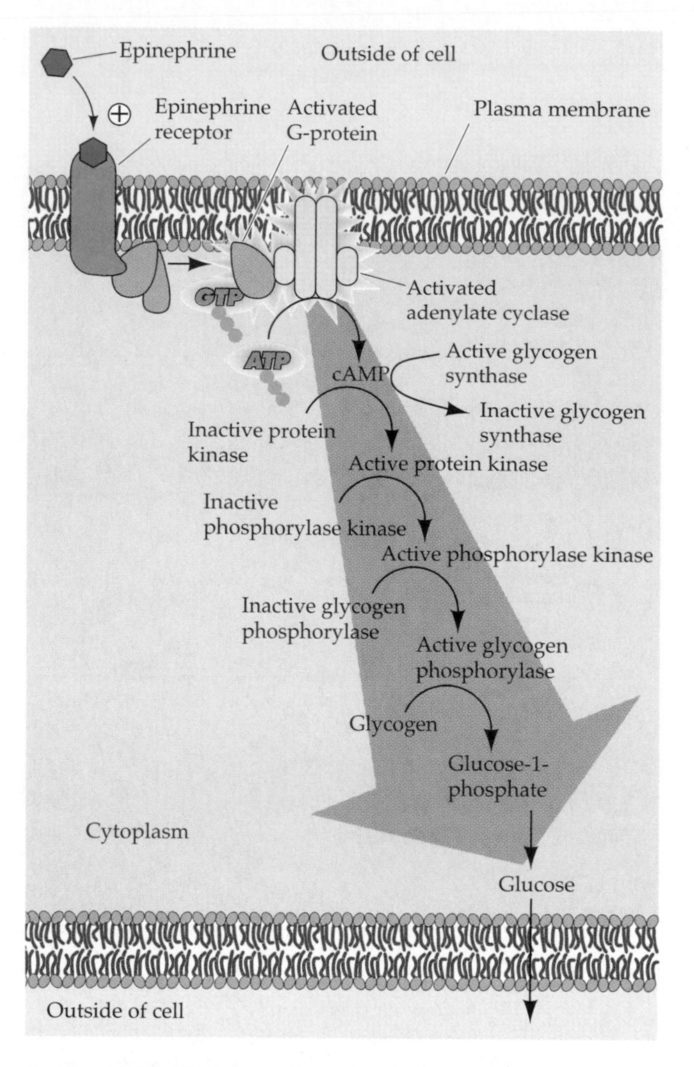

36.17 Epinephrine and cAMP
Epinephrine stimulates production of cAMP in liver cells. In liver cells cAMP then triggers a cascade of events that stimulates breakdown of glycogen to glucose and inhibits glycogen synthesis.

dephosphorylate the three enzymes controlling glycogen metabolism. Glycogen synthase is thus activated while phosphorylase kinase and glycogen phosphorylase are inactivated. Thus the hormone-induced breakdown of glycogen to glucose is reversed when epinephrine is no longer present.

Other Second Messengers

There are other second messengers besides cAMP in animal cells. Two second messengers, **inositol trisphosphate** (IP3) and **diacylglycerol** (DAG), are produced from membrane phospholipids that contain phosphoinositols. When certain water-soluble hormones bind to their receptor proteins, a G-protein is activated that in turn activates the enzyme phospholipase C. Active phospholipase C cleaves phosphatidylinositol to form IP3 and DAG. IP3 enters the cytoplasm while DAG remains in the membrane. DAG activates a protein kinase that is also in the membrane. IP3 causes the release of intracellular calcium ions that can stimulate the protein kinase as well as regulate other cellular functions that depend on calcium ions (Figure 36.18). Some hormones that act through the IP3 and DAG second-messenger system are norepinephrine and vasopressin.

Another second messenger, active in some target cells, is **cGMP** (cyclic guanosine monophosphate), a close chemical analog of cAMP. In many cases, cGMP acts in opposition to cAMP by activating a phosphodiesterase that breaks down cAMP. The effects of

36.18 The IP3 and DAG Second-Messenger System
A G-protein activates the enzyme phospholipase C, which catalyzes the hydrolysis of a membrane phospholipid to form inositol trisphosphate (IP3) and diacylglycerol (DAG). DAG stays in the membrane and IP3 enters the cytoplasm. Both serve as second messengers.

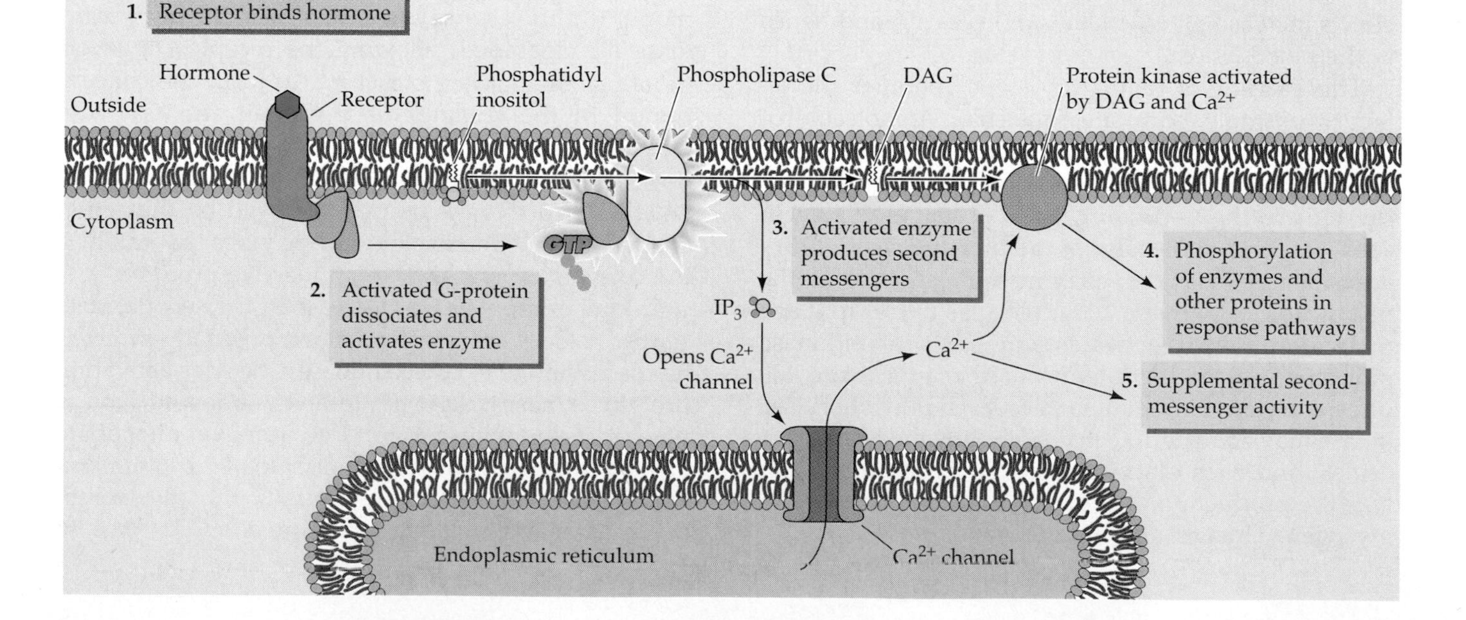

insulin on some target cells are mediated by cGMP; that is, insulin is the first messenger and cGMP the second messenger.

Calcium Ions

Calcium ions (Ca^{2+}) mediate many responses of different kinds of cells. As we shall see in Chapters 38 and 40, Ca^{2+} ions play crucial roles in the functions of both nerve cells and muscle cells. In some cases, such as activating protein kinase C (see Figure 36.18) and controlling some membrane ion channels, Ca^{2+} acts directly. In other situations, however, Ca^{2+} must combine with a calcium-binding protein, of which the most widely distributed is **calmodulin**. Calmodulin is activated by binding with Ca^{2+}, and the active complex can then trigger cell responses including activation of more protein kinases, smooth muscle contraction, microtubule assembly, protein synthesis, and various secretory events. Another calcium-binding protein, troponin, regulates a key reaction in the contraction of the skeletal muscles of vertebrates.

Lipid-Soluble Hormones

Steroid hormones—such as estrogens, progesterone, and the hormones of the adrenal cortex—as well as thyroxine, generally do not react with receptors on the target-cell surface (although it is now known that there are some steroid receptors bound to plasma membranes). These hormones are all lipid-soluble, which, as you may recall from Chapter 5, means that they pass readily through the lipid-rich plasma membrane. They act by stimulating the synthesis of new kinds of proteins through gene activation rather than by altering the activity of proteins already present in the target cells.

Once inside a cell, a lipid soluble hormone binds to a receptor protein in the cytoplasm (Figure 36.19). The presence of a receptor protein is what distinguishes a responsive cell from a nonresponsive cell. A receptor protein is specific for a particular hormone, and it changes shape when it binds its hormone. The hormone–receptor complex associates with acidic chromosomal proteins, and thus with the DNA of the chromosomes. The receptor protein itself cannot bind the acidic chromosomal proteins unless it has already bound a hormone molecule and undergone the necessary change in structure. Once associated with the chromosomal proteins, the hormone activates the transcription of certain genes into messenger RNAs, which are exported to the cytoplasm and translated into specific proteins.

The actions of lipid-soluble hormones are slower and last longer than the actions of water-soluble hormones. When water-soluble hormones bind to their receptors on cell surfaces, some induce changes in membrane permeabilities to ions and some activate

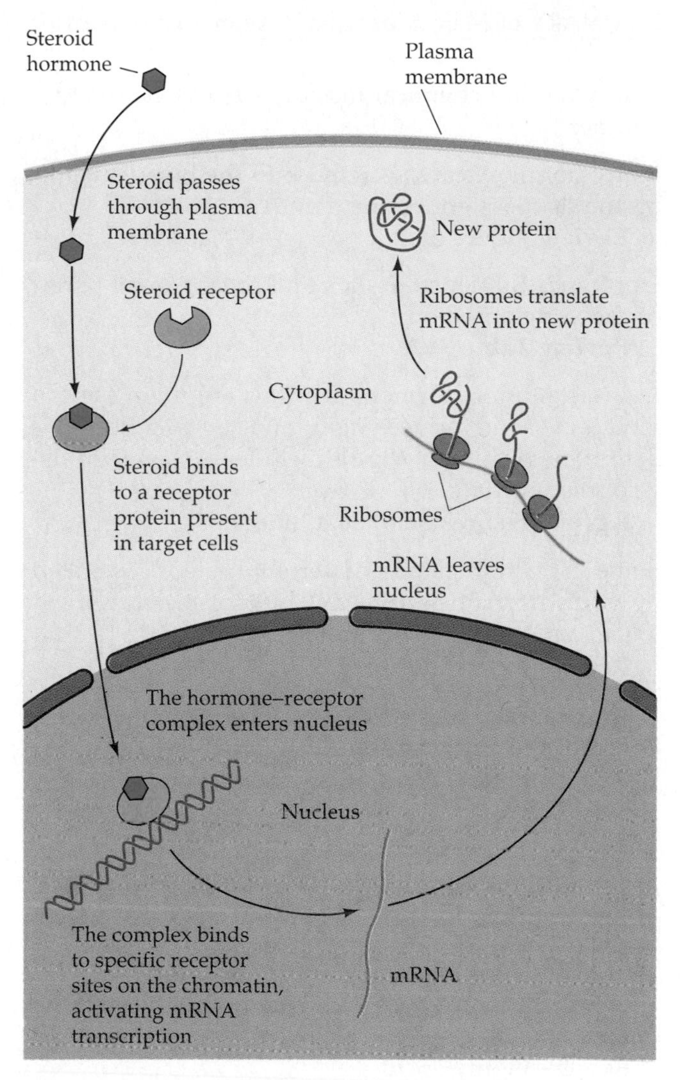

36.19 Action of Lipid-Soluble Hormones
The receptors of lipid-soluble hormones, such as steroids, are inside of cells. These hormones activate gene transcription.

or inactivate enzymes within the cell. In general, these are much more rapid and fleeting responses than the changes in gene expression induced by lipid-soluble hormones.

Hormones in Control and Regulation

In the last chapter we discussed the importance of control and regulation in animal physiology. Control and regulation require information, and that information is coded as electrical signals in the nervous system (see Chapter 38) or as chemical signals in endocrine systems. Throughout the rest of this section, we will see how these two types of information are used in physiological systems. In the next chapter, however, we will build on what we have learned about hormones by studying reproduction, since reproductive systems depend very heavily on hormonal mechanisms for control and regulation.

SUMMARY of Main Ideas about Animal Hormones

Hormones are chemical messages from one cell to another.

Many hormones are secreted into the bloodstream by the ductless endocrine glands.
Review Figure 36.2

Hormones influence tissues and organs in all parts of the vertebrate body.
Review Table 36.1

In insects, brain hormone controls the timing of molt, ecdysone induces molt, and juvenile hormone determines whether a molt includes a change in developmental state.
Review Figures 36.3, 36.4, and 36.5

Some endocrine glands of vertebrates are controlled by tropic hormones from the anterior pituitary, which also secretes several other hormones.
Review Figure 36.6

The posterior pituitary secretes two hormones produced in the hypothalamus.
Review Figure 36.6

The hypothalamus controls the secretion of some anterior pituitary hormones through the production of neurohormones that reach the anterior pituitary through portal blood vessels.
Review Figure 36.8 and Table 36.2

Thyroxine, a hormone produced by the thyroid, influences metabolism and development.
Review Figure 36.10

Parathormone, a hormone from the parathyroid glands, and calcitonin, from the thyroid, control calcium metabolism.
Review Figure 36.11

Insulin and glucagon, hormones from the pancreas, regulate blood sugar.

Hormones of the adrenal medulla stimulate "fight-or-flight" responses.

Hormones of the adrenal cortex control stress responses and salt balance.
Review Figures 36.12 and 36.13

Hormones of the gonads control sexual development, reproduction, and sexual behavior.
Review Figure 36.14

A hormone is either water-soluble or lipid-soluble.

Water-soluble hormones do not cross cell membranes readily, and their receptors project from the surface of the target cell.
Review Figure 36.16

After a water-soluble hormone binds to its receptor, the receptor relays the message to a second messenger within the cell.
Review Figures 36.16 and 36.18

Lipid-soluble hormones pass easily through cell membranes, and their receptors are in the cytoplasm or nucleus.
Review Figure 36.19

cAMP is an important second messenger that can trigger a cascade of intracellular events that amplify the response of the cell to a hormone molecule.
Review Figure 36.17

SELF-QUIZ

1. Which of the following statements is true for all hormones?
a. They are secreted by glands.
b. They have receptors on cell surfaces.
c. They may stimulate different responses in different cells.
d. They target cells distant from their site of release.
e. When the same hormone occurs in different species, it has the same action.

2. The hormone ecdysone
a. is released from the posterior pituitary.
b. stimulates molt and metamorphosis in insects.
c. maintains an insect in larval stages unless brain hormone is present.
d. stimulates secretion of juvenile hormone from the prothoracic glands.
e. keeps the insect exoskeleton flexible to permit growth.

3. The posterior pituitary
a. produces oxytocin.
b. is under the control of hypothalamic releasing neurohormones.
c. secretes tropic hormones.
d. secretes neurohormones.
e. is under feedback control by thyroxine.

4. Growth hormone
a. can cause adults to grow taller.
b. stimulates protein synthesis.
c. is released by the hypothalamus.
d. can be obtained only from cadavers.
e. is a steroid.

5. Both epinephrine and cortisol are secreted in response to stress. Which of the following statements is also *true for both* of these hormones?
   a. They act to increase blood glucose.
   b. The receptors are on the surfaces of target cells.
   c. They are secreted by the adrenal cortex.
   d. Their secretion is stimulated by adrenocorticotropin.
   e. They are secreted into the blood within seconds of the onset of stress.
6. Prior to puberty
   a. the pituitary is secreting luteinizing hormone and follicle-stimulating hormone, but the gonads are unresponsive.
   b. the hypothalamus does not secrete much gonadotropin-releasing hormone.
   c. males can stimulate massive muscle development through a vigorous training program.
   d. testosterone plays no role in development of the male sex organs.
   e. genetic females will develop male genitals unless estrogen is present.
7. Which of the following is *not* true of cyclic AMP?
   a. It is broken down by adenylate cyclase.
   b. It is involved in the chain of events whereby epinephrine stimulates liver cells to break down glycogen.
   c. It is a second messenger mediating intracellular responses to many hormones.
   d. Many of its effects are mediated by protein kinases.
   e. A molecule of cAMP activates only a single protein kinase molecule.
8. Steroid hormones
   a. are all produced by the adrenal cortex.
   b. have only cell surface receptors.
   c. are lipophobic.
   d. act through altering the activity of proteins in the target cell.
   e. act through stimulating the production of new proteins in the target cell.
9. Which is a likely cause of goiter?
   a. The thyroid gland is producing too much parathormone.
   b. Circulating levels of thyrotropin are too low.
   c. An inadequate supply of functional thyroxine.
   d. An oversupply of functional thyroxine.
   e. Too much iodine in the diet.
10. Parathormone
   a. stimulates osteoblasts to lay down new bone.
   b. reduces blood calcium levels.
   c. stimulates calcitonin release.
   d. is produced by the thyroid gland.
   e. is released when blood calcium levels fall.

FOR STUDY

1. Compare the mechanisms of action of peptide and steroid hormones.
2. Explain how both hyperthyroidism and hypothyroidism can cause goiter. Include the roles of the hypothalamus and the pituitary in your answers.
3. Explain the developmental abnormalities that can produce a genetic male with female secondary sexual characteristics. Describe the gonads of such an individual.
4. How did Sutherland's experiments demonstrate that the result of epinephrine combining with a membrane-bounded receptor is the production of a second messenger?
5. How can cAMP working through a protein kinase activate one enzyme while inactivating another enzyme in the same cell?

READINGS

Atkinson, M. A. and N. K. MacLaren. 1990. "What Causes Diabetes?" *Scientific American*, July. This paper reveals how malfunctions of the immune system cause insulin-dependent diabetes.

Berridge, M. J. 1985. "The Molecular Basis of Communication within the Cell." *Scientific American*, October. An authoritative account of second messengers and their roles in biological phenomena.

Bloom, F. E. 1981. "Neuropeptides." *Scientific American*, October. A description of the discovery, synthesis, distribution, and actions of peptides that serve as chemical messengers in the nervous system and as hormones in the body. Focuses on vasopressin, oxytocin, endorphins, enkephalins, and a few others.

Cantin, M. and J. Genest. 1986. "The Heart as an Endocrine Gland." *Scientific American*, February. Interesting account of the discovery and characterization of a hormone half a century after its existence was predicted.

Carafoli, E. and J. T. Penniston. 1985. "The Calcium Signal." *Scientific American*, November. Calcium as a second messenger; the roles of calcium-binding proteins.

Eckert, R., D. Randall and G. Augustine. 1988. *Animal Physiology*, 3rd Edition. W. H. Freeman, San Francisco. An excellent textbook; particularly useful with respect to second messengers and regulatory physiology.

Fernald, R. D. 1993. "Cichlids in Love." *The Sciences*, July/August. A fascinating study of "wimpy" and "macho" behavior among cichlid fish.

Snyder, S. H. 1985. "The Molecular Basis of Communication between Cells." *Scientific American*, October. An overview of the relationships between the nervous and endocrine systems; the focus is on chemical messengers and their molecular biology.

Vander, A. J., J. H. Sherman and D. S. Luciano. 1994. *Human Physiology: The Mechanisms of Body Function*, 6th Edition. McGraw-Hill, New York. Chapter 10 deals specifically with hormonal regulation.

Reproduction: An Essential Feature of Animal Life Like humans, birds reproduce sexually. The cycle of birth, sexual maturation, and reproduction provides a continuous line of genetic information from generation to generation.

37

Animal Reproduction

We are sexually reproducing animals. Usually we think of eggs and sperm as the means for reproducing ourselves, but let's consider another viewpoint. Very early in the life of a new human embryo, during the first cell divisions, arises a population of cells that wanders through the body until the sex organs form, at which time these nomadic cells migrate to the sex organs. Eventually these cells give rise to the eggs and sperm that will produce the next generation. It is as if we the organisms are devices created by our sex cells to reproduce themselves. The sex cells themselves are part of a lineage of cells called the germ line, which is punctuated each generation by meiosis and recombination. This view of the relation between the organism and the sex cells reveals the centrality of reproductive processes in the lives of animals. Because these processes depend extensively on hormonal mechanisms of control and integration, we take up the topic of animal reproduction immediately following our discussion of animal hormones and before we learn about the nervous system.

ASEXUAL REPRODUCTION

Sexual reproduction is a nearly universal trait of animals, but many species can reproduce asexually as well. Offspring produced asexually are genetically identical to one another. Asexual reproduction is highly efficient because there is no mating, which requires energy and involves risks. In addition, all individuals of the population can convert resources into offspring, allowing the population to grow as rapidly as resources permit. However, asexual reproduction does not generate genotypic diversity. An asexually reproducing population does not have a wide variety of genotypes on which natural selection can act as the environment changes. Nevertheless, some animals reproduce asexually.

Budding

A common mode of asexual reproduction in simple multicellular animals is for a new individual to arise as an outgrowth of an older one—a process called **budding.** Some sponges form buds of undifferentiated cells on the outsides of their bodies. These buds grow by mitotic cell division, and the cells undergo differentiation before the buds break away from the parents and become independent sponges. Many freshwater sponges produce internal buds, or **gemmules.** A gemmule consists of several undifferentiated cells. Eventually the gemmules escape from the parent and become free-living individuals, genetically identical to the parent. Budding is part of the life cycle of some cnidarians, such as *Hydra* (Fig-

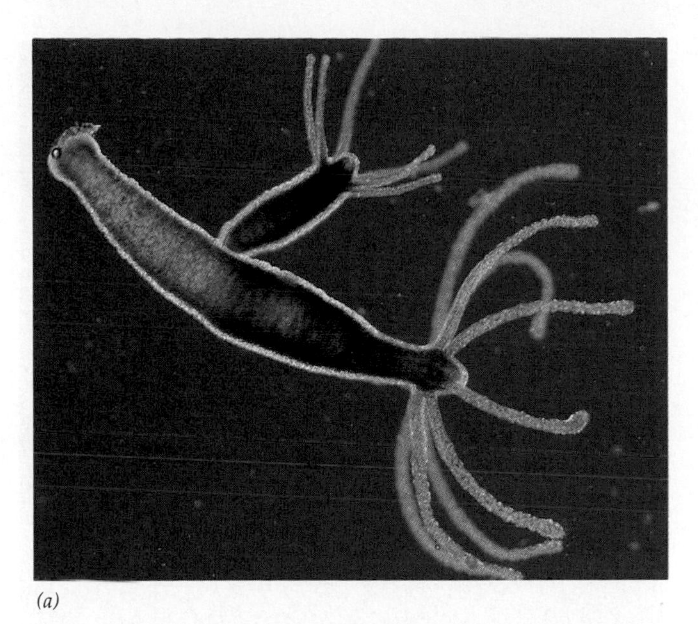

(b)

(c)

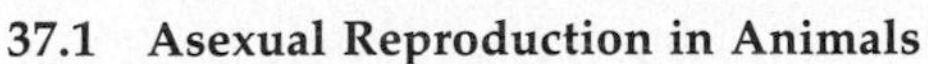

(a)

37.1 Asexual Reproduction in Animals
(a) Budding: A new individual forms as an outgrowth from an adult *Hydra*. *(b)* Regeneration: A single amputated arm from the sea star *Asterias rubens* develops into a new animal. *(c)* Parthenogensis: Aphids can hatch from unfertilized eggs.

ure 37.1*a*). The bud resembles the parent and may grow as large as the parent before it becomes an independent organism.

Regeneration

The cells in sponges and certain cnidarians that initiate budding are **totipotent;** that is, they have the ability to give rise to new, complete organisms. Totipotency is not a characteristic of most cells of most animals (see the discussion on determination and differentiation in Chapter 17). There are notable exceptions, however, in which pieces of animals can develop into whole animals. A dramatic example of such **regeneration** was unwittingly produced by a group of public officials who tried to protect oyster beds by instituting a search-and-destroy mission aimed at sea stars that were preying on the oysters. They "killed" these echinoderm predators by cutting them into pieces and dumping the pieces back into the sea. Sea stars, however, have remarkable abilities of regeneration. If they lose arms they regenerate new ones, and if a severed arm includes a portion of the central disc of the animal's body, it can regenerate into a complete sea star (Figure 37.1*b*). In their attempt to eliminate the sea stars, the public officials created large numbers of pieces capable of regeneration, and the predator population increased enormously.

Regeneration usually follows an animal's being broken by an outside force, but in some species the breakage is a normal event initiated by the animal itself. Certain species of segmented worms (annelids) develop segments with rudimentary heads bearing sensory organs, then they break apart. Each fragmented segment forms a new worm.

Parthenogenesis

Parthenogenesis is a type of asexual reproduction in which the offspring develop from unfertilized eggs (Figure 37.1*c*). Many animals, especially arthropods, reproduce parthenogenetically, as do some species of fish, amphibians, and reptiles. Most species that reproduce parthenogenetically also engage in sexual reproduction or sexual behavior. Aphids, for example, are parthenogenetic in the spring and summer, multiplying rapidly while conditions are favorable. Some of the unfertilized eggs laid in spring and summer develop into male aphids, others into females. As conditions become less favorable, the aphids mate and the females lay fertilized eggs. These eggs do not hatch until the following spring, and they yield only females. Species capable of parthenogenesis frequently switch from asexual to sexual reproduction when environmental conditions change. Parthenogenesis is used when conditions are stable and favorable; sexual reproduction introduces genotypic variability when conditions are changing, stressful, or unpredictable.

In some species parthenogenesis is part of the mechanism that determines sex. For example, in ants and in most species of bees and wasps, females develop from unfertilized eggs and are haploid and males develop from fertilized eggs and are diploid. Most females are sterile workers, but a select few become fertile queens. After a queen mates, she has a supply of sperm that she controls, enabling her to produce either fertilized or unfertilized eggs. Thus the queen determines when and how much of the colony resources are expended on males.

Parthenogenetic reproduction in some species requires a sex act even though this act does not fertilize the egg. The eggs of parthenogenetically reproducing ticks and mites, for example, develop only after the animals have mated, but the eggs remain unfertilized. Some species of beetles have no males at all and can reproduce only parthenogenetically, yet their eggs require sperm to trigger development. These beetles therefore mate with males of closely related, but different, species.

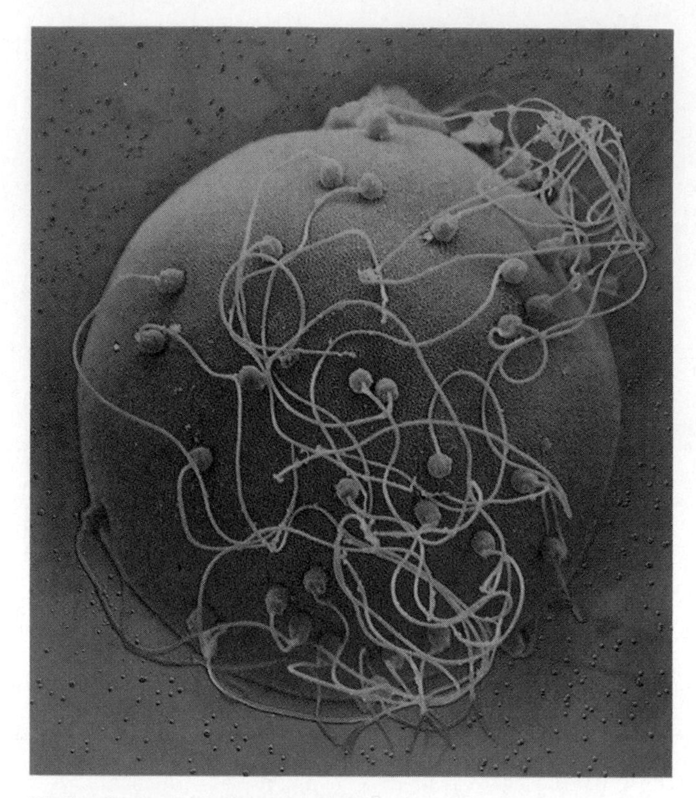

37.2 Gametes Differ in Size
Male gametes—the sperm—are small and motile, propelled by long flagella. The female gamete—the egg—is large and provisioned to nourish the early stages of the embryo's development. In this micrograph of mammalian fertilization, many sperm have attached to a single egg, but only one sperm cell will enter and fertilize it.

SEXUAL REPRODUCTIVE SYSTEMS OF ANIMALS

The enormous genotypic diversity in most sexually reproducing species derives from the independent assortment of chromosomes and the recombination of alleles on those chromosomes. As you know from Chapter 10, an animal has two alleles for each gene, and many pairs of alleles are heterozygous. A sexually reproducing animal packages single alleles from each pair into reproductive cells called **gametes.** Because each gamete receives one or the other allele of a gene, each gamete contains a complete set of genes, with a unique assortment of alleles. Sexual reproduction requires **fertilization**—the fusion of two gametes (almost always from different individuals) to form a **zygote.** The zygote receives half of its alleles from each parent and therefore has a new, unique genotype. Natural selection acts on the genotypic diversity produced by this process. Individuals that have genotypes best suited to environmental conditions are the most likely to survive and produce the largest number of offspring.

Both sexes, female (♀) and male (♂), produce haploid gametes from germ cells. The tiny gametes of males are called **sperm;** they move by beating their flagella. The much larger female gametes are called **eggs,** or **ova** (singular: ovum) and are nonmotile (Figure 37.2). Sperm and eggs are produced in the primary sex organs, the **gonads.** Male gonads are **testes** (singular: testis), and female gonads are **ovaries.** In addition to primary sex organs, most animals (except sponges and cnidarians) have accessory sex organs, including ducts, glands, and structures that deliver and receive gametes. The primary and accessory sex organs of an animal constitute its reproductive system.

Gametogenesis

Gametogenesis is the formation of gametes, and in all animal species except sponges, it takes place in the gonads. As we described in the introduction to this chapter, the gametes derive from a special lineage of cells called the **germ line.** Those cells are not produced by the gonads; they come to reside in the gonads only after the gonads have formed in the embryo. The germ cells are diploid, and they proliferate by mitosis. The cells resulting from the mitotic proliferation of germ cells in the gonads of females are called **oogonia** (singular: oogonium), and those in the gonads of males are called **spermatogonia** (singular: spermatogonium). Meiosis, the next step in gametogenesis, reduces the chromosomes to the haploid number, and the haploid cells mature into sperm and ova. Meiosis is central to the formation of both sperm and ova; you might review the discussion of meiosis in Chapter 9 before reading further.

Spermatogenesis is the process by which sperm form from germ cells. In mammals, sperm are produced in the tubules within the testes. The process begins when the diploid spermatogonia near the wall of a seminiferous tubule increase in size and divide by mitosis to become **primary spermatocytes.** The primary spermatocytes undergo the first meiotic division to form **secondary spermatocytes,** which are haploid. (Recall that the first meiotic division halves the number of chromosomes.) These cells remain connected by bridges of cytoplasm after each division. The second meiotic division produces four haploid **spermatids** for each primary spermatocyte that entered meiosis (Figure 37.3*a*).

Spermatids differ from one another genetically because the random orientation of chromosomes at the first meiotic metaphase shuffles the parental genomes. A given spermatid contains some maternal chromosomes and some paternal chromosomes; the particular combination is a matter of chance. Crossing over during the first meiotic division also contributes to the genetic differences among spermatids.

As the spermatocytes develop into spermatids and the spermatids develop into sperm, they move pro-

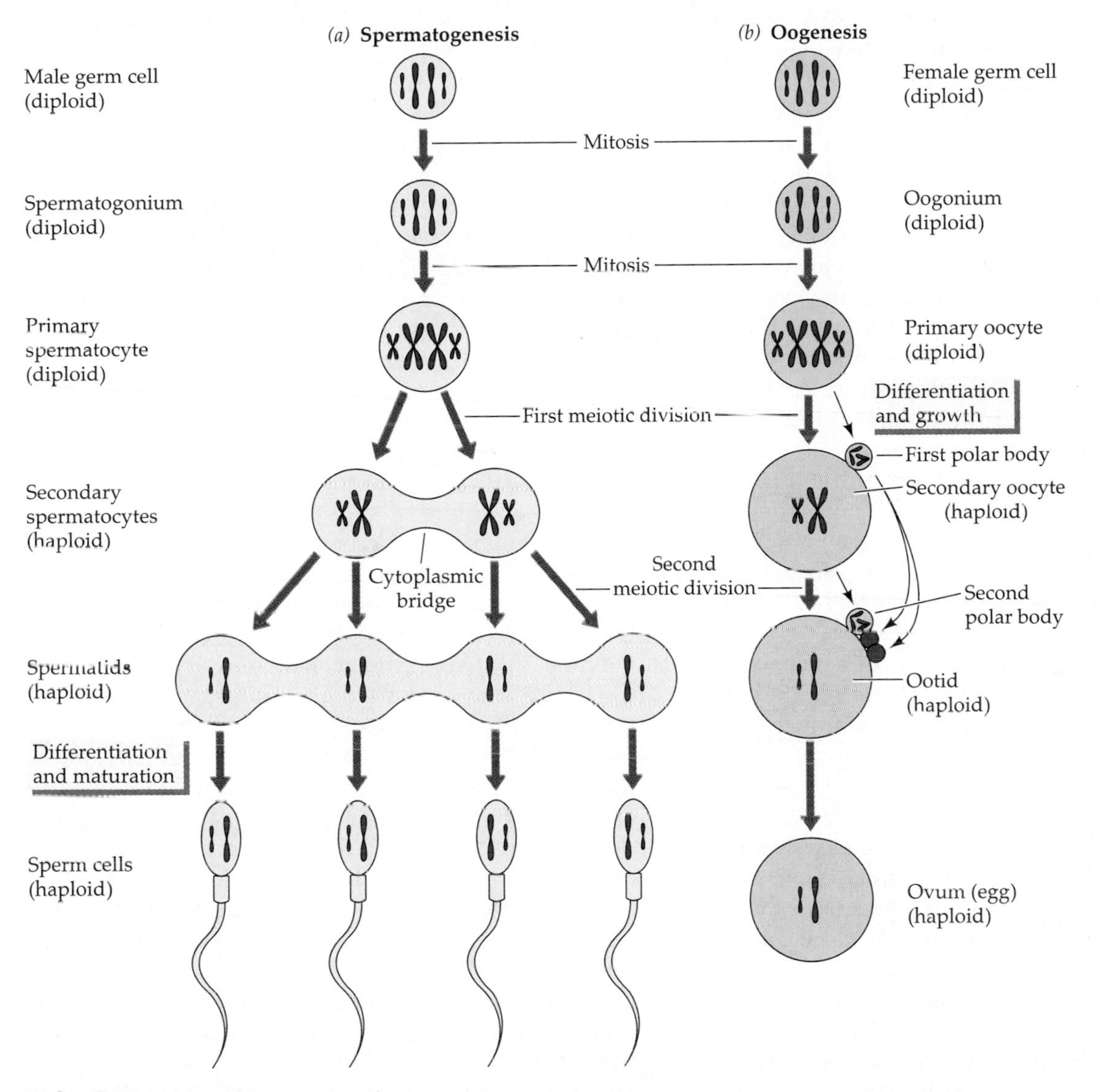

37.3 Gametogenesis
(a) The formation of haploid spermatids from diploid spermatogonia. Spermatids, all of which are different genetically, will differentiate into sperm. A mature mammalian sperm cell is seen in Figure 37.2 and diagrammed in Figure 37.6. *(b)* Diploid oogonia develop into larger primary oocytes that grow and accumulate materials and energy. The first meiotic division produces a haploid secondary oocyte and a small, adjacent, nucleus-containing polar body. The second meiotic division produces another polar body (the first polar body may also divide at this time) and the haploid egg.

gressively from the outermost region of the seminiferous tubule toward the center. The fully differentiated sperm are finally released from the Sertoli cells. The entire process takes about ten weeks. Each day a human male produces about 30 million sperm.

Oogenesis is the process of meiosis and development of the oogonia into eggs. Some oogonia develop into **primary oocytes,** which enter the first meiotic division but arrest in prophase I (Figure 37.3*b*). During this arrest, the primary oocytes enlarge, gaining yolk, ribosomes, cytoplasmic organelles, and energy stores. They accumulate rRNA, mRNA, and tRNA, as well as materials from the blood, and they form follicle cells, which surround them. Many of the lipids and proteins stored by vertebrate ova are made in the liver and transported to the ovaries in the bloodstream. The primary oocyte acquires all of the energy, raw materials, and RNA that the egg needs to survive its first cell divisions after fertilization.

Each month during a human female's fertile years, at least one primary oocyte comes out of its resting stage and matures into an egg. As this primary oocyte resumes meiosis, the nucleus completes its first meiotic division near the surface of the cell. The daughter cells of this division receive a grossly unequal share of the cytoplasm of the primary oocyte. One receives almost all of the cytoplasm and becomes the **secondary oocyte,** and the other receives almost none and forms the **first polar body** (see Figure 37.3*b*). The second meiotic division of the large secondary oocyte is also accompanied by an asymmetric division of the cytoplasm. One daughter cell forms the large, haploid **ootid,** which eventually differentiates into an ovum, and the other forms the **second polar body.** The polar bodies eventually degenerate, so the end result of oogenesis is one very large, haploid ovum that is well provisioned for the rapid divisions of the cleavage stage of development.

Sex Types

Most animals are a distinct sex type—male or female. Species having male and female members are called **dioecious** (from the Greek for "two houses"). By contrast, a single individual of some other species may possess both female and male reproductive systems. Such species are called **monoecious** ("one house") or **hermaphroditic** (from the name of a male, Hermaphroditus, whose body, according to Greek myth, was joined with that of a nymph). Almost all invertebrate groups have hermaphroditic species. An earthworm is an example of a **simultaneous hermaphrodite,** meaning that it is both male and female at the same time. When two earthworms mate, both are fertilized and produce offspring. Some animals are **sequential hermaphrodites,** being male and female at different times in their life cycles. As we learned in Chapter 36, the term hermaphroditism can also be used to describe developmental abnormalities in normally dioecious species that give rise to individuals that have both male and female sex organs.

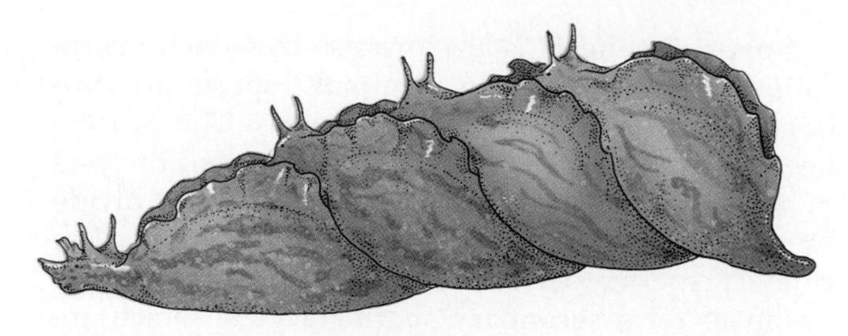

37.4 Hermaphroditic Mating
Although some hermaphroditic species can fertilize themselves, most must mate with another individual; the sea slug *Aplysia* often mates in groups. Here sea slugs form a mating chain in which each animal is functioning as a female for the animal behind and as a male for the animal in front.

Some simultaneous hermaphrodites have a low probability of meeting a potential mate. An example is a parasitic tapeworm. Even though it may be quite large and cause lots of trouble, it may be the only tapeworm in your intestine. Some simultaneous hermaphrodites, such as the tapeworm, can fertilize themselves, but most must mate with another individual (Figure 37.4).

Sequential hermaphroditism confers different advantages on different species. It can reduce the possibility of inbreeding among siblings by making them all the same sex at the same time and therefore incapable of mating with one another. In a species in which only a few males fertilize all the females, sequential hermaphroditism can maximize reproductive success by making it possible for an individual to reproduce as a female until the opportunity arises for it to function as a male. An excellent example is the tropical Pacific fish *Labroides dimidiatus.* All individuals of this species are born female. The population consists of social groups described as harems; each harem consists of three to six females controlled by one male. The male defends the territory of the group from intruders. If the male dies or is removed, the largest, most dominant female in the group changes sex and becomes a functional male, assuming control of the group.

Getting Eggs and Sperm Together

Fertilization can be external or internal. Sexually reproducing animals may release their gametes into the environment, where the meeting of gametes results in fertilization, or the male gametes may be inserted into the female's reproductive tract, where fertilization occurs. Animals that fertilize externally reproduce in aquatic habitats where gametes are not in danger of drying out. External fertilization is the more common pattern among simpler animals, especially those that are sessile.

External fertilization favors the evolution of traits that increase the probability that male and female gametes will meet. One simple adaptation is the production of huge numbers of gametes. A female oyster, for example, may produce 100 million eggs per year, and the number of sperm produced by a male oyster is astronomical. Numbers alone do not guarantee that gametes will meet, however. Also crucial are mechanisms of timing that synchronize the reproductive activities of the males and the females of a population. Seasonal breeders may use photoperiod cues, changes in temperature, or changes in the weather to time their production and release of gametes.

Sexual behavior plays an important role in bringing gametes together. Many species travel great distances to congregate with potential mates and release their gametes at the same time in a suitable environment. An excellent example is the remarkable migration of salmon. These fish hatch and go through juvenile stages in fresh water. They then migrate to the ocean, where they live and grow for three to five years. When finally ready to breed (spawn), they migrate back to the stream in which they were hatched, where they spawn and die.

Because gametes released into a dry environment die quickly, internal fertilization is a major adaptation for terrestrial life. Many aquatic species also practice internal fertilization. A great advantage of internal fertilization is the protection it provides for the early developmental stages of the organism. Animals have evolved an incredible diversity of sexual behaviors and accessory sex organs that facilitate internal fertilization. In general, a tubular structure, the **penis,** enables the male to deposit sperm in the female's accessory sex organ, the **vagina,** or in some species, the **cloaca** (plural: cloacae; a cavity common to the digestive, urinary, and reproductive systems).

Copulation is an act that permits sperm to move directly from the male's reproductive system into the female's reproductive system. The transfer of sperm can also be indirect. The males of some species of mites and scorpions (among the arthropods) and salamanders (among the vertebrates) deposit **spermatophores**—containers filled with their sperm—in the environment. When a female mite finds a spermatophore, she straddles it and opens a pair of plates in her abdomen so that the tip of the spermatophore enters her reproductive tract and allows the sperm to enter. Some female salamanders use the lips of their cloacae to scoop up the portion of the gelatinous spermatophore containing the sperm.

Male squids and spiders play a more active role in spermatophore transfer. The male spider secretes a drop containing sperm into a bit of web; then, with a special structure on a foreleg, he picks up the sperm-containing web and inserts it through the female's genital opening. The male squid uses one special tentacle to pick up a spermatophore and insert it into the female's genital opening. In the process, the tip of his tentacle may break off and remain in the female's body along with the sperm.

Most male insects copulate and transfer spermatophores to the female's vagina through a tubular penis. The genitalia (external parts of the sex organs) of insects often have species-specific shapes that match in a lock-and-key fashion (Figure 37.5). The incredible morphological diversity in genitalia of some groups of species has led to the hypothesis that the lock-and-key fit is a reproductive isolating mechanism (see Chapter 19). This idea is controversial, but at a minimum, the fit between male and female genitalia assures a tight, secure fit between the mating pair during the prolonged period of sperm transfer. The males of some insect species use elaborate structures on their penises to scoop out the female's reproductive tract, removing sperm deposited there by other males. Following this cleaning, a male transfers his own sperm into the tract.

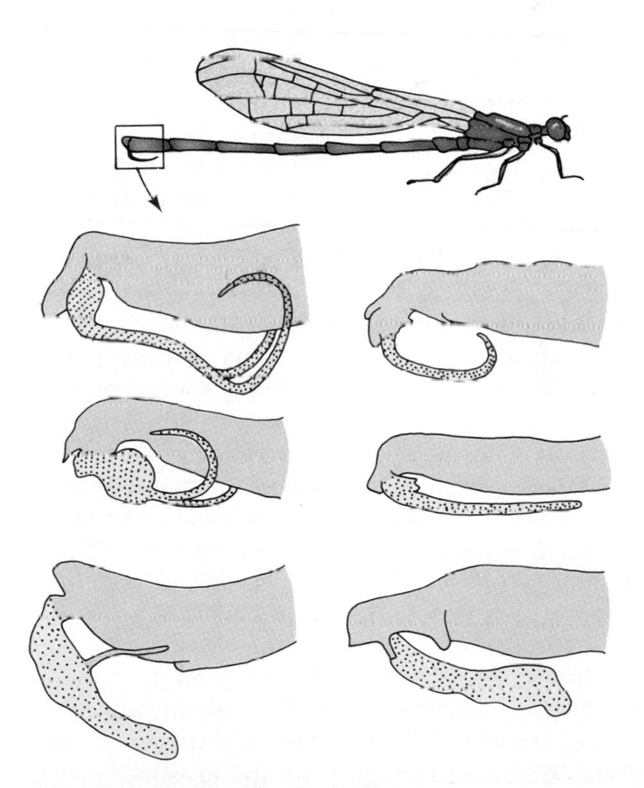

37.5 Species-Specific Penis Shapes
The penises of six species of *Argia*, a genus of damselflies, are remarkably different. Each fits into the corresponding female organ as a key in a lock to facilitate a tight union during a prolonged copulation. The speckled structures are used to clear the female reproductive tract of sperm deposited by other males.

REPRODUCTIVE SYSTEMS IN HUMANS AND OTHER MAMMALS

So far we have seen only a small sampling of the fascinating diversity of animal reproductive systems. We will now look at the mammalian reproductive system in greater depth, using the human as our model. Many of the details are the same for other vertebrates.

The Male

The paired testes of mammals, except those of bats, elephants, and aquatic mammals, are lodged outside the body cavity in a pouch of skin, the **scrotum** (Figure 37.6*a*). Spermatogenesis takes place in most mammals only at a temperature slightly lower than normal body temperature. The scrotum keeps the testes at a temperature optimal for spermatogenesis. Muscles in the scrotum contract in a cold environment, bringing the testes closer to the warmth of the body; in a hot environment they relax, and the testes are suspended farther from the body.

A testis consists of tightly coiled **seminiferous tubules** within which spermatogenesis takes place. The tubule walls are lined with spermatogonia. In going from the tubule wall toward the center, you find germ cells in successive stages of spermatogenesis (see Figure 37.3*a*). Fully differentiated spermatids are shed into the lumen of the tubule. These germ cells are intimately associated with **Sertoli cells**, which nurture them (Figure 37.6*b*). Between the seminiferous tubules are clusters of cells that produce the male sex hormones.

Just after being produced by meiosis, a spermatid bears little resemblance to a sperm. As it differentiates into a sperm, its nucleus becomes compact, its motile flagellum develops into a tail, and most of its cytoplasm is lost. As the head of the sperm forms, it is capped by an **acrosome,** which contains enzymes that will enable the sperm to digest its way into an egg. Between the head and tail of the mature sperm is a midpiece containing two centrioles and mitochondria to provide energy for locomotion (bottom of Figure 37.6). The microtubules that extend from the centrioles into the flagellum have the same pattern as in all typical eukaryotic flagella, the standard 9 + 2 arrangement described in Chapter 4.

From the seminiferous tubules, sperm move into a storage structure called the **epididymis,** where they mature and become motile. The epididymis connects to the **urethra** by a tube called the **vas deferens.** The urethra comes from the bladder, runs through the penis, and opens to the outside of the body at the tip of the **penis**. The urethra is the common duct for the urinary and reproductive systems (see Figure 37.6*a*).

The shaft of the penis is covered with normal skin, but the tip, or **glans penis,** is covered with thinner, more sensitive skin that is especially responsive to sexual stimulation. A fold of skin called the foreskin covers the glans of the human penis. The practice of circumcision removes a portion of the foreskin. There is no strong rationale or justification for circumcision based on health, yet it remains a cultural or religious tradition for many people.

The penis becomes hard and erect during sexual arousal because blood fills shafts of spongy tissue that run the length of the penis (see Figure 37.6*a*). The presence of this blood creates pressure and closes off the vessels that normally drain the penis. Thus, the penis becomes engorged with blood, facilitating insertion into the vagina. Some species of mammals, but not humans, have a bone in the penis; even those species, however, depend on erectile tissue for copulation.

The culmination of the male sex act propels sperm through the vas deferens and the urethra. This process of sperm movement has two steps, **emission** and **ejaculation.** During emission, sperm and the secretions of several accessory glands move into the urethra at the base of the penis. Together, the sperm and these secretions constitute **semen,** the fluid that is ejaculated into the female's vagina. About 60 percent of the volume of the semen is seminal fluid, which comes from the **seminal vesicles.** Seminal fluid is thick because it contains mucus and protein. It also contains fructose, which serves as an energy reserve for the sperm, and modified fatty acids called prostaglandins that stimulate contractions in the female reproductive tract. Another source of secretions is the **prostate gland,** which produces a thin, milky, alkaline fluid. The prostate fluid helps to neutralize the acidity of the urethra and the female reproductive tract to create a favorable environment for the sperm. Prostate fluid also contains a clotting enzyme that works on the protein in the seminal fluid to convert the semen into a gelatinous mass.

Ejaculation, which follows emission, is caused by wavelike contractions of muscles at the base of the penis surrounding the urethra. The rigidity of the erect penis allows these contractions to force the semen through the urethra and out of the body.

The Female

Eggs (ova) are produced and released by the female gonads, the ovaries. The ovaries are paired structures in the lower part of the body cavity. Ovulation releases an egg from the ovary directly into the body cavity. Before it can float away, the released egg is swept into the fringed end of one of the paired tubes called **oviducts** (also known as fallopian tubes). Cilia lining the oviduct propel the egg slowly toward the

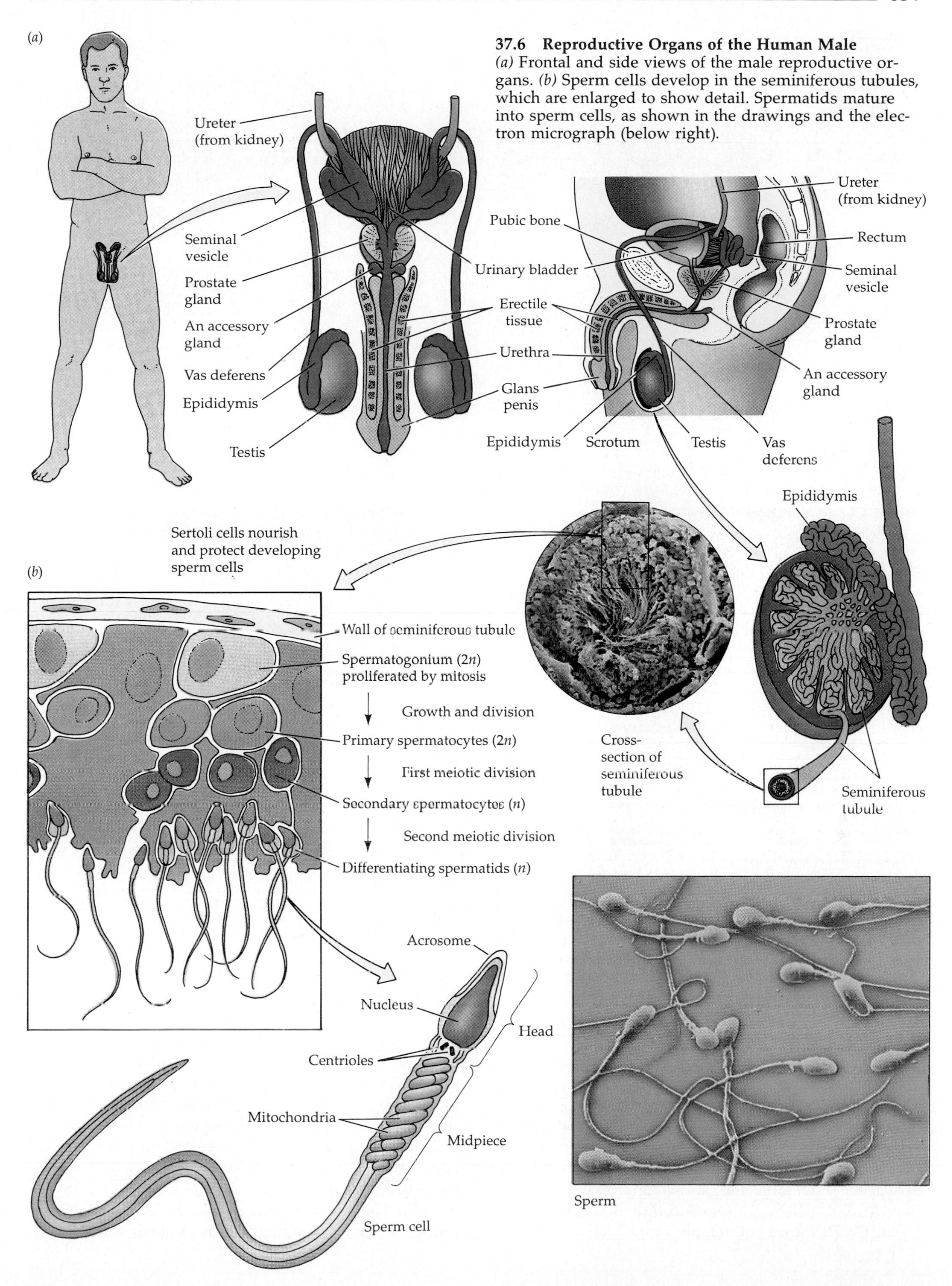

37.6 Reproductive Organs of the Human Male *(a)* Frontal and side views of the male reproductive organs. *(b)* Sperm cells develop in the seminiferous tubules, which are enlarged to show detail. Spermatids mature into sperm cells, as shown in the drawings and the electron micrograph (below right).

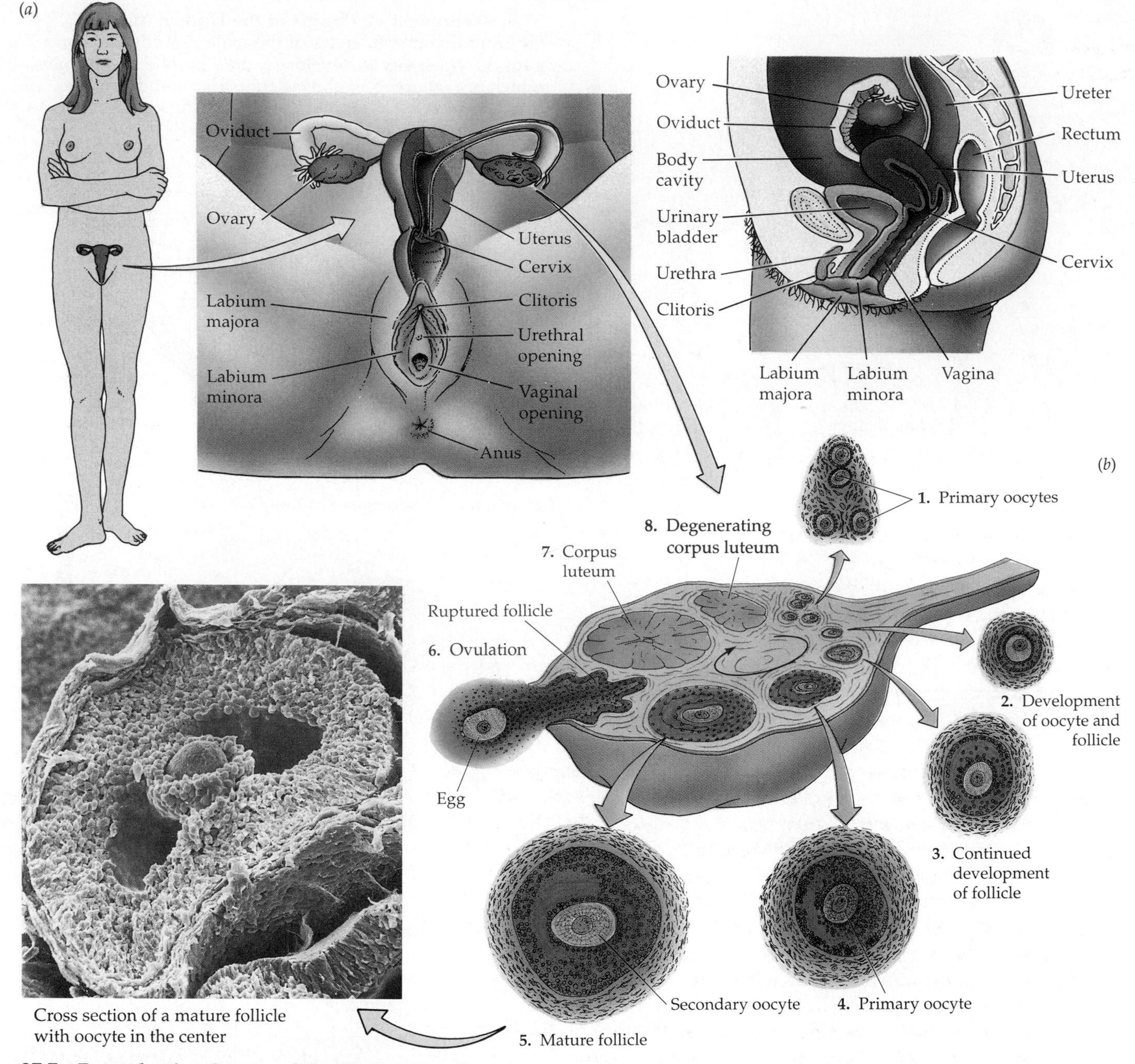

37.7 Reproductive Organs of the Human Female (*a*) Frontal and side views of the female reproductive organs. (*b*) The developing follicle in the ovary. The progression of stages in the ovary is (1–5) development of a follicle, (6) ovulation, and (7–8) growth and degeneration of the corpus luteum. The micrograph shows a mature mammalian follicle; the oocyte is in the center.

uterus, or womb, which is a muscular, thick-walled cavity shaped like an upside-down pear. Babies develop in the uterus. At the bottom of the uterus is an opening, the **cervix,** that leads into the **vagina.** Sperm are ejaculated into the vagina during copulation, and the baby passes through it during birth. Figure 37.7*a* shows the female reproductive organs.

Two sets of skin folds surround the opening of the vagina and the opening of the urethra, through which urine passes. The inner, more delicate folds are the **labia** (singular: labium) **minora** and the outer, thicker folds are the **labia majora.** At the anterior tip of the labia minora is the **clitoris,** a small bulb of erectile tissue that is the anatomical homolog of the penis. The clitoris is highly sensitive and plays an important role in sexual response. The labia minora and the clitoris consist of erectile tissue and become engorged during sexual excitation. The opening of an infant female's vagina is covered by a thin membrane, the hymen, which has no known function. It is eventually ruptured by vigorous physical activity or first intercourse, but it can make first intercourse difficult or painful for the female.

To fertilize an egg, sperm swim from the vagina up through the cervix, the uterus, and most of the oviduct. The egg is fertilized in the upper region of the oviduct. The resulting zygote undergoes its first cell divisions, becoming a **blastocyst,** as it continues to move down the oviduct. When the blastocyst reaches the uterus it implants itself in the uterine lining, the **endometrium.** Some of the tissues of the blastocyst interact with the endometrium to form a structure called the **placenta,** which exchanges nutrients and waste products between the mother's blood and the baby's blood.

The Ovarian Cycle

At birth, a female has about a million primary oocytes in each ovary. By the time she reaches puberty (sexual maturity), she has only about 200,000 primary oocytes in each ovary; the rest have degenerated. During a woman's fertile years, only about 450 of these oocytes will mature completely into eggs and be released. When she is about 50 years old, she reaches **menopause,** the end of fertility. Only a few oocytes are then left in each ovary. Throughout a woman's life, oocytes are degenerating, and no new ones are produced.

A layer of cells surrounds each egg in the ovary. These cells, together with the eggs, constitute the functional unit of the ovary, the **follicle** (Figure 37.7*b*). Between puberty and menopause, 6 to 12 follicles mature within the ovaries of a human female each month. In each of these follicles, the egg enlarges and the surrounding cells proliferate. After about a week one of these follicles is larger than the rest and continues to grow, while the others cease to develop and shrink. In the remaining follicle, the follicular cells nurture the growing egg, supplying it with nutrients and even with macromolecules that it will use in early stages of development if it is fertilized. After two weeks of growth, the follicle ruptures and releases an egg. Following **ovulation,** as this release is called, the follicular cells continue to proliferate and form a mass of endocrine tissue about the size of a marble. This structure, which remains in the ovary, is the **corpus luteum.** It functions as an endocrine gland, producing estrogen and progesterone for about two weeks. It then degenerates unless the egg meets a sperm and is fertilized. We will return to the corpus luteum later in this chapter.

The Menstrual Cycle

Ovulation is part of the regular reproductive cycle in female animals. The human reproductive cycle is called the menstrual cycle because it ends conspicuously with **menstruation,** the sloughing off of the endometrium, the uterine lining. This sloughed-off tissue and blood from the uterine wall are lost through the vagina. The menstrual cycle consists of two coordinated cycles, one in the ovary, which results in the release of an egg each month, and one in the uterus, which prepares the endometrium to receive a blastocyst. The human reproductive cycle has a period of about 28 days or one month (a synonym for menstruation is menses, the Latin word for "months.") Some mammals have shorter ovarian cycles and others have longer ones. Rats and mice have ovarian cycles of about four days; many other mammalian species have only one cycle per year.

Most mammals do not end their cycles with menstruation; instead, the uterine lining is reabsorbed rather than being sloughed off. In these species the reproductive cycle is called the **estrous cycle** because its most striking event is the sexual receptivity of the female at the time of ovulation, called estrus, or "heat." When the female comes into estrus, she actively solicits male attention and may be aggressive to other females. She attracts males by releasing chemical signals as well as through behavior. The human female is unusual among mammals in that she is potentially sexually receptive throughout her reproductive cycle and at all seasons of the year.

Hormonal Control of the Menstrual Cycle

The ovarian and uterine cycles of human females are coordinated and timed by hormones. Gonadotropins secreted by the anterior pituitary are the central elements of this control. Prior to puberty, the secretion of gonadotropins is low, and the ovaries are inactive. At puberty, the hypothalamus increases its release of gonadotropin-releasing hormone, thus stimulating the anterior pituitary to secrete follicle-stimulating hormone and luteinizing hormone (Figure 37.8*a*). In response to these two gonadotropins, ovarian tissue grows and produces estrogen, and the follicles go through early stages of maturation. The rise in estrogen causes the development of secondary sexual characteristics, including the maturation of the uterus. Between puberty and menopause (at which time menstrual cycles cease), the interactions of gonadotropin-releasing hormone, the gonadotropins, and the sex steroids control the reproductive cycle.

Menstruation (menses) marks the beginning of the uterine and ovarian cycles (Figure 37.8*b–d*). A few days before menstruation begins, the anterior pituitary begins to increase its secretion of follicle-stimulating hormone and luteinizing hormone. In response to these gonadotropins, follicles mature in the ovaries and estrogen levels rise slowly. After about a week of growth, all but one of these follicles wither away.

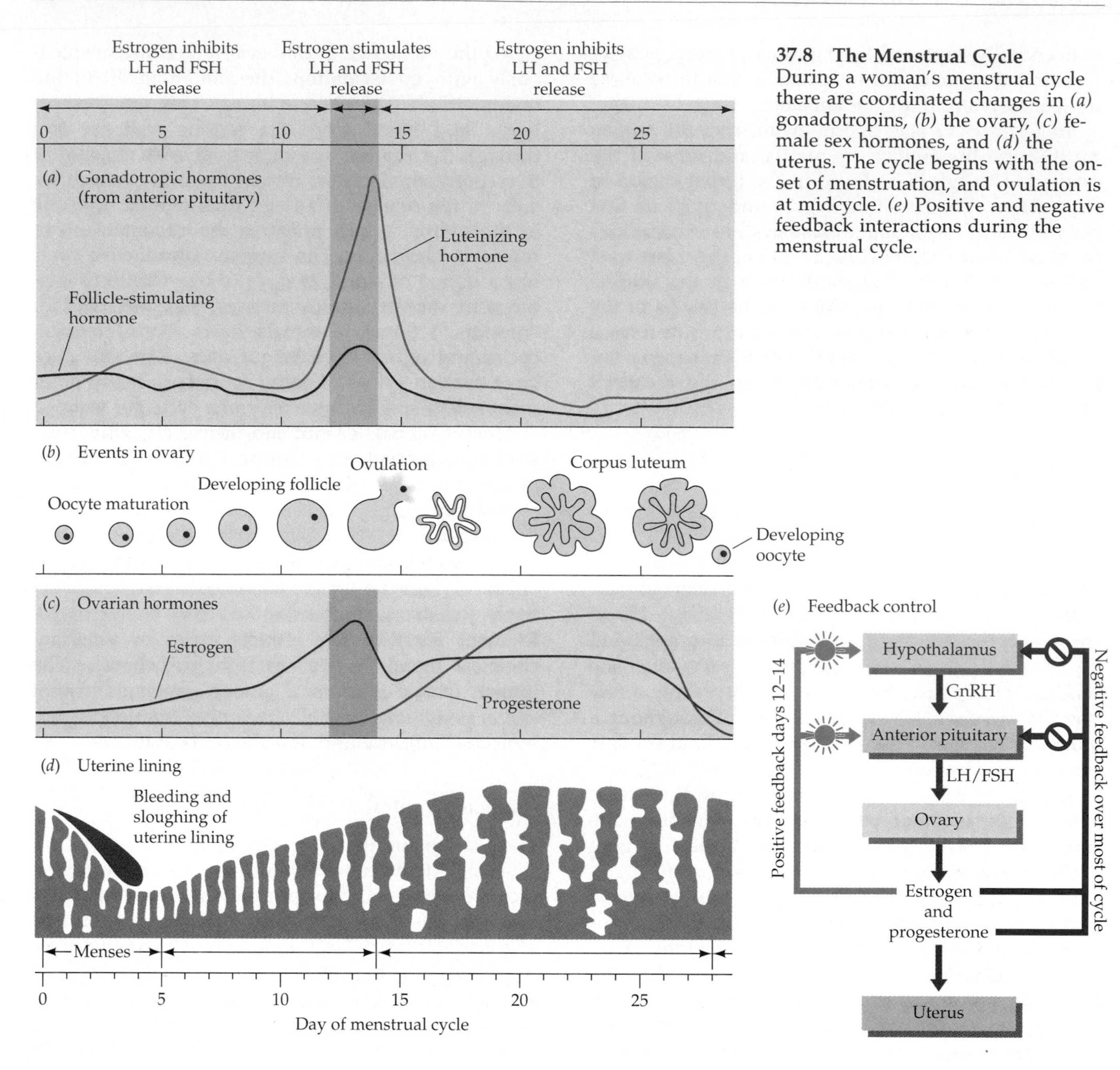

37.8 The Menstrual Cycle During a woman's menstrual cycle there are coordinated changes in *(a)* gonadotropins, *(b)* the ovary, *(c)* female sex hormones, and *(d)* the uterus. The cycle begins with the onset of menstruation, and ovulation is at midcycle. *(e)* Positive and negative feedback interactions during the menstrual cycle.

The one follicle that is still growing secretes increasing amounts of estrogen, stimulating the endometrium to grow. Estrogen exerts a negative feedback on gonadotropin release by the pituitary during the first 12 days of the cycle. Then, on about day 12, estrogen exerts a positive rather than a negative feedback on the pituitary (Figure 37.8*e*). As a result, there is a great surge of luteinizing hormone and, to a lesser extent, follicle-stimulating hormone. The luteinizing hormone surge triggers the mature follicle to rupture and release the egg, and it stimulates the follicle cells to develop into the corpus luteum and to secrete estrogen and progesterone.

Estrogen and especially progesterone secreted by the corpus luteum following ovulation are crucial to the continued development and maintenance of the endometrium. In addition, these sex steroids have negative feedback on the pituitary, inhibiting gonadotropin release and thus preventing new follicles from beginning to mature. If the egg is not fertilized, the corpus luteum degenerates on or about day 26 of the cycle. Without the production of steroids by the corpus luteum, the endometrium sloughs off. The decrease in circulating steroids also relieves the negative feedback on the hypothalamus and pituitary so that gonadotropin-releasing hormone, follicle-stimulating hormone, and luteinizing hormone all increase. The increase in these hormones induces the next round of follicles to develop, and the cycle begins again.

If the egg is fertilized, a zygote is created. The zygote undergoes numerous cell divisions, becoming a blastocyst as it travels down the oviduct. When the blastocyst arrives in the uterus and implants in the endometrium, a new hormone comes into play. A layer of cells covering the blastocyst secretes **human chorionic gonadotropin,** which keeps the corpus luteum functional. These same tissues also produce estrogen and progesterone. Eventually these tissues derived from the blastocyst take over for the corpus luteum. Continued high levels of estrogen and progesterone prevent the pituitary from secreting gonadotropins, and the ovarian cycle ceases for the duration of the pregnancy. This same mechanism is exploited by birth control pills, which contain synthetic hormones resembling estrogen and progesterone that prevent the ovarian cycle through negative feedback to the hypothalamus and pituitary.

Human Sexual Responses

The sexual responses of both women and men consist of four phases: excitement, plateau, orgasm, and resolution. As sexual excitement begins in a woman, her heart rate and blood pressure rise, muscular tension increases, her breasts swell, and her nipples become erect. Her external genitals, including the sensitive clitoris, swell as they become filled with blood, and the walls of the vagina secrete lubricating fluid that facilitates copulation.

As a woman's sexual excitement increases, she enters the plateau phase. Her blood pressure and heart rate rise further, her breathing becomes rapid, and the glans and shaft of the clitoris begin to retract—the greater the excitement, the greater the retraction. The sensitivity that once focused in the clitoris spreads over the external genitals, and the clitoris itself becomes even more sensitive.

Orgasm begins with a contraction of the outer third of the vagina lasting two to four seconds, followed by shorter contractions approximately one second apart. Orgasm may last as long as a few minutes, and, unlike men, some women can experience several orgasms in rapid succession. During the resolution phase, blood drains from the genitals and body physiology returns to close to normal. The resolution phase lasts approximately five to ten minutes after orgasm; if she does not experience orgasm, a woman's resolution phase may take 30 minutes or longer.

The cycle of a man's sexual responses is very similar to that of the woman. The excitement phase is marked by an increase in blood pressure, heart rate, and muscle tension. The penis fills with blood and becomes hard and erect. In the plateau phase, breathing becomes rapid, the diameter of the glans increases, and a clear lubricating fluid oozes from the penis. The testes also swell and the scrotum tightens. Pressure and friction against the nerve endings in the glans and in the skin along the shaft of the penis eventually trigger orgasm. Massive spasms of the muscles in the genital area and contractions in the accessory reproductive organs result in ejaculation. Within a few minutes after ejaculation, the penis shrinks to its normal size, and body physiology returns to resting conditions.

Unlike the sexual response of a female, the male sexual response includes a refractory period immediately after orgasm. During this period, which may last 20 minutes or more, a man cannot achieve a full erection or another orgasm, regardless of the intensity of sexual stimulation. Figure 37.9 shows the male and female response cycles.

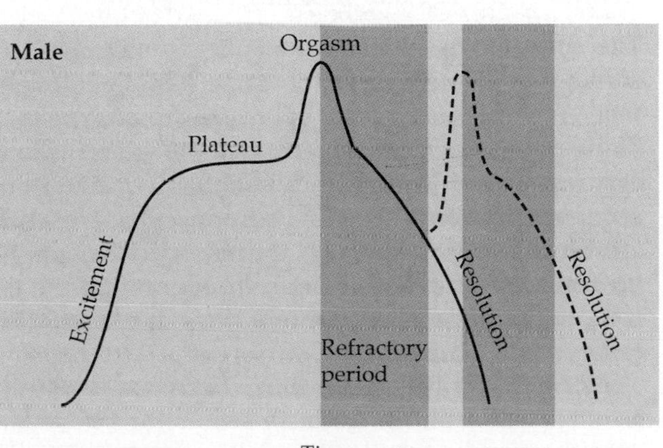

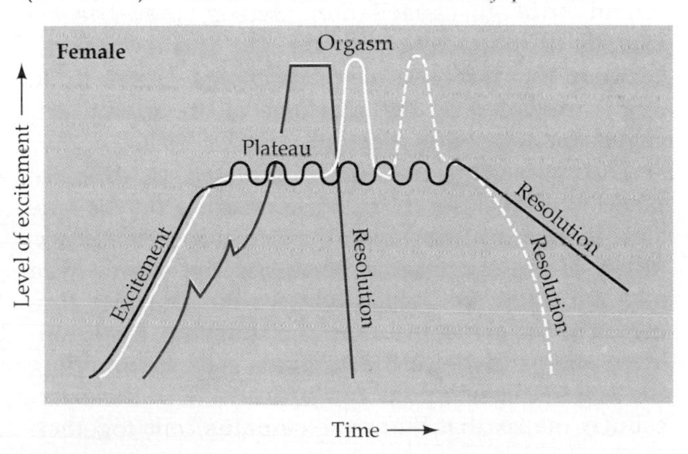

37.9 Human Sexual Responses
The dashed lines show that both males and females may have repeated orgasms, but in the male they are separated by refractory periods during which sexual excitement cannot be maintained. Females have a greater diversity of response cycles, as shown by the three sets of lines. The cycle most similar to that of the male is shown in white. Alternatively, a female may experience sustained multiple orgasms (black curve) or may omit the plateau phase in a surge toward a very intense orgasm (red curve). Females do not have refractory periods.

FERTILIZATION

As you learned in Chapter 17, the union of sperm and egg, or fertilization of the egg, results in a diploid zygote and initiates the development of the embryo. Fertilization is not a single event, but a complex series of processes. It begins with the juxtaposition of sperm and egg, accomplished in most species by sexual behavior. The final distance between sperm and egg must be bridged by the motility of sperm because eggs are universally unable to move. When egg and sperm finally meet, several events take place in sequence: the sperm is activated, the sperm gains access to the plasma membrane of the egg, sperm and egg membranes fuse, and the egg is activated. Egg activation sets up blocks to entry by additional sperm, stimulates the final meiotic division of the egg nucleus, and initiates the first stages of development. The last event of fertilization is fusion of the egg and sperm nuclei to create the diploid nucleus of the zygote. We will now look at each of these steps.

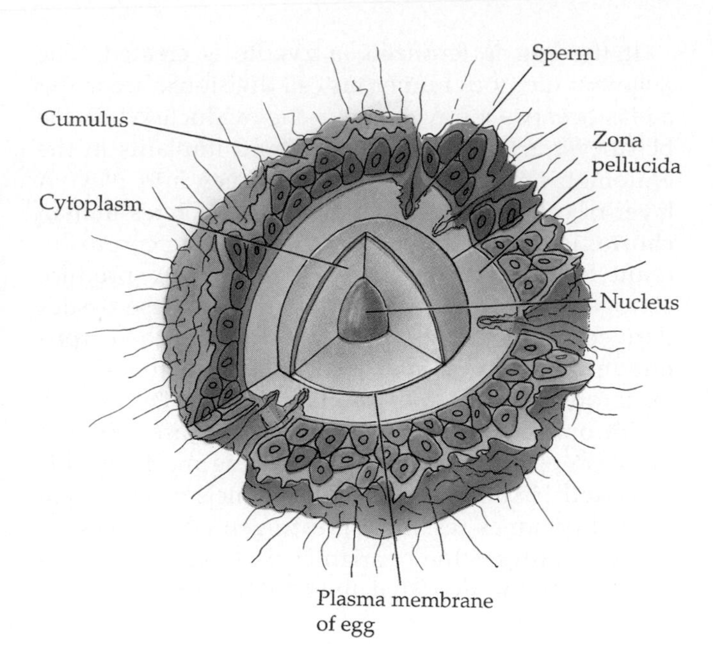

37.10 Barriers to a Sperm Cell
The human egg has several layers that the sperm cell must penetrate to reach the nucleus and fertilize the egg. Both the cumulus and the zona pellucida must be penetrated before the sperm can come into contact and eventually fuse with the egg's plasma membrane.

Activating the Sperm

Mammalian sperm face a formidable task after they are ejaculated into the female's reproductive tract. They must swim up from the vagina, through the uterus, and into the oviducts, where they might find an egg. They are aided in their journey by waves of muscular contractions of the vagina that are part of the female response to sexual stimulation and by stimulation from prostaglandins in the semen. Sperm can reach the upper ends of the oviducts within ten minutes of ejaculation. The mammalian egg, like any other cell, is bounded by a plasma membrane. Immediately surrounding the plasma membrane is a glycoprotein envelope called the **zona pellucida.** Surrounding all of that is a layer called the **cumulus,** consisting of follicle cells in a jelly matrix (Figure 37.10). When sperm are first deposited in the vagina, they are not capable of penetrating all these barriers to fertilize the egg. In the uterine environment, the sperm undergo **capacitation;** that is, they become capable of interacting with the egg and its barriers. Because the response of a capacitated sperm to an egg is mediated by the acrosome of the sperm, it is called the **acrosomal reaction.**

The acrosomal reaction is initiated in different places and at different times depending on the species. In all cases, however, the first step is the breakdown of the membranes bounding the sperm head and the acrosome, which releases the enzymes contained in the acrosome. One such enzyme, **hyaluronidase,** helps disperse the cumulus cells surrounding the egg by digesting the hyaluronic acid in the extracellular matrix that binds the cumulus cells together. Other enzymes released from the acrosome also help disrupt the cumulus. Even though only one sperm fuses with the egg, acrosomal enzymes released from many sperm make the plasma membrane of the egg more accessible.

After the sperm penetrates the cumulus, the sperm head reacts enzymatically with the zona pellucida. The surface of the sperm head contains enzyme molecules and the zona pellucida contains substrate molecules. The enzyme binds to the substrate, linking the sperm to the egg. Acrosomal enzymes then digest a path through the zona pellucida so that the sperm can come into contact and eventually fuse with the plasma membrane of the egg.

Activating the Egg and Blocking Polyspermy

The unfertilized egg is metabolically sluggish, conserving its resources for the early stages of development. The binding of the sperm to the plasma membrane of the egg and the entry of the sperm into the egg activates the egg and initiates a programmed sequence of events. The first responses to fertilization are **blocks to polyspermy,** that is, mechanisms that prevent more than one sperm from entering the egg. If more than one sperm enters the egg, the resulting embryo will probably not survive.

Blocks to polyspermy have been studied intensively in sea urchins. Because sea urchins have large eggs that can be fertilized in dishes of seawater, they are excellent experimental subjects for studying fertilization. Within a tenth of a second after the first sperm enters a sea urchin egg, the egg takes in sodium ions, which changes the electric potential across the egg's plasma membrane. This change prevents the entry of additional sperm and is called the fast block to polyspermy.

There is also a slow block to polyspermy that takes 20 to 30 seconds (Figure 37.11). A sea urchin egg has a membranous structure called a **vitelline envelope,** rather than a zona pellucida, surrounding its plasma membrane. The vitelline envelope is bonded to the plasma membrane and has sperm-binding receptors on its surface. Just under the plasma membrane are cortical vesicles filled with enzymes. The sea urchin egg, like all other animal eggs, contains calcium stored in organelles within the cell. When a sperm enters, the egg releases calcium into its own cytoplasm. The increase in calcium causes the cortical vesicles to fuse with the plasma membrane and release their enzymes, which break the bonds between the vitelline envelope and the plasma membrane. Water then flows by osmosis into the space between the vitelline envelope and the plasma membrane, raising the vitelline envelope away from the plasma membrane to form the **fertilization membrane.** The enzymes from the cortical granules remove the sperm-binding receptors from the surface of the fertilization membrane and cause it to harden, preventing the passage of additional sperm through it.

The release of calcium ions within the egg following fertilization activates the egg metabolically. The pH of the cytoplasm increases, oxygen consumption rises, and protein synthesis increases. The fusion of sperm and egg nuclei does not take place until some time after the sperm enters the egg's cytoplasm—about 1 hour in sea urchins and about 12 hours in mammals. The egg nucleus must complete its second meiotic division before egg and sperm nuclei unite.

Most methods of birth control are focused on events surrounding fertilization. Physical barriers and behavioral changes are used to prevent the meeting of sperm and egg. Hormonal manipulations are employed to disrupt the ovarian cycle and prevent ovulation. Most recently, a chemical means of preventing implantation of the fertilized egg in the endometrium has been developed. Birth control methods and their relative effectiveness are discussed in Box 37.A.

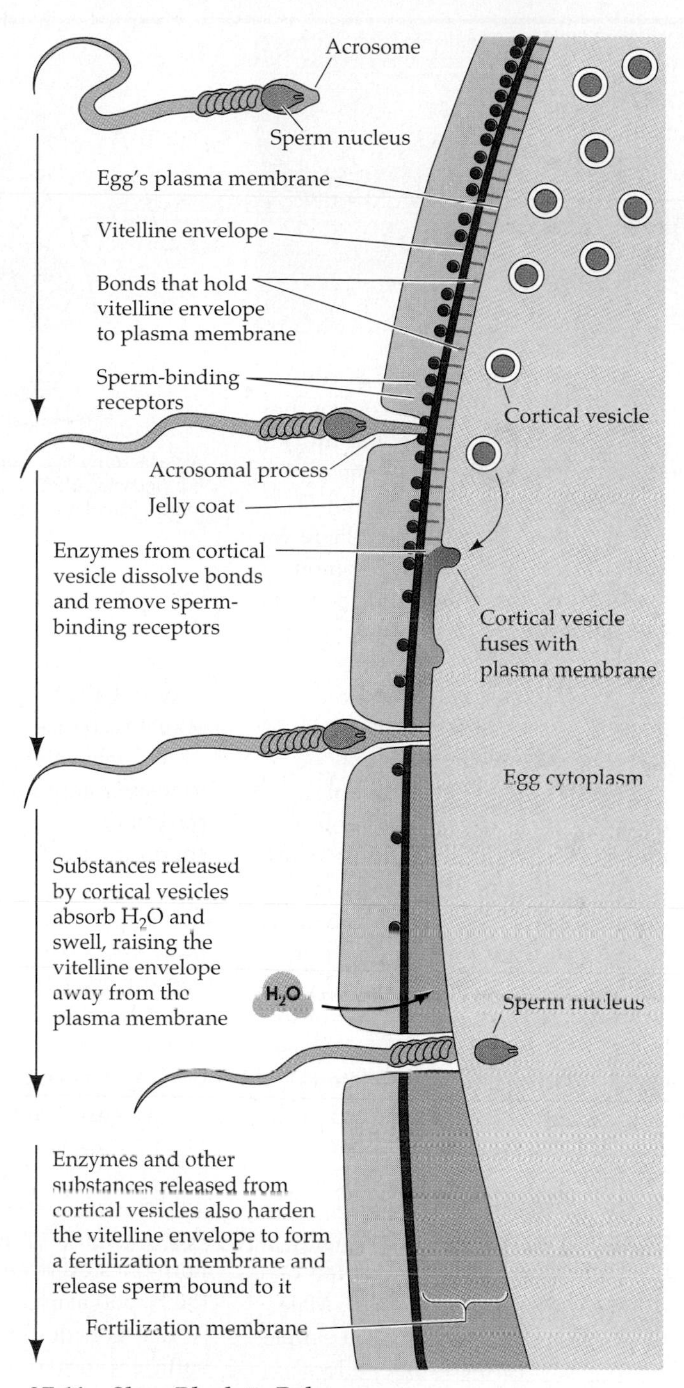

37.11 Slow Block to Polyspermy
As a sea urchin sperm approaches an egg to fertilize it, the contents of cortical vesicles remove the sperm-binding receptors and raise the vitelline envelope, which becomes the fertilization membrane. The fertilization membrane prevents the entry of any other sperm.

CARE AND NURTURE OF THE EMBRYO

After development begins, the embryo requires access to oxygen, removal of carbon dioxide, a continuous source of nutrients, and a suitable physical environment. Two general patterns of care and nurture of the embryo have evolved in animals: oviparity and viviparity.

Oviparity

Oviparous animals lay eggs in the environment; their offspring go through the embryonic stages outside

BOX 37.A

The Technology of Birth Control

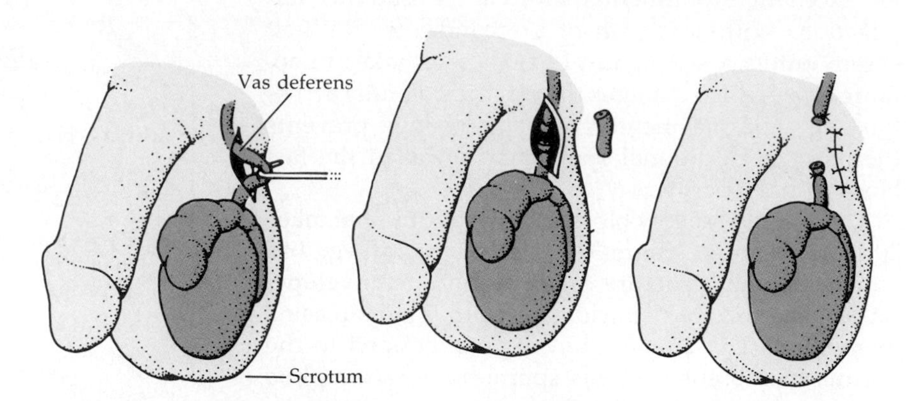

A vasectomy is a minor operation in which each vas deferens is cut and the cut ends are tied closed. The pathway from the testes to the penis is thus interrupted, and the man's ejaculate will not contain sperm.

People use many methods of contraception (birth control) to control the number of their children and the spacing between births. Some of these methods are used by the woman, others by the man. Here we review some of the most common contraception methods and their relative failure rates. Birth control techniques have become very effective, and research continues with the goals of still-greater effectiveness, safety, and convenience.

One of the oldest and simplest methods of contraception is **coitus interruptus,** withdrawal of the penis before ejaculation. The failure rate of this method can be almost as high as 40 percent. Even the few drops of fluid released by the penis during arousal may carry enough sperm to bring about fertilization. Sometimes ejaculation near the vagina allows some sperm to find their way into it, or withdrawal may not be soon enough.

The only certain methods of contraception (at present) are virtually irreversible ones—the **sterilization** of either the man or the woman. Male sterilization by vasectomy is a simple operation performed under a local anesthetic in a doctor's office. As the figure shows, each vas deferens is cut and the cut ends are then tied off. After this minor surgery, the man's ejaculate no longer contains sperm, because sperm cannot pass through the vas deferens after leaving the epididymis, but the operation does not affect his hormone levels or sexual responses.

In female sterilization, the aim is to make it impossible for the egg to travel to the uterus and to block sperm from reaching the egg. The most common method is tubal ligation ("tying the tubes"). A small piece is removed from each oviduct, and the ends of the oviduct are tied off. Alternatively, the oviducts may be burned (cauterized) to seal them off, a process called endoscopy.

The **birth control pill** works by preventing ovulation, so there is no egg to fertilize. The most common pills contain low doses of synthetic estrogens and progesterones. These hormones exert negative feedback on the hypothalamus and the pituitary so that gonadotropin release is not sufficient to permit development of the ovum and the follicle. The ovarian cycle is suspended. On about day 5 of her menstrual cycle, the woman starts taking a pill each day, thus raising her hormone levels. After 20 or 21 days, the pill is discontinued, the lining of the uterus disintegrates, and slight menstrual bleeding occurs. There has been much discussion of negative side effects of oral contraceptives. These side effects include increased risk of blood clot formation, heart attack, and stroke, but they were mostly associated with pills containing higher hormone concentrations than are used in current pills. For pills in use today, there is a very low risk of these side effects, except for women over 35 years who smoke, for whom the risk is significantly greater. Risk of death from using the pill is less than that associated with a full-term pregnancy, and the pill is the most effective method of contraception other than sterilization.

Another highly effective method of contraception (with a failure rate varying from 1 percent to about 7 percent) is the intrauterine device, or **IUD.** The IUD is a small piece of plastic or copper that is inserted in the uterus. The IUD probably works by preventing implantation of the fertilized egg. Complications that can

the mother's body. Eggs are always much larger than sperm because they contain stored nutrients, or yolk, on which the entire course of development depends. Oviparous terrestrial animals, such as reptiles, birds, and insects, coat their eggs with tough, waterproof membranes or shells to keep them from drying out and to protect against predators. The protective coverings of terrestrial eggs must, however, be permeable to oxygen and carbon dioxide.

Oviparous animals may engage in various forms

Common Methods of Birth Control

METHOD	MODE OF ACTION	FAILURE RATE (PREGNANCIES PER 100 WOMEN PER YEAR)
Coitus interruptus	Withdrawal of penis before ejaculation	10–40
Vasectomy	Prevents release of sperm	0.0–0.15
Tubal ligation	Prevents egg from entering uterus	0.0–0.05
"The pill"	Prevents ovulation	0–3
RU486	Prevents development of fertilized egg	0–15
Intrauterine device (IUD)	Prevents implantation of fertilized egg	0.5–6
Condom	Prevents sperm from entering vagina	3–20
Diaphragm/jelly; sponge	Prevents sperm from entering uterus; kills sperm	3–25
Vaginal jelly or foam	Kills sperm; blocks sperm movement	3–30
Rhythm method	Abstinence near time of ovulation	15–35
Douche	Supposedly flushes sperm from vagina	80
(Unprotected)	(No form of birth control)	(85)

arise from its use, including uterine infections and unintended sterility, have led many women to consider other options. Lawsuits against IUD manufacturers and subsequent insurance considerations have resulted in a decline in its use and manufacture in the United States, although it is still widely used in many countries.

Two primary mechanical methods of contraception have been in use for over a century. The **condom** ("rubber," or "prophylactic") is a sheath made of latex or of lamb intestinal material that can be fitted over the erect penis. A condom traps the ejaculate so that sperm do not enter the vagina. Condoms also help prevent the spread of sexually transmitted diseases such as AIDS, syphilis, and gonorrhea. In theory, the use of a condom can be highly effective, with a failure rate near zero; in practice, the failure rate is about 15 percent, because of faulty technique.

The **diaphragm** is a dome-shaped piece of rubber with a firm rim that fits over the woman's cervix and thus blocks sperm from entering the uterus. Smaller than the diaphragm is the **cervical cap,** which fits snugly just over the tip of the cervix. Both are treated first with contraceptive jelly or cream and then inserted through the vagina before intercourse. Failure rates are about the same as for condoms. A device simpler than the diaphragm is the contraceptive vaginal sponge. It is a circular, highly absorbent, polyurethane sponge permeated with a spermicide. Placed in the upper region of the vagina, it blocks, absorbs, and kills sperm. The sponge stays effective for about a day. It is easier to use than the diaphragm but has about the same failure rate.

Used alone, spermicidal foams, jellies, and creams have a failure rate of 25 percent or more. About all that can be said for them is that they are more effective than nothing. Douching (flushing the vagina with liquid after intercourse) is, in spite of popular belief, essentially useless as a method of birth control. Remember that sperm can reach the upper regions of the oviducts ten minutes after ejaculation.

Some people attempt to avoid pregnancy by the **rhythm method.** The couple avoids sex from day 10 to day 20 of the menstrual cycle, when the woman is fertile. The use of a calendar to track the cycle may be supplemented by the basal body temperature method, which is based on the observation that a woman's body temperature drops on the day of ovulation and rises sharply on the day afterward. Other methods of predicting the time of ovulation are under development; significant improvements must be made if the rhythm method's failure rate (between 15 and 35 percent) is to be reduced. Added to the uncertainty of the timing of ovulation is the fact that the ovum remains viable for two to three days and sperm remain viable for up to six days in the female reproductive tract.

A recent addition to birth control technology is a drug, **Ru486,** developed in France. Ru486 is not a contraceptive pill, but a *contragestational* pill. It opposes the actions of progesterone produced by the corpus luteum. Progesterone is essential for maintenance of the uterine lining. If Ru486 is administered (usually with prostaglandins) at the time of the first missed menses after fertilization, it causes the uterine lining to be sloughed off along with the embryo, which is in very early stages of development and implantation.

of protective parental behavior focused on their eggs—nest construction and incubation are good examples—but until the eggs hatch, the embryos are entirely dependent on the nutrients stored in the egg at the time of fertilization. After leaving the protective coverings of the egg, the offspring may receive continuing parental care as it completes its development into a mature organism. Among mammals, only the monotremes—the spiny anteater and the duck-billed platypus—are oviparous (Figure 37.12*a*).

(a)

(b)

37.12 Animals Can be Classified by Where Development Takes Place
(a) Oviparous animals such as the duck-billed platypus lay eggs. *(b)* Marsupials are viviparous even though their infants are born in an extremely premature state. The infants develop in their mother's marsupium, or pouch, as this baby kangaroo (known as a "joey") has done.

Viviparity

Viviparous animals retain the embryo within the mother's body for part of its development. During this time, the embryo depends on nutrients supplied by the mother, not on nutrients stored in the egg. Viviparous animals are said to give birth to "live offspring"—a curious choice of words, because the offspring of oviparous animals are certainly not dead. Most viviparous animals are mammals, and most mammals are viviparous. Viviparous mammals have an enlarged and thickened portion of the female reproductive tract that holds the developing embryo; as you know, this structure is called the uterus.

In marsupials, the order of mammals that includes kangaroos and opossums, the uterus simply holds the embryo and does not have special adaptations to supply it with nutrients. Marsupials are very immature when born. They crawl into a pouch called a marsupium on the mother's belly, attach firmly to the nipple of a mammary gland, and complete their development outside of the mother's body (Figure 37.12*b*).

Mammals other than monotremes and marsupials are called **eutherian mammals.** A distinguishing feature of eutherian mammals is the intimate association of blood supplies of mother and embryo in the placenta. Nutrients pass from mother to embryo and wastes pass from embryo to mother through the placenta. We will discuss the structure of the placenta in the next section.

The eggs of some fishes, amphibians, and reptiles are fertilized internally and then retained within the body of the female until they hatch. The young then leave the mother's body. In such cases, the developing embryos receive all their nutrition from the yolk stored in the eggs, which makes this very different from viviparity. This reproductive pattern is called **ovoviviparity.**

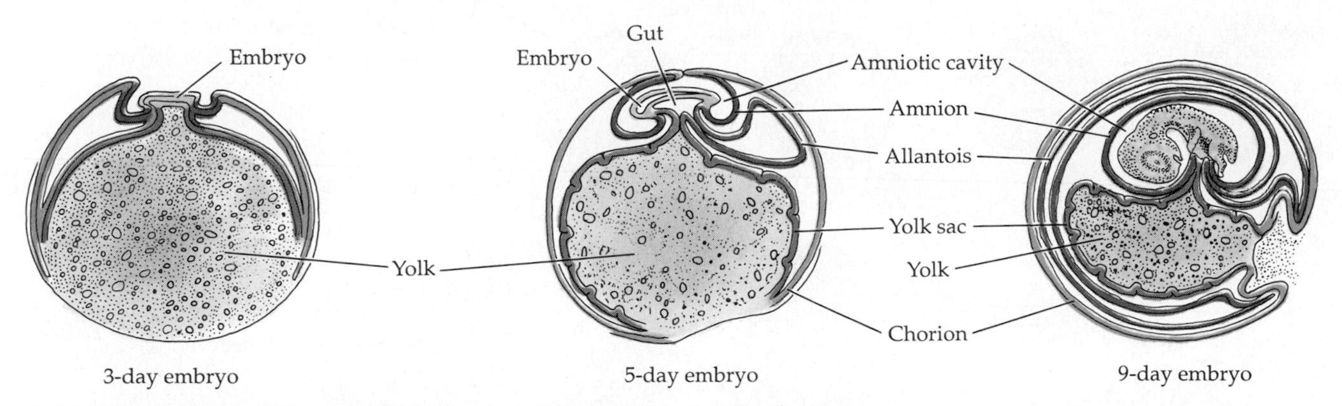

37.13 The Amnion, Allantois, and Chorion: Extraembryonic Membranes These specialized membranes enclose the embryo inside the egg of a chicken. Reptiles and mammals also have these three extraembryonic membranes.

The Extraembryonic Membranes and the Beginning of Development

The embryos of reptiles, birds, and mammals are surrounded by a series of membranes. Anything that reaches the embryo from the environment must pass through these membranes. Figure 37.13 uses the chicken egg to demonstrate how these extraembryonic membranes—amnion, allantois, and chorion—form. The bird embryo starts out as a disc of cells sitting on top of an enormous body of yolk. Early cell division and movements produce the three basic tissue layers of the embryo: the ectoderm, the mesoderm, and the endoderm. Besides making up the body of the embryo, these tissue layers grow out from the embryo to form cavities. The endoderm just over the yolk, along with its associated mesoderm, form a **yolk sac,** which absorbs nutrients from the yolk. Another outgrowth of endoderm and mesoderm becomes the **allantois,** which forms a cavity to receive wastes from the embryo. A growth of ectoderm with associated mesoderm becomes the **amnion,** forming a cavity immediately surrounding the embryo. A more extensive outgrowth of ectoderm and mesoderm becomes the **chorion,** which lines the inside surface of the egg shell.

These extraembryonic membranes are the basic features of the **amniotic egg,** which was a major step in the evolution of reptiles from amphibian ancestors about 300 million years ago. The amniotic egg was the adaptation that freed terrestrial vertebrates from dependence on an aquatic environment for reproduction. Fish or amphibian eggs rapidly dry out if they are exposed to air, but the amniotic egg provides an aqueous environment within which the embryo can develop.

The same extraembryonic membranes found inside birds' eggs also form in mammals. The mammalian blastocyst is a hollow ball of cells that has a central fluid-filled cavity called the blastocoel (Figure 37.14). The first membrane to appear is the chorion; it is apparent by the fifth cell division after fertilization and completely surrounds the blastocyst. It takes more than three days for the human blastocyst to travel down the oviducts to the uterus, where it lives free for the next two to three days. About the sixth day after fertilization, the blastocyst attaches to the lining of the uterus. The chorion plays an important role in implantation by inducing responses in the endometrium. As the blastocyst invades it, the endometrium proliferates and develops more blood vessels. The interaction of the chorion with the wall of the uterus is the beginning of the placenta, which will grow and become the site of exchange of nutrients and wastes between mother and embryo.

A compact inner mass of cells within the blastocyst forms the embryo. As we saw in the bird egg (see Figure 37.13), the amnion surrounds the embryo, creating a fluid filled cavity within which the developing embryo floats. The allantois forms a stalk or cord that connects the embryo with the chorion at the location where the placenta will form. This allantoic stalk becomes the **umbilical cord.** Blood vessels from the embryo grow down the umbilical cord to carry nutrients from and wastes to the placenta. It's easy to see why astronauts taking space walks refer to the cables and the air hoses attaching them to the spacecraft as their umbilical cords.

Cells slough off the embryo and float in the amniotic fluid that bathes the embryo. Later in development, a small sample of the amniotic fluid may be withdrawn with a needle as the first step of a process called **amniocentesis** (Figure 37.15). Some of these cells can be cultured and used for biochemical and genetic analyses that can reveal the sex of the fetus as well as genetic markers for diseases such as cystic fibrosis, Tay–Sachs disease, and Down syndrome.

Amniocentesis usually is not performed until after the fourteenth week of pregnancy, and the tests re-

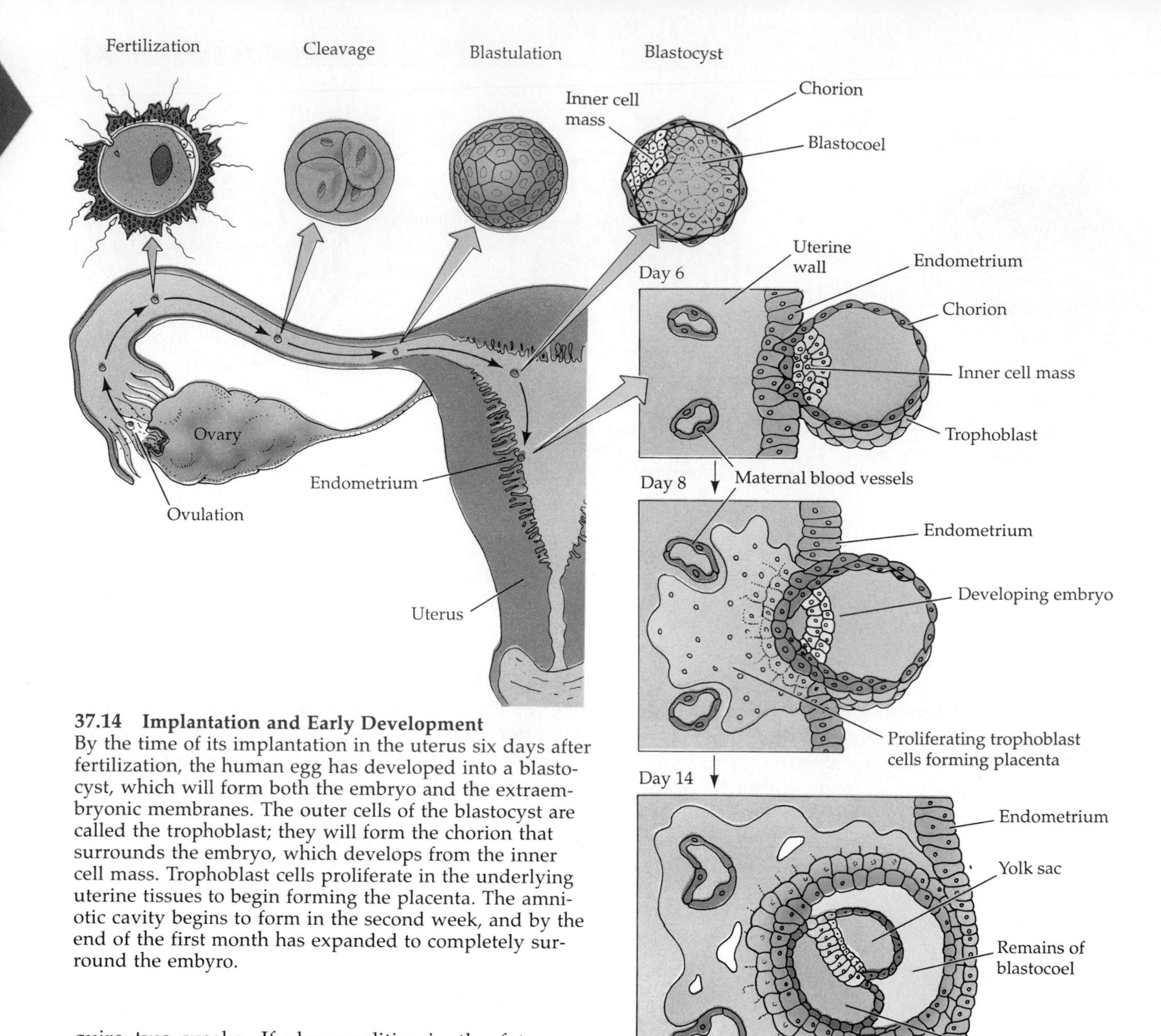

37.14 Implantation and Early Development
By the time of its implantation in the uterus six days after fertilization, the human egg has developed into a blastocyst, which will form both the embryo and the extraembryonic membranes. The outer cells of the blastocyst are called the trophoblast; they will form the chorion that surrounds the embryo, which develops from the inner cell mass. Trophoblast cells proliferate in the underlying uterine tissues to begin forming the placenta. The amniotic cavity begins to form in the second week, and by the end of the first month has expanded to completely surround the embyro.

quire two weeks. If abnormalities in the fetus are detected, termination of the pregnancy at that stage by therapeutic abortion would put the woman's health at greater risk than would an abortion performed earlier. Therefore, a newer technique called **chorionic villus sampling** has been developed. In this test a small sample of the tissue from the surface of the chorion is taken. This test can be done as early as the eighth week of pregnancy, and the results are available in several days.

Pregnancy

Gestation, or pregnancy, is the period from conception (fertilization of egg by sperm) to birth. During gestation the embryo develops in the uterus. In general, the duration of pregnancy in mammals correlates positively with body size; in mice it is about 21 days, in cats and dogs about 60 days, in humans about 266 days, in horses about 330 days, and in elephants about 600 days. In discussing the events

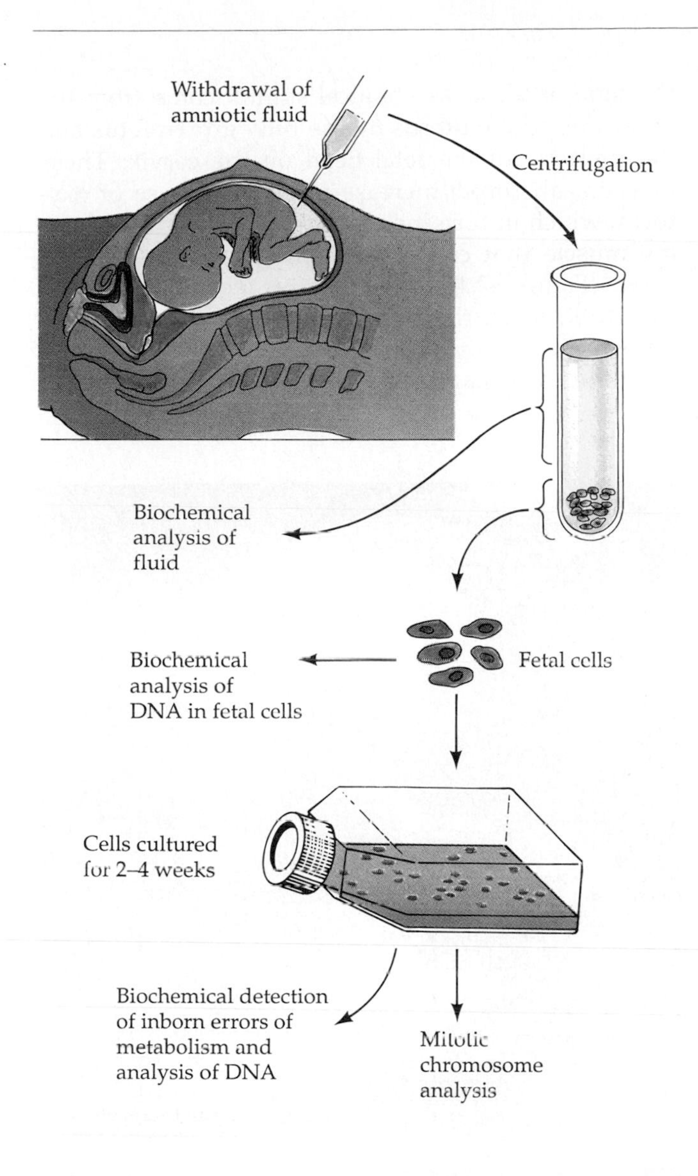

37.15 Amniocentesis Enables Genetic Analysis of the Fetus
Genetic information, including the sex of the fetus, can be gained by amniocentesis—the withdrawal and analysis of a small amount of amniotic fluid. The procedure is usually performed late in the third or early in the fourth month of pregnancy.

of human pregnancy, we divide it into three trimesters of about three months each.

The first trimester begins with fertilization. The blastocyst, as we know, goes through a series of rapid cell divisions, known as cleavage, before implantation. After implantation, the differentiation of tissues and organs we discussed in Chapter 17 begins; the first trimester is the main period of **organogenesis.** The heart begins to beat in week 4, and limbs form by week 8. Figure 37.16*a* shows an embryo midway between these two developmental events. Most organs are present in at least primitive form by the end of the first trimester. Because the first trimester is a time of rapid cell division and differentiation, it is the period during which the embryo is most sensitive to radiation, drugs, and chemicals that can cause birth defects. An embryo can be damaged before the mother even knows she is pregnant. By the end of the first trimester, the embryo appears to be a miniature version of the adult and is called a **fetus.**

Hormonal changes cause major and noticeable responses in the mother during the first trimester, even though the fetus at the end of that time is still so small that it would fit into a teaspoon. Soon after the blastocyst implants, it begins to secrete human chorionic gonadotropin (HCG), the hormone that stimulates the corpus luteum to continue producing estrogen and progesterone. It is HCG that is detected in pregnancy tests. The high levels of estrogen and progesterone prevent menstruation, which would abort the embryo, and exert negative feedback on the hypothalamus and the pituitary, inhibiting the release of follicle-stimulating hormone and luteinizing hormone and preventing a new round of ovulation. Side effects of these hormonal shifts are the well-known symptoms of pregnancy: morning sickness, mood swings, changes in the senses of taste and smell, and swelling of the breasts.

During the second trimester the fetus grows rapidly to about 600 grams, and the mother's abdomen enlarges considerably. The limbs of the fetus elongate, and the fingers, toes, and facial features become well formed (Figure 37.16*b*). Fetal movements are first felt by the mother early in the second trimester, and they become progressively stronger and more coordinated. By the end of the second trimester, the fetus may suck its thumb.

The production of estrogen and progesterone by the placenta increases during the second trimester. The placental tissues do not produce these steroids directly from cholesterol. The tissues receive androgens through the circulation and convert them to estrogen and progesterone. These androgens come from two sources: the adrenal cortex of the mother and the adrenal cortex of the fetus itself. As placental production of these hormones increases, the level of human chorionic gonadotropin and the activity of the corpus luteum decrease. The corpus luteum degenerates by the second trimester, but ovulation and menstruation are still inhibited by the high levels of steroids secreted by the placenta. Along with these hormonal changes, the unpleasant symptoms of early pregnancy usually disappear.

The fetus and the mother continue to grow rapidly during the third trimester. As the fetus approaches its full size, pressure on the mother's internal organs can cause indigestion, constipation, frequent urination, shortness of breath, and swelling of the legs and ankles. Throughout pregnancy the circulatory system of the fetus has been functioning, and as the third trimester approaches its end, other internal organs mature. The digestive system begins to function, the liver stores glycogen, the kidneys produce urine, and the brain undergoes cycles of sleep and waking.

BIRTH

Throughout pregnancy the uterus periodically undergoes slow, weak, rhythmic contractions called Braxton-Hicks contractions. These contractions become gradually stronger during the third trimester and are sometimes called false labor contractions. True labor contractions usually mark the beginning of childbirth, or **parturition.** In some women, however, the first signs of labor are the discharge of the mucous plug that blocks the uterus during pregnancy ("a bloody show") or the rupture of the amnion and the loss of the amniotic fluid ("waters breaking").

Labor

Many factors contribute to the onset of labor. Hormonal and mechanical stimuli increase the contractility of the uterus. Progesterone inhibits and estrogen stimulates contractions of uterine muscle. Toward the end of the third trimester the estrogen–progesterone ratio shifts in favor of estrogen. Oxytocin stimulates uterine contraction; its secretion by the pituitaries of both mother and fetus increases at the time of labor. Mechanical stimuli come from the stretching of the uterus by the fully grown fetus and the pressure of the fetal head on the cervix. These mechanical stimuli increase pituitary release of oxytocin, which in turn increases the activity of the uterine muscle that causes even more pressure on the cervix (Figure 37.17). This positive feedback converts the weak, slow, rhythmic contractions of the uterus into stronger labor contractions.

In the early stage of labor, the contractions of the

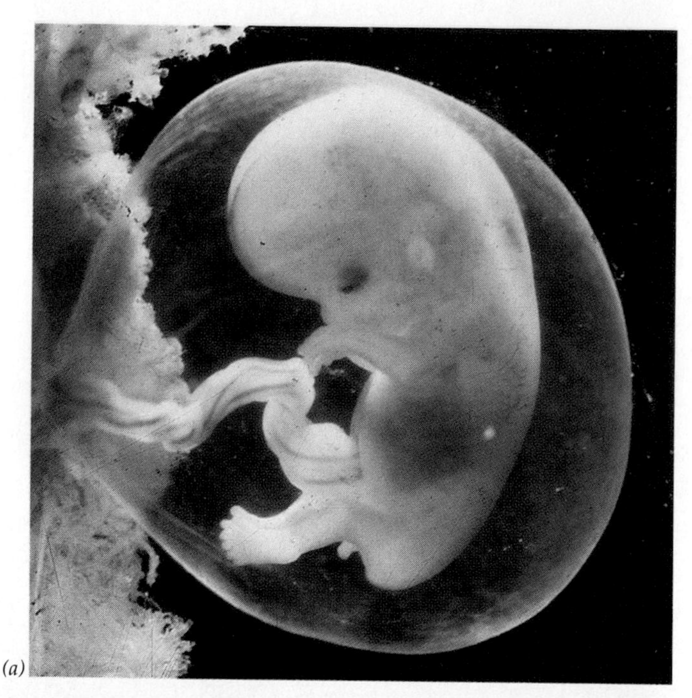

(a)

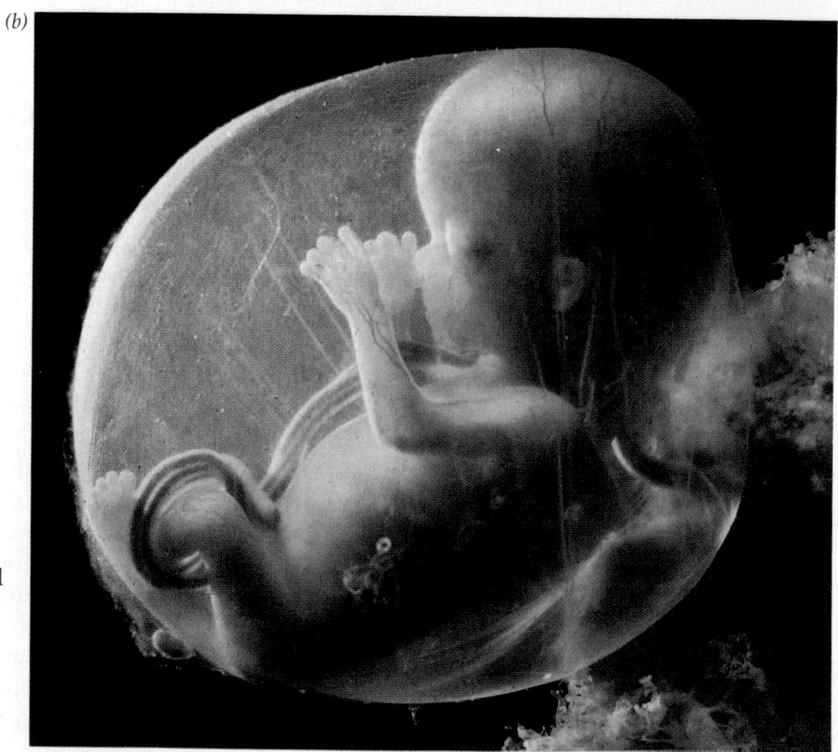

(b)

37.16 A Human Embryo
(a) The first trimester of pregnancy is a period of rapid cell division and differentiation; the organs and body structures of this six-week-old embryo are forming rapidly. *(b)* At four months the fetus moves freely within its protective amniotic membrane. The fingers and toes are fully formed.

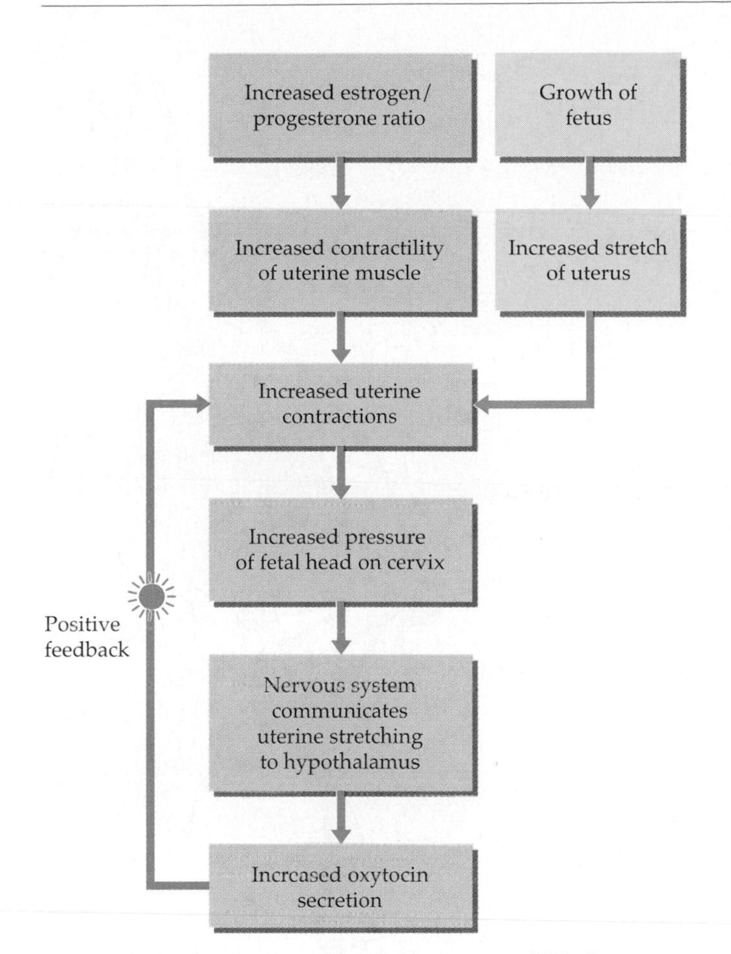

37.17 Increasing Oxytocin Production and Birth
Release of oxytocin by the pituitary gland increases uterine contractions during labor and birth. Note the positive feedback loop.

uterus are 15 to 20 minutes apart, and each lasts 45 to 60 seconds. During this time the contractions pull the cervix open until it is large enough to allow the baby to pass through. This stage of labor lasts an average of 12 to 15 hours in a first pregnancy and 8 hours or less in subsequent ones. Gradually the contractions become more frequent and more intense.

Delivery

In the second stage of labor, which begins when the cervix is fully dilated, the baby's head moves into the vagina and becomes visible from the outside. The usual head down position of the baby at the time of delivery comes about when the fetus shifts its orientation during the seventh month. If the fetus fails to reorient head down, the birth is more difficult. Passage of the fetus through the vagina is assisted by the woman's bearing down with her abdominal and other muscles to help push the baby along. Once the head and shoulders of the baby clear the cervix, the rest of its body eases out rapidly, but it is still connected to the placenta in the mother by the umbilical cord. This second stage of labor may take as little as a minute, or up to half an hour or more in a first pregnancy.

As soon as the baby clears the birth canal, it can start breathing and become independent of its mother's circulation. The umbilical cord may then be clamped and cut. The segment still attached to the baby dries up and sloughs off in a few days, leaving behind its distinctive signature, the belly button—more properly called the umbilicus. The detachment and expulsion of the placenta and fetal membranes takes from a few minutes to an hour, and may be accompanied by uterine contractions.

A **caesarian section** is the surgical extraction of the baby from the uterus (the term comes from Julius Caesar, who was supposedly born this way). It may be necessary for a number of reasons: if the fetus is large and the mother's pelvis small, if the first stage of labor lasts too long, if the cervix fails to dilate sufficiently, or if there is a sudden threat to the health of the baby or the mother.

Lactation

Throughout pregnancy, the high circulating levels of estrogen and progesterone cause the mammary glands of the breasts to develop in preparation for lactation (the secretion of milk). Prolactin secretion from the anterior pituitary also increases progressively during pregnancy, but its effect is countered by estrogen and progesterone, which inhibit the production of milk. Just before birth, the breasts may secrete a few milliliters of fluid each day. This fluid, called **colostrum,** contains little fat, and its rate of production is very low. With expulsion of the placenta, the estrogen and progesterone levels in the mother's bloodstream fall rapidly, and within a few days the well-developed mammary glands are producing milk. This milk does not flow readily into the ducts of the breasts, however. If it did, it would dribble out continuously, rather than just flowing out when the baby suckled. Oxytocin plays an important role in controlling lactation (Figure 37.18*a*). When the baby suckles, stimulation of the breast causes the release of oxytocin from the posterior pituitary. In about 30 seconds this oxytocin reaches the breasts and stimulates contraction of the muscle cells that surround the milk-secreting cells. The milk is thereby "let down," or ejected, into the ducts of the breasts (Figure 37.18*b*).

Oxytocin also has a role just after the delivery of the baby. If the newborn baby is placed at the mother's breast, even though the breast cannot deliver more than colostrum at this time, the suckling of the infant causes oxytocin release, which stimulates continued uterine contractions that help to expel the placenta and inhibit bleeding from the uterine wall.

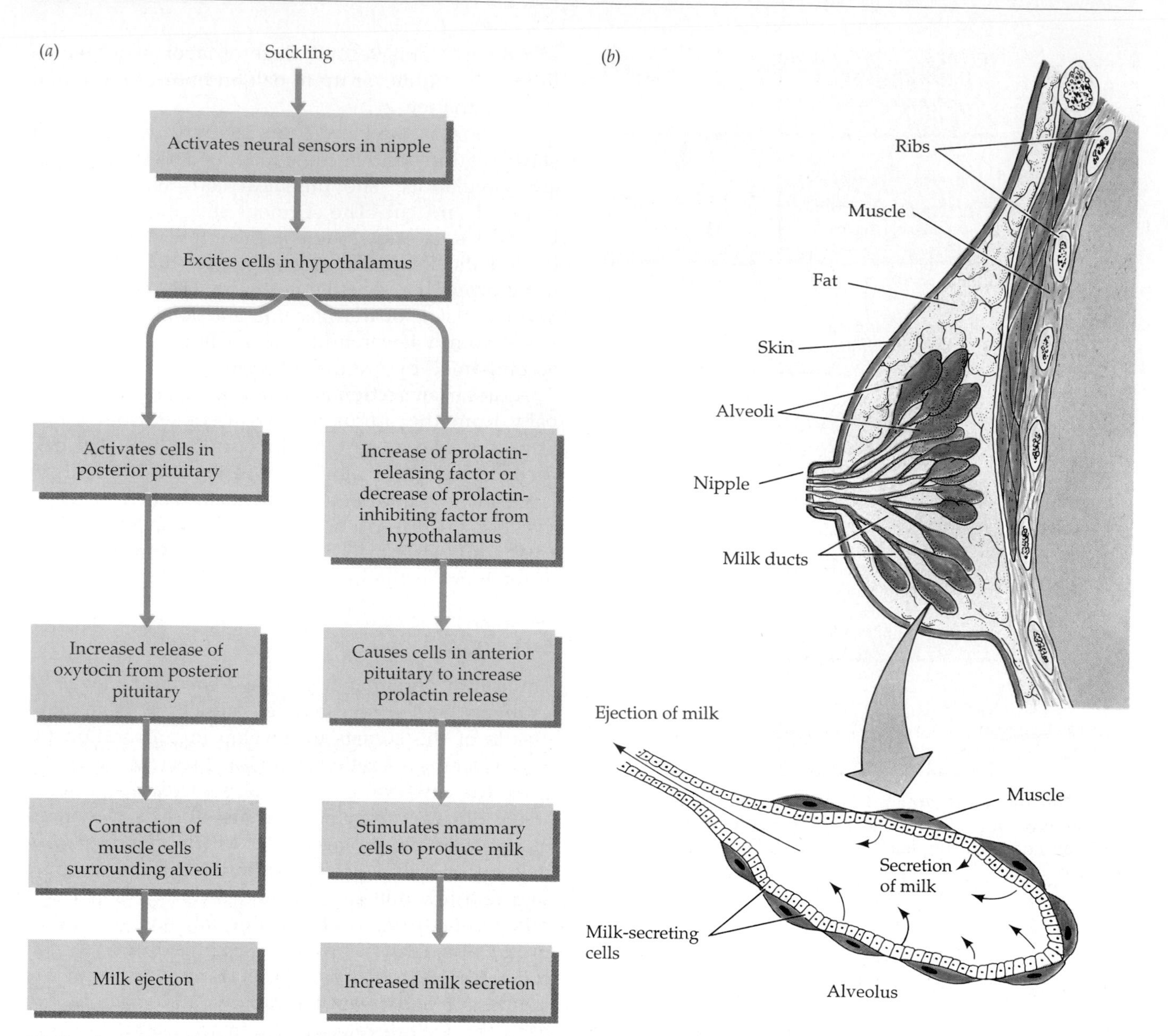

37.18 Lactation is Controlled by Hormones
(a) The role of hormones in lactation. *(b)* The anatomy of the mammary gland. The alveoli are reservoirs for milk.

SUMMARY of Main Ideas about Animal Reproduction

Some simple animals reproduce asexually by budding, regeneration, or parthenogenesis, but most animals reproduce sexually.

Sex produces genotypic diversity through recombination.

The male testes and the female ovaries are the internal reproductive organs in which gametes form.

Both spermatogenesis and oogenesis require mitotic proliferation of primary germ cells, meiosis, and maturation of gametes.
Review Figure 37.3

The reproductive systems of humans include the gonads where gametogenesis takes place, ducts that deliver mature gametes to the site of fertilization, structures for copulation, and in the female, a structure, the uterus, that provides an environment for the developing fetus.
Review Figures 37.6 and 37.7

Hormones control the ovarian and menstrual cycles of the human female so that an ovarian follicle matures each month and releases an ovum at a time when the uterus is prepared to receive it if it becomes fertilized.
Review Figure 37.8

Sexual behavior brings eggs and sperm together, and it involves strong physiological responses and conscious sensations.
Review Figure 37.9

As sperm and egg interact in fertilization, the sperm penetrates the egg's outer layers and binds to the egg plasma membrane. The egg reacts to prevent multiple sperm entry and becomes activated to complete meiosis and begin development.
Review Figures 37.10 and 37.11

Within the human uterus, the placenta, which includes tissues from the mother and from the embryo, supplies the embryo with oxygen and nutrients and removes its waste products.
Review Figure 37.14

Human childbirth is assisted by a positive feedback hormonal mechanism in which stretching of the uterine wall stimulates oxytocin release, which in turn causes uterine muscles to contract.
Review Figure 37.17

Oxytocin and prolactin help control lactation following childbirth.
Review Figure 37.18

SELF-QUIZ

1. Match each of the following modes of asexual reproduction with the statement or description that characterizes it. (Each letter may be used more than once, and more than one letter may apply to each statement.)
   a. Budding
   b. Regeneration
   c. Parthenogenesis
   (i) A form of asexual reproduction that usually follows an animal being broken by an external force, but it can also be initiated by the animal itself.
   (ii) Many freshwater sponges produce clusters of undifferentiated cells which eventually "escape" the parent and become free-living organisms genetically identical to the parent.
   (iii) Offspring develop from unfertilized eggs.
   (iv) The process requires totipotent cells.
   (v) Species that reproduce this way may also engage in sexual reproduction.
2. A species in which the individual possesses both male and female reproductive systems is termed (choose all that apply)
   a. dioecious.
   b. parthenogenetic.
   c. hermaphroditic.
   d. diploid.
   e. monoecious.
3. The major advantage of internal fertilization is that
   a. it ensures paternity.
   b. it permits the fertilization of many gametes.
   c. it reduces the incidence of destructive competitive interactions between the members of a group.
   d. it results in the formation of a stable pair-bond between mates.
   e. it allows the developing organism to enjoy a greater degree of protection during the early phases of development.
4. Which one of the following statements about oocytes is true?
   a. At birth, the human female has produced all the oocytes she will ever produce.
   b. At the onset of puberty, ovarian follicles produce new ones in response to hormonal stimulation.
   c. At the onset of menopause, the human female stops producing them.
   d. They are produced by the human female throughout adolescence.
   e. Those produced by the female are stored in the seminiferous tubules.
5. Spermatogenesis and oogenesis differ in that
   a. spermatogenesis produces gametes with greater energy stores than those produced by oogenesis.
   b. spermatogenesis produces four equally functional diploid cells per meiotic event and oogenesis does not.
   c. oogenesis produces four equally functional haploid cells per meiotic event and spermatogenesis does not.
   d. spermatogenesis produces many gametes with meager energy reserves, whereas oogenesis produces relatively few, well-provisioned gametes.
   e. in humans, spermatogenesis begins before birth, whereas oogenesis does not start until the onset of puberty.
6. The acrosome of the sperm
   a. carries genetic information.
   b. provides energy for movement.
   c. carries the enzymes that facilitate fertilization.
   d. induces ovulation.
   e. prevents polyspermy.
7. During oogenesis in mammals, the second meiotic division occurs
   a. after capacitation.
   b. after implantation.
   c. before ovulation.
   d. before the acrosomal reaction.
   e. after a sperm enters the egg.
8. One of the major differences between the sexual response cycles in human males and females is
   a. the increase in blood pressure in males.
   b. the increase in heart rate in females.
   c. the presence of a refractory period in females after orgasm.
   d. the presence of a refractory period in males after orgasm.
   e. the increase in muscle tension in males.

9. Which of the following membranes is part of the embryonic contribution to placenta formation?
   a. Amnion
   b. Chorion
   c. Uterine membrane
   d. Fertilization membrane
   e. Zona pellucida

10. Contractions of muscles in the uterine wall and in the breasts are stimulated by
   a. progesterone.
   b. estrogen.
   c. prolactin.
   d. oxytocin.
   e. human chorionic gonadotropin.

FOR STUDY

1. Compare and contrast spermatogenesis and oogenesis in terms of the products of and the timetable for each process.
2. Describe how the events during sperm activation and egg activation lead to successful fertilization.
3. Describe the mechanisms controlling lactation.
4. Ovarian and uterine events in the month following ovulation differ depending on whether or not fertilization occurs. Describe the differences and explain their hormonal controls.
5. Explain how positive feedback plays a role in birth.

READINGS

Beaconsfield, P., G. Budwood and R. Beaconsfield. 1980. "The Placenta." *Scientific American*, August. Describes implantation and development of the organ that is the intermediary between fetus and mother. The many functions of the placenta are described.

Epel, D. 1977. "The Program of Fertilization." *Scientific American*, November. The initial events in the sperm–ovum interaction.

Gilbert, S. F. 1994. *Developmental Biology*, 4th Edition. Sinauer Associates, Sunderland, MA. This excellent text on animal development includes chapters on fertilization, as well as on the germ line and gametogenesis.

Johnson, M. H. and B. J. Everitt. 1988. *Essential Reproduction*, 3rd Edition. Blackwell Scientific, Oxford. A concise and comprehensive technical account of the biology of gametogenesis, fertilization, and pregnancy.

Katchadourian, H. A. 1989. *Fundamentals of Human Sexuality*, 5th Edition. Saunders, Philadelphia. An introductory text that covers the anatomy and physiology of sex and reproduction in the first two chapters and then discusses developmental, behavioral, and social aspects of sex.

Short, R. V. 1984. "Breast Feeding." *Scientific American*, April. Breast feeding has hormonal consequences that have contraceptive effects. Trends toward bottle feeding in many developing nations may be causing rises in their rates of population growth.

Wassarman, P. M. 1988. "Fertilization in Mammals." *Scientific American*, December. Examines the molecular and cellular events that surround the fusion of sperm and egg.

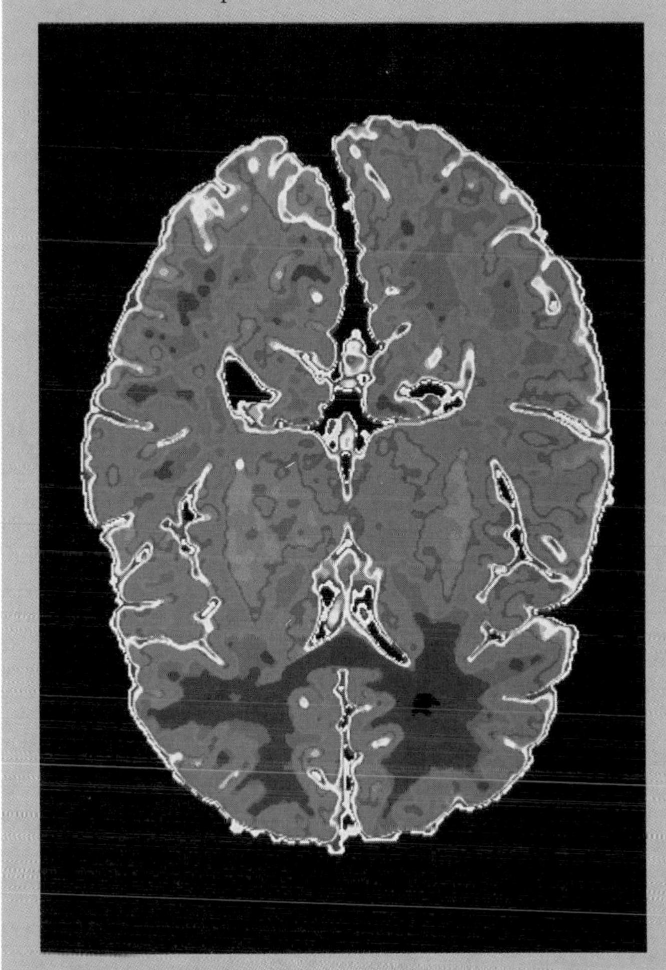

The Human Brain
This horizontal section has been colored by computer to show the different brain regions. The front of the brain is at the top.

38 Neurons and the Nervous System

The human brain weighs about 1.5 kilograms, is mostly water, and has the consistency and color of vanilla custard. The complexity of this small mass of tissue, however, exceeds that of any other known matter. The work of the brain is to process and store information and to control the physiology and behavior of the body. The brain is constantly receiving information from all the senses, integrating and interpreting that information, and generating commands to the muscles and organs of the body. The brain senses the need to act, decides on the appropriate action, orchestrates it, initiates and coordinates it, monitors it, and remembers it. The essence of individuality and personality resides in the brain. You could imagine remaining yourself after replacing any organ of your body except your brain.

Some of the actions controlled by the brain are conscious, or voluntary; others, such as the functions of the heart, lungs, and gut are involuntary. What is remarkable is that so many voluntary and involuntary actions are all going on simultaneously. Every second of its life, the brain processes thousands of bits of information. The brain does not function alone, however; it is part of an essential system—the nervous system—of a human or other animal. Like the hormones we considered in Chapter 36, the nervous system provides communication among cells. In this chapter we discuss nervous systems, particularly the structure and function of the human nervous system, as well as how nerve cells communicate with each other.

COMMUNICATION AND COMPLEXITY

A nervous system uses **sensors** (see Chapter 39), such as the eyes and ears, to transduce (convert) stimuli into messages that it can process. To cause behavior or physiological responses, the nervous system must communicate messages to **effectors**, such as muscles and glands (see Chapter 40). Together the brain and spinal cord make up the **central nervous system**, and nerve cells that carry information to and from the central nervous system make up the **peripheral nervous system**.

The information that flows through the nervous system consists of electrical and chemical messages. The electrical messages are nerve impulses. A nerve impulse is a rapid change in electrical charge across a small portion of the plasma membrane of the nerve cell. The nerve impulse travels along the membrane of a nerve cell, and that can be a long distance in the body because nerve cells have long extensions. When the nerve impulse reaches a point where the nerve cell makes contact with another nerve cell, or a muscle or gland cell, a chemical message is released that communicates across the gap between the two cells. In most cases the nerve impulse cannot spread from cell to cell without the intervention of a chemical message.

Throughout the animal kingdom, nervous systems vary in complexity, ranging from the complex nervous system of humans to the simple nerve nets of cnidarians, which seem to do little more than detect food or danger and cause tentacles to retract and the body to constrict (Figure 38.1). In all cases, however, nervous systems control behavior and the functions of the body. In general, the more complex the behavior and physiological capabilities of a species, the larger is its nervous system. Sometimes size belies capacity, however. Consider, for example, the nervous systems of small spiders that have programmed within them the thousands of precise movements necessary to construct a beautiful web without any prior experience. One of the greatest challenges of biology is to understand how the human brain functions, but it would be a major breakthrough even to understand a much simpler nervous system. Much progress in neurobiology (the science that studies nervous systems) has come from research on the simpler nervous systems of invertebrates. In this chapter we will frequently use the human nervous system as our model, but our knowledge comes from research on a long list of animal species. The way nerve cells function is almost identical in animals as different as squids and humans.

38.1 Nervous Systems Vary in Complexity
As we compare animals that have increasingly complex sensory and behavioral abilities, we find (moving clockwise from the sea anemone to the earthworm) information processing increasingly centralized in ganglia (collections of nerve cells) or in a brain. The brain and spinal cord in the human constitute the central nervous system, which communicates with the body through nerves that make up the peripheral nervous system.

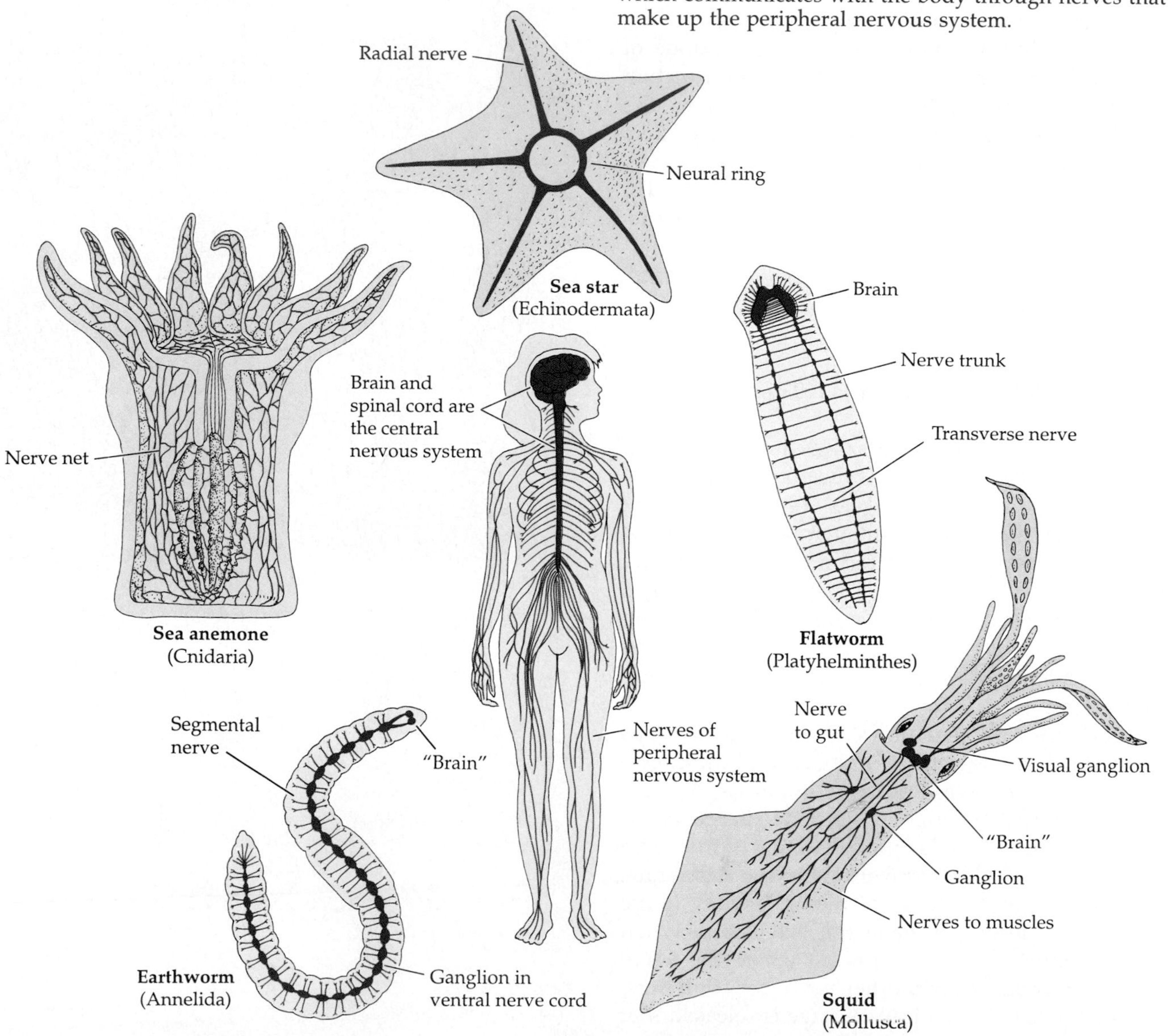

CELLS OF THE NERVOUS SYSTEM

The functions of the brain depend on the properties of its cells. **Neurons** are the cells that make it possible for the brain and the rest of the nervous system to transmit and integrate information. The important property of neurons is that their plasma membranes can generate electrical signals called nerve impulses, or **action potentials**. Their plasma membranes can also conduct these electrical signals rapidly from one location on a cell to the most distant reaches of that cell—a distance that can be more than a meter for some neurons. Where a neuron contacts another neuron or a muscle or gland cell, special structures called synapses transmit the message carried by the action potential in one neuron to the next cell. Much of neurobiology focuses on the structure and function of neurons, but they are not the only type of cell in the nervous system. In fact, there are more **glial cells** than neurons in the brain. As we will see shortly, glial cells do not generate or conduct action potentials; they have supporting roles for neurons.

Neurons

Most neurons have four regions: a cell body, dendrites, an axon, and axon terminals (Figure 38.2), but

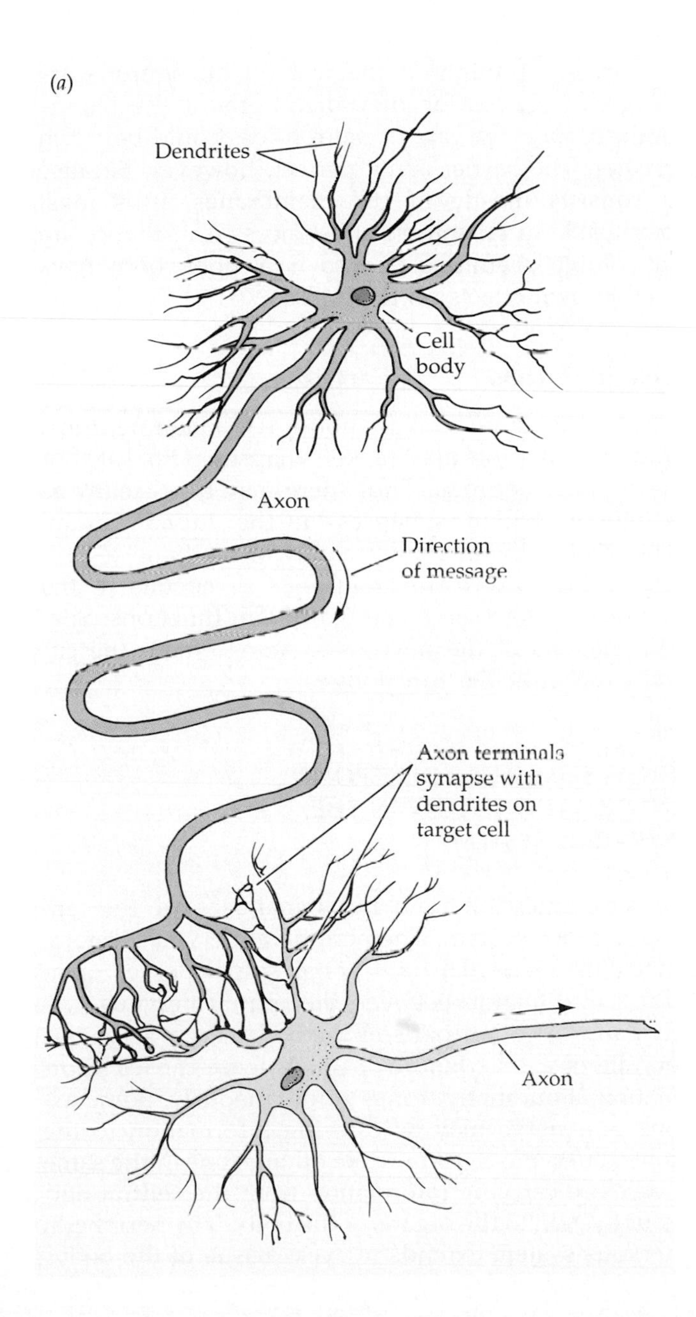

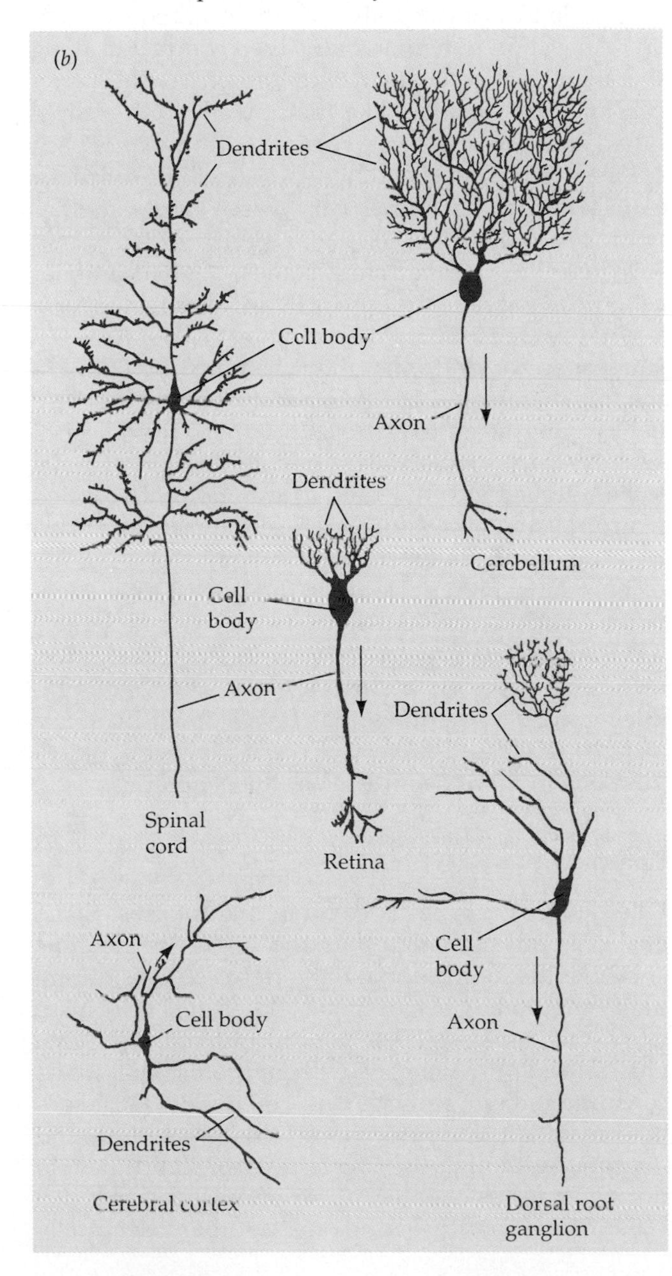

38.2 Neurons
(a) A generalized diagram of a neuron includes a cell body, dendrites that collect input, an axon that conducts action potentials, and axon terminals that make synapses with the target cells (red cell). *(b)* Neurons that are specific to various parts of the body.

the variation in different types of neurons is considerable. The **cell body** contains the nucleus and most of the cell's organelles. Many projections may sprout from the cell body. Most nerve-cell projections are bushlike **dendrites** (from the Greek *dendron* = "tree"), which bring information from other neurons or sensory cells to the neuron's cell body. In most neurons one projection is much longer than the others and is called the **axon**. Axons usually carry information away from the cell body. The length of the axon varies greatly in different types of neurons—as does the degree of branching of the dendrites. The axons of some neurons are remarkably long. The cell body of the neuron that causes your little toe to flex is located in the spinal cord in the middle of your back, and its axon goes all the way down your leg to the muscles in the little toe. Axons are the "telephone lines" of the nervous system. Information received by the dendrites can influence the cell body to generate an action potential that is then conducted along the axon to the cell that is its target. At the target cell, the axon divides, like the frayed end of a rope, into a spray of fine nerve endings. At the tip of each of these tiny nerve endings is a swelling called an **axon terminal** that comes very, very close to another neuron, a muscle cell, or a gland cell.

Where an axon terminal comes close to another cell, the membranes of both cells are modified to form a **synapse** (see Figure 38.2*a*). In most cases there is a small space or cleft only about 25 nm wide between the two membranes at the synapse. An action potential arriving at an axon terminal that forms such a synapse causes molecules stored in the axon terminal to be released into the cleft. These molecules, called neurotransmitters, diffuse across the cleft and bind to receptors on the postsynaptic membrane. The site of neurotransmitter release is the presynaptic membrane. Most individual neurons make and receive thousands of synapses. A neuron integrates information (synaptic inputs) from many sources by producing action potentials that travel down its single axon to target cells. We will discuss synaptic transmissions in more detail later in the chapter.

Glia

Like neurons, glial cells come in many forms. Some glial cells physically support and orient the neurons and help the neurons make the right contacts during their embryonic development. Other glial cells insulate axons, as we will see later in this chapter. Some glial cells supply neurons with nutrients; still others consume foreign particles and cell debris. Glial cells help maintain the proper ionic environment around neurons. Although they have no axons and they do not generate or conduct nerve impulses, some glial cells communicate with one another electrically through a special type of contact called a **gap junction**, a connection that enables ions to flow between cells (see Chapter 5). Gap junctions can also exist between neurons and between muscle cells. In these cases action potentials can cross between the cells without being converted into chemical messages.

Glial cells called **astrocytes** (because they look like stars) create the **blood–brain barrier**. Blood vessels throughout the body are very permeable to many chemicals, some of which are toxic. The brain is better protected from toxic substances than most tissues of the body because of the blood–brain barrier. Astrocytes form this barrier by surrounding the smallest, most permeable blood vessels in the brain. Protection of the brain is crucial because, unlike other tissues of the body, the brain cannot recover from damage by generating new cells. Shortly after birth, neurons in the brain cease cell division, and at that time we have the greatest number of neurons we will ever have in our lives. Throughout the rest of life neurons are progressively lost as they die. Without the blood–brain barrier, the rate of neuron loss could be much greater. The barrier is not perfect, however. Because it consists mostly of cell membranes, it is most permeable to fat-soluble substances. Anesthetics are fat-soluble chemicals and so is alcohol; both have well-known effects on the brain.

Cells in Circuits

The human brain contains about 100 billion neurons, and each neuron may make synapses with 1,000 or more other neurons. Thus there may be as many as a million billion synapses in the human brain. Therein lies the incredible ability of the brain to process information. The thousands of circuits in the nervous system serve many different functions. Specific regions of the nervous system contain the circuits for particular functions.

FROM STIMULUS TO RESPONSE: INFORMATION FLOW IN THE NERVOUS SYSTEM

In vertebrates the brain and spinal cord are the central nervous system. The peripheral nervous system, made up of cranial nerves and spinal nerves, conducts information between the central nervous system and the other parts of the body. Each **nerve** is a bundle of axons (Figure 38.3). A nerve carries information about many things simultaneously. Some axons in a nerve may be carrying information to the central nervous system while other axons in the same nerve are carrying information from the central nervous system to the organs of the body. The peripheral nervous system extends to every tissue of the body.

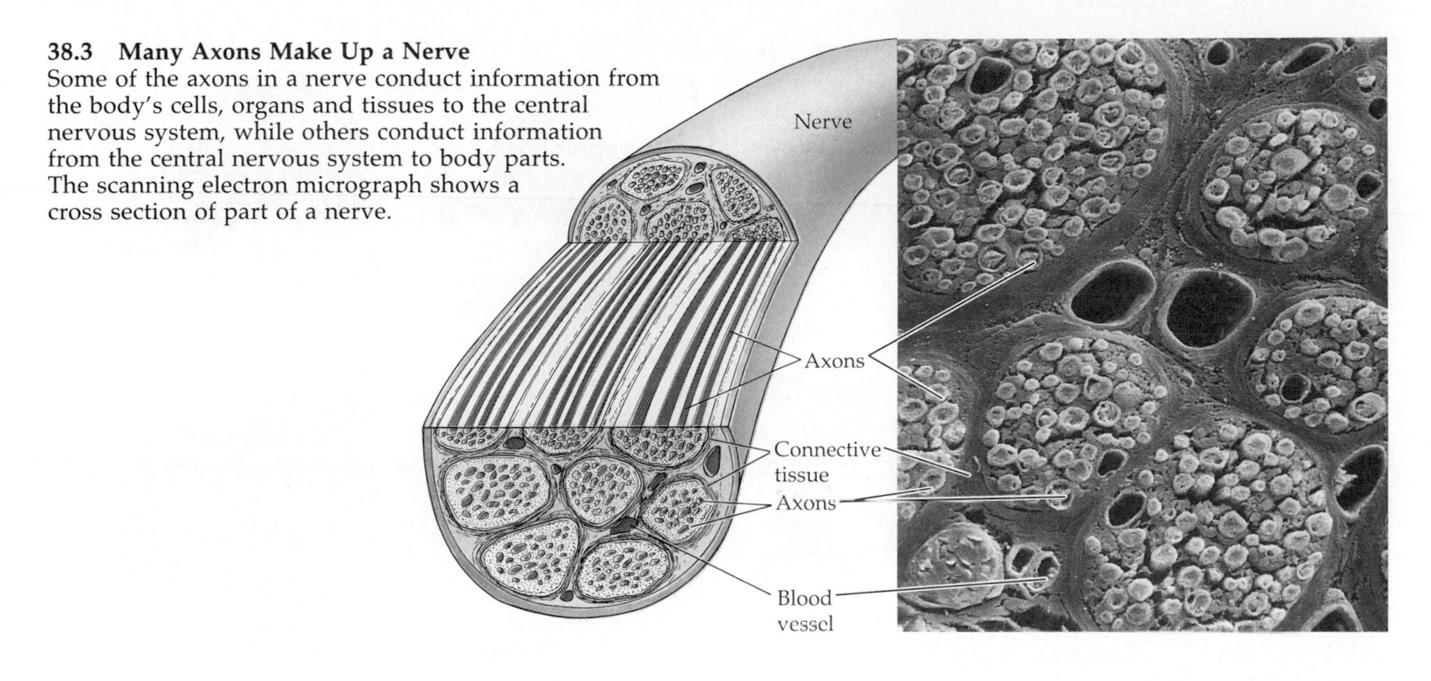

38.3 Many Axons Make Up a Nerve
Some of the axons in a nerve conduct information from the body's cells, organs and tissues to the central nervous system, while others conduct information from the central nervous system to body parts. The scanning electron micrograph shows a cross section of part of a nerve.

Conceptually, the peripheral nervous system has two sides: an **afferent** side that brings information to the central nervous system, and an **efferent** side that carries information away from the central nervous system (Figure 38.4). The afferent side has a division that carries information from our conscious senses and a division that carries information of which we are unaware. We are consciously aware of vision, hearing, touch, taste, pain, balance, and the position of the limbs of our body. We are not consciously aware of most physiological conditions, such as our blood pressure, body temperature, blood sugar level, and oxygen supply. The efferent side of the peripheral nervous system also has two divisions: a voluntary division that executes our conscious movements and an involuntary, or autonomic, division that controls involuntary physiological functions. You must also keep in mind that hormones provide information to the central nervous system, and that neurohormones are important output messages. To translate this conceptual blueprint into an understanding of the anatomical structures of the nervous system, we will first consider their development.

38.4 Organization of the Nervous System
The peripheral nervous system (colored backgrounds) carries information both to and from the central nervous system.

Development of the Vertebrate Nervous System

The nervous system of a vertebrate begins as a hollow tube of neural tissue. The tube runs the length of the embryo on its dorsal side. At the head end of the embryo, this neural tube forms three swellings that become the basic divisions of the brain: the hindbrain, the midbrain, and the forebrain. The rest of the neural tube forms the spinal cord. The cranial and spinal nerves, which are the peripheral nervous system, sprout from the neural tube and grow throughout the embryo.

Each of the three regions in the embryonic brain develops into several structures in the adult brain. From the hindbrain come the **medulla**, the **pons**, and the **cerebellum**. The medulla is continuous with the spinal cord, the pons is in front of the medulla, and the cerebellum is an outgrowth of the pons. The medulla and pons control some physiological functions, such as breathing and circulation. The cerebellum orchestrates and refines behavior patterns.

From the embryonic midbrain come structures that

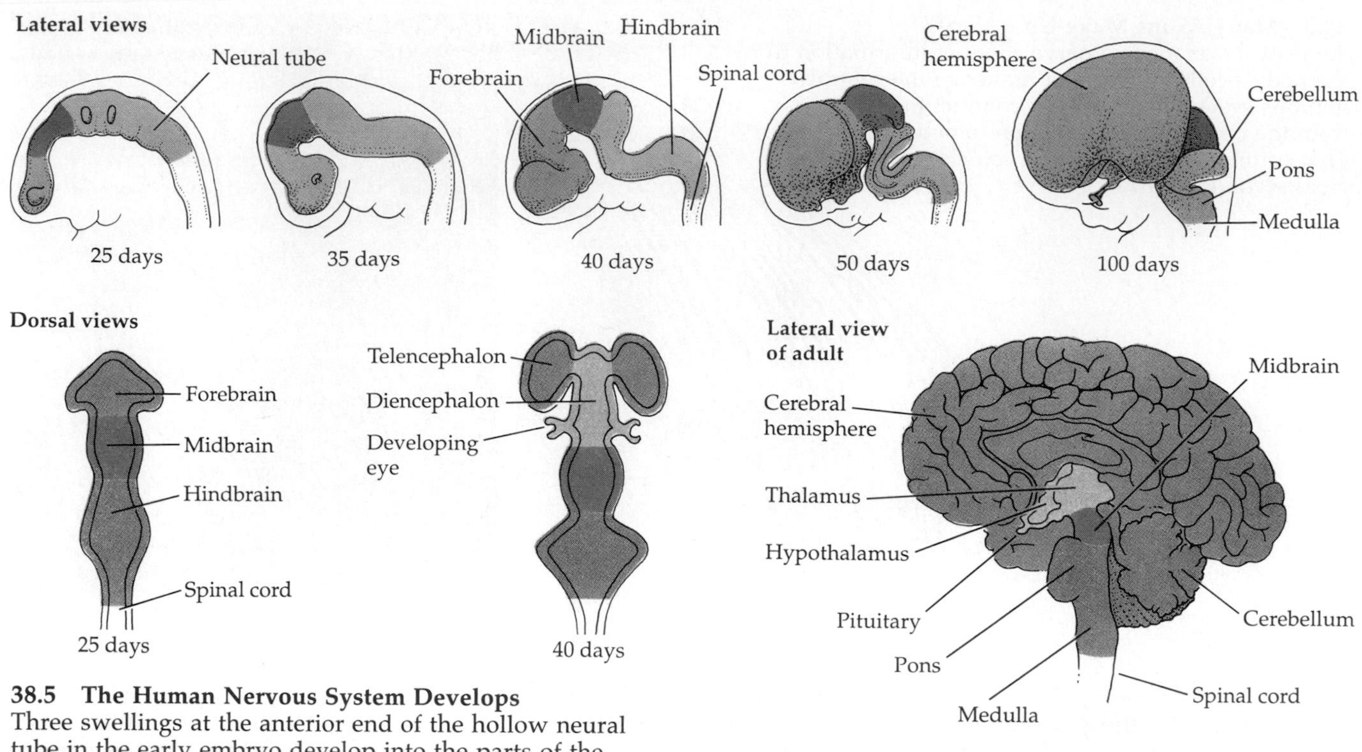

38.5 The Human Nervous System Develops
Three swellings at the anterior end of the hollow neural tube in the early embryo develop into the parts of the adult brain. The final view is an adult brain section (cut down the brain midline).

process aspects of visual and auditory (hearing) information. The embryonic forebrain develops into the **diencephalon** and the surrounding **telencephalon**, which consists of two **cerebral hemispheres**, also referred to collectively as the **cerebrum**. In mammals the telencephalon plays major roles in sensory perception, learning, memory, and conscious behavior. The diencephalon forms the **thalamus**—the final relay station for sensory information going to the telencephalon—and the **hypothalamus**, which regulates many physiological functions and biological drives (see Chapters 35 and 36).

Figure 38.5 shows how the hollow neural tube of the human embryo develops into hindbrain, midbrain, and forebrain. Note the somewhat linear arrangement of the structures. Information moves up and down this linear neural axis. A communication from the spinal cord to the telencephalon travels through the medulla, pons, midbrain, and diencephalon. These four structures are referred to collectively as the **brain stem**. In general, more-primitive and autonomic (involuntary) functions are located farther down the neural axis, and more-complex and evolutionarily advanced functions are higher on the neural axis.

As we go up the vertebrate phylogenetic scale from fish to mammals, the telencephalon increases in size, complexity, and importance. The forebrain dominates the nervous systems of mammals; when it is damaged, severe impairment or even coma results. A shark, by contrast, can swim almost normally with its telencephalon removed.

The Spinal Cord

The spinal cord conducts information between the brain and the organs of the body, and it processes and integrates information. A cross section of the spinal cord reveals a central area of gray matter in the shape of a butterfly surrounded by an area of white matter (Figure 38.6). The gray matter contains the cell bodies of the spinal neurons, and the white matter contains the axons that conduct information up and down the spinal cord. Spinal nerves leave the spinal cord at regular intervals on each side. Each spinal nerve has two roots, one connecting with the **dorsal horn** of the gray area, the other connecting with the **ventral horn** of the gray area. Each spinal nerve carries both afferent information from the sense organs and efferent information to the muscles and glands of the body. Afferent information enters the spinal cord through the dorsal roots of the nerves, and efferent information leaves the spinal cord through the ventral roots.

Information entering the dorsal horn can be transmitted to neurons that will carry it to the brain, to neurons that send commands directly out to muscles, or to cells called **interneurons** that reside entirely in

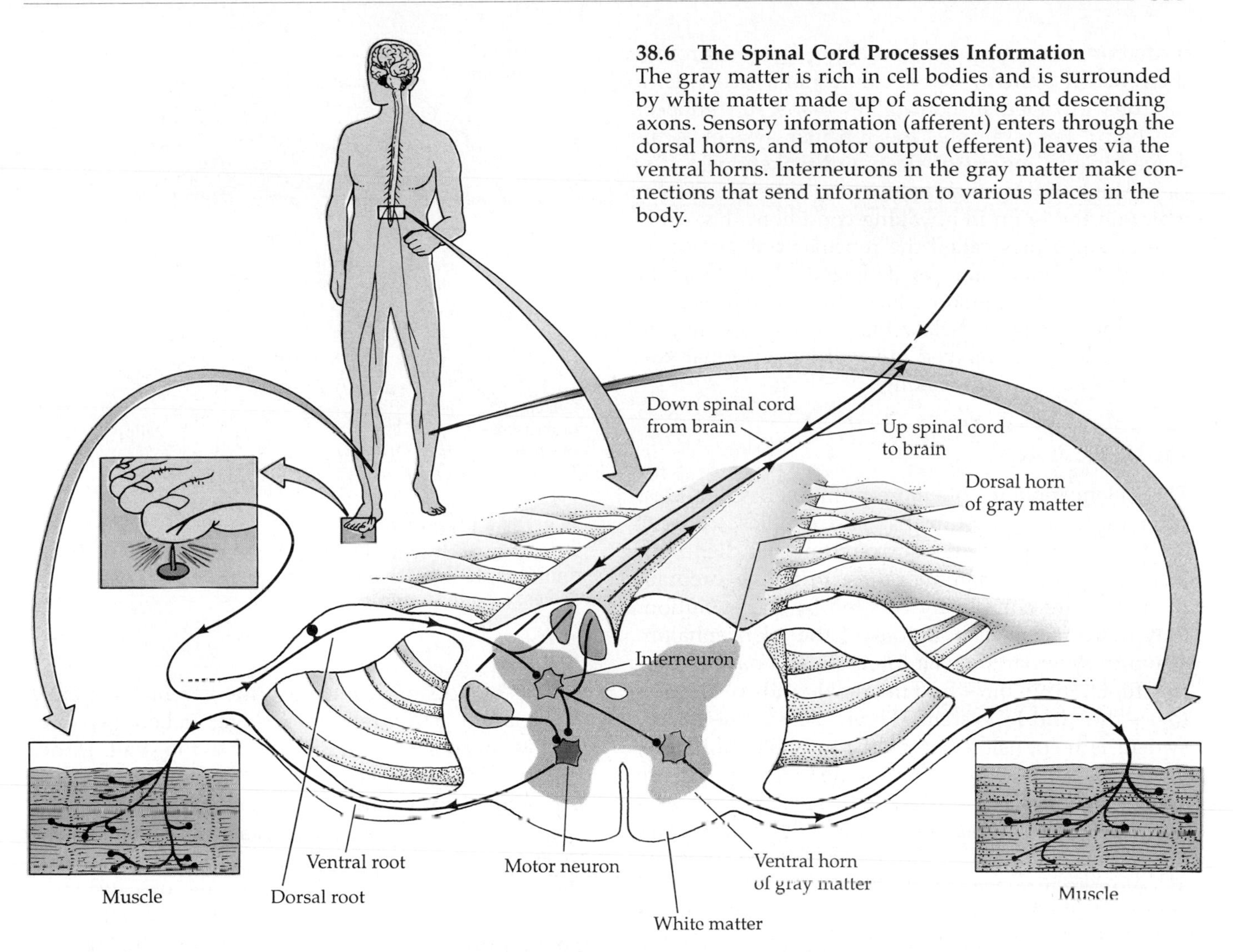

38.6 The Spinal Cord Processes Information
The gray matter is rich in cell bodies and is surrounded by white matter made up of ascending and descending axons. Sensory information (afferent) enters through the dorsal horns, and motor output (efferent) leaves via the ventral horns. Interneurons in the gray matter make connections that send information to various places in the body.

the gray matter of the spinal cord. Interneurons connect with efferent neurons in the ventral horns and communicate with other spinal neurons up and down the spinal axis to amplify the response and to generate more-complex motor patterns. The spinal cord processes and integrates a lot of information. For example, when you step on a tack, spinal circuits coordinate the rapid pulling back that is carried out by many muscles on both sides of your body. Spinal circuits can also generate repetitive motor patterns such as those of walking.

The Reticular System

The **reticular system** extends throughout the core of the medulla, pons, and midbrain (Figure 38.7). In-

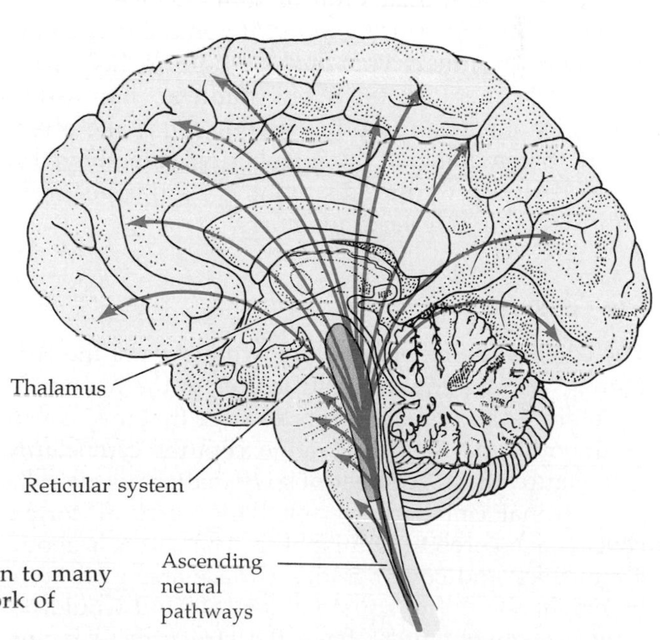

38.7 The Reticular System
Neurons in the core of the brain stem relay afferent information to many brain centers, and neuronal activity within this reticular network of neurons (blue) controls levels of alertness and sleep.

formation coming up the neural axis passes through the reticular system, where connections are made to neurons involved in controlling many functions of the body. For example, the reticular system is involved in the control of sleep and wakefulness. Because high levels of activity in the reticular system maintain the brain in a waking condition, this structure is sometimes called the reticular activating system. If the brain stem is damaged at midbrain or higher levels, the person is likely to remain in a coma, but if the damage is below the reticular system, the person may be paralyzed but will have normal patterns of sleeping and waking.

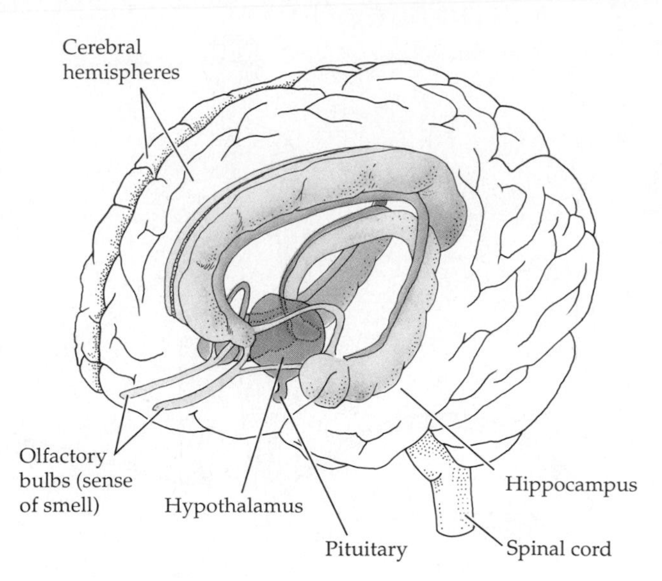

38.8 The Limbic System
Structures deep within the cerebral hemispheres and surrounding the hypothalamus control aspects of motivation, drives, emotions, and memory.

The Limbic System

The telencephalon of more-primitive vertebrates, such as fishes, amphibians, and reptiles, consists only of a few structures surrounding the diencephalon. In birds and mammals these primitive forebrain structures are completely covered by the evolutionarily more recent elaborations of the telencephalon. The primitive parts of the forebrain still have important functions in birds and mammals and are referred to as the **limbic system** (Figure 38.8). The limbic system is responsible for basic physiological drives, instincts, and emotions. Within the limbic system are areas that when stimulated with small electric currents can cause intense sensations of pleasure, pain, or rage. If a rat is given the opportunity to stimulate its own pleasure centers by pressing a switch, it will ignore food, water, and even sex, pushing the switch until it is exhausted. Pleasure and pain centers in the limbic system are believed to play roles in learning and in physiological drives.

One part of the limbic system, the **hippocampus**, is necessary in humans for the transfer of short-term memory to long-term memory. If you are told a new telephone number, you may be able to hold it in short-term memory for a few minutes, but within half an hour it is forgotten unless you make a real effort to remember it. Remembering something for more than a few minutes requires its transfer from short-term to long-term memory.

The Cerebrum

The cerebral hemispheres, the two halves of the cerebrum, are the dominant structures in the mammalian brain. In humans they are so large that they cover all the other parts of the brain except the cerebellum (see Figure 38.5). A sheet of gray matter (tissue rich in neuronal cell bodies) called the **cerebral cortex** covers each cerebral hemisphere. The cortex is about 4 mm thick and covers a total surface area over both hemispheres of one square meter. Since it would be rather inconvenient to have flat structures a meter square on top of our heads, the cerebral cortex is convoluted, or folded, into ridges called gyri and valleys called sulci so that it fits into the skull. Under the cerebral cortex is white matter, made up of the axons that connect the cell bodies in the cortex with each other and with other areas of the brain. The human cerebral cortex contains about 80 percent of the nerve cell bodies in the entire nervous system.

The cerebrum is divided into four lobes: the **temporal, frontal, parietal**, and **occipital** lobes (Figure 38.9*a*). Specific regions of each lobe govern certain functions (Figure 38.9*b*). The temporal lobe processes auditory (hearing) information and is responsible for language. An area of cortex on the underside of the temporal lobe is specialized to recognize faces. If this part of your temporal lobe were damaged, you would remember people's names but would not be able to match the names with the correct faces. The occipital lobe processes visual information.

The frontal and the parietal lobes are separated by a deep valley called the **central sulcus**. The strip of parietal lobe cortex just behind the central sulcus is the **primary somatosensory cortex**. This area receives information through the thalamus about touch and pressure sensations. The whole body surface is represented in this strip of cortex, with the head at the bottom and the legs at the top (Figure 38.10). Areas of the body that have many sensory neurons and are capable of making fine distinctions in touch (such as the lips and the fingers) have disproportionately large representation. If a very small area of the somatosensory cortex is stimulated electrically, the subject reports feeling specific sensations, such as touch, from a very specific part of the body.

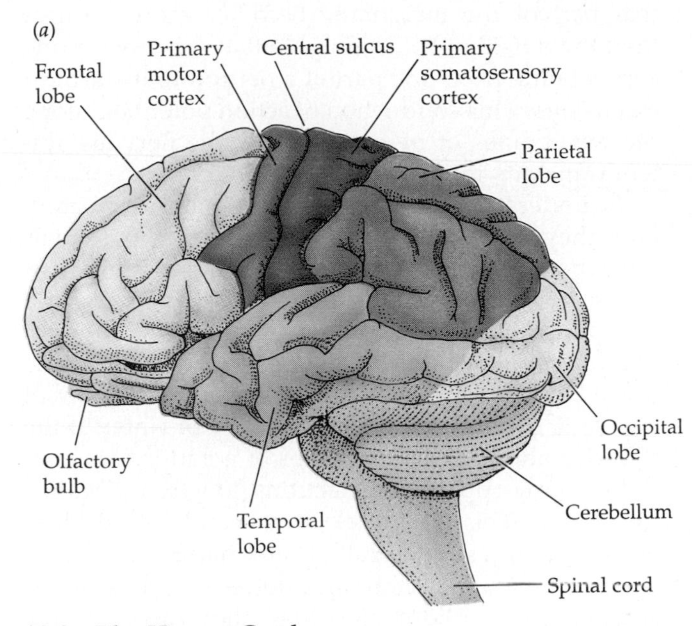

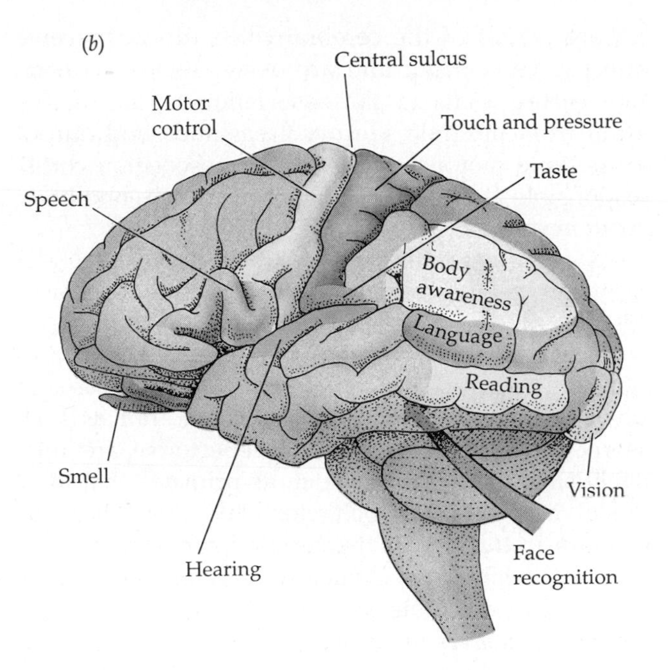

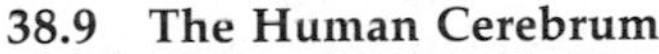

38.9 The Human Cerebrum
The highly convoluted halves of the cerebrum, viewed here from the left cerebral hemisphere, cover most of the other structures of the brain. *(a)* Each cerebral hemisphere is divided into four lobes. *(b)* Different functions are located in particular areas of these lobes.

A strip of the frontal lobe cortex just in front of the central sulcus is the **primary motor cortex**. The cells in this region have axons that extend to muscles in specific parts of the body. Once again, parts of the body are represented in the primary motor cortex, with the head region on the lower side and the lower part of the body in the top areas; fine motor control has the greatest representation. If a very small region of the frontal lobe is electrically stimulated, the response is the twitch of a muscle, but not a coordinated, complex behavior.

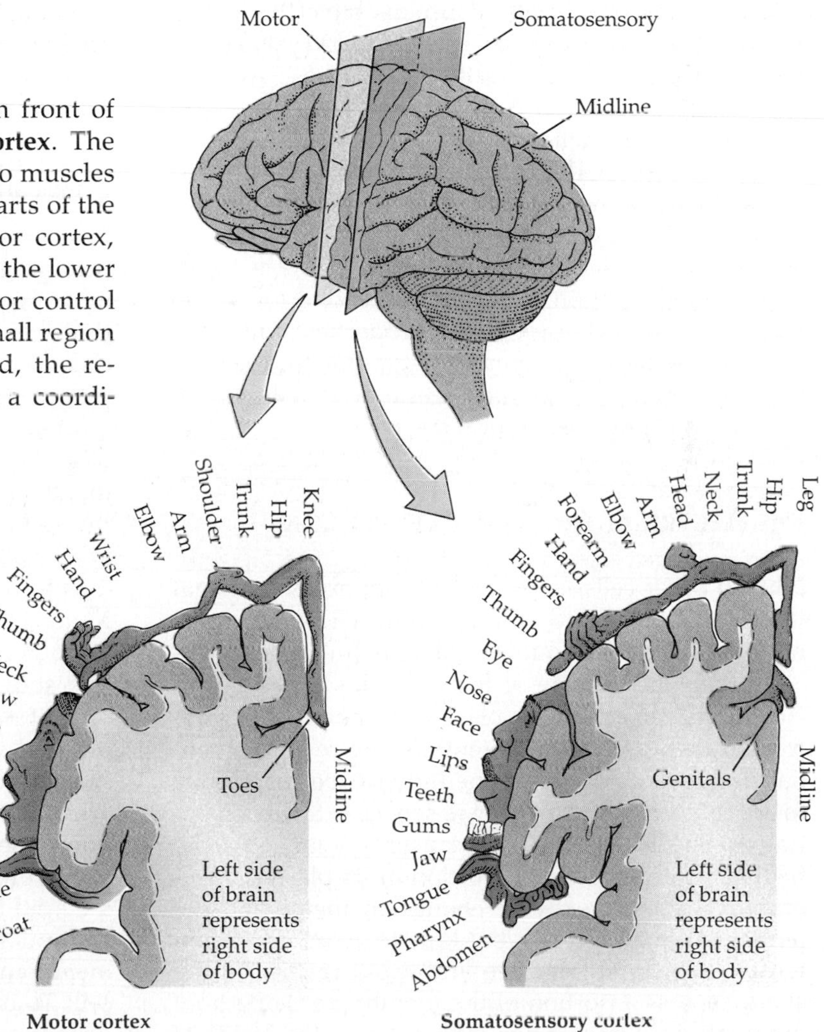

38.10 The Body Is Represented in the Primary Motor and Somatosensory Cortexes
Cross sections through the primary somatosensory cortex and primary motor cortex can be represented as maps of the human body. Body parts are shown in relation to the brain area devoted to them. The left side of the body is represented in the right cerebral cortex, and the right side of the body is represented in the left cerebral cortex.

Large areas of the cerebral cortex do not receive direct sensory input and are designated as **association cortex**. Cells in the association cortex receive input from multiple sensory areas and send output to multiple motor areas. Areas of association cortex responsible for reading, language, speech, and body awareness are indicated in Figure 38.9*b*.

As we mentioned earlier, the size of the telencephalon relative to the rest of the brain increases substantially as we go from fish to amphibian, to reptile, to birds and mammals. Even when we consider only mammals, the cerebral cortex increases in size and complexity when going from animals such as rodents, whose behavioral repertoires are relatively simple, to animals such as primates that have much more complex behavior. The most dramatic increase in the size of the cerebral cortex took place during the last several million years of human evolution. The incredible intellectual capacities of *Homo sapiens* are the result of the enlargement of the cerebral cortex. Humans do not have the largest brains in the animal kingdom; elephants, whales, and porpoises have larger brains in terms of mass. If we compare brain size to body size, however, humans and dolphins top the list. Humans have the largest ratio of brain size to body size, and they have the most highly developed cerebral cortexes. Another feature of the cerebral cortex that reflects increasing behavioral and intellectual capabilities is the ratio of association cortex to primary sensory and motor cortexes. Humans have the largest relative amount of association cortex.

Now that we have some appreciation for the whats and wheres of brain functions, it is time to turn to the hows. We will examine the properties of neurons that enable them to respond to stimulation, to create action potentials, to conduct action potentials, and to process and integrate information.

THE ELECTRICAL PROPERTIES OF NEURONS

Like all other cells, neurons have an excess of negative electrical charge within them. The inside of a neuron is usually about 70 millivolts (mV) more negative than the outside of the cell. This difference in electric charge across the plasma membrane of a neuron is called its **resting potential**. The resting potential provides a means for neurons to be responsive to specific stimuli. A neuron is sensitive to any chemical or physical factor that causes a change in the resting potential across a portion of its plasma membrane. The most extreme change in membrane potential is the electrical event known as an **action potential**, which is a sudden and rapid reverse in the charge across a portion of the membrane. For a brief moment, only one or two milliseconds, the inside of that part of the membrane becomes more positive than the outside. An action potential can move along a membrane from one part of a neuron to its farthest extensions. This conduction of action potentials along the membranes of neurons is how the nervous system transmits information.

To understand how resting potentials are created, how they are perturbed, and how action potentials are generated and conducted along membranes, it is necessary to know a little about electricity, ions, and the special ion-channel proteins in the membranes of neurons. Voltage is the tendency for electrons to move between two points. Voltage is to the flow of electrons what pressure is to the flow of water. If the negative and the positive poles of a battery are connected by a copper wire, electrons flow from negative to positive. This flow of electrons can be used to do work. As you may recall from Chapter 2, electric charges cross cell membranes not as electrons but as charged ions. The major ions that carry electric charges across the membranes of neurons are sodium (Na^+), chloride (Cl^-), potassium (K^+), and calcium (Ca^{2+}). It is also important to remember that ions with opposite charges attract each other. With these basics of bioelectricity in mind, we can ask how the resting potential of the membrane is maintained, and how the flows of ions through membrane channels are turned on and off to generate action potentials.

Pumps and Channels

Like those of all other cells, the membranes of neurons are lipid bilayers that are rather impermeable to ions. The membranes of neurons, however, contain three classes of proteins that give neurons their special electrical properties. These proteins act as pumps, channels, and receptors.

Membrane pumps use energy to move ions or other molecules against their concentration gradients. The major pump in neuronal membranes is the sodium–potassium pump that we learned about in Chapter 5. The action of this pump expels Na^+ ions from the inside of the cell, exchanging them for K^+ ions from outside the cell. The pump expels about three Na^+ ions for every two K^+ ions it brings in. The sodium–potassium pump keeps the concentration of K^+ inside the cell greater than that of the external medium, and the concentration of Na^+ inside the cell less than that of the external medium. The concentration differences established by the pump mean that K^+ would diffuse out of the cell and Na^+ would diffuse into the cell if the ions could cross the lipid bilayer. By itself, the unequal distribution of K^+ and Na^+ ions on the two sides of the plasma membrane does not create the resting potential of the cell. To explain the resting potential, we must introduce ion channels.

Ion channels are pores formed by proteins in the lipid bilayer. These water-filled pores allow ions to pass through. They are selective—that is, they may allow only one type of ion to pass through; thus there are potassium channels, sodium channels, chloride channels, and calcium channels. Ions can move in either direction through a channel. Most ion channels of neurons behave as if they contain a "gate" that opens to allow ions to pass under some conditions, but closes under other conditions. **Voltage-gated** channels open or close in response to the voltage across the membrane, and **chemically gated** channels open or close depending on the presence or absence of a specific chemical that binds to the channel protein or to a separate receptor that in turn alters the channel protein. Both voltage-gated and chemically gated channels play important roles in neuronal functions, as we will see, but first we will see how nongated channels, channels that are always open, are responsible for maintaining the resting potential in neurons and other cells.

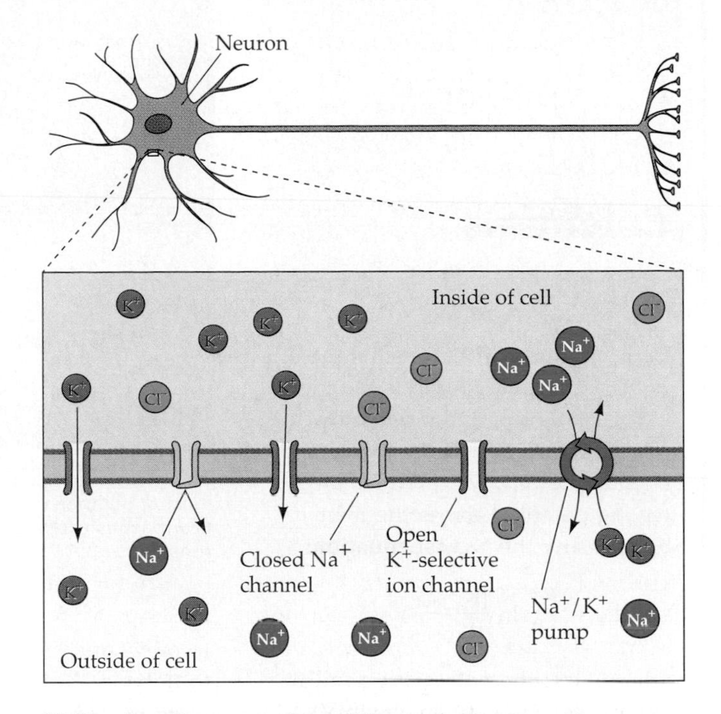

38.11 Ion Pumps and Ion Channels Produce the Resting Potential
The sodium–potassium pumps of a neuron establish ion concentration gradients across its plasma membrane. Under resting conditions more K^+ channels are open in the membrane than are Na^+ channels. Because of the diffusion of K^+ ions out of the cell, the membrane has an excess of positive charges on the outside and negative charges on the inside. This difference between the two charges is the resting potential.

The Resting Potential

Potassium channels are the most common type of open channel in the plasma membranes of neurons. These channels make neurons much more permeable to K^+ than to any other ions. As Figure 38.11 shows, this characteristic is what explains the resting potential. Because the neuron's plasma membrane is permeable to K^+, and because the sodium–potassium pump keeps the concentration of K^+ inside the cell much higher than that outside the cell, K^+ tends to diffuse out of the cell. If a K^+ ion leaves the cell, it leaves behind an unmatched negative charge. The tendency of the K^+ ions to diffuse out of the cell through the open channels thus causes the inside of the cell to become negatively charged in comparison to the external medium. Negative charges attract positive charges, such as those carried by K^+ ions. Eventually, a balance is established between the tendency for K^+ ions to diffuse down their concentration gradient to the outside of the cell and the attraction of the unmatched negative charges to pull them back inside the cell. The resting potential is the voltage difference that creates that balance. If you know the K^+ concentrations inside and outside the cell, you can calculate the resting potential of the membrane (Box 38.A).

Changes in Membrane Potentials

Because K^+ ions tend to leave the cell, the plasma membrane is polarized—that is, regions of unequal electric charge are separated, resulting in a resting potential. If a stimulus perturbs this resting potential, and the inside of the cell becomes *more* negative, the membrane is **hyperpolarized**. If the inside of the cell becomes *less* negative, the membrane is **depolarized**. Changes in the gated channels can cause a membrane to hyperpolarize or depolarize. Think, for example, what would happen if some sodium channels in the cell membrane opened. Na^+ ions would diffuse into the cell because of their higher concentration on the outside of the membrane, and they would also be attracted into the cell by the excess negative charge. As a result of the entry of Na^+ ions, the membrane would become depolarized in comparison to its resting condition.

There are also gated Cl^- channels in neuronal membranes, and the concentration of Cl^- ions is greater in the extracellular fluid than in the intracellular fluid. What would happen to membrane polarity if some of these Cl^- channels opened? The opening and closing of ion channels resulting in changes in the polarity of the membrane is the basic mechanism by which neurons respond to electrical, chemical, or other stimuli.

What good does it do for a neuron to undergo a change in its resting membrane potential at a particular location? Can that information be passed on to

BOX 38.A

Calculating Membrane Potentials

This stellate ganglion from a squid is almost 6 millimeters at its widest point. Giant axons extend from the huge neurons.

If we assume that the membrane of a cell is permeable to only one type of ion, such as K^+, and if we know the concentration of that ion inside and outside the cell, we can calculate the resting potential across the membrane using the **Nernst equation:**

$$E_K = \frac{RT}{zF} \ln \frac{[K^+]_{out}}{[K^+]_{in}}$$

where E_K is the **potassium equilibrium potential** in millivolts (mV). This is the potential difference across the membrane that will prevent a net diffusion of K^+ ions down their concentration gradient. R is the universal gas constant, T is the absolute temperature, z is the number of charges carried by each ion, F is the Faraday constant, and $\ln [K^+]_{out}/[K^+]_{in}$ is the natural logarithm of the K^+ concentration difference.

This equation may seem complex, but think of it this way: When the cell is in equilibrium, the amount of work necessary to move a K^+ ion into the cell against its concentration gradient equals the amount of work necessary to move a positive electric charge out of the cell against the voltage difference. Rather than consider just the tiny amounts of work associated with moving single ions or charges, we calculate in terms of moles of ions or charges. The work to move one mole of K^+ ions is equal to (RT) $(\ln [K^+]_{out}/[K^+]_{in})$, and the work to move one mole of positive charges is equal to $E_K zF$.

Using the Nernst equation is easier if we assume a temperature of 20°C (293 K), combine all constants, change to base 10 logarithms, and convert the equation to the following form:

$$E_K = (58 \text{ mV}) \left(\ln \frac{[K^+]_{out}}{[K^+]_{in}} \right)$$

For example, consider the huge neuron from a squid, shown in the photograph. The concentration of K^+ outside the neuron is 20 mM, whereas the concentration of K^+ inside the neuron is 400 mM. The base 10 logarithm of 20/400 is −1.3. Plugging this number into the simplified Nernst equation gives us a potassium equilibrium potential of −75 mV.

This calculated value is not too far from the measured resting potential in this cell, −60 mV. The Nernst equation can be used to calculate the equilibrium potential for any ion to which the membrane is permeable.

Why is the calculated potassium equilibrium potential using the Nernst equation only close to, and not the same as, the resting potential recorded from this cell using an electrode? In reality, the membrane is somewhat permeable to other ions in addition to K^+. Those other ions also influence the membrane potential, but less so because the membrane is much less permeable to these ions than it is to K^+ ions. Another equation, the Goldman equation, includes all of the ions to which the membrane is permeable, along with the relative permeabilities of the membrane to those ions. When the Goldman equation is used, the calculated membrane potential matches the measured potential extremely well. The Goldman equation for the squid axon would be:

$$V_m = \frac{RT}{F} \ln \frac{P_K[K^-]_{out} + P_{Na}[Na^+]_{out} + P_{Cl}[Cl^-]_{in}}{P_K[K^-]_{in} + P_{Na}[Na^+]_{in} + P_{Cl}[Cl^-]_{out}}$$

The P's in the equation stand for relative membrane permeabilities; for the resting squid neuron they are $P_K : P_{Na} : P_{Cl} = 1 : 0.04 : 0.45$. When these values are used to calculate V_m, the resting potential for the squid neuron, the result is −60 mV, the same as the measured resting potential.

Nernst Equilibrium Potentials of Ions in and around Squid Neurons

ION	CONCENTRATION IN CYTOPLASM (mM)	CONCENTRATION IN EXTRACELLULAR FLUID (mM)	NERNST POTENTIAL (mV)
K^+	400	20	−75
Na^+	50	440	+55
Cl^-	52	560	−60

other parts of the cell? A local perturbation of membrane potential causes electric currents to flow, and the flow spreads the change in membrane potential. The nerve cell is a poor conductor of electricity, however, and the change in membrane polarity diminishes and disappears before it gets very far from the site of stimulation. Communication of a stimulus by the flow of electric current is useful only over very short distances. As we will see, however, electric current is an important part of the mechanisms by which synapses and sensory stimuli generate action potentials, and it is also involved in the propagation of an action potential along an axon.

Action Potentials

Action potentials enable neurons to convey information over long distances with no loss of the signal. An action potential is a sudden and major change in membrane potential that lasts for only one or two milliseconds. It is conducted along the axon of a neuron at speeds up to 100 m/s, which is equivalent to running the length of a football field in one second. Using an electrode—a fine, electrically insulated wire or a very thin glass pipette containing a solution that conducts electric charges—we can record very tiny, local, electrical events that occur across cell membranes. If we place the tips of a pair of electrodes on the two sides of the membrane of a resting axon and measure the voltage difference, it is about −70 mV (Figure 38.12). If these electrodes are exposed to an action potential traveling down the axon, they register a rapid change in membrane potential, from −70 mV to about +40 mV. The membrane potential rapidly returns to its resting level of −70 mV as the action potential passes by. At any location along this axon we could insert another pair of electrodes and record the same action potential. The height of the action potential does not change as it travels along the axon. The action potential is an all-or-nothing, self-regenerating event.

Voltage-gated sodium channels are primarily responsible for the action potential. At a normal membrane resting potential these channels are mostly closed. If a stimulus or a synaptic input makes the membrane less negative—that is, depolarizes it—these Na^+ channels have a higher probability of flipping open briefly—for less than a millisecond. Because of the action of the sodium–potassium pump, Na^+ concentration is much higher outside the axon than inside, so whenever the sodium channels open in a part of the membrane, Na^+ ions from the outside enter the cell at that location. The entering Na^+ ions make the inside of the membrane more positive. Eventually a membrane potential is reached at which so many Na^+ channels open that the membrane potential suddenly rises to a very positive value.

The opening of the Na^+ channels causes the rise of the action potential—what neurobiologists call the spike. What causes the return of the membrane to resting potential? The main reason for the drop is that, after opening briefly, the sodium channels close and remain inactive for a few milliseconds. This is long enough for the membrane to return to resting potential. Some axons also have voltage-gated potassium channels. Because these channels open more slowly than the sodium channels and stay open longer, they help return the voltage across the membrane to its resting level by allowing K^+ ions to carry excess positive charges out of the cell.

The behavior of the voltage-gated sodium channels can be explained by assuming that they have two voltage-sensitive gates, an activation gate and an inactivation gate. Under resting conditions the activation gate is closed and the inactivation gate is open. Depolarization of the membrane to threshold causes both gates to change state, but the activation gate responds faster. As a result, the channel is open for the passage of Na^+ ions for a brief period of time between the opening of the activation gate and the closing of the inactivation gate. The inactivation gates remain closed for a few milliseconds before they spontaneously open again, thus explaining why the membrane has a **refractory period** (a period during which it cannot act) before it can fire another action potential. When the inactivation gates finally open, the activation gates are closed and the membrane is poised to respond once again to a depolarizing stimulus by firing another action potential.

The difference in concentration of Na^+ ions across the plasma membrane of neurons is the "battery" that drives the action potential. How rapidly does the battery run down? It might seem that a substantial number of Na^+ and K^+ ions would have to cross the membrane for the membrane potential to go from −70 mV to +40 mV, and back to −70 mV again. In fact, only about one Na^+ (or K^+) ion in 10 million actually moves through the channels during the passage of an action potential. Thus the effect of a single action potential on the concentration ratios of Na^+ or K^+ is very small. Even hundreds of action potentials barely change the concentration differences of Na^+ and K^+ on the two sides of the membrane. Thus it is not difficult for the sodium–potassium pump to keep the "battery" charged, even when the cell is generating many action potentials every second.

Propagation of the Action Potential

How does an action potential move over long distances? When one part of an axon fires an action potential, the adjacent regions of membrane also become depolarized because of the spread of local electric current. Such movements of ions depolarize the

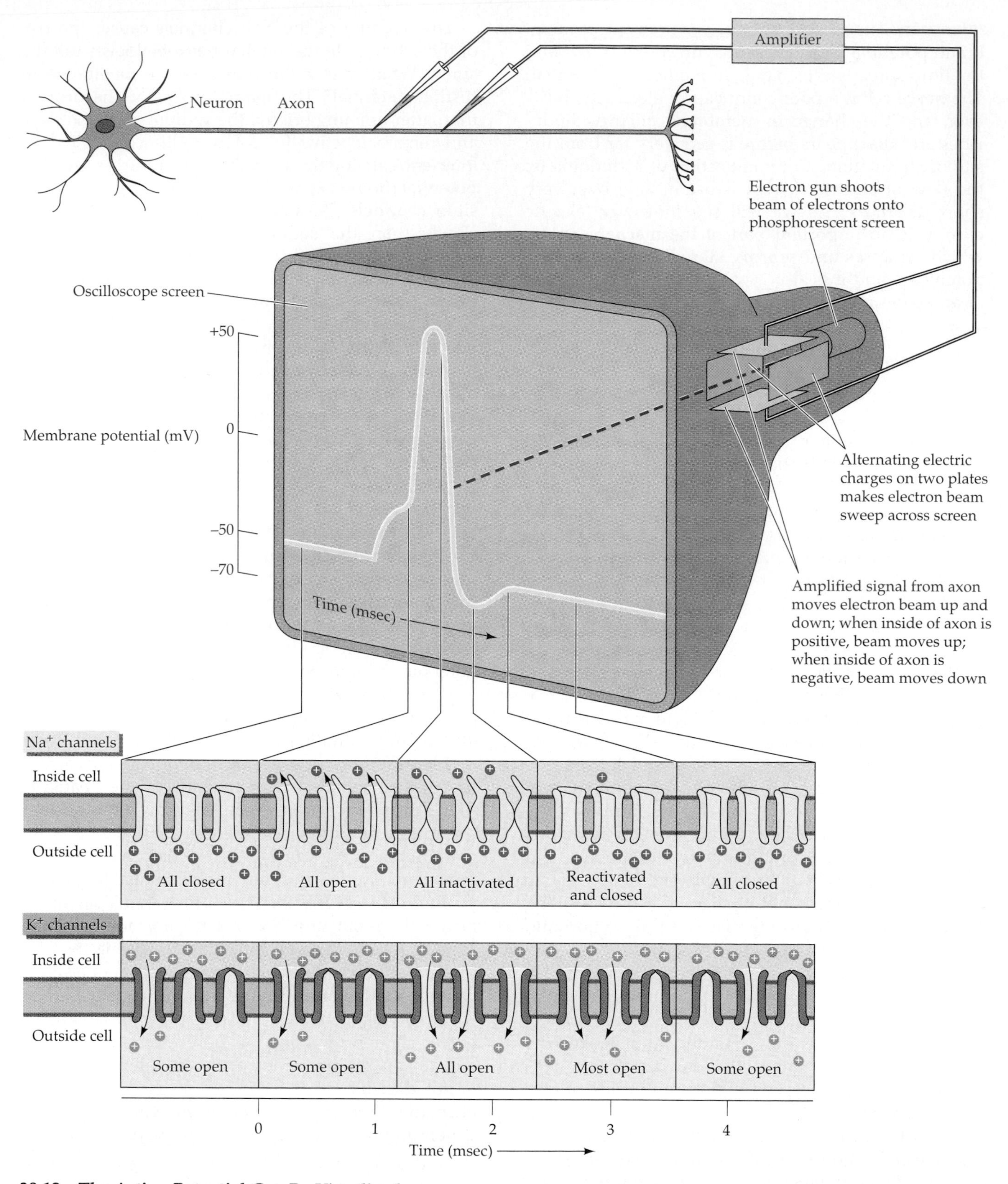

38.12 The Action Potential Can Be Visualized on an Oscilloscope
A pair of electrodes detects an action potential as a voltage change across the membrane of an axon. The signal from the electrodes is amplified and fed into an oscilloscope. A beam of electrons sweeps across the screen in a set period of time. That beam is deflected up if the signal from the electrodes is positive and down if the signal is negative. Thus the action potential is seen on the screen as a change in membrane potential through time. The action potential is created in the axon by voltage-gated Na^+ and K^+ channels opening and closing, as depicted at the bottom of the figure. The membrane potential at any given time depends on which and how many channels are open.

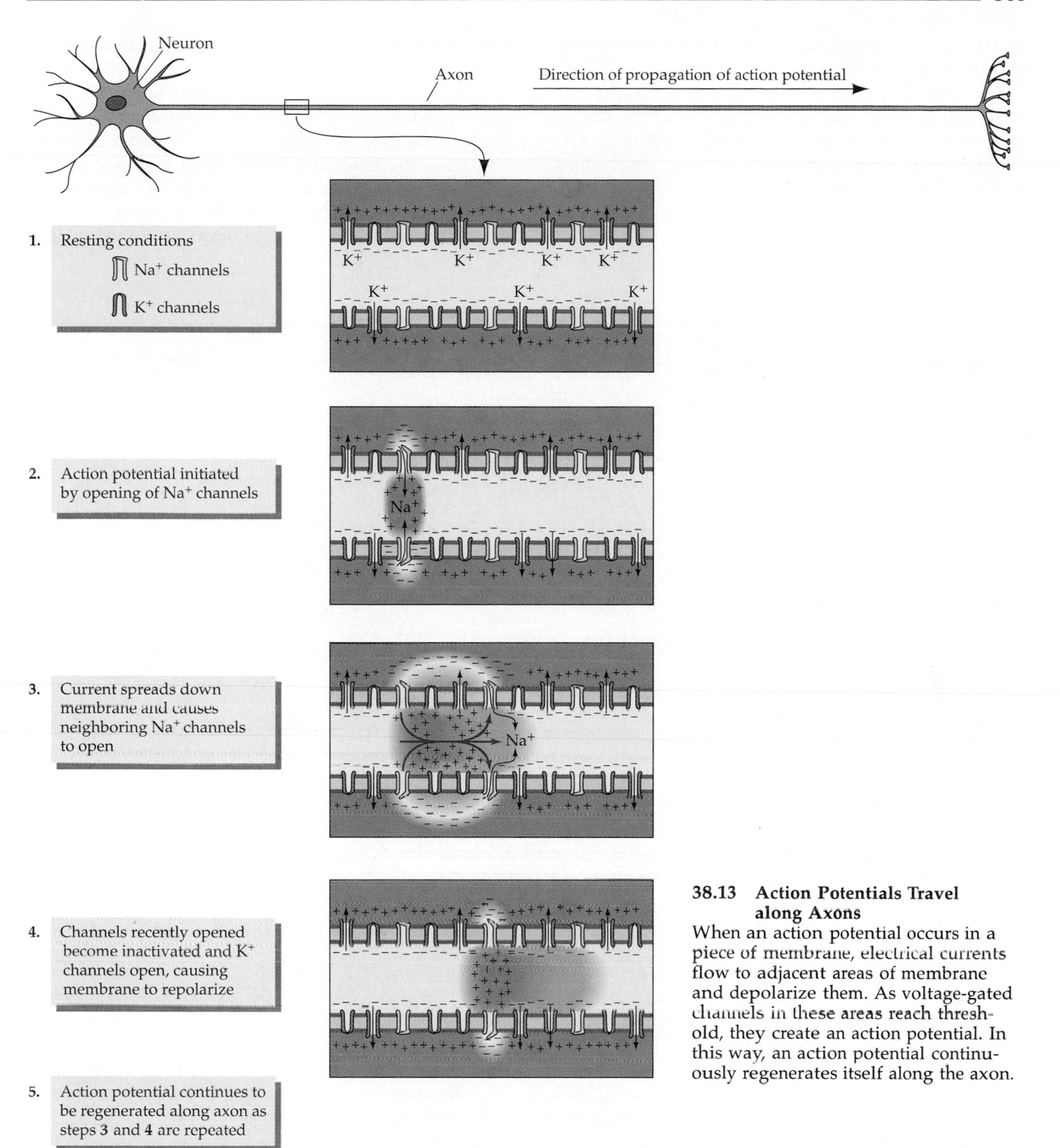

38.13 Action Potentials Travel along Axons When an action potential occurs in a piece of membrane, electrical currents flow to adjacent areas of membrane and depolarize them. As voltage-gated channels in these areas reach threshold, they create an action potential. In this way, an action potential continuously regenerates itself along the axon.

region of the membrane adjacent to that experiencing the action potential (Figure 38.13). Outside the axon, positive ions flow rapidly toward the depolarized region, attracted by the negative charge balance there. Inside the cell, positive charges move *away* from the depolarized region, where they are more abundant, and *toward* the adjacent, more negative regions. The net result for the membrane is a tendency to repolarize at the point of the existing action potential and to depolarize at the adjacent regions. If the depolarization of an adjacent region of membrane brings it to the threshold level that causes massive opening of sodium channels, an action potential is generated. Because an action potential always brings the area of membrane adjacent to it to threshold, it is self-regenerating and propagates itself along the axon. The action potential propagates itself in only one direction; it cannot reverse itself because the part of the membrane it came from is undergoing its refractory period.

The action potential does not travel along all axons at the same speed. Action potentials travel faster in large-diameter axons than in small-diameter axons. Among invertebrates, the axon diameter determines the rate of conduction, and axons that transmit messages involved in escape behavior are very thick. The axons that enable squid to escape predators are almost a millimeter in diameter. These giant axons were the most important experimental material used in the classic studies in neurophysiology that produced basic discoveries about action potentials and their conduction. The British physiologists A. L. Hodgkin and A. F. Huxley performed most of these studies in the 1940s and 1950s. Their work led to our understanding of electrical events in the membrane and to more-recent techniques that have enabled us to understand these events in terms of individual ion channels (Box 38.B).

BOX 38.B

Studying the Electrical Properties of Membranes with Voltage Clamps and Patch Clamps

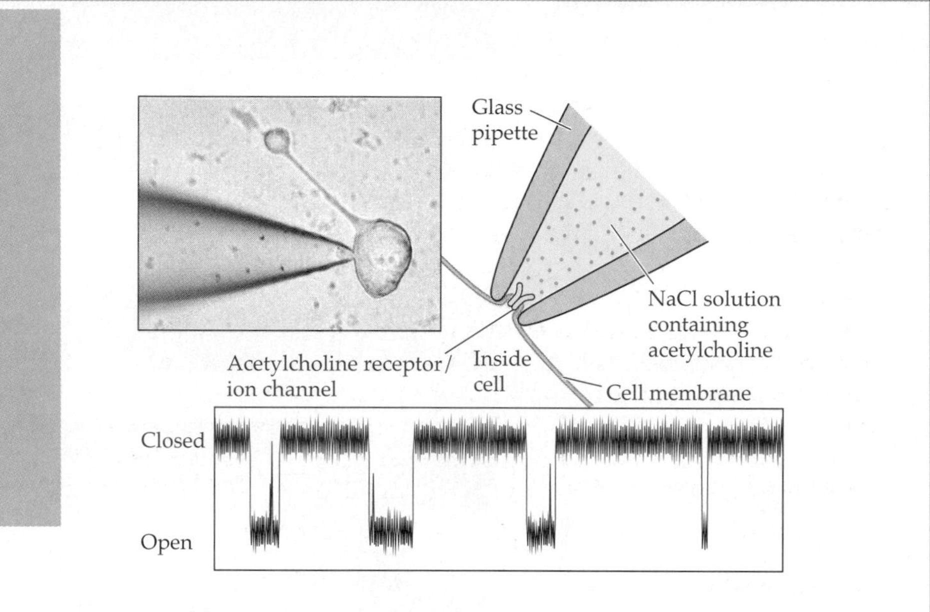

The large size of the squid axon made it possible for Hodgkin and Huxley to develop the technique of **voltage clamping** to study the electrical properties of the axonal membranes. Two fine electrodes were placed within the axon. One of these electrodes was used to measure voltage, the other to pass electric current into the cell to control (or clamp) the voltage across the cell membrane. Monitoring the change in the current necessary to maintain the voltage clamp at a particular level gave a direct measure of the ionic currents across the membrane. Thus, when the membrane was depolarized by an action potential, it was possible to measure the changing electric currents due to Na^+ ions and K^+ ions flowing across the membrane. Hodgkin and Huxley were able to determine the separate contributions of the Na^+ and K^+ ions by changing the concentrations of these ions inside or outside the cell. Later, poisons and drugs that block specific channels were used in voltage clamp experiments. For example, the puffer fish poison, tetrodotoxin, blocks the sodium channel that is responsible for the rise in the action potential, and tetraethylammonium ions block the potassium channel that helps the membrane repolarize following the action potential. Voltage clamp studies led to the first explanations of action potentials in terms of the properties of ion channels, but those ion channels were only postulated because there were no means of observing them directly.

A technique called **patch clamping**, developed in the 1980s by Bert Sakmann and Erwin Neher, made it possible to study single ion channels in membranes and therefore the molecular basis for the electrical properties of membranes revealed by voltage clamping experiments. In patch clamping (shown in the figure), a polished glass pipette is placed in contact with the cell membrane. Slight suction makes a seal between the pipette and the patch of membrane under the tip of the pipette. Movements of ions, and therefore electric charges, through channels in the patch of membrane can be recorded through the pipette. The solution filling the pipette determines the ion concentrations on the outside of the patch and may also contain chemicals to bind to receptors. If the patch is torn loose from the cell, the ion concentrations on the interior side of the membrane can also be changed. A patch may contain only one or a few ion channels; thus the electrical recording from that patch can show individual channels opening and closing. Neher and Sakmann received the Nobel prize in 1991.

In nervous systems more complex than those of invertebrates, increasing the speed of action potentials by increasing the diameter of axons would result in enormous nerves. The optic nerve from each of our eyes contains about one million axons. If we used plain, simple axons to build optic nerves that conduct information from the eye to the brain as fast as ours do, each optic nerve would have to be about the diameter of the eyeball itself. Other groups of axons in the brain, the spinal cord, and the peripheral nervous system would be equally unwieldy. Evolution has increased propagation velocity in vertebrate axons in a different way. The axons of vertebrates are insulated with membranous wrappings produced by specialized glial cells called **Schwann cells**. The membranous wrapping is called **myelin**. It is the myelin that gives the light, shiny appearance to "white matter," which is any nervous system tissue containing mostly axons. The myelination of an axon is not continuous. At regular intervals called **nodes of Ranvier** the bare axon is exposed (Figure 38.14*a*).

How do the nodes of Ranvier and the myelin insulation of the axon increase propagation velocity? At the nodes of Ranvier an axon can fire action potentials, but in the adjacent regions of the axon that are insulated with myelin, electric charges cannot accumulate or cross the membrane. Therefore, the local depolarization caused by the action potential at a node causes electric current to flow to the next place in the membrane where charges can accumulate and cross—the next node of Ranvier. Electric current flows very fast compared to how long ion channels take to open and close. Thus the depolarization caused by an action potential at one node spreads

(*a*) Myelination

Schwann cell

Node of Ranvier

Wrapping of Schwann cell membranes around axon

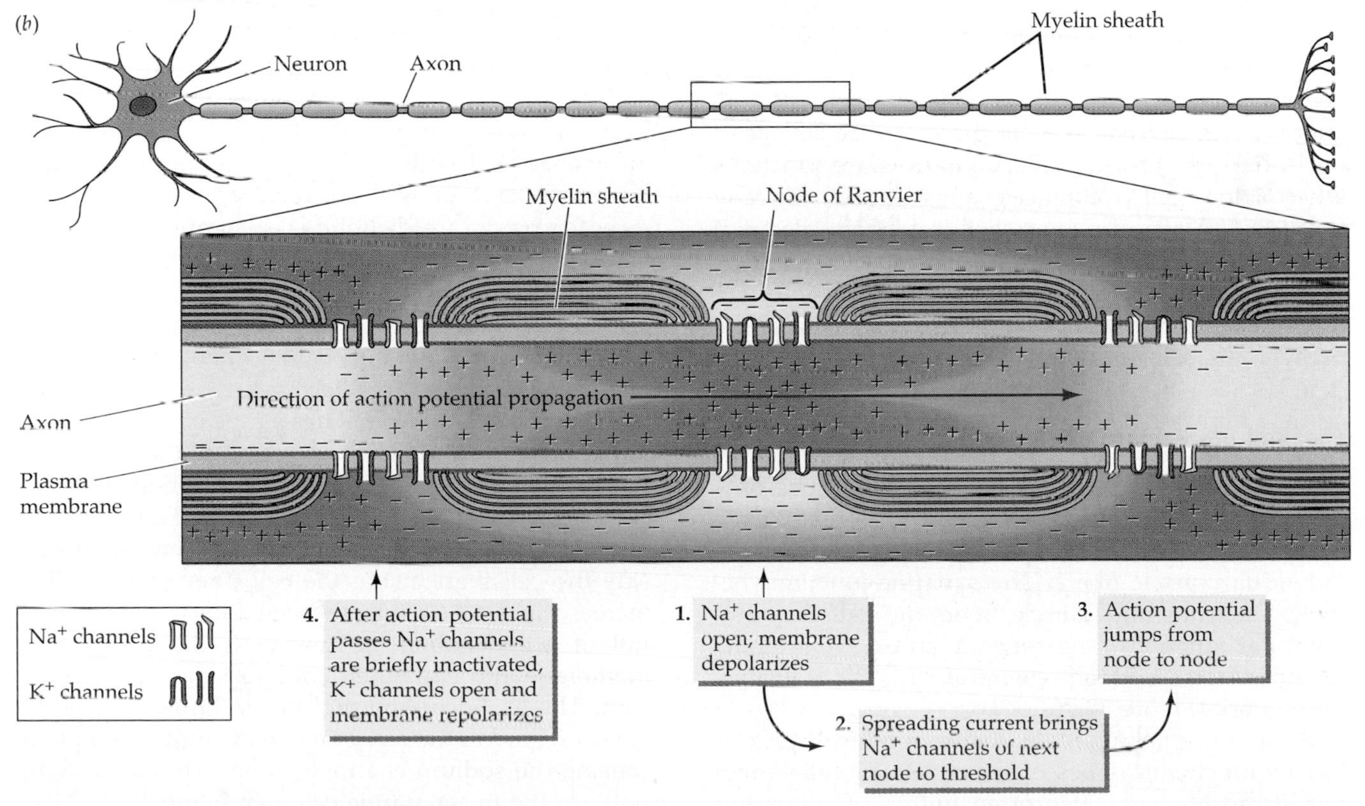

38.14 Saltatory Action Potentials
(*a*) Some axons are myelinated by the wrappings of Schwann cell membranes. (*b*) Action potentials can occur only at gaps in the myelin wrap (nodes of Ranvier), but electric currents created by an action potential can flow to adjacent nodes and depolarize them. As a result, the action potential is conducted down the axon by jumping from node to node.

very rapidly to the next node. When that node reaches threshold, it fires an action potential, and so forth, with the action potentials jumping from node to node down the axon (Figure 38.14*b*). This form of impulse propagation is called **saltatory** (jumping) **conduction** and is much quicker than continuous impulse conduction down an unmyelinated axon.

You have probably experienced the difference in the velocity at which action potentials travel down myelinated and unmyelinated axons. If you touch a very hot or very cold object, you experience a sharp pain before you sense whether the object is hot or cold. With the sensing of the temperature also comes a burning pain different from the first sharp pain. Sensory axons carrying sharp-pain sensation are myelinated, but most axons carrying information about temperature, as well as axons carrying the sensation of burning, aching pain, are unmyelinated. As a result, you know that something is wrong before you know what it is or how bad it is. The unmyelinated temperature and burning-pain axons conduct impulses at only 1 to 2 meters per second, whereas the myelinated sharp-pain axons conduct impulses at velocities up to 5 or 6 m/s. The largest myelinated axons in the human nervous system conduct impulses at velocities up to 120 meters per second.

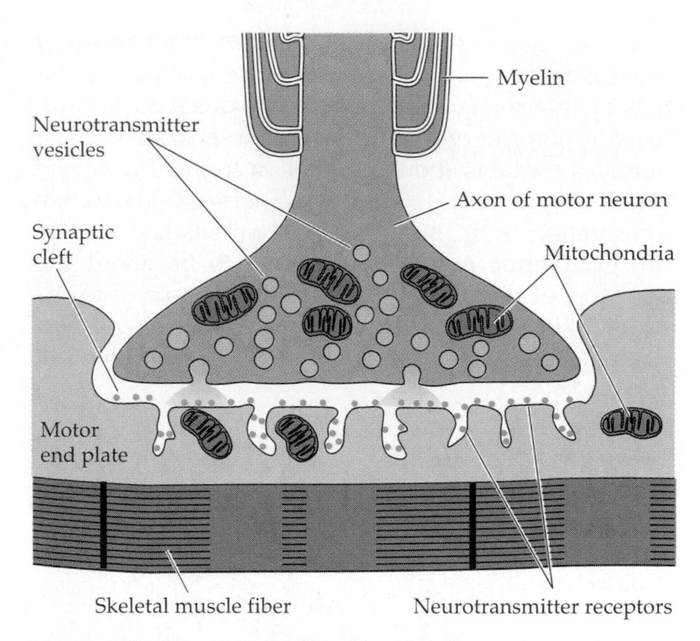

38.15 The Neuromuscular Junction Is a Chemical Synapse
A motor neuron communicates chemically with muscle cells at the neuromuscular junction when transmitter (green) crosses the synaptic cleft.

SYNAPTIC TRANSMISSION

The most remarkable abilities of nervous systems stem from the interactions of neurons. These interactions process and integrate information. Our nervous systems can orchestrate complex behaviors, deal with complex concepts, and learn and remember because large numbers of neurons interact with one another. The mechanisms for these interactions lie in the synapses between cells. **Synapses** are junctions where one cell influences another cell directly through the transfer of a chemical or an electrical message. Most synaptic transmissions are chemical, and we will focus first on those. Chemical information crosses synapses in one direction only, from the **presynaptic cell** to the **postsynaptic cell**.

The Neuromuscular Junction

A motor neuron that innervates a muscle has only one axon, but that axon can branch into many axon terminals that form synaptic junctions with many individual muscle fibers. The synaptic junctions between neurons and muscle fibers are called neuromuscular junctions, and they are an excellent model for how fast, excitatory chemical synaptic transmissions work (Figure 38.15).

Axon terminals contain many spherical vesicles filled with chemical messenger molecules called **neurotransmitters**. The neurotransmitter of all motor neurons that innervate vertebrate skeletal muscles is acetylcholine. The postsynaptic membrane is part of the muscle cell's plasma membrane, but it is slightly modified in the area of the synapse and is called a **motor end plate**. The modification that makes a patch of membrane a motor end plate is the presence of acetylcholine receptor molecules. The receptors function as chemically gated channels that allow both Na^+ and K^+ ions to pass through. Since the resting membrane is already fairly permeable to K^+ ions, the major change that occurs when these channels open is the movement of Na^+ ions into the cell. When a receptor binds acetylcholine, a channel opens and Na^+ ions move into the cell, making the cell more positive inside.

The transmission of a chemical message across a neuromuscular junction begins when an action potential arrives at the axon terminal. The plasma membrane of the axon terminal has a type of voltage-gated ion channel found nowhere else on the axon: the voltage-gated calcium channel. The action potential causes the calcium channels to open (Figure 38.16). Because Ca^{2+} ions are in greater concentration outside the cell than inside the cell, they rush in. The increase in Ca^{2+} inside the cell causes the vesicles full of acetylcholine to fuse with the presynaptic membrane and eject their contents into the synaptic cleft. The acetylcholine molecules diffuse across the cleft and bind to the receptors on the motor end plate, causing the sodium channels to open briefly and depolarize the postsynaptic cell membrane.

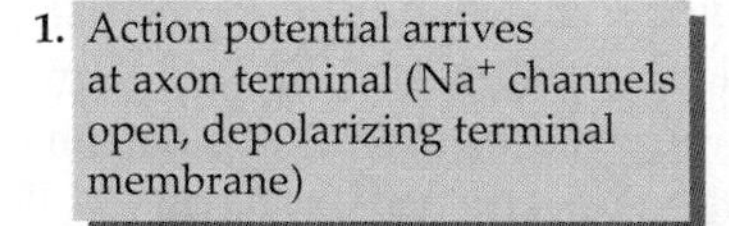

2. Depolarization of terminal membrane causes voltage-gated Ca^{2+} channels to open. Ca^{2+} enters cell and triggers fusion of transmitter vesicles with presynaptic membrane

3. Neurotransmitter molecules diffuse across synaptic cleft and bind to receptors on post-synaptic membrane. Activated receptors open chemically gated Na^+ channels and depolarize postsynaptic membrane

4. Neurotransmitter is broken down and taken back up to be reused in acetylcholine synthesis

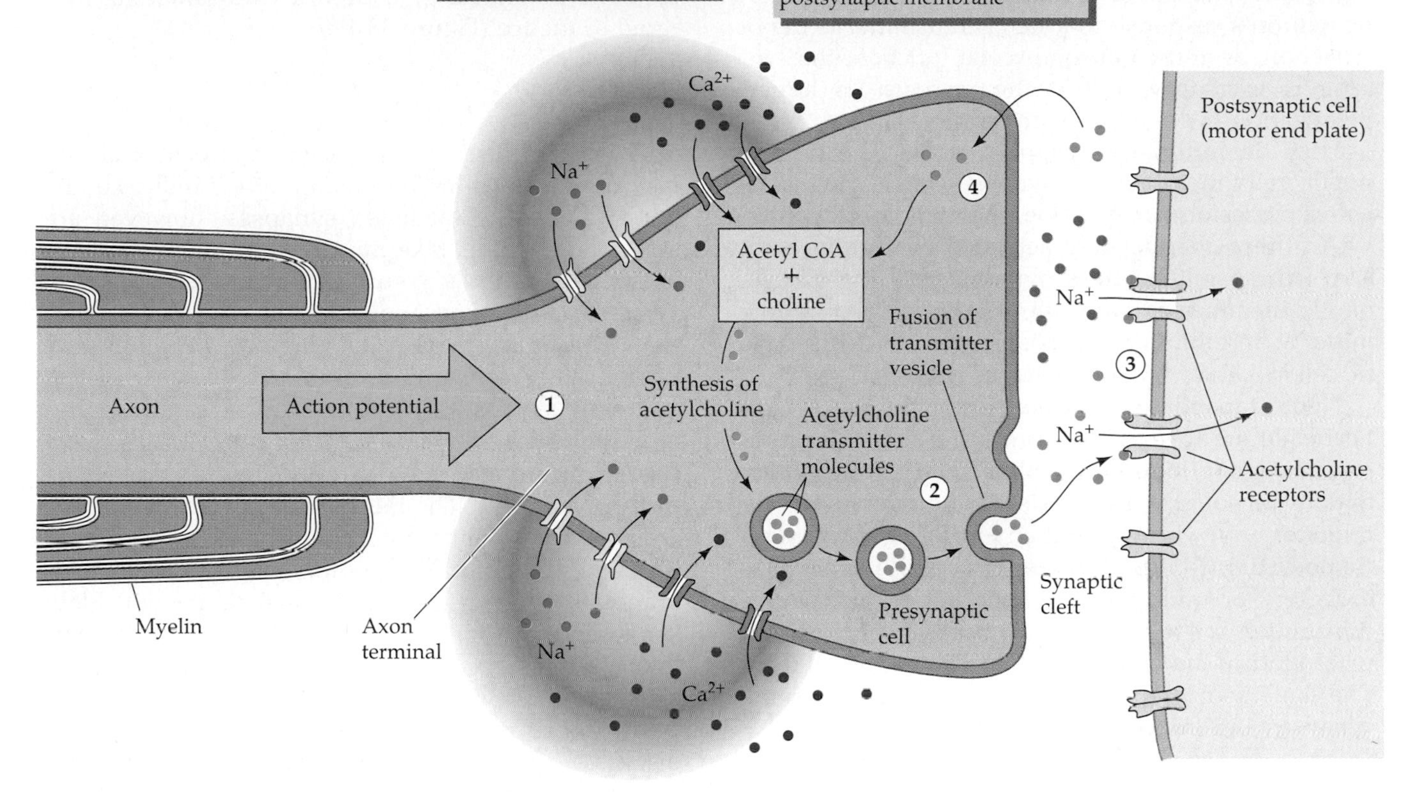

38.16 Synaptic Transmission Begins with the Arrival of an Action Potential

Events in the Postsynaptic Membrane

The postsynaptic membranes differ from the presynaptic membranes in an important way: Because motor end plates have very few voltage-gated sodium channels, they do not fire action potentials. This is true not only of motor end plates, but also of dendrites and of most regions of nerve cell bodies. The binding of neurotransmitter to receptors and the resultant opening of chemically gated ion channels perturbs the resting potential of the postsynaptic membrane. This local change in membrane potential spreads to neighboring regions of the plasma membrane of the postsynaptic cell. Eventually, the spreading depolarization may reach an area of membrane that does contain voltage-gated channels. The entire plasma membrane of a skeletal muscle fiber, except for the motor end plates, has voltage-gated sodium channels. If the axon terminal of a presynaptic cell releases sufficient amounts of neurotransmitter to depolarize a motor end plate enough to bring the surrounding membrane to threshold, action potentials are fired in those areas of membrane. These action potentials are then conducted throughout the muscle fiber's system of membranes, causing the fiber to contract. (We'll learn about the coupling of muscle membrane action potentials and contraction of muscle fibers in Chapter 40.)

How much neurotransmitter is enough? Neither a single acetylcholine molecule nor the contents of an entire vesicle (about 10,000 acetylcholine molecules) is enough to bring the plasma membrane of a muscle cell to threshold. A single action potential in an axon terminal, however, releases about 100 vesicles, which is enough to fire an action potential in the muscle fiber and cause it to twitch.

Excitatory and Inhibitory Synapses

In vertebrates, the synapses between motor neurons and skeletal muscle are always excitatory; that is, motor end plates always respond to acetylcholine by depolarizing. There are several different kinds of synapses between neurons, however. Recall that a given neuron may have many dendrites. Axon terminals from many other neurons may make synapses with those dendrites and with the cell body. The axon terminals of different presynaptic neurons may store and release different neurotransmitters, and membranes of the dendrites and cell body of a postsynaptic neuron may have receptors to a variety of neurotransmitters. Thus a given postsynaptic neuron can receive various chemical messages. If the postsynaptic neuron's response to a neurotransmitter is depolarization, as at the neuromuscular junction, the synapse is excitatory, but if the response is hyperpolarization the synapse is inhibitory.

How do inhibitory synapses work? The postsynaptic cells in inhibitory synapses have chemically gated potassium or chloride channels as receptors. When these channels are activated by binding with a neurotransmitter, they hyperpolarize the postsynaptic membrane. Thus the release of neurotransmitter at an inhibitory synapse makes the postsynaptic cell *less* likely to fire an action potential.

Neurotransmitters that depolarize the postsynaptic membrane are excitatory and bring about an **excitatory postsynaptic potential** (EPSP). Neurotransmitters that hyperpolarize the postsynaptic membrane are inhibitory; they bring about an **inhibitory postsynaptic potential** (IPSP). However, whether a synapse is excitatory or inhibitory depends not on the neurotransmitter but on the postsynaptic receptors—on what kind of ion channels the postsynaptic cell has. The same neurotransmitter can be excitatory at some synapses and inhibitory at others.

Summation

Individual neurons "decide" whether or not to fire action potentials by summing excitatory and inhibitory postsynaptic potentials. This summation ability of neurons is the major mechanism by which the nervous system integrates information. Each neuron may receive 10,000 or more synaptic inputs, yet it has only one output—an action potential in a single axon. All the information contained in the thousands of inputs a neuron receives is reduced to the rate at which that neuron generates action potentials in its axon. For most neurons the critical area for "decision making" is the **axon hillock**, the region of the cell body at the base of the axon. The plasma membrane of the axon hillock is not insulated by glia and has many voltage-gated channels. Excitatory and inhibitory postsynaptic potentials from anywhere on the dendrites or the cell body spread to the axon hillock. If the resulting combined potential depolarizes this area of membrane to threshold, the axon fires an action potential. Because postsynaptic potentials decrease as they spread from the site of the synapse, all postsynaptic potentials do not have equal influences on the axon hillock. A synapse at the end of a dendrite has less influence than a synapse on the cell body near the axon hillock.

Excitatory and inhibitory postsynaptic potentials can be summed over space or over time. Spatial summation adds up the simultaneous influences of synapses from different sites on the postsynaptic cell (Figure 38.17*a*). Temporal summation adds up postsynaptic potentials generated at the same site in a rapid sequence (Figure 38.17*b*).

Other Synapses

Synapses that use chemically gated ion channels are fast; their actions happen within a few milliseconds. Some chemically mediated synapses, however, are slow; their actions take hundreds of milliseconds or even many minutes. Neurotransmitters at these slow synapses activate second-messenger systems, rather than directly controlling ion channels in the postsynaptic cell. Presynaptic events are the same in fast and slow synapses, but when the neurotransmitter of a slow synapse binds to a receptor, it activates a second messenger such as cAMP (cyclic adenosine monophosphate). The mechanisms of slow synapses are therefore similar to the mechanisms of certain hormones that bind to receptors in the plasma membranes of their target cells (see Chapter 36). Slow synapses may open ion channels, influence membrane pumps, activate enzymes, and induce gene expression.

All the neuron-to-neuron synapses that we have discussed up to now are between the axon terminals of a presynaptic cell and the cell body or dendrites of a postsynaptic cell. Synapses can also form between the axon terminals of one cell and the axon of another cell. Such a synapse can modulate how much neurotransmitter the second cell releases in response to action potentials traveling down its axon. We refer to this mechanism of regulating synaptic strength as **presynaptic excitation** or **presynaptic inhibition**.

Electrical synapses, or gap junctions, are completely different from chemical synapses because they directly couple neurons electrically. At gap junctions, the presynaptic and postsynaptic cell membranes are separated by a space of 2 to 3 nm, but the membrane proteins of the two neurons form connexons—molecular tunnels that bridge the two cells—through which ions and small molecules can readily pass (see Figure 5.8). Electrical transmission across gap junctions is very fast and can proceed in either direction; that is, stimulation of either neuron can result in an action potential in the other. Gap junctions are less common in the complex nervous systems of vertebrates than they are in the simple nervous systems of invertebrates, for two very important reasons. First, electrical continuity between neurons does not allow summation of synaptic inputs, which is what enables complex nervous systems to integrate information. Second, an effective electrical synapse requires a large area of contact between the presynaptic and postsynaptic cells. This condition rules out the possibility of thousands of synaptic inputs to a single neuron—which is the norm in complex nervous systems.

Neurotransmitters and Receptors

At present more than 25 neurotransmitters are recognized, and more will surely be discovered. No others are as thoroughly understood as **acetylcholine**, the neurotransmitter at all synapses between motor neurons and skeletal muscles (in the voluntary division of the peripheral nervous system). Acetylcholine and norepinephrine are the neurotransmitters of the other efferent division of the peripheral nervous system, the autonomic, which controls involuntary physiological functions. These two neurotransmitters also play roles in the central nervous system, but they constitute only a small percentage of the neurotransmitter content of the brain.

The workhorse neurotransmitters of the brain are simple amino acids. Glutamic acid and aspartic acid are excitatory, whereas glycine and gamma-aminobutyric acid (GABA) are inhibitory. Another important group of neurotransmitters is the monoamines, which are derivatives of amino acids. They include dopamine and norepinephrine (derivatives of tyrosine) and serotonin (a derivative of tryptophan). A number of peptides also function as neurotransmitters. A very exciting recent discovery is that two gases, carbon monoxide and nitric oxide, are used by neurons as intercellular messengers.

A neurotransmitter may have several different types of receptors in different tissues and may induce different actions. For example, acetylcholine has two well-known receptor types, called **muscarinic** and **nicotinic** because of other compounds that also bind to them. Nicotine, the active ingredient in tobacco, binds to acetylcholine receptors in the skeletal mus-

38.17 The Postsynaptic Membrane Integrates Information
Postsynaptic potentials can be summed over space or time. When the sum exceeds a threshold, action potentials are generated. *(a)* Spatial summation occurs when several postsynaptic potentials arrive at the axon hillock simultaneously. *(b)* Temporal summation means that postsynaptic potentials created at the same site in rapid succession can also be summed.

Dendrites
S1
S2
S3
S4
Excitatory synapses
Axon hillock
Axon
Neuron
Action potential

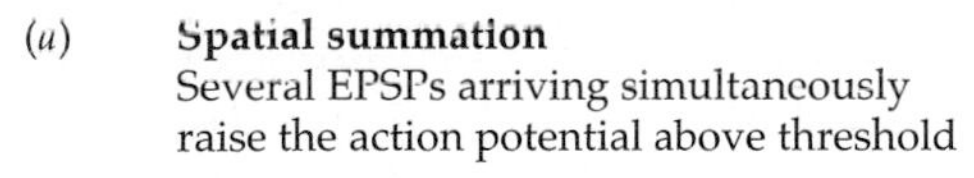
(*a*) **Spatial summation**
Several EPSPs arriving simultaneously raise the action potential above threshold

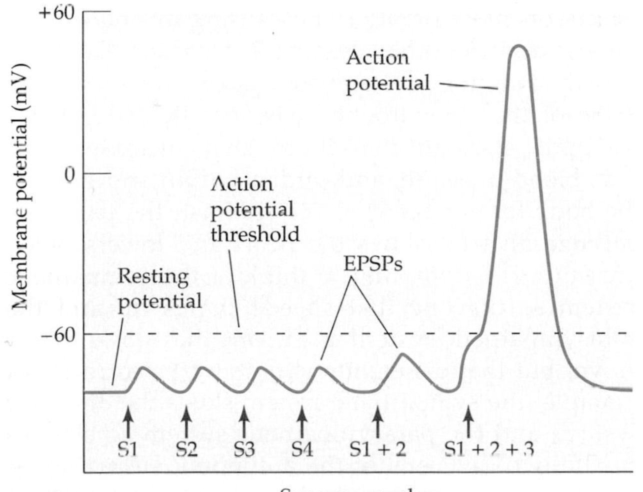

(*b*) **Temporal summation**
Two or more EPSPs in rapid succession raise the action potential above threshold

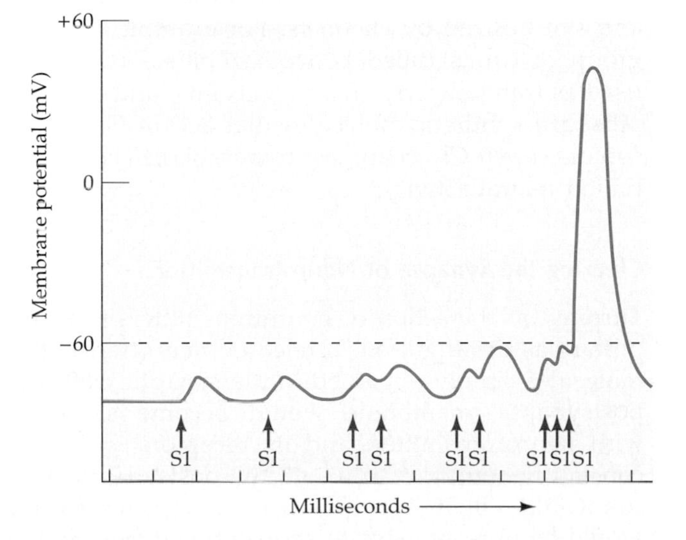

cles, but not to those in heart muscle or in the autonomic nervous system. Muscarine, a compound found in the deadly poisonous mushroom *Amanita muscaria,* binds to the acetylcholine receptors in heart muscle and in the autonomic nervous system, but not to those in skeletal muscle. Both types of acetylcholine receptors are found in the central nervous system, where nicotinic receptors tend to be excitatory and muscarinic tend to be inhibitory. These receptors are the reason that smoking tobacco has behavioral and physiological effects and is addictive and why a number of cultures around the world have used *Amanita* mushrooms as hallucinogenic drugs.

The drug **curare**, extracted from the bark of a South American plant and used by native peoples to make poisoned darts and arrows, binds to nicotinic receptors but does not activate them. Therefore, skeletal muscles in an animal poisoned by curare cannot respond to motor neuron activation. The animal goes into flaccid (relaxed) paralysis and dies because it stops breathing. Curare is used medically to treat severe muscle spasms and to prevent muscle contractions that would interfere with surgery. Another compound, **atropine**, which is extracted from the plant *Atropa belladonna,* binds to muscarinic receptors and prevents acetylcholine from activating them. Atropine is used medically to increase heart rate, decrease secretions of digestive juices, and decrease spasms of the gut. Most people have encountered atropine; it is what the eye doctor uses to dilate the pupils for eye examinations. In the past atropine was used cosmetically to make the eyes look big and dark—hence the plant's species name, belladonna, meaning "beautiful lady." Of course, these beautiful ladies could not see very well.

The ability of compounds extracted from plants and animals to bind to certain neurotransmitter receptors is the basis for neuropharmacology, the study and development of drugs that influence the nervous system. Natural products are still an important source of drugs, but today many drugs are designed and synthesized by chemists. For example, a major group of drugs called benzodiazepines, which are used as tranquilizers, muscle relaxants, and sleeping pills, are synthetic molecules that act on GABA receptors, open Cl^- channels, hyperpolarize cells, and inhibit neural activity.

Clearing the Synapse of Neurotransmitter

Turning off the action of neurotransmitters is as important as turning it on. If released neurotransmitter molecules simply remained in the synaptic cleft, the postsynaptic membrane would become saturated with neurotransmitter, and its receptors would be constantly bound. As a result, the postsynaptic neuron would remain hyperpolarized or depolarized and would be unresponsive to short-term changes in the presynaptic neuron. Thus neurotransmitter must be cleared from the synaptic cleft shortly after it is released by the axon terminal.

Neurotransmitter action is terminated in one of several ways. First, enzymes may destroy the neurotransmitter. For example, acetylcholine is rapidly destroyed by the enzyme **acetylcholinesterase**, which is present in the synaptic cleft in close association with the acetylcholine receptors on the postsynaptic membrane. Some of the most deadly nerve gases that were developed for chemical warfare work by inhibiting acetylcholinesterase. As a result, acetylcholine lingers in the synaptic clefts, causing the victim to die of spastic muscle paralysis. Some agricultural insecticides, such as malathion, also inhibit acetylcholinesterase and can poison farm workers if used without safety precautions. Second, neurotransmitter may simply diffuse away from the cleft. Third, neurotransmitter may be taken up via active transport by nearby cell membranes. Each of these mechanisms—enzymatic destruction, diffusion, and active transport—can clear the synaptic cleft so that a new, discrete signal can pass through the synapse.

NEURONS IN CIRCUITS

Because neurons can interact in the complex ways we have just discussed, networks of neurons can process and integrate information. Next we will examine networks in different parts of the nervous system.

The Autonomic Nervous System

The autonomic nervous system controls the organs and organ systems of the body by influencing the activities of glands and involuntary muscles. There are two divisions of the autonomic nervous system, the **sympathetic** and the **parasympathetic**. These two divisions work in opposition to each other in their effects on most organs, one causing an increase in activity and the other causing a decrease. The best-known autonomic nervous system functions are those of the sympathetic division called the "fight-or-flight" mechanisms—those that increase heart rate, blood pressure, and cardiac output and prepare the body for emergencies. By contrast, the parasympathetic division slows the heart and lowers blood pressure. It is tempting to think of the sympathetic system as the one that speeds things up and the parasympathetic system as the one that slows things down, but that distinction is not always correct. For example, the sympathetic system slows the digestive system, and the parasympathetic system accelerates it. The two divisions of the autonomic nervous system are easily distinguished by their anatomy, their neurotransmitters, and their actions (Figure 38.18).

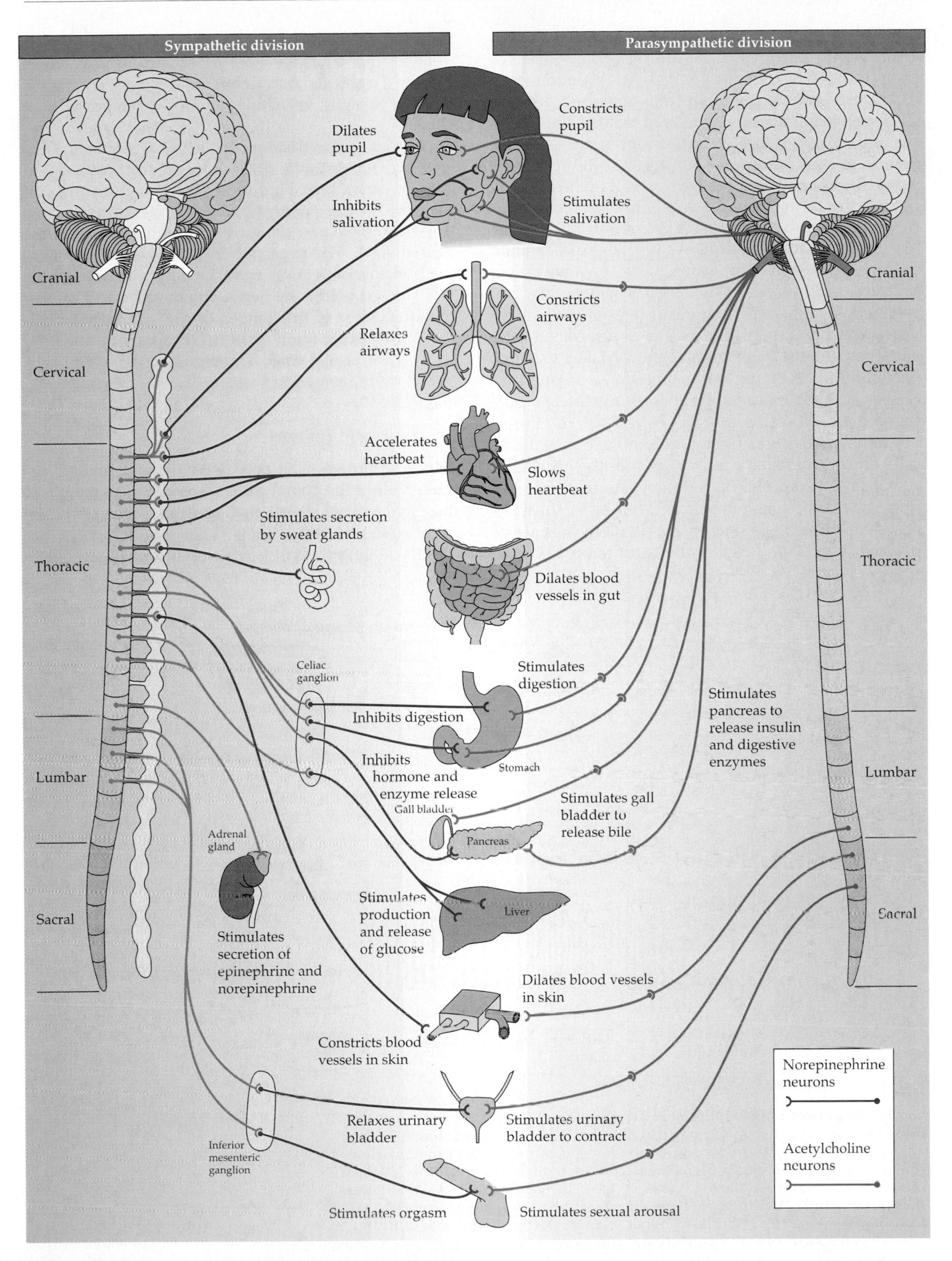

38.18 Organization of the Autonomic Nervous System

Both divisions of the autonomic nervous system are efferent pathways of the central nervous system. Each autonomic efferent begins with a neuron that has its cell body in the brain stem or spinal cord and uses acetylcholine as its neurotransmitter. These cells are called **preganglionic neurons** because the second neuron in each autonomic output pathway resides in a **ganglion** (a collection of neuron cell bodies) that is outside the central nervous system. The second neuron in an autonomic nervous system output pathway is a **postganglionic neuron** because its axon extends from the ganglion. The axons of the postganglionic cells end on the cells of the target organs.

The postganglionic neurons of the sympathetic division use norepinephrine as their neurotransmitter, and those of the parasympathetic division use acetylcholine as their neurotransmitter. In organs that receive both sympathetic and parasympathetic input, the target cells respond in opposite ways to norepinephrine and acetylcholine. A region of the heart called the pacemaker, which generates the heartbeat, provides an example. Norepinephrine increases the firing rate of pacemaker cells and causes the heart to beat faster. Acetylcholine decreases the firing rate of pacemaker cells and causes the heart to beat slower. In another example, norepinephrine causes muscle cells in the digestive tract to hyperpolarize, which slows digestion. Acetylcholine depolarizes muscle cells in the gut, which accelerates digestion (Figure 38.19).

Anatomy also distinguishes the sympathetic from the parasympathetic division of the autonomic nervous system (see Figure 38.18). The preganglionic neurons of the parasympathetic division come from the brain stem and the lowest (sacral) segment of the spinal cord. The preganglionic neurons of the sympathetic division come from the upper regions of the spinal cord below the neck—the thoracic and lumbar regions. Most of the ganglia of the sympathetic nervous system are lined up in two chains, one on either side of the spinal cord. The parasympathetic ganglia are close to, sometimes sitting on, the target organs.

Monosynaptic Reflexes

Much information is processed through neural circuits within the spinal cord. The simplest example of a spinal neural circuit that controls behavior is the **monosynaptic reflex loop**. This type of reflex depends on neural circuits made up of a sensory neuron

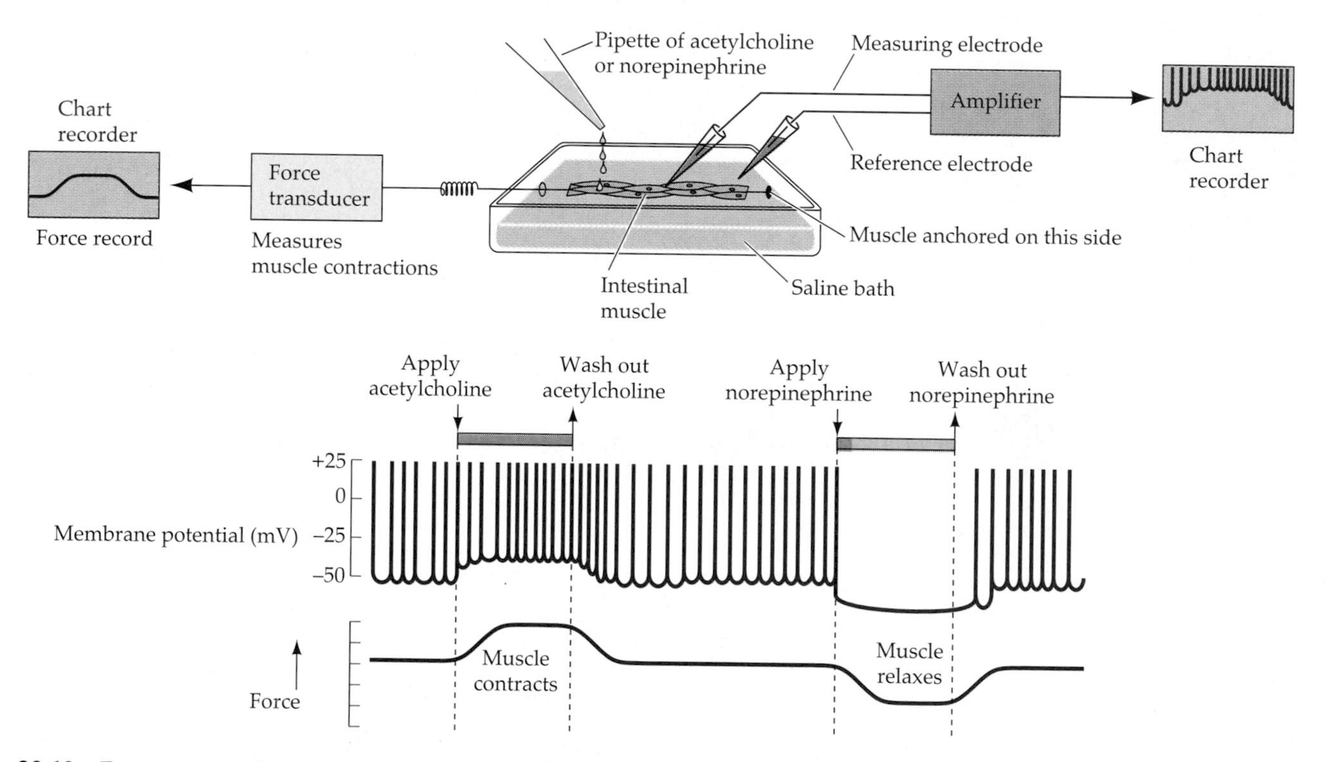

38.19 Responses to Postganglionic Transmitter
In this experiment, a strip of intestinal muscle is mounted in a saline bath that allows electric current to flow so that the force of contractions of the muscle can be measured. An electrode records action potentials in a muscle cell. When acetylcholine is dripped onto the muscle, the cells depolarize, fire action potentials more rapidly, and increase their force of contraction. Norepinephrine, on the other hand, causes the cells to hyperpolarize, decrease their rate of firing, and decrease their force of contraction.

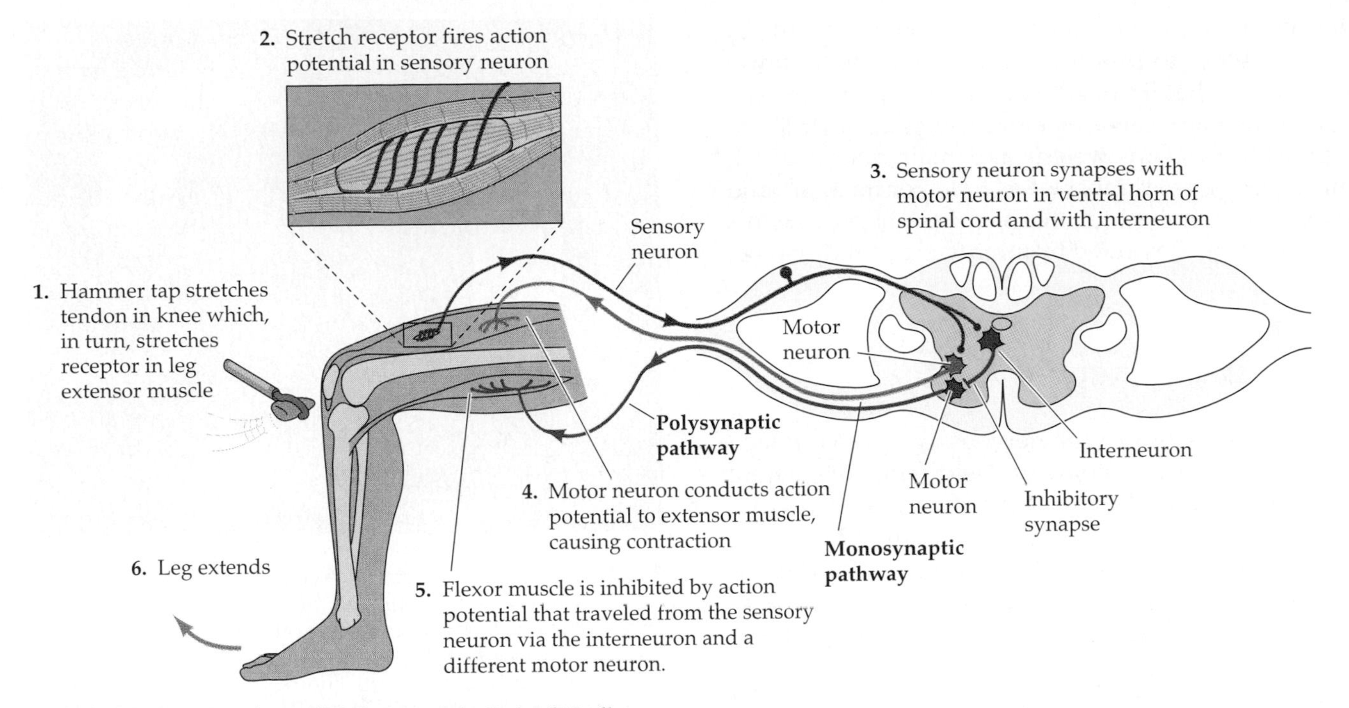

38.20 Monosynaptic and Polysynaptic Spinal Reflexes The knee-jerk reflex is an example of a *monosynaptic* reflex loop in action. Because muscles work in antagonistic pairs, however, one must relax while the other contracts. Thus the knee-jerk reflex is accompanied by a parallel reflex that involves spinal interneurons and therefore more than one synapse—that is, it is *polysynaptic*.

and a motor neuron with just one synapse (hence the term *mono*synaptic) between them. It is common in a visit to a physician to have your reflexes tested by being struck below the kneecap with a rubber mallet. The response is a rapid, involuntary extension of the leg—the knee-jerk reflex. Similar reflexes can be elicited by sharp blows to tendons of the wrist, elbow, ankle, and other joints. Figure 38.20 shows how the knee-jerk reflex works.

The mallet blow on the tendon causes a quick stretch of the muscle attached to that tendon. Within the muscle are modified muscle fibers wrapped in connective tissue. These **muscle spindles** are stretch sensors that activate sensory neurons when they are stretched. The number of nerve impulses per second carried by the sensory neuron signals the degree of stretch. The cell body of the sensory neuron from the muscle spindle is in a ganglion on the dorsal root of the spinal cord, and the axon of this neuron extends all the way to the gray matter of the ventral horn of the spinal cord. There the sensory fiber branches and forms synapses with motor neurons for the same muscle from which the sensory neuron originated. Each motor neuron sends impulses along its axon, which leaves the spinal cord through the ventral root. The axon of the motor neuron synapses on the stretched muscle, causing it to contract. The function of this reflex loop is to adjust the contraction in the muscle to changing loads. An increased load on the limb stretches the muscle, and the stretch reflex returns it to the desired position by increasing the strength of contraction of the muscle.

Even though the knee-jerk reflex is involuntary, you are aware of being struck on the knee. This means the information also travels to your brain. Branches of the sensory neuron form synapses with interneurons in the dorsal horn of the spinal cord. These interneurons send axons up the dorsal white matter tracts to the thalamus and on to the cerebral cortex. Motor commands from the cerebral cortex descend the spinal cord in other white matter tracts to form synapses with the same motor neurons involved in the reflex loop. Thus the same muscle can be controlled both by involuntary reflexes and by conscious commands.

Polysynaptic Reflexes

Most involuntary reflex circuits include more than one synapse. A number of interneurons in the central nervous system are necessary for more-complex responses. For example, a limb can move in opposite directions because muscles work in pairs. One muscle of a pair is an extensor and the other is a flexor. At the same time a flexor contracts, the extensor must relax, or the limb cannot move. A polysynaptic reflex is added to the monosynaptic circuit controlling the stretch reflex to cause relaxation of the opposing muscle (see Figure 38.20).

Much more complex polysynaptic reflexes are responsible for coordinated escape movements. If you step on a tack, many muscles in your foot and leg work together to produce a coordinated withdrawal response of that limb, while muscles on the opposite side of the body cause extension of your other leg to support your body weight and maintain your balance. The large number of muscle contractions and relaxations required for this sequence of movements are initially orchestrated by interneurons in the spinal cord.

HIGHER BRAIN FUNCTIONS

Very few functions of the nervous system have been worked out to the point of identifying the neural circuits that underlie them. Brain processes responsible for phenomena such as thought, perception, memory, and learning are extremely complex. Nevertheless, neurobiologists using a wide range of techniques are making considerable progress in understanding some of the neural mechanisms involved in these higher functions of the nervous system. For example, there have been rapid advances in understanding how the nervous system processes visual information, which we will discuss in the next chapter. The remainder of this chapter focuses on several complex aspects of brain and behavior that present challenges. Neurobiologists want to learn how individual neurons function and how neurons interact in circuits to produce these behaviors.

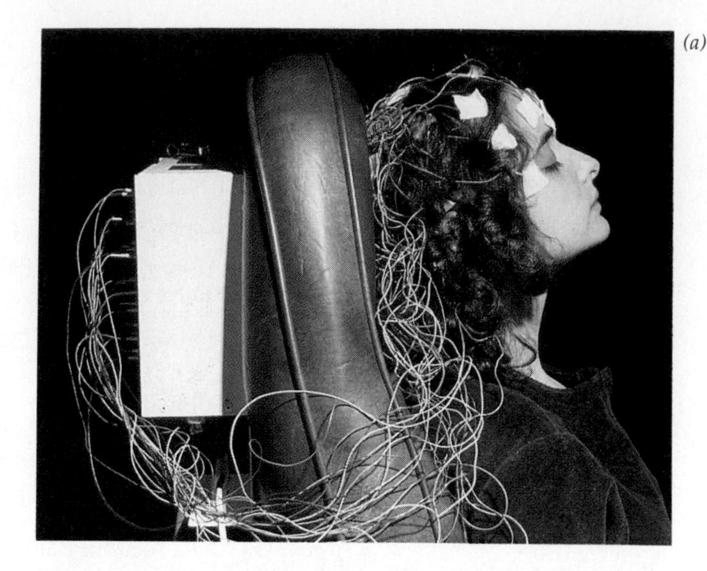

38.21 Patterns of Electrical Activity in the Cortex Characterize Stages of Sleep
(a) Electrical activity of the cerebral cortex is detected by electrodes placed on the scalp and recorded on moving chart paper by a polygraph. The resulting record is the electroencephalogram (EEG). *(b)* The pattern of the EEG is different in wakefulness and in different stages of sleep. *(c)* During a night we cycle through the different stages of sleep. We experience our deepest slow-wave sleep (stage 4) during the first half of the night. We dream during REM sleep, which usually occurs in four or five episodes during the night. We often awaken briefly after REM sleep.

Sleeping and Dreaming

A dominant feature of our behavior is the daily cycle of being asleep and awake. All birds and mammals sleep, and probably all other vertebrates sleep as well. We spend one-third of our lives sleeping, yet we do not know why or how. We do know, however, that we need to sleep. Loss of sleep impairs alertness and performance. Most people in our society—certainly most college students—are chronically sleep-deprived. Many of the accidents and serious mistakes that endanger lives can be attributed to impaired alertness due to sleep loss. Yet insomnia (difficulty in falling asleep) is one of the most common medical complaints. Thus it is important to learn more about the neural control of sleep.

A common tool of sleep researchers is the **electroencephalogram** (EEG), a record of the electrical activity occurring primarily in the cerebral cortex (Figure 38.21). Sleep researchers also record the electrical activity of one or more skeletal muscles as an **electromyogram** (EMG) on the same moving chart. EEG and EMG patterns reveal the transition from being awake to being asleep; they also reveal that there are different states of sleep. Mammals other than humans have two major sleep states: **slow-wave sleep** and rapid eye-movement sleep, commonly referred to as **REM sleep**. In humans we characterize sleep as REM sleep or non-REM sleep. Human non-REM sleep can be further divided into four stages, and only the two deepest stages are considered true slow-wave sleep. When you fall asleep at night you enter non-REM sleep and progress from stage 1 to stage 4, with stages 3 and 4 being deep, restorative, slow-wave-sleep. After this first episode of non-REM sleep, you enter an episode of REM sleep. Throughout the night, you have four or five cycles of non-REM/REM sleep (see Figure 38.21). About 80 percent of human sleep is non-REM sleep and 20 percent is REM sleep.

Vivid dreams and nightmares occur during REM sleep, which gets its name from the jerky movements the eyeballs make during this state. The most remarkable feature of REM sleep is that commands from the brain almost completely paralyze the skeletal muscles. Occasional muscle twitches break through the paralysis, as in a dog that appears to be trying to run in its sleep. If you look closely at a sleeping dog when its legs and paws are twitching, you will be able to see the rapid eye movements. The function of muscle paralysis during REM sleep is probably to prevent the acting out of dreams. Sleepwalking occurs during non-REM sleep.

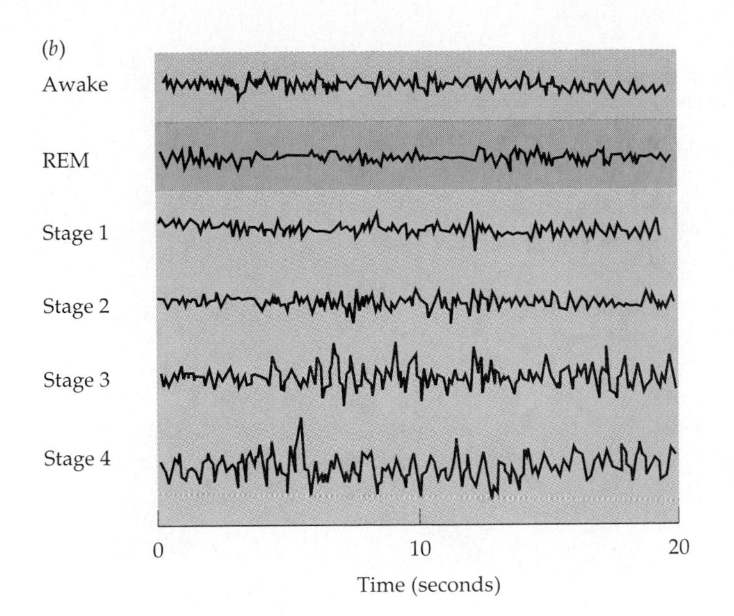

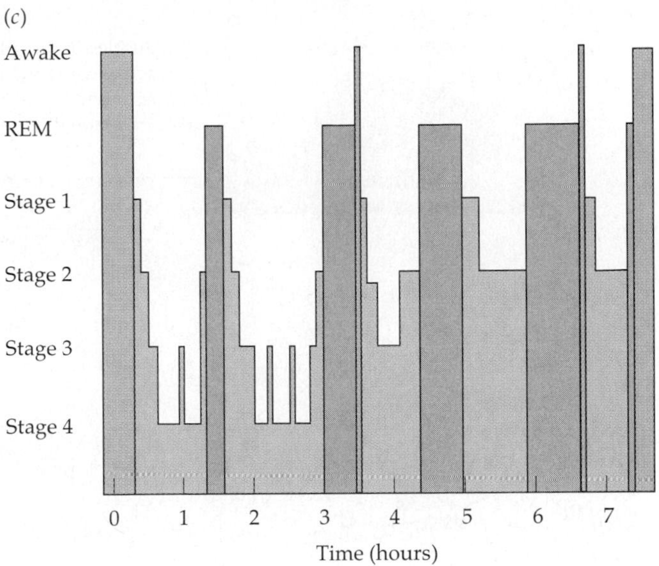

Learning and Memory

Learning is the modification of behavior by experience. **Memory** is the ability of the nervous system to retain what is learned and what is experienced. Even very simple animals can learn and remember, but these two abilities are developed the most in humans. Language, culture, artistic creativity, and scientific progress are made possible by these abilities. Consider the amount of information associated with learning a language. The capacity of human memory and the rate at which items can be retrieved are remarkable features of the nervous system. Is it possible to understand these phenomena in terms of the cells and molecules that make up the brain?

Habituation and sensitization are simple forms of learning that can be studied in all nervous systems, from the nerve nets of cnidarians to the human central nervous system. **Habituation** is learning to ignore a repeated stimulus that conveys little information. **Sensitization** is learning to be especially aware of a stimulus that conveys important information. For example, humans can habituate to noisy, busy, crowded environments, ignoring most of the barrage of sensory information they receive from those environments. Hearing your name spoken in a crowded, noisy room, however, immediately gets your attention.

Habituation and sensitization can be understood as processes that rely on individual synapses. The synaptic basis of habituation and sensitization has been studied extensively by Eric Kandel and his colleagues at Columbia University. The animal they use in their experiments is a marine mollusk called a sea slug or a sea hare (*Aplysia californica*). Because the sea slug does not have a shell, it is very vulnerable to predators. When the sea slug is undisturbed, its siphon is extended to take in water to ventilate its gill membranes. If the siphon is touched, the animal withdraws it (Figure 38.22*a*). If the siphon is touched repeatedly, the animal habituates to the stimulus and no longer withdraws it. The researchers found that the siphon-withdrawal reflex depends on sensory neurons and motor neurons with only one synapse between them. By studying the characteristics of that synapse, they learned that habituation was due to a decrease in the amount of neurotransmitter released by the axon terminals of the presynaptic cells. This reduced release of neurotransmitter could last a considerable time and was not due simply to fatigue or depletion of neurotransmitter.

The researchers also studied sensitization. To sensitize the sea slug, they applied mild electric stimulation to its tail just before gently touching its siphon. Now the animal responded to the touch of its siphon with a much more vigorous withdrawal. Study of the synapse in the withdrawal reflex pathway showed that sensitization was due to the action of a third neuron that formed synapses with the axon terminals of the sensory cell. This sensitizing neuron caused the axon terminal of the sensory neuron to release more transmitter in response to each action potential (Figure 38.22*b*).

Another form of learning that is widespread among animal species is **associative learning**, in which two unrelated stimuli are linked to the same response. The simplest example of associative learning is the **conditioned reflex**, discovered by the Russian physiologist Ivan Pavlov. Pavlov was studying control of digestive functions in dogs and observed that a dog salivates at the sight or smell of food—a simple autonomic reflex. He discovered that if he rang a bell just before the food was presented to the dog, after a few trials the dog would salivate at the sound of the bell, even if no food followed. The

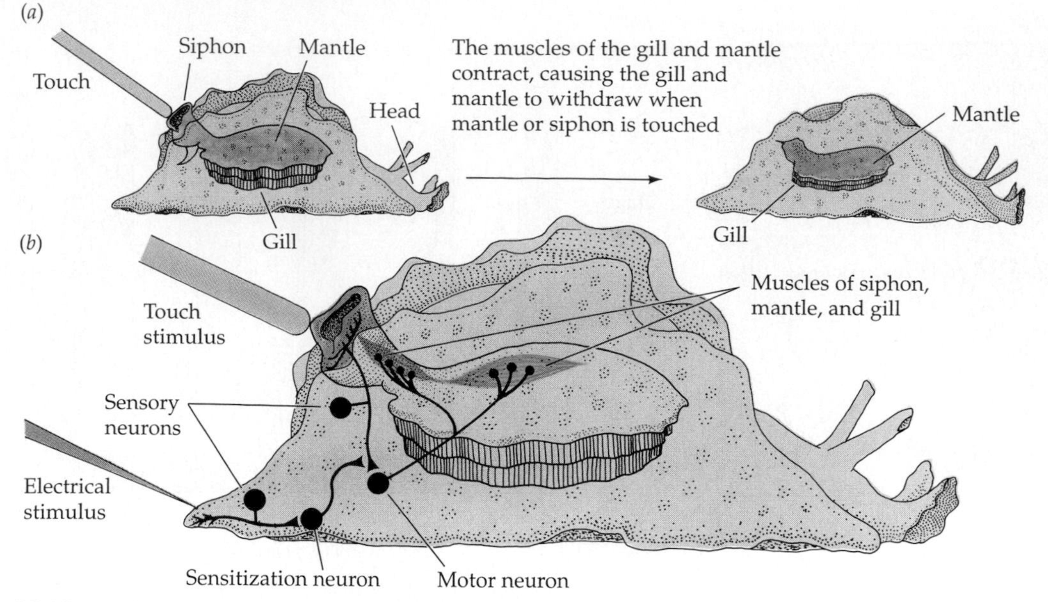

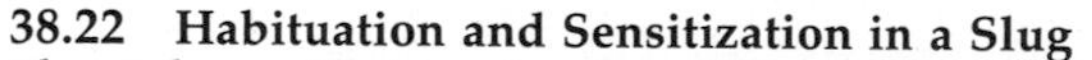

38.22 Habituation and Sensitization in a Slug
The siphon-withdrawal reflex of the sea slug *Aplysia,* shown in *(a),* can be overcome by habituation. *(b)* The opposite effect, sensitization, can also be achieved. Applying a mild electrical shock to the animalUs tail just prior to touching the siphon intensifies the siphon-withdrawal reflex. Habituation and sensitization can be understood in terms of the neurons and synapses diagrammed here.

salivation reflex was conditioned to be associated with the sound of a bell, which normally is unrelated to feeding and digestion. This simple form of learning has been studied extensively in efforts to understand its underlying neural mechanisms.

Attempts to treat human diseases have sometimes led to increases in scientific understanding. Epilepsy is a disorder characterized by uncontrollable increases in neural activity in specific parts of the brain. The resulting epileptic fits can endanger the afflicted individual. At one time, serious cases of epilepsy were sometimes treated by destroying the part of the brain from which the surge of activity originated. To find the right area, the surgery was done under local anesthesia and different regions of the brain were electrically stimulated with fine electrodes while the patient described the resulting sensations. When some regions of association cortex were stimulated, patients sometimes reported vivid memories, as if reliving events from the past. Such observations were the first evidence that memories have anatomical locations in the brain and exist as properties of neurons and networks of neurons. Because the destruction of a small area of the brain does not completely erase a memory, however, it is postulated that memory is a function distributed over many brain regions and that a memory may be stimulated via many different routes.

You should be able to recognize several forms of memory from your own experience. There is **immediate memory** for events that are happening now. Immediate memory is almost perfectly photographic, but it lasts only seconds. **Short-term memory** contains less information, but it lasts longer—on the order of 10 to 15 minutes. If you are introduced to a group of new people, you may remember most of their names for 5 or 10 minutes, but will have forgotten them in an hour or so if you have not repeated them, written them down, or used them in a conversation lasting longer than the round of introductions. Repetition, use, or reinforcement by something that gets your attention (such as the title President or Queen) facilitates the transfer of short-term memory to **long-term memory**, which can last for days, months, or years.

Knowledge about neural mechanisms for the transfer of short-term memory to long-term memory has come from observations of patients who have lost parts of the limbic system, notably the hippocampus. A famous case is that of H. M., who had his hippocampus removed on both the left and right sides of his brain in an effort to control severe epilepsy. Following his surgery, H. M. could not transfer any information to long-term memory. If someone were introduced to him, had a conversation with him, and then left the room for an hour, when that person returned, he or she was unknown to H. M., and it was as if the previous conversation had never taken place. H. M. had normal memory for events that happened before his surgery, but he could remember post-surgery events for only 10 or 15 minutes.

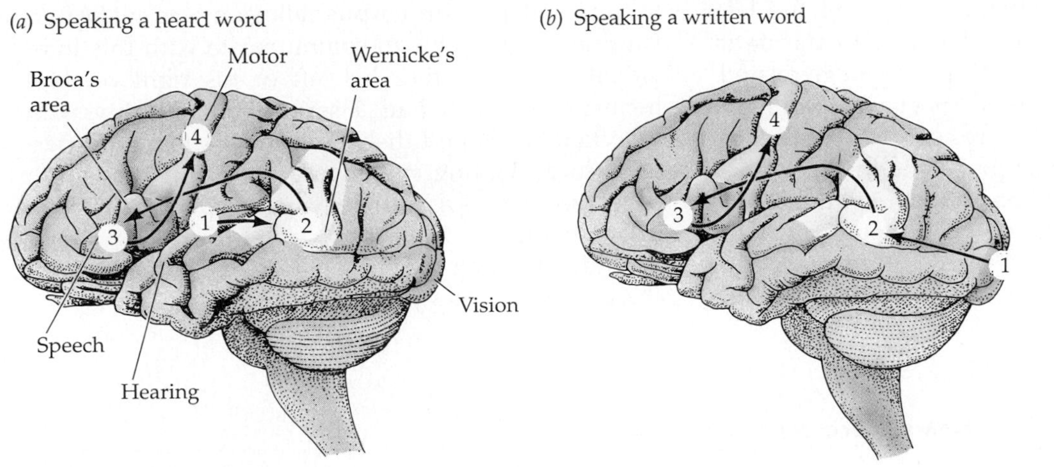

38.23 Language Areas
Different regions of the left cerebral cortex participate in the use of *(a)* spoken and *(b)* written language.

Language, Lateralization, and Human Intellect

No aspect of brain function is as fundamentally related to consciousness and intellect as is language. Therefore, studies of the brain mechanisms underlying the acquisition and use of language are extremely interesting to neuroscientists. A curious fact of language abilities is that they are located mostly in one cerebral hemisphere—the left hemisphere in 97 percent of all people. This phenomenon is referred to as the **lateralization** of language functions. Some of the most fascinating research on this subject has been done by Roger Sperry and his colleagues at the California Institute of Technology; Sperry received the 1981 Nobel prize in medicine for this work. The two cerebral hemispheres are connected by a white-matter tract called the **corpus callosum**. In one severe form of epilepsy, bursts of action potentials travel from hemisphere to hemisphere across the corpus callosum. Cutting the tract eliminates the problem, and patients function quite normally following the surgery.

Experiments have revealed interesting deficits in the language abilities of these "split-brain" patients, however. With the connections between the two hemispheres cut, the knowledge or experience of the right hemisphere can no longer be expressed in language, nor can language be used to communicate with the right hemisphere. Sensory input from the right hand goes to the left cerebral hemisphere, and sensory input from the left hand goes to the right cerebral hemisphere. If a split-brain patient is blindfolded and a familiar tool is placed in his or her right hand, the patient can identify the tool and describe its use. If the tool is placed in the left hand, however, the patient can use the tool correctly, but cannot name it or describe its use. In split-brain individuals, the right hemisphere has lost access to the language abilities that reside predominantly in the left hemisphere. Language is said to be *lateralized* to the left hemisphere.

The brain mechanisms of language in the left hemisphere have been the focus of much research. The experimental subjects are persons who have suffered damage to some region of the left hemisphere and are left with one of many forms of **aphasias**, deficits in the abilities to use or understand words. The known language areas of the left hemisphere are shown in Figure 38.23*a*. Broca's area, in the frontal lobe just in front of the motor cortex, is essential for speech. Damage to Broca's area results in halting, slow, unclear speech or even complete loss of speech, but the patient can still read and understand language. In the temporal lobe, close to its border with the occipital lobe, is Wernicke's area, which is more involved with sensory than with motor aspects of language. Damage to Wernicke's area can cause a person to lose the ability to speak sensibly while retaining the abilities to form the sounds of normal speech and to imitate its rhythm. Moreover, such a patient cannot understand spoken or written language. Near Wernicke's area is the angular gyrus, which is believed to be essential for integration of spoken and written language.

Normal language ability depends on the flow of information between various areas of the left cerebral cortex (Figure 38.23). Input from spoken language travels from the primary auditory (hearing) cortex to Wernicke's area. Input from reading language travels from the primary visual cortex to the angular gyrus to Wernicke's area. Commands to speak are formulated in Wernicke's area and travel to Broca's area and from there to the primary motor cortex. Damage to any one of these areas or the pathways between them can result in aphasia.

Young children can recover remarkably from even severe damage to the left cerebral hemisphere because lateralization of language abilities to the left hemisphere has not fully developed, and the right hemisphere can take over all of the language-related functions. One such patient in Sperry's split-brain studies produced provocative results with respect to the relationship between language and intellect. This patient had language functions in both hemispheres. Because of left-brain damage in childhood, his right hemisphere had developed language functions, and over time his left hemisphere had recovered. Then, to treat epilepsy, his corpus callosum was cut. Afterward it was possible to communicate with this individual through either his left or his right cerebral hemisphere. Each had a separate personality with individual likes and dislikes. Each responded differently to evaluating events and projecting plans into the future. It was as if there were two persons housed in one brain.

Understanding the brain will be one of the greatest challenges in biology for many years to come.

SUMMARY of Main Ideas about Neurons and the Nervous System

Nervous systems process and integrate information received from sensors and communicate commands to effectors.

Nervous systems are composed of cells called neurons, which have cell bodies, dendrites, and axons.
Review Figure 38.2

Neurons communicate with other cells at synaptic junctions.
Review Figures 38.15 and 38.16

A nerve is a bundle of many axons carrying information to and from the central nervous system.
Review Figure 38.3

The brain and spinal cord are the central nervous system, and the cranial and spinal nerves are the peripheral nervous system.
Review Figure 38.4

From a hollow tube the vertebrate nervous system develops a hindbrain, a midbrain, and a forebrain.
Review Figure 38.5

Different regions of the central nervous system have different functions.

The spinal cord receives and sends information.
Review Figure 38.6

The reticular system controls sleeping and waking.
Review Figure 38.7

The limbic system functions in emotion, drive, instinct, and memory.
Review Figure 38.8

The convoluted cerebral hemispheres are the dominant structures of the human brain, and regions within them are responsible for motor functions and particular kinds of sensory information.
Review Figures 38.9, 38.10, and 38.23

The resting potential of a neuron is the difference prevailing most of the time between the electric charge on the inside of the plasma membrane and that on the outside.
Review Figure 38.11

Action potentials are rapid reverses in charge across portions of the plasma membrane.
Review Figure 38.12

Action potentials are self-regenerating, all-or-nothing events that transmit information by traveling down axons.
Review Figure 38.13

Myelinated axons propagate action potentials faster than other axons do.
Review Figure 38.14

Synaptic inputs added together generate action potentials when the sum exceeds a threshold.
Review Figure 38.17

Neurons arranged in circuits control the activities of most organs of the body.
Review Figures 38.18 and 38.20

The two states of human sleep are REM sleep and non-REM sleep, and non-REM sleep can be divided into four stages based on electrical activity in the brain.
Review Figure 38.21

Habituation and sensitization are simple forms of learning and memory that can be studied in the less complex nervous systems of invertebrates.
Review Figure 38.22

Human use of language involves communication between different areas of the cortex.
Review Figure 38.23

SELF-QUIZ

1. In the nervous system, the *most* abundant cell type is the
   a. motor neuron.
   b. sensory neuron.
   c. preganglionic parasympathetic neuron.
   d. glial cell.
   e. preganglionic sympathetic neuron.

2. Within the nerve cell, information moves from
   a. dendrite to cell body to axon.
   b. axon to cell body to dendrite.
   c. cell body to axon to dendrite.
   d. axon to dendrite to cell body.
   e. dendrite to axon to cell body.

3. Which of the following statements is *not* true?
   a. Sensory afferents carry information of which we are consciously aware.
   b. Visceral afferents carry information about physiological functions of which we are not consciously aware.
   c. The voluntary motor division of the efferent side of the peripheral nervous system executes conscious movements.
   d. The cranial nerves and spinal nerves are parts of the peripheral nervous system.
   e. Afferent and efferent axons never travel in the same nerve.

4. Which of the following statements is *not* true?
   a. In the spinal cord, the white matter contains the axons that conduct information.
   b. The limbic system is involved in basic physiological drives, instincts, and emotions.
   c. The limbic system consists of primitive forebrain structures.
   d. Most nerve cell bodies in the human nervous system are contained within the limbic system.
   e. In humans, a part of the limbic system is necessary for the transfer of short-term memory to long-term memory.

5. Which of the following statements accurately describes an action potential?
   a. Its magnitude *increases* along the axon.
   b. Its magnitude *decreases* along the axon.
   c. All action potentials in a single neuron are of the same magnitude.
   d. During an action potential the transmembrane potential of a neuron remains constant.
   e. It permanently shifts a neuron's transmembrane potential away from its resting value.

6. A neuron that has just fired an action potential cannot be immediately restimulated to fire a second action potential. The short interval of time during which restimulation is not possible is called
   a. hyperpolarization.
   b. the resting potential.
   c. depolarization.
   d. repolarization.
   e. the refractory period.

7. The rate of propagation of an action potential depends on
   a. whether or not the axon is myelinated.
   b. the axon's diameter.
   c. whether or not the axon is insulated by glial cells.
   d. the cross-sectional area of the axon.
   e. all of the above

8. The binding of neurotransmitter to the postsynaptic receptors in an inhibitory synapse results in
   a. depolarization of the transmembrane potential.
   b. generation of an action potential.
   c. hyperpolarization of the transmembrane potential.
   d. increased permeability of the membrane to sodium ions.
   e. increased permeability of the membrane to calcium ions.

9. Whether a synapse is excitatory or inhibitory depends on
   a. the type of neurotransmitter.
   b. the presynaptic terminal.
   c. the size of the synapse.
   d. the nature of the postsynaptic neurotransmitter receptors.
   e. the concentration of neurotransmitter in the synaptic space.

10. The part of the brain that differs the most in complexity between mammals and amphibians is
    a. the midbrain.
    b. the forebrain.
    c. the cerebellum.
    d. the limbic system.
    e. the hippocampus.

FOR STUDY

1. Compare and contrast the two divisions of the autonomic nervous system. Emphasize distinctions with respect to their anatomical organization, neurotransmitters used, and general effects on the functions of specific organ systems.

2. Outline the development of the vertebrate nervous system. Where on the neural axis are the more evolutionarily primitive and advanced functions located?

3. Describe the electrochemical and structural elements involved in the establishment and maintenance of the neuron's transmembrane resting potential.

4. Describe the processes and structures involved in (a) the initiation and propagation of an action potential, and (b) synaptic transmission.

5. Define and describe the synaptic basis for habituation and sensitization in *Aplysia*.

READINGS

Camhi, J. M. 1984. *Neuroethology: Nerve Cells and the Natural Behavior of Animals*. Sinauer Associates, Sunderland, MA. This advanced text covers the properties and functions of neurons but emphasizes aspects relevant to sensory abilities of animals and the neural control of behavior. Examples are taken from a wide variety of vertebrates and invertebrates.

Kandel, E. R., J. H. Schwartz and T. M. Jessell. 1991. *Principles of Neural Science*, 3rd Edition. Elsevier, New York. A very thorough, advanced text in neurobiology.

Llinas, R. R. 1988. *The Biology of the Brain: From Neurons to Networks*, and Llinas, R. R. 1990. *The Workings of the Brain: Development, Memory, and Perception*. W. H. Freeman, New York. These two volumes are a rich collection of articles on aspects of neuroscience that were published in *Scientific American* since 1976. They provide a broad yet selectively in-depth survey of modern neurobiology.

Nicholls, J. G., A. R. Martin and B. G. Wallace. 1992. *From Neuron to Brain*, 3rd Edition. Sinauer Associates, Sunderland, MA. An advanced text on cellular neurobiology.

Shepherd, G. M. 1994. *Neurobiology*, 3rd Edition. Oxford University Press, New York. A comprehensive advanced text that covers the full range of neurobiology from molecular mechanisms to human behavior.

Thompson, R. F. 1985. *The Brain: An Introduction to Neuroscience*. W. H. Freeman, New York. A well-written and easy-to-understand introductory text on neuroscience, covering topics from membrane events to the neural basis of behavior.

"Angela, you *look* great. I *hear* that you and Carl are going to the Khyber Pass for dinner."

"Yes, you raved so much about the wonderful *aromas* and complex, spicy *flavors* that we couldn't resist. After all, you have good *taste*, my friend."

"You'll also like the *feel* of the place. The tables are low and you sit on *soft* cushions. The walls are covered with *brightly colored* hangings. Stay in *touch*, I'd like to know how you liked it."

"I just hope the food isn't too *hot*, or you'll *hear* from me sooner than you think."

We cannot discuss anything for very long without using words that refer to our senses. Our senses are the window through which we view the world, and the world is what our senses tell us it is. Because different species look through different sensory windows, their views of the world differ. Dogs do not see color, but they have keener senses of hearing and smell than humans do. As you gaze at a beautiful sunset, your dog may be sniffing around at your feet and pricking up its ears as it hears small animals moving in the underbrush. Bees can see patterns on flowers that reflect ultraviolet light; we cannot. In environments that are totally dark to us, some snakes can "see" the infrared radiation emitted by bodies warmer than the environment. Bats can use reflected sound to avoid obstacles and catch small insects. The sounds bats emit are extremely intense, but they are beyond our range of hearing. In murky waters, the duck-billed platypus uses its sensitive bill to feel and taste food items, and electric fishes detect other fishes by the electric fields they create. How the environment "looks" to any animal depends on what information that animal receives from its sensors.

A Keen Sense of Smell Put to Work
A search dog trained to sniff out drugs, alcohol, and guns patrols school lockers.

39 Sensory Systems

WHAT IS A SENSOR?

Sensors are cells of the nervous system that transduce (convert) physical or chemical stimuli into signals that are transmitted to other parts of the nervous system for processing and interpretation. Most sensors are modified neurons, but some are other types of cells closely associated with neurons. Sensors are specialized for specific types of stimuli. In this chapter we will examine chemosensors, which respond to specific molecular structures; mechanosensors, which respond to mechanical forces; and photosensors, which respond to light. In general, the sensor possesses a membrane protein that detects the stimulus and responds by altering the flow of ions across the cell membrane (Figure 39.1). The resulting change in membrane potential causes the sensor either to fire action potentials itself or to secrete neurotransmitter

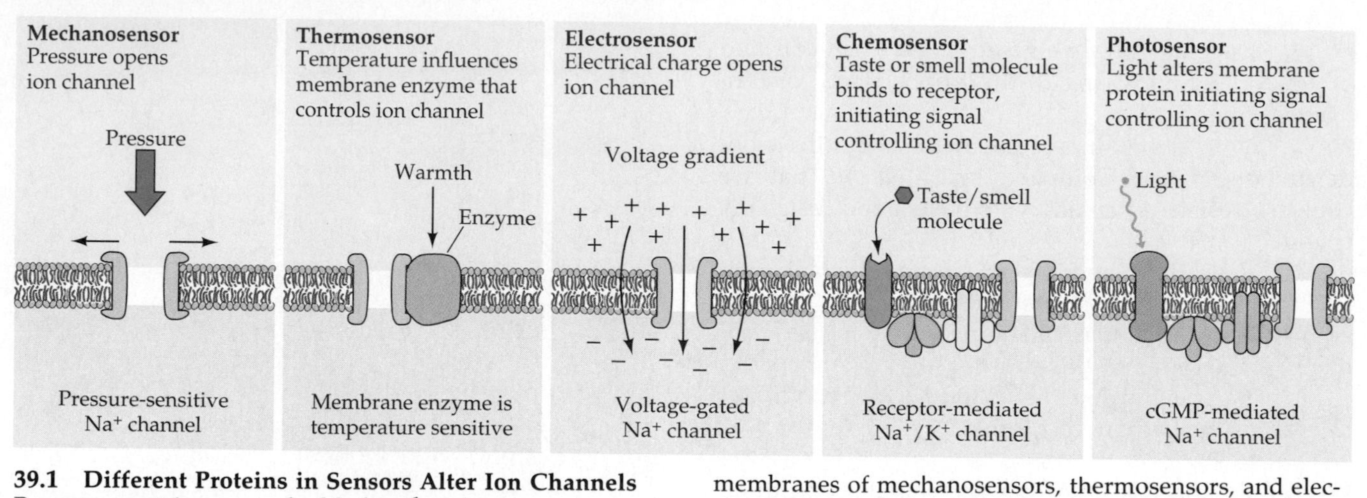

39.1 Different Proteins in Sensors Alter Ion Channels Receptor proteins are embedded in the plasma membranes of sensors. Sensory stimuli such as pressure or warmth modify the receptor proteins, which in turn modify ion channels. The receptor proteins in the plasma membranes of mechanosensors, thermosensors, and electrosensors are themselves the ion channels. The receptor proteins in the plasma membranes of chemosensors and photosensors initiate biochemical cascades that eventually open or close ion channels.

onto an associated cell that fires action potentials. Ultimately the stimulus is transduced into the universal message of the nervous system—action potentials (see Chapter 38).

Sensation

If the messages derived from all sensors are the same, how can we perceive different sensations? Sensations such as temperature, itch, pressure, pain, light, smell, and sound differ because the messages from sensors arrive at different places in the central nervous system. Action potentials arriving in the visual cortex are interpreted as light, in the auditory cortex as sound, in the olfactory bulb as smell, and so forth. A small patch of skin on your arm contains sensors that increase their firing rates when the skin is warmed and others that increase activity when the skin is cooled. Other types of sensors in the same patch of skin respond to touch, movement of hairs, irritants such as mosquito bites, and pain from cuts or burns. As we learned in Chapter 38, these sensors in your arm transmit their messages through axons that enter the central nervous system through the dorsal horn of the spinal cord. The synapses made by those axons in the central nervous system and the subsequent pathways of transmission determine whether the stimulation of the patch of skin on your arm is perceived as warmth, cold, pain, touch, itch, or tickle.

The specificity of sensory circuits is dramatically illustrated in persons who have had a limb or part of a limb amputated. Although the sensors from that region are gone, the axons that came from those sensors to the spinal cord may remain. If those axons are stimulated, the person feels specific sensations as if they were coming from the limb that is no longer there—a phenomenon known as a phantom limb.

The messages from some sensors communicate information about internal conditions in the body, but we may not be consciously aware of that information. The brain receives continuous information about such things as body temperature, blood sugar, blood carbon dioxide and oxygen, arterial pressure, muscle tensions, and positions of limbs. All this information is important for maintaining homeostasis, but thankfully we don't have to think about it—if we did, we would have no time to think about anything else. Sensors produce information that the nervous system can use, but that information does not always result in conscious sensation.

Sensory Organs

Some sensors are assembled with other types of cells into sensory organs such as eyes, ears, and noses that enhance the ability of the sensors to collect, filter, and amplify stimuli. For example, a photosensor detects electromagnetic radiation of only a particular range of wavelengths and therefore filters out radiation of other wavelengths. This filtering is the basis for color vision, and the specificity of photosensors explains why some insects can see ultraviolet light and some snakes can see infrared radiation. In some simple organisms photosensors sense only the presence of light, but in more-complex animals, photosensors are combined with other cell types into eyes. We'll learn how eyes collect light and focus it onto sheets of photosensors so that patterns of light can be detected. The basis of vision is the ability of the eye to filter available light information for patterns and colors.

Similarly, we'll see that the sense of hearing depends on mechanosensors, but the accessory structures that constitute the ear make it possible to amplify low levels of sound and filter it so that it also conveys directional information. Some sensory organs can reduce the level of stimulus energy that reaches sensors. For example, the pupillary reflex of vertebrate eyes varies the amount of light falling on the photosensors, and tiny muscles in ears can dampen the energy from loud sounds before it reaches the sensitive mechanosensors.

Sensory Transduction

In this chapter we will examine several sensor types and the sensory organs with which they are associated. A general question in each case is, how does the sensor transduce stimulus energy into action potentials? Although different for different sensors, the details of sensory transduction all fit into a general pattern. Figure 39.1 illustrated the first steps of sensory transduction: A receptor protein is activated by a specific stimulus. The activated receptor protein opens or closes specific ion channels in the plasma membrane of the sensor by one of several mechanisms. The receptor protein may be part of the ion channel and by changing its conformation may open or close the channel directly. Alternatively, the activated receptor protein may set in motion intracellular events that eventually affect the ion channel. Figure 39.2 reviews these first steps of sensory transduction and outlines the subsequent steps.

The opening or closing of ion channels in response to a stimulus changes the sensor's membrane potential, which is called the **receptor potential**. Such changes in membrane potential can spread electrotonically over short distances, but to travel long distances in the nervous system receptor potentials must be converted into action potentials. Interestingly, the intracellular events involved in converting the original, stimulus-induced alteration of the ion channels into the generation of action potentials can amplify the signal. In other words, the energy in the output of the sensor can be much greater than the energy in the stimulus.

Receptor potentials produce action potentials in two ways: by generating action potentials within the sensors or by causing the release of neurotransmitter that induces an associated neuron to generate action potentials. In the first case, the sensor has a region of plasma membrane with voltage-gated sodium channels. A receptor potential in such a cell is called a **generator potential** because it generates action potentials by causing the voltage-gated Na^+ channels to open.

A good example of generator potentials is found in stretch sensors of crayfish (Figure 39.3). By placing

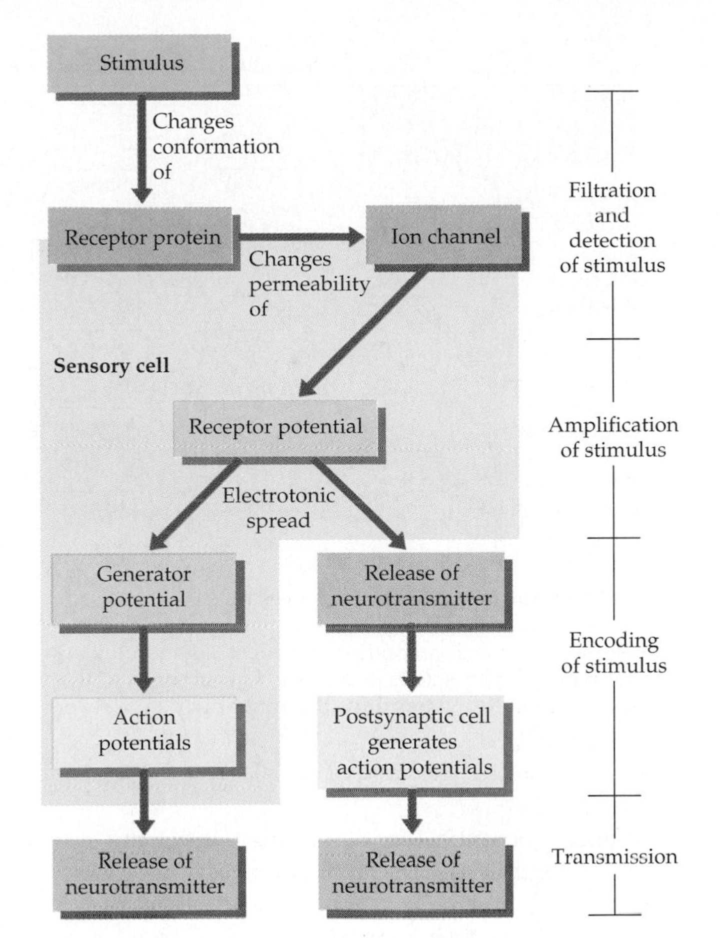

39.2 Sensory Transduction Involves Several Steps The detection of a stimulus and subsequent change in ion channels was described in Figure **39.1.** Sensors process and amplify the stimulus, either producing action potentials or releasing neurotransmitters that induce associated neurons to produce action potentials.

an electrode in the cell body of the crayfish stretch sensor, we can record the changes in the receptor potential that result from stretching the muscle to which the dendrites of the cell are attached. These changes in receptor potential become a generator potential at the base of the sensor's axon, where there are voltage-gated Na^+ channels. Action potentials generated here travel down the axon to the central nervous system. The rate at which the axon fires action potentials depends on the magnitude of the generator potential, which in turn depends on how much the muscle is stretched.

In sensors that do not fire action potentials, the spreading receptor potential reaches a presynaptic patch of plasma membrane and induces the release of neurotransmitter. The neurotransmitter can then activate ligand-gated ion channels on a postsynaptic membrane and cause the postsynaptic cell to fire action potentials. An excellent example is the photosensor that we will study in some detail later in this chapter. In either case, the stimulus is transduced

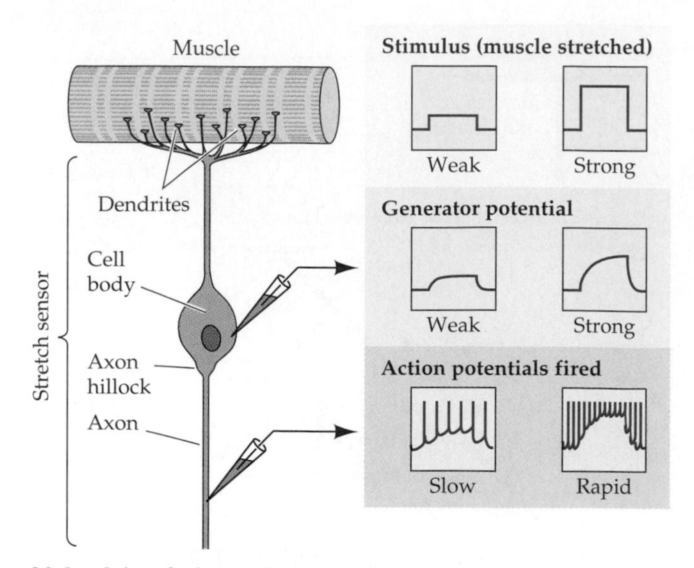

39.3 Stimulating a Sensor Produces a Generator Potential
The stretch sensor of a crayfish produces a generator potential when the muscle is stretched. The strength of the generator potential determines the rate at which axon potentials are fired.

into action potentials, and the intensity of the stimulus is coded by the frequency of action potentials.

Changing Sensitivity

An important characteristic of many sensors is that they can stop being excited by a stimulus that initially caused them to be active. In other words, they adapt to the stimulus. **Adaptation** enables an animal to ignore background or unchanging conditions while remaining sensitive to changes or to new information. (Note that this use of "adaptation" is different from its application in an evolutionary context.) When you dress, you feel each item of clothing touch your skin, but the sensation of clothes touching your skin is not constantly on your mind throughout the day. You are immediately aware, however, when a seam rips, your shoe comes untied, or someone touches your back ever so lightly. The ability of animals to discriminate between important and unimportant stimuli is partly due to the fact that some sensors adapt; it is also due to information processing by the central nervous system. Some sensors adapt very little or do so slowly; examples are pain sensors and sensors for balance.

In the rest of this chapter we will learn how sensory systems gather and filter stimuli, transduce stimuli into action potentials, and transmit action potentials to the central nervous sytem, as well as how the central nervous sytem processes that information to yield perceptions about the internal and external worlds. Sensors are also important components of autonomic regulatory mechanisms, as we will see in subsequent chapters.

CHEMOSENSORS

Animals receive information about chemical stimuli through **chemosensors**, which respond to specific molecules in the environment. Chemosensors are responsible for smell, taste, and the monitoring of aspects of the internal environment such as the level of carbon dioxide in the bloodstream. Chemosensitivity is universal among animals. A colony of corals responds to a small amount of meat extract in the seawater around it by extending bodies and tentacles and searching for food. A single amino acid can stimulate this response. If, however, an extract from an injured individual of the colony is released into the water, the colony members retract their tentacles and bodies to avoid danger. Humans have similar reactions to chemical stimuli. Upon smelling freshly baked bread we salivate and feel hungry, but we gag and retch when we smell rotting meat. Information from chemosensors can cause powerful behavioral and autonomic responses.

Chemosensation in Arthropods

Arthropods use chemical signals to attract mates. These signals, called **pheromones**, demonstrate the sensitivity of chemosensory systems. The female silkworm moth releases a pheromone called bombykol from a gland at the tip of her abdomen. The male silkworm moth has sensors for this molecule on his antennae (Figure 39.4). Each feathery antenna carries about 10,000 bombykol-sensitive hairs, and each hair has a dendrite of a sensor cell at its core. A single molecule of bombykol is sufficient to activate a dendrite and generate action potentials in the antennal nerve that transmits the signal to the central nervous system. When approximately 200 hairs per second are activated, the male flies upwind in search of the female. Because of the male's high degree of sensitivity, the sexual message of a female moth is likely to reach any male within a downwind area stretching over several kilometers. Because the rate of firing in the male's sensory nerves is proportional to bombykol concentrations in the air, the male can follow a concentration gradient and home in on the emitting female.

Many arthropods have chemosensory hairs, with each hair containing one or more specific types of sensor. For example, crabs and flies have chemosensory hairs on their feet; they taste potential food by stepping in it. These hairs have sensors for sugars, amino acids, salts, and other molecules (Figure 39.5). After a fly steps in a drop of sugar water and tastes

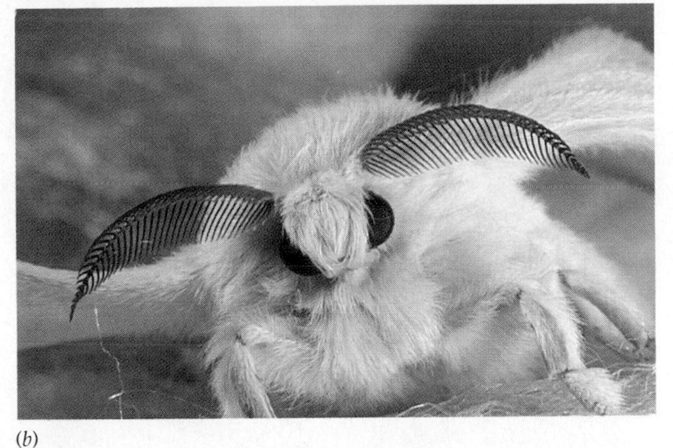

(*a*) (*b*)

39.4 A Scent That Travels Several Kilometers
Mating in silkworms of the genus *Bombyx* is coordinated by a chemical attractant. *(a)* The female moth releases the attractant from a gland at the tip of her abdomen. *(b)* From as far away as several kilometers, a male moth detects this chemical attractant in the air passing over his antennae, which are covered with chemosensitive hairs.

it, its proboscis (a tubular feeding structure) extends and it feeds. Potential food items stimulate extension of the proboscis; other substances do not.

Olfaction

The sense of smell, called **olfaction**, is another form of chemosensation. In vertebrates, the smell sensors are neurons embedded in a layer of epithelial cells at the top of the nasal cavity (Figure 39.6). The axons of smell sensors project to the olfactory bulb of the brain, and their dendrites end in olfactory hairs that project to the surface of the nasal epithelium. A protective layer of mucus covers the epithelium. Molecules must diffuse through this mucus to get to the receptors on the olfactory hairs. When you have a cold or an attack of hay fever, the amount of mucus increases and the epithelium swells. With this in mind, study Figure 39.6 and you will easily understand why you lose your sense of smell at those times. A dog has up to 40 million nerve endings per square centimeter of nasal epithelium, many more than humans do. Although humans have a sensitive olfactory system, we are unusual among mammals in that we depend more on vision than on olfaction (we tend to join bird-watching societies more often than mammal-smelling societies). Whales and porpoises have no olfactory sensors and hence no sense of smell.

How does an olfactory sensor transduce the structure of a molecule into action potentials? A molecule that triggers an olfactory sensor is called an odorant molecule. Odorant molecules bind to receptors on

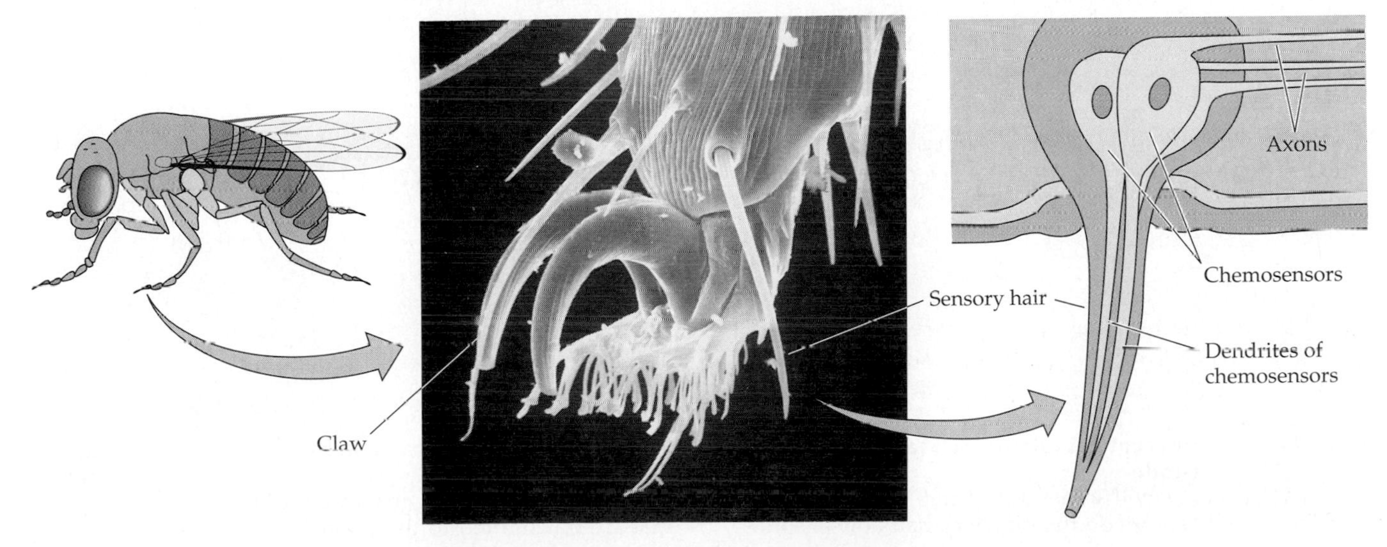

39.5 Tasting with the Feet
Using sensory hairs on their feet, flies such as this fruit fly (*Drosophila melanogaster*) can identify a potential food source by stepping in it.

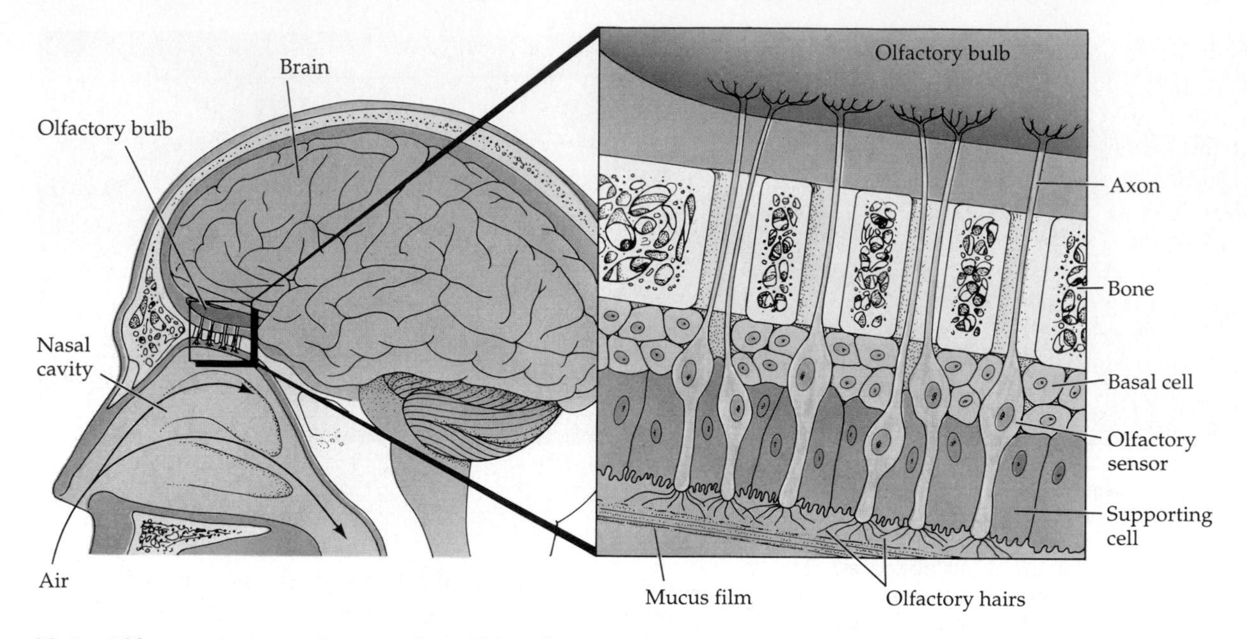

39.6 Olfactory Sensors Communicate Directly with the Brain
The sensors of the human olfactory system are embedded in the tissue lining the nasal cavity. The sensors' receptors are on the olfactory hairs. The axons of the sensors project to the olfactory bulb of the brain.

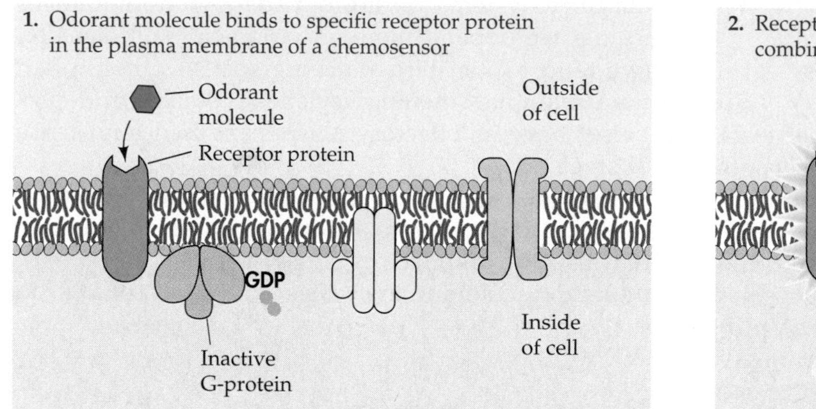

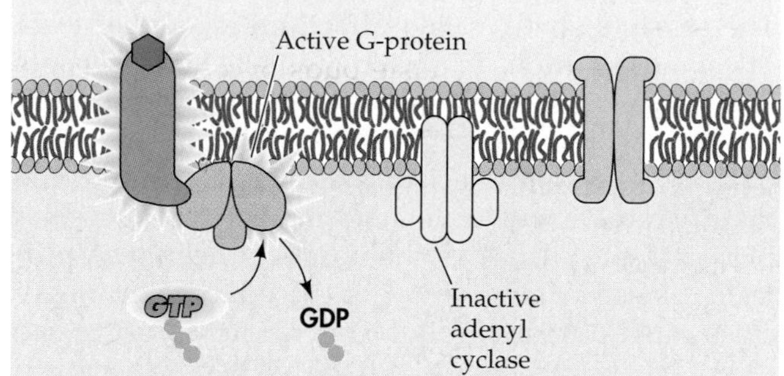

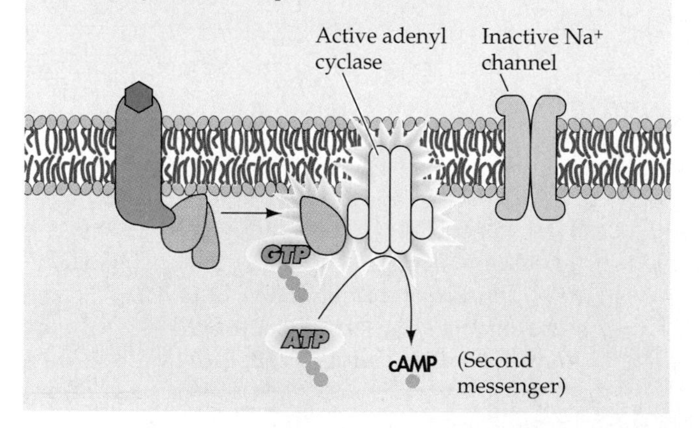

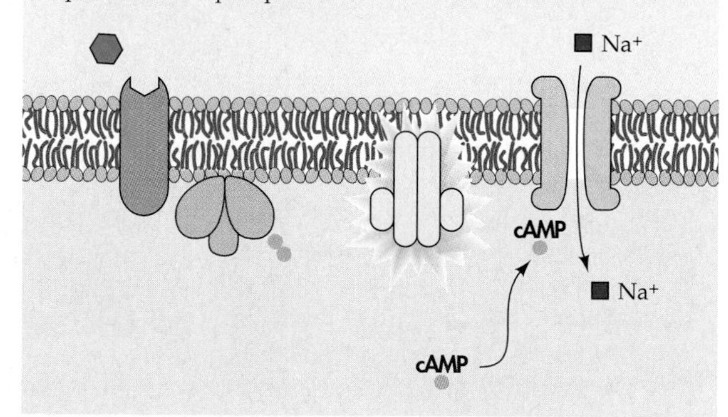

39.7 Olfactory Receptor Proteins Activate the cAMP Cascade
In order for us to smell something, odorant molecules must bind to receptors on the olfactory hairs of sensors and initiate the cAMP cascade by activating a G-protein. The cascade amplifies the signal so that a single odorant molecule can cause the generation of action potentials that are transmitted to the brain.

the olfactory hairs of the sensors. Olfactory receptors are specific for particular odorant molecules, and they work like a lock-and-key mechanism does. If a "key" (an odorant molecule) fits the "lock" (the receptor), a G-protein is activated, which in turn activates an enzyme (adenyl cyclase, for example) that causes an increase of a second messenger in the cytoplasm of the sensor. The second messenger binds with sodium channel proteins in the sensor's plasma membrane and opens the channels, causing an influx of Na^+. The sensor thus depolarizes to threshold and fires action potentials (Figure 39.7).

The olfactory world has an enormous number of "keys"—molecules that produce distinct smells. Are there a correspondingly large number of "locks"—receptor proteins? Indeed there are. Researchers have recently discovered an enormous family of genes that code for olfactory receptor proteins.

How does the sensor signal the intensity of a smell? It responds in a graded fashion to the concentration of odorant molecules: The more odorant molecules that bind to receptors, the more action potentials that are generated and the greater the intensity of the perceived smell.

Gustation

The sense of taste, called **gustation**, in humans and other vertebrates depends on clusters of sensors called **taste buds**. The taste buds of terrestrial vertebrates are in the mouth, but some fish have taste buds in the skin that enhance their ability to taste their environment. Some fish living in murky water are sensitive to small amounts of amino acids in the water around them and can find food without the use of vision. The duck-billed platypus, a monotreme mammal, has similar abilities because it has taste buds on the sensitive skin of its bill. What is a taste bud and how does it work?

A taste bud is a cluster of many taste sensors. A human tongue has approximately 10,000 or so taste buds. The taste buds are embedded in the epithelium of the tongue or are found on the raised papillae of the tongue. Look at your tongue in a mirror; the papillae make it look fuzzy. Each papilla has many taste buds. A taste bud's outer surface has a pore that exposes the tips of the taste sensors (Figure 39.8). Microvilli (tiny hairlike projections) increase the surface areas of the sensors where their tips converge at the taste pore. Taste sensors, unlike olfactory sensors, are not neurons. At their bases, taste sensors form synapses with dendrites of sensory neurons.

Gustation begins at receptors in the membranes of the microvilli. As with olfactory transduction, receptors on the sensors bind molecules, and the binding causes changes in the membrane polarity of the sensors. Because the taste sensors are not neurons, however, they do not fire action potentials. Instead, they release neurotransmitter onto the dendrites of the sensory neurons. The sensory neurons respond to the neurotransmitter by firing action potentials that are conducted to the central nervous system. The tongue does a lot of hard work, so its epithelium is shed and replaced at a rapid rate. Taste buds last only a few days before they are replaced, but the sensory neurons associated with them live on, always forming new synapses as new taste buds form.

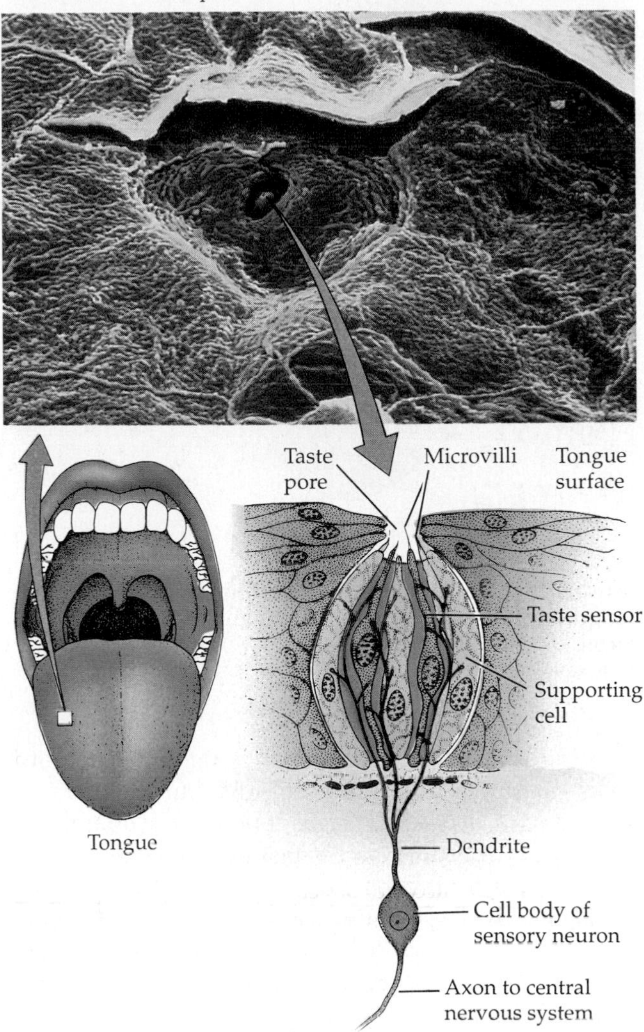

39.8 Taste Buds Are Clusters of Taste Sensors
The scanning electron micrograph of a papilla on the tongue shows the taste pore, where microvilli of the sensors that make up the taste bud come into contact with stimuli. The drawing to the right is a cross section of a taste bud. Note that the taste sensors are not neurons.

You may have heard it said that humans can perceive only four tastes: sweet, salty, sour, and bitter. Particular regions of the tongue have taste buds responsible for these general categories of taste, but the regions overlap to a large extent (Figure 39.9). You can map your own tongue by dipping toothpicks

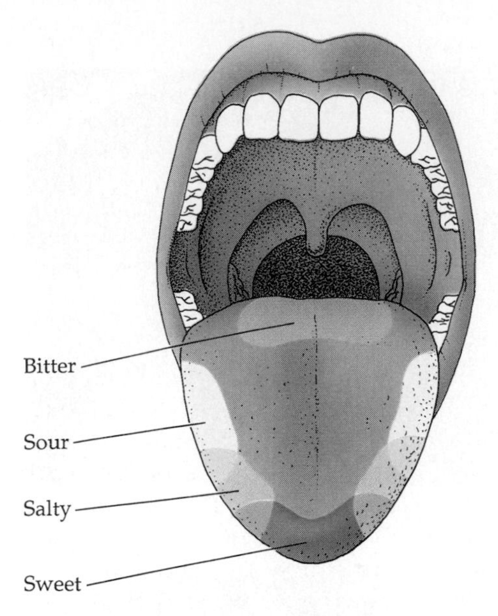

39.9 A Taste Map
Sensors for sweet, salty, sour, and bitter in the human tongue reside in specific regions that overlap in some areas.

in different solutions and then touching the toothpicks to different regions of the surface of your tongue. In actuality, taste buds can distinguish among a variety of sweet-tasting molecules and a variety of bitter-tasting molecules. The full complexity of the chemosensitivity that enables us to enjoy the subtle flavors of food comes from the combined activation of gustatory and olfactory sensors; hence you lose some of your sense of "taste" when you have a cold.

Why does a snake continually sample the air with its forked tongue darting in and out? If the snake, like us, tasted only sweet, salty, sour, and bitter with its tongue, it would not get much useful information by tasting the air. The forks of the snake's tongue fit into cavities in the roof of its mouth that are richly endowed with olfactory sensors. The tongue samples the air and presents the sample directly to these sensors. Thus the snake is really using its tongue to smell its environment, not to taste it. Why doesn't the snake use the flow of air to and from its lungs as we do to smell the environment? Air flow to and from the lungs is slow and intermittent in reptiles, but the tongue can dart in and out many times in a second. It is a quick source of olfactory information.

MECHANOSENSORS

Mechanosensors are specialized cells sensitive to mechanical forces that distort their membranes. A variety of mechanosensors in the skin are responsible for the perception of touch, pressure, and tickle. Stretch sensors in muscles, tendons, and joints give information about the position of the parts of the body in space and the forces acting on them. Stretch sensors in the walls of blood vessels signal blood pressure. "Hair" cells with extensions that are sensitive to being bent are incorporated into mechanisms for hearing and mechanisms for signaling the body's position with respect to gravity. Physical distortion of a mechanosensor's plasma membrane causes ion channels to open and alters the resting potential of the cell; this change leads to the generation of action potentials (see Figure 39.3). The rates of action potentials in the sensory nerves tell the central nervous system the strengths of the stimuli that are exciting the mechanosensor.

Touch and Pressure

Objects touching our skin generate varied sensations because our skin is packed with diverse mechanosensors (Figure 39.10). The outer layers of skin, especially hairless skin such as lips and fingertips, contain many whorls of nerve endings enclosed in connective tissue capsules. These very sensitive mechanosensors are called **Meissner's corpuscles**, and they respond to objects that touch the skin even lightly. Meissner's corpuscles adapt very rapidly, however. That is one reason why you roll a small object between your fingers, rather than holding it still, to discern its shape and texture. As you roll it,

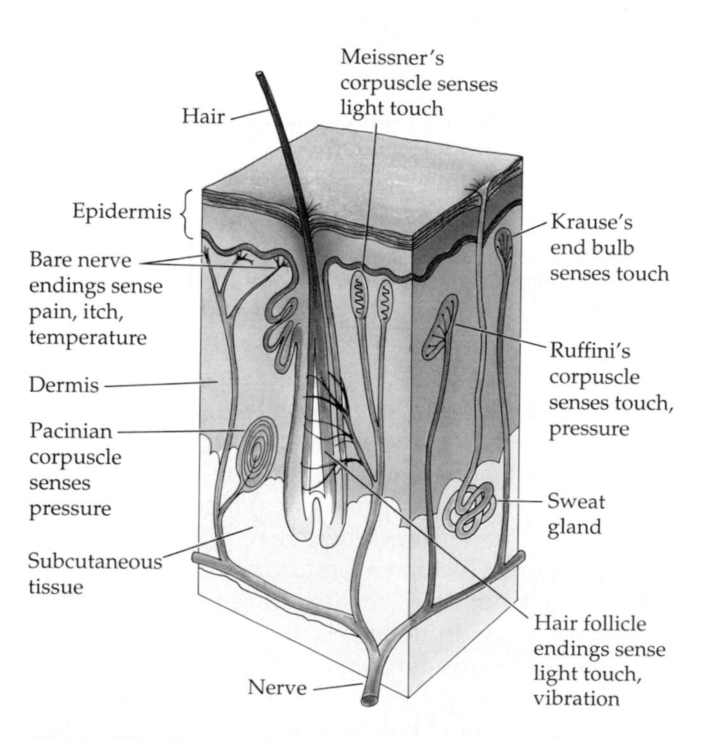

39.10 The Skin Feels Many Sensations
Even a very small patch of skin contains a diversity of sensors that send information to the brain.

you continue to stimulate sensors anew. When you hold an object still, the sensors originally activated adapt. Try it. Also in the outer regions of the skin are **expanded-tip tactile sensors** of various kinds. They differ from Meissner's corpuscles in that they adapt only partially and slowly. They are useful for providing steady-state information about objects that continue to touch the skin.

The density of the tactile sensors varies across the surface of the body. A two-point discrimination test demonstrates this fact. If you lightly touch someone's back with two toothpicks, you can determine how far apart the two stimuli have to be before the person can distinguish whether he or she was touched with one or two points. The same test applied to the person's lips or fingertips reveals a finer spatial discrimination; that is, the person can identify as separate two stimuli that are closer together than those on the back.

Deep in the skin, extensions of neurons wrap around hair follicles. When the hairs are displaced, those neurons are stimulated. Also deep within the skin is another type of mechanosensor, the **Pacinian corpuscles**. These sensors look like onions because they are made up of concentric layers of connective tissue cells encapsulating an extension of a sensory neuron. Pacinian corpuscles respond especially well to vibrations applied to the skin, but they adapt rapidly to steady pressure. The connective tissue capsule is important in the adaptation of these sensors. An initial pressure distorts the corpuscle and the membrane of the neuron at its core, but the layers of the capsule rapidly rearrange to redistribute the force, thus eliminating the distortion of the membrane of the neuron.

Stretch Sensors

An animal receives information from **stretch sensors** about the position of its limbs and the stresses on its muscles and joints. These mechanosensors are activated by being stretched. They continuously feed information to the central nervous system, and that information is essential for the coordination of movements. We encountered one important type of stretch sensor, the muscle spindle, when we discussed the monosynaptic reflex loop in Chapter 38. Muscle spindles are embedded in connective tissue within skeletal muscles and consist of modified muscle fibers that are innervated in the center with extensions of sensory neurons. Whenever the muscle is stretched, the spindle cells are also stretched, and the neurons transmit action potentials to the central nervous system. Earlier in this chapter (see Figure 39.3), we learned how crayfish stretch sensors transduce physical force into action potentials.

Another stretch sensor is found in tendons and ligaments. It is called the **Golgi tendon organ**. Its role is to provide information about the force generated by a contracting muscle. When a contraction becomes too forceful, the information from the Golgi tendon organ feeds into the spinal cord, inhibits the motor neuron, and causes the contracting muscle to relax, thus protecting the muscle from tearing.

Hair Cells

Hair cells are mechanosensors that are not neurons. From one surface they have projections called **stereocilia** that look like a set of organ pipes. When these stereocilia are bent, they alter receptor proteins in the hair cell's plasma membrane. When the stereocilia of some hair cells are bent in one direction, the receptor potential becomes more negative, and when they are bent in the opposite direction, it becomes more positive. When the receptor potential becomes more positive, the hair cells release neurotransmitter to the sensory neurons associated with them, and the sensory neurons send action potentials to the brain.

Hair cells are found in the lateral line sensory system of fishes. The lateral line consists of a canal just under the surface of the skin that runs down each side of the fish. The canal has numerous openings to the external environment. Many structures called cupulae project into the lateral line canal. Each cupula contains hair cells whose stereocilia are embedded in gelatinous (jellylike) material. Movements of water in the lateral line canal move the cupulae and stimulate the hair cells (Figure 39.11). Thus the lateral line provides information about movements of the fish through the water as well as information about the moving objects, such as predators or prey, that cause pressure waves in the surrounding water.

Many invertebrates have equilibrium organs called **statocysts** that use sensory hairs to signal the position of the animal with respect to gravity. In the case of the lobster, the statocyst is a chamber lined with hollow, nonliving hairs made of chitin. Each hair receives the dendrite of a sensory neuron. In the center of the statocyst is a dense **statolith** consisting of grains of sand (Figure 39.12). Due to gravity, the statolith stimulates the sensory hairs that are lowest, as determined by the position of the animal. When a scientist replaced the statoliths of lobsters with iron filings and held a magnet over the animals, they swam upside down. When he held the magnet to their sides, they swam on their sides. The behavior of the lobsters proved the role of the statoliths.

Vertebrates also have equilibrium organs. The mammalian inner ear, for example has two organs of equilibrium that also use hair cells to detect the position of the body with respect to gravity. In the next section we will examine the structure of the ear. For the moment it is enough to know that the inner ear

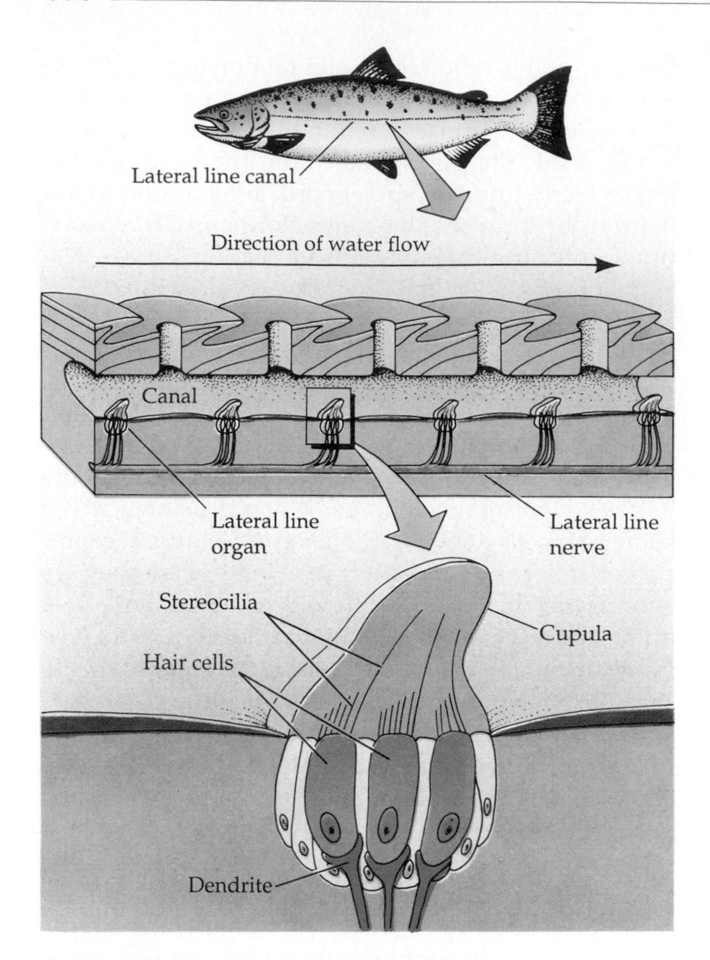

39.11 The Lateral Line System Contains Mechanosensors
Hair cells in the lateral line organs of a fish detect movement of the water around the animal, giving the fish information about its own movements and the movements of objects nearby.

contains three **semicircular canals** at right angles to one another (Figure 39.13). Each semicircular canal has a swelling called an **ampulla**, which contains a group of hair cells with their stereocilia embedded in a gelatinous cupula. The canals are filled with fluid. As an animal's head changes position, the fluid in its semicircular canals moves, puts pressure on the cupulae, and bends the stereocilia of the hair cells. The second equilibrium organ is found in the vestibule. This organ, the **vestibular apparatus**, has two chambers whose function is like that of the statocysts of invertebrates. Hair cells line the floors of the chambers; their stereocilia are embedded in a layer of gelatinous material. On top of this layer are **otoliths** (literally, "ear stones"), which are granules of calcium carbonate. As the head moves, gravity pulls on the dense otoliths, which bend the stereocilia of the hair cells.

Auditory Systems

The stimuli that animals perceive as sound are pressure waves. Auditory systems use mechanosensors to transduce pressure waves into action potentials. These systems include special structures to gather sound, direct it to the sensors, and amplify it. Human hearing provides good examples of these aspects of auditory systems. The organs of hearing are the ears. The two prominent structures on the sides of our heads usually thought of as ears are the **ear pinnae**. The pinna of an ear collects sound waves and directs them into the auditory canal leading to the actual hearing apparatus in the middle ear and the inner ear. If you have seen a rabbit or a horse change the orientation of its ear pinnae to focus on a particular sound, then you have witnessed the role of ear pinnae in hearing.

The human ear is diagrammed at progressively higher levels of magnification in Figure 39.14. The eardrum, or **tympanic membrane**, covers the end of the auditory canal. The tympanic membrane vibrates in response to pressure waves traveling down the auditory canal. The chamber of the middle ear, an air-filled cavity, lies on the other side of the tympanic membrane. The middle ear is open to the throat at

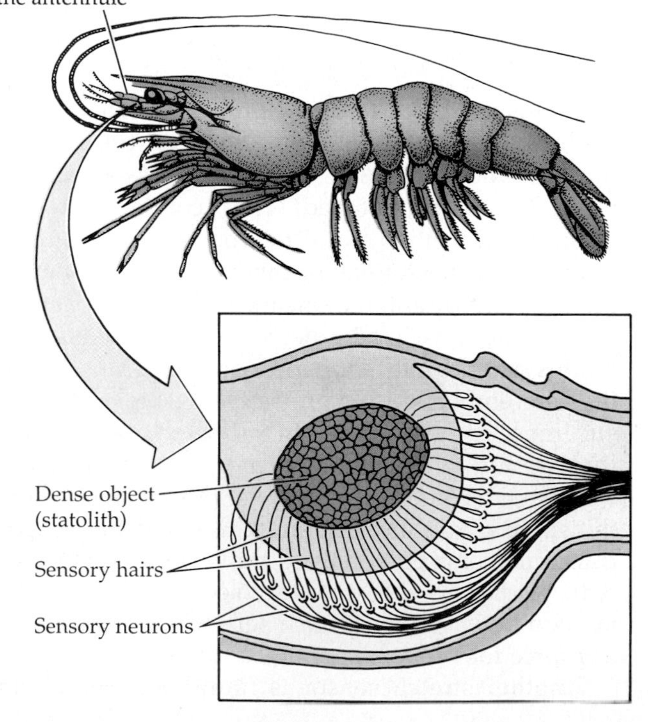

39.12 How a Lobster Knows Which Way Is Up
The statocyst is a sense organ found in many invertebrates. The force of gravity moves statoliths within the sensory hair-lined statocyst, giving the animal information about its position.

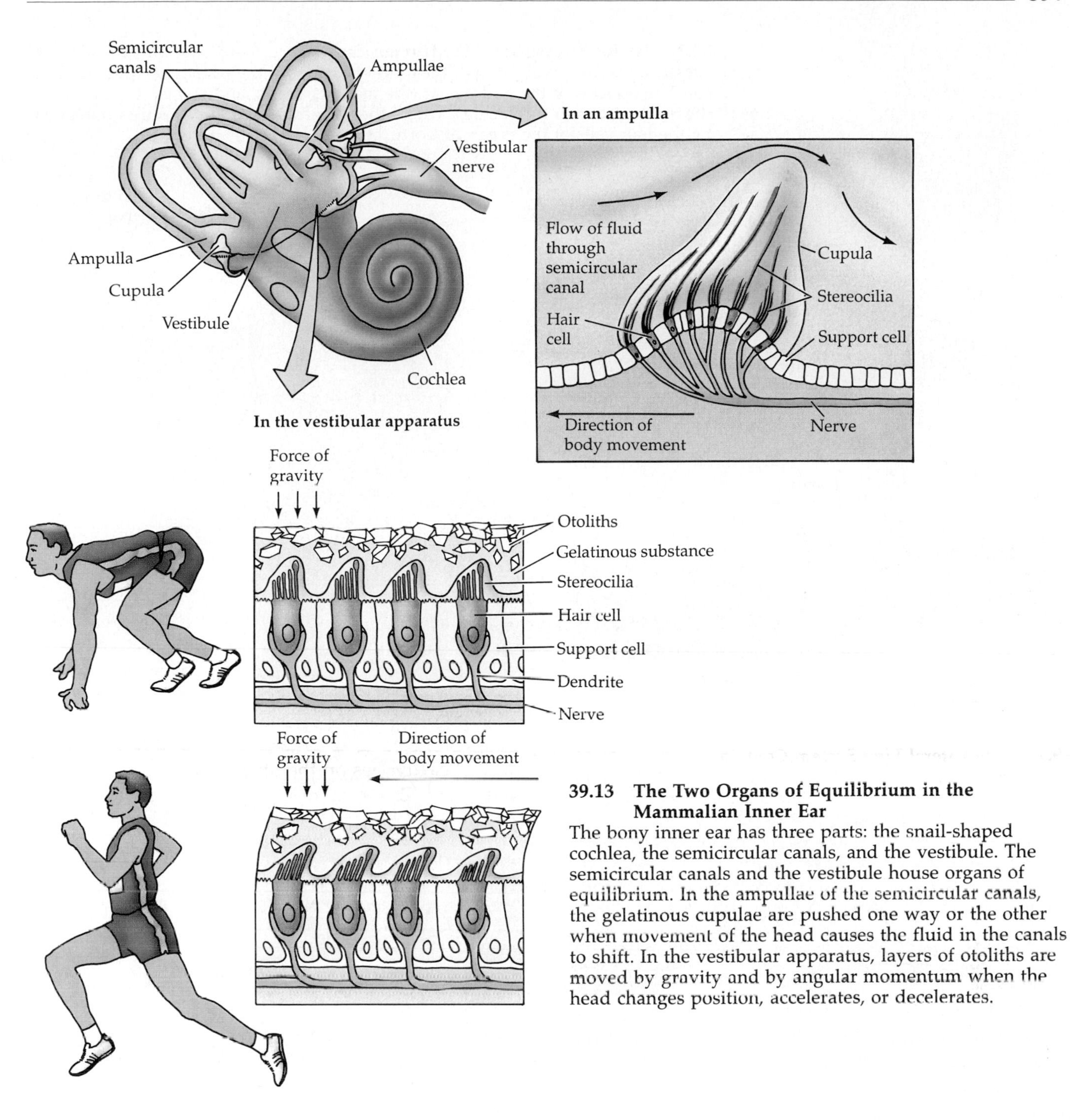

39.13 The Two Organs of Equilibrium in the Mammalian Inner Ear
The bony inner ear has three parts: the snail-shaped cochlea, the semicircular canals, and the vestibule. The semicircular canals and the vestibule house organs of equilibrium. In the ampullae of the semicircular canals, the gelatinous cupulae are pushed one way or the other when movement of the head causes the fluid in the canals to shift. In the vestibular apparatus, layers of otoliths are moved by gravity and by angular momentum when the head changes position, accelerates, or decelerates.

the back of the mouth through the eustachian tube. Because air flows through the **eustachian tube**, pressure equilibrates between the middle ear and the outside world. When you have a cold or hay fever, the tube becomes blocked by mucus or by tissue swelling, and you have difficulty "clearing your ears," or equilibrating the pressure in the middle ear with the outside air pressure. Then the flexible tympanic membrane bulges in or out, dampening your hearing and sometimes causing earaches.

The middle ear contains three delicate bones called the **ear ossicles**, individually named the **malleus** (hammer), **incus** (anvil), and **stapes** (stirrup). The ossicles transmit the vibrations of the tympanic membrane to the fluid-filled inner ear, where they will be transduced into action potentials. The leverlike action of the ossicles amplifies the vibrations about twentyfold. The malleus is attached to the center of the tympanic membrane and at the other end of the chain of ossicles, the stapes is attached to a smaller membrane called the **oval window**, which covers an opening into the inner ear. The incus serves as a pivot point. When the tympanic membrane moves in, the lever action of the ossicles pushes the stapes, and the

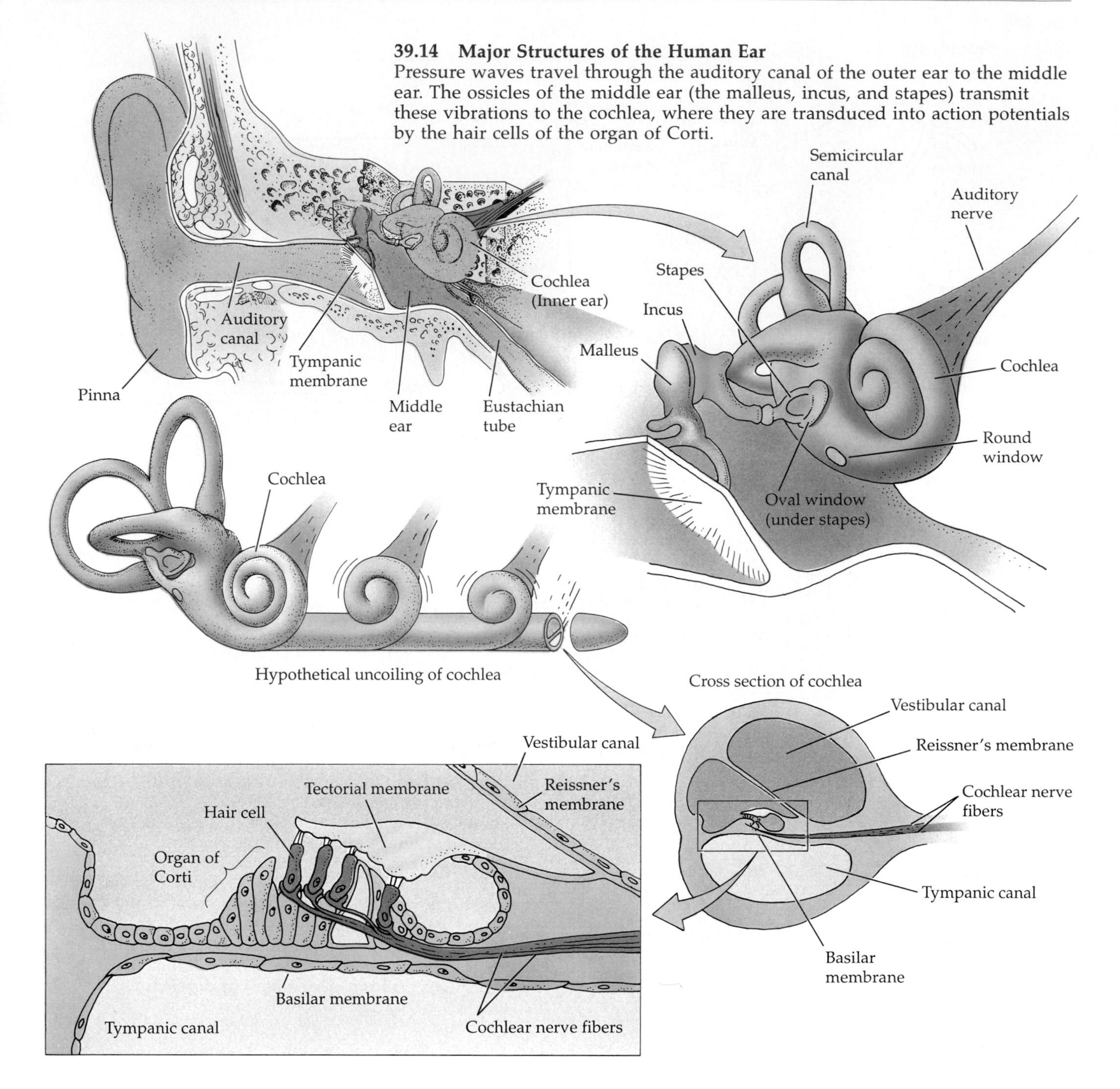

39.14 Major Structures of the Human Ear
Pressure waves travel through the auditory canal of the outer ear to the middle ear. The ossicles of the middle ear (the malleus, incus, and stapes) transmit these vibrations to the cochlea, where they are transduced into action potentials by the hair cells of the organ of Corti.

oval window bulges into the inner ear. When the tympanic membrane moves out, the stapes and the oval window are also pulled out. In this way, pressure waves in the auditory canal are converted into pressure waves in the fluid-filled inner ear.

Pressure waves are transduced into action potentials in the inner ear. The inner ear is a long, narrow, coiled chamber called the **cochlea** (from the Latin and Greek words for "snail" or "shell"). The cross section of the cochlea in Figure 39.14 reveals that it is composed of three parallel canals separated by two membranes: Reissner's membrane and the basilar membrane. Sitting on the basilar membrane is the **organ of Corti**, the apparatus that transduces pressure waves into action potentials in the auditory nerve—the nerve that conveys information from the ear to the brain. The organ of Corti contains hair cells whose stereocilia are in contact with an overhanging, rigid shelf called the tectorial membrane. Whenever the basilar membrane flexes, the tectorial membrane bends the hair cell stereocilia. As a consequence, the hair cells depolarize or hyperpolarize, altering the rate of action potentials transmitted to the brain by their associated sensory neurons.

What causes the basilar membrane to flex, and how does this mechanism distinguish sounds of different frequencies? In Figure 39.15 the cochlea is shown uncoiled to make it easier to understand its structure and function. To simplify matters we have left out Reissner's membrane, thus combining the upper and the middle canals into one upper canal. The purpose of Reissner's membrane is to contain a specific aqueous environment for the organ of Corti separate from the aqueous environment in the rest of the cochlea. This role is important for the nutrition of the sensitive organ of Corti, but it has nothing to do with the transduction of sound waves. The simplified diagram of the cochlea in Figure 39.15 reveals two additional features that are important for its function. First, the upper and lower chambers separated by the basilar membrane are joined at the end of the cochlea farthest from the oval window, making one continuous canal that folds back on itself. Second, just as the oval window is a flexible membrane at the beginning of the cochlea, the **round window** (see also Figure 39.14) is a flexible membrane at the end of the long cochlear canal.

Air is highly compressible, but fluids are not. Therefore, a sound pressure wave can travel through air without much displacement of the air, but a sound pressure wave in fluid causes displacement of the fluid. Imagine holding a screen-door spring slightly stretched between your two hands. Someone could grab the spring in the center and move it back and forth without moving its ends—it is compressible. Now imagine holding a broomstick in the same way. If someone grabs its middle and moves it back and forth, obviously the ends will move too—the broomstick is incompressible.

How does this comparison of springs and broomsticks relate to the inner ear? When the stapes pushes the oval window in, the fluid in the upper canal of the cochlea is displaced. Think about what happens if the oval window moves in very slowly. The cochlear fluid displacement travels down the upper canal, round the bend, and back through the lower chamber. At the end of the lower canal the displacement is absorbed by the round window membrane's bulging outward. Now what happens if the oval window vibrates in and out rapidly? The waves of fluid displacement do not have enough time to travel all the way to the end of the upper canal and back through the lower canal. Instead, they take a shortcut by crossing the basilar membrane, causing it to flex. The more rapid the vibration, the closer to the oval and round windows the wave of displacement will flex the basilar membrane. Thus different pitches of sound will flex the basilar membrane at different locations and activate different sets of hair cells (see Figure 39.15). This ability of the basilar membrane to respond to vibrations of different frequencies is enhanced by its structure. Near the oval and round windows (the proximal end) it is narrow and stiff, but it gradually gets wider and more flexible toward the opposite (distal) end. So it is easier for the proximal basilar membrane to resonate with high frequencies and for the distal basilar membrane to resonate with lower frequencies. A complex sound made up of many frequencies will distort the basilar membrane at many places simultaneously and activate a unique subset of hair cells.

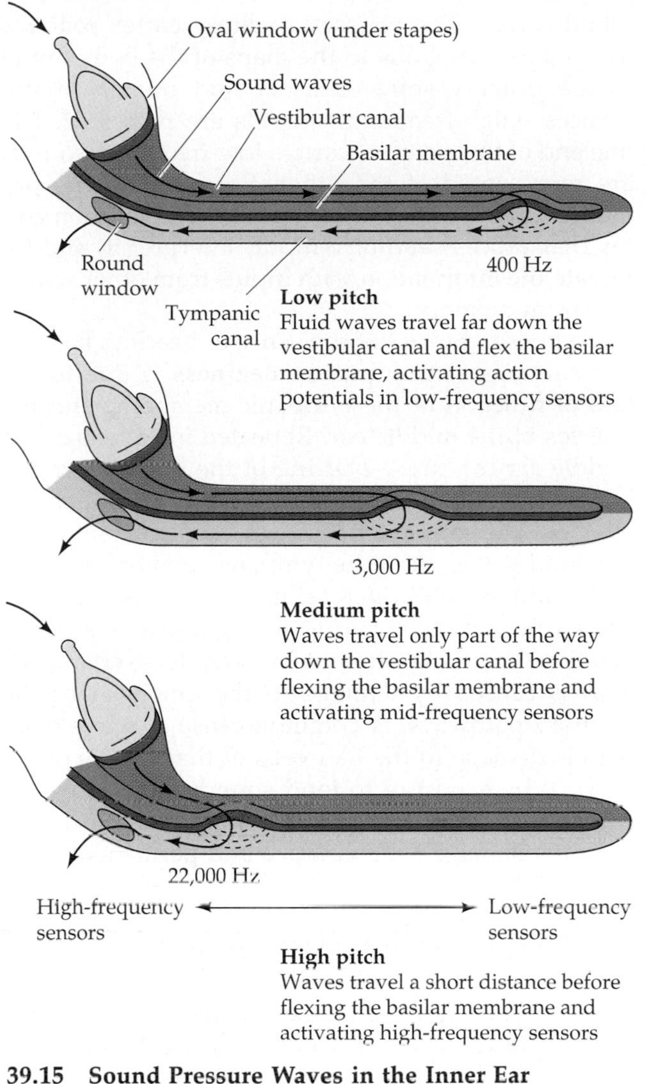

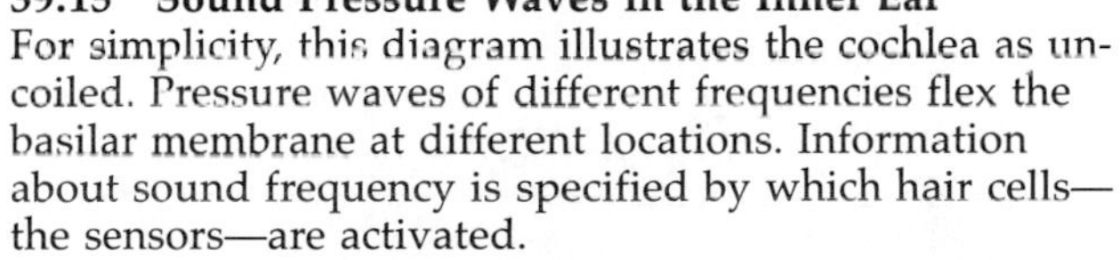

39.15 Sound Pressure Waves in the Inner Ear
For simplicity, this diagram illustrates the cochlea as uncoiled. Pressure waves of different frequencies flex the basilar membrane at different locations. Information about sound frequency is specified by which hair cells—the sensors—are activated.

Action potentials generated by the mechanosensors at different places along the organ of Corti travel to the brain stem along the auditory nerve. The auditory pathways make several synapses in the brain stem and send off collateral fibers to the reticular activating system. That is why sudden or loud noises wake us up or get our attention. Eventually the auditory pathways reach the temporal lobes of the cerebral cortex. The primary auditory cortex contains tone maps analogous to the maps of the body found in the primary somatosensory and primary motor cortices. High-frequency sounds are represented at one end of this patch of cortex, low-frequency sounds are represented at the other. Surrounding the primary auditory cortex are the areas of association cortex that process auditory input, interpret it, and integrate the information with inputs from other senses and from memory.

Deafness, the loss of the sense of hearing, has two general causes. **Conduction deafness** is due to the loss of function of the tympanic membrane and the ossicles of the middle ear. Repeated infections of the middle ear can cause scarring of the tympanic membrane and stiffening of the connections between the ossicles. The consequence is less-efficient conduction of sound waves from the tympanic membrane to the oval window. With increasing age, the ossicles progressively stiffen, resulting in a gradual loss of the ability to hear high-frequency sounds. **Nerve deafness** is caused by damage to the inner ear or the auditory pathways. A common cause of nerve deafness is damage to the hair cells of the delicate organ of Corti by exposure to loud sounds such as jet engines, pneumatic drills, or highly amplified rock music. This damage is cumulative and permanent.

PHOTOSENSORS AND VISUAL SYSTEMS

Sensitivity to light—photosensitivity—confers upon the simplest animals the ability to orient to the sun and sky and gives more-complex animals instantaneous detailed information about objects in the environment. It is not surprising that simple and complex animals can sense and respond to light. What is remarkable is that across the entire range of animal species evolution has conserved the same basis for photosensitivity: the molecule rhodopsin. In this section we will learn how rhodopsin responds when stimulated by light energy, how that response is transduced into neural signals, and how the brains of vertebrates process those signals to result in vision. We will also examine the structures of eyes, the organs that gather and focus light energy onto photosensitive cells. Finally, we will learn how the brain uses action potentials from the retina to create our mental image of the visual world.

Rhodopsin

Photosensitivity depends on the ability of a molecule to absorb photons of light and to respond by changing its conformation. The molecule that does this in the eyes of all animals is **rhodopsin** (Figure 39.16). Rhodopsin consists of a protein, **opsin**, which by itself does not absorb light, and a light-absorbing group, **11-*cis* retinal**. The light-absorbing group is cradled in the center of the opsin, and the entire rhodopsin molecule sits in the plasma membrane of a photosensor cell. When the 11-*cis* retinal absorbs a photon of light energy, its shape changes into a different isomer of retinal—all-*trans* retinal. This conformational change puts a strain on the bonds between retinal and opsin, and the two components

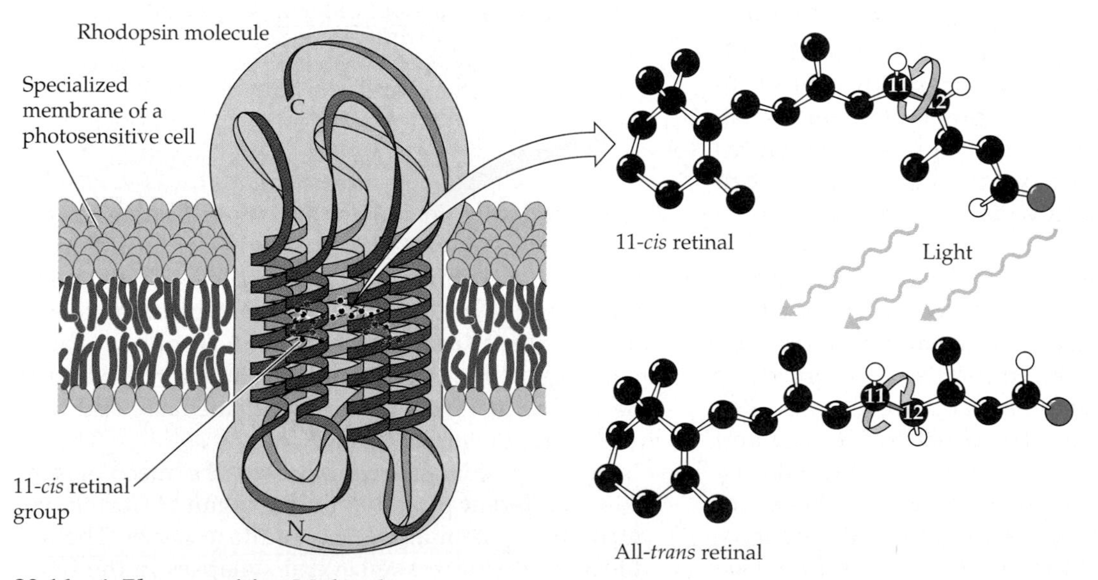

39.16 A Photosensitive Molecule
Rhodopsin is a transmembrane protein (opsin) that contains a light-responsive group (11-*cis* retinal). When 11-*cis* retinal absorbs a photon of light energy, it changes shape, becoming all-*trans* retinal, which is not responsive to light. The molecule returns spontaneously to the 11-*cis* conformation.

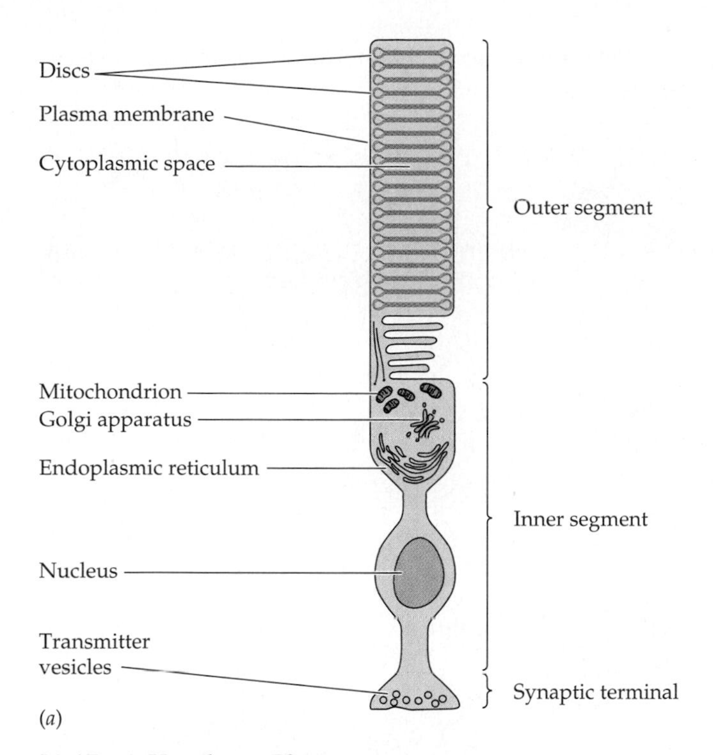

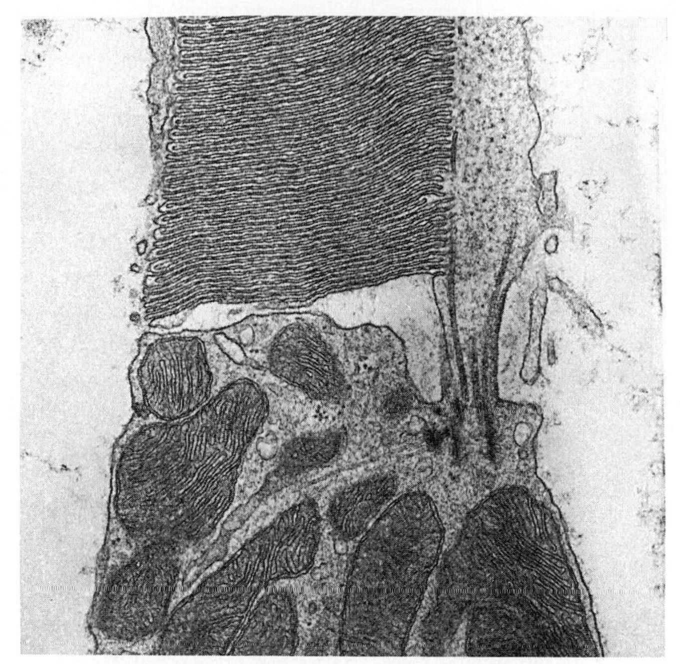

39.17 A Vertebrate Photosensor
(a) The rod cell of the vertebrate retina is a neuron modified for photosensitivity. The membranes of a rod cell's discs are densely packed with rhodopsin. *(b)* A transmission electron micrograph of a section through a photosensor.

break apart. As a result of this disassociation of retinal and opsin, referred to as bleaching, the molecule loses its photosensitivity. The retinal spontaneously returns to its 11-*cis* isomer and recombines with opsin to become, once again, photosensitive rhodopsin.

How does the light-induced change in rhodopsin's conformation transduce light into a cellular response? As retinal goes from the 11-*cis* to the all-*trans* forms, its interactions with opsin pass through several unstable intermediate stages. One stage is known as photoexcited rhodopsin because it triggers a cascade of reactions that changes ion flows, producing the alteration of membrane potential that is the photosensor's response to light. Let's explore these events of transduction in more detail.

The rhodopsin molecule sits in the membrane of a photosensor. How does this molecule communicate to the cell that it has absorbed a photon? And how does the sensor then communicate to the nervous system that its rhodopsin molecules are receiving light? To answer these questions, we must see how photosensors respond to light. A good example of a photosensor cell is a vertebrate **rod cell**, which is a modified neuron (Figure 39.17). A dense layer of photosensor cells at the back of the eye forms the **retina**, which, as we will see, is the structure that transduces the visual world into the language of the nervous system. Each rod cell in the retina has an inner segment and an outer segment. The inner segment contains the usual organelles of a cell and has a synaptic terminal at its base where the cell communicates with other neurons. The outer segment is highly specialized and contains a stack of discs. These discs form by the invagination (folding inward) and pinching off of the plasma membrane. The membranous discs are densely packed with rhodopsin; their function is to capture photons of light passing through the rod cell.

To see how the rod cell responds to light, we can penetrate a single rod cell with an electrode. Through this electrode we can record the receptor potential of the rod cell in the dark and in the light (Figure 39.18). From what we have learned about other types of sensors, we might expect stimulation of the photosensor with light would make its receptor potential less negative, but the opposite is true. When the rod cell is kept in the dark, it already has a very high resting potential in comparison with other neurons. In fact, the plasma membrane of the rod cell is fairly permeable to Na^+ ions, so these positive charges are continually entering the cell. When a light is flashed on the dark-adapted rod cell, its receptor potential becomes more negative—it hyperpolarizes. The rod cell itself *does not* generate action potentials. The rod cell does change its rate of neurotransmitter release, however, as its membrane polarity changes. Later we will learn how other cells in the visual pathway respond to neurotransmitter released from the photosensors so that information is communicated to the brain in the form of action potentials.

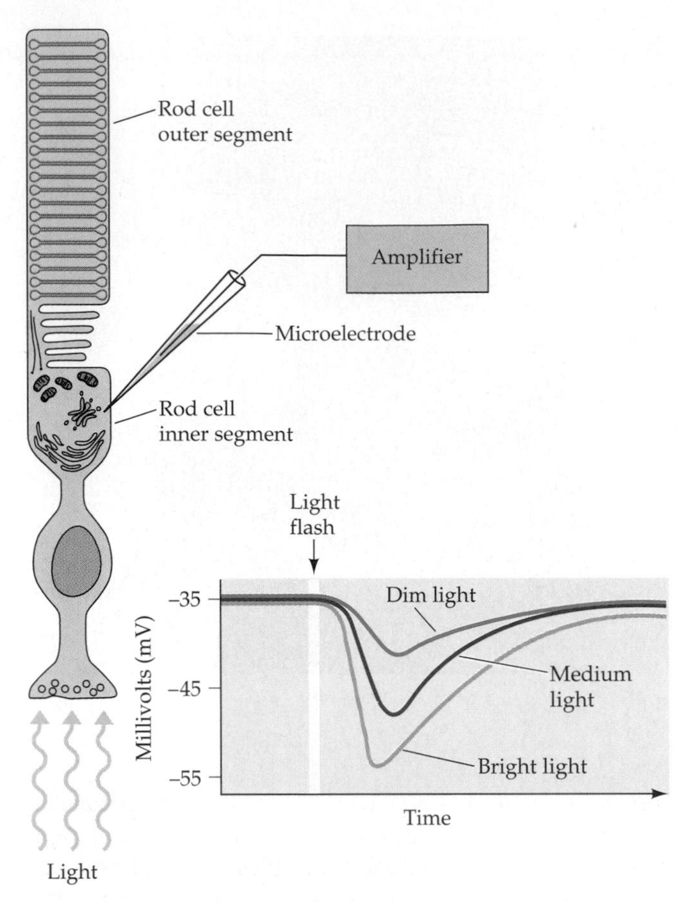

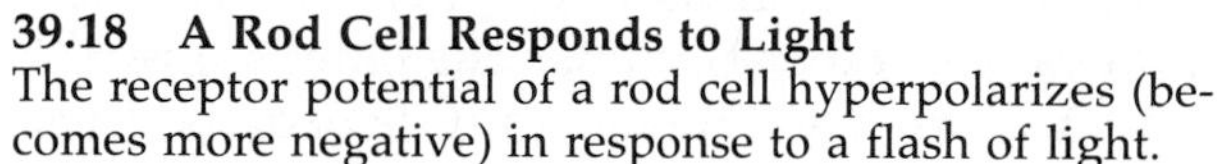

39.18 A Rod Cell Responds to Light
The receptor potential of a rod cell hyperpolarizes (becomes more negative) in response to a flash of light.

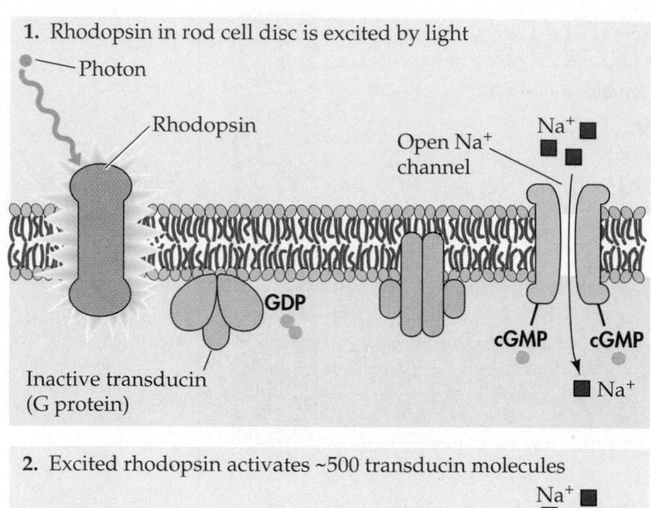

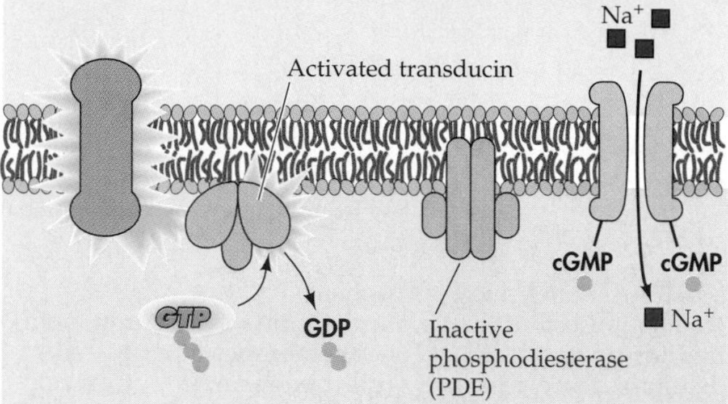

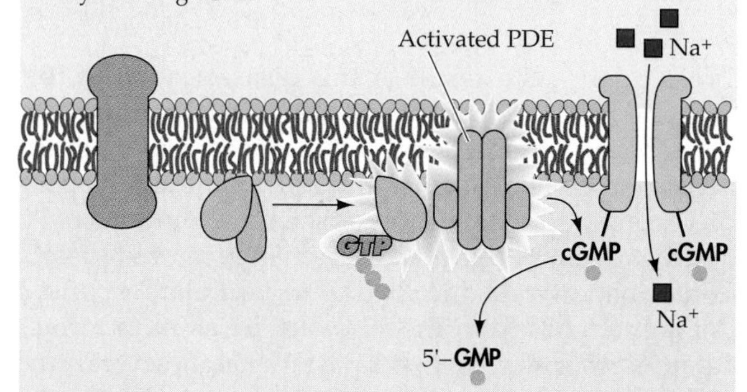

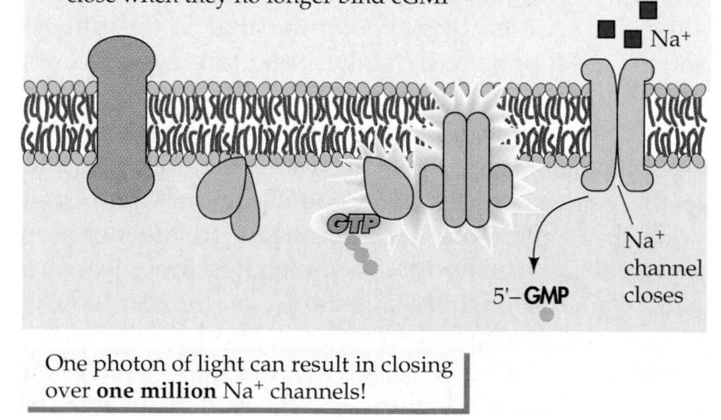

How does the absorption of light by rhodopsin hyperpolarize the rod cell? When rhodopsin is excited by light, it initiates a cascade of events. The photoexcited rhodopsin combines with and activates another protein, a G-protein called transducin. Activated transducin in turn activates a phosphodiesterase. Active phosphodiesterase converts cyclic GMP (cGMP) to 5′-GMP. What was that cGMP doing before it was converted to 5′-GMP? It was holding open the sodium channels and keeping the cell depolarized. As cGMP is destroyed, the sodium channels close, and the cell hyperpolarizes. This may seem like a roundabout way of doing business, but its significance is its enormous amplification ability. Each molecule of photoexcited rhodopsin can activate about 500 transducin molecules, thus activating about 500 phosphodiesterase molecules. The catalytic prowess of a molecule of phosphodiesterase is great; it can hydrolyze over 4,000 molecules of cGMP per second. The bottom line is that a single photon of light can cause the closing of over a million sodium channels and thereby change the rod cell's receptor potential (Figure 39.19). Now let's see how photosensors work in animals.

39.19 Light Absorption Closes Na^+ Channels
A key player in these events is a G-protein called transducin. Inactive transducin has three subunits, one of which binds GDP. Activated rhodopsin causes the GDP to be replaced by GTP, and the transducin molecule splits. The subunit with GTP can activate a phosphodiesterase (PDE) molecule. Active phosphodiesterase degrades cGMP bound to the Na^+ channel and causes Na^+ channels to close.

Visual Systems of Invertebrates

Flatworms, simple multicellular animals, obtain directional information about light from photosensitive cells organized into eye cups (Figure 39.20). The eye cups are bilateral structures, and each is partially shielded from light by a layer of pigmented cells lining the cup. Because the openings of the eye cups face in opposite directions, the photosensors on the two sides of the animal are unequally stimulated unless the animal is facing directly toward or away from a light source. Using directional information about light sources, the flatworm moves away from light.

Arthropods (such as crustaceans, spiders, and insects) have evolved **compound eyes** that provide them with information about patterns or images in the environment. Each compound eye consists of many optical units called **ommatidia** (singular: ommatidium). The number of ommatidia in a compound eye varies from only a few in some ants, to 800 in fruit flies, to 10,000 in some dragonflies. Each ommatidium has a lens structure that directs light onto photosensors called retinula cells. Flies have seven elongated retinula cells in each ommatidium. The inner borders of the retinula cells are covered with microvilli that contain rhodopsin and constitute a light trap. Since the microvilli of the different retinula cells overlap, they appear to form a central rod, called a rhabdom, down the center of the ommatidium. Axons from the retinula cells communicate with the nervous system (Figure 39.21). Because each ommatidium of a compound eye is directed at a slightly different part of the visual world, only a crude, or perhaps broken, image of the visual field can be communicated from the compound eye to the central nervous system.

Vertebrate and Cephalopod Eyes

Vertebrates and cephalopod mollusks have evolved eyes with exceptional abilities to form images of the visual world. These eyes operate like cameras, and considering that they evolved independently of each other, their high degree of similarity is remarkable (Figure 39.22). The vertebrate eye is a spherical, fluid-filled structure bounded by a tough connective tissue layer called the sclera. At the front of the eyeball, the sclera forms the transparent **cornea** through which light enters the eye. Just inside the cornea is the pigmented **iris**, which gives the eye its color. The important function of the iris is to control the amount of light that reaches the photosensors at the back of the eyeball, just as the diaphragm of the camera controls the amount of light reaching the film. The central opening of the iris is the **pupil**. The iris is under control of the autonomic nervous system. In bright light the iris constricts and the pupil is very small, but as light levels fall, the iris relaxes and the pupil enlarges.

Behind the iris is the crystalline lens, which helps focus images on the photosensitive layer—the ret-

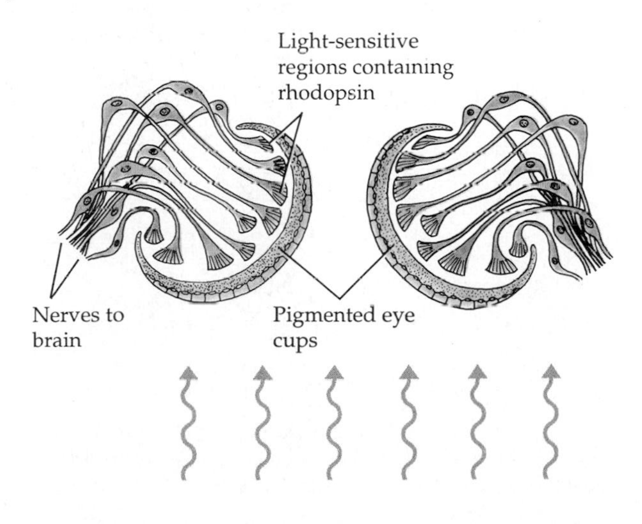

39.20 A Simple Photosensory System Gives Directional Information
Although flatworms do not "see" as we understand it, the eye cups of this flatworm enable it to move away from a light source to an area where it may be less visible to predators.

39.21 Eyes of an Insect
(a) The compound eyes of a horsefly; each eye contains hundreds of ommatidia. *(b)* Each ommatidium focuses light on a rhabdom consisting of overlapping, light-sensitive plasma membranes of a few photosensors.

(a)

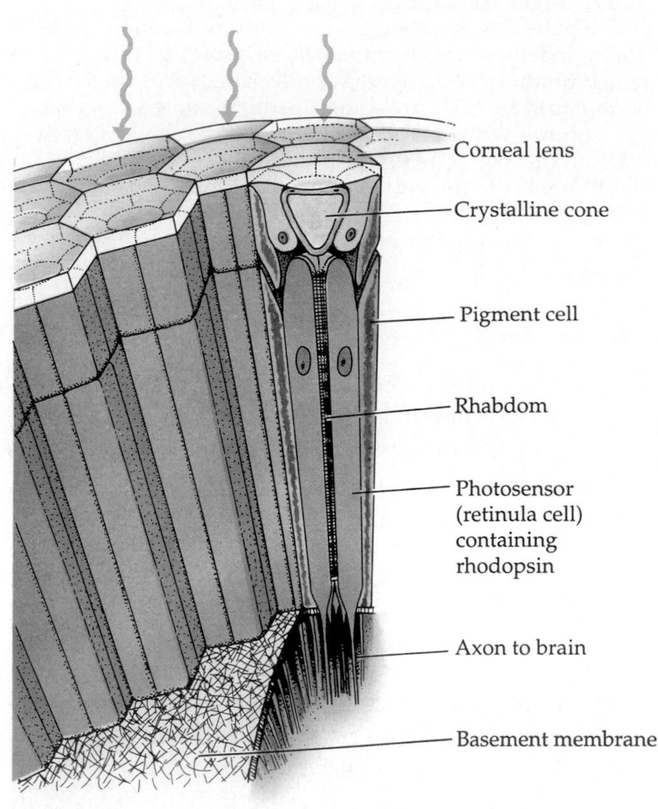

(b)

ina—at the back of the eye. The cornea and the fluids of the eye chambers also help focus light on the retina, but the lens is responsible for the ability to accommodate—to focus on objects at various locations in the near visual field. To focus a camera on objects close at hand, you must adjust the distance between the lens and the film. Fishes, amphibians,

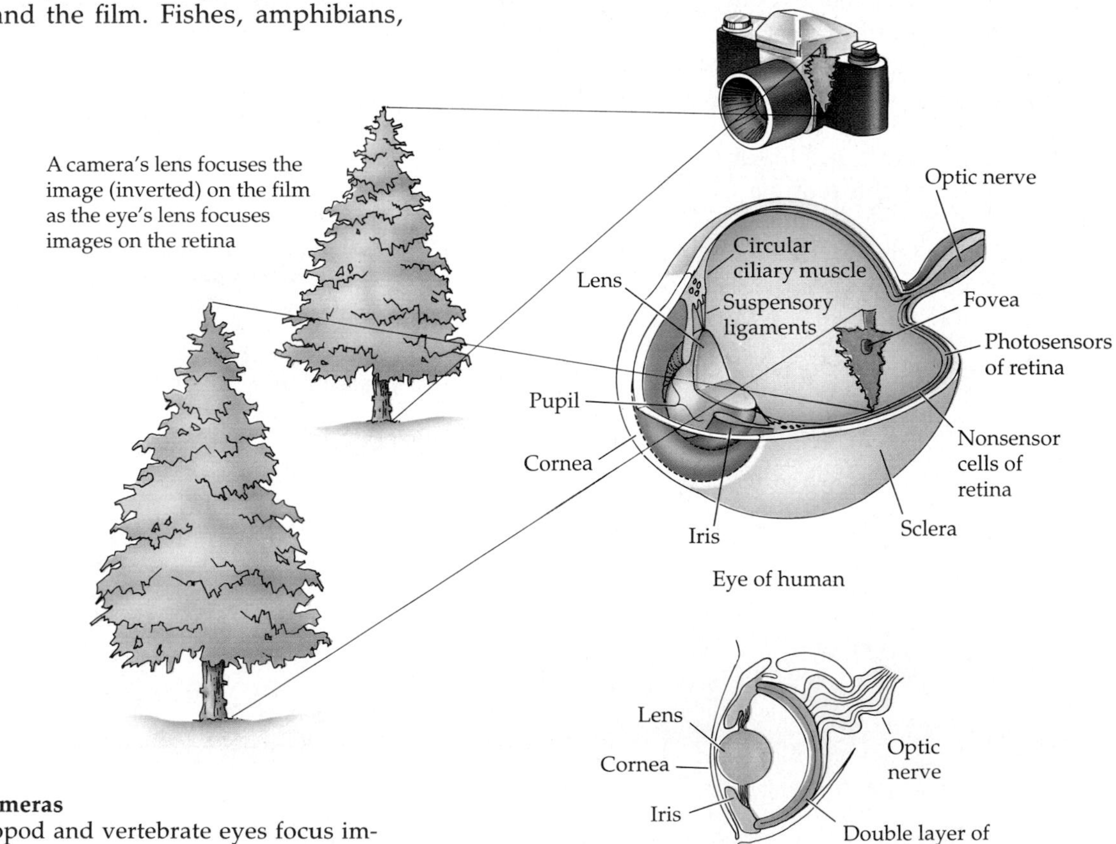

39.22 Eyes Like Cameras
The lenses of cephalopod and vertebrate eyes focus images on layers of photosensors, just as a camera lens focuses images on film.

and reptiles accommodate in a similar manner, moving the lenses of their eyes closer to or farther from their retinas. Mammals and birds use a different method; they alter the shape of the lens.

The lens is contained in a connective tissue sheath that tends to keep it in a spherical shape, but it is also suspended by suspensory ligaments that pull it into a flatter shape. Circular muscles called the ciliary muscles counteract the pull of the suspensory ligaments and permit the lens to round up. With the ciliary muscles at rest, the flatter lens has the correct optical properties to focus distant images on the retina, but not close images. Contracting the ciliary muscles rounds up the lens, changing its light-bending properties to bring close images into focus (Figure 39.23). As we get older, our lenses become less elastic, and we lose the ability to focus on objects close at hand without the help of corrective lenses. Prolonged concentration on small, close objects (such as the type on these pages) tires and strains the eyes by overworking the ciliary muscles.

The Vertebrate Retina

The retina is an extension of the brain. During development, neural tissue grows out from the brain to form the retina. In addition to a layer of photosensors, the retina includes layers of cells that process the visual information from the photosensors and transmit it to the brain in the form of action potentials in the optic nerves. A curious feature of the anatomy of the retina is that the outer segments of the photosensors are all the way at the back of the retina so light must pass through all the layers of retinal cells before reaching the place where photons are captured by rhodopsin. Later we will examine in detail how the cells of the retina process information, but first let's describe some general features of retinal organization.

The density of photosensors is not the same across the entire retina. Light coming from the center of the field of vision falls on an area of the retina called the **fovea**, where the density of sensors is the highest. The human fovea has about 160,000 sensors per square millimeter. A hawk has about 1 million sensors per square millimeter of fovea, making its vision about eight times sharper than ours. In addition, the hawk has two foveas in each eye. One fovea receives light from straight ahead; the other receives light from below. Thus while the hawk is flying, it sees both its projected flight path and the ground below, where it might detect a mouse scurrying in the grass.

The fovea of a horse is a long, vertical patch of retina. The horse's lens is not good at accommodation, but it focuses distant objects that are straight ahead on one part of this long fovea and close objects that are below the head on another part of the fovea. When horses are startled by an object close at hand, they pull their heads back and rear up to bring the object into focus on the close-vision fovea.

Where blood vessels and the bundle of axons going to the brain pass through the back of the eye, there is a blind spot on the retina. You are normally not aware of your blind spot, but you can find it. Stare straight ahead, holding a pencil in your outstretched hand so that the eraser is in the center of your field of vision. While continuing to stare straight ahead, slowly move the pencil to the side until the eraser disappears. When this happens, the light from the eraser is focused directly on your blind spot.

Until now we have referred to only one type of photosensor, the rod cell. There are, however, two major types of photoreceptors, both named for their shapes: rod cells and **cone cells** (Figure 39.24). A human retina has about 3 million cones and about 100 million rods. Rod cells are more sensitive to light

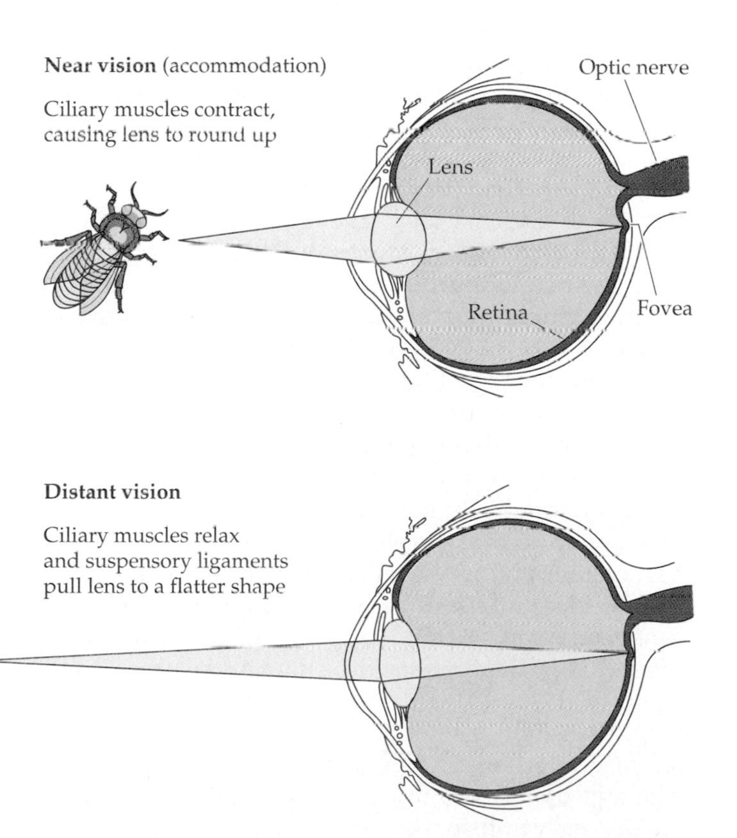

39.23 Coming into Focus
Mammals and birds focus their eyes on close objects by changing the shape of the lens.

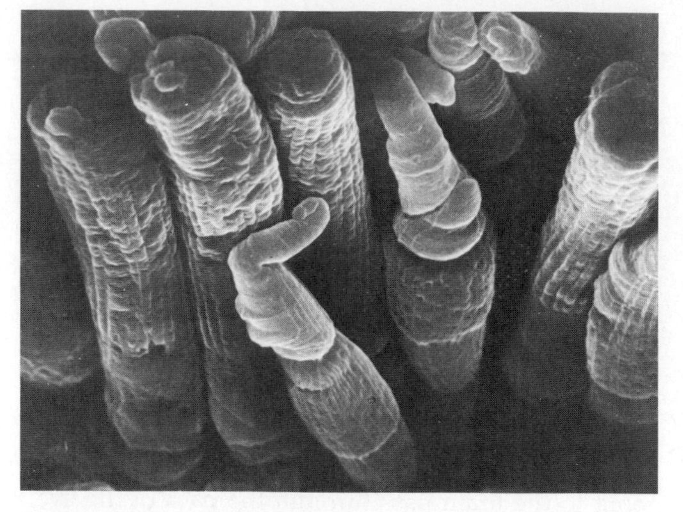

39.24 Rods and Cones
This scanning electron micrograph of photosensors in the retina of a mud puppy, an amphibian, shows cylindrical rods and tapered cones.

but do not contribute to color vision. Cones are responsible for color vision but are less sensitive to light. Cones are also responsible for our sharpest vision because, even though there are many more rods than cones in human retinas, our foveas contain mostly cones.

Because cones have low sensitivity to light, they are of no use at night. At night our vision is not very sharp, and we see only in black and white. You may have trouble seeing a small object such as a keyhole at night when you are looking straight at it—that is, when its image is falling on your fovea. If you look a little to the side, so that the image falls on a rod-rich area of retina, you can see the object better. Astronomers looking for faint objects in the sky learned this trick a long time ago. Animals that are nocturnal (such as flying squirrels) may have only rods in their retinas and have no color vision. By contrast, some animals that are active only during the day (such as chipmunks and ground squirrels) have only cones in their retinas.

How do cone cells enable us to see color? There are at least three kinds of cone cells, each possessing slightly different types of opsin molecules. Because different cone cells have different opsin molecules, they differ in the wavelengths of light they absorb best. Although the retinal group is the light absorber, its molecular interactions with opsin tune its spectral sensitivity. Some opsins cause retinal to absorb most efficiently in the blue region, some in the green, and some in the red (Figure 39.25). Intermediate wavelengths of light excite these classes of cones in different proportions. Recently the genes coding for the different opsins of humans were identified: one for blue-sensitive opsin, one for red-sensitive opsin, and several for green-sensitive opsin.

The human retina is organized into five layers of neurons that receive visual information and process it before sending it to the brain (Figure 39.26). Strangely enough, as mentioned earlier, the layer of photosensors is all the way at the back of the retina, so light has to traverse all the other layers before reaching the rods and cones. The disc-containing ends of the rods and cones are partly buried in a layer of pigmented epithelium that absorbs photons not captured by rhodopsin and prevents any backward scattering of light that might decrease visual sharpness. Nocturnal animals, such as cats, have a highly reflective layer, called the tapetum, behind the photosensors. Photons not captured on their first pass through the photosensors are reflected back, thus increasing visual sensitivity (but not sharpness) in low-light conditions. The reflective tapetum is what makes cats' eyes appear to glow in the dark.

A first step in investigating how the human retina processes visual information is to study how its five layers of neurons are interconnected and how they influence one another. The neurons at the back of the retina are the photosensors. As we know, the photosensors hyperpolarize in response to light, but they do not generate action potentials. The cells at the front of the retina are ganglion cells. Ganglion cells fire action potentials, and their axons form the optic nerves that travel to the brain. The photosensors and ganglion cells are connected by bipolar cells. The changes in membrane potential of rods and cones in response to light alter the rate at which the rods and cones release neurotransmitter at their synapses with the bipolar cells. Like rods and cones, bipolar cells do not fire action potentials. In response to neurotransmitter from the photosensors, the membrane potentials of bipolar cells change, altering the rate at which they release neurotransmitter onto ganglion cells. The ganglion cells generate action poten-

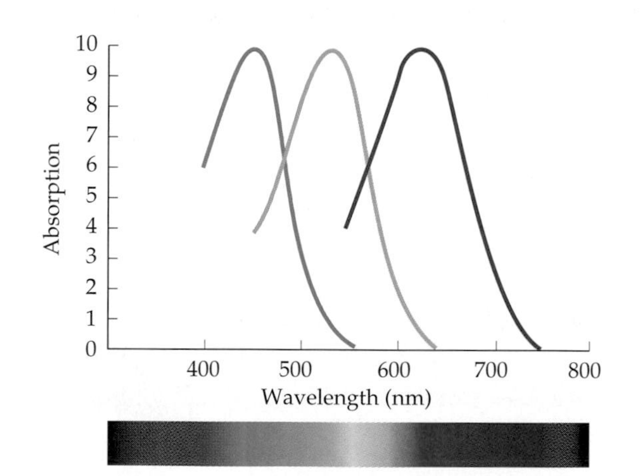

39.25 Absorption Spectra of Cone Cells
Human color vision is based on three kinds of cone cells. Each absorbs a different band of wavelengths most effectively.

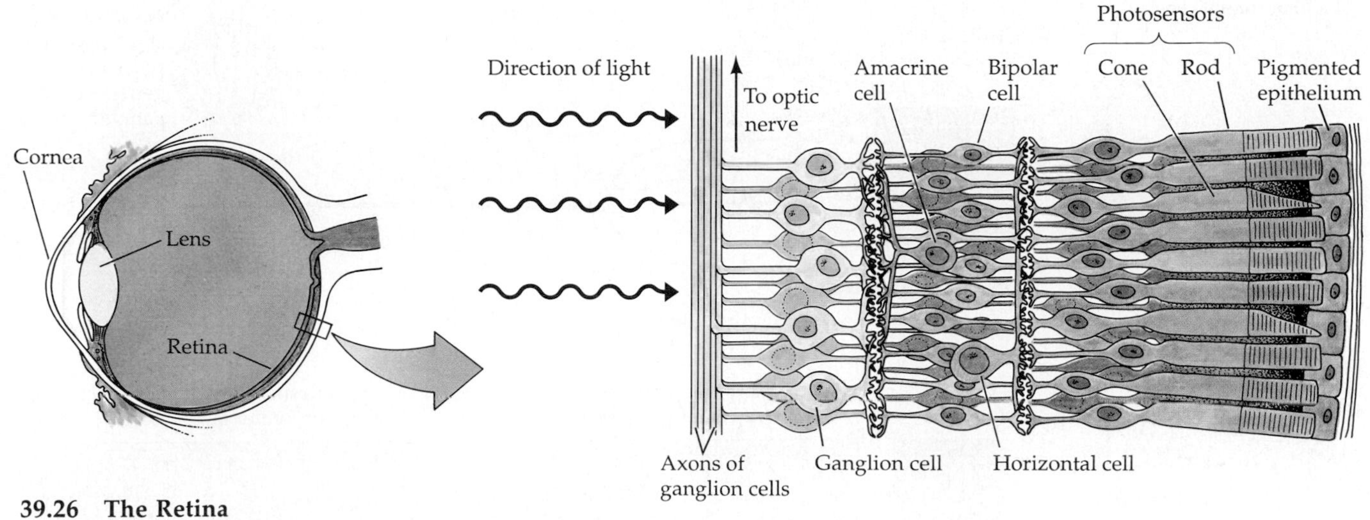

39.26 The Retina
Light travels through layers of transparent neurons—ganglion, amacrine, bipolar, and horizontal cells—and is absorbed by discs in the rods and cones (the photosensory layer) at the back of the retina. The visual information is then processed forward through several layers of neurons, with final convergence on ganglion cells that send their axons to the brain.

tials, and the rate of neurotransmitter release from the bipolar cells determines the rate at which ganglion cells fire action potentials. Thus the direct flow of information in the retina is from photosensor to bipolar cell to ganglion cell. Ganglion cells send the information to the brain.

What do the other two layers (the horizontal cells and the amacrine cells; see Figure 39.26) do? They communicate laterally across the retina. Horizontal cells connect neighboring pairs of photosensors and bipolar cells. Thus the communication between a photosensor and its bipolar cell can be influenced by the amount of light absorbed by neighboring photosensors. This lateral flow of information sharpens the perception of contrast between light and dark patterns falling on the retina. Amacrine cells connect neighboring pairs of bipolar cells and ganglion cells. One role of amacrine cells is to adjust the sensitivity of the eyes according to the overall level of light falling on the retina.

Knowing the paths of information flow through the retina still doesn't tell us how that information is processed. What does the eye tell the brain in response to a pattern of light falling on the retina? One aspect of information processing in the retina is information reduction. There are over 100 million photosensors in each retina, but only about 1 million ganglion cells sending messages to the brain. How is the information from all those photosensors reduced to the messages sent to the brain by the ganglion cells? This question was addressed in some elegant, classic experiments in which scientists used electrodes to record the activity of single ganglion cells in living animals while their retinas were stimulated with spots of light. They found that each ganglion cell has a well-defined **receptive field** that consists of a specific group of photosensors. Stimulating these photosensors with light activates the ganglion cell (Figure 39.27*a*). Information from many photosensors is reduced in this way to a single message.

The receptive fields of ganglion cells are all circular, but the way a spot of light influences the activity of the ganglion cell depends on where in the receptive field it falls. Each ganglion cell's receptive field is divided into two concentric areas, called the center and the surround. There are two kinds of receptive fields, on-center and off-center. Stimulating the center of an on-center receptive field excites the ganglion cell; stimulating the surround inhibits it. Stimulating the center of an off-center receptive field inhibits the ganglion cell, and stimulating the surround excites it (Figure 39.27*b*). Center effects are always stronger than surround effects. The response of a ganglion cell to stimulation of the center of its receptive field depends on how much of the surround area is also stimulated. A small dot of light directly on the center has the maximal effect, a bar of light hitting the center and pieces of the surround has less of an effect, and a large, uniform patch of light falling equally on center and surround has no effect. Ganglion cells communicate information about light-dark contrasts falling on their receptive fields to the brain.

How are receptive fields related to the connections between the neurons of the retina? The photosensors in the center of a ganglion cell's receptive field are connected to that ganglion cell by bipolar cells. The photosensors in the surround send information to the center photosensors and thus to the ganglion cell through the lateral connections of horizontal cells. Thus the receptive field of a ganglion cell is due to a pattern of synapses between photosensors, horizontal cells, bipolar cells, and ganglion cells. The recep-

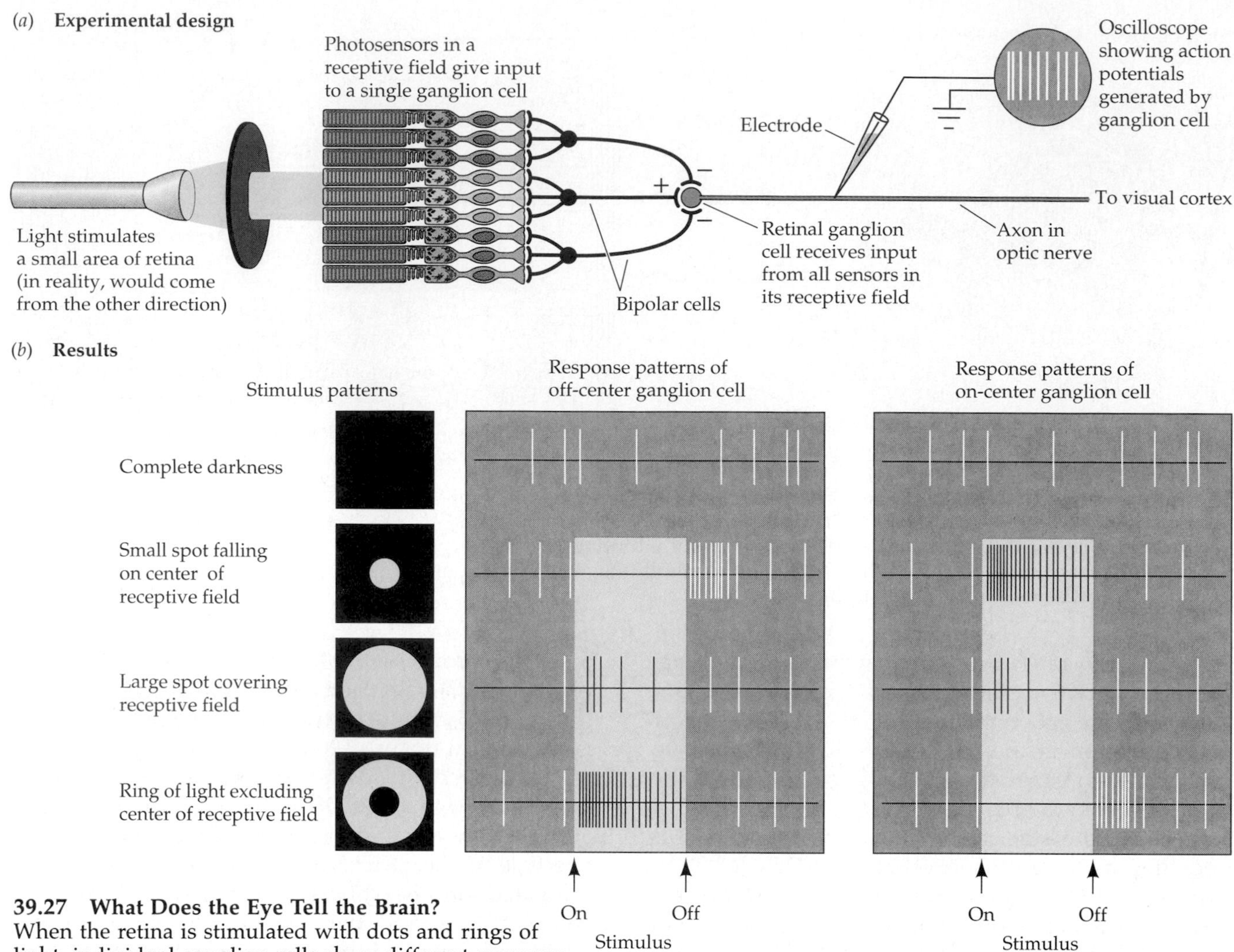

39.27 What Does the Eye Tell the Brain? When the retina is stimulated with dots and rings of light, individual ganglion cells show different response properties. Each ganglion cell is spontaneously active and responds to light shining on a small circular area of retina in its receptive field. Some ganglion cells are stimulated and some are inhibited by a spot of light falling on the receptive field centers. An "on-center" ganglion cell is inhibited by a ring of light falling on the peripheral area of its receptive field. The opposite is true of "off-center" ganglion cells. Cells may show a brief response when the stimulus is turned off.

tive fields of neighboring ganglion cells can overlap greatly; a given photosensor can be connected to several ganglion cells.

The eye sends the brain simple messages about the pattern of light intensities falling on small, circular patches of retina. How does the brain create a mental image of the visual world from this information? David Hubel and Torsten Wiesel of Harvard University tackled this question. Information from the retina is transmitted through the optic nerves to a relay station in the thalamus and then to the brain's visual processing area, in the occipital cortex at the back of the cerebral hemispheres. Hubel and Wiesel recorded the activities of single cells in the visual processing areas of the brains of living animals while they stimulated the animals' retinas with spots and bars of light. They found that, like ganglion cells, cells in the visual cortex have receptive fields—specific areas of the retina that when stimulated by light influence the rate at which the cortical cells fire action potentials.

Hubel and Wiesel learned that cells in the visual cortex have different types of receptive fields. One type of cell, called the **simple cell**, is maximally stimulated by a bar of light with a certain orientation falling on one place on the retina. Simple cells probably receive input from a number of ganglion cells that have their circular receptive fields lined up in a row. Another type of cortical cell, the **complex cell**, is also maximally stimulated by a bar of light with a particular orientation, but the bar may fall anywhere on a large area of retina described as that cell's receptive field. Complex cells seem to receive input from a number of simple cells that share a certain stimulus orientation but that have receptive fields in

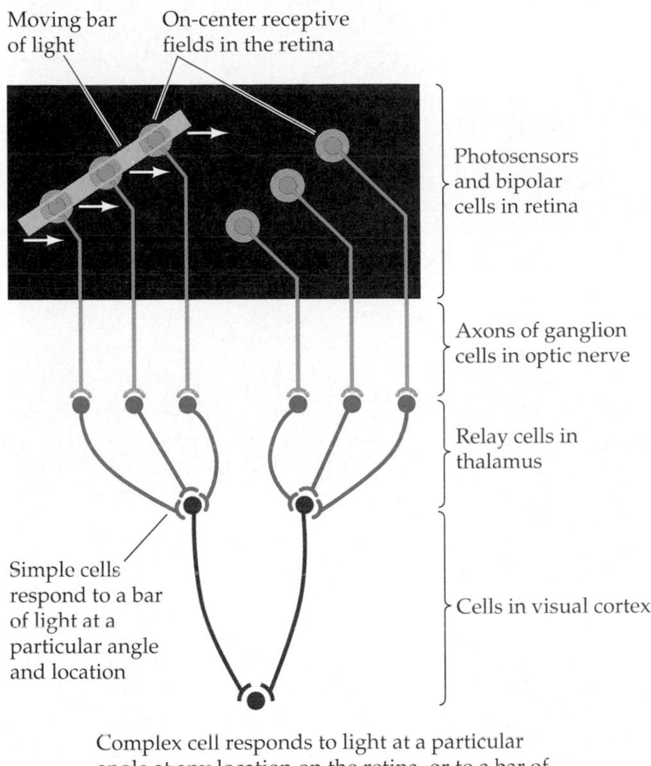

39.28 Receptive Fields of Cells in the Visual Cortex Cells in the visual cortex respond to specific patterns of light falling on the retina. Ganglion cells that project information about circular receptive fields converge on simple cells in the cortex in such a way that simple cells have linear receptive fields. Simple cells project to complex cells in such a way that complex cells can respond to linear stimuli falling on different areas of the retina.

different places on the retina (Figure 39.28). Some complex cells respond best when the bar of light moves in a particular direction, perhaps depending on the combination of on-center and off-center receptive fields.

The concept that emerges from these experiments is that the brain assembles a mental image of the visual world by analyzing edges of light patterns falling on the retina. This analysis is conducted in a massively parallel fashion. Each retina sends 1 million axons to the brain, but there are at least 200 million neurons in the visual cortex. A bit of information from a retinal ganglion cell is received by hundreds of cortical cells, each responsive to a different combination of orientation, position, and even movement of contrasting lines in the pattern of light falling on the retina.

Binocular Vision

How do we see objects in three dimensions? The quick answer is that our two eyes see overlapping, yet slightly different, fields. Turn a typical conical flowerpot upside down and look down at it so that the bottom of the pot is exactly in the center of your overall field of vision. You see the bottom of the pot, and you see equal amounts of the sides and rim of the pot as concentric circles around the bottom. Now if you close your left eye, you see more of the right side and right rim of the pot. With your right eye closed you see more of the left side and left rim of the pot. The discrepancies in the information coming from your two eyes are interpreted by the brain to provide information about the depth and the three-dimensional shape of the flowerpot. If you are blind in one eye, you have great difficulty discriminating distances. Animals whose eyes are on the sides of their heads have nonoverlapping fields of vision and, as a result, poor depth vision, but they can see predators creeping up from behind!

The story of how the brain integrates information from two eyes begins with the paths of the optic nerves. If you look at the underside of the brain, the optic nerves from the two eyes appear to join together just under the hypothalamus and then separate again. The place where they join is called the **optic chiasm**. Axons from the half of each retina closest to your nose cross in the optic chiasm and go to opposite sides of your brain. The axons from the other half of each retina go to the same side of the brain. The result of this division of axons in the optic chiasm is that all visual information from your left visual field (everything left of straight ahead) goes to the right side of your brain, as shown in red in Figure 39.29. All visual information from your right visual field goes to the left side of your brain, as indicated in green. Both eyes transmit information about a specific spot in your right visual field, for example, to the same place in the left visual cortex. How are the two sources of information integrated?

Cells in the visual cortex are organized in columns. These columns alternate: left eye, right eye, left eye, right eye, and so on. Cells closest to the border between two columns receive input from both eyes and are therefore **binocular cells**. Binocular cells interpret distance by measuring the disparity between where the same stimulus falls on the two retinas. What is disparity? Hold your finger out in front of you and look at it, closing one eye and then the other. Your finger appears to jump back and forth because its image falls on a different position on each retina. Repeat the exercise with an object at a distance. It doesn't appear to jump back and forth as much because there is less disparity in the positions of the image on the two retinas. Certain binocular cells respond optimally to a stimulus falling on both retinas

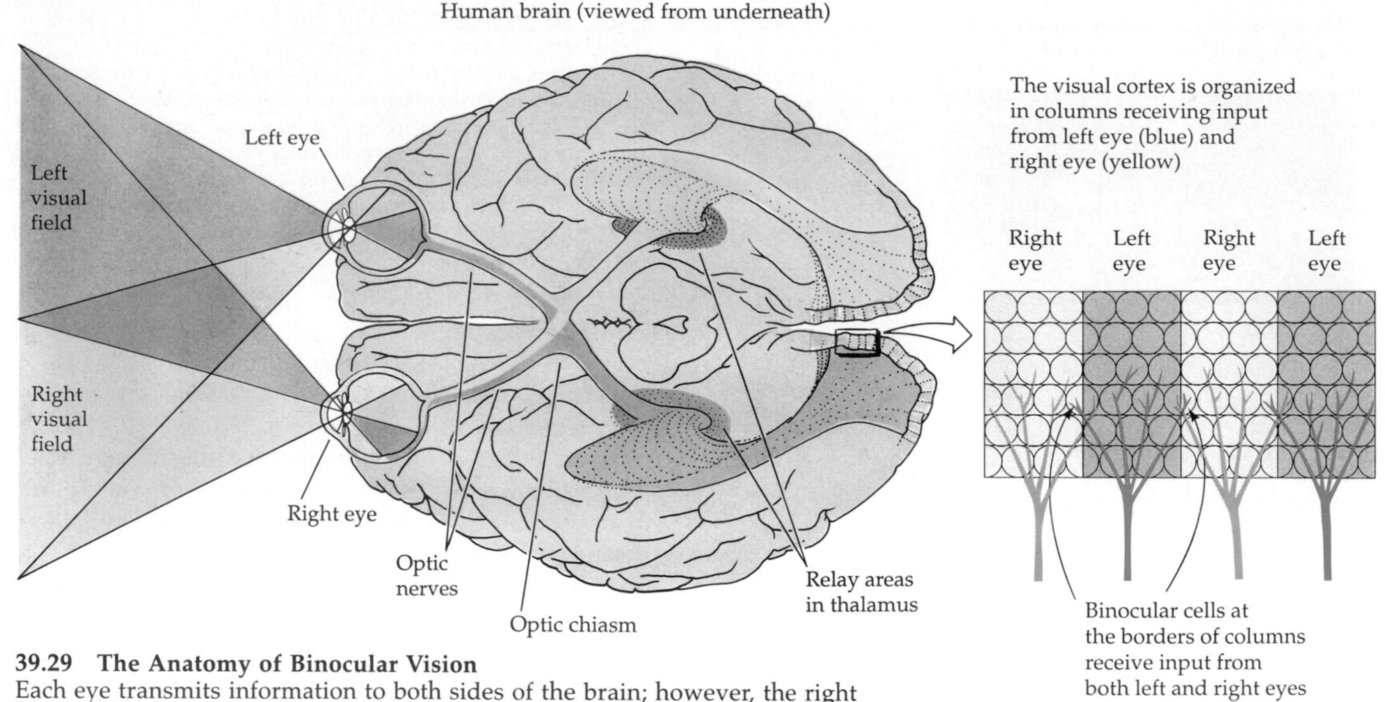

39.29 The Anatomy of Binocular Vision
Each eye transmits information to both sides of the brain; however, the right brain processes all the information from the left visual field and vice versa. The visual cortex sorts visual field information according to whether it comes from the right or left eye. Binocular cells in the visual cortex receive the same visual information from both eyes.

with a particular disparity. Which set of binocular cells is stimulated depends on how far away the stimulus is.

When we look at something, we can detect shape, color, depth, and movement. Where does all this information come together? Is there a single cell that fires only when a red sports car drives by? Probably not. A specific visual experience comes from simultaneous activity in many cells. To add to the complexity, a visual experience is not strictly visual, but is enhanced by information from the other senses and from memory.

OTHER SENSORY WORLDS

After emphasizing the incredible neural complexity of sensory integration, we must recognize that humans make use of only a subset of the information available to us in the environment. Other animals have sensory systems that enable them to use different subsets and different types of information.

Infrared and Ultraviolet Detection

When discussing vision we use the term "visible spectrum," but what we really mean is light visible to us. The human visible spectrum is a very narrow region of the entire, continuous range of electromagnetic radiation in the environment (see Figure 8.6). For example, we cannot see ultraviolet radiation, but many insects can. One of the seven photosensors in each ommatidium of a fruit fly is sensitive to ultraviolet light. The visual sensitivity of many pollinating insects includes the ultraviolet part of the spectrum. Some flowers have patterns that are invisible to us but that show up if we photograph them with film that is sensitive to ultraviolet light (see Figure 48.23). Those patterns provide information to prospective pollinators, but humans are not equipped to receive that information.

At the other end of the spectrum is infrared radiation, which we sense as heat. Other animals extract much more information from the infrared radiation—especially infrared radiation emitted by potential prey. A group of snakes known as pit vipers have pit organs, one just in front of each eye, that can sense infrared radiation (Figure 39.30). In total darkness these snakes can locate a prey item such as a mouse, orient to it, and strike it with great accuracy based on the directional information that comes to them from the warmth of the mouse's body.

Echolocation

Some species emit intense sounds and create images of their environments from the echoes of those sounds. Bats, porpoises, dolphins, and, to a lesser

39.30 **Pit Organs "See" Heat**
The eyelash viper of the Costa Rican rainforest is a pit viper. The "hole" just below and in front of its eye is a pit organ that senses infrared radiation. Such snakes can locate prey in total darkness based on directional information they receive through their pit organs.

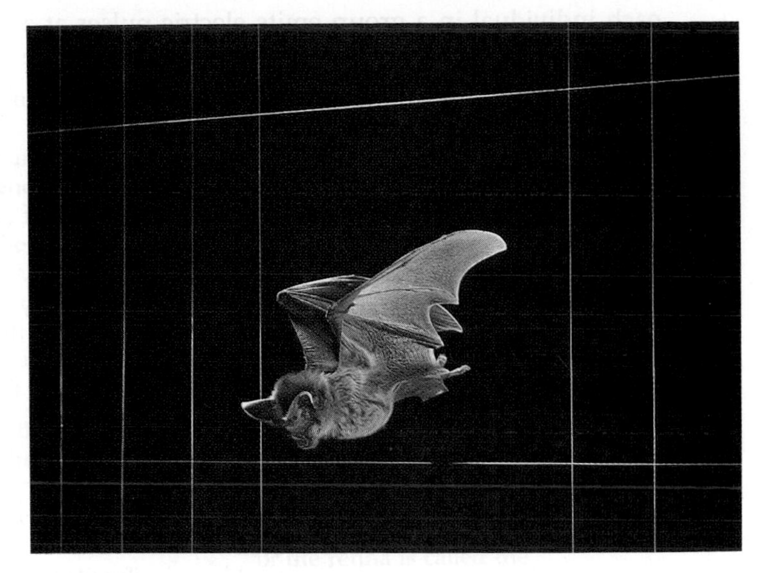

39.31 **Flying in the Dark**
Using echolocation, this bat avoids an obstacle course of thin wires while flying in complete darkness.

extent, whales, are accomplished echolocators. Some species of bats have elaborate modifications of their noses to direct the sounds they emit, as well as impressive ear pinnae to collect the returning echoes. The sounds they emit as pulses (about 20 to 80 per second) are above our range of hearing, but they are extremely loud in contrast to their faint echoes bouncing off small insects. An echolocating bat is similar to a construction worker who is trying to overhear a whispered conversation while using a pneumatic drill. To avoid deafening themselves, bats use muscles in their middle ears to dampen their sensitivity while they emit sounds, then relax them quickly enough to hear the echoes. The ability of bats to use echolocation to "see" their environment is so good that in a totally dark room strung with fine wires, bats can capture tiny flying insects while navigating around the wires (Figure 39.31).

Detection of Electric Fields

We already discussed the mechanosensors in the lateral lines of fishes (see Figure 39.11). The lateral lines of some species, especially ones that live in murky waters (catfish, for example) also contain *electro*sensors. These sensors enable the fish to detect weak electric fields, which can help them locate prey. The use of electrosensors is even more sophisticated in species called electric fishes. These animals have evolved electric organs in their tails that generate a continuous series of electric pulses, creating a weak electric field around their bodies (Figure 39.32). Any objects in the environment, such as rocks, logs, plants, or other fish, disrupt the electric fish's electric field, and the electrosensors of the lateral line detect those disruptions. In some species of electric fish,

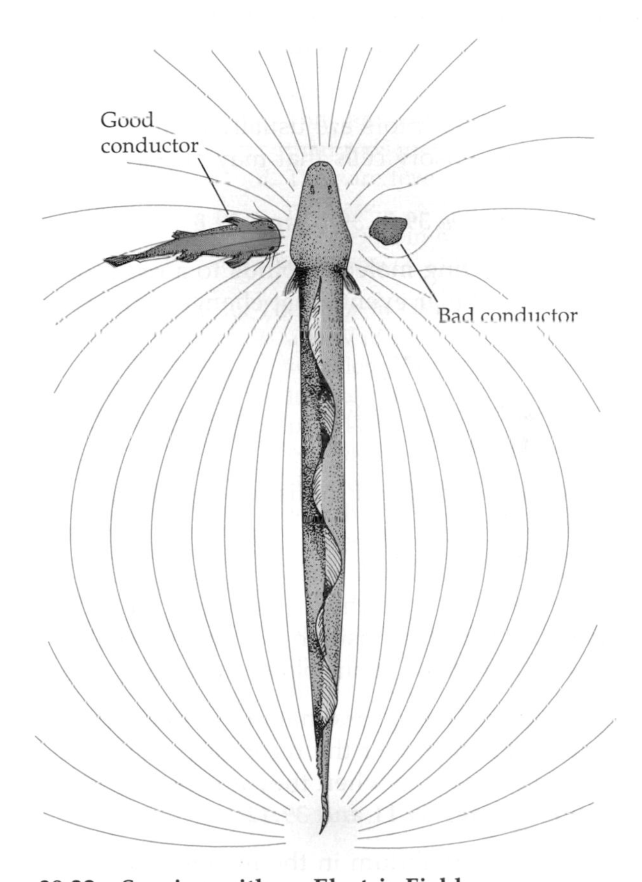

39.32 **Sensing with an Electric Field**
Fish such as the electric eel generate personal electric fields and sense perturbations within their fields. Objects that are poor conductors of electricity perturb the field greatly, whereas good conductors perturb it only slightly or not at all.

4. Describe and contrast two sensory systems that enable animals to "see" in the dark. What problems or limitations are inherent in these systems in comparison with vision?

5. Describe what is meant by a receptor potential and how it functions to encode intensity of stimulus. Use a specific sensor as your example.

READINGS

Camhi, J. M. 1984. *Neuroethology: Nerve Cells and the Natural Behavior of Animals.* Sinauer Associates, Sunderland, MA. This text is particularly good for putting basic neurophysiology in the context of whole animals and their behaviors. Part II deals with the sensory worlds of animals.

Hubel, D. H. 1988. *Eye, Brain, and Vision.* Scientific American Library Series No. 22. W. H. Freeman, New York. A comprehensive and beautifully illustrated book about the neurophysiology and neuroanatomy of vision. It is very readable for the nonexpert, yet it presents the depth and breadth of knowledge and experience of someone who has been a major contributor to this area of research.

Hudspeth, A. J. 1983. "The Hair Cells of the Inner Ear." *Scientific American*, January. The inner ear transduces mechanical forces of pressure waves into action potentials transmitted to the brain. This paper describes the cells that accomplish the transduction and explain how they do it.

Knudsen, E. I. 1981. "The Hearing of the Barn Owl." *Scientific American*, June. The barn owl can use its remarkably precise and sensitive auditory system to locate prey in complete darkness. This paper explains the neurophysiological basis for its extreme accuracy.

Konishi, M. 1993. "Listening with Two Ears." *Scientific American*, April. Continuing the research reported in the article by Knudsen in 1981, this paper shows how the brain of the barn owl integrates signals from its two ears to create a map of its auditory environment.

Koretz, J. F. and G. H. Handelman, 1988. "How the Human Eye Focuses." *Scientific American*, July. In addition to describing in detail how the eye accommodates for distance, this article explains the changes that occur with aging.

Newman, E. A. and P. H. Hartline. 1982. "The Infrared 'Vision' of Snakes." *Scientific American*, March. Some snakes are able to use infrared radiation emitted by objects in their environment to construct a sensory world.

Stryer, L. 1987. "The Molecules of Visual Excitation." *Scientific American*, July. This paper describes the molecular chain of events that transduce photons of light falling on the retina into action potentials that are transmitted to the brain.

Suga, N. 1990. "Biosonar and Neural Computation in Bats." *Scientific American*, June. The use of echolocation by bats to construct a view of their environment requires complex neural processing of information by their auditory systems.

A Spider's Jump Seems Effortless

40

Effectors

Information from sensors is not of much value to an animal unless it can respond. An obvious response is movement, and a fascinating array of adaptations have evolved that enable animals to move. Consider jumping. When bending down to smell a flower you see a spider. Startled by the spider, neural signals from your brain routed through spinal circuits activate muscles in your legs to contract, causing your legs to extend, and you jump. At the same time, the spider jumps in the opposite direction. Did its nervous system also cause muscles to contract and extend its jumping legs? No; the spider doesn't have such muscles. Instead, extracellular fluids were squeezed into its hollow jumping legs, and the increased pressure in the legs caused them to extend and the spider to jump.

While this drama was playing out, a flea sensed your body heat and jumped 20 centimeters onto your leg. The flea's jump involves still a different mechanism, one that works in the same way as a slingshot. The flea is so small and its initial acceleration so great that no muscle can contract fast enough to cause such a movement. At the base of its jumping legs is an elastic material that is compressed by slow muscles while the flea is resting. When a trigger mechanism is released, the elastic material recoils like a slingshot and propels the flea a distance perhaps 200 times its body length.

Another remarkable jumper that combines the mechanisms used by humans with those used by the flea is a kangaroo. As you run faster and faster, the number of strides you take per minute and the energy you expend per minute increases rapidly. Neither effect is true for the kangaroo. When running at speeds from about 5 to 25 kilometers per hour, the kangaroo has the same number of strides per minute and no increase in its metabolic rate. How can this be so? As we will learn in this chapter, muscles are attached to bones by tendons. Like the material at the base of the flea's legs, tendons can be elastic. The kangaroo's tendons stretch when it lands, and their recoil helps power the next jump. To move faster, the kangaroo increases its stride length, thereby increasing the stretch on its tendons each time it lands and the magnitude of the recoil at the initiation of each jump.

HOW ANIMALS RESPOND TO INFORMATION FROM SENSORS

Jumping is one way an animal can respond to information received by its sensors. In the examples of jumping behavior just discussed, muscles generate mechanical forces, and structures such as bones, elastic tissues, exoskeletons, and fluid-filled cavities apply mechanical force to cause behavior. Adaptations

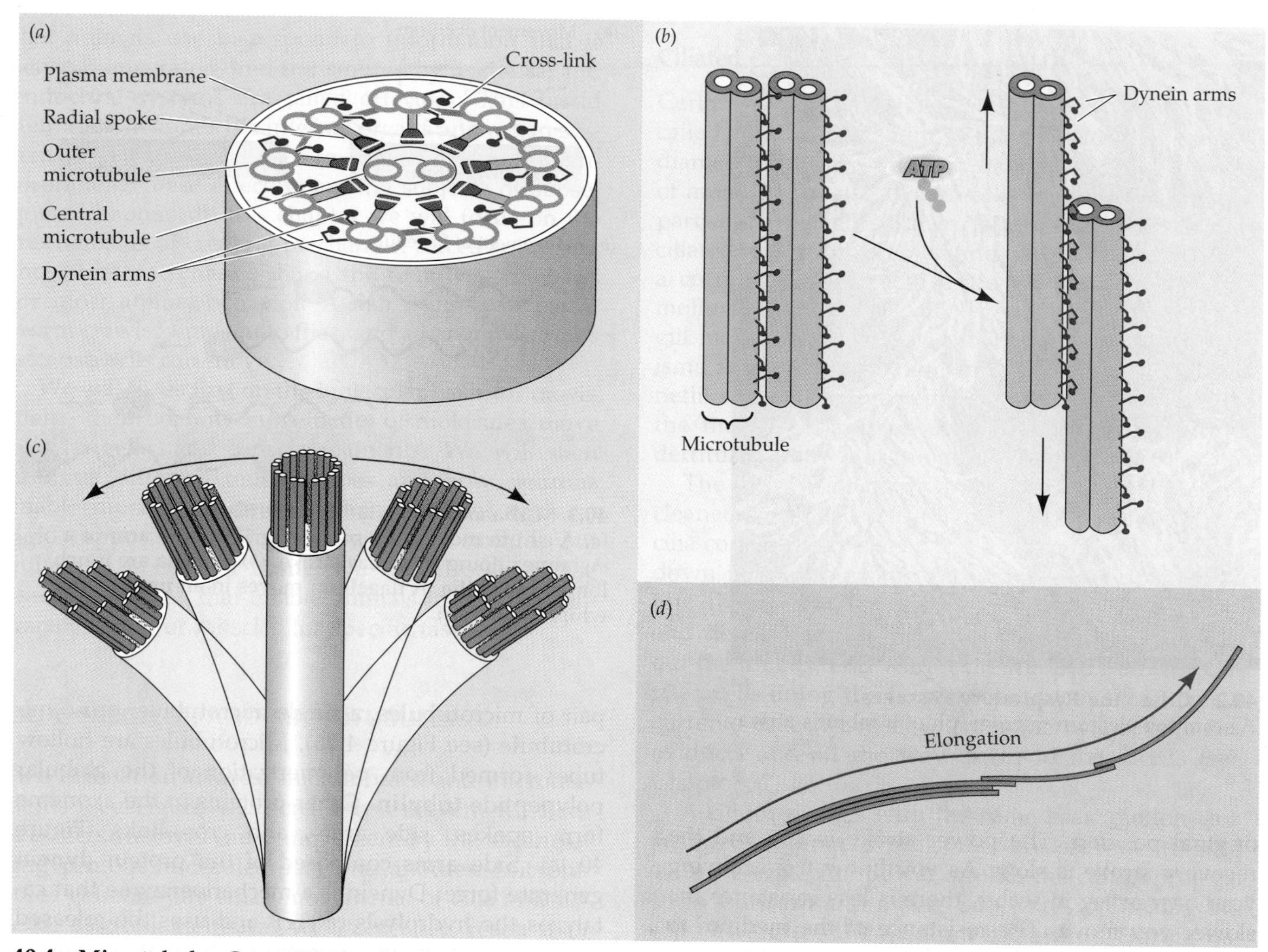

40.4 Microtubules Create Motion by Pushing against Each Other
Cilia and flagella move because of the actions of proteins in the axoneme. *(a)* A cross section of an axoneme. The microtubules occur in doublets that run the length of the axoneme. *(b)* Dynein arms of the doublets generate force by making and breaking cross-links with the neighboring microtubule pair. *(c)* When microtubule pairs try to slide past each other, the axoneme bends because the microtubules are anchored together at the base. *(d)* If all links between microtubules except the dynein arms are eliminated, ATP causes the microtubules to slide past each other, and the axoneme elongates.

the axoneme with enzymes that hydrolyze proteins, however, disrupts the spokes and cross-links, leaving only the microtubules and dynein arms intact. If the isolated microtubules are then exposed to Ca^{2+} and ATP, they row past one another and the whole structure elongates manyfold (Figure 40.4*d*), demonstrating that the forces moving microtubule pairs along one another are the basis for the bending of the intact axoneme. When dynein (and only dynein) is removed from isolated axonemes, the microtubules lose their ATPase activity and their motility. Restoring purified dynein to the axonemes restores ATPase activity and motility.

Microtubules as Intracellular Effectors

Microtubules play important roles in cell movements. As components of the cytoskeleton, microtubules also contribute to the cell's shape. Cells can change shape and move by polymerizing and depolymerizing the tubulin in their microtubules. During mitosis, the spindle that moves chromosomes to the mitotic poles at anaphase forms by the polymerization of tubulin. Another example of microtubule involvement in cell movement is the growth of the axons of neurons in the developing nervous system. Neurons find and make their appropriate connections by sending out long extensions that search for the correct contact cells. If polymerization of tubulin is chemically inhibited, the neurons do not extend. Microtubules are important intracellular effectors for changing cell shape, moving organelles, and enabling cells to respond to the environment.

MICROFILAMENTS AND CELL MOVEMENT

Like microtubules, protein **microfilaments** change cell shape and cause cell movements. The dominant

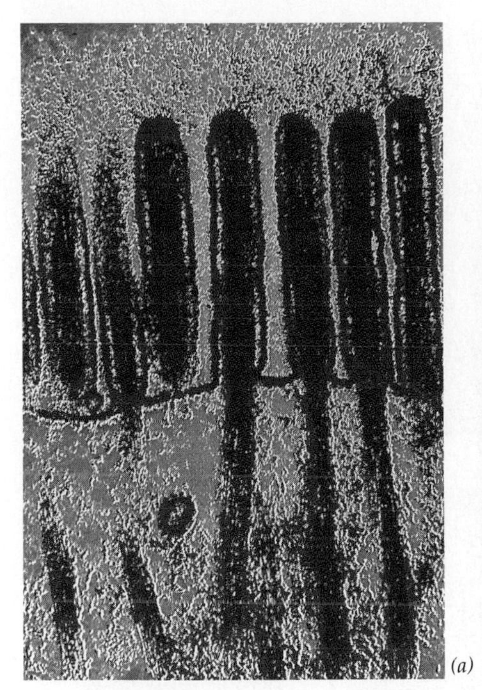

40.5 Cell Projections Supported by Microfilaments
(a) The cells lining the gut have numerous fingerlike projections called microvilli. *(b)* Some mechanosensors, such as these hair cells in the organ of Corti, have stereocilia. Both microvilli and hair cells are stiffened by microfilaments.

microfilament in cells is the protein **actin**, and bundles of cross-linked actin strands form important structural components of cells. For example, the microvilli (tiny projections) that increase the absorptive surface area of the cells lining the gut are stiffened by actin microfilaments (Figure 40.5*a*), as are the stereocilia of the sensory hair cells mentioned in Chapter 39 (Figure 40.5*b*). Like microtubules, actin microfilaments can change the shape of a cell simply by polymerizing and depolymerizing. The projections sent out by phagocytic cells (see Figure 16.2) are an example of this process.

Together with the protein myosin, actin microfilaments generate the contractile forces responsible for many aspects of cell locomotion and changes in cell shape. For example, the contractile ring that divides a cell undergoing mitosis into two daughter cells is composed of actin microfilaments in association with myosin. The mechanisms that many cells employ to engulf materials (phagocytosis and pinocytosis; see Chapter 4) also rely on actin microfilaments and myosin. Nets of actin and myosin beneath the cell membrane change a cell's shape during phagocytosis.

The movement of certain cells in multicellular animals is due to the activity of actin microfilaments and myosin. During development many cells migrate by such **amoeboid movement**, and throughout an animal's life phagocytic cells (see Chapter 16) circulate in the blood, squeeze through the walls of the blood vessels, and wander through the tissues by amoeboid movement. The mechanisms of amoeboid movement have been studied extensively in the protist for which this type of movement was named—the amoeba, which lives in freshwater streams and ponds.

The amoeba moves by extending lobe-shaped projections called pseudopods and then seemingly squeezing itself into them (Figure 40.6). The cytoplasm in the core of the amoeba is relatively liquid and is called **plasmasol**, but just beneath the plasma membrane the cytoplasm is much thicker and is called **plasmagel**. To form a pseudopod, the thick plasmagel in a certain area of the cell thins, allowing a bulge to form. Just under the cell surface, in the plasmagel, is a network of actin microfilaments that interacts with myosin to squeeze plasmasol into the bulge, thus forming a pseudopod. As the network continues to contract, cytoplasm streams in the direction of the pseudopod. Eventually the cytoplasm at the leading edge of the pseudopod converts to gel

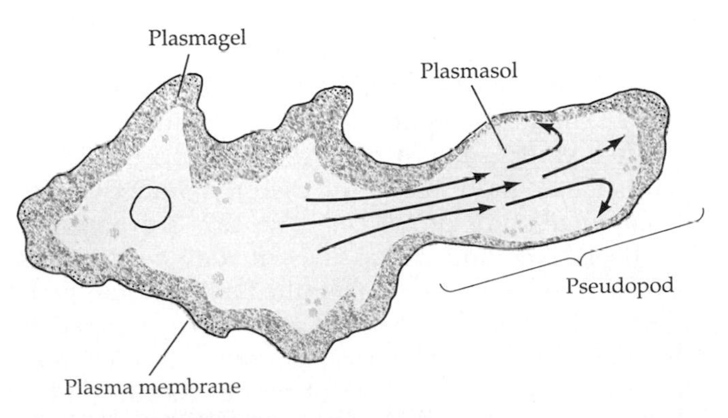

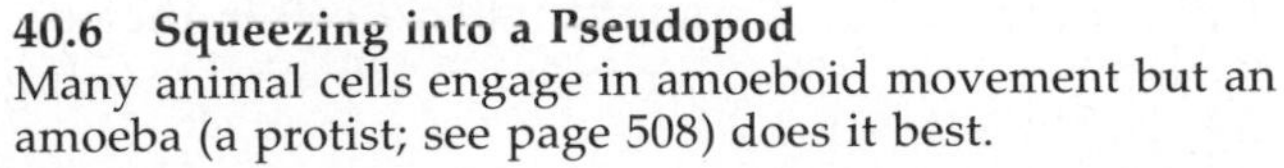

40.6 Squeezing into a Pseudopod
Many animal cells engage in amoeboid movement but an amoeba (a protist; see page 508) does it best.

and the pseudopod stops forming. Thus the basis for amoeboid motion is the ability of the cytoplasm to cycle through sol and gel states and the ability of the microfilament network under the cell membrane to contract and cause the cytoplasmic streaming that pushes out a pseudopod.

MUSCLES

Muscle contraction is the most important effector mechanism that animals have for responding to their environments. All behavioral and most physiological responses depend on muscle cells. Muscle cells are specialized for contraction and have high densities of actin microfilaments and myosin. Such cells are found throughout the animal kingdom. They account for the thrashing movements of nematodes, the expansion–contraction movements of earthworms, the pulsating movements of jellyfish, and the limb movements of arthropods and vertebrates. Muscle cells are found in the walls of blood vessels, guts, bladders, and hearts. Wherever contraction of whole tissues takes place in animals, muscle cells are responsible. In all cases, the molecular mechanism of contraction is the same, but there are many specializations of muscle cells fitting them to the wide variety of functions they serve. We begin our study of muscle by looking at the three types of muscle cells found in vertebrates: smooth muscle, skeletal muscle, and cardiac (heart) muscle.

Smooth Muscle

Smooth muscle provides the contractile forces for most of our internal organs, which are under the control of the autonomic nervous system. Smooth muscle moves food through the digestive tract, controls the flow of blood through blood vessels, and empties the urinary bladder. Structurally, smooth muscle cells are the simplest muscle cells. They are usually long and spindle-shaped, and each cell has a single nucleus. Because the filaments of actin and myosin in smooth muscle are not as regularly arranged as those in the other muscle types, the contractile machinery is not obvious when the cells are viewed under the light microscope (Figure 40.7*a*).

If we study smooth muscle from a particular organ, such as the walls of the digestive tract, we find it has interesting properties. The cells are arranged in sheets, and individual cells in the sheets are in electrical contact with one another through gap junctions. As a result, an action potential generated in the membrane of one smooth muscle cell can spread to all the cells in the sheet of tissue. Another interesting property of a smooth muscle cell is that the resting potential of its membrane is sensitive to being stretched. If the wall of the digestive tract is stretched in one location (such as by receiving a mouthful of food), the membranes of the stretched cells depolarize, reach threshold, and fire action potentials that cause the cells to contract. Thus smooth muscle con-

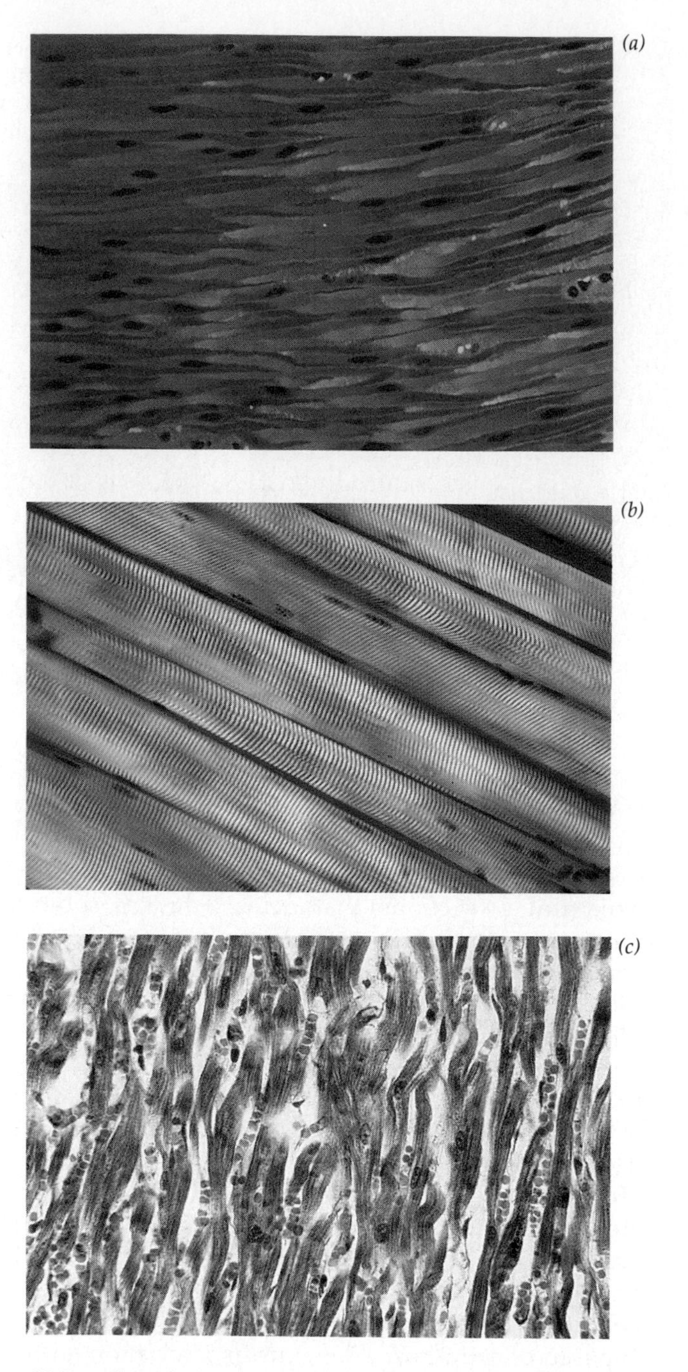

40.7 The Three Types of Vertebrate Muscle Tissue *(a)* Smooth muscle cells are usually arranged in sheets such as those that make up the walls of the stomach and the intestine. The dark structures in this sheet of muscle cells are nuclei. *(b)* Skeletal, or striated, muscle appears to be striped, or banded, because of its highly regular arrangement of contractile filaments. *(c)* Cardiac muscle is also striated, but cardiac muscle fibers branch and create a meshwork that resists tearing or breaking.

tracts after being stretched, and the harder it is stretched, the stronger the contraction. Later in this chapter we will discuss how membrane depolarization triggers contraction.

Skeletal Muscle

Skeletal muscle carries out all voluntary movements, such as running or playing a piano, and generates the movements of breathing. Skeletal muscle is called **striated muscle** because the highly regular arrangement of its actin microfilaments and myosin gives it a striped appearance (Figure 40.7*b*). Skeletal muscle cells, or **muscle fibers**, are large, and they have many nuclei because they develop through the fusion of many individual cells. A muscle such as your biceps (which bends your arm) is composed of many muscle fibers bundled together by connective tissue.

What is the relation between a muscle fiber and the actin and myosin filaments responsible for contraction? Each muscle fiber is composed of **myofibrils**: bundles of contractile filaments made up of actin and myosin (Figure 40.8). Within each myofibril are thin filaments, which are actin microfilaments, and thick filaments, which are composed of myosin. If we cut across the myofibril at certain locations, we see only thick filaments, in other locations only thin filaments, but in most regions of the myofibril, each thick myosin filament is surrounded by six thin actin filaments.

A longitudinal view of a myofibril reveals the striated appearance of skeletal muscle. The band pattern of the myofibril is due to repeating units called **sarcomeres**, which are the units of contraction (Figure 40.8*a*). Each sarcomere is made of overlapping filaments of actin and myosin. As the muscle contracts, the sarcomeres shorten, and the appearance of the band pattern changes.

The observation that the widths of the bands in the sarcomeres change when a muscle contracts led two British biologists, Hugh Huxley and Andrew Huxley, to propose a molecular mechanism of muscle contraction. Let's look at the myofibril's band pattern in detail (Figure 40.8*b*). Each sarcomere is bounded by Z lines that anchor the thin actin filaments. Centered in the sarcomere is the A band, which contains all the myosin filaments. The H zone and the I band, which appear light, are regions where actin and myosin filaments do not overlap in the relaxed muscle. The dark stripe within the H zone is called the M band; it contains proteins that help hold the myosin filaments in their regular hexagonal arrangement. When the muscle contracts, the sarcomere shortens. The H zone and I band become much narrower, and the Z lines move toward the A band as if the actin filaments were sliding into the region occupied by the myosin filaments. This observation led Huxley and Huxley to propose the **sliding-filament theory** of muscle contraction: Actin and myosin filaments slide past each other as the muscle contracts.

To understand what makes the filaments slide, we must examine the structure of actin and of myosin. The myosin molecule consists of two long polypeptide chains coiled together, each ending in a globular head (Figure 40.8*c*). The myosin filament is made up of many myosin molecules arranged in parallel with their heads projecting laterally from one or the other end of the filament (Figure 40.8*d*). The actin filament consists of a helical arrangement of two chains of monomers like two strands of pearls twisted together (Figure 40.8*e*). Twisting around the actin chains are two strands of another protein, **tropomyosin**. The myosin heads have sites that can bind to actin and thereby form bridges between the myosin and the actin filaments. The myosin heads also have ATPase activity; they bind and hydrolyze ATP. The energy released changes the orientation of the myosin head.

These details explain the cycle of events that cause actin and myosin filaments to slide past one another and shorten the sarcomere (Figure 40.8*f*). A myosin head binds to an actin filament. Upon binding, the head changes its orientation with respect to the myosin filament, exerting a force that causes the actin and myosin filaments to slide about 5 to 10 nanometers. Next, the myosin head binds a molecule of ATP, which causes it to release the actin. When the ATP is hydrolyzed, the energy released causes the myosin head to return to its original conformation, in which it can bind again to actin. The hydrolysis of the ATP is like cocking the hammer of a pistol; binding with actin pulls the trigger.

We have been discussing the cycle of contraction in terms of a single myosin head. Don't forget that each myosin filament has many myosin heads at both ends and is surrounded by six actin filaments; thus the contraction of the sarcomere involves a great many cycles of interaction between actin and myosin molecules. That is why when a single myosin head breaks its contact with actin, the actin filaments do not slip backwards.

An interesting aspect of this contractile mechanism is that ATP is needed to break the actin–myosin bonds but not to form them. Thus muscles require ATP to stop contracting. This fact explains why muscles stiffen soon after animals die, a condition known as rigor mortis. Death stops the replenishment of the ATP stores of muscle cells, so the myosin–actin bridges cannot be broken, and the muscles stiffen. Eventually the proteins begin to lose their integrity and the muscles soften. Because these events have regular time courses that differ somewhat for different regions of the body, an examination of the stiffness of the muscles of a corpse sometimes helps a coroner estimate the time of death.

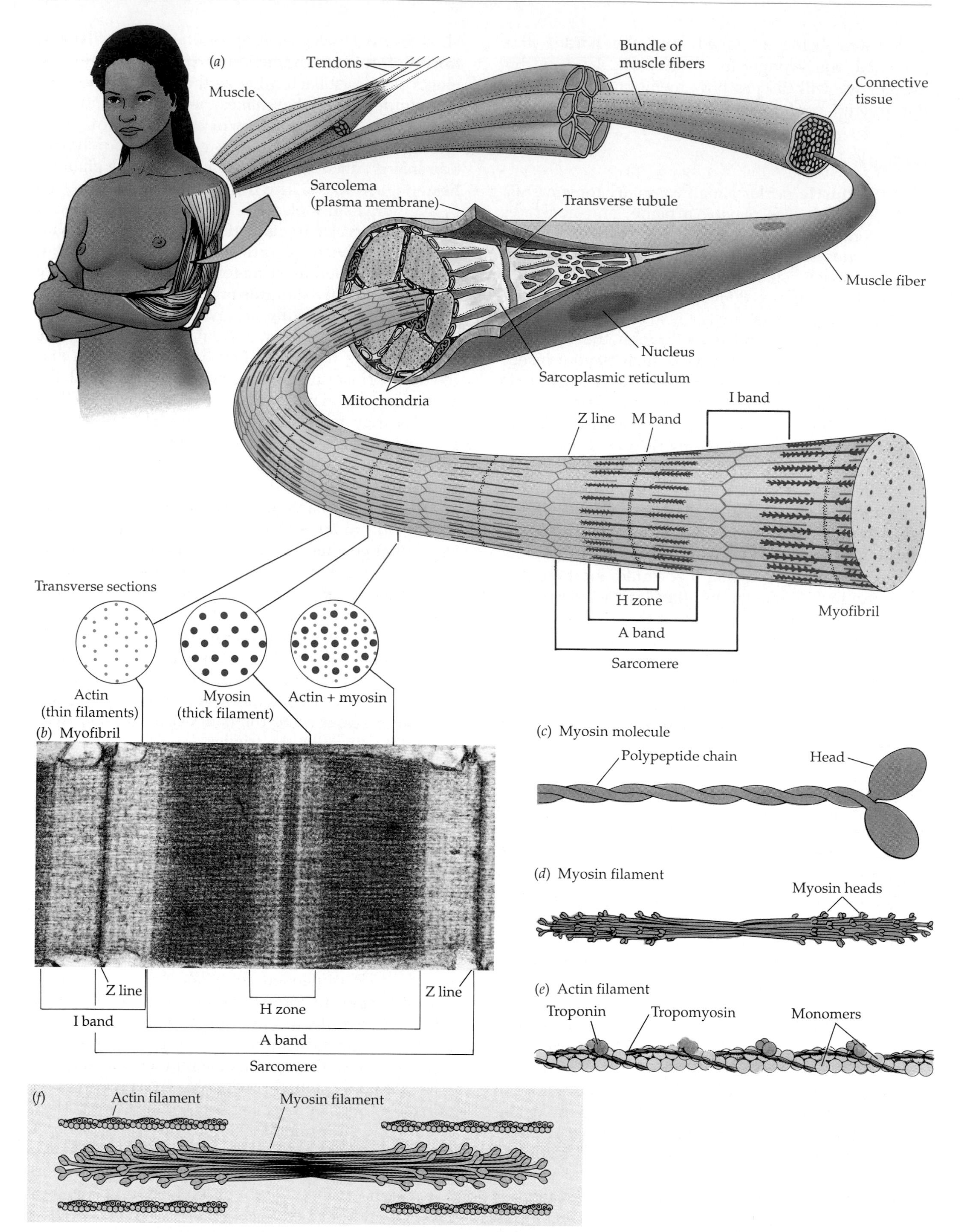
(a)
Tendons
Muscle
Bundle of muscle fibers
Connective tissue
Sarcolema (plasma membrane)
Transverse tubule
Muscle fiber
Nucleus
Sarcoplasmic reticulum
Mitochondria
I band
Z line
M band
H zone
A band
Sarcomere
Myofibril
Transverse sections
Actin (thin filaments)
Myosin (thick filament)
Actin + myosin
(b) Myofibril
Z line
Z line
H zone
I band
A band
Sarcomere
(c) Myosin molecule
Polypeptide chain
Head
(d) Myosin filament
Myosin heads
(e) Actin filament
Troponin
Tropomyosin
Monomers
(f)
Actin filament
Myosin filament

40.8 The Structure of Skeletal Muscle from Tissue Down to Molecules
(*a*) A skeletal muscle is made up of bundles of muscle fibers. Each muscle fiber is a multinucleate cell containing numerous myofibrils, which are highly ordered assemblages of thick myosin and actin filaments. Within each muscle fiber the nuclei, the mitochondria, and the sarcoplasmic reticulum surround the myofibrils. (*b*) The structure of the myofibrils gives muscle fibers their characteristic striated appearance. Where there are only actin filaments the myofibril appears light; where there are both actin and myosin filaments the myofibril appears dark. (*c,d*) Myosin filaments consist of bundles of molecules with long polypeptide tails and globular heads. (*e*) Actin filaments consist of two chains of monomers twisting around each other. Two polypeptide chains of tropomyosin twist around the actin chains. (*f*) Actin and myosin filaments overlap in the sarcomere.

Controlling the Actin–Myosin Interaction

Muscle contractions are initiated by nerve action potentials arriving at the neuromuscular junction. Motor neurons are generally highly branched and can innervate up to 100 muscle fibers each. All the fibers innervated by a single motor neuron are a **motor unit** and contract simultaneously in response to the action potentials fired by that motor neuron. To understand the fine control the nervous system has over the sliding of actin and myosin filaments, we must examine the membrane system of the muscle fiber and some additional protein components of the actin filaments.

Like neurons, vertebrate skeletal muscle fibers are excitable cells: When they are depolarized to a threshold that opens their voltage-gated sodium channels, their plasma membranes generate action potentials, just as the membranes of axons do. The initial depolarization that spreads across the muscle fiber plasma membrane is generated at the neuromuscular junction—the synapse between the motor neuron and the muscle fiber. As we saw in Chapter 38, neurotransmitter from the motor neuron binds to receptors in the postsynaptic membrane, causing ion channels to open. The depolarization of the postsynaptic membrane spreads to the surrounding plasma membrane of the muscle fiber which contains voltage-gated ion channels. When threshold is reached, the plasma membrane fires an action potential that is rapidly conducted to all points on the surface of the muscle fiber.

The plasma membrane of the muscle fiber is continuous with a system of tubules that descends into and branches throughout the cytoplasm (also called the sarcoplasm) of the muscle fiber. These are the transverse tubules, or **T-tubules**, and they communicate with a network of membranes, called the **sarcoplasmic reticulum**, that surrounds every myofibril (see Figure 40.8*a*). The wave of depolarization that spreads over the plasma membrane of the muscle fiber also spreads through the T-tubule system. Calcium pumps in the membranes of the sarcoplasmic reticulum cause this membrane-bounded compartment of the fiber to take up and sequester Ca^{2+}. As a result, there is a high concentration of Ca^{2+} in the sarcoplasmic reticulum and low concentration of Ca^{2+} in the sarcoplasm surrounding the filaments. When a wave of depolarization spreads through the T-tubule system, it induces Ca^{2+} channels in the sarcoplasmic reticulum to open, resulting in the diffusion of Ca^{2+} ions out of the sarcoplasmic reticulum and into the sarcoplasm surrounding the microfilaments. The Ca^{2+} stimulates the interaction of actin and myosin and the sliding of the filaments. How does it do so?

Remember that an actin filament is a helical arrangement of two strands of actin monomers. Lying in the grooves between the two actin strands is the two-stranded protein tropomyosin (Figure 40.9*a*). At regular intervals the filament also includes another globular protein, **troponin**. The troponin molecule has three subunits; one binds actin, one binds tropomyosin, and one binds Ca^{2+}. When Ca^{2+} is sequestered in the sarcoplasmic reticulum, the tropomyosin strands block the sites where myosin heads can bind to the actin. When the T-tubule system depolarizes, Ca^{2+} is released into the sarcoplasm, where it binds to the troponin, changing the shape of the troponin molecule. Because the troponin is also bound to the tropomyosin, this conformational change of the troponin twists the tropomyosin enough to expose the actin–myosin binding sites. This initiates the cycle of making and breaking actin–myosin bridges; the filaments are pulled past one another, and the muscle fiber contracts. When the T-tubule system repolarizes, the calcium pumps remove the Ca^{2+} ions from the sarcoplasm, causing the tropomyosin to return to the position in which it blocks the binding of the myosin heads to the actin strands, and the muscle fiber returns to its resting condition. Figure 40.9*b* reviews the cycle.

Twitches, Graded Contractions, and Tonus

In vertebrate skeletal muscle, the arrival of an action potential at the neuromuscular junction causes an action potential in the muscle fiber. The spread of the action potential through the membrane system of the muscle fiber causes a minimum unit of contraction, called a **twitch**. A twitch can be measured in terms of the tension, or force, it generates (Figure 40.10). If action potentials in the muscle fiber are adequately separated in time, each twitch is a discrete, all-or-none phenomenon. If action potentials are fired more rapidly, however, new twitches are triggered before the filaments have had a chance to return to their

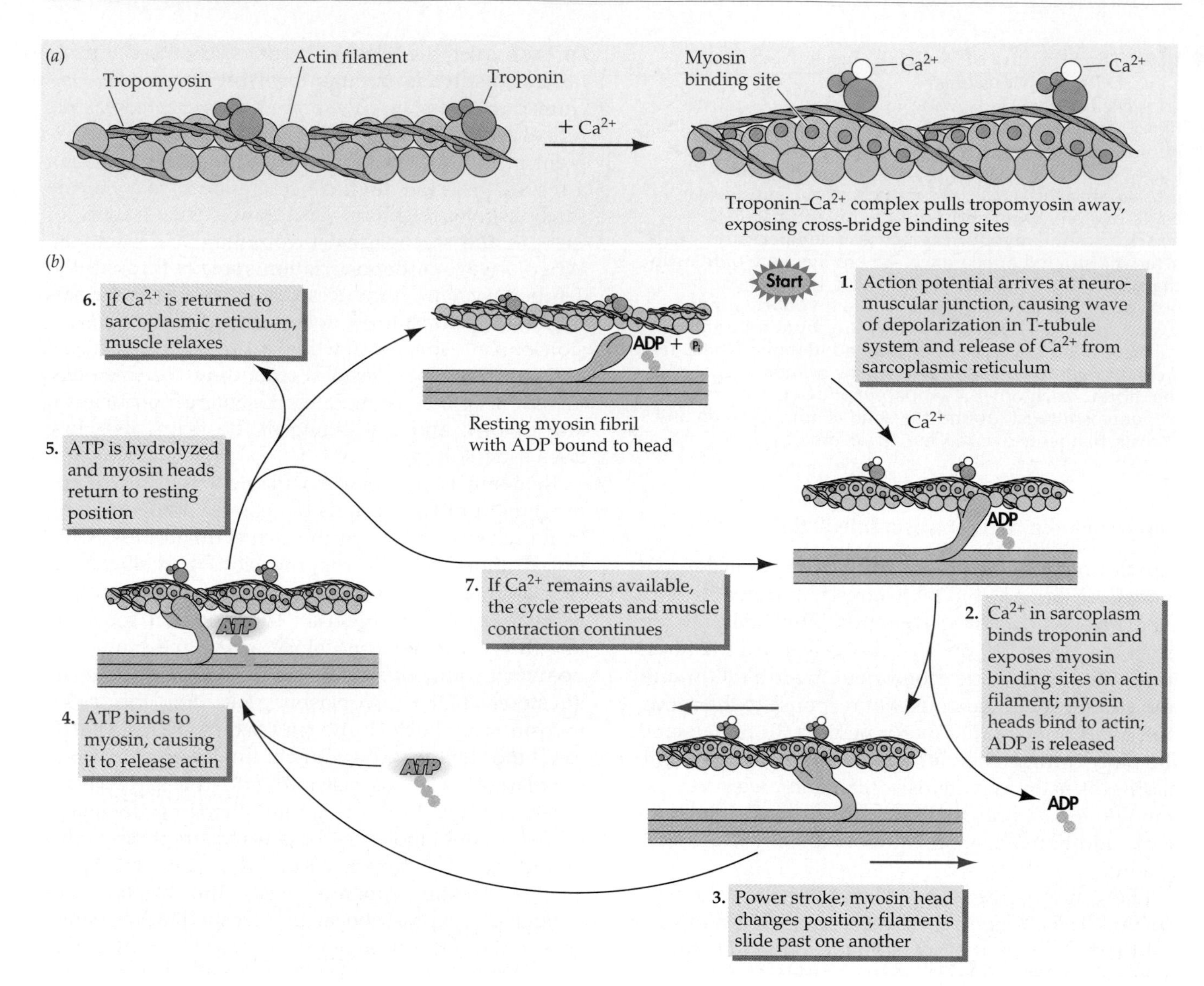

40.9 Tropomyosin and Troponin
Tropomyosin and troponin molecules on actin filaments control the formation of crossbridges. *(a)* When Ca^{2+} binds to troponin it exposes crossbridge binding sites. *(b)* As long as crossbridge sites and ATP are available, the cycle of actin and myosin interactions continues and the filaments slide.

resting condition. As a result, the twitches sum and the tension generated by the fiber increases and becomes more continuous. Thus the individual muscle fiber can show a graded response to increased levels of stimulation by its motor neuron. At high levels of stimulation, the calcium pumps in the sarcoplasmic reticulum can no longer remove Ca^{2+} ions from the sarcoplasm between action potentials, and the contractile machinery generates maximum tension—a condition known as **tetanus**. (Do not confuse this condition with the disease tetanus, which is caused by a bacterial toxin and characterized by spastic contractions of skeletal muscles.)

How long a muscle fiber can maintain a tetanic contraction depends on its supply of ATP. Eventually the fiber will fatigue. It may seem paradoxical that the lack of ATP causes fatigue, since the action of ATP is to break the actin–myosin bonds. But remember that the energy released from the hydrolysis of ATP "recocks" the myosin heads, allowing them to cycle through another power stroke. The situation is like rowing a boat upstream: You cannot maintain your position relative to the stream bank by just holding the oars out against the current; you have to keep rowing.

The ability of a whole muscle to generate different levels of tension depends also on how many muscle fibers in that muscle are activated. Whether a muscle contraction is strong or weak depends both on how many motor neurons to that muscle are firing and on the rate at which those neurons are firing. These two factors can be thought of as spatial and temporal summation, respectively. Both types of summation increase the strength of contraction of the muscle as

a whole. Faster twitching of individual fibers causes temporal summation (Figure 40.10), and an increase in the number of motor units involved in the contraction causes spatial summation. (Remember that a motor unit is all the muscle fibers innervated by a single neuron, and that a single muscle consists of many motor units.)

Many muscles of the body maintain a low level of tension called **tonus** even when the body is at rest. For example, the muscles of our neck, trunk, and limbs that maintain our posture against the pull of gravity are always working, even when we are standing or sitting still. Muscle tonus comes from the activity of a small but changing number of motor units in a muscle; at any one time some of the muscle's fibers are contracting and others are relaxed. Tonus is constantly being readjusted by the nervous system.

Fast- and Slow-Twitch Fibers

Not all skeletal muscle fibers are alike in how they twitch, and a single muscle may contain more than one type of fiber. The two major types of skeletal muscle fibers are **slow-twitch fibers** and **fast-twitch fibers** (Figure 40.11*a*). Slow-twitch fibers are also

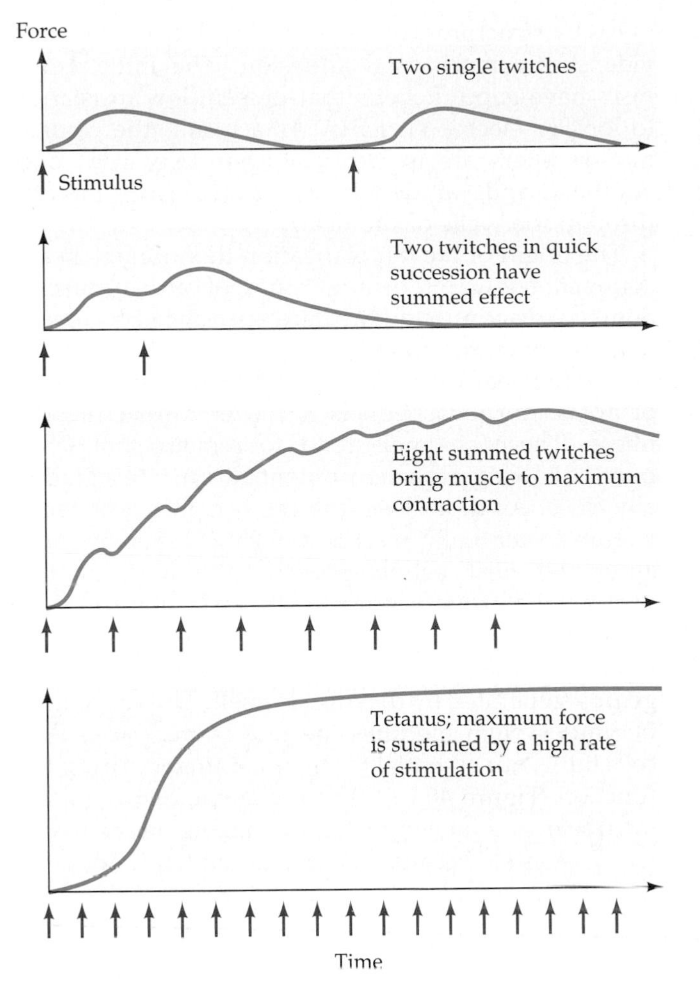

40.10 Twitches and Tetanus
Each red arrow represents a single, brief electric pulse to the nerve that is innervating the muscle. This stimulus elicits a twitch, the minimum unit of contraction of a muscle. Twitches that occur in rapid succession have a summed effect. Tetanus is the maximum state of tension a muscle can achieve.

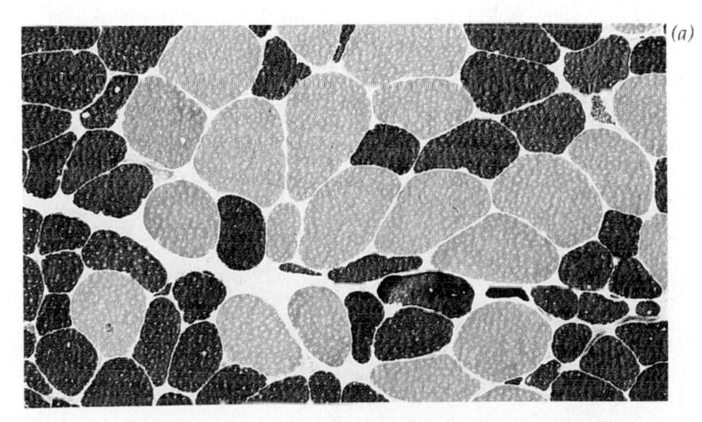

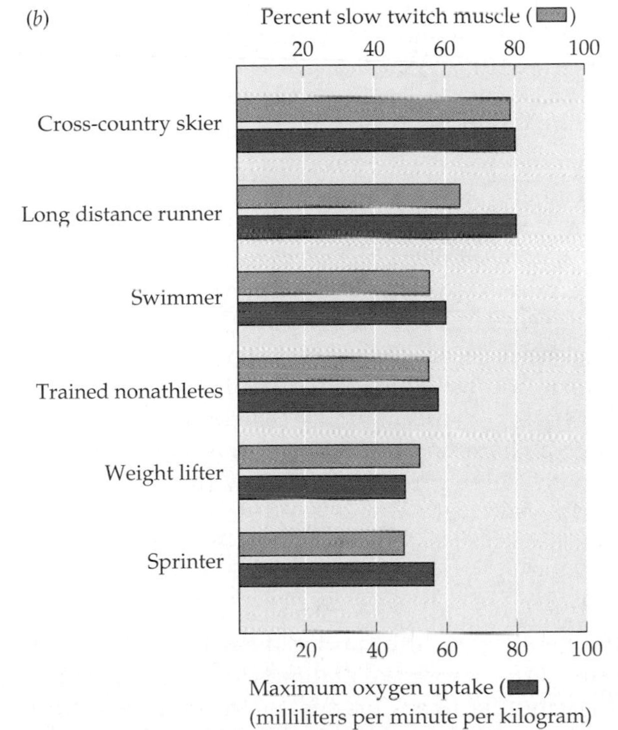

40.11 Two Types of Muscle Fibers are Specialized For Fast or For Sustained Contraction
(*a*) Skeletal muscle consists of fast- and slow-twitch fibers. In this stained micrograph, which is a cross section of a skeletal muscle, slow-twitch fibers are dark and fast-twitch fibers are light. (*b*) World-class athletes in different sports have different distributions of fiber types. Slow-twitch fibers are better adapted for sustained aerobic activity. Fast-twitch fibers can generate maximum tension quickly, but they also fatigue quickly. For comparison, we include here average values for amateur athletes in good condition. The maximum oxygen uptake is a measure of the maximum level of aerobic metabolism the person can sustain for a period of time greater than a few minutes.

called red muscle because they have lots of the oxygen-binding molecule myoglobin, they have lots of mitochondria, and they are well supplied with blood vessels. A single twitch of a slow-twitch fiber produces low tension. The maximum tension a slow-twitch fiber can produce is low and develops slowly, but these fibers are highly resistant to fatigue. Because slow-twitch fibers have substantial reserves of glycogen and fat, their abundant mitochondria can maintain a steady, prolonged production of ATP if oxygen is available. Muscles with high proportions of slow-twitch fibers are good for long-term, aerobic work (work that requires lots of oxygen). Champion long-distance runners, cross-country skiers, swimmers, and bicyclists have leg and arm muscles consisting mostly of slow twitch fibers (Figure 40.11*b*).

Fast-twitch skeletal muscle fibers are also called white muscle because they have fewer mitochondria, little or no myoglobin, and fewer blood vessels than slow-twitch fibers do. The white meat of domestic chickens is composed of fast-twitch fibers. Fast-twitch fibers can develop maximum tension more rapidly than slow-twitch fibers can, and that maximum is greater, but they fatigue rapidly. The myosin of fast-twitch fibers has high ATPase activity, so they can put the energy of ATP to work very rapidly, but they cannot replenish it rapidly enough to sustain contraction for a long time. Fast-twitch fibers are especially good for short-term work that requires maximum strength. Champion weight lifters and sprinters have leg and arm muscles with high proportions of fast-twitch fibers.

What determines the proportion of fast- and slow-twitch fibers in your muscles? The most important factor is your genetic heritage, so there is some truth to the statement that champions are born, not made. To a certain extent, however, you can alter the properties of your muscle fibers through training. With aerobic training, the oxidative capacity of fast-twitch fibers can improve substantially. But a person born with a high proportion of fast-twitch fibers will never become a champion marathon runner, and a person born with a high proportion of slow-twitch fibers will never become a champion sprinter.

Cardiac Muscle

Although striated, the cardiac (heart) muscle of vertebrates differs from skeletal muscle in several ways. Cardiac muscle fibers branch to create a meshwork of contractile elements (see Figure 40.7*c*). Unlike skeletal muscles, which can easily be torn along the length of the muscle fibers, the meshwork of the cardiac muscle cannot be separated in this way; thus the heart walls can withstand high pressures without danger of tearing or forming leaks. Unlike skeletal muscle fibers, each cardiac muscle fiber is a uninucleate cell. Each cell is joined to other cardiac muscle fibers by structures called **intercalated discs** that provide strong mechanical adhesion. The intercalated discs have gap junctions that present low resistance to ions or electric currents. As a result, the cardiac muscle fibers are in electrical continuity with one another, and a depolarization spreads rapidly through the walls of the heart.

The origin of the depolarization that triggers heart contraction is an interesting feature of cardiac muscle. Some cardiac muscle fibers are specialized for pacemaking function; they initiate the rhythmic contraction of the heart. Pacemaking is due to a unique class of potassium ion channels found in cardiac muscle fibers. These channels tend to remain somewhat open following an action potential, but they gradually close. As they close, the cell becomes less negative and eventually reaches threshold, thereby initiating the next action potential. Because of the pacemaking property of cardiac muscle fibers, a heart removed from an animal continues to beat with no input from the nervous system; the heartbeat is **myogenic**—generated by the muscle itself. The autonomic nervous system modifies the rate of the pacemaker cells but is not essential for their continued, rhythmic function (Figure 40.12). The myogenic nature of the heartbeat is a major factor in making heart transplants possible because the implanted heart does not depend on neural connections to beat.

SKELETAL SYSTEMS

Muscles can only contract and relax. Without something rigid to pull against, muscles can do little more

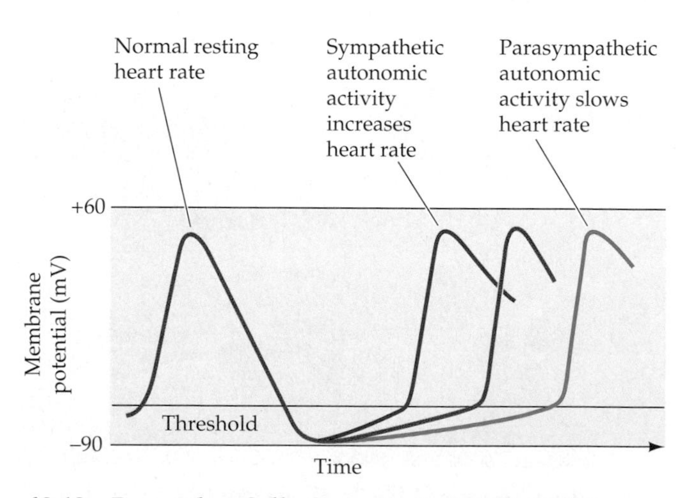

40.12 Pacemaker Cells Generate Action Potentials Spontaneously
The resting potentials of the plasma membranes of pacemaker cells spontaneously depolarize to threshold and fire action potentials. The rate of spontaneous depolarization determines heart rate. Sympathetic stimulation increases the rate of depolarization, whereas parasympathetic stimulation decreases the rate of depolarization.

than lie in a formless mass that twitches and changes shape. **Skeletal systems** provide rigid supports against which muscles can pull, thereby creating directed movements. The three types of skeletal systems found in animals are hydrostatic skeletons, exoskeletons, and endoskeletons.

Hydrostatic Skeletons

The simplest type of skeleton is the **hydrostatic skeleton** of cnidarians, annelids, and many other soft-bodied invertebrates. It consists of a volume of incompressible fluid (water) enclosed in a body cavity surrounded by muscle. When muscles oriented in a certain direction contract, the fluid-filled body cavity bulges out in the opposite direction. The sea anemone has a hydrostatic skeleton (see Figure 26.14). Its body cavity is filled with seawater. To extend its body and its tentacles, the anemone closes its mouth and constricts muscle fibers that are arranged in circles around its body. Contraction of these circular muscles puts pressure on the liquid in the body cavity, and that pressure forces the body and tentacles to extend. If alarmed, the anemone retracts its tentacles and body by contracting muscle fibers that are arranged longitudinally in the body wall and in the long dimension of the tentacles.

The hydrostatic skeletons of some animals have become adapted for locomotion. An annelid such as the earthworm uses its hydrostatic skeleton to crawl (see Figure 26.24). The earthworm's body cavity is divided into many separate segments. The body wall has a muscle layer in which the muscle fibers are arranged in circles around the body cavity, and another muscle layer in which the muscle fibers run lengthwise. A closed compartment in each segment of the worm is filled with fluid. If the circular muscles in a segment contract, the compartment in that segment narrows and elongates. If the lengthwise (longitudinal) muscles of a segment contract, the compartment shortens and bulges outward. Alternating contractions of the the circular and longitudinal muscles create waves of narrowing and widening, lengthening and shortening, that travel down the body of the earthworm. The bulging, short segments serve as anchors as the narrowing, expanding segments project forward, and longitudinal contractions pull other segments forward. Bristles help the widest parts of the body to hold firm against the substrate. The alternating waves of contraction and extension along its body allow the earthworm to make fairly rapid progress through or over the soil (Figure 40.13).

Another type of locomotion made possible by adaptation of the hydrostatic skeleton is the jet propulsion used by the squid and the octopus. Muscles surrounding a water-filled cavity in these cephalopods contract, putting the water under pressure and expelling it from the animal's body. As the water shoots out under pressure, it propels the animal in the opposite direction.

Exoskeletons

An **exoskeleton** is a hardened outer surface to which internal muscles can be attached. Contractions of those muscles cause jointed segments of the exoskeleton to move relative to each other. The simplest

40.13 A Hydrostatic Skeleton and Locomotion
(a) An earthworm's hydrostatic skeleton consists of fluid-filled compartments separated by septa. *(b)* Contractions of circular muscles cause a compartment (and its corresponding segment) to elongate, and contractions of longitudinal muscles cause a compartment (and its segment) to shorten. Alternating waves of elongation and contraction move the earthworm through the soil. Bristles prevent parts of the worm from moving backward as waves of contractions pass by.

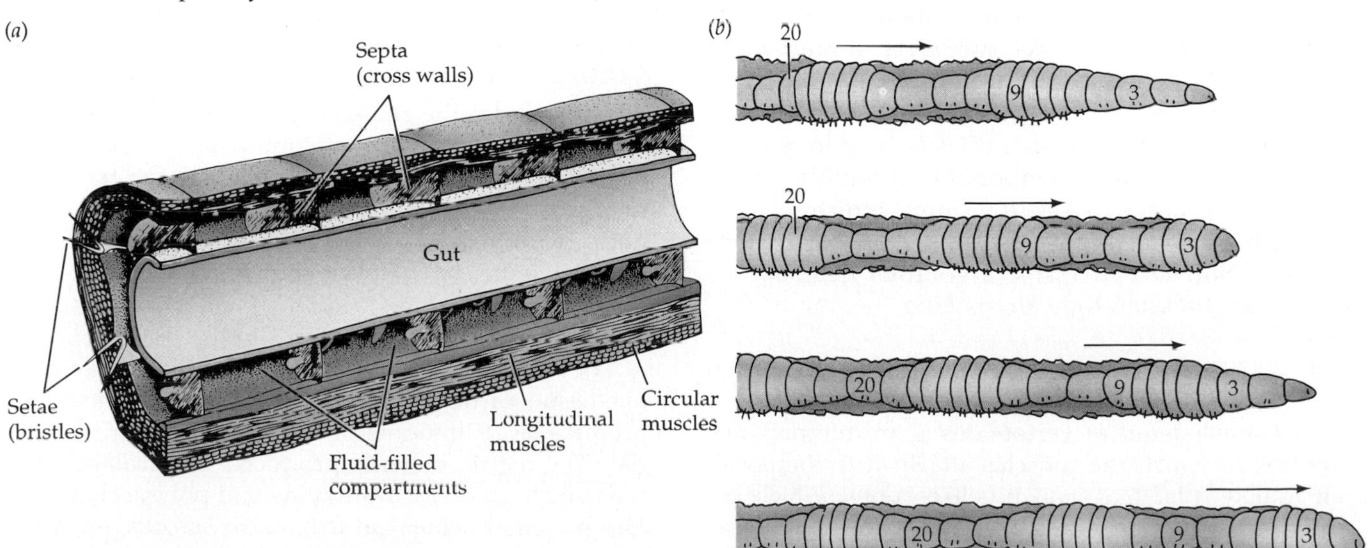

example of an exoskeleton is the shell of a mollusk, which generally consists of just one or two pieces. Some marine bivalves and snails have shells composed of protein strengthened by crystals of calcium carbonate (a rock-hard material). These shells can be massive, affording significant protection against predators. The shells of land snails generally lack the hard mineral component and are much lighter. Molluscan shells can grow as the animal grows, and growth rings are usually apparent on the shells. The soft parts of the molluscan body have a hydrostatic skeleton as well. The hydrostatic skeleton is used in locomotion; the exoskeleton mainly provides protection. (Some scallops, however, swim by opening their shells and snapping them shut—another version of jet propulsion.)

The most complex exoskeletons are found among the arthropods. Plates of exoskeleton, or **cuticle**, cover all the outer surfaces of the arthropod's body and all its appendages. The plates are secreted by a layer of cells just below the exoskeleton. A continuous, layered waxy coating covers the entire body. The skeleton contains stiffening materials everywhere except at the joints, where flexibility must be retained. The layers of cuticle include an outer, thin, waxy epicuticle that protects the bodies of terrestrial arthropods from drying out, and a thicker, inner endocuticle that forms most of the structure. The endocuticle is a tough, pliable material found only in arthropods. It consists of a complex of protein and **chitin**, a nitrogen-containing polysaccharide. In marine crustaceans the endocuticle is further toughened by insoluble calcium salts. The thickness of the exoskeleton varies, forming a very efficient armor. Muscles attached to the inner surfaces of the arthropod exoskeleton move its parts around the joints (Figure 40.14).

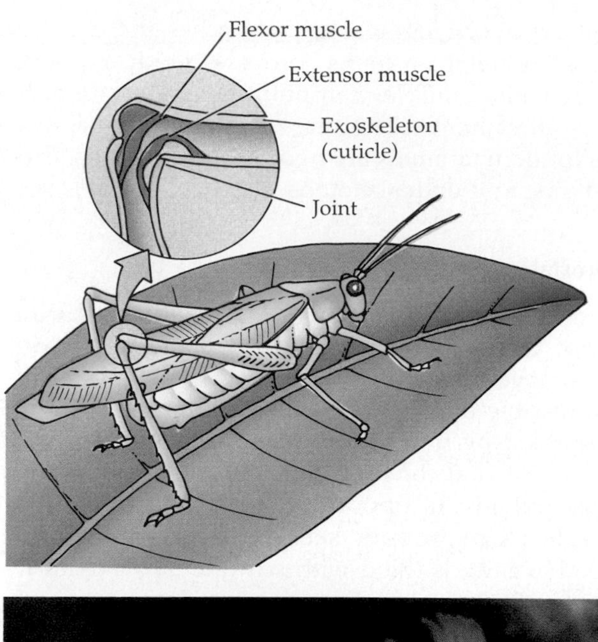

40.14 An Insect's Exoskeleton
Muscles attached to the exoskeleton move parts around flexible joints. The insect in the photograph is a katydid.

An exoskeleton protects all the soft tissues of the animal, but it is itself subject to damage such as abrasion and crushing. The greatest drawback of the arthropod exoskeleton is that it cannot grow. Therefore, if the animal is to become larger, it must **molt**, shedding its exoskeleton and forming a new, larger one. During this process the animal is vulnerable because the new exoskeleton takes time to harden. The animal's body is temporarily unprotected and, without the firm exoskeleton against which its muscles can exert maximum tension, it is unable to move rapidly. Soft-shelled crabs, a gourmet delicacy, are crabs caught when they are molting.

Vertebrate Endoskeletons

The **endoskeleton** of vertebrates is an internal scaffolding to which the muscles attach. It is composed of rodlike, platelike, and tubelike bones, which are connected to each other at a variety of joints that allow a wide range of movements. The human skeleton consists of 206 bones (some of which are shown in Figure 40.15) and is divided into an **axial skeleton**, which includes the skull, vertebral column, and ribs, and an **appendicular skeleton**, which includes the pectoral girdle, the pelvic girdle, and the bones of the arms, legs, hands, and feet. Endoskeletons do not provide the protection that exoskeletons do, but their advantage is that bones continue to grow. Because bones are inside the body, the body can enlarge without shedding its skeleton.

The endoskeleton consists of two kinds of connective tissue: cartilage and bone. Connective tissue cells produce large amounts of extracellular matrix material. The matrix material produced by cartilage cells is a rubbery mixture of proteins and polysaccharides. The principal protein in the matrix is collagen. Col-

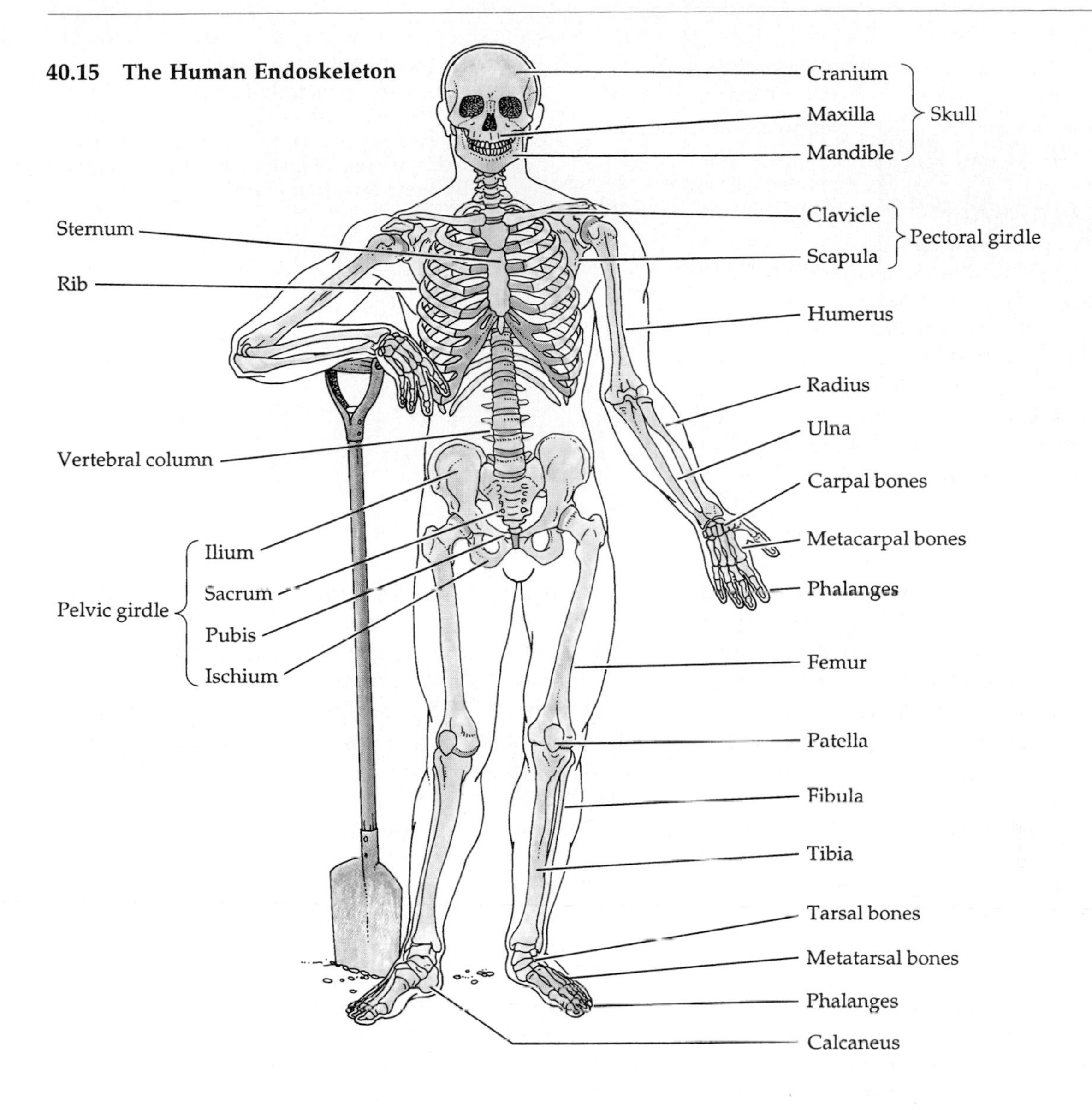

40.15 The Human Endoskeleton

lagen fibers run in all directions through the gel-like matrix and give it the well-known strength and resiliency of "gristle." Cartilage is found in parts of the endoskeleton where both stiffness and resiliency are required, such as on the surfaces of joints, where bones move against each other. Cartilage is also the supportive tissue in stiff but flexible structures such as the larynx (voice box), the nose, and the ears. Sharks and rays are called cartilaginous fishes (see Figure 27.14) because their skeletons are composed entirely of cartilage. In all other vertebrates, cartilage is the principal component of the embryonic skeleton, but over the course of development it is gradually replaced by bone.

Bone consists mostly of extracellular matrix material that contains collagen fibers as well as crystals of insoluble calcium phosphate, which give bone its rigidity and hardness. The skeleton serves as a reservoir of calcium for the rest of the body and is in dynamic equilibrium with soluble calcium in the extracellular fluids of the body. This equilibrium is under hormonal control by calcitonin and parathyroid hormone (see Figure 36.11). If too much calcium is taken from the skeleton, the bones are seriously weakened.

The living cells of bone—called osteoblasts, osteocytes, and osteoclasts—are responsible for the dynamic remodeling of bone structure that is constantly under way. **Osteoblasts** lay down new matrix on bone surfaces. These cells gradually become surrounded by matrix and eventually become enclosed within the bone, at which point they cease laying down matrix but continue to exist within small lacunae (cavities) in the bone. In this state they are called **osteocytes**. In spite of the vast amounts of matrix between them, osteocytes remain in contact with one another through long cellular extensions that run through tiny channels in the bone. Communication between osteocytes is believed to be important in controlling the activities of the cells that are laying down new bone or eroding it away.

The cells that erode or reabsorb bone are the **osteoclasts**. They are derived from the same cell lineage

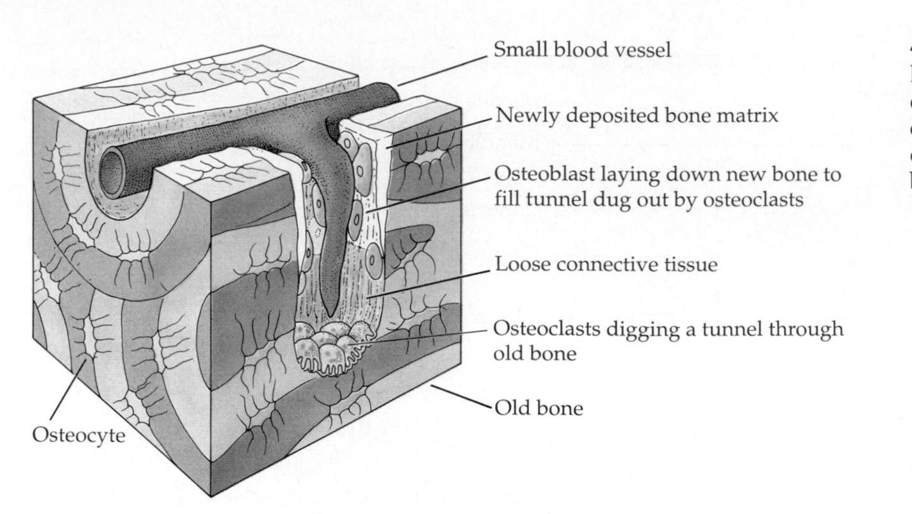

40.16 Renovating Bone
Bones are constantly being remodeled by osteoblasts, which lay down bone, and osteoclasts, which dissolve bone. Osteocytes are osteoblasts that become trapped by their own handiwork.

that produces the white blood cells. Osteoclasts burrow into bone, forming cavities and tunnels. Osteoblasts follow osteoclasts, depositing new bone (Figure 40.16). Thus the interplay of osteoblasts and osteoclasts constantly replaces and remodels the bones. How the activities of these cells are coordinated is not understood, but stress placed on bones provides information used in the process. A remarkable finding in studies of astronauts spending long periods in zero gravity was that their bones decalcified. Conversely, certain bones of athletes can become considerably thicker than they were prior to training or than the same bones in nonathletes. Both thickening and thinning of bones are experienced by someone who has a leg in a cast for a long time. The bones of the uninjured leg carry the person's weight and thicken, while the bones of the inactive leg in the cast thin. The jawbones of people who lose their teeth experience less compressional force during chewing and become considerably remodeled (Figure 40.17).

Types of Bone

Bones are divided into two types, based on how they develop. Membranous bone forms on a scaffolding of connective tissue membrane; cartilage bone forms first as cartilaginous structures and is gradually hardened (ossified) to become bone. The outer bones of the skull are membranous bones; the bones of the limbs are cartilage bones. Cartilage bones can grow throughout ossification. In the long bones of the legs and arms, for example, ossification occurs first at the centers and later at each end (Figure 40.18). Growth can continue until these areas of ossification join. The membranous bones forming the skull cap grow until their edges meet. The fontanel, or "soft spot," on the top of a baby's head is where the skull bones have not yet joined.

The composition of bone may be **compact** (solid and hard) or **cancellous** (having numerous internal cavities that make it appear spongy, even though it is rigid). The architecture of a specific bone depends on its position and function, but most bones have both compact and cancellous regions. The shafts of the long bones of the limbs, for example, are cylinders of compact bone surrounding central cavities that contain the bone marrow, where the cellular elements of the blood are made. The ends of long bones are cancellous (Figure 40.19). The cancellous bone is light in weight because of its numerous cavities, but it is also strong because its internal meshwork constitutes a system of supporting struts and ties. It can withstand considerable forces of compression. The rigid, tubelike shaft can withstand compression as well as bending forces. Architects and nature alike use hollow tubes as lightweight structural elements. In a solid rod that is subjected to a bending force, one side of the rod is compressed while the other side is stretched, and both help to resist the force. Because the center of a solid rod contributes very little to its ability to resist bending, hollowing out a rod reduces its weight but not its strength.

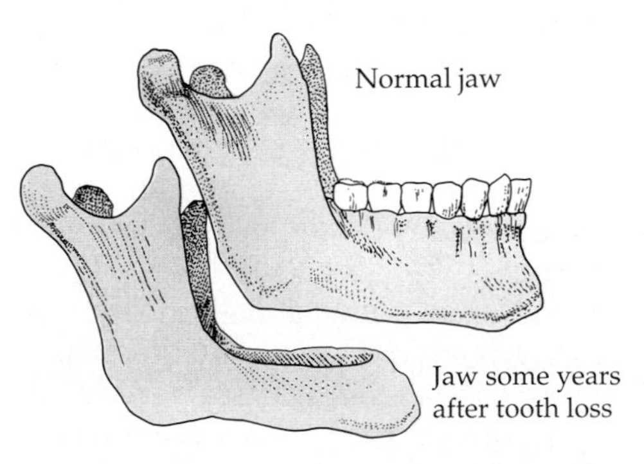

40.17 Tooth Loss and Jawbone Structure
Human jawbone is reabsorbed after tooth loss because of lack of compressional forces.

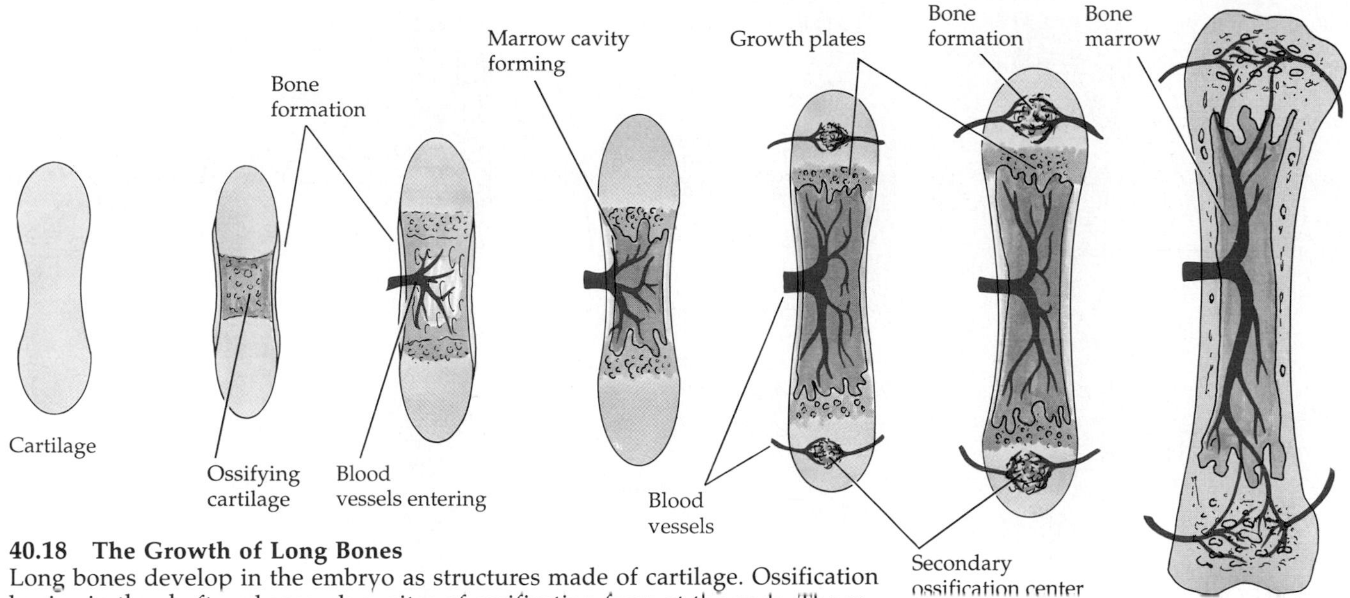

40.18 The Growth of Long Bones
Long bones develop in the embryo as structures made of cartilage. Ossification begins in the shaft and secondary sites of ossification form at the ends. The regions between the ossified parts can grow. Eventually the areas of ossification fuse and elongation of the bone ceases.

Most of the compact bone in mammals is called Haversian bone because it is composed of structural units called **Haversian systems** (Figure 40.20). Each system is a set of thin, concentric bony cylinders, between which are the osteocytes in their lacunae. Through the center of each Haversian system runs a narrow canal containing blood vessels (see Figure 40.16). The osteocytes in one Haversian system connect only with osteocytes in the same system; no channels cross the boundaries (called glue lines) between systems. An important feature of Haversian bone is its resistance to fracturing. If a crack forms in one Haversian system, it tends to stop at the nearest glue line.

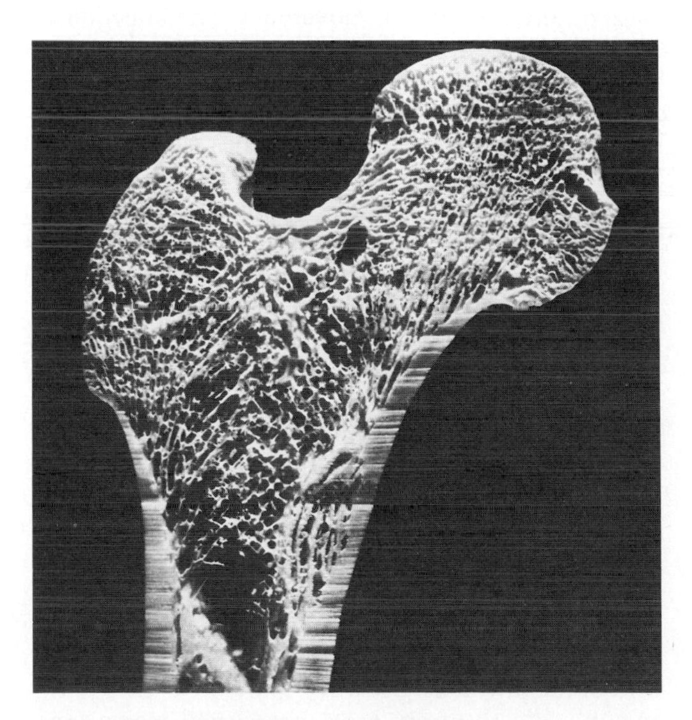

40.19 Internal Architecture of Bone
Bone may have cancellous (spongy) and compact regions. The ends of long bones are cancellous and the shafts are tubes of compact bone.

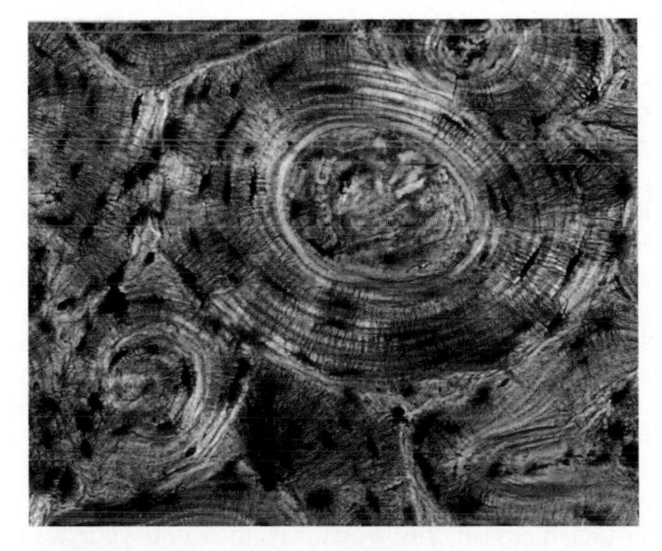

40.20 Haversian Systems in Bone
Osteoblasts lay down bone in layers. In long bones these layers form concentric tubes parallel to the long axis of the bone. At the center of the tube is a canal containing blood vessels and nerves. This micrograph is colored because the bone cross section was illuminated with polarized light.

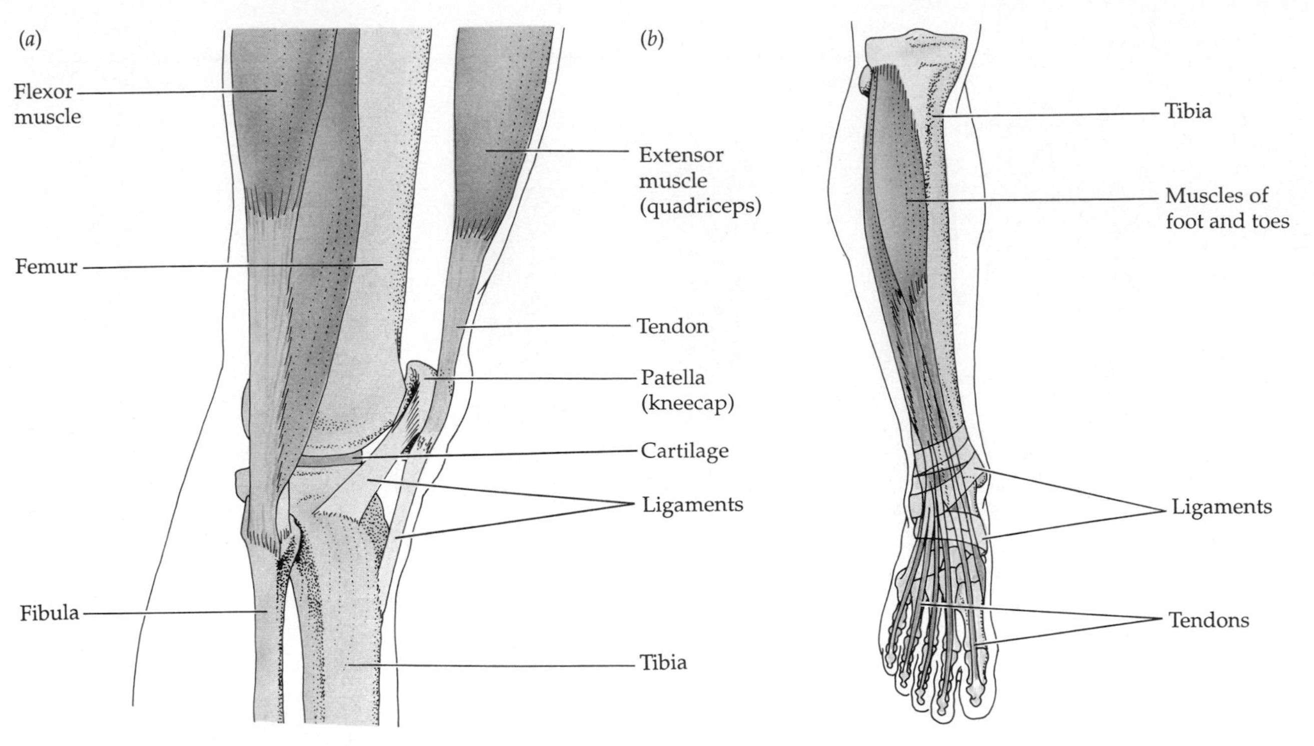

40.21 Joints, Ligaments, and Tendons
(a) This side view of the human knee shows tendons that attach muscle to bone and ligaments that attach bone to bone. Flexor and extensor muscles work antagonistically to operate the joint. *(b)* Tendons connect muscles in the front of the leg to foot and toe bones. These tendons, which pass under strapping ligaments at the front of the ankle, elevate the foot and toes.

Joints and Levers

Muscles and bones work together around **joints**, where two or more bones come together. Since muscles can only contract and relax, they create movement around joints by working in antagonistic pairs: When one contracts, the other relaxes. With respect to a particular joint, such as the knee, we can refer to the muscle that flexes the joint as the **flexor** and the muscle that extends the joint as the **extensor** (Figure 40.21*a*). The bones that meet at the joint are held in place by **ligaments**, which are flexible bands of connective tissue. Other straps of connective tissue, **tendons**, attach the muscles to the bones. In many kinds of joints, only the tendon spans the joint, sometimes moving over the surfaces of the bone like a rope over a pulley. It is the tendon of the quadriceps muscle traveling over the knee joint that is tapped to elicit the knee-jerk reflex.

Ligaments can also hold tendons in place and change the direction of the force they exert. For example, many of the muscles that extend your toes are actually in your lower leg. The tendons from these muscles travel over the front of your ankle, across the upper surfaces of the foot bones, and attach to the bones of your toes. Straps of ligaments over the ankle hold these tendons in place and allow them to bend at a right angle at the ankle (Figure 40.21*b*).

The human skeleton has a wide variety of joints with different ranges of movement. The knee joint is a simple hinge that has almost no rotational movement and can flex in one direction only. At the shoulders and hips are ball-and-socket joints that allow movement in almost any direction. A pivotal joint between the two bones of the forearm where they meet at the elbow allows the smaller bone, the radius, to rotate when the wrist is twisted from side to side. Several kinds of joints permit some rotation, but not in all directions as do the ball-and-socket joints. Examples of these joints are found in the bones of the hands, and they give the hands a wide range of possible movements (Figure 40.22).

Bones around joints and the muscles that work with these bones can be thought of as levers. A lever has a power arm and a load arm that work around a fulcrum (pivot). The ratio of these two arms determines whether a particular lever can exert a lot of force over a short distance or is better at translating force into big or fast movements. Compare the jaw joint and the knee joint (Figure 40.23). The power arm of the jaw is long relative to the load arm, allowing the jaw to apply great pressures over a small distance, such as when you crack a nut with your teeth. The power arm of the lower leg, on the other hand, is short relative to the load arm, so you can run fast, jump high, and deliver swift kicks, but you

40.22 Types of Joints
The designs of joints are similar to mechanical counterparts and enable a variety of movements.

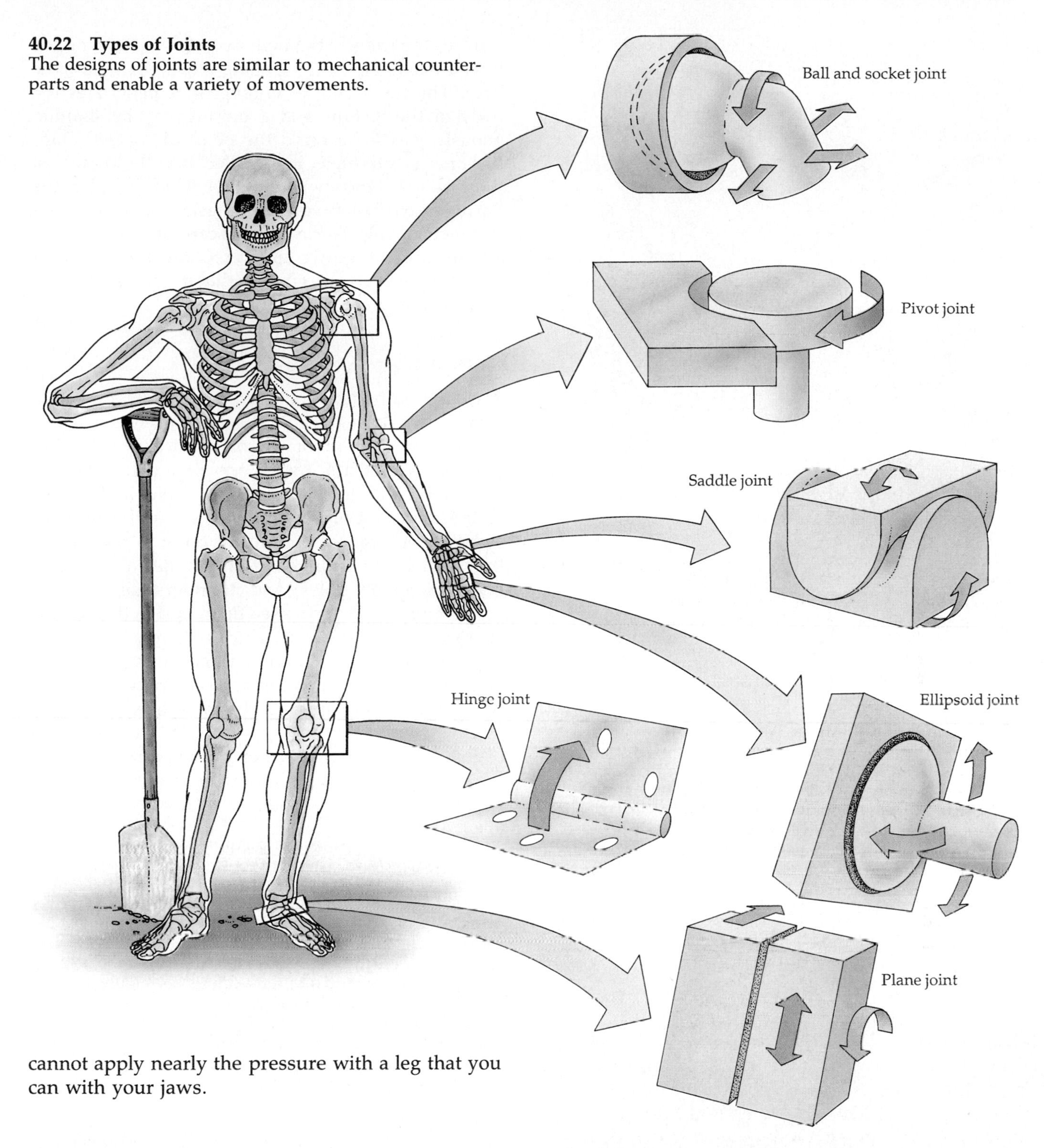

cannot apply nearly the pressure with a leg that you can with your jaws.

OTHER EFFECTORS

Muscles in animals serve almost universally as effectors. Other effectors are more specialized and are not shared by many animal species. Some specialized effectors are used for defense, some for communication, some for capture of prey or avoidance of predators. A discussion of all the effectors animals use would take an entire book, but we can briefly mention a few here to give a sampling of their evolutionary diversity.

Some animals possess highly specialized organs that are fired like miniature missiles to capture prey and repel enemies. **Nematocysts** are elaborate cellular structures produced only by hydras, jellyfish, and other cnidarians. They are concentrated in huge numbers on the outer surface of the tentacles of the animal. Each nematocyst is made up of a slender thread coiled tightly within a capsule, which is armed with a spinelike trigger projecting to the outside (see Figure 26.10). When a potential prey organism

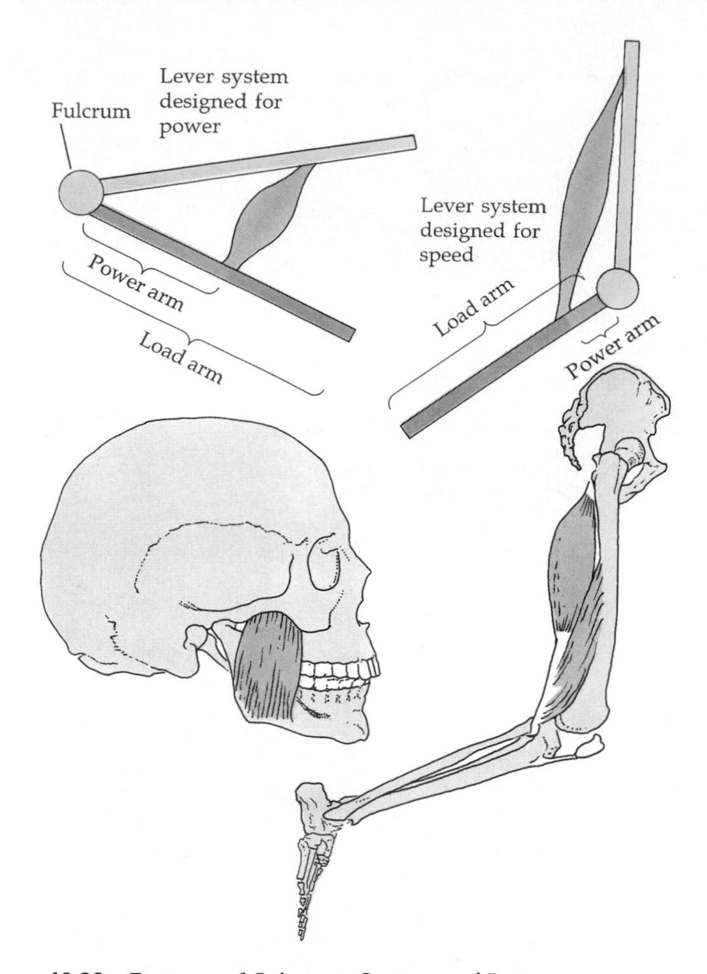

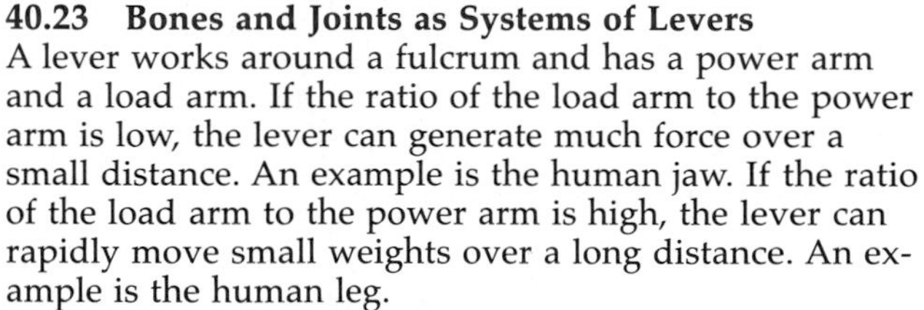

40.23 Bones and Joints as Systems of Levers
A lever works around a fulcrum and has a power arm and a load arm. If the ratio of the load arm to the power arm is low, the lever can generate much force over a small distance. An example is the human jaw. If the ratio of the load arm to the power arm is high, the lever can rapidly move small weights over a long distance. An example is the human leg.

(a)

brushes the trigger, the nematocyst fires, turning the thread inside out and exposing little spines along its base. The thread either entangles or penetrates the body of the victim, and a poison may be simultaneously released around the point of contact. Once the prey is subdued, it is pulled into the mouth of the cnidarian and swallowed. A jellyfish called the Portuguese man-of-war has tentacles that can be several meters long. These animals can capture, subdue, and devour full-grown mackerel, and the poison of their nematocysts is so potent that it can kill a human who gets tangled in the tentacles.

Chromatophores

A change in body color is an effector response that some animals use to camouflage themselves in a particular environment or to communicate with other animals. **Chromatophores** are pigment-containing cells in the skin that can change the color and pattern of the animal. Chromatophores are under nervous system or hormonal control, or both; in most cases they can effect a change within minutes or even seconds. In squids, soles, and flounders, all of which spend much time on the sea floor, and in the famous chameleons (a group of African lizards; see Figure 27.20*b*) and a few other animals, chromatophores enable the animal to blend in with the background on which it is resting and thus be more likely to escape discovery by predators (Figure 40.24*a*). In other kinds of fishes and lizards, a color change sends a signal

40.24 Chromatophores Help Animals Camouflage Themselves
(a) This octopus has adapted its chromatophores so that it is almost indistinguishable from its background of coral, sponges, and other animals. The octopus is the largest element in the photograph; its eyes are partially closed and its head fills much of the lower left quarter of the scene. *(b)* The chromatophores of the octopus are highly elastic. Radiating muscle fibers change them from small spheres of pigment to large sheets of pigment. Different chromatophores have different pigments, so the coloration of the animal depends on which chromatophores are extended and which are contracted.

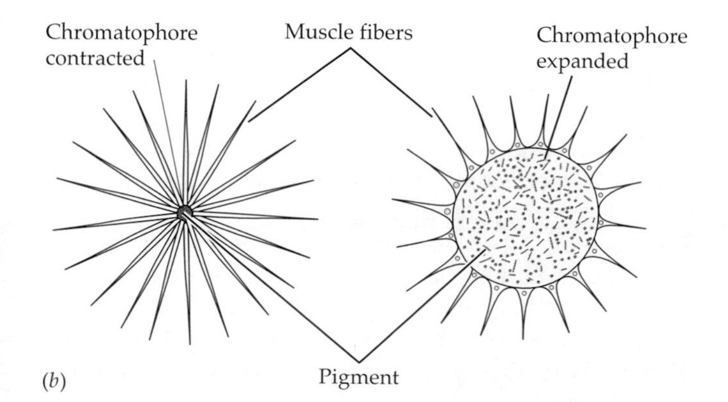

to potential mates and territorial rivals of the same species.

There are three principal types of chromatophore cells. The most common type has fixed cell boundaries, within which pigmented granules may be moved about by microfilaments. When the pigment is concentrated in the center of each chromatophore, the animal is pale; the animal turns darker when the pigment is dispersed throughout the cell. Some other chromatophores are capable of amoeboid movement. They can mold themselves into shapes with a minimal surface area, leaving the tissue relatively pale, or they can flatten out to make the tissue appear darker. Cephalopods have chromatophores that can undergo shape changes as a result of the action of muscle fibers radiating outward from the cell. When the muscles are relaxed, the chromatophores are small and compact and the animal is pale. To darken the animal, the muscles contract and spread the chromatophores over more of the surface (Figure 40.24*b*). Chromatophores with different pigments enable animals to assume different hues or to become mottled to match the background more precisely.

Glands

Glands are effector organs that produce and release chemicals. Some glands produce chemicals that are responsible for communication among animal cells (see Chapter 36). Other glands produce chemicals that are used defensively or to capture prey. Certain snakes, frogs, salamanders, spiders, mollusks, and fish have poison glands. Many of these poisons have proven to be of practical use to humans. The poison dendrotoxin, which certain tribes of the Amazonian jungles use on the tips of their arrows, comes from the skin of a frog (see Figure 27.17*b*) and blocks certain potassium channels. The snake venom bungarotoxin inactivates the neuromuscular acetylcholine receptors. The pufferfish poison tetrodotoxin blocks voltage-gated sodium channels. A poison from a mollusk, conotoxin, blocks calcium channels. There are many such examples, and as you can imagine, they are useful research tools for neurobiologists.

Not all defensive secretions are poisonous. A well-known example is the odoriferous chemical mercaptan, which is sprayed by skunks. Human olfaction is more sensitive to mercaptan than to any other compound. The bombardier beetle (see Figure 48.8*a*) is another spectacular example of an animal that releases an irritating defensive secretion.

The glands that produce and release pheromones, chemicals used in communication among animals—which we mentioned when we discussed chemical sensors in Chapter 39 and which we will discuss further in Chapter 45—are another category of effectors. Still other effector glands produce secretions necessary to facilitate physiological functions. Examples are salivary gland secretions that aid digestion and sweat gland secretions that are an effective means of heat loss.

Sound and Light Producers

The ability to produce sound is an extremely important effector for humans, since speech is one of the most distinctive features of our species. Mammals, birds, and amphibians have evolved a variety of organs that create sound by passing streams of air over structures that vibrate. Insects such as cicadas produce sounds by rubbing together rough surfaces on their appendages. Sound production is by no means a universal effector, however. Most species of animals do not produce sound.

Production of light is even more rare. The classic example of a light-producing animal is the firefly. In an organ at the tip of the firefly's abdomen, an enzyme, luciferase, catalyzes the reaction of a protein, luciferin, with ATP, releasing light energy. The primary function of the light is to attract mates. Bioluminescence is used by a number of plant and animal species (see Box 7.A), and in many cases its function is not known.

Electric Organs

A number of fishes can generate electricity, including the electric eel, the knife fish, the torpedo (a type of ray), and the electric catfish. The electric fields they generate are used for sensing the environment (see Figure 39.32), for communication, and for stunning potential predators or prey. The electric organs of these animals evolved from muscle, and they produce electric potentials in the same general way as nerves and muscles do. Electric organs consist of very large, disc-shaped cells arranged in long rows like stacks of coins. When the cells discharge simultaneously, the electric organ can generate far more current than can nerve or muscle. Electric eels, for example, can produce up to 600 volts with an output of approximately 100 watts—enough to light a row of light bulbs or to temporarily stun a person.

SUMMARY of Main Ideas about Effectors

Effectors enable animals to respond to stimulation from their internal and external environments.

Most effector mechanisms generate mechanical forces that move or change the shapes of cells or whole animals.

Cellular movement comes from two structures: microtubules and microfilaments.

Microtubules and microfilaments depend on long protein molecules that can slide past each other.
Review Figure 40.4

Microtubules move cilia and flagella.
Review Figure 40.3

Microfilaments are responsible for amoeboid movement and for muscle cell contractions.
Review Figure 40.6

The three types of vertebrate muscle are smooth, skeletal, and cardiac.
Review Figure 40.7

Skeletal muscle contracts when actin and myosin filaments in muscle cells slide past each other.
Review Figure 40.8

In a skeletal muscle cell, the microfilaments are organized into myofibrils; myofibrils are made up of units called sarcomeres that contain overlapping myosin and actin filaments.

Myosin has globular heads that bind with actin, change their orientation, and release, causing the filaments to slide past each other and contract the muscle fiber

Action potentials cause the sarcoplasmic reticulum to release calcium ions (Ca^{2+}) into the cytoplasm of the muscle fiber.

Ca^{2+} binds to troponin molecules on the actin filaments and thereby exposes the sites to which myosin can bind.
Review Figure 40.9

In skeletal muscle, a single action potential causes a minimum unit of contraction called a twitch.
Review Figure 40.10

Fast-twitch muscle fibers can generate maximum tension quickly but also fatigue rapidly; slow-twitch fibers generate less tension and do so more slowly but are resistant to fatigue.
Review Figure 40.11

Some cardiac muscle depolarizes spontaneously giving it the ability to serve as a pacemaker.
Review Figure 40.12

Skeletal systems provide rigid supports against which muscles can pull.

Hydrostatic skeletons have fluid-filled cavities that can be squeezed by muscles.
Review Figure 40.13

Exoskeletons are hardened outer surfaces to which internal muscles are attached.
Review Figure 40.14

Endoskeletons are internal, jointed systems of rodlike, platelike, and tubelike rigid supports consisting of bone and cartilage to which muscles attach.
Review Figure 40.15

Bone is continually being remodeled by cells that lay down new bone (osteoblasts) and cells that absorb bone (osteoclasts).
Review Figure 40.16

Bones, which ossify (harden), can grow until centers of ossification meet.
Review Figure 40.18

Muscles and bones work together around joints as systems of levers.
Review Figures 40.22 and 40.23

Tendons connect muscles to bones; ligaments connect bones to each other and help direct the forces generated by muscles by holding tendons in place.
Review Figure 40.21

Other effector organs include nematocysts, chromatophores, glands, and structures that produce sound, light, and electric pulses.

SELF-QUIZ

1. The movement of cilia and flagella is due to
- *a.* polymerization and depolymerization of tubulin.
- *b.* making and breaking of cross-bridges between actin and myosin.
- *c.* contractions of microtubules.
- *d.* changes in conformations of dynein molecules.
- *e.* the spokes of the axoneme using energy of ATP to contract.

2. Smooth muscle differs from both cardiac and skeletal muscle in that
- *a.* it can act as a pacemaker for rhythmic contractions.
- *b.* contractions of smooth muscle are not due to interactions between neighboring microfilaments.
- *c.* neighboring cells can be in electrical continuity through gap junctions.
- *d.* neighboring cells are tightly coupled by intercalated discs.
- *e.* the membranes of smooth muscle cells are depolarized by stretching.

3. Fast-twitch fibers differ from slow-twitch fibers in that
- *a.* they are more common in the leg muscles of champion sprinters than in the same muscles of marathon runners.
- *b.* they have more mitochondria.
- *c.* they fatigue less rapidly.
- *d.* their abundance is more a product of genetics than of training.
- *e.* they are more common in the leg muscles of champion cross-country skiers than in the same muscles of weight lifters.

4. The role of Ca^{2+} in the control of muscle contraction is
- *a.* to cause depolarization of the T-tubule system.
- *b.* to change the conformation of troponin, thus exposing myosin binding sites.
- *c.* to change the conformation of myosin heads, thus causing microfilaments to slide past each other.
- *d.* to bind to tropomyosin and break actin–myosin cross-bridges.
- *e.* to block the ATP binding site on myosin heads, enabling muscle to relax.

5. Which of the following statements about muscle contractions is *not* true?
- *a.* A single action potential at the neuromuscular junction is sufficient to cause a muscle to twitch.
- *b.* Once maximum muscle tension is achieved, no ATP is required to maintain that level of tension.
- *c.* An action potential in the muscle cell activates contraction by releasing Ca^{2+} into the sarcoplasm.
- *d.* Summation of twitches leads to a graded increase in the tension that can be generated by a single muscle fiber.
- *e.* The tension generated by a muscle can be varied by controlling how many of its motor units are active.

6. Which of the following statements about the structure of skeletal muscle is true?
- *a.* The bright bands of the sarcomere are the regions where actin and myosin filaments overlap.
- *b.* When a muscle contracts, the A bands of the sarcomere (dark regions) lengthen.
- *c.* The myosin filaments are anchored in the Z lines.
- *d.* When a muscle contracts, the H bands of the sarcomere (light regions) shorten.
- *e.* The sarcoplasm of the muscle cell is contained within the sarcoplasmic reticulum.

7. The long bones of our arms and legs are strong and can resist both compressional and bending forces because
- *a.* they are solid rods of compact bone.
- *b.* their extracellular matrix contains crystals of calcium carbonate.
- *c.* their extracellular matrix consists mostly of collagen and polysaccharides.
- *d.* they have a very high density of osteoclasts.
- *e.* they consist of lightweight cancellous bone with an internal meshwork of supporting elements.

8. If we compare the jaw joint with the knee joint as lever systems,
- *a.* the jaw joint can apply greater compressional forces.
- *b.* their ratios of power arm to load arm are about the same.
- *c.* the knee joint has greater rotational abilities.
- *d.* the knee joint has a greater ratio of power arm to load arm.
- *e.* only the jaw is a hinged joint.

9. Which of the following statements about skeletons is true?
- *a.* They can consist only of cartilage.
- *b.* Hydrostatic skeletons can be used only for amoeboid locomotion.
- *c.* An advantage of exoskeletons is that they can continue to grow throughout the life of the animal.
- *d.* External skeletons must remain flexible, so they never include calcium carbonate crystals as do bones.
- *e.* Internal skeletons consist of four different types of bones: compact, cancellous, dermal, and cartilage.

10. Chemicals used by neurophysiologists to block voltage-gated sodium channels have come from
- *a.* chromatophores.
- *b.* nematocytes.
- *c.* electric eels.
- *d.* luciferase.
- *e.* poison glands of fish.

FOR STUDY

1. Describe in outline form all the events that occur between the arrival of an action potential at a motor nerve terminal and the contraction of a muscle fiber.
2. How do we know that the basis for the movement of cilia resides in the dynein components of the axoneme?
3. Wombats are powerful digging animals and kangaroos are powerful jumping animals. How do you think the structure of their legs would compare in terms of their designs as lever systems?
4. Maria and Margaret are identical twin sisters. Their mother was an Olympic marathon runner and their father was on the varsity rowing team in college. Maria has become a serious cross-country skier and Margaret has joined the track team as a sprinter. Which one do you think will have the greatest chance of becoming a champion in her sport, and why?
5. If an adolescent breaks a leg bone close to the ankle joint, after the break heals that leg may not grow as long as the other one. Explain why, including an explanation of why the leg grows at all.

READINGS

Alexander, R. M. 1991. "How Dinosaurs Ran." *Scientific American*, April. An example of how physics and engineering approaches can be used to study the effectors of extinct animals.

Cameron, J. N. 1985. "Molting in the Blue Crab." *Scientific American*, May. How an arthropod deals with its exoskeleton in order to grow.

Caplan, A. I. 1984. "Cartilage." *Scientific American*, October. A component of the vertebrate skeletal system, cartilage plays a surprisingly diverse group of roles in the developing and mature animal.

Carafoli, E. and J. T. Penniston. 1985. "The Calcium Signal." *Scientific American*, November. Calcium as a second messenger; calcium in muscle contraction.

Cohen, C. 1975. "The Protein Switch of Muscle Contraction." *Scientific American*, November. Interaction of muscle proteins and calcium ions.

Eckert, R., D. Randall and G. Augustine. 1988. *Animal Physiology*, 3rd Edition. W. H. Freeman, New York. An excellent advanced textbook. Chapter 10 deals with muscle and Chapter 11 with cell motility.

Gans, C. 1974. *Biomechanics: An Approach to Vertebrate Biology*. University of Michigan Press, Ann Arbor. A small classic on the architecture of animals and how their structure is adapted to their environment and lifestyle.

Hadley, N. F. 1986. "The Arthropod Cuticle." *Scientific American*, July. Properties of the exoskeleton of the most abundant and diversified phylum of animals.

Lazarides, E. and J. P. Revel. 1979. "The Molecular Basis of Cell Movement." *Scientific American*, May. Microtubules and microfilaments in action.

Schmidt-Nielsen, K. 1990. *Animal Physiology: Adaptation and Environment*, 4th Edition. Cambridge University Press, New York. Chapter 11 gives a marvelous treatment of skeletons, muscles, and other effectors.

Vander, A. J., J. H. Sherman and D. S. Luciano. 1994. *Human Physiology: The Mechanisms of Body Function*, 6th Edition. McGraw-Hill, New York. Chapter 11 deals with the structure and function of human muscle.

41

Gas Exchange in Animals

Portable Life Support
In space there is no oxygen to breathe. Because humans cannot live without oxygen, astronauts who leave their spaceship take their own atmosphere to support respiration.

The wail that heralds the birth of an infant and brings joy to its parents also initiates the infant's breathing, a process that must continue every minute of its life. Whether awake or asleep, exercising or resting, talking or eating, we must breathe. For brief moments, such as when we swim under water, we can hold our breath, but the urge to breathe mounts rapidly and soon becomes overwhelming. By ignoring a petulant child who threatens to hold its breath until its demands are met, we do not endanger the child's life. Events such as drowning or choking, however, that prevent breathing but are not under a person's own control can lead quickly to death.

BREATHING AND LIFE

Breathing is absolutely coupled with life. We explained the reason for this connection in Chapter 7. Cells need a constant supply of energy to carry out their functions. This energy comes in the form of ATP produced through the oxidation of nutrient molecules. An animal's production of ATP can be sustained only where oxygen gas (O_2) is present. Some cells, such as muscle cells, can survive short periods without O_2 by deriving ATP from glycolysis only and incurring an O_2 debt that has to be paid back later. Brain cells, however, have little capacity to function in the absence of O_2. In the short run, lack of O_2 causes loss of consciousness. As the length of time without O_2 increases, cell functions grind to a halt, cells die, and tissues and organs suffer irreversible damage. Breathing provides the body with the O_2 required to support the energy metabolism of all its cells. Breathing also eliminates one of the waste products of cell metabolism, carbon dioxide (CO_2).

For humans and other large animals, breathing makes possible the exchange of CO_2 and O_2 between the body and the environment. Gas molecules diffuse across membranes that separate the internal environment of the body from the external environment. Some animals, especially small and inactive ones, accomplish gas exchange without breathing because the contact between their gas-exchange membranes and the environment is adequate to support diffusion of O_2 in and diffusion of CO_2 out without their expending energy to move the environment over those membranes, which is what breathing does. Gas exchange is commonly referred to as respiration, as in artificial respiration, but whole-animal gas exchange should not be confused with cellular respiration, even though the two processes are tightly linked.

Animals differ greatly in the rates of gas exchange necessary to support their energy metabolism. At room temperature, a frog consumes about 0.01 liters of O_2 per hour, and a resting human consumes about

15 liters of O_2 per hour. A marathon runner consumes about 150 liters of O_2 per hour during a race. The runner, however, can be outdistanced and outlasted by fish breathing water and by birds flying at very high altitudes, where there is little O_2. We'll learn why in this chapter.

LIMITS TO GAS EXCHANGE

Diffusion is the only means by which respiratory gases are exchanged. There are no active transport mechanisms for respiratory gases. This fact is true for all gas-exchange systems, despite their diverse structures and functions. To help you understand the adaptations of gas-exchange systems that we will cover in this chapter, review the discussion of diffusion in Chapter 5. Because diffusion is strictly a physical phenomenon, it is limited by physical factors such as whether animals breathe air or water.

Breathing Air or Water

Animals that breathe water must have a more efficient gas-exchange system than animals that breathe air. O_2 can be exchanged more easily in air than in water for several reasons. First, the oxygen content of air is much higher than the oxygen content of an equal volume of water. The maximum O_2 content of a rapidly flowing stream splashing over rocks and tumbling over waterfalls is less than 10 ml of O_2 per liter of water. The O_2 content of fresh air is about 200 ml of O_2 per liter of air. Second, O_2 diffuses much more slowly in water than in air. In a still pond, the O_2 content of the water may be zero only a few millimeters below the surface if the water has not been disturbed. Finally, when an animal breathes, it performs work to move air or water over its gas-exchange surfaces. More energy is required to move water than to move air because water is denser and thicker than air.

The slow diffusion of O_2 molecules in water imposes a gas-exchange constraint on air-breathing animals as well as on water-breathing animals. Eukaryotic cells respire in their mitochondria, which are in the cytoplasm—an aqueous medium. Cells are bathed with extracellular fluid—also an aqueous medium. The slowness of O_2 diffusion in water limits the efficiency of O_2 distribution from gas-exchange membranes to the sites of cellular respiration in both

41.1 Keeping in Touch with the Medium
(a) No cell in the leaflike body of this marine flatworm is more than a millimeter away from seawater. *(b)* The same is true of sponges; they have body walls perforated by many channels lined with flagellated cells. These channels communicate with the outside world and with a central cavity. The flagella maintain currents of water through the channels, through the central cavity, and out of the animal. Every cell in the sponge is very close to the respiratory medium. *(c)* The gills of this newt project like a feathery fringe and provide a large surface area for gas exchange. Blood circulating through the gills comes into close contact with the respiratory medium.

(a)

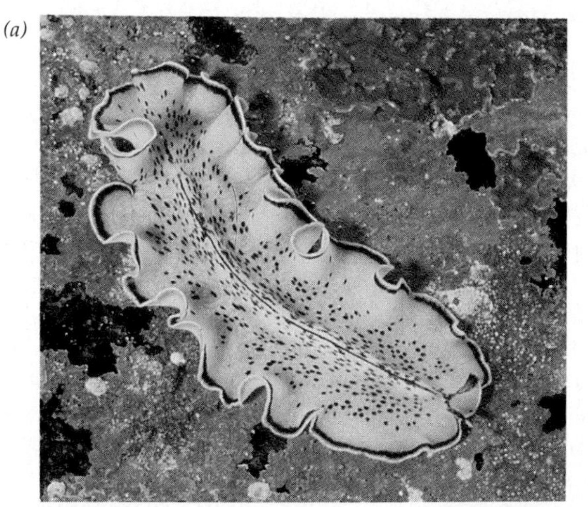

(b)

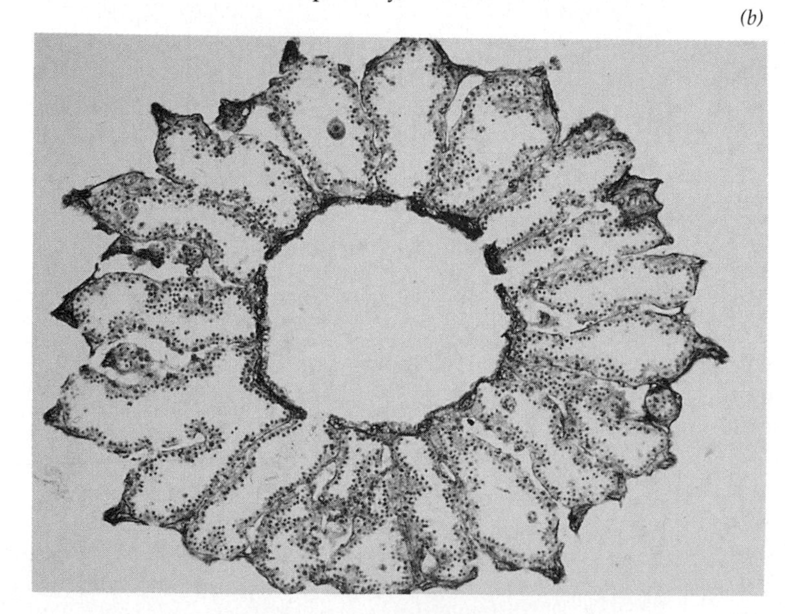

(c)

air-breathing and water-breathing animals. In water, a supply of O_2 that is adequate to support the metabolism of a typical animal cell can be obtained only if the diffusion path is shorter than about 1 millimeter. Therefore, in an animal that lacks an internal system for transporting gases, no cell may be more than about 1 millimeter from the outside world. This is a severe size limit, but one way to accommodate it and still grow bigger is to have a flat, leaflike body plan, which is common among simple invertebrates (Figure 41.1*a*). Another way is to have a very thin body built around a central cavity through which water circulates (Figure 41.1*b*). Otherwise, specialized structures are required to provide an increased surface area for diffusion, and an internal circulatory system is needed to carry gases to and from these exchange structures (Figure 41.1*c*).

Temperature

Temperature is a crucial factor influencing gas exchange in animals that breathe water. Almost all water breathers are ectotherms. The body temperatures of aquatic ectotherms are closely tied to the temperature of the water around them. As the temperature of the water rises, so does their body temperature, and because of Q_{10} effects (see Chapter 35), energy expenditure and oxygen demand rise exponentially. But warm water holds less gas than cold water does. (Just think of what happens when you open a warm bottle of beer or soda.) Thus aquatic ectotherms are in a double bind: As the temperature of their environment goes up, so does their demand for O_2; but the availability of O_2 in their environment goes down (Figure 41.2). If the animal performs work to move water across its gas-exchange surfaces (as fish do, for example), the energy the animal must expend increases as water temperature rises. Thus as water temperature goes up, the water breather must extract more O_2 from the environment, or it must decrease its energy expenditures for activities other than breathing.

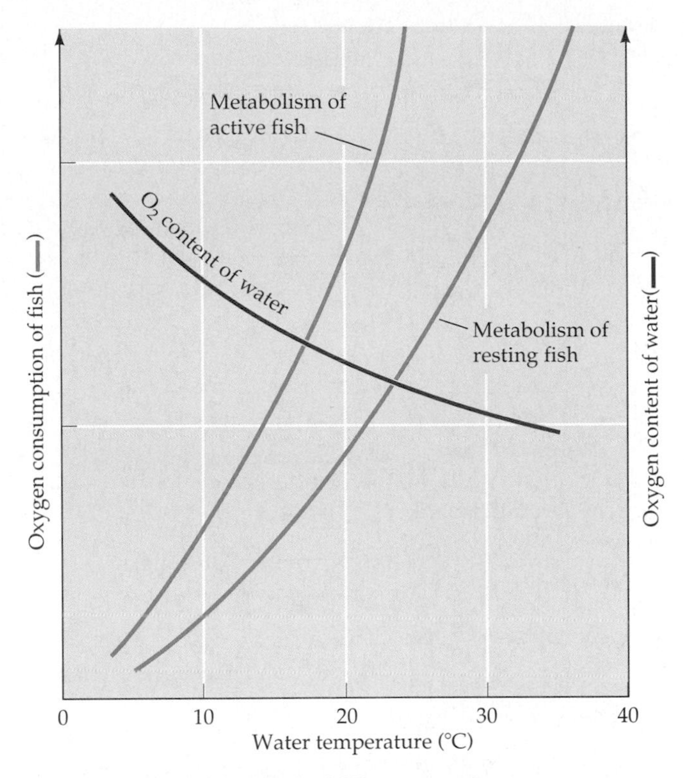

41.2 The Double Bind of Water Breathers
As water temperature increases, so do the body temperatures of water-breathing ectotherms, and therefore their oxygen needs increase. However, warm water carries less oxygen in solution than does cold water.

Altitude

Just as a rise in temperature reduces the supply of O_2 available for aquatic animals, an increase in altitude reduces the O_2 supply for air breathers. The amount of O_2 in the atmosphere decreases with increasing altitude. One way of expressing the concentration of gases in air and in water is by their **partial pressures**. At sea level, the pressure exerted by the atmosphere is the equivalent to that produced by a column of mercury 760 mm high. We therefore say that the **barometric pressure** (atmospheric pressure) is 760 mm of mercury (Hg). Because dry air is 20.9 percent O_2, the partial pressure of oxygen (P_{O_2}) at sea level is 20.9 percent of 760 mm Hg, or about 159 mm Hg. As you go higher in elevation, there is less and less air above you, so barometric pressure declines. At an altitude of 5,300 meters, barometric pressure is only half as much as it is at sea level, so the P_{O_2} at that altitude is only 80 mm Hg. At the summit of Mount Everest (8,848 meters) the P_{O_2} is only about 50 mm Hg or roughly one-third what it is at sea level. Remember that diffusion of O_2 into the body is dependent on O_2 concentration differences between the air and the body fluids, so the drastically reduced O_2 concentration in the air at a high altitude constrains O_2 uptake. This low O_2 concentration is why mountain climbers who venture to the heights of Mount Everest breathe O_2 from pressurized bottles they carry with them (Figure 41.3).

Carbon Dioxide Exchange with the Environment

Respiratory gas exchange is a two-way street. CO_2 diffuses out of the body as O_2 diffuses in. Given the same concentration gradient, CO_2 and O_2 molecules diffuse at about the same rate whether in air or in water. However, the concentration gradients for diffusion of O_2 and CO_2 across gas-exchange membranes are generally not the same. The concentration of CO_2 in the atmosphere is so low, and its solubility

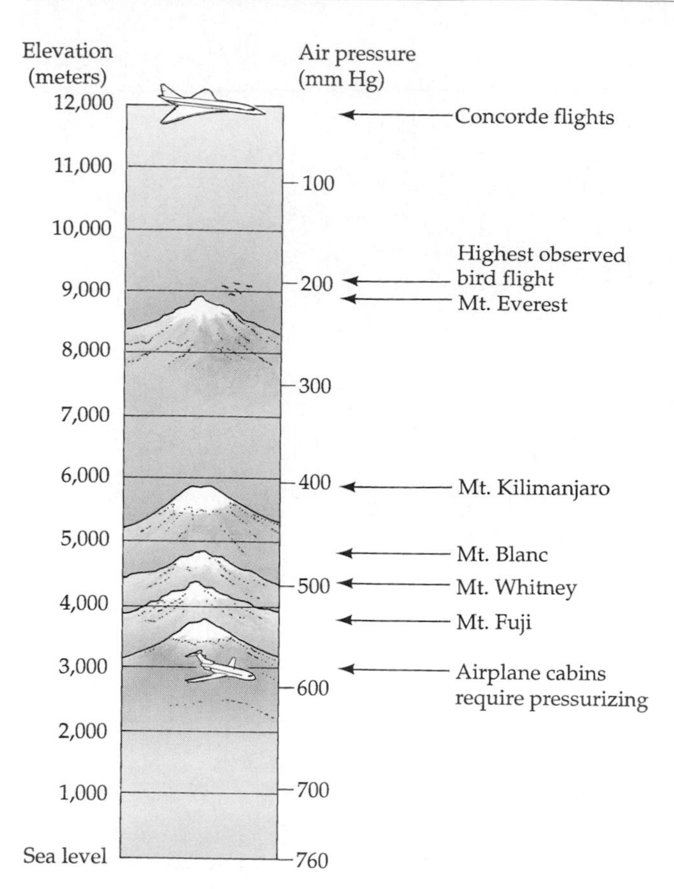

41.3 Scaling Heights
The oxygen content of the atmosphere decreases with altitude. Therefore, airplane cabins must be pressurized and mountain climbers must carry pressurized containers of oxygen. Birds, however, have been observed flying over even the highest peaks.

in the aquatic environment is so high, that diffusion of CO_2 from an animal is usually not a problem. Transporting CO_2 from where it is produced in the body to where it diffuses into the environment, however, can be a limiting factor in gas exchange and hence metabolism.

RESPIRATORY ADAPTATIONS

Animals have evolved a great diversity of adaptations to maximize their rates of gas exchange. All of these adaptations, however, work through influencing a few physical parameters that are described by a simple equation called **Fick's law of diffusion**:

$$Q = DA \frac{C_1 - C_2}{L}$$

Fick's law describes the rate, Q, at which a substance diffuses between two locations. D is the diffusion coefficient, which is a characteristic of a particular substance diffusing in a particular medium at a particular temperature. For example, perfume has a higher D than does motor oil, and substances diffuse faster in air than in water. A is the cross-sectional area over which the substance is diffusing. C_1 and C_2 are the concentrations of the substance at two locations, and L is the distance between those locations. Therefore, $(C_1 - C_2)/L$ is a concentration gradient. Animals can maximize D for the respiratory gases by using air rather than water for the gas-exchange medium whenever possible. All other adaptations for maximizing respiratory gas exchange must influence the surface area for exchange or the concentration gradient across that surface area.

Surface Area

There are many anatomical adaptations that maximize specialized body surface areas over which gases can diffuse (Figure 41.4). **External gills** are highly branched and folded elaborations of the body surface that provide a large surface area for gas exchange. They consist of thin, delicate membranes that minimize the path length traversed by diffusing molecules of O_2 and CO_2 (see Figure 41.1*c*). Because external gills are vulnerable to damage and are tempting morsels for carnivorous organisms, it is not surprising that protective body cavities for gills have evolved. Many mollusks, arthropods, and fishes have **internal gills** in such cavities.

Like water-breathers, air-breathing animals have adapted by increasing their surface area for gas exchange, but their structures are quite different from gills. First, gas-exchange surfaces in air breathers

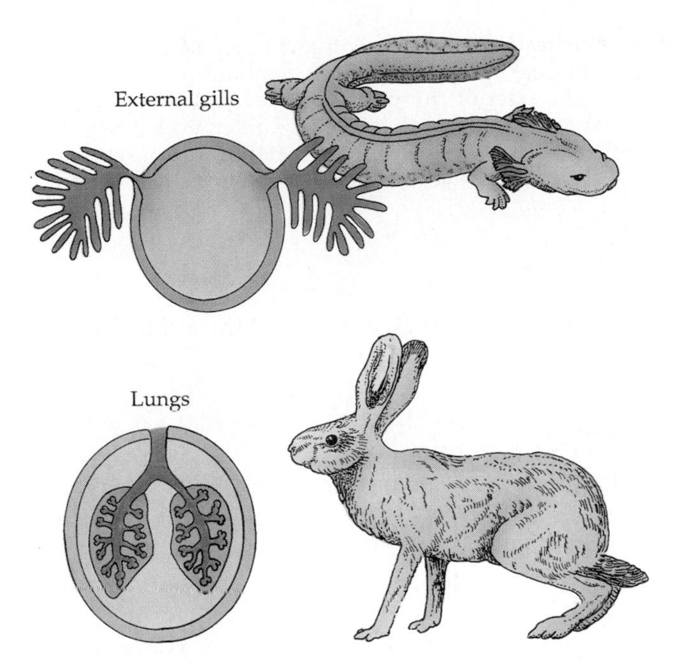

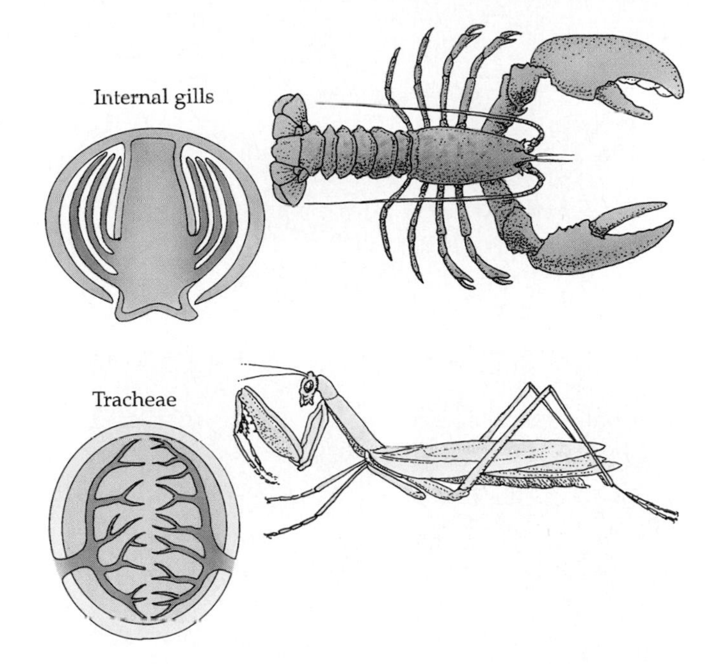

41.4 Gas-Exchange Organs
Increased surface area for the diffusion of respiratory gases is a common feature. Gills are adaptations for breathing water; lungs and tracheae are adaptations for breathing air.

must be in moist internal cavities to prevent drying out. Second, surface elaborations such as gills work only in water because without water for support, they collapse and stick together like the pages of a wet magazine and thus lose effective surface area. That is why a fish suffocates in air in spite of the much higher O_2 concentration. The gas-exchange structures, or **lungs**, of most air-breathing vertebrates are highly divided, elastic air sacs, which we describe in greater detail later in the chapter. The gas-exchange structures of insects are highly branched systems of air-filled tubes that branch through all the tissues of the insect's body. In both cases the surface areas for gas exchange are greatly enhanced by their division into many small units.

Ventilation and Perfusion

Fick's law of diffusion points to another way besides increasing surface area for increasing respiratory gas exchange. Animals can maximize the concentration gradients for the respiratory gases across the gas-exchange membranes in several ways. First, gill and lung membranes can be very thin so that the path length for diffusion (L) is small. Second, the environmental side of the exchange surfaces can be exposed to fresh respiratory medium (air or water) with the highest possible O_2 concentration and the lowest possible CO_2 concentration. Third, the opposite conditions—the lowest possible O_2 concentration and highest possible CO_2 concentration—can be maintained on the internal sides of the exchange surfaces. Mechanisms that move substances over the gas-exchange surfaces are important for maximizing concentration gradients.

External gills are exposed to fresh respiratory medium as they wave around in the environment. Gas-exchange surfaces that are enclosed in body cavities, however, must be **ventilated**; that is, the animal must move fresh respiratory medium over internal gills or lungs. Breathing consists of the movements that ventilate gills or lungs. **Perfusion** is the movement of blood across the internal side of the gas-exchange membranes. Blood carries O_2 away as it diffuses across from the environmental side, and it brings CO_2 to the exchange surfaces so that it can diffuse in the opposite direction.

An animal's **gas-exchange system** is made up of its gas-exchange surfaces and the mechanisms it uses to ventilate and perfuse those surfaces. The following sections describe four gas-exchange systems. First we will look at the unique gas-exchange system of insects. Then we will describe two remarkably efficient systems: fish gills and bird lungs. Finally, we will discuss mammalian lungs, which in comparison to fish gills and bird lungs are a relatively inefficient gas-exchange system.

Insect Respiration

Respiratory gases diffuse through air most of the way to and from every cell of an insect's body. This diffusion is achieved through a system of air tubes, or **tracheae**, that open to the outside environment through holes called **spiracles** in the sides of the

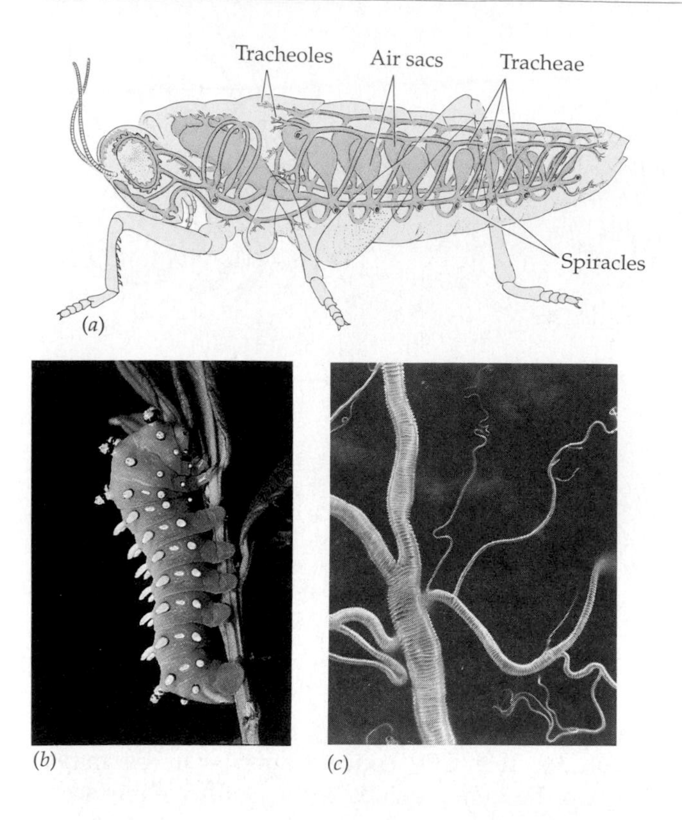

41.5 The Tracheal Gas-Exchange System of Insects
(a) The tracheal system extends throughout the body and opens to the exterior through spiracles. *(b)* The spiracles of the larva of a silk moth look like golden eyes down the side of the animal. *(c)* A scanning electron micrograph of part of the tracheal system shows tracheoles and air capillaries.

abdomen. The tracheae branch into even finer tubes, or tracheoles, until they end in tiny **air capillaries** (Figure 41.5). In the insect's flight muscle and other highly active tissue, no mitochondrion is more than a few micrometers away from an air capillary. Because the diffusion rate of oxygen is about 300,000 times higher in air than in water, air capillaries enable insects to supply oxygen to their cells at high rates. Many insects metabolize at extremely high rates, but this relatively simple gas-exchange system is well able to provide them with the oxygen they need. The rate of diffusion in insect tracheae and air capillaries is limited, however, by the small diameter and by the length of these dead-end airways, so insects are relatively small animals.

Some species of bugs that dive and stay under water for long periods make use of an interesting variation on diffusion. These bugs carry with them a bubble of air. A small bubble may not seem like a very large reservoir of oxygen, yet these bugs can stay under water almost indefinitely with their small air tanks. The secret has to do with the partial pressure of O_2 in the bubble. When the bug dives, the air bubble contains about 80 percent nitrogen and 20 percent O_2. As the insect consumes the O_2 in its bubble, the bubble shrinks, but its nitrogen concentration increases, and its O_2 concentration decreases. When the partial pressure of O_2 in the bubble falls below the partial pressure of O_2 in the surrounding water, O_2 diffuses into the bubble. For many of these small bugs, the rate of O_2 diffusion into the bubble is enough to meet the O_2 demand of the animal while it is under water.

Fish Gills

The internal gills of fishes are marvelously adapted for gas exchange. They offer a large surface area for gas exchange between blood and water. They are supported by five or six bony **gill arches** on either side of the fish between the mouth cavity and the protective **opercular flaps** (Figure 41.6*a*). Water flows into the fish's mouth and out from under the opercular flaps. Each gill arch is lined with hundreds of leaf-shaped gill filaments arranged in two columns. These columns of gill filaments point toward the opercular opening, which is the direction of water flow (Figure 41.6*b*). The tips of the gill filaments of adjacent arches interlock. The upper and lower flat surfaces of each gill filament have rows of evenly spaced folds, or **lamellae**, which greatly increase the gill surface area. The surface area of the lamellae is the site of gas exchange. The interlocking network of gill filaments and lamellae directs the flow so that practically all water that passes across the gills comes into close contact with the gas-exchange surfaces.

The flow of blood perfusing the inner surfaces of the lamellae is unidirectional because of the arrangement of the **afferent blood vessels**, which bring blood to the gills, and the **efferent blood vessels** which take blood away from the gills (Figure 41.6*c*). Blood flow through the lamellae is in the opposite direction to the water flow over the lamellae. Such **countercurrent flow** makes gas exchange much more efficient than parallel flow (Box 41.A). Countercurrent exchange is an important principle in a number of different physiological systems.

The very delicate structure of the lamellae minimizes the path length for diffusion of gases between blood and water. Blood travels in blood vessels through the gill arches and the gill filaments, but in the lamellae the blood flows between the two surfaces of the lamellae as a sheet not much more than one red blood cell thick. The surfaces of the lamellae consist of highly flattened epithelial cells with almost no cytoplasm, so the water and the red blood cells are separated by little more than 1 or 2 μm.

Besides a large surface area and a short diffusion path length, what more can be done to maximize the

41.6 Internal Gills in Fish Enable Countercurrent Exchange
(a) Opercular flaps protect and help ventilate the gills. *(b)* Each gill arch supports two rows of gill filaments; each filament is folded into many thin, flat lamellae that are the gas-exchange surfaces. O_2 diffuses from water into the blood. *(c)* Blood flow through the lamellae is countercurrent to the flow of water over the lamellae.

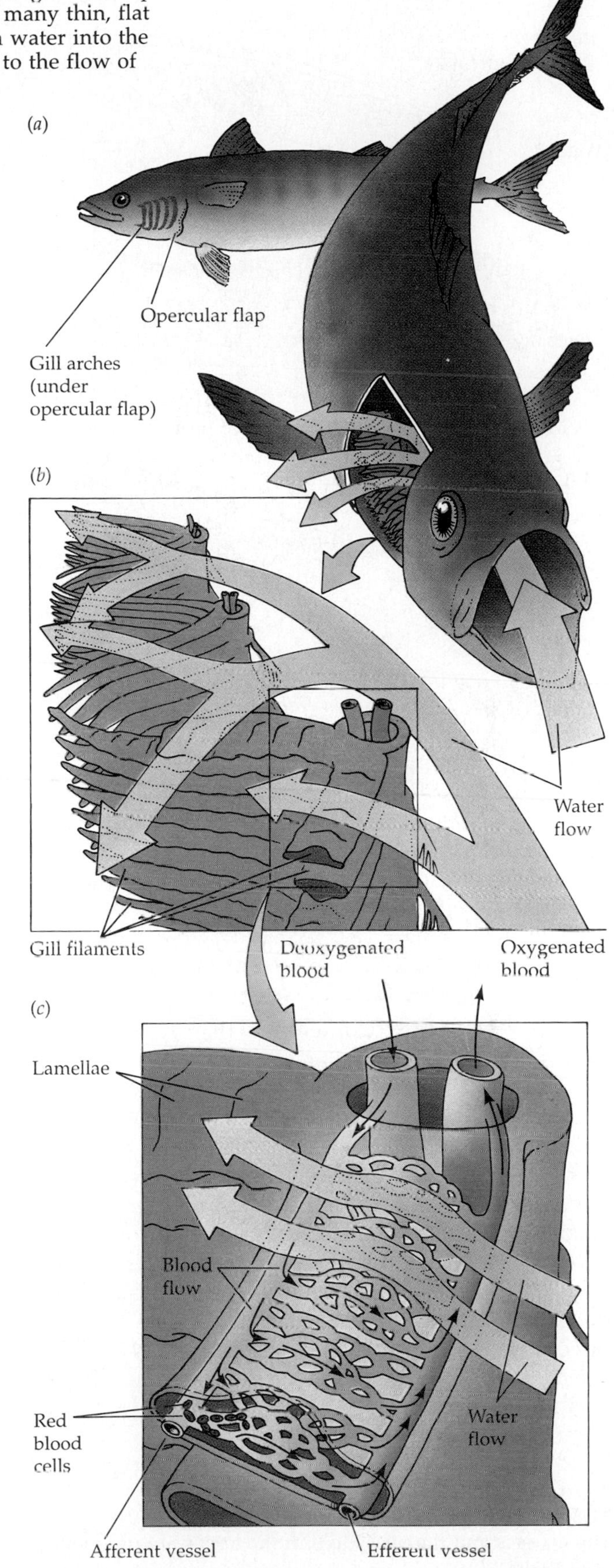

rate of diffusion? The concentration difference of O_2 between water and blood can be maximized. Fish accomplish this task by ventilating the external surface and perfusing the internal surface of the lamellae. A constant flow of water moving over the gills maximizes the O_2 concentration on the external surfaces. On the internal side, the circulation of blood minimizes the concentration of O_2 by sweeping the O_2 away as rapidly as it diffuses across.

Most fishes ventilate the external surfaces of their gills by means of a two-pump mechanism that maintains a unidirectional and constant flow of water over the gills. The closing and contracting of the mouth cavity acts as a **positive-pressure pump**, pushing water over the gills. The opening and closing of the opercular flaps acts as a **negative-pressure pump**, or suction pump, pulling water over the gills. Because these pumps are slightly out of phase, they maintain an almost continuous flow of water across the gills (Figure 41.7).

In summary, fish can extract an adequate supply of O_2 from meager environmental sources by maximizing the surface area for diffusion, minimizing the path length for diffusion, and maximizing oxygen extraction efficiency by constant, unidirectional, countercurrent flow of blood and water over the opposite sides of the gas-exchange surfaces.

Bird Lungs

Birds can sustain extremely high levels of activity for much longer periods than mammals can, and they can do so at very high altitudes, where mammals cannot even survive because the oxygen content of the air is so low. Yet the lungs of a bird are smaller than those of a mammal of a similar size. A unique feature of birds is that in addition to lungs they have **air sacs** at several locations in their bodies. The air sacs connect with one another, with the lungs, and with air spaces in some of the bones of the bird. Even though the air sacs as well as the lungs receive inhaled air, the air sacs are not gas-exchange surfaces. If a sample of air or pure oxygen is tied off in an air sac, its composition does not change rapidly, as it would if the O_2 were diffusing into the blood and CO_2 were diffusing into the air sac.

The anatomy of the bird lung is unique among air-breathing vertebrates. As in other air-breathing vertebrates, air enters and leaves the system through a

(*a*) Mouth open, mouth cavity expanding, opercular flaps closed, opercular cavity expanding

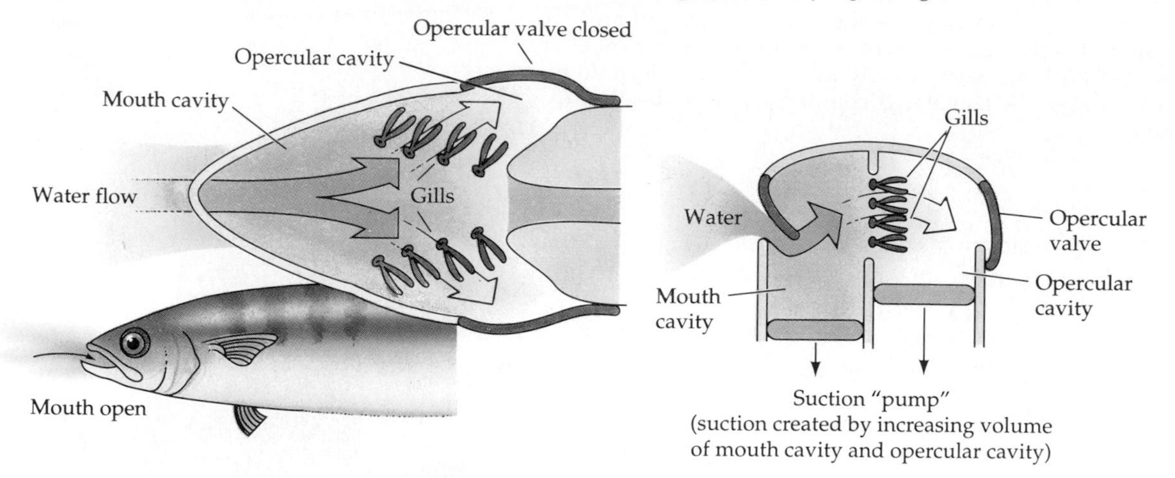

(*b*) Mouth closed, mouth cavity contracting, opercular flaps opening, opercular cavity contracting

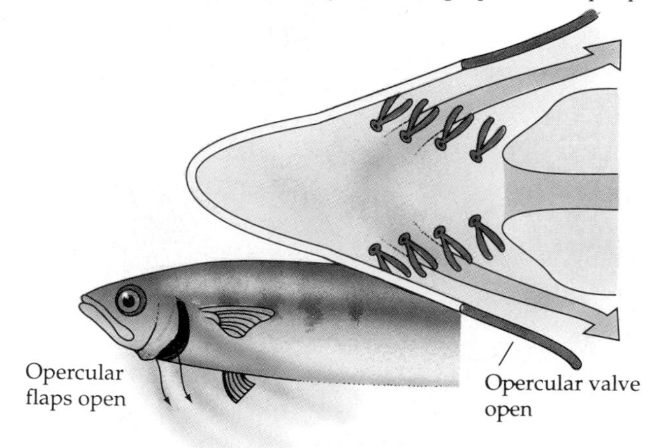

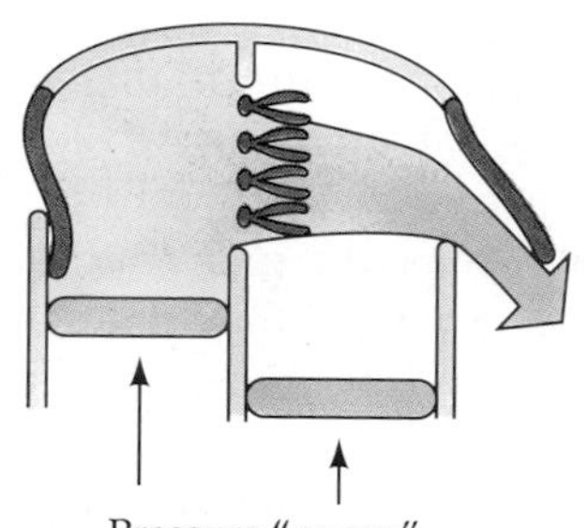

41.7 Two Pumps Maintain Constant Water Flow
The mouth cavity (a positive-pressure pump) and the opercular cavity (a negative-pressure pump) work together to ventilate fish gills.

trachea (commonly known as the "windpipe"), which divides into smaller airways called **bronchi** (singular, *bronchus*). In air-breathing vertebrates other than birds, the bronchi generate trees of branching airways that become finer and finer until they dead-end in clusters of microscopic air sacs, where gases are exchanged. In bird lungs, however, there are no dead-ends, so air can flow completely through the lungs. The bronchi distribute air to the air sacs and to the lungs (Figure 41.8). In the lungs the bronchi divide into tubelike **parabronchi** that guide the unidirectional flow through the lungs (Figure 41.9). Air capillaries off the parabronchi increase the surface area for gas exchange.

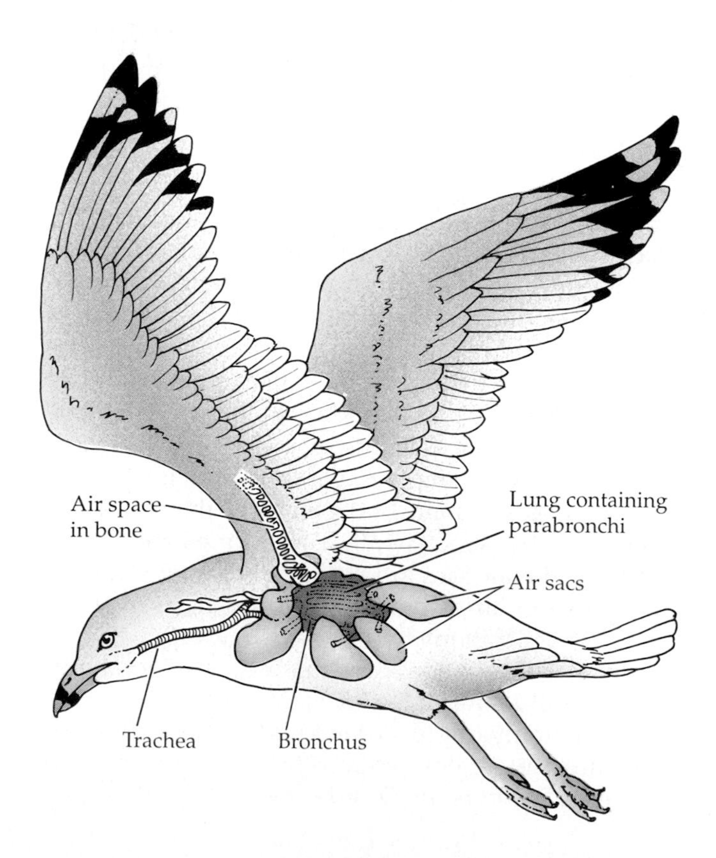

41.8 The Respiratory System of a Bird
The air sacs and the air spaces in the bones are unique to bird anatomy.

BOX 41.A

Countercurrent Exchangers

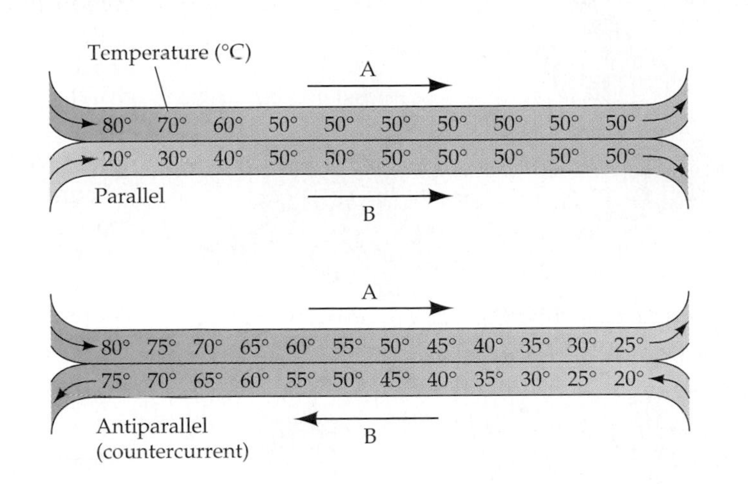

Consider two pipes, side by side, in which parallel streams of water flow in the same direction. In pipe A, the water that enters has a temperature of 80°C; in B, the water that enters has a temperature of 20°C. If we assume that heat may be exchanged between the two pipes but may not be lost to the environment, then what happens is clear: Heat is transferred from A to B until the temperatures in the two pipes are identical—approximately 50°C in this example. The second law of thermodynamics (see Chapter 6) tells us that heat cannot flow spontaneously from a cooler to a warmer system; therefore, once both pipes reach 50°C, no further heat can be transferred from A to B.

Suppose now that we make a minor change; we have the water flow in opposite directions (antiparallel) in pipes A and B. As before, heat flows from A to B, but in this case the transfer of heat is much more complete—instead of about half of the heat being transferred, almost all of it passes from A to B. As water flows through pipe B, it gets hotter and hotter, but it is always cooler than the water in pipe A. As water approaches the end of pipe A, it is relatively cool, but it is still warmer than that in pipe B, so it continues to give up heat. Thus heat transfer occurs along the entire region of overlap of the two pipes. The only difference between these two examples is that in the first case the flow is parallel and in the second case antiparallel. The antiparallel system is usually called a **countercurrent exchanger**.

Countercurrent systems are common in the animal kingdom. In fish gills the efficiency of the oxygenation of blood is maximized by having blood flow in the direction opposite to that of the oxygen-bearing water (see Figure 41.6*c*). When the water leaves the gill, it has lost much of its oxygen, and the blood has become maximally oxygenated. If blood flowed in the same direction as the water, oxygen exchange would be much less complete. In Chapter 44 we will see the operation of a countercurrent exchange system in the vertebrate kidney—in a structure called the loop of Henle—allowing the formation of a steep salt concentration gradient within the kidney. In Chapter 35 we encountered heat exchangers in "hot fish" (see Figure 35.17*b*) and in the extremities of desert animals (see Figure 35.20*a*).

Another unusual feature of bird lungs is that in comparison to mammalian lungs they expand and contract relatively little during a breathing cycle. To make things even more puzzling, the bird lungs contract during inhalation and expand during exhalation!

The puzzle of how birds breathe was solved when researchers used very small gas sensors placed at different locations in the air sacs and airways to follow the path of air flow. They discovered that the air sacs can be divided into an anterior group and a posterior group, and that these two groups act as bellows to maintain a continuous, unidirectional flow of air through the lungs (Figure 41.10). When the bird inhales, the fresh air coming in through the trachea goes primarily to the posterior air sacs and into the portion of the lungs closest to these air sacs. Simultaneously, air that was in the lungs flows into the anterior air sacs. When the bird exhales, the air in the posterior air sacs flows through the lungs to the anterior air sacs and continues out through the trachea. Thus the flow of air through the bird respiratory system during a breathing cycle is in and out through the trachea but unidirectional through the posterior air sacs, through the parabronchi of the lungs, and through the anterior air sacs. The advantages of this unique gas-exchange system are similar to those of fish gills. Because air from the outside flows unidirectionally and practically continuously over the gas-exchange surfaces, the concentration of O_2 on the environmental side of those surfaces is maximized. Furthermore, the unidirectional flow of air through the system makes possible a pattern of

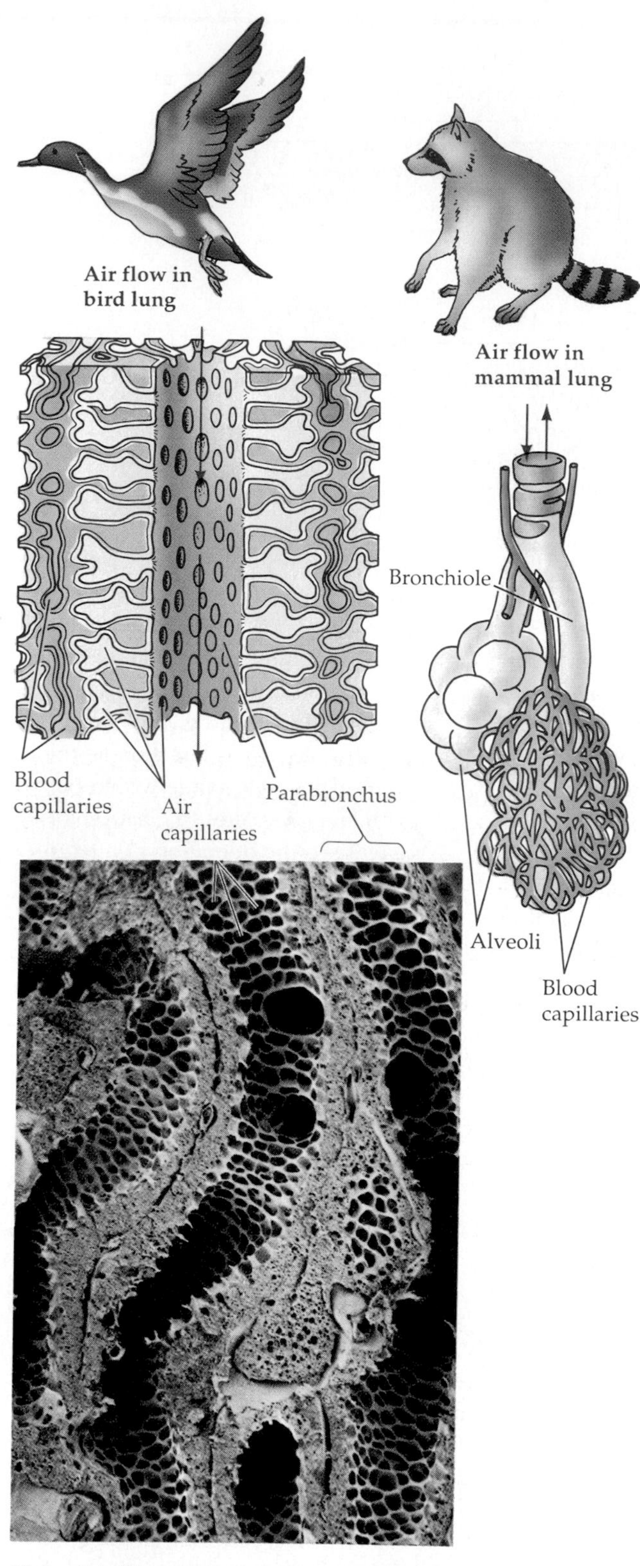

41.9 Air Flow Through Bird Lungs Is Constant and Unidirectional
The gas-exchange surfaces of mammals are alveoli, which are blind sacs, so air flow must be tidal. The gas-exchange surfaces of birds are air capillaries branching off the parabronchi that run through the lungs; these structures are shown in the scanning electron micrograph.

blood flow to minimize the O_2 concentration on the internal side of the exchange surfaces. However, in birds, the flow appears to be crosscurrent (at right angles) rather than countercurrent to the airflow.

It is now clear how birds can fly over Mount Everest. A bird supplies its gas-exchange surfaces with a continuous flow of fresh air that has an oxygen concentration close to that of the ambient air. Even though the P_{O_2} of the ambient air is only slightly above the P_{O_2} of the blood, diffusion of O_2 from air to blood can take place.

Next we will see why mammals would not be able to fly over Mount Everest—even if they could fly!

Breathing in Mammals

Vertebrate lungs have their origins in outpocketings of the digestive tract (Figure 41.11). At the beginning of their evolution, lungs were dead-end sacs, and

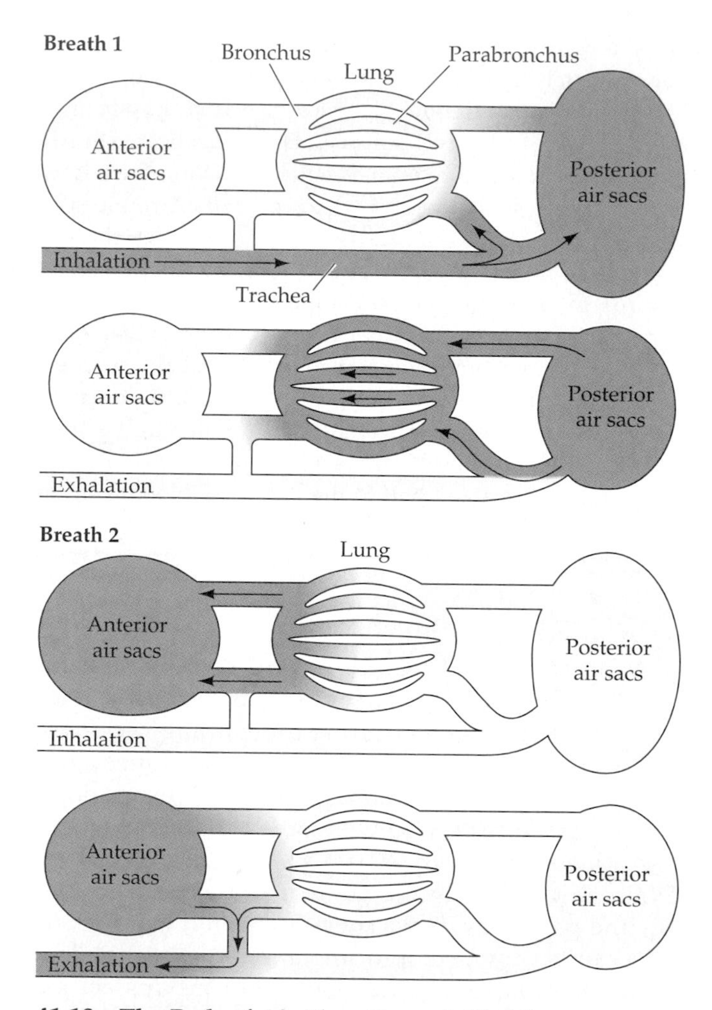

41.10 The Path of Air Flow through Bird Lungs
The fresh air a bird takes in with one breath (green) will travel through the lungs in one direction, from the posterior to the anterior air sacs. Two cycles of inhalation and exhalation are required for the air to travel the full length of the bird's respiratory tract.

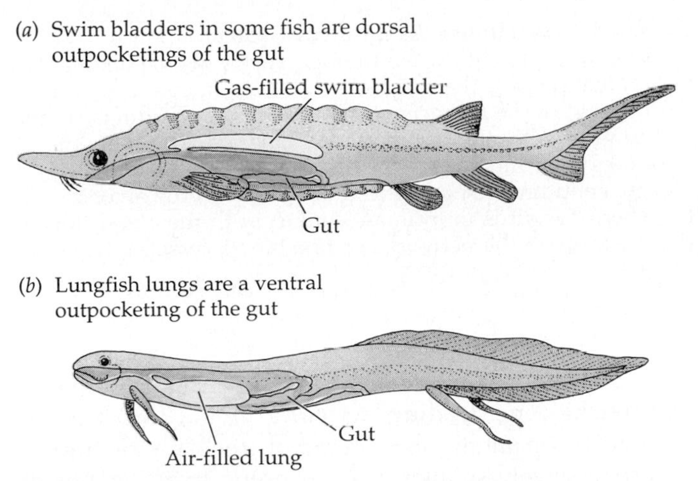

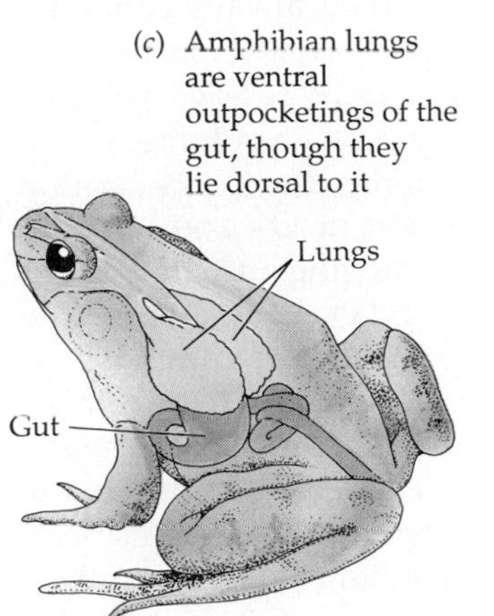

41.11 Lung Evolution
(a) Outpocketings of the digestive tract evolved into swim bladders in some fish. In the lungfish *(b)* and terrestrial vertebrates such as amphibians *(c)*, gut outpockets evolved into lungs.

they remain so today in all air-breathing vertebrates except birds. Because lungs are dead-end sacs, ventilation cannot be constant and unidirectional, but must be tidal: Air comes in and then flows out by the same route. A spyrometer shows how we use our lung capacity in breathing (Figure 41.12). When we are at rest, the amount of air that our normal breathing cycle moves per breath is called the **tidal volume** (about 500 ml for an average human adult). We can breathe much more deeply and inhale more air than our resting tidal volume, and the additional volume of air we can take in above normal tidal volume is our **inspiratory reserve volume**. Conversely, we can forcefully exhale more air than we normally do during a resting exhalation. This additional amount of air that can be forced out of the lungs is the **expiratory reserve volume**. Even after the most extreme exhalation possible, however, some

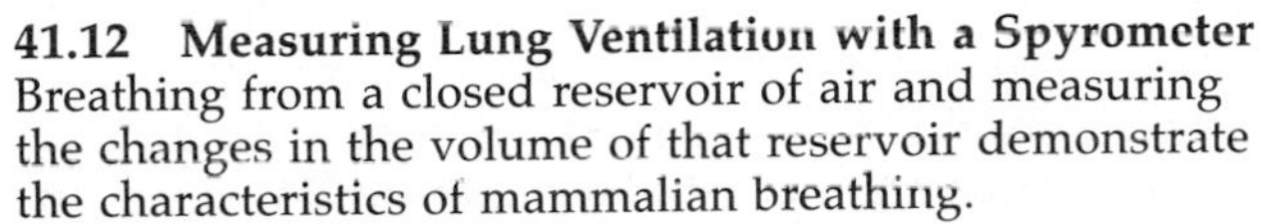
41.12 Measuring Lung Ventilation with a Spyrometer
Breathing from a closed reservoir of air and measuring the changes in the volume of that reservoir demonstrate the characteristics of mammalian breathing.

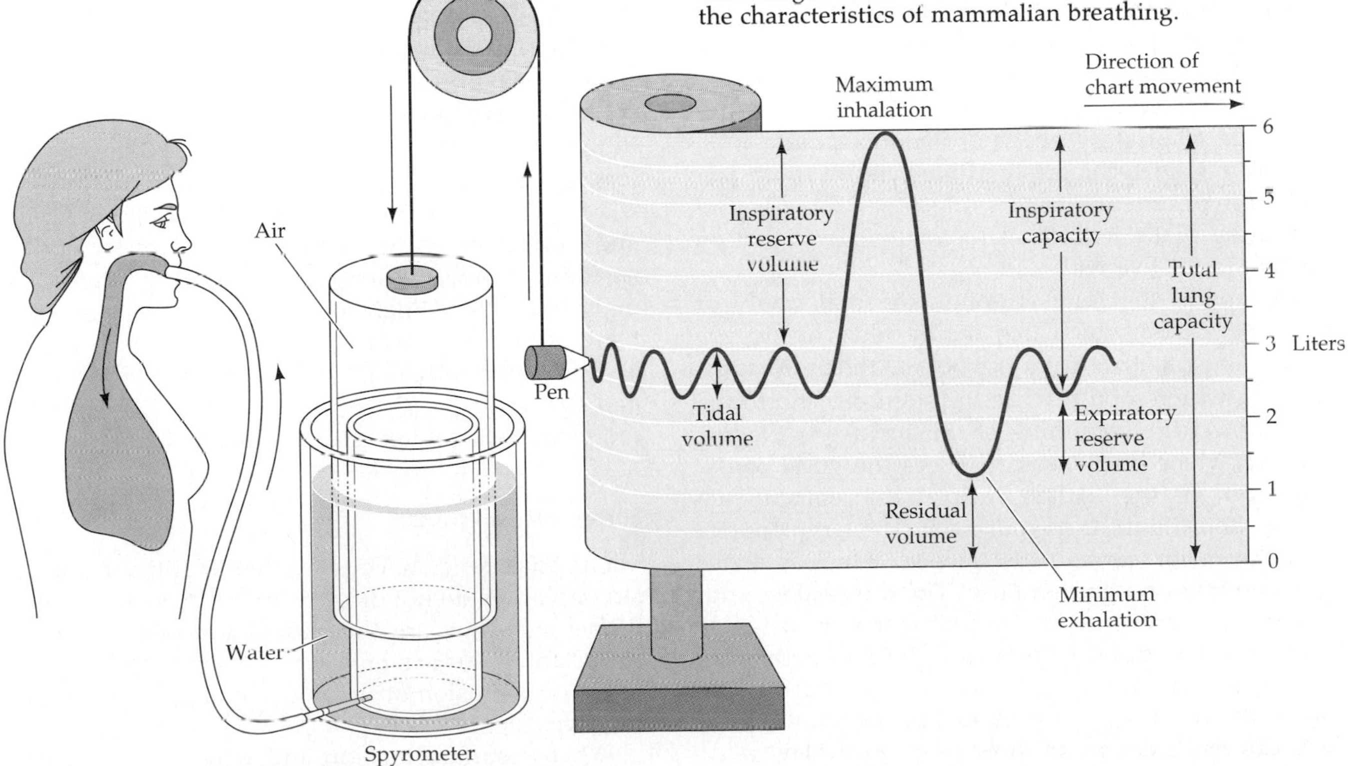

air remains in the lungs. The lungs and airways cannot be collapsed completely; they always contain a **residual volume**. The **total lung capacity** is the sum of the residual volume, expiratory reserve volume, tidal volume, and inspiratory reserve volume.

Tidal breathing severely limits the concentration difference driving the diffusion of O_2 from air into the blood. Fresh air is not moving into the lungs during half of the respiratory cycle; therefore the average O_2 concentration of air in the lungs is less than it is in the air outside the lungs. The incoming air also mixes with the stale air that was not expelled by the previous exhalation. The lung volume that is not ventilated with fresh air is **dead space**. This dead space consists of the residual volume and, depending on the depth of breathing, some or all of the expiratory reserve volume. The scale in Figure 41.12 tells us that a tidal volume of 500 ml of fresh air mixes with up to 2,000 ml of stale moist air before reaching the gas-exchange surfaces. When the P_{O_2} in the ambient air is 150 mm Hg, the P_{O_2} of the air that reaches the gas-exchange surfaces is only about 100 mm Hg. By contrast, the P_{O_2} in the water bathing the lamellae of the fish gills or in the air flowing through the air capillaries of the bird lung is the same as the P_{O_2} in the outside water or air.

As well as reducing the concentration difference, tidal breathing reduces the efficiency of gas exchange in another way. It does not allow countercurrent gas exchange between air and blood. Because the air enters and leaves the gas-exchange structures by the same route, there is no anatomical way that blood can flow parallel to and countercurrent to the air flow.

Mammalian lungs possess some interesting and important design features that maximize the rate of gas exchange: an enormous surface area and a very short path length for diffusion. Mammalian lungs serve the respiratory needs of mammals quite well, considering the ecologies and lifestyles of these animals. Environmental factors other than low O_2 concentration make it difficult for mammals to live on top of Mount Everest!

Air enters the lungs through the oral cavity or nasal passages, which join in the pharynx (Figure 41.13). The pharynx gives rise both to the esophagus, through which food reaches the stomach, and to the airways. At the beginning of the airways is the **larynx**, or "voice box," which houses the vocal cords. The larynx is the "Adam's apple" that you can see and feel on the front of your neck. The larynx opens into the major airway, the trachea, which is about the diameter of a garden hose. The thin walls of the trachea are prevented from collapsing by rings of cartilage that support them as air pressure changes during the breathing cycle. If you run your fingers down the front of your neck just below your larynx, you can feel a couple of these rings of cartilage.

41.13 The Human Respiratory System ▶
The lungs lie within the thoracic cavity, which is bounded by the ribs and the diaphragm. Pleural membranes line the part of the thoracic cavity containing the lungs, so the lungs are actually in the pleural cavities. Air enters the lungs from the oral cavity or nasal passages via the trachea and bronchi, and eventually reaches the alveoli. There, the air is in intimate contact with the blood flowing through the networks of fine blood vessels surrounding the alveoli.

The trachea branches into two slightly smaller bronchi, one leading to each lung. The bronchi branch repeatedly to generate a treelike structure of progressively smaller airways going to all regions of the lungs. Structurally, each of these bronchi is a smaller version of the trachea; they all have supporting cartilage rings. As the branching of the bronchial tree continues to produce still smaller airways, the cartilage supports eventually disappear, marking the transition to **bronchioles**. The branching continues until the bronchioles are less than the diameter of a pencil lead, at which point the tiny, thin-walled air sacs called **alveoli** begin. Alveoli resemble clusters of grapes on a system of stems (see Figure 41.13). The "stems" are the bronchioles, which have about six more branch points. Finally, terminal bronchioles end in alveoli. The alveoli are the sites of gas exchange. Because the airways only conduct the air to and from the alveoli, their volume is physiological dead space. If you trace an airway from the primary bronchus leaving the trachea down to the very last terminal bronchiole, you pass about 23 branching points. Thus there are 2^{23} terminal bronchioles—a very large number. The number of alveoli is even larger, about 300 million in the human lungs. Even though each alveolus is very small, the combined surface area for diffusion of respiratory gases is about 70 square meters, or the size of a badminton court.

Each alveolus consists of very thin cells. Between and surrounding the alveoli are networks of the smallest of blood vessels, also made up of exceedingly thin cells. Where blood vessel meets alveolus there is very little space between them (see Figure 41.13), so the diffusion path length between the air and the blood is only 2 μm. Even the diameter of a red blood cell is much greater—about 7 μm.

Surfactant and Mucus

Mammalian lungs have two other adaptive features, although they do not directly influence gas-exchange properties. They are the production of mucus and surfactant. A surfactant is any substance that reduces the surface tension of the liquid lining the insides of the alveoli.

What is surface tension and why do we need to

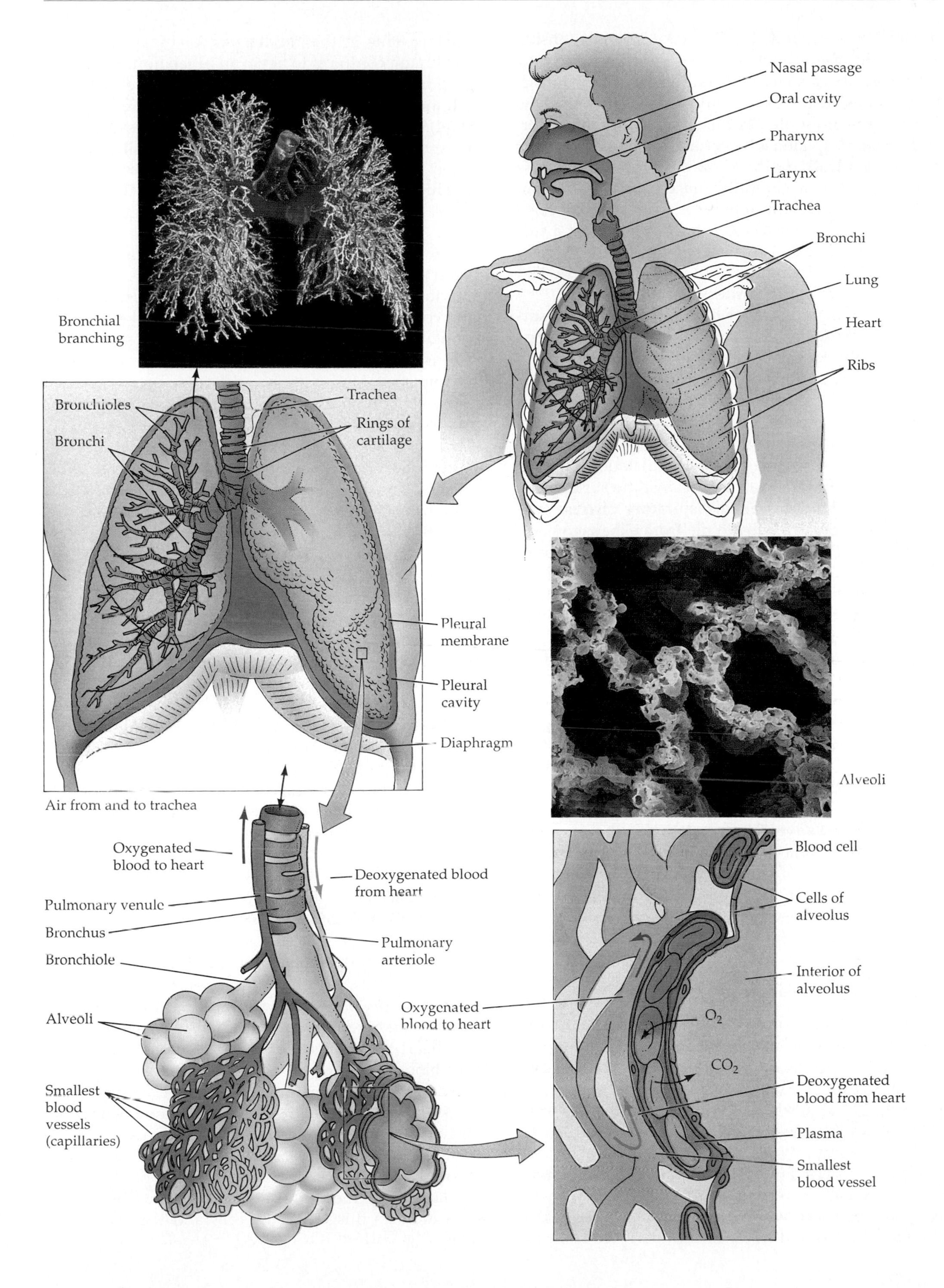
Nasal passage
Oral cavity
Pharynx
Larynx
Trachea
Bronchi
Lung
Heart
Ribs
Bronchial branching
Bronchioles
Bronchi
Trachea
Rings of cartilage
Pleural membrane
Pleural cavity
Diaphragm
Alveoli
Air from and to trachea
Oxygenated blood to heart
Deoxygenated blood from heart
Pulmonary venule
Bronchus
Pulmonary arteriole
Bronchiole
Alveoli
Smallest blood vessels (capillaries)
Blood cell
Cells of alveolus
Interior of alveolus
Oxygenated blood to heart
O_2
CO_2
Deoxygenated blood from heart
Plasma
Smallest blood vessel

consider it as we study lung function? Surface tension arises from the attractive (cohesive) forces between the molecules of a liquid. At the surface of the liquid, these cohesive forces are unbalanced and give the surface the properties of an elastic membrane. Surface tension explains why certain insects called water striders can walk on the surface of water (see Figure 2.18) and why a carefully placed razor blade can "float." A surfactant interferes with the cohesive forces that create surface tension. Detergent is a surfactant; when added to water it causes the floating razor blade to sink and can make walking on water difficult for the water strider.

The thin, aqueous layer lining the alveoli has surface tension, which can make inflation of the lungs very difficult. Surface tension in the alveoli normally is reduced by surfactant molecules produced by certain cells in the alveoli. If a baby is born more than a month prematurely, its alveoli may not be producing surfactant. Such a baby has great difficulty breathing because an enormous inhalation effort is required to stretch the alveoli against the surface tension. This condition, called **respiratory distress syndrome**, may cause a baby to die from exhaustion and suffocation. The common treatment is to put the baby on a respirator to assist its breathing and to give the baby hormones to speed its lung development. A new approach, applying surfactant to the lungs in an aerosol, is very promising.

Many cells lining the airways produce a sticky mucus that captures bits of dirt and microorganisms as they are inhaled. The mucus must be continually cleared from the airways, however; the beating of cilia lining the airways accomplishes this task (see Figure 40.2). The cilia move the mucus with its trapped debris up toward the pharynx, where it is swallowed. This phenomenon, called the mucus escalator, can be adversely affected by inhaled pollutants. Smoking one cigarette can immobilize the cilia of the airways for hours. Hacking, or smoker's cough, results from the need to clear the obstructing mucus from the airways when the mucus escalator is out of order.

Mechanics of Ventilation

As Figure 41.13 shows, the lungs are suspended in the **thoracic cavity**, which is bounded on the top by the shoulder girdle, on the sides by the rib cage, and on the bottom by a domed sheet of muscle, the **diaphragm**. The thoracic cavity is lined on the inside by the **pleural membranes**, which divide it into right and left **pleural cavities**. Because the pleural cavities are closed spaces, any effort to increase their volume creates negative pressure—suction—inside them. Negative pressure within the pleural cavities causes the lungs to expand as air flows into them from the outside. This is the mechanism of inhalation. The diaphragm contracts to begin an inhalation. This contraction pulls the diaphragm down, increasing the volume of the thoracic and pleural cavities (Figure 41.14). As pressure in the pleural cavities becomes more negative, air enters the lungs. Exhaling begins when the contraction of the diaphragm ceases, the diaphragm relaxes and moves up, and the elastic recoil of the lungs pushes air out through the airways.

The diaphragm is not the only muscle that increases the volume of the thoracic cavity. Between the ribs are **intercostal muscles**, and one set of these intercostal muscles expands the thoracic cavity by lifting the ribs up and outward. When heavy demands are placed on the respiratory system, such as during strenuous exercise, the intercostal muscles and the diaphragm contract together and increase the volume of air inhaled.

Inhalation is always an active process, with muscles contracting; exhalation is usually passive, with muscles relaxing. However, there is another set of intercostal muscles that can be called into play for forceful exhalations. Place your hands on your ribs and abdomen while breathing shallowly, then deeply. Feel which muscles are active during inhalation and which are active during exhalation.

When the diaphragm is at rest between breaths, the pressure in the pleural cavities is still slightly negative. This slight suction keeps the alveoli partially inflated. If the thoracic wall is punctured, by a knife wound for example, air leaks into the pleural cavity, and the pressure from this air causes the lung to collapse. If the hole in the thoracic wall is not sealed, the breathing movements of diaphragm and intercostal muscles pull air into the pleural cavity rather than into the lung, and ventilation of the alveoli in that lung ceases.

TRANSPORT OF RESPIRATORY GASES BY THE BLOOD

The circulatory system is the subject of the next chapter, but since two of the substances it transports are the respiratory gases (O_2 and CO_2), we must mention aspects of it here. The circulatory system uses a pump (the heart) and a network of blood vessels to transport blood and the substances it carries around the body. As O_2 diffuses across the gas-exchange surfaces into the vessels, the circulating blood sweeps it away. This internal perfusion of the gas-exchange surfaces minimizes the concentration of O_2 on the internal side and promotes the diffusion of O_2 across the surface at the highest possible rate. The blood then delivers this O_2 to the cells and tissues of the body.

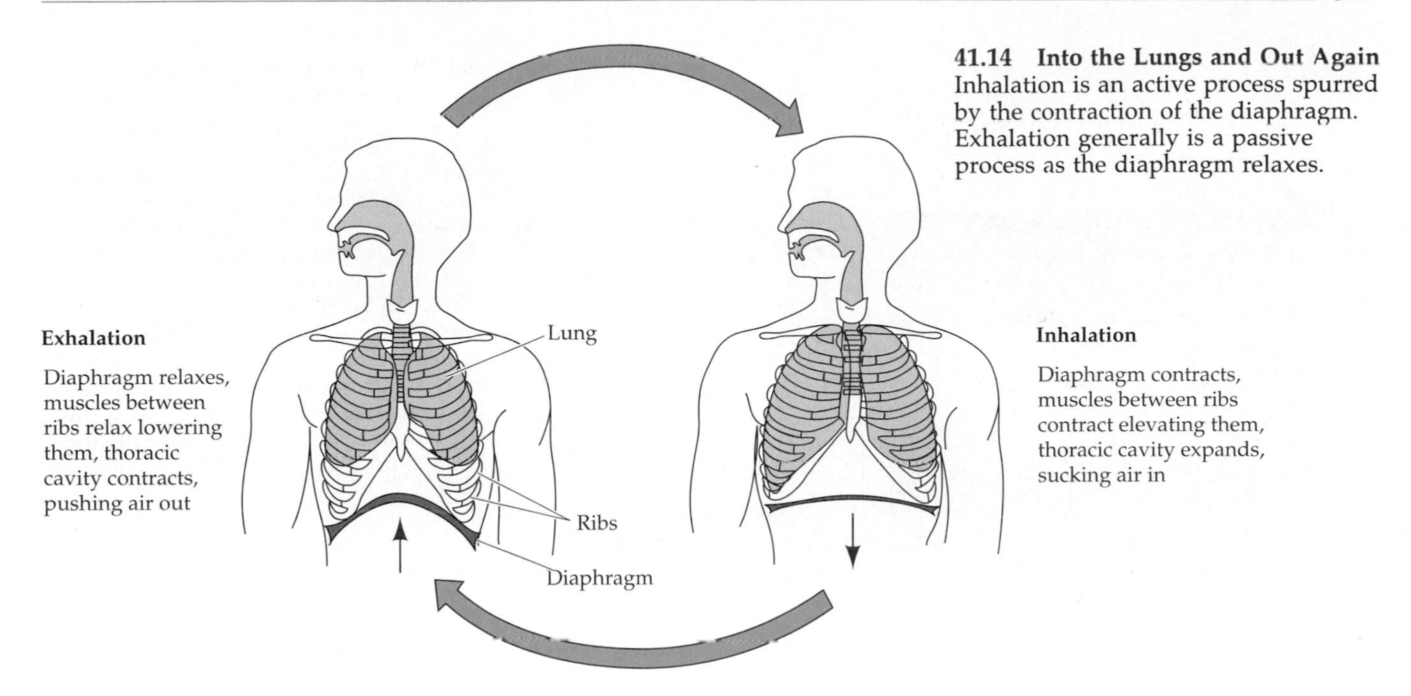

41.14 Into the Lungs and Out Again Inhalation is an active process spurred by the contraction of the diaphragm. Exhalation generally is a passive process as the diaphragm relaxes.

The liquid part of the blood, the **blood plasma**, carries some O_2 in solution, but the ability of the blood to pick up and transport O_2 would be quite limited if plasma were the only means available. Blood plasma carries about 0.3 ml of oxygen per 100 ml. To support the O_2 needs of a person at *rest*, the heart would have to pump about 5,000 liters of blood plasma *per hour* (enough to fill the gas tanks of about 100 cars). Fortunately, the blood also contains **red blood cells** (see page 293), which are red because they are loaded with the oxygen-binding pigment hemoglobin. Hemoglobin increases the capacity of the blood to transport oxygen by about 60-fold.

Hemoglobin

Red blood cells contain enormous numbers of hemoglobin molecules. Hemoglobin is a protein consisting of four polypeptide subunits (see Figure 3.20). Each of these polypeptides surrounds a heme group—an iron-containing ring structure that can reversibly bind a molecule of O_2. As O_2 diffuses into the red blood cells, it binds to hemoglobin. Once O_2 is bound, it cannot diffuse back across the cell membrane. By mopping up O_2 molecules as they enter the red blood cells, hemoglobin maximizes the concentration difference driving the diffusion of O_2 into the red blood cells. In addition, hemoglobin enables the red blood cells to carry a large amount of O_2 for use by the tissues of the body.

The ability of hemoglobin to pick up or release O_2 depends on the concentration or partial pressure of O_2 in its environment. When the P_{O_2} of the blood plasma is high, as it usually is in the lung capillaries, each molecule of hemoglobin can carry its maximum load of four molecules of O_2. As the blood circulates through capillary beds elsewhere in the body, it encounters lower P_{O_2}'s. At these lower P_{O_2}'s the hemoglobin releases some of the O_2 it is carrying. The lower the P_{O_2} of the environment, the more O_2 is released from hemoglobin to diffuse out of the red blood cells and into the tissues. The relation between P_{O_2} and the amount of O_2 bound to hemoglobin is not linear, however. This relationship is described by a sigmoid (S-shaped) curve (Figure 41.15), and it is an important property of hemoglobin.

Remember that the hemoglobin molecule consists of four subunits, each of which can bind one molecule of O_2. At very low P_{O_2}'s, one subunit will bind an O_2 molecule. As a result, the shape of this subunit changes, causing an alteration in the quarternary structure of the whole hemoglobin molecule (see Chapter 3). This structural change makes it easier for the other subunits to bind a molecule of O_2; that is, their O_2 affinity is increased. Only small increases in the P_{O_2} cause most hemoglobin molecules to pick up a second and then a third molecule of O_2. The influence of the binding of O_2 by one subunit on the binding affinity of the other subunits is called **positive cooperativity**, because binding of the first molecule makes binding of the second easier, and so forth. After the binding of the third molecule of O_2 a large increase in P_{O_2} is required for the fourth subunits of all the hemoglobin molecules to be loaded because there are fewer and fewer available binding sites as more and more of the hemoglobin molecules become fully saturated.

The significance of the interactions of the hemoglobin subunits that result in the sigmoid shape of the hemoglobin–oxygen binding curve in Figure

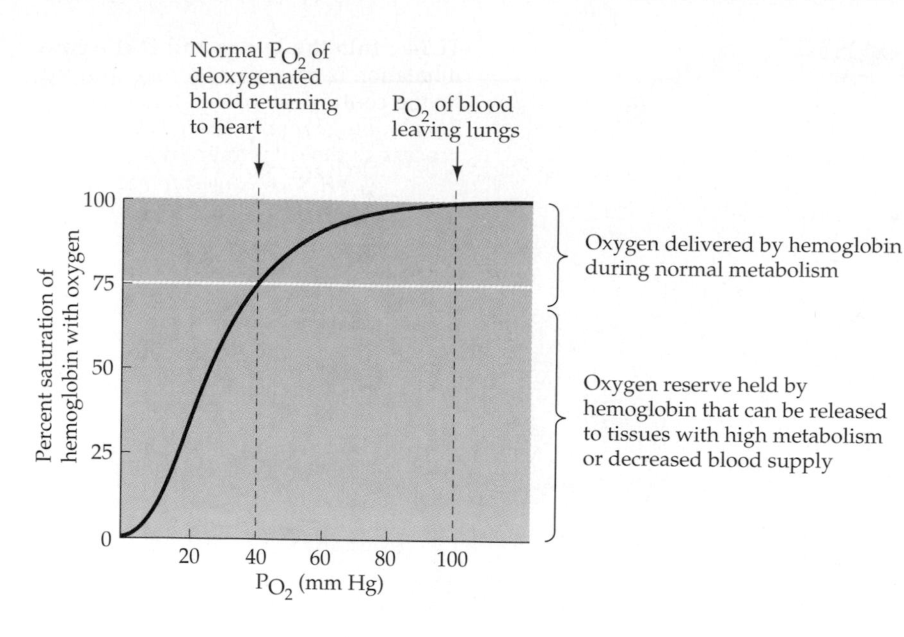

41.15 The Binding of Oxygen to Hemoglobin Depends on the Concentration of Oxygen in the Blood Plasma
As the blood flows through the lungs, the partial pressure of oxygen in the plasma reaches about 100 mm Hg. At this P_{O_2}, each hemoglobin molecule can bind four molecules of oxygen and thus can be 100 percent saturated with oxygen. In body tissues, the P_{O_2} in the plasma may be at about 40 mm Hg. At this lower oxygen concentration, each molecule of hemoglobin can bind only three molecules of oxygen, so the hemoglobin will be only 75 percent saturated.

41.15 is best appreciated by considering the dynamics of unloading the O_2 in the tissues. The P_{O_2} that normally exists in the alveoli of the lungs is about 100 mm Hg, and at this P_{O_2} the hemoglobin is 100 percent saturated (each molecule of hemoglobin carrying four molecules of O_2). The P_{O_2} in mixed venous blood is usually about 40 mm Hg. Thus the hemoglobin returning to the heart from the body is still about 75 percent saturated. That means that most hemoglobin molecules drop only one of their four O_2 molecules as they circulate through the body.

This system may seem very inefficient for delivery of oxygen to the tissues, but it is actually extremely adaptive. When a tissue becomes oxygen-starved and its local P_{O_2} falls below 40 mm Hg, the hemoglobin flowing through that tissue will drop much more of its oxygen load with only small additional decreases in P_{O_2}. The steep portion of the sigmoid hemoglobin–oxygen binding curve comes into play when tissue P_{O_2} falls below the normal 40 mm Hg. Thus the cooperative oxygen-binding property of hemoglobin is very effective in making O_2 available to the tissues precisely when and where it is most needed.

Myoglobin

Muscle cells have their own oxygen-binding molecule, **myoglobin**. Myoglobin consists of just one polypeptide chain associated with an iron-containing ring structure that can bind one molecule of oxygen (see Figure 3.19). Myoglobin has a higher affinity for O_2 than hemoglobin does (Figure 41.16), so it picks up and holds oxygen at P_{O_2}'s at which hemoglobin is releasing its bound O_2. Myoglobin provides a reserve of oxygen for the muscle cells for times when metabolic demands are high and blood flow is interrupted as contracting muscles constrict blood vessels. When hemoglobin has no more O_2 to give up, and tissue P_{O_2} falls even lower, myoglobin releases its bound O_2. Diving mammals such as seals that can remain active under water for many minutes have high concentrations of myoglobin in their muscles. Muscles called upon for extended periods of work frequently have more myoglobin than muscles that are used for short, intermittent periods. This is one of the reasons for the difference in appearance of the "white" and "dark" meat of chickens and turkeys. These birds are not long-distance fliers, and their flight muscles (the white meat) have little myoglobin. Ducks and geese, however, come from distinguished lineages of long-distance fliers. Their flight muscles have much myoglobin, as well as more mitochondria and more blood vessels, and thus appear dark.

Regulation of Hemoglobin Function

The various factors that influence the oxygen-binding properties of hemoglobin also influence oxygen delivery to tissues. For example, there are variations in the chemical composition of the polypeptide chains that form the hemoglobin molecule. The normal hemoglobin of adult humans has two each of two kinds of polypeptide chains—two α chains and two β chains—and has the oxygen-binding characteristics shown in Figure 41.16. Before birth, the fetus has a different form of hemoglobin consisting of two α chains and two γ chains. The chemical composition of fetal hemoglobin enables fetal blood to pick up O_2 from maternal blood when both are at the same P_{O_2}. Fetal hemoglobin thus has an oxygen-binding curve that plots to the left of the adult curve. This difference between maternal and fetal hemoglobin facilitates the transfer of O_2 from the mother's blood to the blood of the fetus in the placenta.

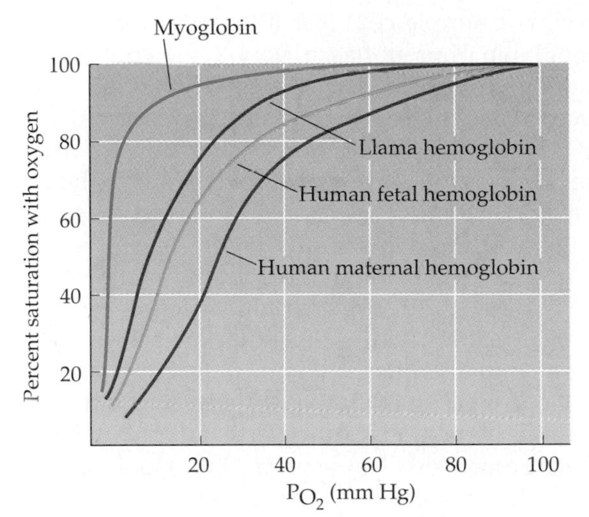

41.16 Oxygen-Binding Adaptations
Llamas are used as pack animals in the high Andes Mountains because they are so well adapted to high altitudes, where the partial pressure of oxygen is low. Evolution has adapted the oxygen-binding properties of different hemoglobins and of myoglobin. Llama hemoglobin has such a high affinity for oxygen that it is 100 percent saturated even at the low P_{O_2}'s found in the Andes. The higher oxygen affinity of fetal hemoglobin in comparison to maternal hemoglobin enables fetal blood to pick up oxygen from maternal blood when both are at the same P_{O_2}. Myoglobin can serve as an oxygen reservoir because it remains 100 percent saturated until the P_{O_2} falls so low that hemoglobin has given up most of its oxygen.

Llamas and vicuñas are mammals native to high altitudes in the Andes mountains of South America (see Figure 41.16). The hemoglobins of these animals, like those of the human fetus, must pick up O_2 in an environment with a low P_{O_2}. In the animal's natural habitat, over 5,000 meters above sea level, the P_{O_2} is below 85 mm Hg, and the P_{O_2} in their lungs is about 50 mm Hg. The hemoglobins of llamas and vicuñas have oxygen-binding curves much to the left of the curves of hemoglobins of most other mammals—in other words, they can become saturated with O_2 at lower P_{O_2}'s than those of other animals can.

The oxygen-binding properties of normal adult hemoglobin are influenced by physiological conditions. The influence of pH on the function of hemoglobin has been well studied and is known as the **Bohr effect**. As the pH of the blood plasma falls, the oxygen-binding curve shifts to the right (Figure 41.17). This shift means that the hemoglobin will release more O_2 to the tissues. Where does hemoglobin encounter a decreased pH as it circulates through the body? In tissues with very high metabolic rates the pH is reduced by the release of acidic metabolites such as lactic acid, fatty acids, and CO_2, which combines with water to form carbonic acid. Because of the Bohr effect, hemoglobin releases more of its bound oxygen in these tissues—another way that O_2 is supplied where and when it is most needed.

Diphosphoglyceric acid is a normal intermediate metabolite that plays an important role in regulating hemoglobin function. The mature mammalian red

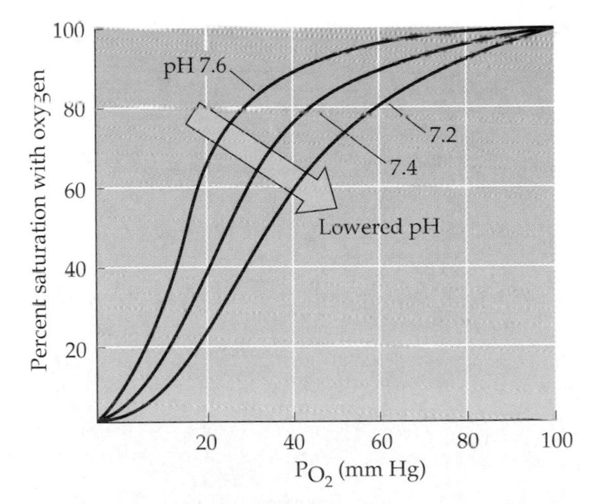

41.17 The Oxygen-Binding Properties of Hemoglobin Can Change
Changes in pH affect the oxygen-binding capacity of hemoglobin. Lowering pH shifts the binding curve to the right; more oxygen is then being released to the tissues.

blood cell is a simple cell. It is little more than a sac of hemoglobin, but it has a very high content of diphosphoglyceric acid. The concentration of diphosphoglyceric acid in red blood cells increases in response to exercise and during acclimation to high altitude. Diphosphoglyceric acid reversibly combines with deoxygenated hemoglobin and changes its shape so that it has a lower affinity for O_2. The result is that at any P_{O_2}, hemoglobin releases more of its bound O_2 than it otherwise would. In other words, diphosphoglyceric acid shifts the oxygen-bindng curve of mammalian hemoglobin to the right.

The llama and the human employ opposite adjustments of hemoglobin function as adaptations for life at high altitudes. The llama's hemoglobin has a left-shifted oxygen-binding curve, which means that it can become 100 percent saturated with O_2 at the low P_{O_2}'s at high altitude. As a consequence, the llama's tissues must operate at a lower P_{O_2}. By contrast, human hemoglobin acquires, through acclimation, a right-shifted oxygen-binding curve. Human hemoglobin never becomes fully saturated with O_2 at high altitude, but more of the O_2 carried by that hemoglobin is released to the tissues.

Transport of Carbon Dioxide from the Tissues

Delivering O_2 to the tissues is only half of the respiratory function of the blood. The blood also must take metabolic waste products away from the tissues. Because we are concerned with respiratory gases in this chapter, the metabolic waste product we will consider is carbon dioxide. CO_2 is highly soluble and readily diffuses through cell membranes, moving from its site of production in a cell into the blood, where its concentration is lower. Very little CO_2 is transported by the blood in this dissolved form, however. Most CO_2 produced by tissues is transported to the lungs in the form of the **bicarbonate ion**, HCO_3^-. How and where CO_2 becomes HCO_3^-, is transported, and then is converted back to CO_2 is an interesting story.

When CO_2 dissolves in water, some of it slowly reacts with the water molecules to form carbonic acid (H_2CO_3), some of which then dissociates into a proton (H^+) and a bicarbonate ion (HCO_3^-). This sequence of events is expressed as follows:

$$CO_2 + H_2O \rightleftharpoons H_2CO_3 \rightleftharpoons H^+ + HCO_3^-$$

In the blood plasma, the reaction between CO_2 and H_2O goes too slowly to have much of an effect. It is different, however, in the red blood cells, where the enzyme **carbonic anhydrase** speeds up the conversion of CO_2 to H_2CO_3. The newly formed carbonic acid dissociates, and the resulting bicarbonate ion diffuses back out into the plasma (Figure 41.18). This action of carbonic anhydrase in the red blood cells creates a sink for CO_2, thus facilitating the diffusion of CO_2 from tissue cells to plasma to red blood cells. Most CO_2 is transported by the blood as bicarbonate ions in the plasma. Some CO_2 is also carried in chemical combination with deoxygenated hemoglobin as **carboxyhemoglobin**.

41.18 Carbon Dioxide Is Transported as Bicarbonate Ions
In tissues, CO_2 diffuses from cells into plasma and into the red blood cells. In the red blood cells, CO_2 is rapidly converted to bicarbonate ions because carbonic anhydrase is present. Bicarbonate ions leave red blood cells in exchange for chloride ions. In the lungs, these processes are reversed. Some CO_2 combines with hemoglobin (Hb).

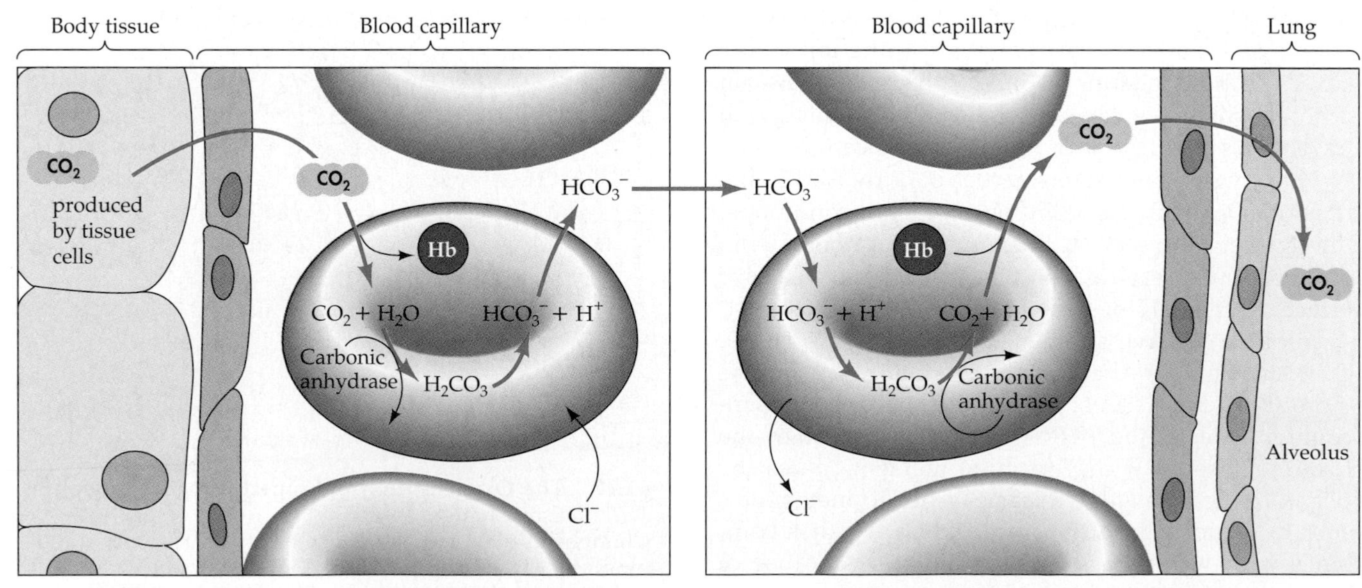

In the lungs, the reactions involving CO_2 and bicarbonate are reversed. CO_2 diffuses from the blood plasma into the air in the alveoli and is exhaled. Breathing keeps CO_2 concentrations in the alveoli low, so CO_2 diffuses from red blood cells to the plasma. The loss of CO_2 from the red blood cells shifts the equilibrium between CO_2 and bicarbonate. As the HCO_3^- in the red blood cells is converted back to CO_2, more HCO_3^- moves into the red blood cells from the plasma. Remember that an enzyme like carbonic anhydrase only speeds up a reversible reaction; it does not determine its direction. Direction is determined by concentrations of reactants and products (see Chapter 6).

REGULATION OF VENTILATION

We must breathe every minute of our lives. We don't worry about our need to breathe or even think about it very often because breathing is an autonomic function of the nervous system. The breathing pattern easily adjusts itself around other activities such as speech and eating. Most impressively, our breathing rates change to match the metabolic demands of our bodies. Now we will learn how the regular respiratory cycle is generated and how it is controlled so that we get the oxygen we need and eliminate the carbon dioxide we produce as our levels of activity change.

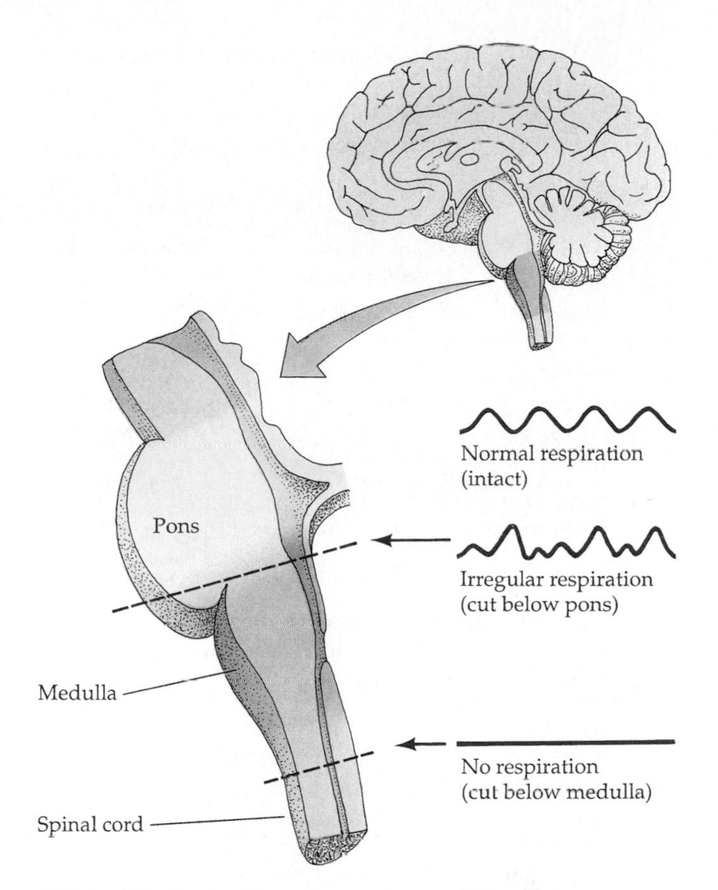

41.19 The Brain Stem Generates and Controls Breathing Rhythm
Severing the brain stem at different levels reveals that the basic breathing rhythm is generated in the medulla and modified by neurons in or above the pons.

The Ventilatory Rhythm

Breathing is an involuntary function. The complex, coordinated movements of the diaphragm and other muscles do not require conscious thought. Automatically, the central nervous system maintains a ventilatory rhythm and modifies its depth and frequency to meet the demands of the body for O_2 supply and CO_2 elimination. The ventilatory rhythm ceases if the spinal cord is severed in the neck region, showing that the rhythm is generated in the brain. If the brain stem is cut just above the medulla, the segment of the brain stem just above the spinal cord, a crude ventilatory rhythm remains (Figure 41.19).

Groups of neurons within the medulla increase their firing rates just before an inhalation begins. As more and more of these neurons fire, and fire faster and faster, the inhalation muscles contract. Suddenly the neurons stop firing, the inhalation muscles relax, and exhalation begins. Exhalation is usually a passive process that depends on the elastic recoil of the lung tissues. When respiratory demand is high, however, as during strenuous exercise, exhalation neurons in the medulla increase their firing rates and accelerate the ventilatory rhythm by adding an active component to the exhalation phase of the cycle. Brain areas above the medulla can also modify the ventilatory rhythm to accommodate speech, ingestion of food, coughing, and emotional states.

As respiratory demands increase, the activities of the inhalation neurons also increase, thus contributing to greater depth of inhalation. An override reflex, however, prevents the ventilatory muscles from overdistending and damaging the lung tissue. This reflex is named the Hering–Breuer reflex (after the two physiologists who discovered it). It begins with stretch sensors in the lung tissue. When stretched, these sensors send impulses via the vagus nerve that inhibit the inhalation neurons.

Matching Ventilation to Metabolic Needs

When the partial pressure of O_2 and the partial pressure of CO_2 in the blood change, the respiratory rhythm changes to return these values to normal levels. An early experimental approach to understanding gas exchange in humans addressed the reasonable expectation that the blood P_{O_2}, or P_{CO_2}, or both, should provide feedback to the respiratory cen-

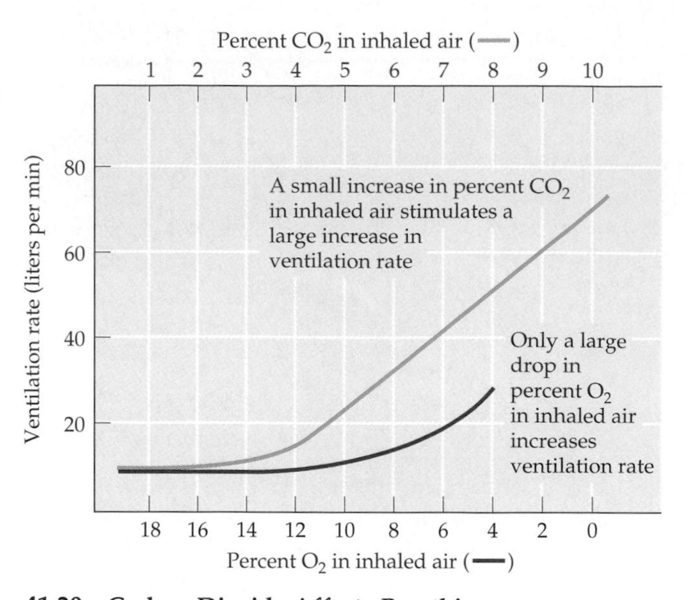

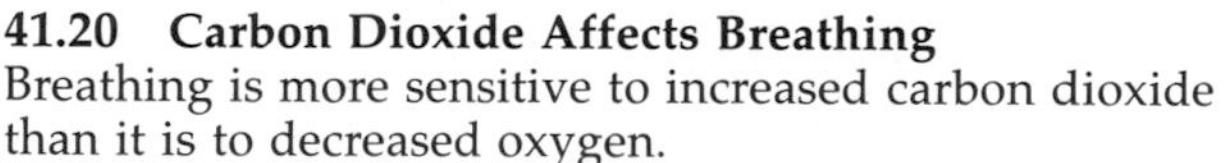
41.20 Carbon Dioxide Affects Breathing
Breathing is more sensitive to increased carbon dioxide than it is to decreased oxygen.

ters in the brain. Dramatic and disastrous insight regarding this expectation was provided by three French physiologists in 1875. They wanted to investigate the physiological effects of breathing low concentrations of O_2. Sophisticated gas pumps and pressure chambers did not exist in 1875, so the three decided to go up in a balloon to very high altitudes and observe the effects of the rarefied atmosphere on one another. They noted no ill effects and continued to throw out ballast, going higher than 8,000 meters. Then all three became unconscious. The balloon finally descended on its own, and one of the physiologists regained consciousness to find his two colleagues dead. This infamous flight of the balloon *Zenith* is tragic proof that the human body is not very good at sensing its own need for O_2.

Humans and other mammals are very sensitive, however, to increases in the P_{CO_2} of the blood, whether caused by energy demands or by the composition of the air breathed. If you re-breathe a small volume of air, thereby increasing the P_{CO_2} of that reservoir, your breathing becomes deeper and more rapid, and you become anxious and agitated. You react the same way even if pure O_2 is continually released into the reservoir to keep its P_{O_2} constant. Typical ventilatory responses to changes in blood P_{O_2} and P_{CO_2} are shown in Figure 41.20.

It makes sense that CO_2 rather than O_2 is the dominant feedback stimulus for ventilation. As we have seen, animals have evolved respiratory systems and hemoglobin properties that work to keep the blood that is leaving the gas-exchange surfaces fully saturated with O_2 over a broad range of alveolar P_{O_2}'s and metabolic rates. Normal fluctuations in metabolism and ventilation have very little effect on the maximum amount of O_2 carried by the blood. By contrast, small changes in metabolism and alveolar P_{CO_2} do influence the concentration of CO_2 in the blood. Changes in blood P_{CO_2} are a much finer index of energy demands and respiratory performance than is the O_2 content of the blood.

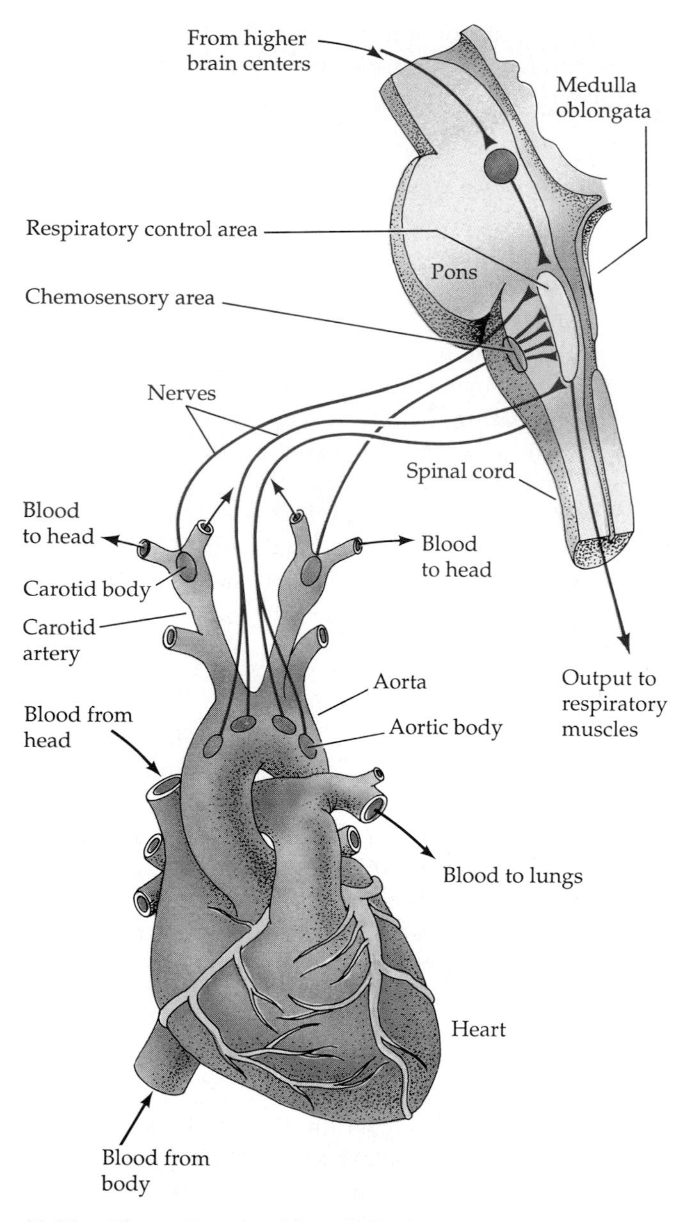

41.21 Chemosensors Sense Changes in Gas Concentrations
Chemosensors on large blood vessels leaving the heart are sensitive to the partial pressure of oxygen in the blood; other chemosensors on the surface of the medulla are sensitive to the partial pressure of carbon dioxide in the blood. The body uses information from these chemosensors to match breathing rate to metabolic demand.

Where are gas concentrations in the blood sensed? The major site of CO_2 sensitivity is an area on the ventral surface of the medulla, not far from the groups of neurons that generate the ventilatory rhythm. Sensitivity to the O_2 concentration of the blood resides in nodes of tissue on the large blood vessels leaving the heart, the aorta and the carotid arteries (Figure 41.21). These carotid and aortic bodies receive enormous supplies of blood, and they contain chemosensory nerve endings. If the blood supply to these structures decreases, or if the P_{O_2} of the blood falls dramatically, the chemosensors are activated and send impulses to the respiratory centers. Although we are not very sensitive to changes in blood P_{O_2}, the carotid and aortic bodies can stimulate increases in ventilation during exposure to very high altitude or when blood volume or blood pressure are very low.

SUMMARY of Main Ideas about Gas Exchange in Animals

Respiratory gas-exchange systems of animals promote diffusion of O_2 from the environment into the body and diffusion of CO_2 from the body to the environment.

Maximizing the surface area for diffusion of respiratory gases is an adaptation that takes many forms in different species, including flat body shapes, external gills, internal gills, lungs, and both air-filled and water-filled channels that branch throughout the body.
Review Figures 41.1 and 41.4

In comparison to air-breathing animals, water breathers are under the constraints of lower O_2 content in their environment, decreasing O_2 availability but increasing O_2 need as temperature rises, and the fact that more work is required to move water than to move air over gas-exchange surfaces.
Review Figure 41.2

The O_2 content of air decreases as barometric pressure decreases.
Review Figure 41.3

The concentration gradient for diffusion of respiratory gases is maximized by ventilating the environmental side of exchange membranes with air or water and perfusing the internal side with blood.

The tracheal gas-exchange system of insects supports the high metabolic rate of insect tissue by distributing O_2 to cells through air capillaries.
Review Figure 41.5

Respiratory systems of fish and birds are highly efficient because of adaptations that enable continuous, unidirectional flow over their gas-exchange surfaces.
Review Figures 41.6, 41.7, 41.8, 41.9 and 41.10

Countercurrent flow of water and blood on opposite sides of respiratory gas-exchange surfaces augments the efficiency of respiratory gas exchange in fish.
Review Figure 41.6 and Box 41.A

The lungs of amphibians, reptiles, and mammals require tidal ventilation, which limits their gas-exchange effiency by limiting the concentration gradients for respiratory gas diffusion.
Review Figures 41.12 and 41.14

The alveoli of lungs present large surface areas for gas exchange.
Review Figure 41.13

Hemoglobin binds O_2, carries it in the bloodstream, and releases it where it is needed.
Review Figures 41.15, 41.16, and 41.17

CO_2 produced in tissues leaves the body via the blood and the lungs.
Review Figure 41.18

The breathing rhythm is generated in the medulla and modified by inputs from other brain regions.
Review Figure 41.19

CO_2 concentration in the blood is a major feedback signal controlling the rate of breathing; O_2 concentration has a lesser effect.
Review Figure 41.20

Chemosensors on the suface of the medulla detect CO_2 and chemosensors on blood vessels leaving the heart detect O_2 concentration.
Review Figure 41.21

SELF-QUIZ

1. Which of the following statements is *not* true?
   a. Respiratory gases are exchanged by diffusion only.
   b. Oxygen has a lower rate of diffusion in water than in air.
   c. The oxygen content of water falls as the temperature of water rises, all other things being equal.
   d. The amount of oxygen in the atmosphere decreases with increasing altitude.
   e. Birds have evolved active transport mechanisms to augment their respiratory gas exchange.

2. Which of the following statements about the respiratory system of birds is *not* true?
   a. Respiratory gas exchange does not occur in the air sacs.
   b. The respiratory system of birds can achieve more complete exchange of O_2 from air to blood than that of humans.
   c. Air passes through birds' lungs in only one direction.
   d. The gas-exchange surfaces in bird lungs are the alveoli.
   e. A breath of air remains in the bird respiratory system for two breathing cycles.

3. In a countercurrent exchange system
   a. the fluids in two tubes flow in opposite directions.
   b. gas exchange between two streams is less complete than if they flowed in parallel.
   c. a gas and a liquid flow in parallel tubes.
   d. the materials in solution move against the current.
   e. the fluids in two tubes are funneled into a final common vessel.

4. In the human respiratory system
   a. the lungs and airways are completely collapsed after a forceful exhalation.
   b. the average O_2 concentration of air inside the lungs is always less than it is in the air outside the lungs.
   c. the P_{O_2} of the blood leaving the lungs is greater than the P_{O_2} of the exhaled air.
   d. the amount of air that is moved per breath during normal, at-rest breathing is termed the total lung capacity.
   e. oxygen and carbon dioxide are actively transported across the alveolar–capillary membranes.

5. Which of the following statements about the human respiratory system is *not* true?
   a. During inhalation there is a negative pressure in the space between the lung and the thoracic wall.
   b. Smoking one cigarette can immobilize the cilia lining the airways for hours.
   c. The respiratory control center in the medulla responds more strongly to changes in arterial O_2 concentration than to changes in arterial CO_2 concentrations.
   d. Without surfactant, the work of breathing is greatly increased.
   e. The diaphragm contracts during inhalation and relaxes during exhalation.

6. The hemoglobin of a human fetus
   a. is the same as that of an adult.
   b. has a higher affinity for O_2 than that of an adult.
   c. has only two protein subunits instead of four.
   d. is supplied by the mother's red blood cells.
   e. has a lower affinity for O_2 than that of an adult.

7. As blood flows through an active muscle, it
   a. becomes saturated with oxygen.
   b. takes up only a small amount of oxygen.
   c. unloads more of its oxygen than it would in a resting muscle.
   d. tends to decrease the partial pressure of oxygen in the muscle tissues.
   e. is denatured.

8. Most carbon dioxide is carried in the blood
   a. in red blood cell cytoplasm.
   b. dissolved in the plasma.
   c. in the plasma as bicarbonate ions.
   d. bound to plasma proteins.
   e. in red blood cells bound to hemoglobin.

9. Myoglobin
   a. binds O_2 at P_{O_2}'s at which hemoglobin is releasing its bound O_2.
   b. has a lower affinity for O_2 than hemoglobin does.
   c. consists of four polypeptide chains, just as hemoglobin does.
   d. provides an immediate source of O_2 for muscle cells at the onset of activity.
   e. can bind four O_2 molecules at once.

10. When the level of CO_2 in the blood becomes greater than the set operating range,
   a. the rate of respiration decreases.
   b. the pH of the blood rises.
   c. the respiratory centers become dormant.
   d. the rate of respiration increases.
   e. the blood becomes more alkaline.

FOR STUDY

1. Compare and contrast the respiratory systems of birds, fish, and humans. Why can birds and fish outperform mammals in environments where the concentration of O_2 is low?

2. What does the following chemical equation represent and how does it relate to gas exchange in human lungs and in active tissues?

$$CO_2 + H_2O \rightleftharpoons H_2CO_3 \rightleftharpoons H^+ + HCO_3^-$$

3. Describe and contrast the adaptations of llamas and humans for gas exchange at high altitude.

4. Describe, in terms of the structure of the human respiratory system, how air is brought into and then expelled from the lungs.

5. Workers A and B must inspect two large gas-storage tanks. Tank 1 contains 100 percent N_2 (nitrogen gas); tank 2 contains 100 percent CO_2. The tanks are *not* flushed out before inspection. Worker A goes into tank 1, becomes unconscious, and dies. Worker B goes into tank 2, feels strangely short of breath, and leaves the tank while feeling somewhat dizzy. Explain why worker A died whereas worker B lived.

READINGS

Eckert, R., D. Randall and G. Augustine. 1988. *Animal Physiology: Mechanisms and Adaptations*, 3rd Edition. W. H. Freeman, New York. An outstanding textbook of animal physiology. Chapter 14 covers gas exchange.

Feder, M. E. and W. W. Burggren. 1985. "Skin Breathing in Vertebrates." *Scientific American*, November. Not all breathing involves lungs or gills. This article discusses adaptations possessed by many vertebrates for gas exchange through the skin.

Perutz, M. F. 1978. "Hemoglobin Structure and Respiration." *Scientific American*, December. An authoritative article on hemoglobin structure and function by the man who received the Nobel prize for his work on this subject.

Schmidt-Nielsen, K. 1971. "How Birds Breathe." *Scientific American*, December. Describes the complex adaptations of the avian respiratory system.

Schmidt-Nielsen, K. 1990. *Animal Physiology: Adaptation and Environment*, 4th Edition. Cambridge University Press, New York. An outstanding textbook that emphasizes the comparative approach.

Vander, A. J., J. H. Sherman and D. S. Luciano. 1994. *Human Physiology: The Mechanisms of Body Function*, 6th Edition. McGraw-Hill, New York. Chapter 15 deals with human respiration.

Heart Attack!
Hospital emergency rooms deal with the aftermath of sudden heart attacks. Here an emergency team attempts to stimulate a failed heart to resume beating.

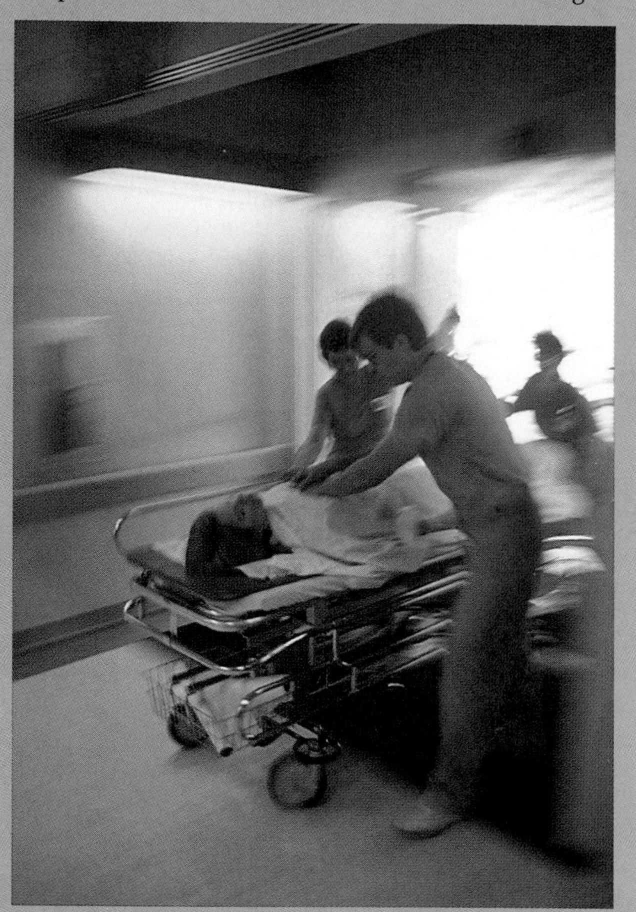

42

Internal Transport and Circulatory Systems

Sweating, severe chest pain, fainting, 911, paramedics, flashing lights, emergency room, intensive care—heart attack. We all will experience this sequence of events directly or indirectly. More than one-third of the deaths in the United States are due to heart failure, and many such deaths are those of a sudden nature that we call heart attacks. Why is it so serious when the heart fails? The answer is that all organs of the body depend on the heart.

The needs of all the cells in the body of an animal are served by the internal fluid environment (see Chapter 35). Cells take up nutrients from the fluid that bathes them, and they release their waste products into that fluid. The activities of the cells change the internal fluid environment in ways that are not healthy for the cells themselves. The organs return different aspects of the internal environment to optimal levels. The lungs take in oxygen and eliminate carbon dioxide, the digestive tract takes in nutrients, the liver controls nutrient levels and eliminates toxic compounds, and the kidneys control salt concentrations and eliminate toxic wastes. The cells of each organ depend on the activities of all the other organs, and only through a system of transport, or circulation, can the activities of each organ influence the internal environment of the entire body. So when the heart stops, transport stops, the internal environment deteriorates, and cells get sick and die.

TRANSPORT SYSTEMS

A heart is not all there is to a transport system; it is just the pump. In addition, there must be a vehicle (blood) to transport materials through the system and a series of conduits (blood vessels) through which the materials can be pumped around the body. Heart, blood, and vessels together constitute a circulatory system, also known as a cardiovascular system (from the Greek *kardia* = "heart" and the Latin *vasculum* = "small vessel"). In this chapter we will explore adaptations of animals that directly serve the needs of their cells by bringing them nutrients and removing their wastes. Although this chapter focuses on cardiovascular systems, the simplest transport systems are not vascular; they do not use vessels. Some do not even use a pump to move things around. But all carry out the essential task of transport systems: transporting substances to and away from every cell of the bodies of animals.

Gastrovascular Cavities

A circulatory system is unnecessary if all the cells of an organism are close enough to the external environment that nutrients, respiratory gases, and wastes

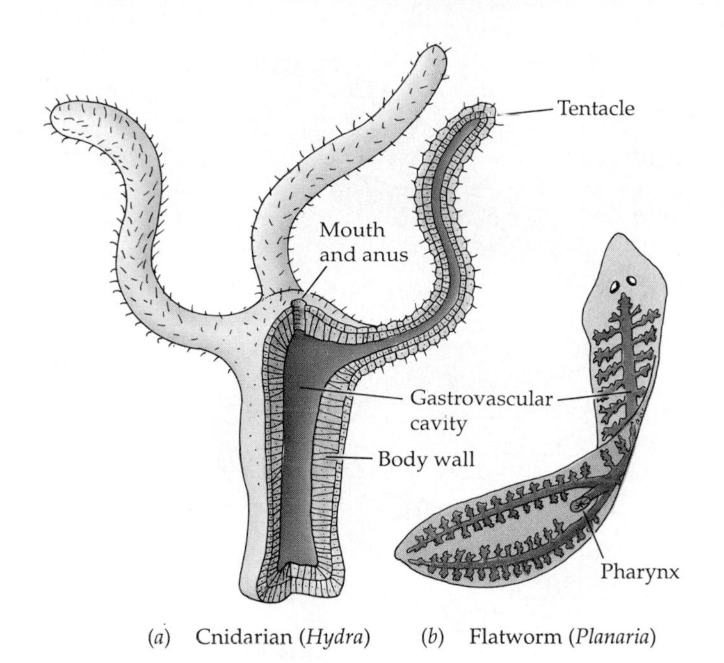

42.1 Gastrovascular Cavities
Gastrovascular cavities in animals without circulatory systems serve the metabolic needs of the innermost cells of the body. *(a)* The gastrovascular cavity of a cnidarian, *Hydra*, extends into the tentacles. No cell of the hydra is more than one cell away from either the gastrovascular cavity or the external medium. *(b)* The gastrovascular cavity of a flatworm, *Planaria*, extends into all regions of the animal's flattened body.

can diffuse directly between the cells and the outside environment. Small aquatic invertebrates have various structures and shapes that permit direct diffusional exchange. The hydra, a cnidarian, is a good example (see Figure 26.12). This aquatic animal is cylindrical and only two cell layers thick. Each of the hydra's cells contacts water that is either surrounding the animal or circulating through its **gastrovascular cavity**, a dead-end sac that serves both for digestion ("gastro") and transport ("vascular") (Figure 42.1*a*). The cells of some other invertebrates are served by diffusion from highly branched gastrovascular systems. Flattened body shapes minimize the diffusion path length—the distance that molecules have to diffuse between cells and the external environment (Figure 42.1*b*). A central gastrovascular system cannot, however, serve the needs of larger animals with many layers of cells. Transport in such animals requires a circulatory system. All terrestrial animals require circulatory systems because none of their cells are served by an external medium; all of their cells must be served by the interstitial fluids.

Open Circulatory Systems

In the simplest circulatory systems the interstitial fluid is simply squeezed through intercellular spaces as the animal moves. In these **open circulatory systems** there is no distinction between interstitial fluid and blood. Usually a muscular pump, or heart, assists the distribution of the fluid throughout the tissues. The contractions of the heart propel the interstitial fluid through vessels leading to different regions of the body, but the fluid leaves those vessels to trickle through the tissues and eventually to return to the heart. In the arthropod shown in Figure 42.2*a* the fluid returns to the heart through valved holes called **ostia** when the heart relaxes. In the mollusk in Figure 42.2*b*, open vessels aid in the return of interstitial fluid to the heart.

Closed Circulatory Systems

In a **closed circulatory system** some components of the blood never leave the vessels. Blood circulates through the vascular system, pumped by one or more muscular hearts. The system keeps the circulating blood separate from the interstitial fluid.

A simple example of a closed circulatory system is

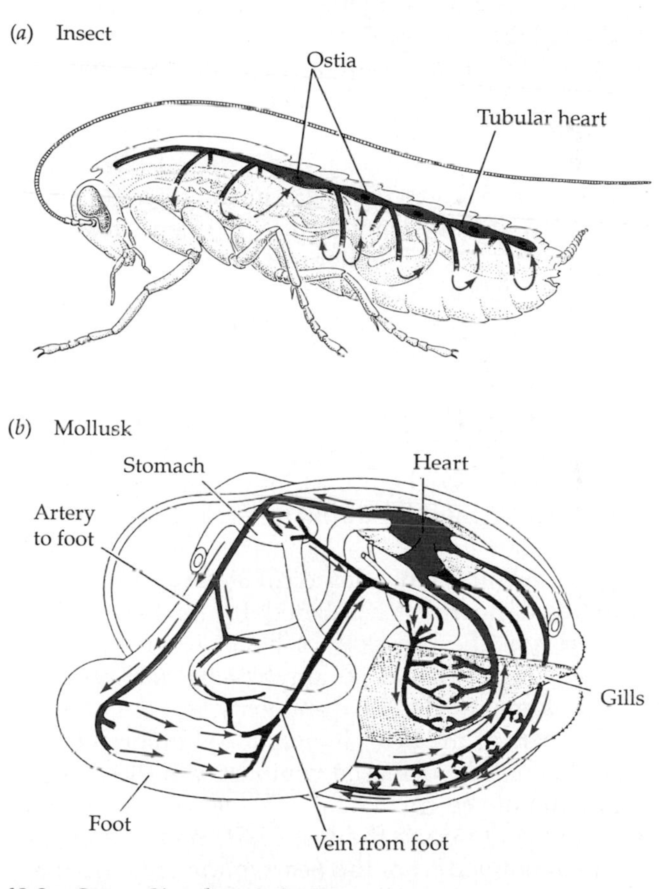

42.2 Open Circulatory Systems
The open circulatory systems of an arthropod *(a)* and a mollusk *(b)*. In both, blood is pumped by a tubular heart and directed to regions of the body through vessels that open into interstitial spaces. In arthropods the blood, or hemolymph, reenters the heart through the ostia, and in the mollusk a system of vessels drains the interstitial spaces and returns the blood to the heart.

that of the common earthworm, an annelid (see Figure 26.24). One large blood vessel on the ventral side of the earthworm carries blood from its anterior end to its posterior end. In each segment of the worm, smaller vessels branch off and transport the blood to even smaller vessels in the tissues of that segment. Here, respiratory gases, nutrients, and metabolic wastes diffuse between the blood and the interstitial fluids. The blood then flows into larger vessels that lead into a single large vessel on the dorsal side of the worm. The dorsal vessel carries the blood from the posterior to the anterior end. Five pairs of vessels connect the large dorsal and ventral vessels in the anterior end, thus completing the circuit (Figure 42.3). The dorsal vessel and the five pairs of connecting vessels serve as hearts for the earthworm; their contractions keep the blood circulating. The direction of circulation is determined by one-way valves in the dorsal vessel and in the five pairs of connecting vessels.

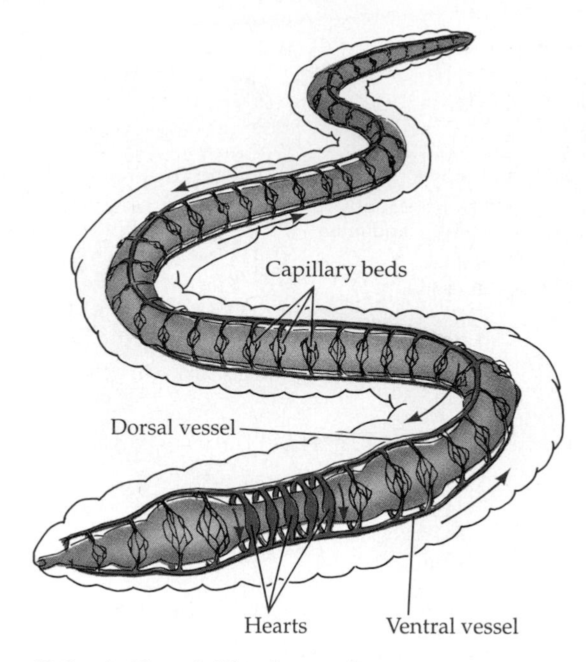

42.3 A Closed Circulatory System
In a closed circulatory system, blood is confined to the blood vessels, kept separate from the interstitial fluid, and is pumped by one or more muscular hearts. The earthworm, with large dorsal and ventral blood vessels and a branching network of smaller vessels, exemplifies this system.

Closed circulatory systems have several advantages over open ones. First, closed systems can deliver oxygen and nutrients to the tissues and carry away metabolic wastes more rapidly than open systems can. Second, closed systems can direct blood to specific tissues. Third, cellular elements and large molecules that function within the vascular system can be kept within it; examples are red blood cells and large molecules that help in the distribution of hormones and nutrients. Overall, closed circulatory systems can support higher levels of metabolic activity, especially in larger animals. How then do highly active insect species achieve high levels of metabolic output with their open circulatory systems? The key to answering this question is to remember something you learned in the previous chapter: Insects do not depend on their circulatory systems for respiratory gas exchange (see Figure 41.5); oxygen diffuses directly to their muscles through tracheae and air capillaries.

All vertebrates have closed circulatory systems and chambered hearts. Chambered hearts have valves that prevent the backflow of blood when the heart contracts. From fishes to amphibians to reptiles to birds and mammals, the complexity and the number of chambers of the heart increase. An important consequence of this increased complexity is the gradual separation of the circulation into two circuits, one to the lungs and one to the rest of the body. In fishes, blood is pumped from the heart to the gills and then to the tissues of the body and back to the heart. In birds and mammals, blood is pumped from the heart to the lungs and back to the heart in the **pulmonary circuit**, and from the heart to the rest of the body and back to the heart in the **systemic circuit**. We will see how the separation of the circulation into two circuits improves the efficiency and capacity of the circulatory system.

The vascular system includes **arteries** that carry blood away from the heart, and **veins** that carry blood back to the heart. **Arterioles** are small arteries, and **venules** are small veins. **Capillaries** are very small, thin-walled vessels that connect arterioles and venules. Exchanges between the blood and the interstitial fluid take place only across capillary walls.

Circulatory Systems of Fishes

Fish hearts have two chambers. A less muscular chamber called the **atrium** receives blood from the body and pumps it into a more muscular chamber, the **ventricle**, which then pumps the blood to the gills, where gas exchange takes place. Blood leaving the gills collects in a large dorsal artery, the **aorta**, which distributes it to smaller arteries and arterioles leading to all the organs and tissues of the body. In the tissues the blood flows through capillary beds, then collects in venules and veins, and is eventually returned to the heart (Figure 42.4*a*). Most of the pressure imparted to the blood by the contraction of the ventricle is dissipated by the high resistance of the many tiny, narrow spaces the blood flows through in the gills. As a result, the blood entering the aorta

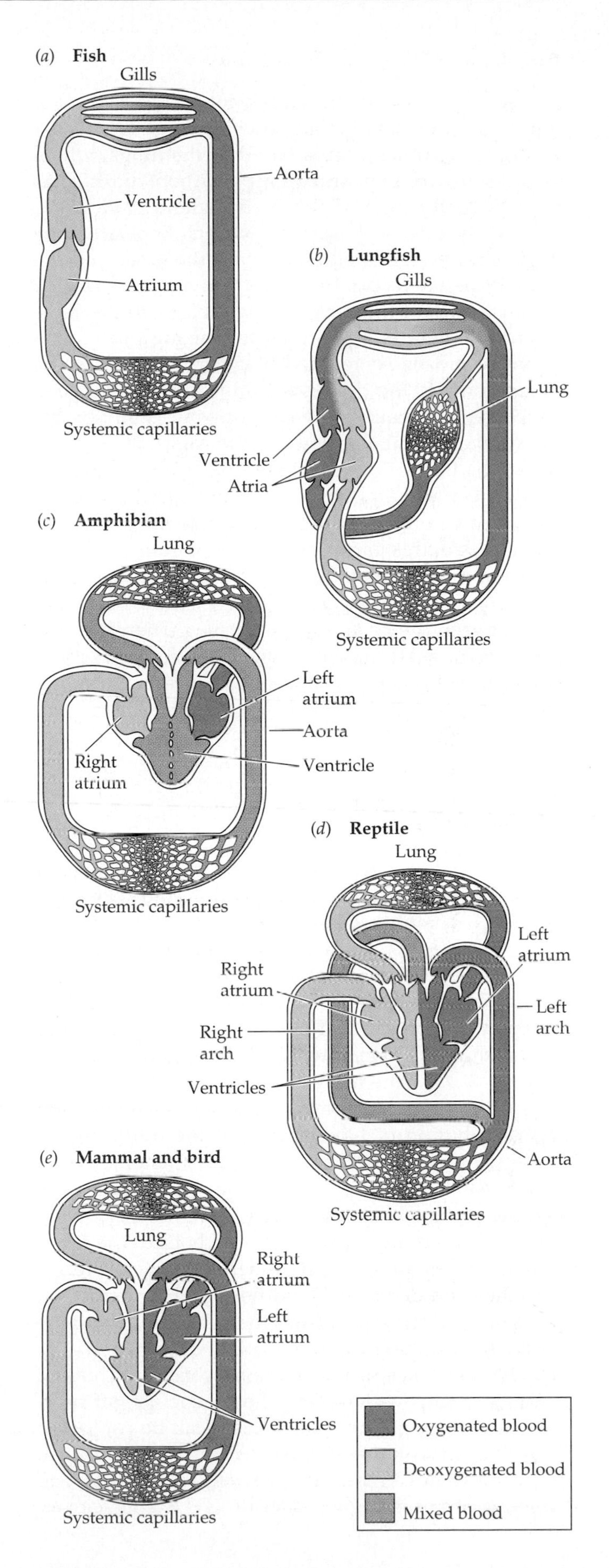

42.4 Vertebrate Circulatory Systems
All vertebrates have closed circulatory systems. *(a)* Fishes have a heart with two chambers: a single atrium and a single ventricle. *(b)* The lungfish heart has two atria, one receiving oxygenated blood from the lung and one deoxygenated blood from the body. *(c)* The pulmonary and systemic circuits are partially separated in adult amphibians. The heart is three-chambered, with two atria and one ventricle. *(d)* The ventricle of the reptilian heart is partially divided by a septum to direct the flow of oxygenated blood to the body and deoxygenated blood to the lungs. An advantage of this incomplete division of the ventricle is that most of the blood can be directed to the systemic circuit when the animal is not breathing. *(e)* Birds and mammals have four-chambered hearts. Their pulmonary and systemic circuits are totally separate.

of the fish is under low pressure, limiting the ability of the fish circulatory system to supply the tissues with oxygen and nutrients. This limitation on arterial blood pressure does not seem to hamper the performance of many rapidly swimming species, such as tuna and marlin.

An important evolutionary step, however, is reflected in the circulatory systems of African lungfish. These fish are periodically exposed to water with low oxygen content or to situations in which their aquatic environment dries up. Their adaptation for dealing with these conditions is an outpocketing of the gut that serves as a lung (see Figure 41.12*b*). This lung contains many thin-walled blood vessels, so deoxygenated blood flowing through those vessels can pick up oxygen from air gulped into the lung. What blood vessels serve this new organ? The last pair of gill arteries is modified to carry blood to the lung, and a new vessel carries oxygenated blood from the lung back to the heart. In addition, two other gill arches have lost their gill filaments, and their blood vessels deliver blood from the heart directly to the dorsal aorta (Figure 42.4*b*). Because a few of the gill arches retain gill filaments, African lungfish can breathe either air or water.

In the evolution of vertebrate circulatory systems, the lungfish reveals the transition step leading to separate pulmonary and systemic circuits. Adaptations of the lungfish heart are also evolutionarily important. Unlike the hearts of other fish, the lungfish heart has a partially divided atrium with the left side receiving oxygenated blood from the lungs and the right side receiving deoxygenated blood from the other tissues. The two bloodstreams stay mostly separate as they flow through the ventricle and the large vessel leading to the gill arches, so the oxygenated blood goes to the gill arteries leading to the dorsal aorta, and the deoxygenated blood goes to the arches with functional gill filaments and to the lung.

Circulatory Systems of Amphibians

Pulmonary and systemic circulation are partially separated in adult amphibians such as frogs and toads. A single ventricle pumps blood to the lungs, where it picks up oxygen and dumps carbon dioxide, as well as to the rest of the body, where it picks up carbon dioxide and dumps oxygen. Separate atria receive the oxygenated blood from the lungs and the deoxygenated blood from the body (Figure 42.4*c*). Because both of these atria deliver blood to the same ventricle, there is a potential for mixing of the oxygenated and deoxygenated blood in which case the blood going to the tissues would not carry a full load of oxygen. In reality, however, mixing is limited because anatomical features of the ventricle tend to direct the flow of deoxygenated blood from the right atrium to the pulmonary circuit and the flow of oxygenated blood from the left atrium to the aorta. The advantage of this partial separation of pulmonary and systemic circulation is that the high resistance of the capillary beds of the gas-exchange organ no longer lies between the heart and the tissues. Therefore, the amphibian heart delivers blood to the aorta, and hence to the body, at a higher pressure than can the fish heart, which pumps the blood through the gills first.

Circulatory Systems of Reptiles and Crocodilians

Turtles, snakes, and lizards have three-chambered hearts, while crocodilians (crocodiles and alligators) have four-chambered hearts. It is an oversimplification, however, to say that turtles, snakes, and lizards have three-chambered hearts. They have two separate atria and a ventricle that is partially divided by complex anatomical features that probably influence the mixing of oxygenated and deoxygenated blood (Figure 42.4*d*). The most important and unusual feature of their hearts, however, is the ability to alter the distribution of blood going to the lungs and to the rest of the body. Because reptiles generally have lower metabolic rates than birds and mammals, they can get along without breathing continuously. When they are breathing, blood is routed both to the lungs and to the rest of the body. When they aren't breathing, they can decrease blood flow to their lungs and send most of the blood from the heart directly to the body. They apparently redirect blood flow by changing the resistance in the pulmonary circuit. Think of a single pump with several hoses coming off of it. The flow through individual hoses can be controlled by valves that alter resistance. When a snake, lizard, or turtle is not breathing, the resistance in its pulmonary circuit is higher than in its systemic circuit, and most of the blood pumped out of the ventricle goes to the body.

Have crocodilians, with their four-chambered hearts, lost this adaptation to control distribution of blood to lungs and body? Not really. All fish, amphibians, and reptiles have two aortas leaving their ventricle. In crocodilians, one aorta comes from the right ventricle and the other from the left ventricle, but they are connected to each other soon after they leave the heart. Because the ventricles are completely separate, they can generate different pressures when the heart contracts. When the animal is breathing, the pressure in the left ventricle and its aorta is higher than the pressure in the right ventricle. This higher pressure is communicated to the aorta leaving the right ventricle, where it creates a back pressure on the valve between the right ventricle and its aorta. Thus when the animal is breathing, all of the blood pumped out of its right ventricle goes to the lungs. When the animal is not breathing, the pressures in the ventricles equalize, and blood from both ventricles flows into the aortas, increasing the proportion of blood going to the body.

Crocodilians have sophisticated adaptations that enable them (like mammals and birds) to maintain a systemic blood pressure that is higher than the pulmonary blood pressure when they are breathing, but enables them (like reptiles) to redirect pulmonary blood to the systemic circuit when they are not breathing. From the circulatory perspective, it is no wonder that reptiles and crocodilians have been such a successful group of animals for such a long time.

Circulatory Systems of Birds and Mammals

Birds and mammals have four-chambered hearts and fully separated pulmonary and systemic circuits (Figure 42.4*e*). Several advantages arise from this design. First, since oxygenated and deoxygenated bloodstreams cannot mix, the systemic circuit is always receiving arterial blood with the highest oxygen content. Second, respiratory gas exchange is maximized because the blood with the lowest oxygen content and highest CO_2 content is sent to the lungs. Third, because the systemic and the pulmonary circuits are completely separate, they can operate at different pressures. Why is this important? Mammalian and bird tissues have high nutrient demands and thus a very high density of the smallest vessels, the capillaries. Many small vessels present lots of resistance to the flow of blood. Therefore, higher pressure is required to maintain adequate blood flow in the systemic circuit. The pulmonary circuit does not have such a large number of capillaries and such a high resistance, so it doesn't require such high pressure.

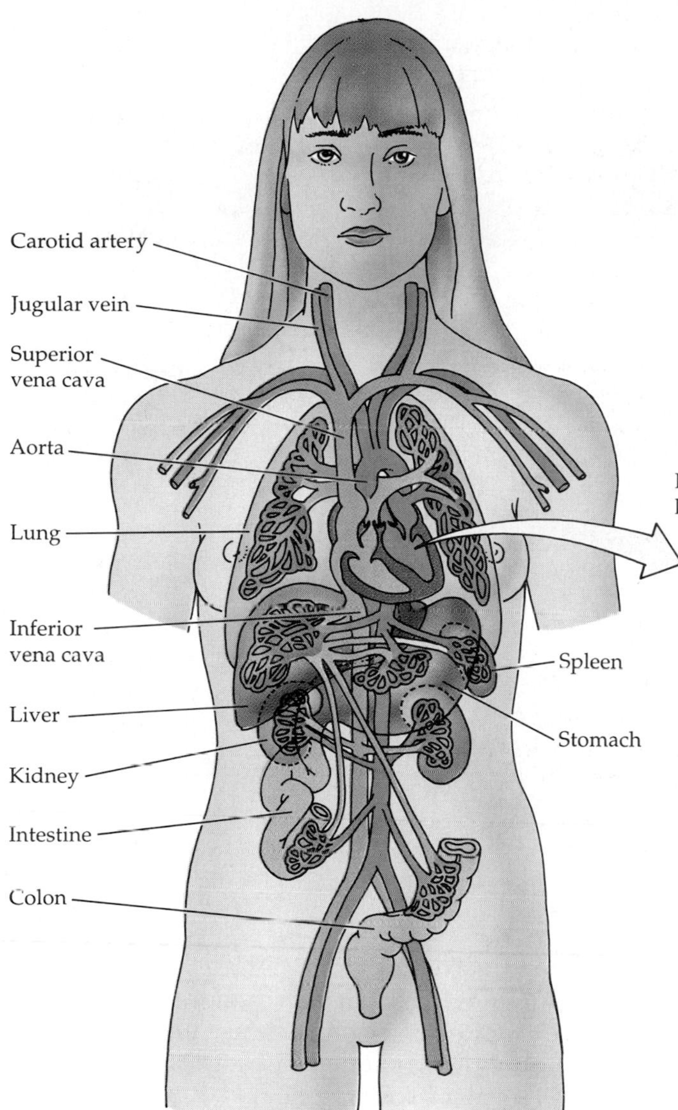

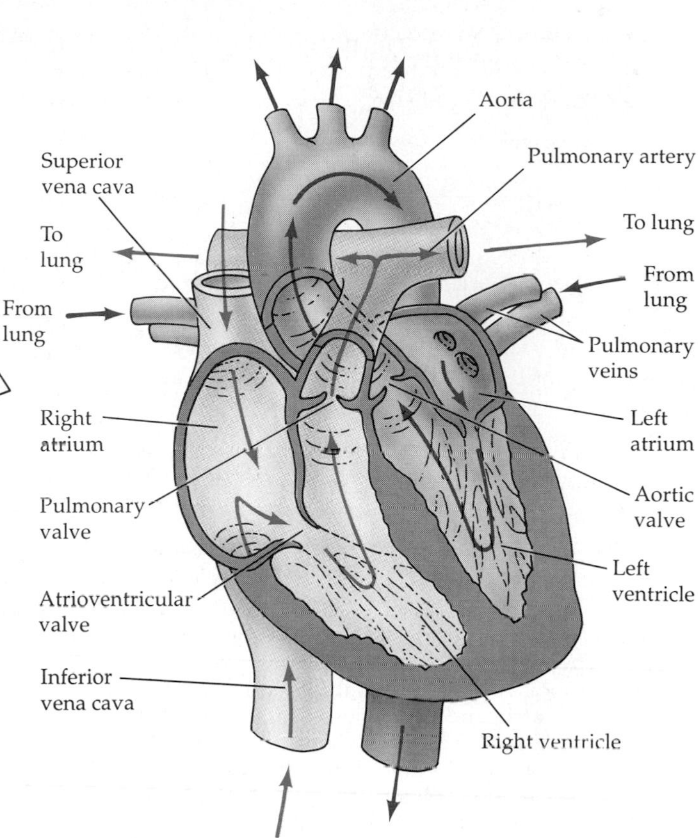

42.5 The Human Heart and Circulation
Deoxygenated blood from the tissues of the body enters the right atrium and flows through an atrioventricular valve into the right ventricle. The right ventricle pumps the blood into the pulmonary circuit, from which it returns to the left atrium and flows into the left ventricle through an atrioventricular valve. The left ventricle pumps blood into the systemic circuit. The atrioventricular valves prevent blood from flowing back into the atria when the ventricles contract. Pulmonary and aortic valves prevent blood from flowing back into ventricles from the arteries when the ventricles relax.

THE HUMAN HEART

Structure and Function

Like all mammalian hearts, the human heart has four chambers, two atria and two ventricles (Figure 42.5). The atrium and ventricle on the right side of your body are called the right atrium and right ventricle. The atrium and ventricle on the left side of your body are called the left atrium and left ventricle. Each atrium pumps blood into its respective ventricle, and the ventricles pump blood into arteries. The right ventricle pumps blood through the pulmonary circuit, and the left ventricle pumps blood through the systemic circuit. Valves between the atria and ventricles, the **atrioventricular valves**, prevent backflow of blood into the atria when the ventricles contract. The **pulmonary valves** and the **aortic valves** positioned between the ventricles and the arteries prevent the backflow of blood into the ventricles.

Let's follow the circulation of the blood through the heart. The right atrium receives blood from the **superior vena cava** and the **inferior vena cava**, large veins that collect blood from the upper and lower body, respectively. The veins of the heart itself also drain into the right atrium. The right ventricle pumps blood into the **pulmonary artery**, which transports it to the lungs. The **pulmonary veins** return the oxygenated blood from the lungs to the left atrium, from which it enters the left ventricle. The walls of the left ventricle are powerful muscles that contract around the blood with a wringing motion starting from the

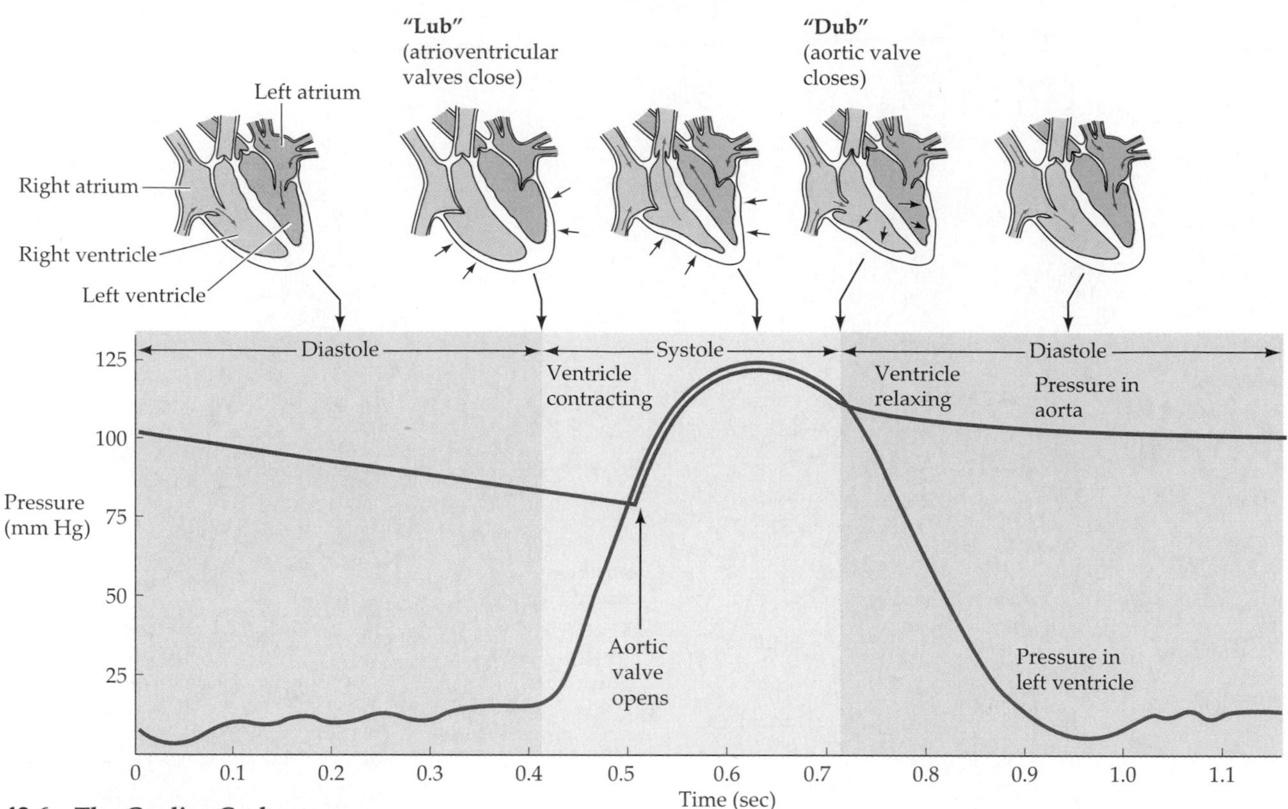

42.6 The Cardiac Cycle
The rhythmic contraction (systole) and relaxation (diastole) of the atria and ventricles is called the cardiac cycle. During diastole the heart fills with blood. Just before systole, the atria contract and maximize the filling of the ventricles. When ventricles contract, the atrioventricular valves shut (first heart sound, "lub") and pressure in the ventricles builds up until the aortic and pulmonary valves open. Blood then leaves the ventricles. At the end of systole the ventricles relax, pressure in the ventricles falls because pressure is now greater in the aorta, and the aortic and pulmonary valves slam shut (second heart sound, "dub"). Pressure in the ventricles continues to fall until the atrioventricular valves open and the heart refills.

bottom. When pressure in the left ventricle is high enough to push open the aortic valve, the blood rushes into the aorta to begin its circulation throughout the body and eventually back to the right atrium. Note in Figure 42.5 how much more massive the left ventricle is than the right ventricle. Because there are many more arterioles and capillaries in the systemic circuit than in the pulmonary circuit, the resistance is higher in the systemic circuit and the left ventricle must squeeze with greater force than the right, even though both are pumping the same volume of blood.

The pumping of the heart—contraction of the two atria followed by contraction of the two ventricles—is the **cardiac cycle**. Contraction of the ventricles is called **systole** and relaxation of the ventricles is called **diastole** (Figure 42.6). The sounds of the cardiac cycle, the "lub-dub" heard through a stethoscope placed on the chest, are created by the slamming shut of the heart valves. As the ventricles begin to contract, the atrioventricular valves close ("lub"), and when the ventricles begin to relax, the pressure in the aorta and pulmonary artery causes the aortic and pulmonary valves to bang shut ("dub"). Defective valves produce the sounds of heart murmurs. For example, if an atrioventricular valve is defective, blood will flow back into the atria with a "whoosh" sound following the "lub."

The cardiac cycle can be felt in the pulsation of arteries such as the one that supplies blood to your hand. You can feel your pulse by placing two fingers from one hand lightly over the wrist of the other hand just below the thumb. During systole, blood surges through the arteries of your arm and hand and you can feel that as a pulsing of the artery in your wrist.

You can measure blood-pressure changes associated with the cardiac cycle in the large artery in your arm by using an inflatable pressure cuff called a sphygmomanometer and a stethoscope (Figure 42.7). When the inflation pressure of the cuff exceeds maximum (systolic) blood pressure in the artery, blood flow in the artery stops. As the pressure in the cuff is gradually released, a point is reached when blood pressure at the peak of systole is greater than the pressure in the cuff. At this point, a little blood squirts through the closed artery and the artery slams

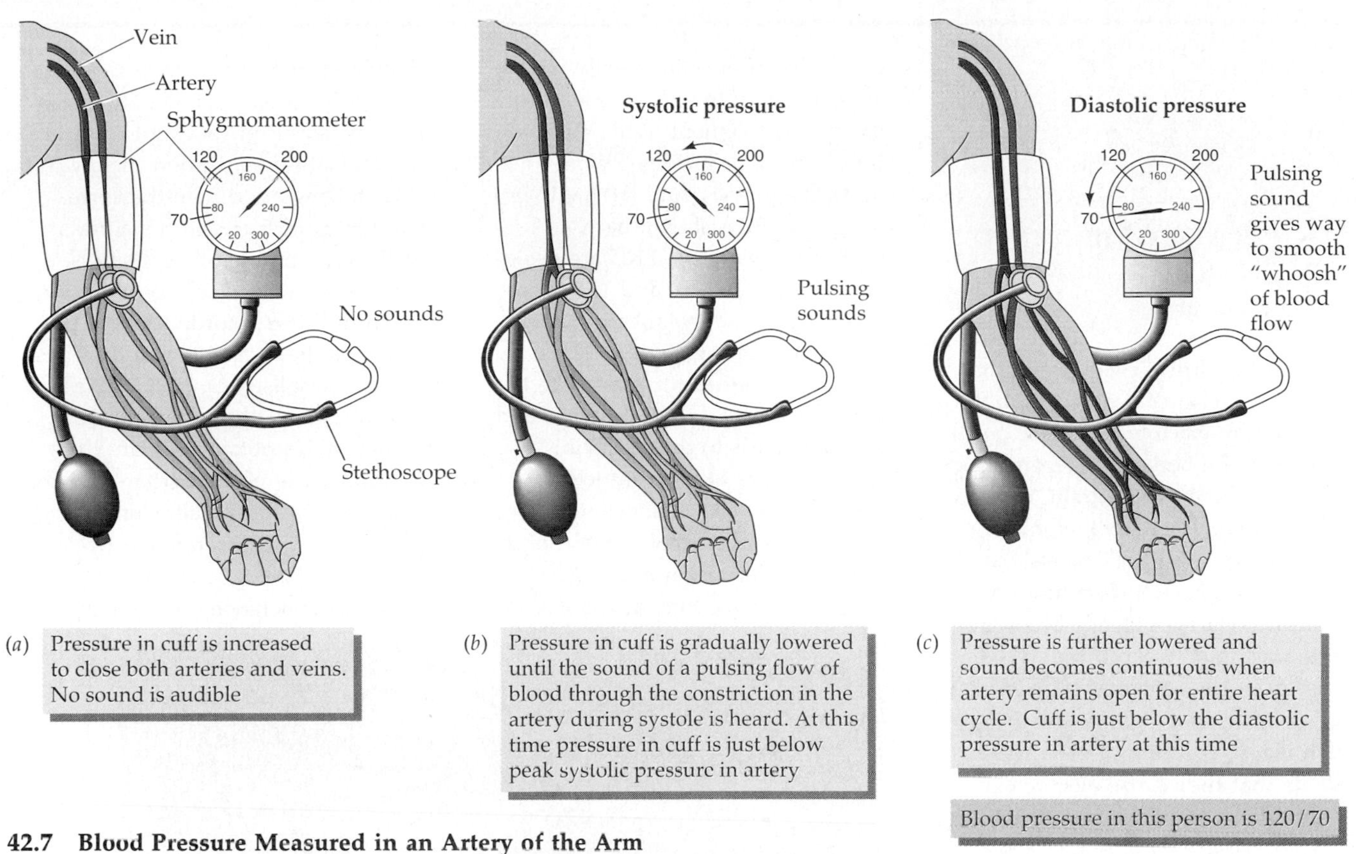

42.7 Blood Pressure Measured in an Artery of the Arm
Blood pressure in a major artery can be measured with an inflatable pressure cuff called a sphygmomanometer.

shut, producing a sound that can be heard through a stethoscope applied to the arm. The pressure at which these slamming sounds are first heard is the systolic blood pressure. As the cuff pressure is reduced even more, the slamming sounds gradually disappear to be replaced by a "whoosh" sound as the blood flow in the artery becomes more continuous. The pressure at which slamming sounds are no longer heard is the diastolic blood pressure. In a conventional blood-pressure reading, the systolic value is placed over the diastolic value. Normal values for a young adult might be 120 mm of mercury (Hg) during systole and 80 mm Hg during diastole, or 120/80.

Cardiac Muscle and the Heartbeat

As we saw in Chapter 40, some unique properties help cardiac muscle function as an effective pump. One such property is that the cardiac muscle cells are in electrical continuity with one another. Special junctions called gap junctions enable action potentials to spread rapidly from cell to cell. (Recall what we learned about gap junctions in Chapters 5 and 38.) Because a spreading action potential stimulates contraction, large groups of cardiac muscle cells contract in unison; this coordinated contraction is important for pumping blood. The massive electrical events associated with the heartbeat can be measured on the body surface as an electrocardiogram or EKG (Box 42.A).

Another important property of cardiac muscle cells is that some of them have the ability to initiate action potentials and therefore can contract spontaneously without stimulation from the nervous system. These cells stimulate neighboring cells to contract, thereby acting as pacemakers. The important characteristic of a pacemaker cell is that its resting membrane potential gradually becomes less negative (depolarizes) until it reaches the threshold voltage for initiating an action potential (see Figure 40.12). The nervous system controls the heartbeat (speeds it up or slows it down) by influencing the rate at which pacemaker cells undergo their gradual depolarization between action potentials.

Under normal circumstances the pacemaker activity of the heart originates from modified cardiac muscle cells located at the junction of the superior vena cava and right atrium, in the **sinoatrial node** (Figure 42.8). An action potential spreads from the sinoatrial node across the atrial walls, causing the two atria to contract in unison. There are no gap junctions, how-

BOX 42.A

The Electrocardiogram

During the cardiac cycle, electrical events in the cardiac muscle can be recorded by placing electrodes on the surface of the body. The recording is called an **electrocardiogram**, or EKG (*E*KG because the Greek word for heart is *kardia*, but ECG is also used). The EKG is an important tool for diagnosing heart problems. The action potentials that sweep through the muscles of the atria and the ventricles prior to their contraction are such massive, localized electrical events that they cause electric currents to flow outward from the heart to all parts of the body. Electrodes placed on the surface of the body at different locations—usually on the wrists and ankles—detect those electric currents at different times because the heart is positioned asymmetrically in the chest cavity. The appearance of the EKG depends on the exact placement of the electrodes used for the recording. Placing them on the right wrist and left ankle produced the EKG shown here.

The waves of the EKG are designated as P, Q, R, S, and T. P corresponds to the depolarization and contraction of the atrial muscles; Q, R, and S together correspond to the depolarization of the ventricles; and T corresponds to the relaxation and repolarization of the ventricles. Where is the electrical event corresponding to the repolarization of the atria? Repolarization of the atria occurs at the same time as the massive depolarization of the ventricles, so the QRS complex completely masks the smaller electrical event resulting from atrial repolarization. Below the EKG tracing are drawn the corresponding pressures in the aorta as well as the timing of the heart sounds.

From EKGs recorded after a person has a heart attack, cardiologists (heart specialists) can determine which regions of the heart were damaged. To obtain such an EKG, electrodes are positioned around the heart on the chest wall. Comparing EKGs from the different electrodes tells the cardiologist which region of the heart is behaving abnormally.

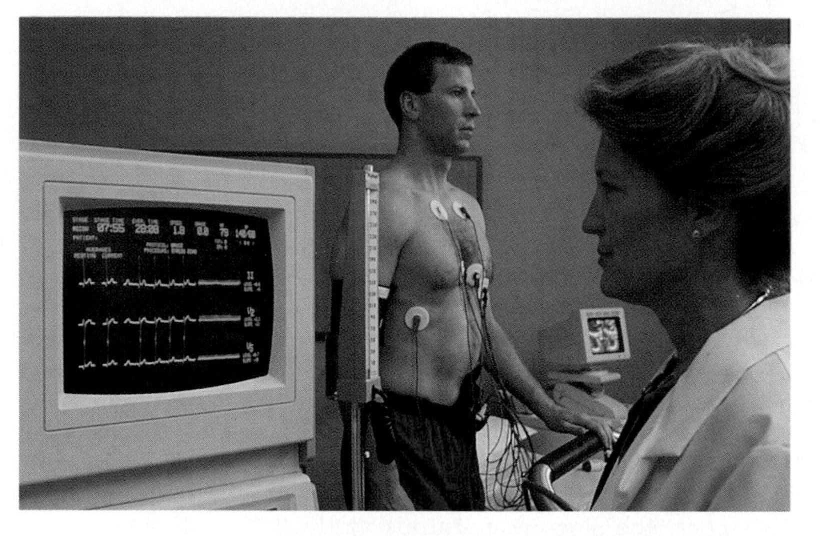

The electrocardiogram is used to monitor heart function during an exercise tolerance test.

A normal electrocardiogram (EKG)

The sounds heard through a stethoscope occur at the beginning and end of systole

Some abnormal EKGs

Besides detecting rhythmic irregularities in heartbeat (arrhythmias), EKGs can detect damage to the heart muscle (infarctions) or decreased blood supply to the heart muscle (ischemias) by changes in the size and shape of the EKG curves

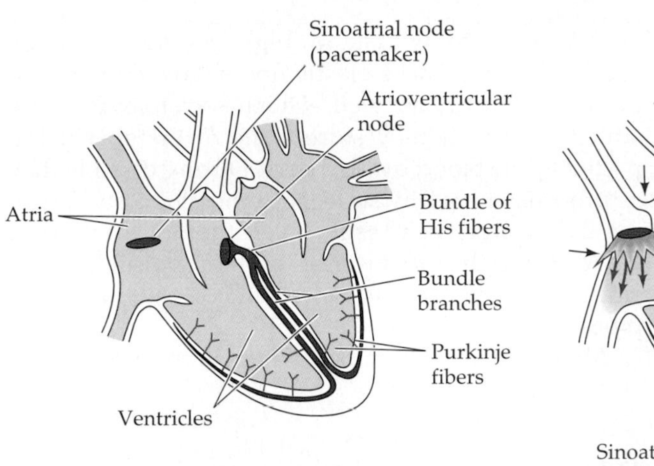

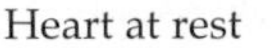

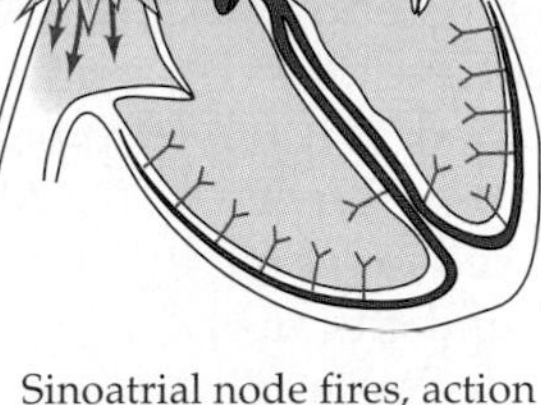

Sinoatrial node fires, action potentials spread through atria which contract

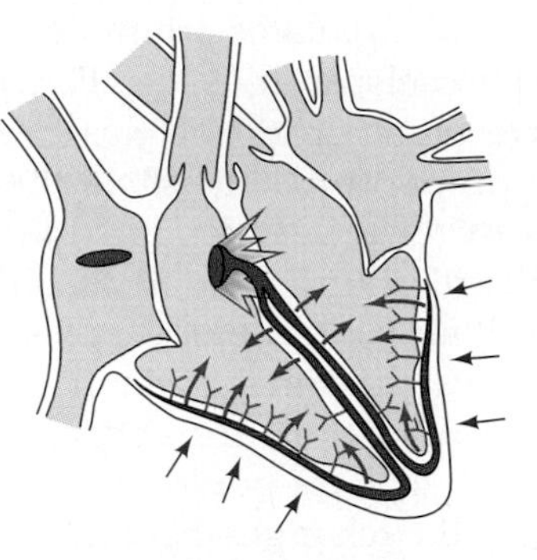

Atrioventricular node fires, sending impulses along conducting fibers; ventricles contract

42.8 The Heartbeat
Pacemaker cells in the sinoatrial node initiate action potentials that spread through the walls of the atria, causing them to contract. Because the walls of the ventricles are not in electrical continuity with the atrial muscle tissue, the action potentials initiated in the pacemaker must pass through the atrioventricular node. When cells of the atrioventricular node fire action potentials, they spread rapidly through the bundle of His and Purkinje fibers to all regions of the ventricular muscle, causing it to contract.

ever, between the atria and the ventricles. The action potential initiated in the atria passes to the ventricles through another node of modified cardiac muscle cells, the **atrioventricular node**. The atrioventricular node passes the action potential on to the ventricles via modified muscle fibers called the **bundle of His**. The bundle of His divides into right and left bundle branches, which connect with **Purkinje fibers** that branch throughout the ventricular muscle.

The timing of the spread of the action potential from atria to ventricles is important. The atrioventricular node imposes a short delay in the spread of the action potential from atria to ventricles. Then the action potential spreads very rapidly throughout the ventricles, causing them to contract. Thus the atria contract before the ventricles do, so the blood passes progressively from the atria to the ventricles to the arteries.

Control of the Heartbeat

The activity of pacemaker cells, and therefore the heart rate, is altered by acetylcholine and norepinephrine released by the autonomic nervous system (Figure 42.9; see also Figure 40.12). Parasympathetic nerves can release acetylcholine onto the sinoatrial and atrioventricular nodes. Acetylcholine slows the pace of action potential generation, thereby slowing the heartbeat. Overactivity of the parasympathetic system can even lead to fainting, which is referred to as a **vagal reaction** because parasympathetic fibers reach the heart from the brain via a nerve called the vagus nerve. A vagal reaction can be stimulated by deeply felt grief or by having blood withdrawn from a vein.

The opposite effect—speeding up of the heartbeat—is stimulated when sympathetic nerves release norepinephrine onto the cells of the sinoatrial and atrioventricular nodes. An increase in sympathetic nervous system activity also elevates the level of epinephrine (adrenaline) in the blood, which contributes to the excitatory effect on the heart. Norepinephrine and epinephrine also strengthen the contractions of the cardiac muscle cells so that more blood is ejected per beat.

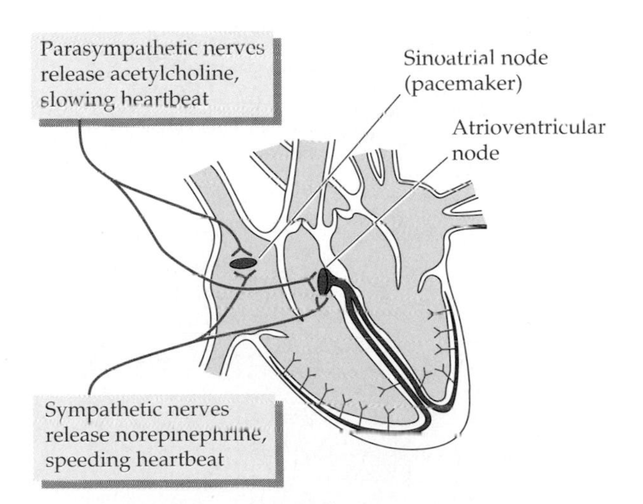

42.9 Changing the Heartbeat through the Autonomic Nervous System
Parasympathetic nerves slow the pacemaker; sympathetic nerves speed up the pacemaker.

THE VASCULAR SYSTEM

The blood circulates around the body in a system of blood vessels: arteries, capillaries, and veins. Arteries receive blood from the heart, and accordingly they have properties that enable them to withstand high pressure. The arteries are important in controlling the distribution of blood to different organs and in controlling central blood pressure. Veins have characteristics that enable them to return blood to the heart at low pressure and also to serve as a blood reservoir. The properties of capillaries make them the site of all exchanges between the blood and the internal environment. It is important to understand how the structure of the different vessels supports their functions.

Arteries and Arterioles

Blood pressure is highest in the vessels that carry blood away from the heart—the arteries and arterioles—and their structure reflects this fact. The walls of the large arteries have many elastic fibers that enable them to withstand high pressures (Figure 42.10, top left). These elastic fibers have another important function as well. During systole they are stretched and thereby store some of the energy imparted to the blood by the heart. During diastole they return this energy by squeezing the blood and pushing it forward. As a result, even though the flow of blood through the arterial system pulsates, it is smoother than it would be through a system of rigid pipes.

Abundant smooth muscle fibers in the arteries and

42.10 Anatomy of Blood Vessels
The anatomical characteristics of blood vessels match their functions. Arteries have lots of elastic fibers and muscle fibers (top left). They must withstand high pressures. Arterioles have muscle fibers that control blood flow to different capillary beds. The total cross sectional area of capillaries is larger than for any other class of vessels and they are more permeable, thus suiting them for their function of exchange of nutrients and wastes with the extracellular fluids. Because veins (top right) operate under low pressure, they have valves to prevent backflow of blood. As veins get larger they have more elastic fibers, enabling them to accommodate changing volumes of blood.

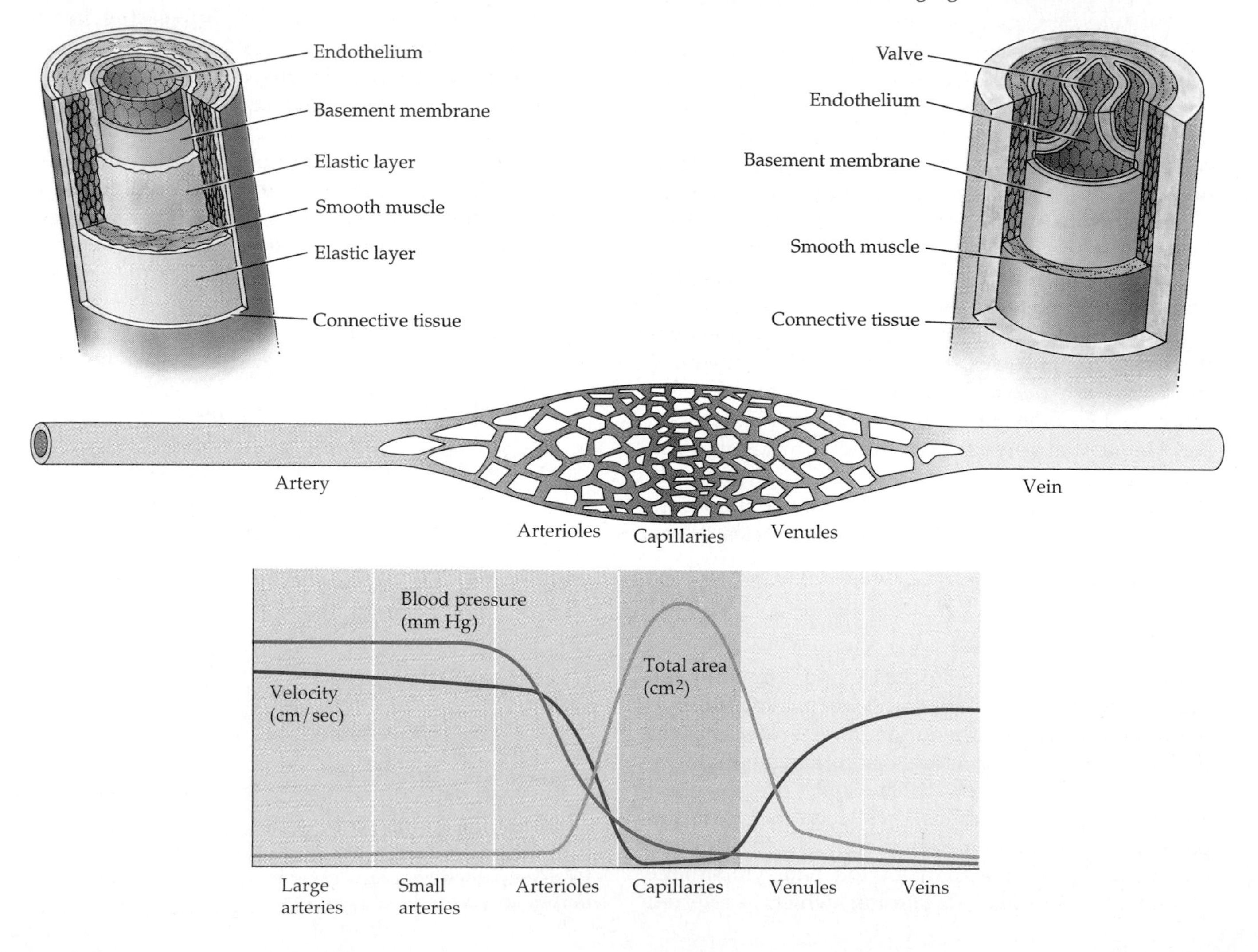

BOX 42.B

Vascular Disease

Vascular disease is by far the largest single killer in the developed Western world; it is responsible for about half the deaths each year. The immediate cause of most of these deaths is heart attack or stroke, but those events are the end result of a disease called **atherosclerosis** (hardening of the arteries) that begins many years before symptoms are detected. Hence atherosclerosis is called the silent killer. What is atherosclerosis, and how can it be prevented?

Healthy arteries have a smooth internal lining of endothelial cells (see Figure 42.10). This lining can be damaged by chronic high blood pressure, smoking, a high-fat diet, and other causes. Fatty deposits called **plaque** begin to form at sites of endothelial damage. First the endothelial cells at the damaged site swell and proliferate; then they are joined by smooth muscle cells migrating from below. Lipids, especially cholesterol, are deposited in these cells so that the plaque becomes fatty. Fibrous connective tissue invades the plaque, and along with deposits of calcium makes the artery wall less elastic; this process is what gives us the terms artherosclerosis and hardening of the arteries.

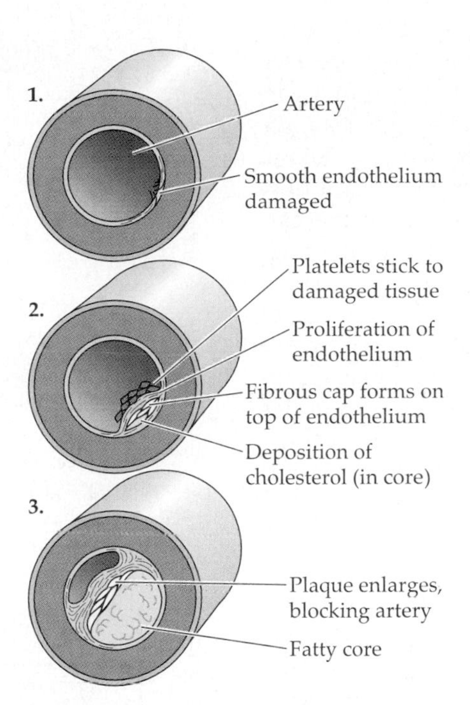

The development of atherosclerotic plaque in an artery.

The growing plaque narrows the artery and causes turbulence in the blood flowing over it. Blood platelets, which are discussed later in this chapter, stick to the plaque and initiate the formation of a blood clot, called a **thrombus**, that further blocks the artery. The blood supply to the heart itself flows through the **coronary arteries**. These arteries are highly susceptible to atherosclerosis; as they narrow, blood flow to the heart muscles decreases. Chest pains and shortness of breath during mild exertion are symptoms of this condition. A person with atherosclerosis is at high risk of forming a thrombus in a coronary artery. Such a **coronary thrombosis** can totally block the vessel, causing a heart attack, or **coronary infarction**. A piece of a thrombus breaking loose, called an **embolus,** is likely to travel to and become lodged in a vessel of smaller diameter, blocking its flow (an **embolism**). Arteries already narrowed by plaque formation are likely places for an embolus to lodge. If an embolism occurs in an artery in the brain, the cells fed by that artery die. This is called a stroke. The specific damage resulting from a stroke, such as memory loss, speech impairment, or paralysis, depends on the location of the blocked artery.

The most important solution to vascular disease is prevention, not treatment. The risk factors for developing atherosclerosis are: high-fat and high-cholesterol diet, smoking, a sedentary lifestyle, **hypertension** (high blood pressure), obesity, certain medical conditions such as diabetes, and genetic predisposition. There is not much you can do about the genes you inherit or about some forms of diabetes, but the other risk factors can be avoided, thus decreasing the significance of genetic predisposition. It is never too early to take steps to prevent atherosclerosis. Many American children are overweight, and up to 25 percent may have cholesterol levels that are too high. Many teenagers already have well-developed plaques in their arteries. Changes in diet and behavior can prevent and reverse these trends and help to fend off the silent killer.

arterioles contract and expand the diameter of those vessels. These changes in diameter alter the resistance of the vessels, which controls the flow of blood. By increasing and decreasing the resistance of these vessels, neural and hormonal mechanisms control the distribution of blood to different tissues of the body and control central blood pressure. The arteries and arterioles are called the **resistance vessels** because their resistance varies. Diseases of the arteries cause about half of the human deaths each year in developed countries (Box 42.B).

Capillaries

Beds of capillaries connect arterioles to venules. No cell of the body is more than a couple of cell diameters away from a capillary. The needs of cells are served by the exchange of materials between blood and interstitial fluid. This exchange takes place across the capillary walls. It is possible because capillaries have thin, permeable walls and because blood flows through them slowly under very low pressure (see Figure 42.10). To anyone who has played with a garden hose, it may seem strange that big arteries have high pressure and fast flow and that when the blood flows into the small capillaries the pressure and flow decrease. When you restrict the diameter of the garden hose by placing your thumb over the opening, the pressure in the hose increases, which in turn increases the velocity of the water spraying out of the hose. This puzzle is resolved by one more piece of information. Arterioles branch into so many capillaries that the total cross-sectional area of capillaries is much greater than that of any other class of vessels. Even though each capillary is so small that the red blood cells pass through in single file (Figure 42.11), each arteriole gives rise to such a large number of capillaries that together they have a much greater capacity for blood than do the arterioles. An analogy is a fast-flowing river dividing up into many small rivulets flowing across a flat, broad delta. Each rivulet may be small and its flow sluggish, yet all together they accommodate all of the water poured into the delta by the river.

Exchange in Capillary Beds

The walls of capillaries are permeable to water and small molecules, but not to large molecules such as proteins. Blood pressure therefore tends to squeeze water and small molecules out of the capillaries and into the surrounding interstitial spaces. This process is filtration. The large molecules that cannot cross the capillary wall create an osmotic potential (also called osmotic pressure) that tends to draw water back into the capillary.

Blood pressure is highest on the arterial side of a capillary bed and steadily decreases as the blood flows to the venous side. Therefore, more water is squeezed out of the capillaries on the arterial side of the bed. The osmotic potential pulling water back into the capillary gradually becomes the dominant force as the blood flows toward the venous side of the bed. The interactions of the two opposing forces—the blood pressure versus the osmotic potential—determines the net flow of water (Figure 42.12).

The balance between blood pressure and osmotic potential changes if the blood pressure in the arterioles and the permeability of the capillary walls change. An example of such a change is associated with the inflammation that accompanies injuries to the skin or allergic reactions. The inflamed area becomes hot and red because blood flow to the area

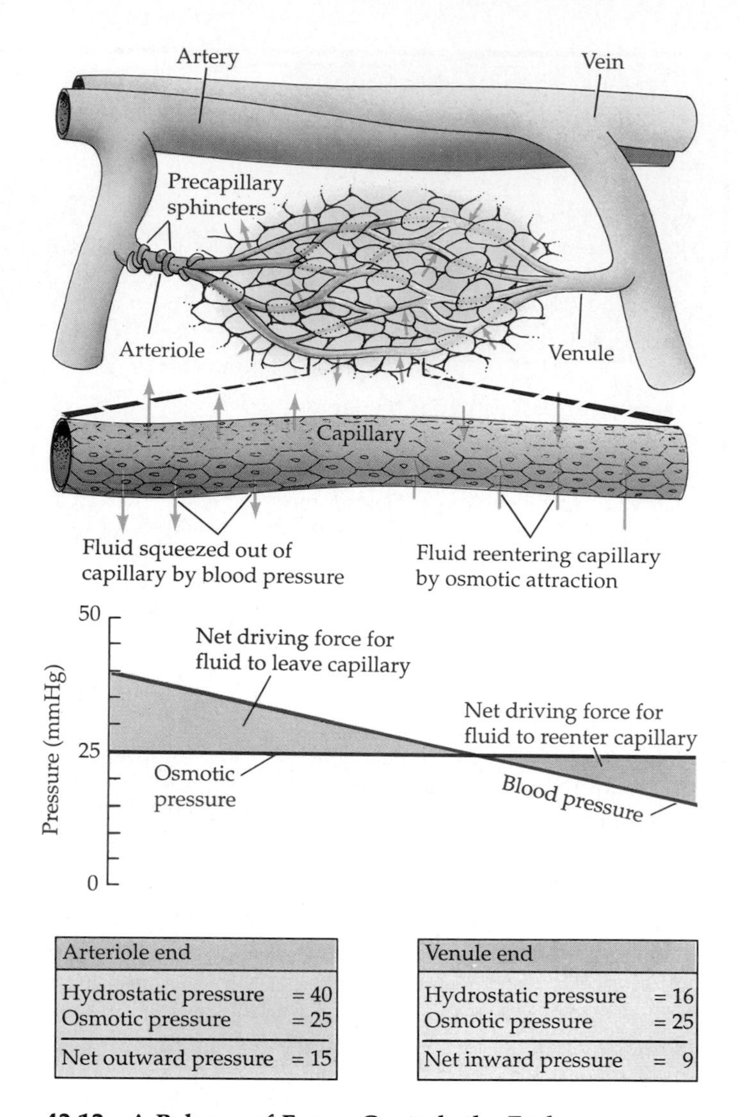

Arteriole end	
Hydrostatic pressure	= 40
Osmotic pressure	= 25
Net outward pressure	= 15

Venule end	
Hydrostatic pressure	= 16
Osmotic pressure	= 25
Net inward pressure	= 9

42.12 A Balance of Forces Controls the Exchange of Fluids between Blood Vessels and Interstitial Space
Fluids are squeezed out of capillaries by blood pressure and pulled back in by osmotic pressure created by large molecules that cannot leave the capillaries.

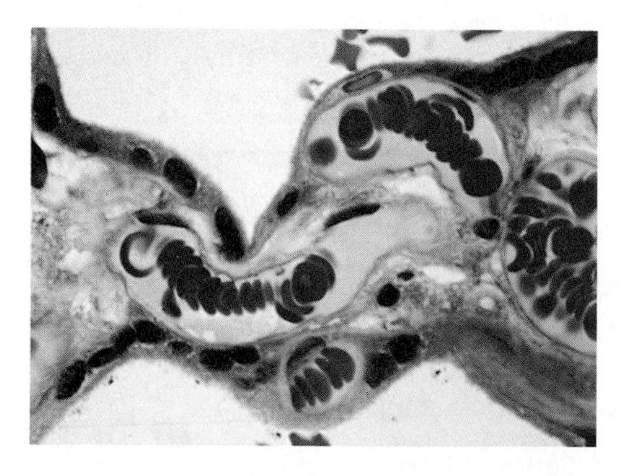

42.11 A Narrow Lane
Red blood cells—the disclike structures—pass through capillaries slowly and in single file.

increases. The inflamed tissue also swells. The major cause of these events is a chemical called **histamine** that is released mainly by certain white blood cells flowing through the damaged tissue. Histamine makes blood vessels expand, thus increasing blood flow to the area and increasing pressure in the capillaries. Because histamine also increases the permeability of the capillaries and venules, more water leaves the capillaries and venules, and the tissue swells due to the accumulation of interstitial fluids, a condition known as **edema**. The use of drugs called **antihistamines** can alleviate inflammation and allergic reactions.

The loss of water from the capillaries increases if the osmotic potential of the blood decreases, as is seen in the disease kwashiorkor. This disease is caused by severe protein starvation. When the body has no amino acids available for the synthesis of essential proteins, it begins to break down its own blood proteins. Thus fewer molecules are available in the blood to maintain the osmotic potential that pulls water back into the capillaries. The result is that interstitial fluids build up, swelling the abdomen and the extremities (Figure 42.13).

Whether specific small molecules cross a capillary wall depends on the architecture of the capillary, the type of substance, and the concentration difference between the blood and the interstitial fluid. Capillary walls are membranous, and lipid-soluble substances pass freely across a membrane from the area of higher concentration to that of lower concentration (see Chapter 5). Consider what happens in the capillary beds of skeletal muscle tissue. Because the concentration of oxygen is high in the blood coming from the arteriole but very low in active skeletal muscle tissue, oxygen readily moves from the blood into the muscle. At the same time, carbon dioxide rapidly moves into the blood because its concentration is high in the working muscle but low in the blood. The concentrations of these gases in the blood thus change rapidly as the blood travels through the capillary beds.

Small molecules in the blood generally can pass through the capillary walls, but the capillaries in different tissues are differentially selective to the sizes of molecules that can pass through them. In all capillaries, O_2, CO_2, glucose, lactate, and small ions such as Na^+ and Cl^- can cross. In the capillaries of the brain, not much else can cross unless it is a lipid-soluble substance such as alcohol; this high selectivity of brain capillaries is known as the blood–brain barrier (see Chapter 38). In other tissues the capillaries are much less selective and even have pores to permit the passage of large molecules. Such capillaries are found in the digestive tract, where nutrients are absorbed, and in the kidneys, where wastes are filtered. Some capillaries have large gaps that permit the movement of even larger substances. These capillaries are found in the bone marrow, spleen, and liver. Substances can move across many capillary walls by endocytosis (see Chapter 4).

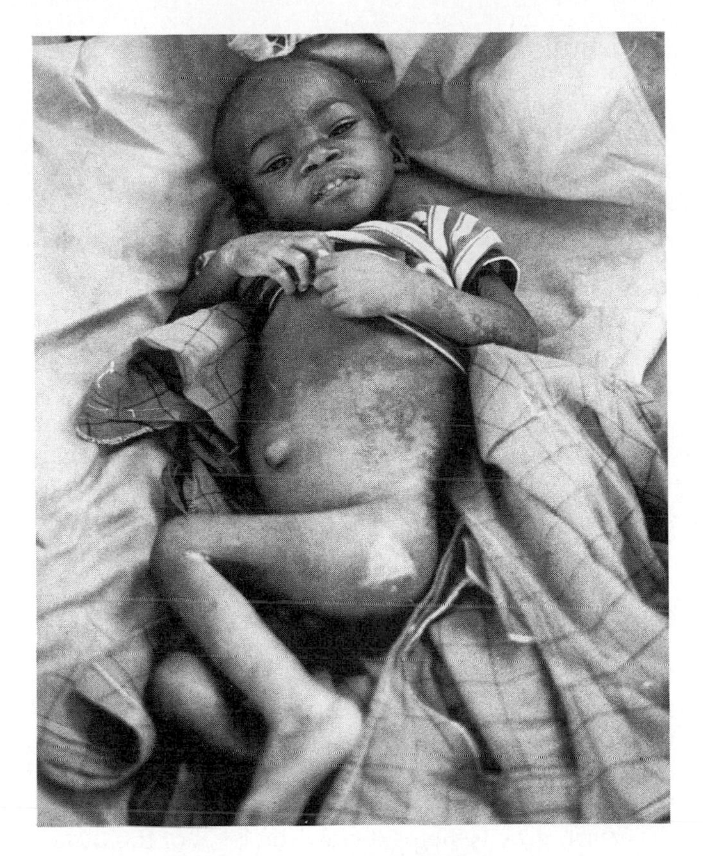

42.13 Kwashiorkor, "The Rejected One"
A child may show the symptoms of protein malnutrition soon after breast-feeding is terminated if the new diet lacks protein. Swollen abdomen, face, hands, and feet due to edema and spindly limbs are hallmarks of serious protein starvation. The limbs are spindly because the body breaks down muscle tissue, as well as blood proteins, to obtain needed amino acids.

The Lymphatic System

The interstitial fluid that accumulates outside the capillaries contains water and small molecules but no red blood cells and less protein than is in blood. A separate system of vessels—the **lymphatic system**—returns the interstitial fluid to the blood. After entering the lymphatic vessels, the interstitial fluid is called **lymph**. Fine lymphatic capillaries merge progressively into larger and larger vessels and end in a major vessel—the **thoracic duct**—which empties into the superior vena cava returning blood to the heart (see Figure 16.3). Lymphatic vessels have one-way valves that keep the lymph flowing toward the thoracic duct. The propelling force moving the lymph is pressure on the lymphatic vessels from the contractions of nearby skeletal muscles.

Mammals and birds have lymph nodes along the major lymphatic vessels. Lymph nodes are an important component of the defensive machinery of the body. They are a major site of lymphocyte production and of the phagocytic action that removes microorganisms and other foreign materials from the circulation. The lymph nodes also act as mechanical filters. Particles become trapped there and are digested by the phagocytes that are abundant in the nodes. Lymph nodes swell during infection. Some of them, particularly those on the side of the neck or in the armpit, become noticeable at such times. The nodes also trap metastasized cancer cells, that is, those that have broken free of the original tumor. Because such cells may start additional tumors, surgeons often remove the neighboring lymph nodes when they excise a malignant tumor.

Venous Return

Blood flows back to the heart through the veins, but what propels it? The pressure of the blood flowing from capillaries to venules is extremely low, so it cannot be the beating of the heart that propels blood through the veins. If the veins are above the level of the heart, gravity helps; below the level of the heart, blood must be moved against the pull of gravity. In actuality, blood tends to accumulate in veins, and the walls of veins are more expandable than the walls of arteries. As much as 80 percent of the total blood volume may be in the veins at any one time. Veins are called **capacitance vessels** because of their high capacity to store blood.

Blood must be returned from the veins so that circulation can continue. If too much blood remains in the veins, then too little blood returns to the heart, and thus too little blood is pumped to the brain; a person may faint as a result. Fainting is self-correcting because a fainting person falls, thereby changing from the position in which gravity caused blood to accumulate in the lower body. There are means other than fainting, however, by which blood is moved from the tissues back to the heart.

The most important of the forces that propel venous and lymphatic return from the regions of the body below the heart is the milking action caused by skeletal muscle contraction around the vessels. As muscles contract, the vessels are squeezed and the blood moves through them. Blood flow might be temporarily obstructed during a prolonged muscle contraction, but with relaxation of the muscles the blood is free to move again. Within the veins are valves that prevent the backflow of blood. Thus whenever a vein is squeezed, blood is propelled forward because the valves prevent it from flowing backward. In this way blood is gradually pushed toward the heart (Figure 42.14). As we already noted, the lymphatic vessels have similar valves.

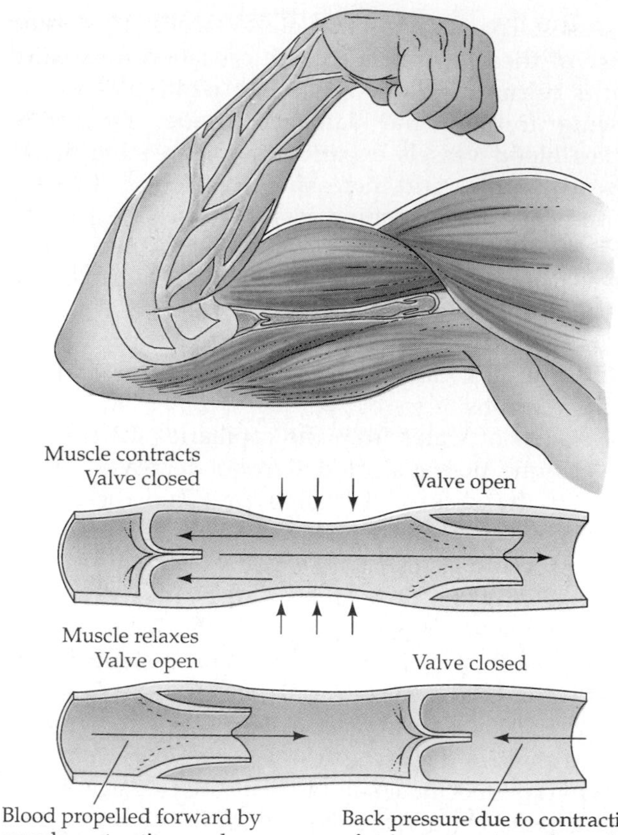

42.14 One-Way Flow
Contractions of skeletal muscles squeeze the veins. This squeezing moves the blood in the veins toward the heart because of one-way valves that prevent backflow.

People who must stand still for prolonged periods, thereby accumulating blood in the veins of the lower body, appreciate the role of muscle contraction in venous return. The guards at Buckingham Palace in England, for instance, must shift their weight and contract their leg muscles periodically to prevent the extreme lowering of blood flow to the heart that causes fainting. Gravity causes edema as well as blood accumulation in veins. The back pressure that builds up in the capillaries when blood accumulates in the veins shifts the balance between blood pressure and osmotic potential so that there is a net movement of fluid into the interstitial spaces. This is why you have trouble putting your shoes back on after you sit for a long time with your shoes off, such as on an airline flight. In persons with very expandable veins, the veins may become so stretched that the valves can no longer prevent backflow. This condition produces varicose (swollen) veins. Draining these veins is highly desirable and can be aided by wearing support hose and periodically elevating the legs above the level of the heart.

During exercise, the milking action of muscles speeds blood toward the heart to be pumped to the lungs and then to the respiring tissues. As an animal runs, its legs act as auxiliary vascular pumps, returning blood to the heart from the veins of the lower body. As a greater volume of blood returns to the heart it contracts more forcefully, and its pumping action becomes more effective. This strengthening of the heartbeat is due to a property of cardiac muscle fibers referred to as the **Frank–Starling law**: If the fibers are stretched, as they are when the volume of returning blood increases, they contract more forcefully. This principle holds (within a certain range) whenever venous return increases, by any mechanism.

The actions of breathing also help return venous blood to the heart. The ventilatory muscles create suction that pulls air into the lungs (see Chapter 41), and this suction also pulls blood and lymph toward the chest, increasing venous return to the right atrium.

Some smooth muscle in the walls of the veins moves venous blood back to the heart by constricting the veins and moving the blood forward. These muscles are rare in most of the veins and are totally absent from lymphatic vessels in humans. They do not play a major role in venous return. However, in the largest veins closest to the heart, smooth muscle contraction at the onset of exercise can suddenly increase venous return and stimulate the heart in accord with the Frank–Starling law, thus increasing cardiac output.

THE BLOOD

We have considered the circulation of the blood in detail without looking closely at the blood itself. Blood is a tissue; it has cellular elements suspended in an aqueous medium of specific, yet complex, composition. The cells of the blood can be separated from the aqueous medium, called **plasma**, by centrifugation (Figure 42.15). If we take a 100-milliliter sample of blood and spin it in a centrifuge, all the cells move to the bottom of the tube, leaving the straw-colored, clear plasma on top. The **packed cell volume** or **hematocrit**, is the percentage of the blood volume made up by cells. Normal hematocrit is about 38 percent for women and 42 percent for men, but the values can vary considerably. They are usually higher, for example, in people living and doing heavy work at high altitude. We will consider next the three classes of cellular elements in blood: the red blood cells (erythrocytes); the white blood cells (leukocytes); and the platelets, which are pinched-off fragments of cells.

Red Blood Cells

Most of the cells in the blood are **erythrocytes**, or red blood cells. The function of red blood cells is to transport the respiratory gases (see Chapter 41). There are about 5 million red blood cells per milliliter of blood but only 5,000 to 10,000 white blood cells in the same volume. Red blood cells form from special cells called

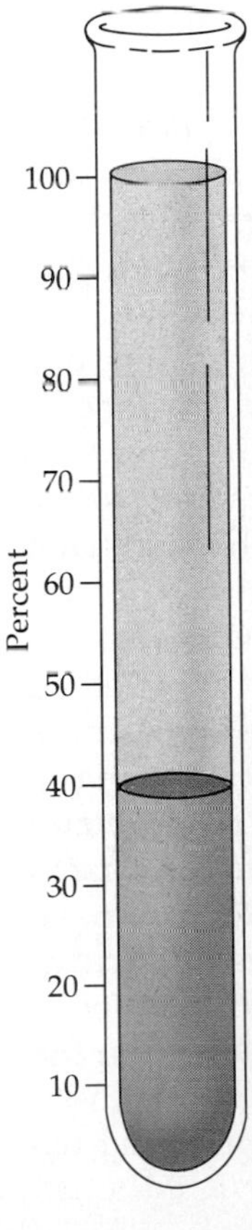

Withdraw blood from arm, place in test tube, and centrifuge

Percent: 100, 90, 80, 70, 60, 50, 40, 30, 20, 10

Plasma portion

	Water	Salts	Plasma proteins	Transported by blood
Components	Water	Salts: Sodium, potassium, calcium, magnesium, chloride, bicarbonate	Plasma proteins: Albumin, Fibrinogen, Immunoglobulins	Transported by blood: Nutrients (e.g. glucose, vitamins); Waste products of metabolism; Respiratory gases (O_2 and CO_2); Hormones; Heat
Functions	Solvent	Osmotic balance, pH buffering, regulation of membrane potentials	Osmotic balance, pH buffering, clotting, immune responses	

Cellular portion (hematocrit)

	Erythrocytes	Leukocytes	Platelets
Components	Erythrocytes (red blood cells)	Leukocytes (white blood cells): Basophil, Eosinophil, Neutrophil, Lymphocyte, Monocyte	Platelets
Number per mm^3 of blood	5–6 million	5,000–10,000	250,000–400,000
Functions	Transport oxygen and carbon dioxide	Destroy foreign cells, produce antibodies; roles in allergic responses	Blood clotting

42.15 The Composition of Blood
Blood consists of a complex aqueous solution, numerous cell types, and cell fragments.

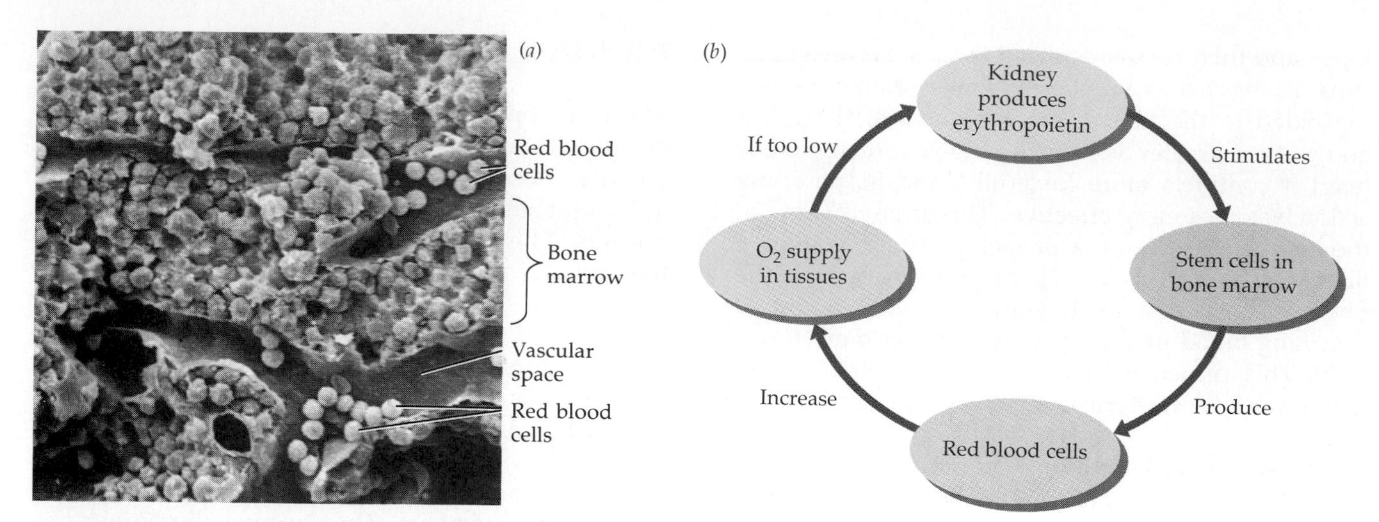

42.16 Red Blood Cells Form in the Bone Marrow
(a) In this scanning electron micrograph of a section of bone, the marrow is surrounded by fine blood vessels. As new red blood cells mature, they squeeze through the endothelium lining the vessels and enter the blood. *(b)* Erythropoietin stimulates stem cells in the bone marrow to produce red blood cells.

stem cells. Stem cells are found in the bone marrow, particularly in the ribs, breastbone, pelvis, and vertebrae (Figure 42.16*a*). Red blood cell production is controlled by a hormone, **erythropoietin**, which is released by cells in the kidney in response to insufficient oxygen (Figure 42.16*b*).

Erythropoietin stimulates stem cells to produce red blood cells. Under normal conditions your bone marrow produces about 2 million red blood cells every second. The developing red blood cells divide many times while still in the bone marrow, and during this time they are producing hemoglobin. When the hemoglobin content of a red blood cell approaches about 30 percent, its nucleus, endoplasmic reticulum, Golgi apparatus, and mitochondria begin to break down. This process is almost complete when the new red blood cell squeezes through pores in the endothelial walls of blood vessels and enters the circulation. Each red blood cell circulates for about 120 days and is then broken down. The iron from its hemoglobin molecules is recycled to the bone marrow. Mature red blood cells are biconcave flexible discs packed with hemoglobin. Their shape gives them a large surface area for gas exchange, and their flexibility enables them to squeeze through the capillaries (see Figures 41.16 and 42.11).

White Blood Cells

Leukocytes, or white blood cells, defend the body against infection (see Chapter 16). Some leukocytes search for and destroy foreign cells; some are phagocytes that consume bacteria, debris, and even dead or damaged cells from our own bodies; and some manufacture antibodies. Leukocytes squeeze through capillary walls and spend a great deal of time outside the vascular system. They move about by amoeboid motion and travel to sites of infection and cell damage by following cues from chemicals released by dead or sick cells.

Platelets and Blood Clotting

Platelets bud off from large cells in the bone marrow. A platelet is just a tiny fragment of a cell with no nucleus, but it is packed with enzymes and chemicals necessary for its function of sealing leaks in the blood vessels—that is, clotting the blood. When a vessel is damaged, collagen fibers are exposed. When a platelet encounters collagen fibers, it is activated. It swells, becomes irregularly shaped and sticky, and releases chemicals that activate other platelets and initiate the clotting of blood. The sticky platelets form a plug at the damaged site, and the subsequent clotting forms a stronger patch on the vessel.

The clotting of blood requires many steps and many **clotting factors**. The absence of any one of these factors can cause excessive bleeding and thus can be lethal. Because the liver produces most of the clotting factors, liver diseases such as hepatitis and cirrhosis can result in excessive bleeding. The sex-linked trait hemophilia (see Chapter 10) is an example of a genetic inability to produce one of the clotting factors. Blood clotting factors participate in a cascade of steps that activate other substances circulating in the blood. The cascade begins with cell damage and platelet activation and continues with the conversion of an inactive circulating enzyme, **prothrombin**, to its active form, **thrombin**. Thrombin causes circulating protein molecules called **fibrinogen** to polymerize and form **fibrin** threads. The fibrin threads form the meshwork that clots the blood cells, seals the vessel, and provides a scaffold for the formation of scar tissue (Figure 42.17).

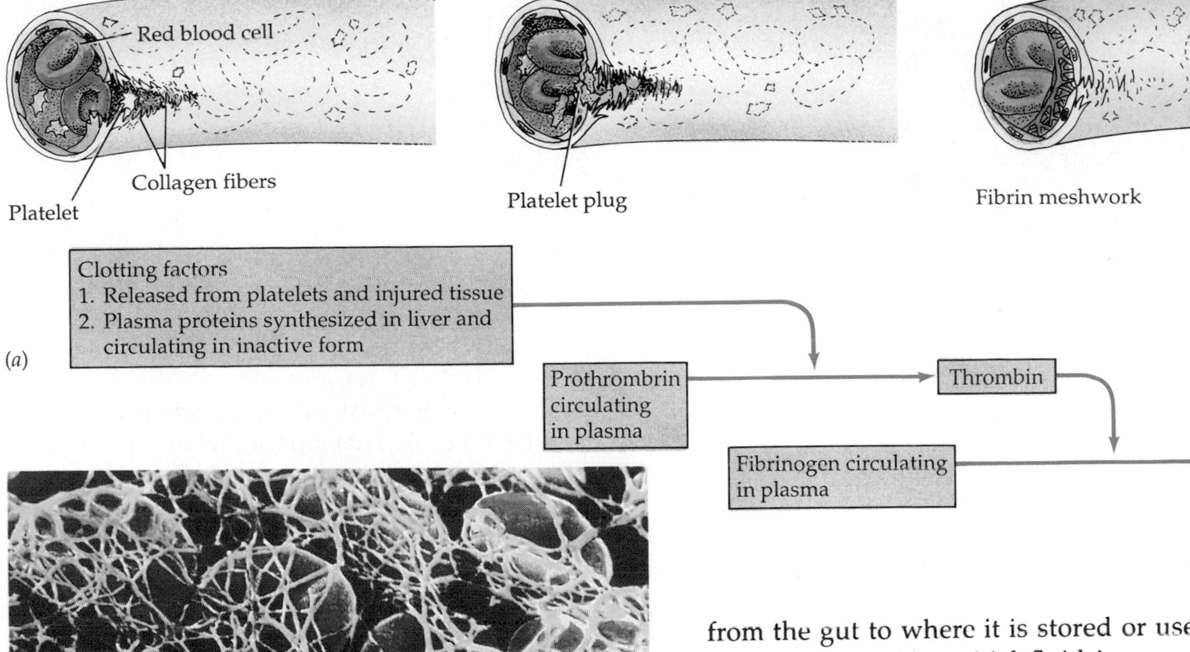

42.17 Blood Clotting
(a) Damage to a blood vessel initiates a cascade of events that produces a fibrin meshwork. *(b)* As the fibrin forms, red blood cells become enmeshed and form a clot, as this scanning electron micrograph shows.

Plasma

Plasma is a complex solution of gases, ions, nutrient molecules, and proteins. Most of the ions are Na^+ and Cl^- (hence the salty taste of blood), but many other ions are also present. Nutrient molecules in plasma include glucose, amino acids, lipids, cholesterol, and lactic acid. The circulating proteins have many functions. We have just noted proteins that function in blood clotting; others of interest include albumin, which is largely responsible for the osmotic potential in capillaries that prevents a massive loss of water from plasma to interstitial spaces; antibodies (the immunoglobulins); hormones; and various carrier molecules, such as **transferrin**, which carries iron from the gut to where it is stored or used. Plasma is very similar to interstitial fluid in composition, and most of its components move readily between these two fluid compartments of the body. The main difference between the two fluids is the higher concentration of proteins in the plasma.

CONTROL AND REGULATION OF CIRCULATION

The circulatory system is controlled and regulated at many levels. Every tissue requires an adequate supply of blood that is saturated with O_2, that carries essential nutrients, and that is relatively free of waste products. The nervous system cannot monitor and control every capillary bed in the body. Instead, each bed regulates its own blood flow through **autoregulatory mechanisms** that cause the arterioles supplying the bed to constrict or dilate.

The autoregulatory actions of every capillary bed in every tissue influence the pressure and composition of the arterial blood leaving the heart. For example, if many arterioles suddenly dilate, allowing blood to flow through many more capillary beds, arterial blood pressure falls. If all the newly filled capillary beds contribute metabolic waste products to the blood, the concentration of wastes in the blood returning to the heart increases. Thus events in all capillary beds throughout the body produce combined effects on arterial blood pressure and composition. The nervous and endocrine systems respond to changes in arterial blood pressure and composition by changing breathing, heart rate, and blood distribution to match the metabolic needs of the body.

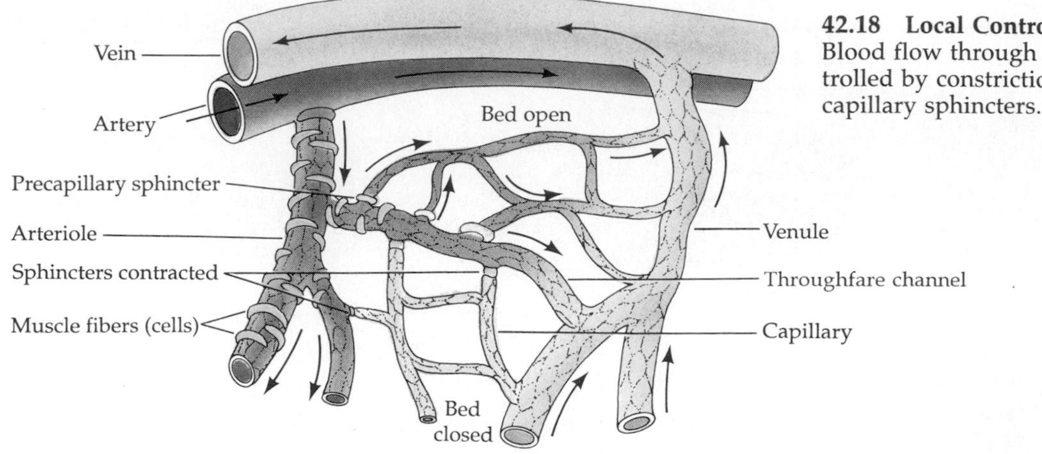

42.18 Local Control of Blood Flow Blood flow through a capillary bed is controlled by constriction of arterioles and precapillary sphincters.

Autoregulation

The autoregulatory mechanisms that adjust the flow of blood to a tissue are part of the tissue itself, but they can be influenced by the nervous system and certain hormones. The amount of blood that flows through a capillary bed is controlled by the degree of contraction of the smooth muscles of the arteries and arterioles feeding that bed: As the muscles contract, they constrict the vessels, thereby decreasing the flow. The flow of blood in a typical capillary bed is diagrammed in Figure 42.18. Blood flows into the bed from an arteriole. Smooth-muscle "cuffs," or precapillary sphincters, on the arteriole can completely shut off the supply of blood to the capillary bed. When the precapillary sphincters are relaxed and the arteriole is open, the arterial blood pressure pushes blood into the capillaries.

Autoregulation depends on the sensitivity of the smooth muscle to the composition of its chemical environment. Low O_2 concentrations and high CO_2 concentrations cause the smooth muscle to relax, thus increasing the supply of blood, which brings in more O_2 and carries away CO_2. Increases in the concentration of products of metabolism other than CO_2, such as lactate, hydrogen ions, potassium, and adenosine, also promote increased blood flow by this mechanism. Hence activities that increase the metabolism of a tissue also increase the blood flow to that tissue.

Control and Regulation by the Endocrine and Nervous Systems

The same smooth muscles of arteries and arterioles that respond to autoregulatory stimuli also respond to signals from the endocrine and central nervous systems. Most arteries and arterioles are innervated by the autonomic nervous system, particularly the sympathetic division. Most sympathetic neurons release norepinephrine, which causes the smooth muscle fibers to contract, thus constricting the vessels and increasing their resistance to blood flow. An exception is found in skeletal muscle, where specialized sympathetic neurons release acetylcholine and cause the smooth muscles of the arterioles to relax and the vessels to dilate, causing more blood to flow to the muscle.

Hormones also can cause arterioles to constrict. Epinephrine has actions similar to those of norepinephrine; it is released from the adrenal medulla during massive sympathetic activation—the fight-or-flight response. Angiotensin, produced when blood pressure to the kidneys falls, causes arterioles to constrict. Vasopressin, released by the posterior pituitary, has similar effects (Figure 42.19). These hormones influence arterioles located for the most part in peripheral tissues (extremities) or in tissues whose functions need not be maintained continuously (such as the gut). By reducing blood flow in those arterioles, the hormones increase the central blood pressure and blood flow to essential organs such as the heart, brain, and kidneys.

The autonomic nervous system activity that controls heart rate and constriction of blood vessels originates in cardiovascular centers in the medulla of the brain stem. Many inputs converge on this central integrative network and influence the commands it issues via parasympathetic and sympathetic fibers (Figure 42.20). Of special importance is information about changes in blood pressure from stretch sensors in the walls of the great arteries—the aorta and the carotid arteries. Increased activity in the stretch sensors indicates rising blood pressure and inhibits sympathetic nervous system output. As a result, the heart slows and arterioles in peripheral tissues dilate. If pressure in the great arteries falls, the activity of the stretch sensors decreases, stimulating sympathetic output. Increased sympathetic output causes the heart to beat faster and the arterioles in peripheral tissues to constrict. When arterial pressure falls, the change in stretch-sensor activity also causes the hypothalamus to release vasopressin, which helps to

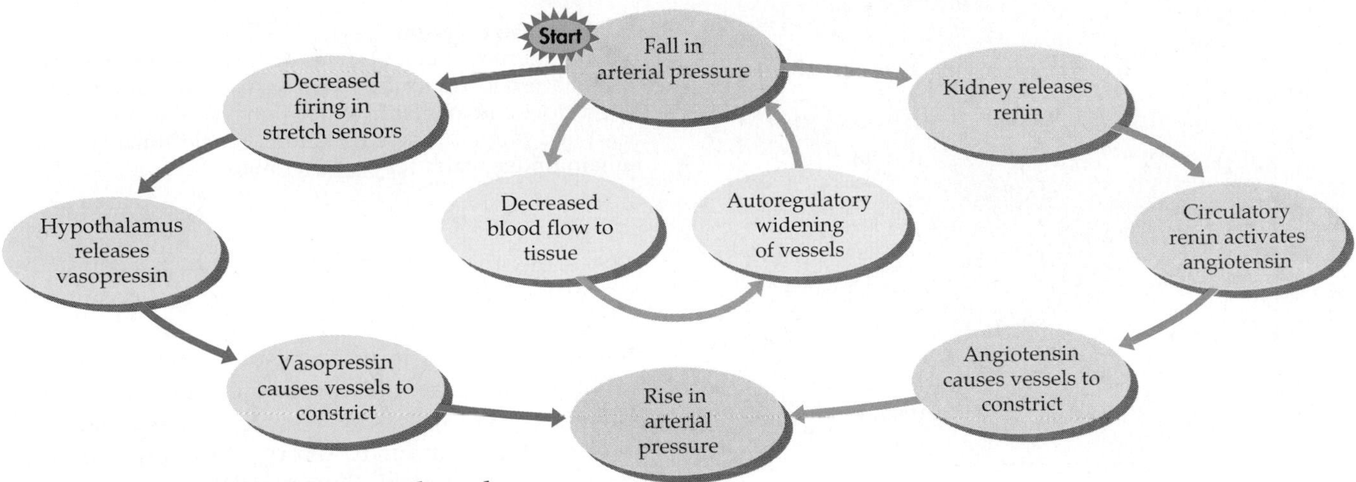

42.19 Controlling Blood Pressure through Vascular Resistance
A fall in arterial pressure reduces blood flow to tissues, resulting in local accumulation of metabolic wastes. This change in the extracellular environment stimulates autoregulatory opening of the arteries that would lead to a further fall in central blood pressure if this were not prevented by the negative feedback mechanisms shown in this diagram, which work through promoting constriction of arteries in less essential tissues.

increase blood pressure by stimulating peripheral arterioles to constrict.

Other information that causes the medullary regulatory system to increase heart rate and blood pressure comes from the carotid and aortic bodies (see Figure 41.23). These nodules of modified smooth muscle tissue are chemosensors that respond to inadequate O_2 supply. If arterial blood flow slows or the O_2 content of the arterial blood falls drastically, these sensors are activated and send signals to the regulatory center.

The regulatory center also receives input from other brain areas. Emotions and the anticipation of intense activity, such as at the start of a race, can cause the center to increase heart rate and blood pressure. A reflex that slows the heart is the so-called diving reflex, which is highly developed in marine mammals (Figure 42.21). Humans also have a diving reflex, which causes the heart to beat more slowly when the face is immersed in water.

A question we can ask about any physiological system is, "What is being regulated and how?" In the respiratory system (see Chapter 41), it is primarily the CO_2 concentration of the blood, and to a lesser extent the O_2 concentration, that is regulated by changes in the depth and frequency of breathing. Regulation in the circulatory system is more complex. The blood flow to individual tissues is regulated by local, autoregulatory mechanisms that cause dilation of local arterioles and precapillary sphincters when the tissue needs more oxygen or has accumulated

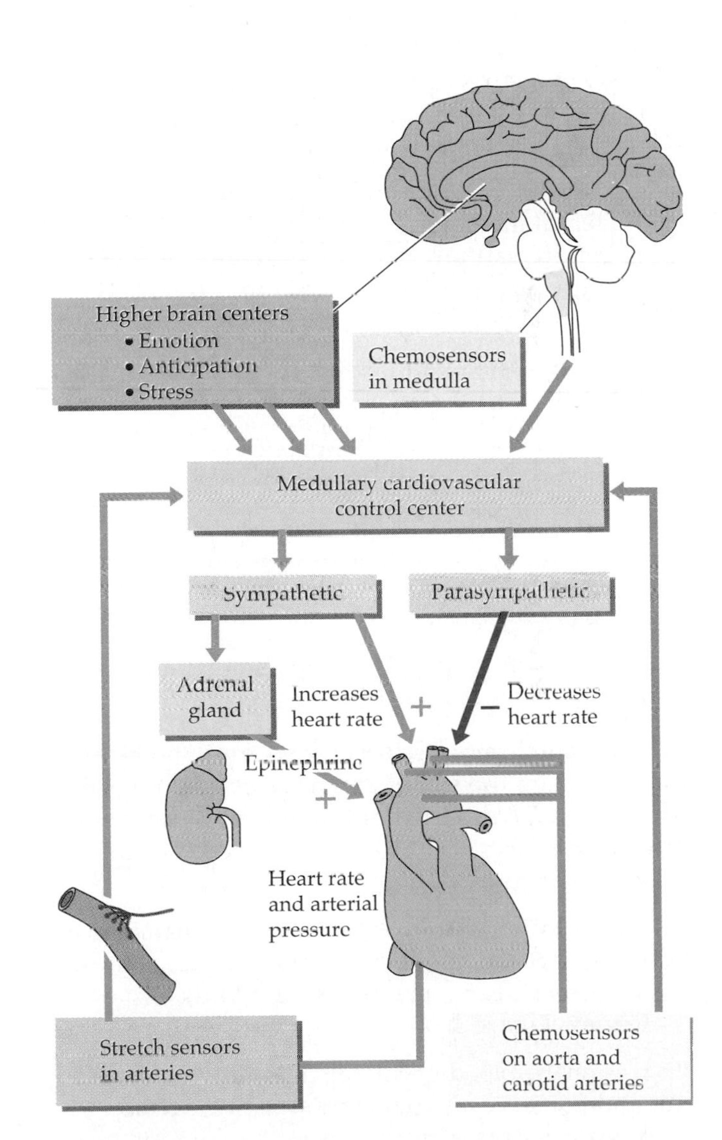

42.20 Regulating Blood Pressure
The autonomic nervous system controls heart rate in response to information about blood pressure and blood composition that is integrated by regulatory centers in the medulla.

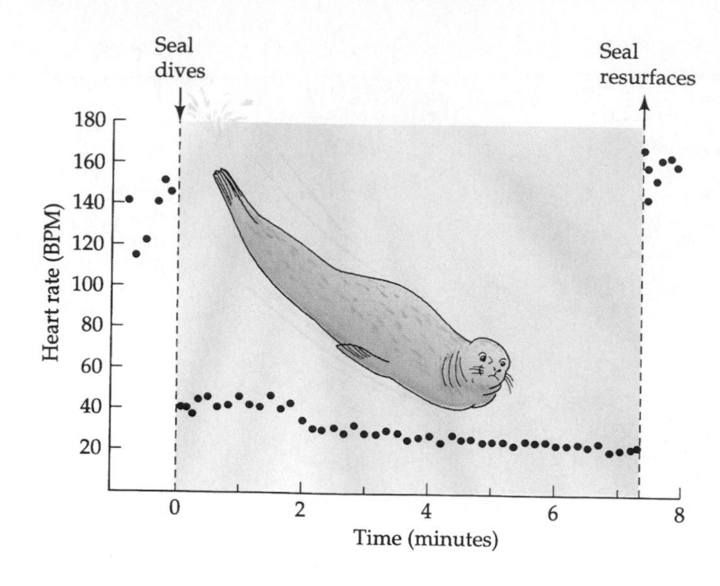

42.21 Master Divers
When a marine mammal dives, its heart rate slows. In addition, arteries to most organs constrict so that almost all blood flow and available oxygen goes to the animal's heart and brain. These adaptations enable some seals to remain under water for up to an hour.

wastes. As more blood flows into such tissues, the central blood pressure falls and the composition of the blood returning to the heart reflects the exchanges that occur in those tissues. Changes in central blood pressure and composition are sensed and both endocrine and central nervous system responses are activated in order to return blood pressure and composition to normal. Thus circulatory functions are matched to the regional and overall needs of the body.

SUMMARY of Main Ideas about Internal Transport and Circulatory Systems

The metabolic needs of the cells of very small animals are met by direct exchange of materials with the external medium. The metabolic needs of cells of larger animals are met by a circulatory system that transports nutrients, respiratory gases, and metabolic wastes.

Circulatory systems can be open or closed.

In open circulatory systems the blood or hemolymph leaves vessels and percolates through tissues.
Review Figure 42.2

In closed circulatory systems the blood is contained in a system of vessels.
Review Figure 42.3

The circulatory systems of vertebrates consist of a heart and a closed system of vessels.

Arteries and arterioles carry blood from the heart; capillaries are the site of exchange between blood and interstitial fluids; venules and veins carry blood back to the heart.
Review Figure 42.10

The vertebrate heart evolved from two chambers in fishes to three in amphibians and reptiles, and to four in crocodilians, mammals, and birds.
Review Figure 42.4

In mammals, blood circulates through two circuits: the pulmonary circuit and the systemic circuit.
Review Figure 42.5

The cardiac cycle has two components: systole, in which the ventricles contract and the "lub" sound is heard; and diastole, in which the ventricles relax and the "dub" sound is heard
Review Figure 42.6

A pacemaker (the sinoatrial node) control the cardiac cycle by initiating a wave of depolarization in the atria, which is conducted to the ventricles through the atrioventricular node.
Review Figure 42.8

The autonomic nervous system controls the pacemaker.
Review Figure 42.9

The measurement of blood pressure using a sphygmomanometer and a stethoscope is based on the cardiac cycle.
Review Figure 42.7

Capillary beds are the site of fluid exchange between the blood and the interstitial fluids.

The exchange of fluids between the blood and interstitial fluids is determined by the balance between blood pressure and osmotic potential in the capillaries.
Review Figure 42.12

Blood can be divided into a plasma portion (water, salts, and proteins) and a cellular portion (red blood cells, white blood cells, and platelets).
Review Figure 42.15

Blood cells form in the bone marrow.

Red blood cells transport oxygen.
Review Figure 42.16

White blood cells defend the body from foreign substances.

Platelets (cell fragments), along with circulating proteins, are involved in clotting responses.
Review Figure 42.17

Blood flow through capillary beds is controlled by local conditions, hormones, and the autonomic nervous system.
Review Figures 42.18 and 42.19

Heart rate is controlled by the autonomic nervous system, which responds to information about blood pressure and blood composition that is integrated by regulatory centers in the brain stem.
Review Figure 42.20

SELF-QUIZ

1. An open circulatory system is characterized by
a. the absence of a heart.
b. the absence of blood vessels.
c. blood with a composition different from that of interstitial fluid.
d. the absence of capillaries.
e. a higher pressure circuit through gills than to other organs.

2. Which of the following statements about vertebrate circulatory systems is *not* true?
a. In fish, oxygenated blood from the gills returns to the heart through the left atrium.
b. In mammals, deoxygenated blood leaves the heart through the pulmonary artery.
c. In amphibians, deoxygenated blood enters the heart through the right atrium.
d. In reptiles, the blood in the pulmonary artery has a lower oxygen content than the blood in the aorta.
e. In birds, the pressure in the aorta is higher than the pressure in the pulmonary artery.

3. Which of the following statements about the human heart is true?
a. The walls of the right ventricle are thicker than the walls of the left ventricle.
b. Blood flowing through atrioventricular valves is always deoxygenated blood.
c. The second heart sound is due to the closing of the aortic valve.
d. Blood returns to the heart from the lungs in the vena cava.
e. During systole the aortic valve is open and the pulmonary valve is closed.

4. Pacemaker actions of cardiac muscle
a. are due to opposing actions of norepinephrine and acetylcholine.
b. are localized in the bundle of His.
c. depend on the gap junctions between cells that make up the atria and those that make up the ventricles.
d. are due to spontaneous depolarization of the plasma membranes of some cardiac muscle cells.
e. result from hyperpolarization of cells in the sinoatrial node.

5. Blood flow through capillaries is slow because
a. lots of blood volume is lost from the capillaries.
b. the pressure in venules is high.
c. the total cross-sectional area of capillaries is larger than that of arterioles.
d. the osmotic pressure in capillaries is very high.
e. red blood cells are bigger than capillaries and must squeeze through.

6. How are lymphatic vessels like veins?
a. Both have nodes where they join together into larger common vessels.
b. Both carry blood under low pressure.
c. Both are capacitance vessels.
d. Both have valves.
e. Both carry fluids rich in plasma proteins.

7. The production of red blood cells
a. ceases if the hematocrit falls below normal.
b. is stimulated by erythropoietin.
c. is about equal to the production of white blood cells.
d. is inhibited by prothrombin.
e. occurs in bone marrow before birth and in lymph nodes after birth.

8. Which of the following does *not* increase blood flow through a capillary bed?
a. High concentrations of CO_2
b. High concentrations of lactate and hydrogen ions
c. Histamine
d. Vasopressin
e. Increase in arterial pressure

9. The clotting of the blood
a. is impaired in hemophiliacs because they don't produce platelets.
b. is initiated when platelets release fibrinogen.
c. involves a cascade of factors produced in the liver.
d. is initiated by leukocytes forming a meshwork.
e. requires conversion of angiotensinogen to angiotensin.

10. Autoregulation of blood flow to a tissue is due to
a. sympathetic innervation.
b. the release of vasopressin by the hypothalamus.
c. increased activity of baroreceptors.
d. chemosensors in carotid and aortic bodies.
e. the effect of local environment on arterioles.

FOR STUDY

1. How is cardiac output increased at the beginning of a race? Include the Frank–Starling law in your answer.
2. The final stages of alcoholism involve loss of liver function and accumulation of fluids in extremities and the abdominal cavity. Explain how these two consequences of alcoholism are related.
3. A sudden and massive loss of blood results in a fall in blood pressure. Describe several mechanisms that help return blood pressure to normal.
4. You can describe the cycle of events in a ventricle of the heart by a graph that plots the pressure in the ventricle on the y axis and the volume of blood in the ventricle on the x axis. What would such a graph look like? Where would the heart sounds occur on this graph? How would the graph differ for the left and the right ventricles?
5. Why doesn't diastolic blood pressure fall to zero between heartbeats? Why does systolic blood pressure increase with *(a)* sympathetic activity, *(b)* increased venous return, and *(c)* age?

READINGS

Eckert, R., D. Randall and G. Augustine. 1988. *Animal Physiology: Mechanisms and Adaptations*, 3rd Edition. W. H. Freeman, New York. An outstanding textbook of animal physiology, with excellent coverage of the circulatory system.

Golde, D. W. and J. C. Gasson. 1988. "Hormones That Stimulate the Growth of Blood Cells." *Scientific American*, July. Now made by recombinant DNA methods, hemopoietins promise to transform the practice of medicine.

Robinson, T. F., S. M. Factor and E. H. Sonnenblick. 1986. "The Heart as a Suction Pump." *Scientific American*, July. A new proposal concerning the filling of the heart, along with interesting general information on cardiac muscle and the connective tissues of the heart.

Scholander, P. F. 1963. "The Master Switch of Life." *Scientific American*, December. Delightful, classic description of the discovery of the diving adaptations of marine mammals.

Vander, A. J., J. H. Sherman and D. S. Luciano. 1994. *Human Physiology: The Mechanisms of Body Function*, 6th Edition. McGraw-Hill, New York. Chapter 14 deals with circulation.

Zapol, W. M. 1987. "Diving Adaptations of the Weddell Seal." *Scientific American*, June. These breath-holding master divers provide an opportunity to study the diving reflex.

Zucker, M. 1980. "The Functioning of Blood Platelets." *Scientific American*, June. The role of platelets in blood clotting.

43 Animal Nutrition

Shark!
To this great white shark, we are food.

"Shark!" That single word can send chills down the spines of ocean bathers and surfers around the world. Why? Because these predators can take us as prey. The image of the great white shark is that of a veritable eating machine. It can be more than 6 meters long and can take a single bite over half a meter in diameter. It attacks at considerable speed and in the final moments of its approach opens its enormous mouth, turns its head upward, and embeds rows of triangular teeth into its prey. The shark then shakes violently to rip away a huge chunk of flesh by the sawing actions of its teeth. The attack can be so violent that some of its teeth are left in the carcass. The loss of teeth is no problem, however, because new teeth continually form and advance forward as if on conveyer belts to take the place of those that are lost.

Now imagine a shark three times bigger than the great white—18 meters long, with a mouth 2 meters wide. Could such a terrifying animal exist only in horror films? No. These are the dimensions of whale sharks, which are not terrifying at all. These enormous creatures are gentle, slow swimmers that divers have been known to hitch rides on by holding onto a fin or a tail. Their huge mouths take in thousands of tons of water per hour, which they filter through their gills to extract the millions of small organisms that are their food. When feeding, the contents of a whale shark's stomach may weigh half a ton or more. What animals eat and how they eat it are some of their most distinguishing characteristics.

DEFINING ORGANISMS BY WHAT THEY EAT

Animals must eat to stay alive, and in a sense they are what they eat. Animals are **heterotrophs**: They derive both their energy and their structural molecules from their food. **Autotrophs** (most plants, some bacteria, and some protists), on the other hand, can trap solar energy through photosynthesis and use it to synthesize all of their structures from inorganic materials. Almost all life runs on solar energy, but heterotrophs receive theirs indirectly, via the autotrophs. Photosynthetic autotrophs provide the nutritional foundation for all other known ecosystems, and heterotrophs have evolved an enormous diversity of adaptations for exploiting that continual source of life (Figure 43.1).

Heterotrophic nutritional lifestyles span a great range. **Saprophytes**, such as fungi, simply absorb organic molecules from dead organisms. Rather than feeding passively, as saprophytes do, some animals actively feed on the remains of dead organisms. **Detritivores**, such as earthworms, process environmental deposits containing organic matter for their nutri-

(a)

(b)

(c)

(d)

43.1 A Focus on the Consumers
Heterotrophs have evolved an amazing range of adaptations for exploiting sources of energy. *(a)* The manatee is a herbivore whose source of food is aquatic vegetation in tropical and subtropical rivers and lagoons. *(b)* The long bill of this Australian spiny-cheeked honeyeater, a fluid feeder, enables it to harvest the tiny amounts of nectar in individual flowers. *(c)* The red file clam of the South Pacific islands obtains food from the ocean water it constantly filters through its system. *(d)* The carnivorous polar bear is a fearsome predator. A strong swimmer, it feeds on fish and seals from the frigid Arctic Ocean; its feet, with hairy soles, are adapted to moving swiftly and surely across the ice packs, so it is equally able to bring down prey, including birds and mammals, on land.

tional needs. All animals that feed on other living organisms can be considered **predators**. **Herbivores** are predators that prey on plants; **carnivores** prey on other animals. **Omnivores**, including humans, prey on both plants and animals. **Filter feeders**, such as clams and blue whales, prey on small organisms by filtering them out of the environmental medium. We are only too familiar with **fluid feeders**, which include mosquitoes, aphids, and leeches, as well as birds that feed on plant nectar. In this chapter we will examine some of the anatomical adaptations for this diversity of nutritional lifestyles.

The statement "We are what we eat" is true only in a limited sense. Although some parasitic animals can absorb all the basic nutrients they need from their environment, most animals must break down the complex molecules in their food into simple units such as amino acids, fatty acids, and sugars. These simplest nutrient units can then be used for synthesis of new molecules or metabolized as an energy source. The breakdown of food molecules is **digestion**. The cells of some animals, such as sponges, engulf particles of food and digest them intracellularly, but most animals process their food through **extracellular digestion**, in a digestive cavity called a gut. The gut of simple animals, such as flatworms and jellyfish, is a saclike structure with only one opening to the environment. More-complex animals have a tubular gut with a separate entrance and exit for food. Such animals ingest (take in) food through a mouth, where it may be broken up. The food then passes through the tubular digestive tract, where digestive enzymes break it down. Throughout this process the food is really outside the body because it has not crossed any cell membranes. Digestion occurs *outside* the cells. The products of digestion are then absorbed into the body and taken up by the cells.

In this chapter we will first consider the nutrients that organisms require and why. Then we will look at how animals procure nutrients, and once ingested, how nutrients are processed, digested, and absorbed. Finally we will learn how the body regulates its traffic in molecules used for metabolic fuel.

NUTRIENT REQUIREMENTS

Animals need food to supply the energy for metabolism and the carbon skeletons that they cannot synthesize but must have in order to build larger organic molecules. Mineral nutrients, such as iron and calcium, that animals require to build functional and structural molecules also come from food. The diet also provides complex organic molecules called vitamins that are needed in small quantities as cofactors for enzymes and for other purposes.

Nutrients as Fuel

In Chapters 6 and 7 we learned that energy in the chemical bonds of food molecules is transferred to the high-energy phosphate bonds of ATP. ATP then provides energy for active transport, biosynthesis of molecules, degradation of molecules, muscle contraction, and other work. Because they are never 100 percent efficient, these energy conversions produce heat as a by-product. Even the useful energy conversions eventually are reduced to heat, as molecules that were synthesized are broken down and energy of movement is dissipated by friction. In time, all the energy that is transferred to ATP from the chemical bonds of food molecules is released to the environment as heat. It is convenient, therefore, to talk about the energy requirements of animals and the energy content of food in terms of a measure of heat energy: the **calorie**. A calorie is the amount of heat necessary to raise the temperature of one gram of water one degree centigrade. Since this value is a tiny amount of energy in comparison to the energy requirements of many animals, physiologists commonly use the **kilocalorie** (kcal) as a unit (1,000 calories = 1 kcal). Nutritionists also use the kilocalorie as a standard unit of energy, but they traditionally refer to it as the **Calorie**, which is always capitalized to distinguish it from the single calorie. A person on a diet of 1,000 Calories per may consume up to 1,000 kcal/day. Such confusion of terms is unfortunate, but we live with it. (It should be noted that physiologists are gradually abandoning the calorie as an energy unit as they switch to the International System of Units. In this system the basic unit of energy is the joule; 1 calorie = 4.184 joules.)

The metabolic rate of an animal (see Chapter 35) is a measure of the overall energy needs that must be met by an animal's ingestion and digestion of food. The components of food that provide energy are fats, carbohydrates, and proteins. Fat yields 9.5 kcal per gram when it is metabolically oxidized, carbohydrate yields 4.2 kcal per gram, and protein yields about 4.1 kcal per gram. The basal metabolic rate of a human is about 1,300 to 1,500 kcal a day for an adult female and 1,600 to 1,800 kcal a day for an adult male. Physical activity adds to this basal energy requirement. Some equivalences of food, energy, and exercise are shown in Figure 43.2.

Although the cells of the body use energy continuously, most animals do not eat continuously. Humans generally eat several meals a day, a lion may eat once in several days, a boa constrictor may eat once a month, and hibernating animals may go five to six months without eating. Therefore, animals must store fuel molecules that can be released as needed between meals. Carbohydrate is stored in liver and muscle cells as glycogen (see Chapter 3), but the total glycogen stores are usually not more than the equivalent of a day's energy requirements. Fat is the most important form of stored energy in the bodies of animals. Not only does fat have the highest energy content per gram, but it can be stored with little associated water, making it more compact. If migrating birds had to store energy as glycogen rather than fat to fuel long flights, they would be too heavy to fly! Protein is not used to store energy, although body protein can be metabolized as an energy source as a last resort.

If an animal takes in too little food to meet its needs for metabolic energy, it is **undernourished** and must make up the shortfall by metabolizing some of the molecules of its own body. Consumption of self for fuel begins with the storage compounds glycogen and fat. Protein loss is minimized for as long as pos-

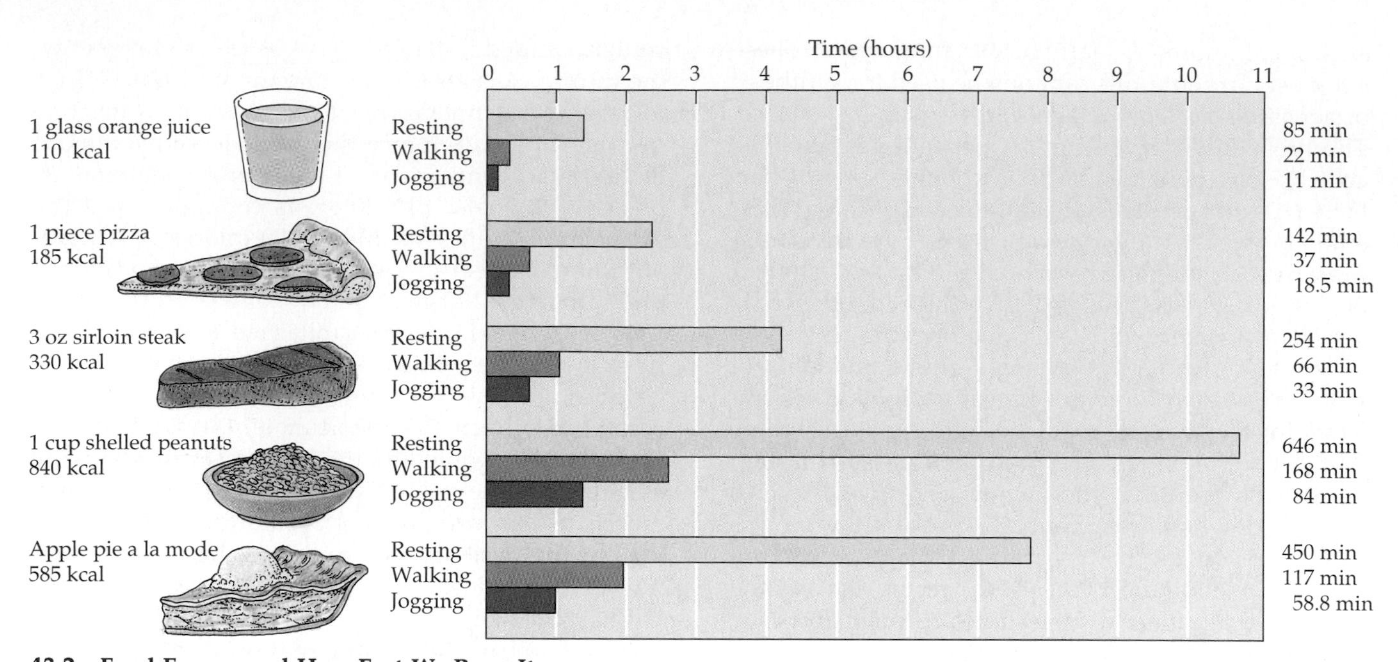

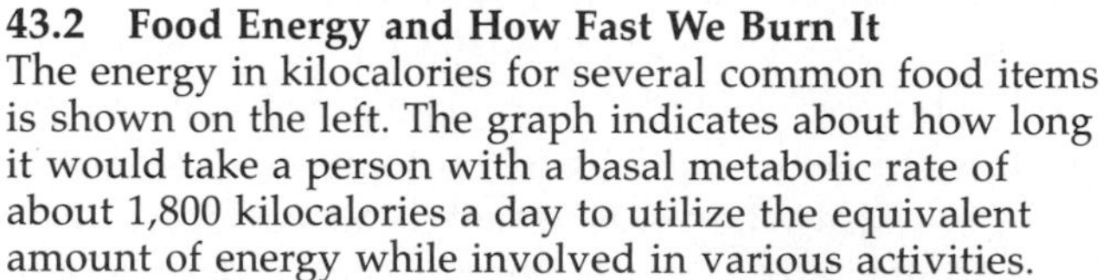
43.2 Food Energy and How Fast We Burn It
The energy in kilocalories for several common food items is shown on the left. The graph indicates about how long it would take a person with a basal metabolic rate of about 1,800 kilocalories a day to utilize the equivalent amount of energy while involved in various activities.

sible, but eventually the body has to use its own proteins for fuel. The breakdown of body proteins impairs body functions and eventually leads to death. Blood proteins are among the first to go, resulting in loss of fluid to the interstitial spaces (edema; see Chapter 42). Muscles atrophy (waste away) and eventually even brain protein is lost. Figure 43.3 shows the course of starvation. Undernourishment is rampant among people in underdeveloped and war-torn nations, and a billion people—one-fifth of the world's population—are undernourished. (Ironically, one cause of life-threatening undernourishment in Western developed nations is a self-imposed starvation called **anorexia nervosa** that results from a psychological aversion to body fat.)

When an animal consistently takes in more food than it needs to meet its energy demands, it is **overnourished**. The excess nutrients are stored as increased body mass. First, glycogen reserves build up; then additional dietary carbohydrate, fat, and protein are converted to body fat. In some species, such as hibernators, seasonal overnutrition is an important adaptation for surviving periods when food is unavailable. In humans, however, overnutrition can be a serious health hazard, increasing the risk of high blood pressure, heart attack, diabetes, and other disorders. A common clay building brick weighs about 5 pounds, so a person who is 50 pounds overweight is constantly carrying around the equivalent of ten bricks. That alone is quite a strain on the heart, but in addition, each extra pound of body tissue includes miles of additional blood vessels through which the heart must pump blood. Obesity is a health hazard, but so are poorly planned fad or crash diets that can lead to malnutrition (discussed in the next section). People spend billions of dollars every year on schemes to lose weight, even though all they need to do is follow a simple rule: Take in fewer calories than your body burns, but maintain a balanced diet.

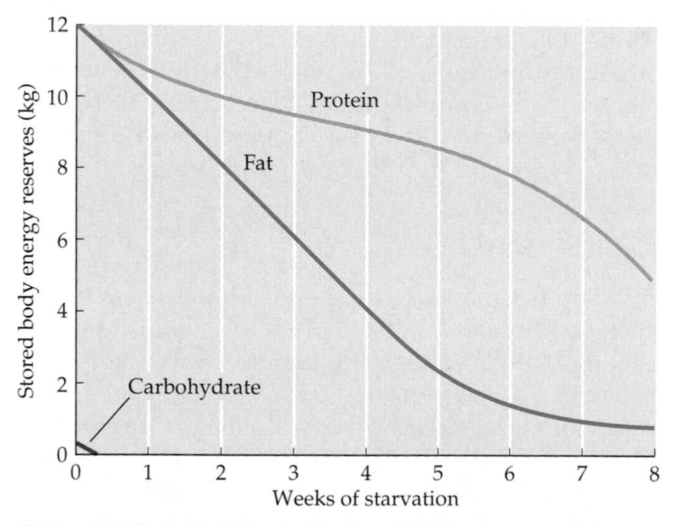

43.3 Depletion of Body Energy Reserves during Starvation
The carbohydrate reserves of our bodies are meager and are depleted by only a single day without food intake. Our major energy reserve is fat; even a person of average body weight has enough fat to survive four or five weeks without food. When most body fat has been exhausted, the only remaining fuel is protein, which is lost at an accelerating rate, with serious consequences and often death.

Nutrients as Building Blocks

Every animal requires certain basic organic molecules (carbon skeletons) that it cannot synthesize for itself but must have in order to build the complex organic molecules needed for life. An example of a required carbon skeleton is the acetyl group (Figure 43.4). Animals cannot make acetyl groups from carbon, oxygen, and hydrogen molecules; they obtain acetyl groups by metabolizing carbohydrates, fats, or proteins. From these acquired acetyl groups, animals create a wealth of other necessary compounds, including fatty acids, steroid hormones, electron carriers for cellular respiration, certain amino acids, and, indirectly, legions of other compounds. The three major classes of nutrients—carbohydrates, fats, and proteins—provide both the energy and the carbon skeletons for biosynthesis.

Because the acetyl group can be derived from the metabolism of virtually any food, it is unlikely ever to be in short supply for an animal with an adequate food supply. Other carbon skeletons, however, are derived from more-limited sources, and an animal can suffer a deficiency of these materials even if its caloric intake is adequate. This state of deficiency is called **malnutrition**. Amino acids, the building blocks of protein, are a good example of such substances. Humans obtain amino acids by digesting protein from food and absorbing the resulting amino acids. The body then synthesizes its own protein molecules, as specified by its DNA, from these dietary amino acids. Another source of amino acids is the breakdown of existing body proteins, which are in constant turnover as the tissues of the body undergo normal remodeling and renewal.

Animals can synthesize some of their own amino acids by taking carbon skeletons synthesized from acetyl or other groups and transferring to them amino groups (—NH_2) derived from other amino acids. Most animals, however, cannot synthesize all the amino acids they need. Each species has certain **essential amino acids** that must be obtained from food. Different species have different essential amino acids and, in general, herbivores have fewer essential amino acids than do carnivores. If an animal does not take in one of its essential amino acids, its protein synthesis is impaired. Think of protein synthesis as typing a story. If the typewriter is missing a key, the story either comes to a stop or has an error in it wherever the letter represented by that key is needed. In protein synthesis, the story usually comes to a stop and a functional protein is not produced.

Humans require eight essential amino acids in their diet: isoleucine, leucine, lysine, methionine, phenylalanine, threonine, tryptophan, and valine (see Table 3.1). All eight are available in milk, eggs, or meat; however, no plant food contains all eight. A strict vegetarian thus runs a risk of protein malnutrition. An appropriate dietary *mixture* of plant foods, however, supplies all eight essential amino acids (Figure 43.5). Wheat, corn, rice, and other grains are deficient in lysine and isoleucine but are well stocked with most of the others. Beans, lentils, and other legumes have lots of lysine but are low in methionine and tryptophan. Eating only grains or only beans would lead to a serious deficiency of one or more amino acids. If grains and beans are eaten together, however, the diet includes all the essential amino acids.

In general, grains are complemented by legumes or by milk products; legumes are complemented by grains and by seeds and nuts. Long before the chemical basis for this complementarity was understood, societies with little access to meat learned appropriate

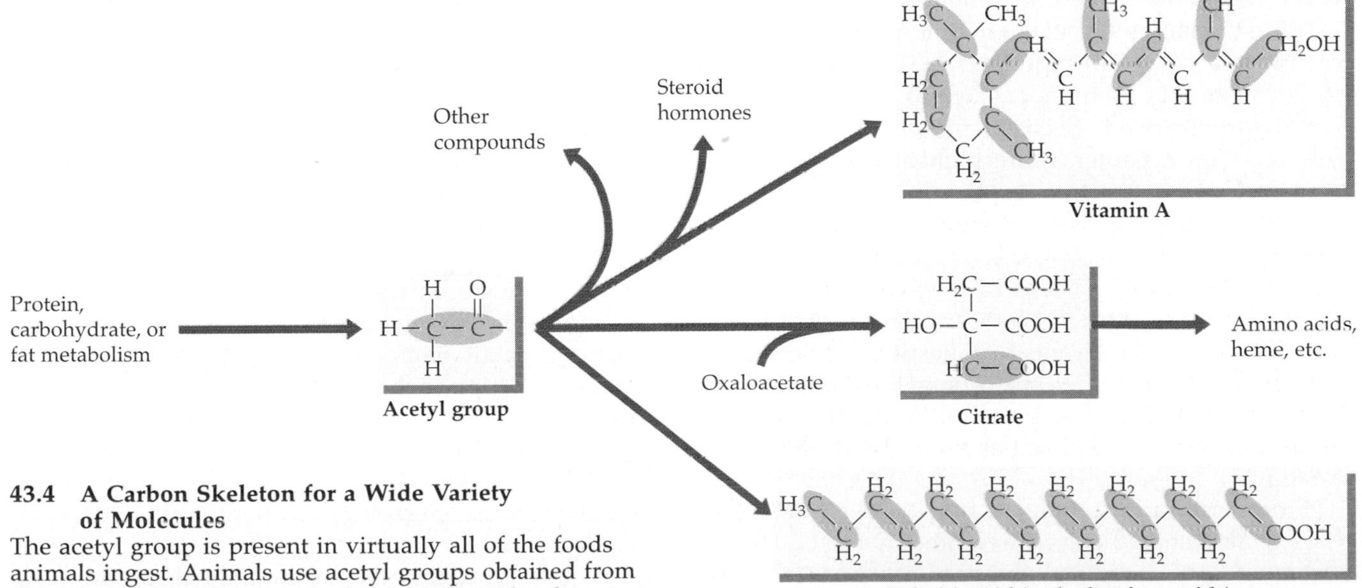

43.4 A Carbon Skeleton for a Wide Variety of Molecules
The acetyl group is present in virtually all of the foods animals ingest. Animals use acetyl groups obtained from their food to build more-complex organic molecules.

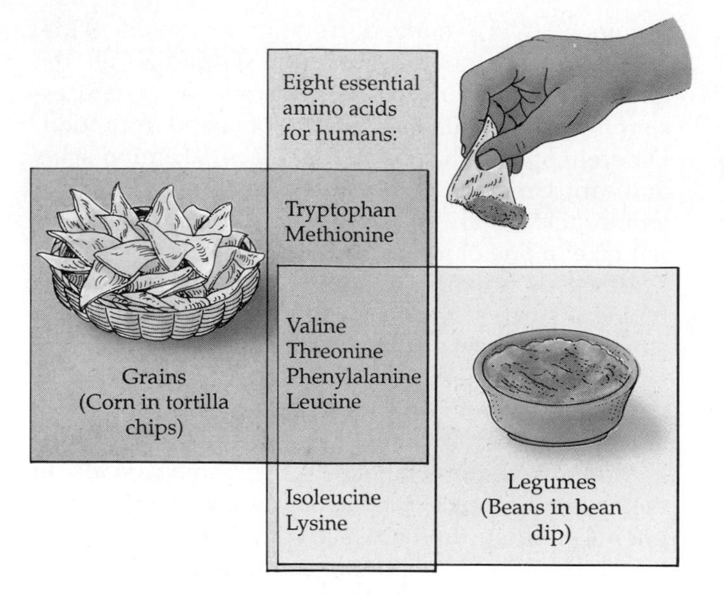

43.5 A Strategy for Vegetarians
By combining cereal grains and legumes, a vegetarian can obtain all eight essential amino acids.

dietary practices through trial and error. Many Central and South American peoples traditionally ate beans with corn, and the native peoples of North America complemented their beans with squash. Remember that we do not retain great stores of free amino acids in our bodies, yet we synthesize proteins continuously. It makes little nutritional sense to eat grains one day and beans the next; they must be eaten together for proper amino acid balance. Excess amino acids are burned for fuel, converted to fat, or excreted as waste.

Why are dietary proteins completely digested to their constituent amino acids before being used by the body? Wouldn't it be more energy-efficient to reuse some dietary proteins directly? There are several reasons why ingested proteins are not used "as is." First, macromolecules such as proteins are not readily taken up through plasma membranes, but their constituent monomers (such as amino acids) are readily transported. Second, protein structure and function (see Chapter 3) are highly species-specific. A protein that functions optimally in one species might not function well in another species. Third, foreign proteins entering the body directly from the gut would be recognized as invaders and be attacked by the immune system (see Chapter 16). Most animals avoid these problems by digesting food proteins extracellularly and then absorbing the amino acids into the body. The new proteins formed from these amino acids are recognized as "self" by the immune system.

From acetyl units obtained from food we can synthesize almost all the lipids required by the body, but we must have a dietary source of essential fatty acids—notably linoleic acid. Essential fatty acids are necessary components of membrane phospholipids, and a deficiency can lead to problems such as infertility and impaired lactation.

Mineral Nutrients

Table 43.1 lists the principal mineral elements required by animals. Certain species require additional elements. Elements required in large amounts are known as **macronutrients**; elements required in only tiny amounts are called **micronutrients**. Some essential elements are required in such minute amounts that deficiencies are never observed, but these elements are nevertheless essential.

Animals need calcium and phosphorus in great quantity. Calcium phosphate is the principal structural material in bones and teeth. Muscle contraction, nerve function, and many other intracellular functions in animals require calcium. Phosphorus is an integral component of nucleic acids. We learned in Chapter 7 about the role of phosphate groups in biological energy transfers. Sulfur is part of the structure of two amino acids and is therefore found in almost all proteins. Other essential compounds also contain sulfur. Iron is the oxygen-binding atom in both hemoglobin and myoglobin, the oxygen-carrying proteins in vertebrate blood and muscle. In addition, iron undergoes redox reactions in some of the electron-carrying proteins of cellular respiration. A number of mineral nutrients—among them magnesium, manganese, zinc, and cobalt—act as cofactors for enzymes. Potassium, sodium, and chloride ions are particularly important in the osmotic balance of tissues and in the electrical properties of membranes, including resting potentials and action potentials.

Animals require large amounts of both sodium and chloride ions. Because plants contain few of these ions, herbivores may travel considerable distances to natural salt licks. Ranchers and game wardens frequently supply salt licks for animals that do not have access to natural sources.

Specific requirements for individual elements are different in different species. In vertebrates, copper is essential in trace amounts for certain enzymes to function properly. For example, hemoglobin synthesis requires copper, even though copper is not part of the hemoglobin molecule. In numerous invertebrate species, however, copper is part of the respiratory pigment hemocyanin, and those animals require more copper than vertebrates do.

Vitamins

Another group of essential nutrients is the **vitamins**. Like essential amino acids and fatty acids, vitamins are organic compounds that an animal cannot make

TABLE 43.1
Mineral Elements Required by Animals

ELEMENT	SOURCE IN HUMAN DIET	MAJOR FUNCTIONS
Macronutrients		
Calcium (Ca)	Dairy foods, eggs, green leafy vegetables, whole grains, legumes, nuts	Found in bones and teeth; blood clotting; nerve and muscle action; enzyme activation
Chlorine (Cl)	Table salt (NaCl), meat, eggs, vegetables, dairy foods	Water balance; digestion (as HCl); principal negative ion in fluid around cells
Magnesium (Mg)	Green vegetables, meat, whole grains, nuts, milk, legumes	Required by many enzymes; found in bones and teeth
Phosphorus (P)	Dairy foods, eggs, meat, whole grains, legumes, nuts	Found in nucleic acids, ATP, and phospholipids; bone formation; buffers; metabolism of sugars
Potassium (K)	Meat, whole grains, fruits, vegetables	Nerve and muscle action; protein synthesis; principal positive ion in cells
Sodium (Na)	Table salt, dairy foods, meat, eggs, vegetables	Nerve and muscle action; water balance; principal positive ion in fluid around cells
Sulfur (S)	Meat, eggs, dairy foods, nuts, legumes	Found in proteins and coenzymes; detoxification of harmful substances
Micronutrients		
Chromium (Cr)	Meat, dairy foods, whole grains, dried beans, peanuts, brewer's yeast	Glucose metabolism
Cobalt (Co)	Meat, tap water	Found in vitamin B_{12}; formation of red blood cells
Copper (Cu)	Liver, meat, fish, shellfish, legumes, whole grains, nuts	Found in active site of many redox enzymes and electron carriers; production of hemoglobin; bone formation
Fluorine (F)	Most water supplies	Resistance to tooth decay
Iodine (I)	Fish, shellfish, iodized salt	Found in thyroid hormones
Iron (Fe)	Liver, meat, green vegetables, eggs, whole grains, legumes, nuts	Found in active sites of many redox enzymes and electron carriers, hemoglobin, and myoglobin
Manganese (Mn)	Organ meats, whole grains, legumes, nuts, tea, coffee	Activates many enzymes
Molybdenum (Mo)	Organ meats, dairy foods, whole grains, green vegetables, legumes	Found in some enzymes
Selenium (Se)	Meat, seafood, whole grains, eggs, chicken, milk, garlic	Fat metabolism
Zinc (Zn)	Liver, fish, shellfish, and many other foods	Found in some enzymes and some transcription factors; insulin physiology

for itself but that are required for its normal growth and metabolism (Box 43.A). Most vitamins function as coenzymes or parts of coenzymes and are required in very small amounts compared with essential amino acids and fatty acids that have structural roles. The list of required vitamins varies from species to species. For example, ascorbic acid (vitamin C) is not a vitamin for most mammals because they can make it themselves. Primates (including humans), however, do not have this ability, so ascorbic acid is a vitamin. If we do not get it in our diet, we develop the disease known as scurvy (Box 43.B). There are 13 such compounds that humans cannot synthesize in sufficient quantities (Table 43.2). They are divided into two groups: Water-soluble vitamins and fat-soluble vitamins.

Water-soluble vitamins (the B complex and vitamin C) play roles in both vertebrates and invertebrates. The B vitamins are coenzymes or parts of coenzymes. The B vitamin niacin, for example, we have encountered already under another name, nicotinamide (see Chapter 7). It is the portion of NAD (nicotinamide adenine dinucleotide) and NADP that undergoes oxidation and reduction in the respiratory chain and in other key redox systems in all living things. Riboflavin (vitamin B_2) similarly is the site of oxidation and reduction in the respiratory chain intermediates FAD (flavin adenine dinucleotide) and FMN (flavin mononucleotide). Vitamin C (ascorbic acid) has a number of functions, among them an essential role in the formation of the structural protein collagen. Collagen is a fibrous protein that is a

TABLE 43.2
Vitamins in the Human Diet

VITAMIN	SOURCE	FUNCTION	DEFICIENCY SYMPTOMS
Water-Soluble			
B_1, thiamin	Liver, legumes, whole grains, yeast	Coenzyme in cellular respiration	Beriberi, loss of appetite, fatigue
B_2, riboflavin	Dairy foods, organ meats, eggs, green leafy vegetables	Coenzyme in cellular respiration (in FAD and FMN)	Lesions in corners of mouth, eye irritation, skin disorders
Niacin (nicotinamide, nicotinic acid)	Meat, fowl, liver, yeast	Coenzyme in cellular metabolism (in NAD and NADP)	Pellagra, skin disorders, diarrhea, mental disorders
B_6, pyridoxine	Liver, whole grains, dairy foods	Coenzyme in amino acid metabolism	Anemia, slow growth, skin problems, convulsions
Pantothenic acid	Liver, eggs, yeast	In acetyl CoA	Adrenal problems, reproductive problems
Biotin	Liver, yeast, bacteria in gut	In coenzymes	Skin problems, loss of hair
B_{12} cobalamin	Liver, meat, dairy foods, eggs	Coenzyme in formation of nucleic acids and proteins, and in red blood cell formation	Pernicious anemia
Folic acid	Vegetables, eggs, liver, whole grains	Coenzyme in formation of heme and nucleotides	Anemia
C, ascorbic acid	Citrus fruits, tomatoes, potatoes	Aids formation of connective tissues; prevents oxidation of cellular constituents	Scurvy, slow healing, poor bone growth
Fat-Soluble			
A, retinol	Fruits, vegetables, liver, dairy foods	In visual pigments	Night blindness, damage to mucous membranes
D, calciferol	Fortified milk, fish oils, sunshine	Absorption of calcium and phosphorus	Rickets
E, tocopherol	Meat, dairy foods, whole grains	Muscle maintenance, prevents oxidation of cellular components	Anemia
K, menadione	Intestinal bacteria, liver	Blood clotting	Blood-clotting problems (in newborns)

major constituent of bone, cartilage, tendons, ligaments, and skin. The water-soluble compounds that are vitamins for humans are essential to all animals. Some species, however, can make some of those compounds in sufficient quantity that they do not require the compounds in their diet.

Fat-soluble vitamins—vitamins A, D, E, and K—have diverse functions. Vitamin A (retinol) is a precursor of retinal, the visual pigment in our eyes. Vitamin D (calciferol) regulates the absorption and metabolism of calcium. Although vitamin D may be obtained in the diet, it can also be produced in human skin by the action of ultraviolet wavelengths of sunlight on certain lipids already present in the body. Thus vitamin D is only a vitamin for individuals with inadequate exposure to the sun, such as people living in cold climates where clothing usually covers most of the body and where the sun may not shine for long periods of time.

The need for vitamin D may have been an important factor in the evolution of skin color. Human races adapted to equatorial and low latitudes have dark skin pigmentation as a protection against the damaging effects of ultraviolet radiation. These peoples generally have extensive skin areas exposed to the sun on a regular basis, so adequate synthesis of vitamin D occurs in their skins. In general, races that became adapted to higher latitudes lost dark skin pigmentation. Presumably, lighter skin facilitates vitamin D production in the relatively small areas of skin exposed to sunlight during the short days of winter. An exception to the correlation between latitude and skin pigmentation is the Inuit peoples of the Arctic. These dark-skinned people obtain plenty of vitamin D from the large amounts of fish oils in their diets; for them, exposure to sunlight is not a factor in obtaining this vitamin.

Vitamin E is poorly understood. Its principal function may be to protect unsaturated fatty acids in cellular membranes from oxidation. Vitamin K functions

BOX 43.A

Beriberi and the Vitamin Concept

Beriberi is Sinhalese (a language of Sri Lanka) for "extreme weakness." This disease of humans is found wherever unbalanced diets are common. It became particularly prevalent in Asia in the nineteenth century, when it became standard practice to mill rice to a high, white polish and discard the hulls that are present in brown rice. There are several forms of beriberi, with different symptoms, but the heart is generally adversely affected.

In 1897 Christiaan Eijkman, working in what is now Indonesia, discovered that chickens developed beriberi-like symptoms when fed a diet of polished rice. During the next decade, other investigators found that people in Malaysia on a polished rice diet developed beriberi, whereas those who ate brown rice did not. Finally, in 1912, Polish-born Casimir Funk showed that pigeons with beriberi could be cured of their symptoms by feeding them a concentrate of rice polishings—the hulls that were discarded to make the rice more "appealing." He went on to suggest that beriberi and some other diseases are dietary in origin, and that they result from deficiencies in specific substances, for which he coined the term *vitamines* because he mistakenly thought that all these substances were amines vital for life. In 1926 thiamin (vitamin B_1)—the substance lost in the rice milling process—was the first vitamin to be isolated in pure form; in 1936 its structure was determined and it was synthesized for the first time.

Eijkman's and Funk's investigations with birds were the first in a series of experiments on animals that established that diseases can result from dietary deficiencies. Before these experiments, all diseases were thought to be caused by microorganisms.

in blood clotting following an injury and hence plays a crucial role in the protection of the body. The fat-soluble vitamins generally are required by vertebrates but not by invertebrates.

When water-soluble vitamins are ingested in excess of bodily needs, they are simply eliminated in the urine. (This is the fate of much of the vitamin C that people take in excessive doses.) The fat-soluble vitamins, however, accumulate in body fat and may build up to life-threatening levels if taken in excess. (Inuit peoples generally do not eat polar bear liver, which is unusually high in fat-soluble vitamin A.)

BOX 43.B

Scurvy and Vitamin C

In 1498 Vasco da Gama sailed around the Cape of Good Hope. In the process, he lost 100 of his 160 crew members to a disease that was becoming all too familiar on ocean voyages lasting several months. The symptoms of **scurvy** include general debility, hemorrhaging and decay of skin and flesh, bleeding gums and loss of teeth, and finally, death.

More than a century later, the British physician James Lind found that shipboard scurvy could be prevented by having sailors eat citrus fruit or sauerkraut. The effects were dramatic. In 1760 one British naval hospital treated 1,754 cases of scurvy. Beginning in 1795, the Royal Navy required all sailors to take a daily ration of lemon juice—and in 1806 the same hospital treated exactly one case of scurvy. The navy switched from lemons to limes in 1865, and since that time British sailors have been referred to as "limeys."

In 1907 researchers induced scurvy in guinea pigs by giving them a diet of dried hay and oats, with no fresh plant material. By supplementing that diet with various foods, the researchers were able to determine which foods prevented scurvy. At last, in 1932, the anti-scurvy factor was isolated from lemon juice. The anti-scurvy factor is ascorbic acid, a compound identified in plant extracts six years earlier by Albert Szent-Györgyi and now commonly called vitamin C.

Vitamin deficiency diseases are rare among primitive societies living according to long-established tradition, but they frequently occur when new habits are thrust upon people by civilization or technology. Scurvy and other deficiency diseases may also arise when people are cut off from their normal diets by such things as ocean voyages or imprisonment, or when armies are on campaign or cities are under siege.

Nutritional Deficiency Diseases in Humans

In humans, chronic shortage of a nutrient produces a characteristic deficiency disease. If the deficiency is not remedied, death may follow. An example is kwashiorkor. Kwashiorkor results from protein deficiency, which causes swelling of the extremities, distension of the abdomen (see Figure 42.13), immune system breakdown, degeneration of the liver, mental retardation, and other problems.

Shortage of any of the vitamins results in specific deficiency symptoms (see Table 43.2). Two deficiency diseases, beriberi and scurvy, were discussed in Boxes 43.A and 43.B. **Pellagra**, which results from a deficiency of the B vitamin niacin, is a common and severe disease in many poor areas. It also occurs frequently in conjunction with chronic alcoholism. Its symptoms include diarrhea, itching skin, abdominal pain, and other problems. Vitamin D deficiency decreases the absorption and use of calcium, leading to softening of the bones and a distortion of the skeleton. This deficiency disease is known as **rickets**. Vitamin B_{12} (cobalamin) is produced by microorganisms that live in our intestines and use the cobalt in our diet. Cobalamin is present in all foods of animal origin. Plants neither use nor produce vitamin B_{12}, and a strictly vegetarian diet (not supplemented by vitamin pills) can lead to **pernicious anemia**, the B_{12} deficiency disease.

Inadequate mineral nutrition can also lead to deficiency diseases. Iodine, for example, is a constituent of the hormone thyroxin, which is produced in the thyroid gland. If insufficient iodine is obtained in the diet, the thyroid gland grows larger in an attempt to compensate for the inadequate production of thyroxin. The swelling that results is called a **goiter** (see Figure 36.9). Goiters are common in mountain areas such as the Andes of South America because of low iodine levels in the soil and hence in the crops grown there. Goiters were once common in Switzerland and in the Great Lakes area of the United States, but the problem was solved by adding small amounts of iodine to table salt or to drinking water.

ADAPTATIONS FOR FEEDING

The ways an animal acquires its nutrients and its adaptations for doing so are frequently its most distinguishing characteristics. The role that a species plays in nature is described as its ecological niche, and its feeding specializations and adaptations are major dimensions of that ecological niche (see Chapter 48). The crucial adaptations that enable a species to exploit a particular source of nutrition are frequently physiological and biochemical. For example, the Australian koala eats nothing but leaves of eucalyptus trees. Eucalyptus leaves are tough, low in nutrient content, and loaded with pungent, toxic compounds that evolved to protect the trees from predators. Yet the koala's gut can digest and detoxify the leaves and absorb all of the nutrients the animal needs from this formidably specialized diet. The feeding adaptations that are most obvious to us, however, are the anatomical and behavioral features that animals use to acquire and ingest their food.

Food Acquisition by Carnivores

The predatory behaviors of many carnivores are legendary. One need only call to mind the hunting skills of hawks, wolves, or any member of the cat family. Carnivores have evolved stealth, speed, power, large jaws, sharp teeth, and strong gripping appendages. A cheetah, for instance, first stalks its prey stealthily from downwind, aided by its natural camouflage. When close enough, it dashes after the prey at speeds as fast as 110 kilometers per hour. It then brings the prey down with its sharp, powerful claws and teeth. Carnivores also have evolved remarkable means of detecting prey. Bats use echolocation, pit vipers sense infrared radiation from the warm bodies of their prey, and certain fishes detect electric fields created in the water by their prey (see Chapter 39).

Adaptations for killing and ingesting prey are diverse and specialized. These adaptations can be especially important when the prey species are capable of inflicting damage on the predators. Many species of snakes take relatively large prey that are well equipped with sharp teeth and claws. A snake may strike with poisonous fangs and immobilize its prey before ingesting it. A boa or python kills its prey by squeezing it with coils of its powerful body. To swallow large prey, a snake's lower jaw disengages from its joint with the skull. The tentacles of jellyfish, corals, squid, and octopus, the long, sticky tongues of frogs and chameleons, and the webs of spiders are other examples of fascinating adaptations for capturing and immobilizing prey.

Because some prey items are impossible for a predator to ingest, digestion is sometimes accomplished externally. Sea stars evert their stomachs (turn them inside out) and digest their molluscan prey while they are still in their shells (Figure 43.6). Spiders usually prey on insects with indigestible exoskeletons. The spider can inject its prey with digestive enzymes and then suck out the liquefied contents, leaving behind the empty exoskeletons frequently seen in old spider webs.

Food Acquisition by Herbivores

Herbivores obtain food less dramatically than predators do. Cows or sheep graze in grassy meadows

43.6 Inside-Out Digestion
This sea star is eating two mollusks. It is holding them with its arms as the tissue from its everted stomach digests them.

while caterpillars munch steadily on leaves. Some herbivores have striking adaptations for feeding, such as the trunk (a flexible, gripping nose) of the elephant, the long neck of the giraffe, or the wing design that enables the hovering flight hummingbirds use to gather nectar swiftly from large numbers of flowers.

Behavior that can almost be described as agricultural is an adaptation of some herbivores. There are species of termites and tropical leaf-cutter ants that prepare and tend subterranean fungus gardens (Figure 43.7). Individuals of these species forage for plant material, which they bring into their nests and process into a spongy comb structure of undigested cellulose. By depositing their feces into the comb they innoculate it with fungal spores that send fungal mycelia throughout the comb. The mycelia produce enzymes that break down the comb's cellulose, providing nutrients for the fungus which then forms fruiting bodies. The insects eat the fruiting bodies, as well as the older comb riddled with mycelia. Through this symbiotic relationship, the fungus helps the insects derive nutrition from the forage they collect. There are also termite species with symbiotic protists in their guts that provide a similar service of producing enzymes that break down cellulose.

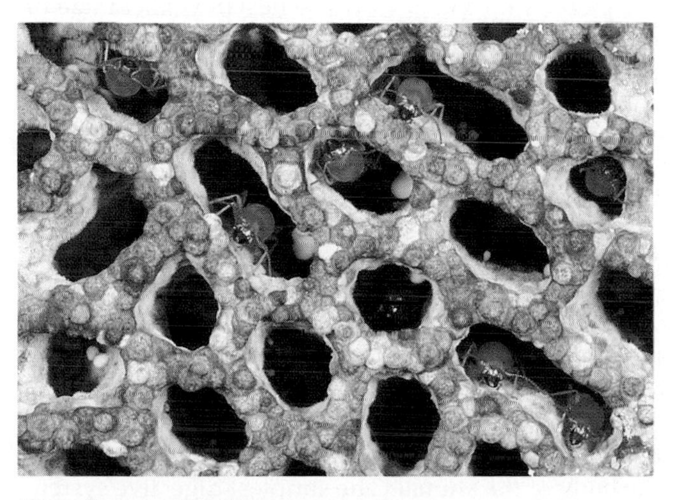

43.7 Fungus Gardens
These termites harvest plant matter, which they cannot digest, and spread it on the walls of their underground chambers. The fungus that grows in these carefully prepared beds helps break down the woody matter for the termites and also serves as food.

Food Acquisition by Filter Feeders

Filter feeders strain out particles or small organisms suspended in water. Most sessile (stationary) aquatic animals, such as sponges, corals, barnacles, and bivalve mollusks, feed in this way. There are also mobile filter feeders, including the baleen whales, tadpoles, mosquito larvae, hundreds of fishes (such as herring, sardines, menhaden, and as we learned at the beginning of the chapter, whale sharks), and even some birds. All filter feeders have some way of passing great volumes of water through a filterlike device at the front end of the gut. Flamingos sift through water and mud with their grooved bills, capturing insect larvae, worms, seeds, bacteria, and other matter. Many stationary filter feeders employ mucus to extract particles from water. Oysters, clams, and other bivalves draw water over their gills, where sticky mucus traps food particles. Their gills are densely covered with cilia that convey the mucus and its trapped particles toward the mouth, where coarse matter is rejected and the rest allowed to enter.

The most impressive filter feeders are the baleen whales, the largest of which (in fact, the largest animal ever to live on this planet) is the blue whale (*Balaenoptera musculus*). A blue whale may grow to a length of 30 meters—the equivalent of several school buses placed end to end. Its tongue is the size of an elephant and its heart is the size of a small car, yet the blue whale eats tiny crustaceans called krill, which it filters from seawater. If you run your tongue along the roof of your mouth you will feel ridges. These ridges are greatly enlarged in baleen whales, forming **baleen plates** with fringed edges (Figure 43.8). The whale approaches a swarm of krill and gulps it in, along with thousands of liters of water. As it closes its mouth, its tongue forces the seawater out through the fringe of the baleen plates, leaving the krill behind. Krill are very abundant in cold, nutrient-rich waters and make it possible for the whale to support its metabolic rate of about a million kilocalories per day.

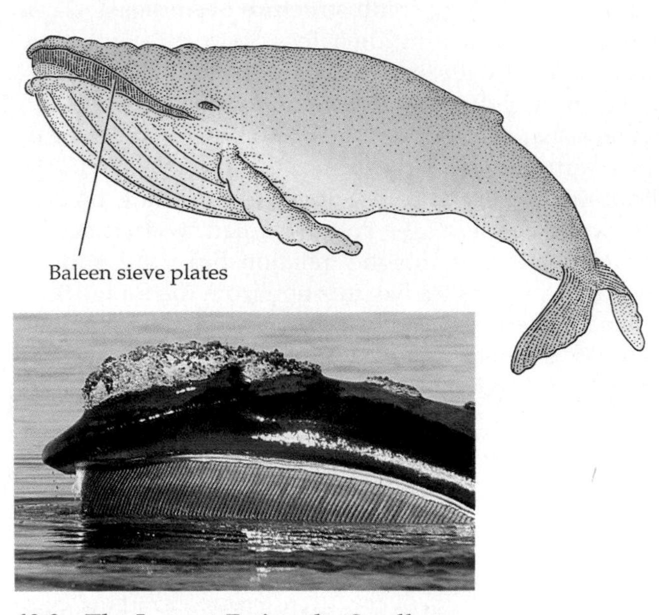

43.8 The Largest Eating the Smallest
Visible in this feeding whale are the baleen plates that hang from the roof of its mouth and filter small animals from the huge volumes of water that pass through them. The front of its snout is encrusted with barnacles, which are also filter feeders.

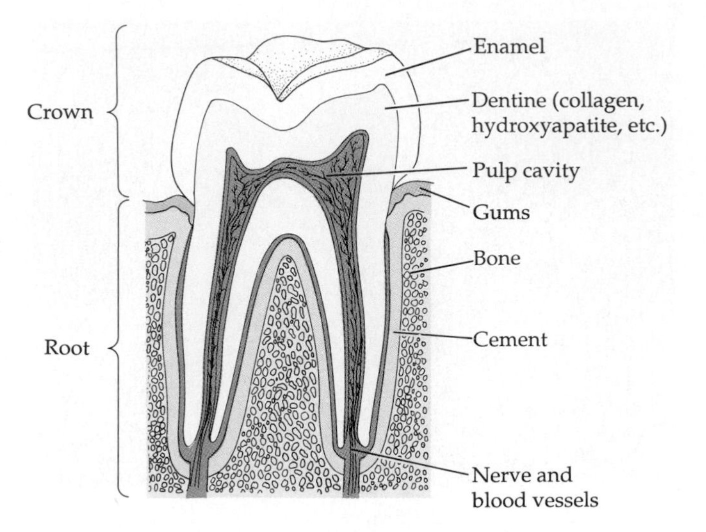

43.9 A Mammalian Tooth Has Three Layers
A section through a mammalian tooth shows that the crown above the gums is covered with hard enamel. Below the enamel is a thick dentine layer that extends into the skull to form the root, and inside the dentine is the pulp cavity containing the tooth's supply of blood vessels and nerves. The tooth is held in its bony socket in the skull by a fairly soft "cement." The teeth of other vertebrates, such as lizards, are not entrenched in bony sockets and consequently must be replaced frequently.

Vertebrate Teeth

Many vertebrate species have distinctive teeth. Teeth are adapted for the acquisition and initial processing of specific types of foods, and because they are one of the hardest structures of the body, they remain in the environment long after the animals die. Paleontologists use teeth to identify animals that lived in the distant past and to understand their behavior.

All mammalian teeth have a general structure consisting of three layers (Figure 43.9). An extremely hard material called **enamel**, composed principally of calcium phosphate, covers the crown of the tooth. Both the crown and the root contain a layer of a bony material called **dentine**, within which is a pulp cavity containing blood vessels, nerves, and the cells that produce the dentine. The shapes and organization of mammalian teeth, however, can be very different, since they are adapted to specific diets (Figure 43.10). In general, incisors are teeth used for cutting, chopping, or gnawing; canines are teeth used for stabbing, ripping, and shredding; and molars and premolars (the cheek teeth) are used for shearing, crushing, and grinding.

The highly varied diet of humans is reflected by our multipurpose set of teeth, as is common among omnivores. Children first develop a set of 20 "milk teeth," which are also called deciduous teeth because they are lost and replaced by permanent teeth. The permanent teeth of adults include four upper and four lower incisors for biting, and two upper and two lower canines for tearing. Behind the canines, on each side of the jaw, are four upper and four lower premolars and six upper and six lower molars for crushing and grinding. This is a total of 32 teeth before the dentist starts extracting them. The last set of molars, or the rearmost tooth on each side of the upper and lower jaw, usually does not erupt through the gums until the person reaches 18 years of age or older. These "wisdom teeth" frequently present problems because the jaws may be too small to accommodate them.

DIGESTION

Most animals digest food extracellularly. Animals take food into a body cavity that is continuous with the outside environment and then secrete digestive enzymes into that cavity. The enzymes act on the food, reducing it to nutrient molecules that can be absorbed by the cells lining the cavity. Only after they are absorbed by the cells are the nutrients within the body of the animal. The simplest digestive system is a gastrovascular cavity that connects to the outside world through a single opening. After a cnidarian captures a prey with its stinging nematocysts (see Figure 26.10), its tentacles cram the prey into the gastrovascular cavity (see Figure 42.1*a*) where en-

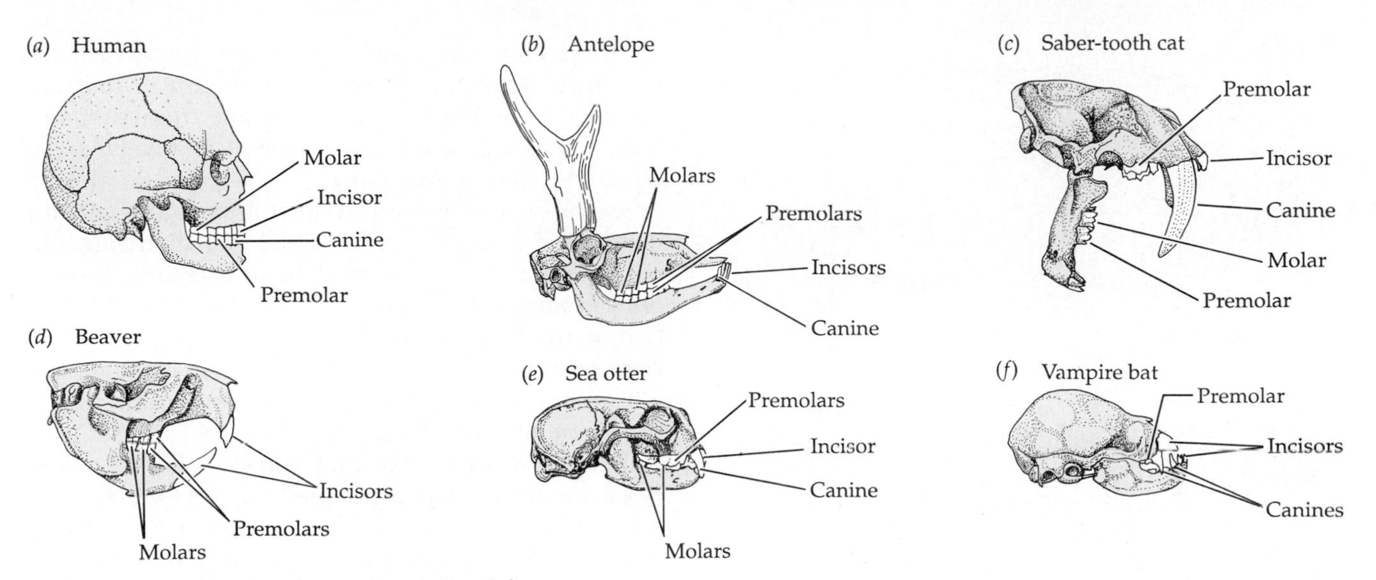

43.10 Mammalian Teeth Are Specialized for Different Diets
(a) A human is an omnivore and has a generalized set of teeth. *(b)* The pronghorn antelope grazes on tough grasses, herbs, and shrubs. Its upper jaw has no incisors or canines, and in the lower jaw they are far forward and are used to tear the leaves off plants. The antelope's cheek teeth are highly adapted for grinding coarse vegetation. *(c)* Carnivores such as the extinct saber-toothed cat have greatly enlarged canine teeth for gripping, killing, and tearing their prey. The incisors are used for scraping muscle off of bone, and the cheek teeth are used for shearing flesh and crushing bones. *(d)* Rodents such as the beaver have enlarged incisors that they use for gnawing. The beaver cuts down and strips the bark from trees with its incisors. The incisor teeth of rodents grow continuously to compensate for wear and tear. The beaver has no canines, but its cheek teeth are well adapted for grinding fibrous plant material. *(e)* The sea otter has well-developed cheek teeth for crushing the shells of marine invertebrates. *(f)* The upper incisors and canines of the vampire bat are triangular and sharp. They are used to make incisions in the skin of large mammals so the bat can drink their blood.

zymes partially digest it. Extracellular digestion in cnidarians is supplemented by intracellular digestion: Cells lining the gut take in some small food particles by endocytosis.

The gastrovascular cavity of flatworms is more complex than that of cnidarians, but it also has only one opening through which food enters and waste products exit. Where the gastrovascular cavity meets the mouth, it narrows to form a tubular **pharynx**. This muscular structure can be pushed out through the mouth during feeding. In addition, extensive branching of the gastrovascular cavity increases the efficiency of absorption of nutrient molecules; the branches increase the surface area through which absorption can occur (Figure 43.11).

Some multicellular animals have no digestive systems. Many of these are internal parasites such as tapeworms. They live in an environment so rich in already digested nutrient molecules that they just absorb them directly into their cells.

Tubular Guts

The guts of all animal groups other than sponges, cnidarians, and flatworms are tubular, with an opening at either end. A mouth takes in food; solid digestive wastes, or feces, are excreted through an anus. Different regions in the tubular gut are specialized for particular functions (Figure 43.12). Remember as we discuss these regions that all locations within the tubular gut are really outside the body of the animal.

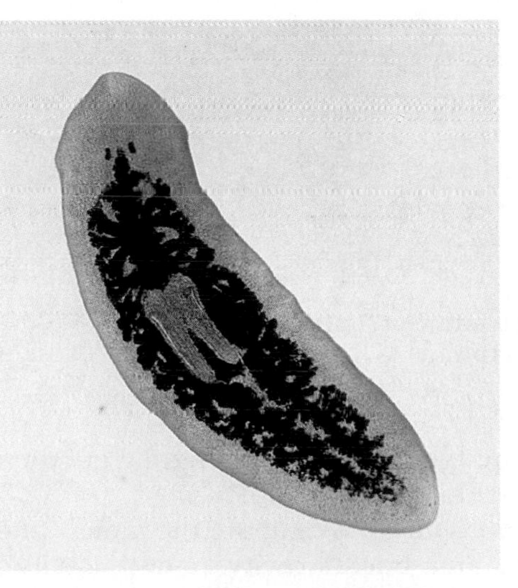

43.11 Digestive Tract of a Flatworm
The gastrovascular system of this flatworm has been stained to reveal all of its branches. This system communicates to the outside through the muscular pharynx visible in the center of the animal.

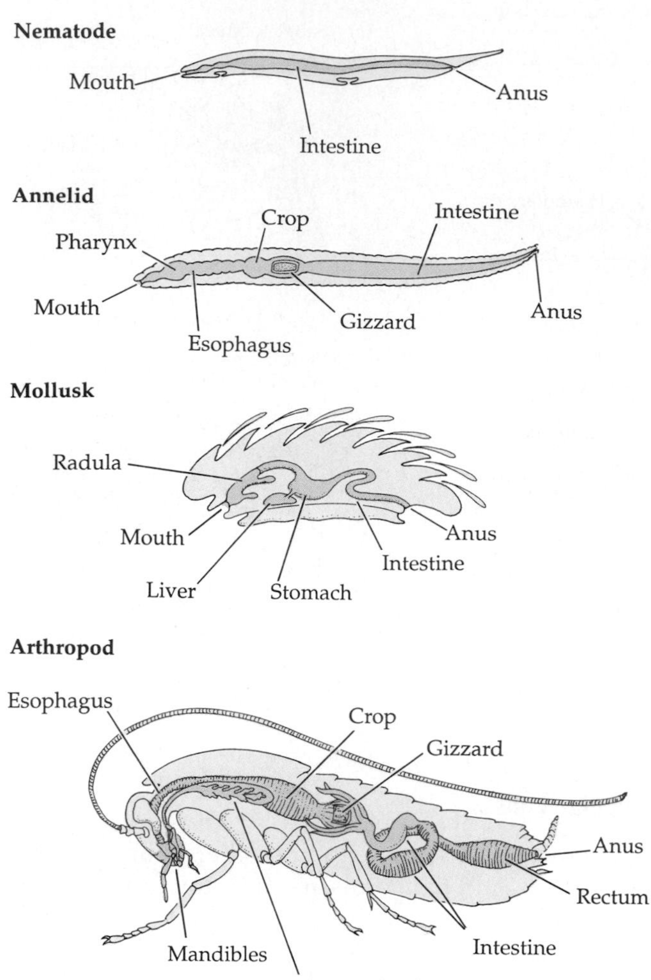

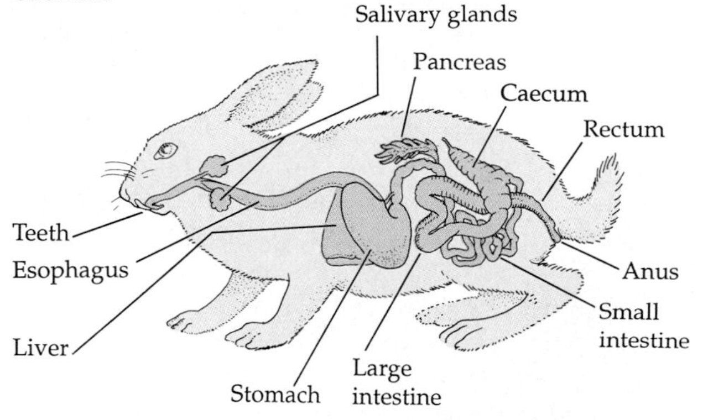

43.12 Compartments Specialized for Digestion and Absorption

Only by crossing the membranes lining the gut does a nutrient molecule enter the body.

At the anterior end of the gut are the mouth (the opening itself) and **buccal cavity** (mouth cavity). Food may be broken up by teeth (in some vertebrates), by the radula in snails, or by mandibles (in insects), or somewhat further along the gut by structures such as the gizzards of birds and earthworms, where muscular contractions of the gut grind the food together with small stones. Some animals simply ingest large chunks of food with little or no fragmentation. **Stomachs** and **crops** are storage chambers, enabling animals to ingest relatively large amounts of food and digest it at leisure. Food may or may not be digested in such a storage chamber, depending on the species. Food delivered into the next section of gut, the **midgut** or **intestine**, is well minced and well mixed. Most digestion and absorption occurs here. Specialized glands such as the pancreas in mammals secrete digestive enzymes into the intestine, and the gut wall itself secretes other digestive enzymes. The **hindgut** recovers water and ions and stores feces so that they can be released to the environment at an appropriate time or place. A muscular **rectum** near the anus assists in the expulsion of undigested wastes, the process of **defecation**.

Within the hindguts of many species are colonies of bacteria that live in cooperation (symbiosis) with their hosts. The bacteria obtain their own nutrition from the food passing through the host's gut while contributing to the digestive processes of the host. Members of the leech genus *Hirudo* produce no enzymes that can digest the proteins in the blood they suck from vertebrates; however, a colony of gut bacteria produces the enzymes necessary to break down those proteins into amino acids, which are subsequently used by both the leeches and the bacteria. Many animals obtain vitamins from the bacteria in their hindguts. Herbivores such as rabbits, cattle, termites, and cockroaches depend on microorganisms in their guts for the digestion of cellulose. In some, specialized regions of their guts may even serve as microbial fermenters. An example is the caecum of the rabbit (see Figure 43.12).

In many animals, the parts of the gut that absorb nutrients have evolved extensive surface areas for absorption. The earthworm has a long, dorsal infolding of its intestine, called the typhlosole (Figure 43.13*a*), that provides extra absorptive surface area. The shark's intestine has a spiral valve, forcing food to take a longer path and thus encounter more absorptive surface (Figure 43.13*b*). In many vertebrates the wall of the gut is richly folded, with the individual folds bearing legions of tiny fingerlike projections called **villi**. The villi in turn have microscopic projections called microvilli, on the cells that line their surfaces (see Figure 43.14). Microvilli present an enormous surface area for the absorption of nutrients.

Digestive Enzymes

Protein, carbohydrate, and fat macromolecules are broken down into their simplest monomeric units by digestive enzymes. All of these enzymes cleave the chemical bonds of macromolecules through a reaction that adds a water molecule at the site of cleavage;

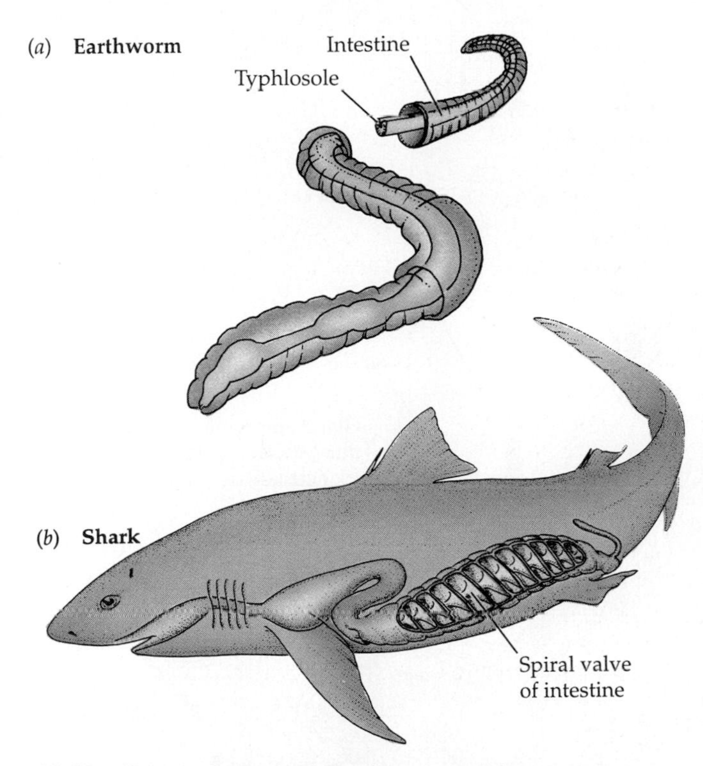

43.13 Greater Intestinal Surface Area Means More Absorption
(a) In earthworms: the adaptation is a typhlosole: a simple longitudinal infolding of the intestinal wall. *(b)* Sharks have evolved a spiral valve that increases the surface area of the intestine.

hence they are generally called **hydrolytic enzymes**. Examples of hydrolytic cleavage are the breaking of the bonds between adjacent amino acids of a protein or peptide (see Figure 6.14) and the breaking of the bonds between adjacent glucose units in starch (see Figure 6.7).

Digestive enzymes are classified according to the substances they hydrolyze: carbohydrases hydrolyze carbohydrates; proteases, proteins; peptidases, peptides; lipases, fats; and nucleases, the nucleic acids. The prefixes *exo-* (outside) and *endo-* (within) indicate where the enzyme cleaves the molecule. Thus an endoprotease hydrolyzes a protein at an internal site along the polypeptide chain, and an exoprotease snips away amino acids at the ends of the molecule (see Figures 6.16 and 6.17).

How can an organism produce enzymes to digest biological macromolecules without digesting itself? The answer is that the digesting, as you know, is usually done *outside* the animal. The gut is simply a tunnel through the animal; food in the gut is outside the body and hence can receive treatment (such as high acidity or potent enzymes) that would be intolerable within a cell or a tissue. (In the cases in which digestion is intracellular, such as in cnidarians, the hydrolytic reactions are localized within food vacuoles.) Most digestive enzymes are produced in an inactive form, known as a **zymogen**. Thus they do not act on the cells that produce them. When secreted into the gut, the zymogens are activated, sometimes by exposure to a different pH but more often by the action of another enzyme.

The gut itself is not digested by activated enzymes because it is protected by a covering of mucus, a slimy material secreted by special cells in the lining of the gut. The mucus also lubricates the gut and protects it from abrasion. If mucus production is inadequate, digestive enzymes or stomach acid can act upon the gut, producing sores called **ulcers**. Insects rely on a different trick to prevent digestion of the gut lining. Within the gut they secrete a thin tube of chitin, a modified polysaccharide (see Figure 3.13) that is also found in the insect's protective exoskeleton. The chitin tube encloses the food and enzymes and protects the gut from abrasion and self-digestion.

STRUCTURE AND FUNCTION OF THE VERTEBRATE GUT

The separate compartments that have specific functions in the digestive tracts of vertebrates are all part of a continuous tube that runs from mouth to anus. The specific functions must be coordinated so that they occur in proper sequence and at appropriate rates. Let's take a tour of the vertebrate gut to see how structure and function work together to move food through the gut and bring about its sequential digestion and the absorption of nutrients.

The Tissue Layers

The cellular architecture of the tube that forms the vertebrate gut follows a common plan throughout. Four major layers of different cell types form the wall of the tube (Figure 43.14). These layers differ somewhat from compartment to compartment, but they are always present. Starting in the cavity, or **lumen**, of the gut, the first tissue layer is the **mucosa**. Nutrients are absorbed across the membranes of the mucosal cells; in some regions of the gut, those membranes have many folds that increase their surface area. Mucosal cells also have secretory functions. Some secrete mucus that lubricates the food and protects the walls of the gut. Others secrete digestive enzymes, and still others in the stomach secrete hydrochloric acid (HCl). At the base of the mucosa are some smooth muscle cells and just outside the mucosa is the second layer of cells, the **submucosa**. Here we find the blood and lymph vessels that carry absorbed nutrients to the rest of the body. The submucosa also contains a network of nerves; these neurons are both sensory (responsible for stomach

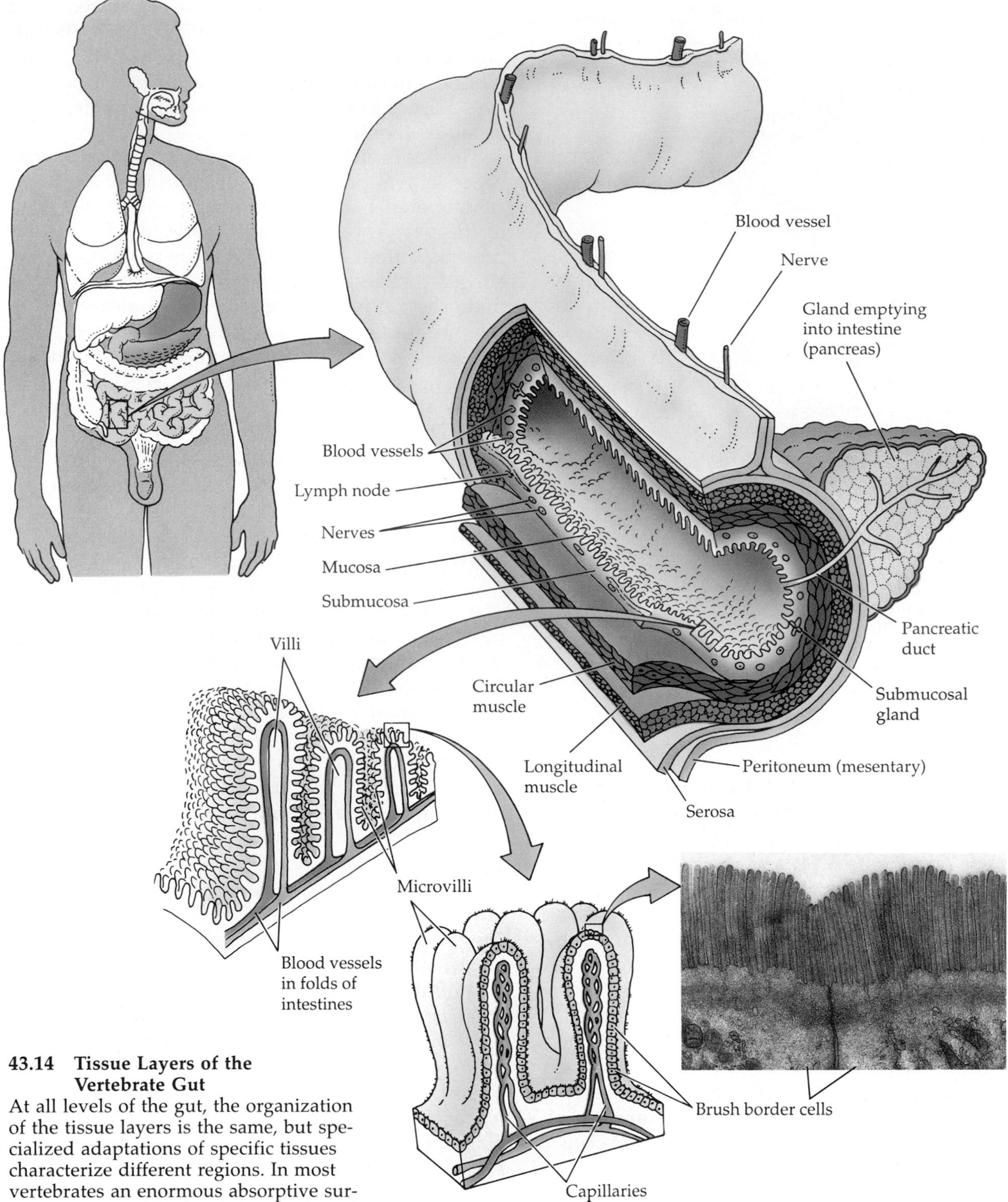

43.14 Tissue Layers of the Vertebrate Gut
At all levels of the gut, the organization of the tissue layers is the same, but specialized adaptations of specific tissues characterize different regions. In most vertebrates an enormous absorptive surface is achieved by the sheer length of the tubular small intestine and the folding of its linning. Finally, the thin, fingerlike microvilli that cover the villi increase absorptive surface area enormously.

aches!) and regulatory (controlling the various secretory functions of the gut).

External to the submucosa are two layers of smooth muscle cells responsible for the movements of the gut. Because the cells of the innermost layer are oriented *around* the gut, this layer is called **circular muscle**. It constricts the lumen. The cells of the outermost layer of the tube's wall, the **longitudinal mus-**

cle, are arranged along the length of the gut. When this layer contracts, the gut shortens. Between these two layers of muscle is a network of nerves that controls the movements of the gut, coordinating the different regions with one another.

Surrounding the gut is a fibrous coat called the **serosa**. Like other abdominal organs, the gut is also covered and supported by a tissue, the **peritoneum**.

Movement of Food in the Gut

Food entering the mouths of most vertebrates is chewed and mixed with the secretions of salivary glands. The muscular tongue then pushes the mouthful, or bolus, of food toward the back of the mouth cavity. By making contact with the soft tissue at the back of the mouth, the food initiates a complex series of neural reflex actions known as swallowing. Stand in front of a mirror and gently touch this tissue at the back of your mouth with the eraser of your pencil or with a cotton swab. You may gag slightly, but you will also experience an uncontrollable urge to swallow. Swallowing involves many muscles doing a variety of jobs that propel the food through the **pharynx** and into the **esophagus** without allowing any of it to enter the windpipe (trachea) or nasal passages (Figure 43.15).

Once the food is in the esophagus, peristalsis takes over and pushes the food toward the stomach. **Peristalsis** is a wave of smooth muscle contraction that moves progressively down the gut from the pharynx toward the anus. The smooth muscle of the gut contracts in response to being stretched. Swallowing a bolus of food stretches the upper end of the esophagus, and this stretching initiates a wave of contraction that slowly pushes the contents of the gut toward the anus. Peristalsis can occasionally run in the opposite direction, however. When your stomach is very full, pressure on your abdomen or a sudden movement can push some stomach contents into the lower end of your esophagus. This can initiate peristaltic movements that bring the acidic, partially digested food into your mouth. When you vomit, contractions of the abdominal muscles explosively force stomach contents out through the esophagus. Prior to vomiting, waves of reverse peristalsis can even bring the contents of the upper regions of the intestine back into the stomach to be expelled.

The movement of food from the stomach into the esophagus is normally prevented by a thick ring of circular smooth muscle at the junction of the esophagus and the stomach. This ring of muscle, called a **sphincter**, is normally constricted. Waves of peristalsis cause it to relax enough to let food pass through. Sphincter muscles are found elsewhere in the digestive tract as well. The pyloric sphincter governs the passage of stomach contents into the intestine. Another important sphincter surrounds the anus.

Digestion in the Mouth and Stomach

In addition to physically disrupting food, the mouth initiates the digestion of carbohydrates through the

43.15 Swallowing and Peristalsis
Food pushed to the back of the mouth triggers the swallowing reflexes. Once food enters the esophagus, peristalsis propels it to the stomach.

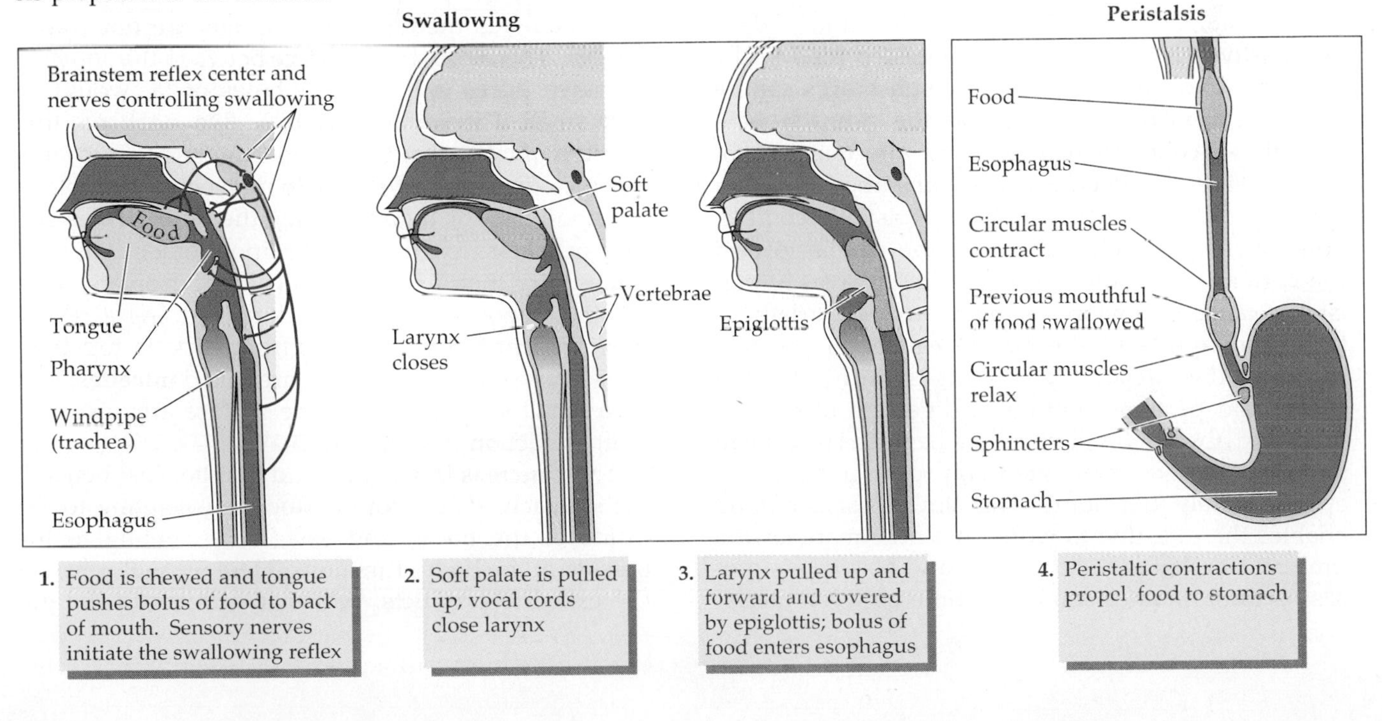

action of the enzyme amylase, which is secreted with the saliva and mixed with the food as it is chewed. Amylase is a carbohydrase; it hydrolyzes the bonds between the 6-carbon sugar units that make up the long-chain starch molecules. The action of amylase is what makes a piece of bread or cracker taste sweet if you chew it long enough.

Most vertebrates can rapidly consume a large volume of food, but digesting that food is a long, slow process. The stomach stores the food devoured in the course of a meal and continues breaking it down physically. The secretions of the stomach kill microorganisms taken in with the food and begin the digestion of proteins. The major enzyme produced by the stomach is an endopeptidase called **pepsin**. Pepsin is secreted as the zymogen **pepsinogen** by cells in the **gastric pits**—deep folds in the stomach lining. Other cells in the gastric pits produce hydrochloric acid, and still others near the openings of the gastric pits and throughout the stomach mucosa secrete mucus.

Hydrochloric acid (HCl) maintains the stomach fluid (the gastric juice) at a pH between 1 and 3. This low pH activates the conversion of pepsinogen to pepsin. This conversion is amplified as the newly formed pepsin activates other pepsinogen molecules, a process called **autocatalysis**. Hydrochloric acid also provides the right pH for the enzymatic action of pepsin. In addition, the low pH helps dissolve the intercellular substances holding the ingested tissues together. Breakdown of the ingested tissues exposes more food surface area to the action of digestive enzymes. Mucus secreted by the stomach mucosa coats and protects the walls of the stomach from being eroded and digested by the HCl and pepsin.

Contractions of the muscles in the walls of the stomach churn its contents, thoroughly mixing them with the stomach secretions. The acidic, fluid mixture of digestive juices and partially digested food in the stomach is called **chyme**. A few substances can be absorbed from the chyme across the stomach wall, including alcohol (hence its rapid effects), aspirin, and caffeine, but even these are absorbed in rather small quantities in the stomach. Peristaltic contractions of the stomach walls push the chyme toward the bottom end of the stomach. These waves of peristalsis cause the pyloric sphincter to relax briefly so that little squirts of the chyme can enter the first region of the intestine, where digestion of carbohydrates and proteins continues, digestion of fats begins, and absorption of nutrients begins. The human stomach empties itself gradually over a period of approximately four hours. This slow passage of food enables the intestine to work on a little material at a time and prolongs the digestive and absorptive processes throughout much of the time between meals.

The Small Intestine, Gall Bladder, and Pancreas

Although the **small intestine** takes its name from its diameter, it is really a very large organ and is the site of the major events of digestion and absorption. The small intestine of an adult human is more than 6 meters long; its coils fill much of the lower abdominal cavity (Figure 43.16). Because of its length, and because of the folds, villi, and microvilli of its lining, its inner surface area is enormous: about 550 square meters, or roughly the size of a tennis court. Across this surface the small intestine absorbs the nutrient molecules from food. The small intestine has three sections: The initial section—the **duodenum**—is the site of most digestion; the **jejunum** and the **ileum** carry out 90 percent of the absorption of nutrients.

Digestion requires many specialized enzymes as well as several other secretions. Two accessory organs that are not part of the digestive tract—the liver and the pancreas—provide many of these enzymes and secretions. The liver synthesizes a substance called **bile** from cholesterol. Bile emulsifies fats just as soap emulsifies grease on your clothes or hands. Bile secreted from the liver flows through the **hepatic duct**. A side branch of the hepatic duct delivers bile to the gallbladder, where it is stored until it is needed to assist in fat digestion (Figure 43.17). Below the branching point to the gallbladder, the hepatic duct is called the **common bile duct**. When undigested fats enter the duodenum, the gallbladder releases bile, which flows down the common bile duct and enters the duodenum.

To understand the role of bile in fat digestion, think of the oil in salad dressing; it is not soluble in water (it is hydrophobic) and tends to aggregate together in large globules. The enzymes that digest fat, the **lipases**, are water-soluble and must do their work in an aqueous medium, but the fats are not water-soluble. Therefore, the interface between the aqueous digestive juices and large globules of fat would be very small if it weren't for bile. Bile stabilizes tiny droplets of fat so that they cannot aggregate into large globules. One end of each bile molecule is soluble in fat (lipophilic, or hydrophobic); the other end is soluble in water (hydrophilic, or lipophobic). Bile molecules bury their lipophilic ends in fat droplets, leaving their lipophobic ends sticking out. As a result, they prevent the fat droplets from sticking together. These very small fat particles are called **micelles**, and their small size maximizes the surface area exposed to lipase action (see Figure 43.19).

The **pancreas** is a large gland that lies just beneath the stomach. It has both endocrine (secreting to the inside of the body) and exocrine (secreting to the outside of the body) functions. Here we will consider the exocrine products, which it delivers to the gut

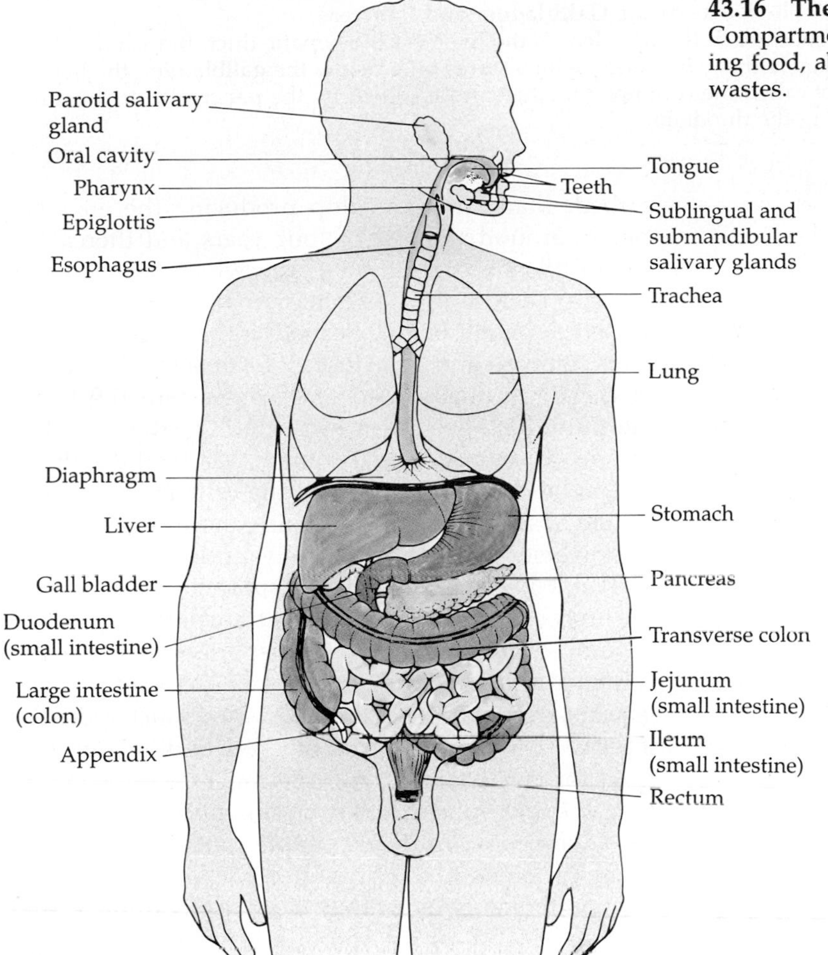

43.16 The Human Gut
Compartments within this long tube specialize in digesting food, absorbing nutrients, and storing and expelling wastes.

through the pancreatic duct. The pancreatic duct joins the common bile duct before entering the duodenum (see Figure 43.17). The pancreas produces a host of digestive enzymes (Table 43.3). As in the stomach, these enzymes are released as zymogens; otherwise they would digest the pancreas and its ducts before they ever reached the duodenum. Once in the duodenum, one of these inactive enzymes, **trypsinogen**, is activated by **enterokinase**, which is produced by cells lining the duodenum (Figure 43.18). Active **trypsin** can cleave other trypsinogen molecules to release even more active trypsin (another example of autocatalysis). Similarly, trypsin acts on the other zymogens secreted by the pancreas and releases their active enzymes. The mixture of zymogens produced by the pancreas can be very dangerous if the pancreatic duct is blocked or if the pancreas is injured by an infection or a severe blow to the abdomen. A few trypsinogen molecules spontaneously converting to trypsin can initiate a chain reaction of enzyme activation that digests the pancreas in a very short period of time, destroying both its endocrine and exocrine functions.

The pancreas produces, in addition to digestive enzymes, a secretion rich in bicarbonate ions (HCO_3^-). Bicarbonate ions neutralize the pH of the chyme that enters the duodenum from the stomach. This neutralization is essential because intestinal enzymes function best at a neutral or slightly alkaline pH.

Absorption in the Small Intestine

Only the smallest products of digestion pass through the mucosa of the small intestine and into the blood and lymphatic vessels that lie in the submucosa. The final digestion of proteins and carbohydrates that produces these absorbable products takes place among the microvilli. The mucosal cells with microvilli produce dipeptidase, which cleaves larger peptides into tripeptides, dipeptides, and individual amino acids that the cells can absorb. These cells also produce the enzymes maltase, lactase, and sucrase, which cleave the common disaccharides into their constituent, absorbable monosaccharides—glucose, galactose, and fructose. Disaccharides are not ab-

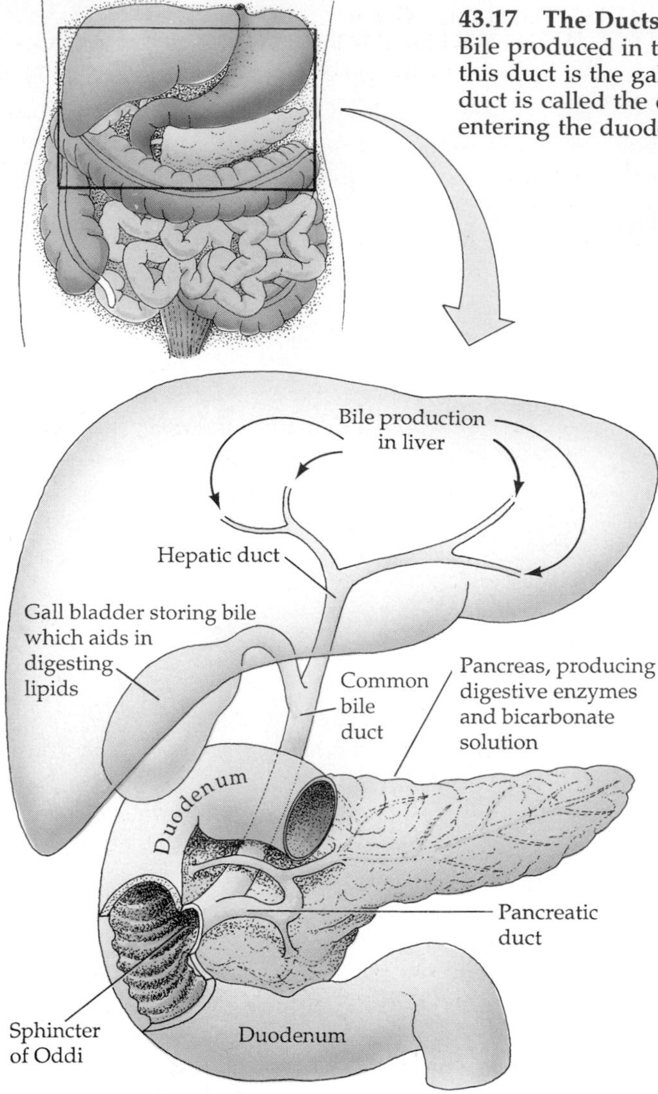

43.17 The Ducts of the Gallbladder and Pancreas
Bile produced in the liver leaves the liver via the hepatic duct. Branching off this duct is the gallbladder, which stores bile. Below the gallbladder, the hepatic duct is called the common bile duct and is joined by the pancreatic duct before entering the duodenum.

sorbed. Many humans stop producing the enzyme lactase around the age of four years and thereafter have difficulty digesting lactose, which is the sugar in milk. Lactose is a disaccharide and cannot be absorbed without being cleaved into its constituent units, glucose and galactose. If a substantial amount of lactose is unabsorbed and passes into the large intestine, its metabolism by bacteria in the large intestine causes abdominal cramps, gas, and diarrhea.

The mechanisms by which the cells lining the intestine absorb nutrient molecules and inorganic ions are diverse and not completely understood. Many inorganic ions are actively transported into or across the mucosa. For example, active transporters exist for sodium, calcium, and iron. Transporters also exist for certain classes of amino acids and for glucose and galactose, but curiously their activity is much reduced if active sodium transport is blocked. Sodium diffuses from the gut contents into the mucosal cells and is then actively transported from the mucosal cells into the submucosa. To diffuse into a mucosal cell, a sodium ion binds to a carrier molecule in the mucosal cell membrane, which also binds a nutrient molecule such as glucose or an amino acid. The diffusion of the sodium ion, driven by a concentration difference, therefore drives the absorption of the nutrient molecule. This mechanism is called **sodium cotransport** (see Figure 5.15).

TABLE 43.3
Sources and Functions of the Major Digestive Enzymes of Humans

ENZYME	SOURCE	ACTION	SITE OF ACTION
Salivary amylase	Salivary glands	Starch → Maltose	Mouth
Pepsin	Stomach	Proteins → Peptides; autocatalysis	Stomach
Pancreatic amylase	Pancreas	Starch → Maltose	Small intestine
Lipase	Pancreas	Fats → Fatty acids and glycerol	Small intestine
Nuclease	Pancreas	Nucleic acids → Nucleotides	Small intestine
Trypsin	Pancreas	Proteins → Peptides; activation of zymogens	Small intestine
Chymotrypsin	Pancreas	Proteins → Peptides	Small intestine
Carboxypeptidase	Pancreas	Peptides → Peptides and amino acids	Small intestine
Aminopeptidase	Small intestine	Peptides → Peptides and amino acids	Small intestine
Dipeptidase	Small intestine	Dipeptides → Amino acids	Small intestine
Enterokinase	Small intestine	Trypsinogen → Trypsin	Small intestine
Nuclease	Small intestine	Nucleic acids → Nucleotides	Small intestine
Maltase	Small intestine	Maltose → Glucose	Small intestine
Lactase	Small intestine	Lactose → Galactose and glucose	Small intestine
Sucrase	Small intestine	Sucrose → Fructose and glucose	Small intestine

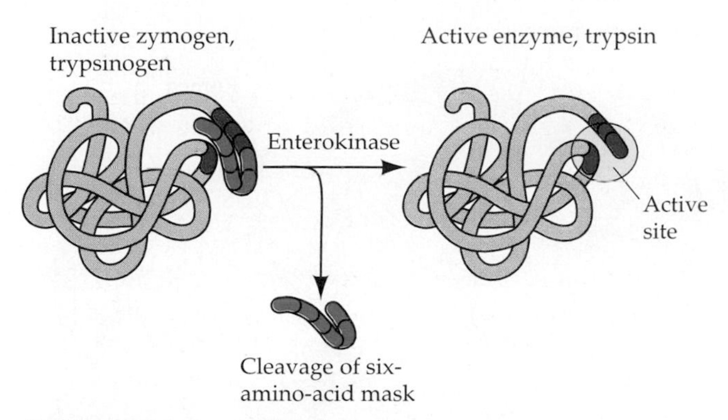

43.18 Zymogen Activation
Powerful digestive enzymes often exist as inactive zymogens until their catalytic activity is required. The zymogen trypsinogen, for example, is secreted from the pancreas; when it reaches the small intestine, the enzyme enterokinase cleaves a chain of six amino acids that masked the active site, transforming trypsinogen into the active digestive enzyme trypsin.

The absorption of the products of fat digestion is much simpler (Figure 43.19). Lipases break down fats into fatty acids and monoglycerides, which are lipid-soluble and are thus able to dissolve in the membranes of the microvilli and diffuse into the mucosal cells. Once in the cells, they are resynthesized into triglycerides, combined with cholesterol and phospholipids, and coated with protein to form water-soluble **chylomicrons**, which are really little particles of fat. The chylomicrons pass into the lymphatic vessels in the submucosa and into the bloodstream through the thoracic duct. Following a meal rich in fats, the chylomicrons can be so abundant in the blood that they give it a milky appearance.

The bile that emulsifies the fats is not absorbed along with the monoglycerides and the fatty acids, but is recycled back and forth between the gut contents and the microvilli. Finally, in the ileum, bile is actively reabsorbed and returned to the liver via the bloodstream. Bile is synthesized in the liver from cholesterol. Cholesterol is also synthesized in the liver, and additionally comes from our diets.

Remember that high cholesterol levels contribute to arterial plaque formation and therefore to cardiovascular disease (see Box 42.B). The body has no way of breaking down excess cholesterol, so high dietary intake or high levels of synthesis create problems. One major way cholesterol leaves the body is through the elimination of unreabsorbed bile in the feces. The rationale for including certain kinds of fiber in our diet is that the fiber binds the bile, decreases its reabsorption in the ileum, and thus helps to lower body cholesterol.

The Large Intestine

Peristalsis gradually pushes the contents of the small intestine into the large intestine, or **colon**. The rate of peristalsis is controlled so that food passes through the small intestine slowly enough for digestion and absorption to be complete, but quickly enough to ensure an adequate supply of nutrients for the body. The material that enters the colon has had most of its nutrients removed, but it contains a lot of water and inorganic ions. The colon reabsorbs water and ions, producing semisolid feces from the slurry of indigestible materials it receives from the small intestine. If too much water is reabsorbed, constipation

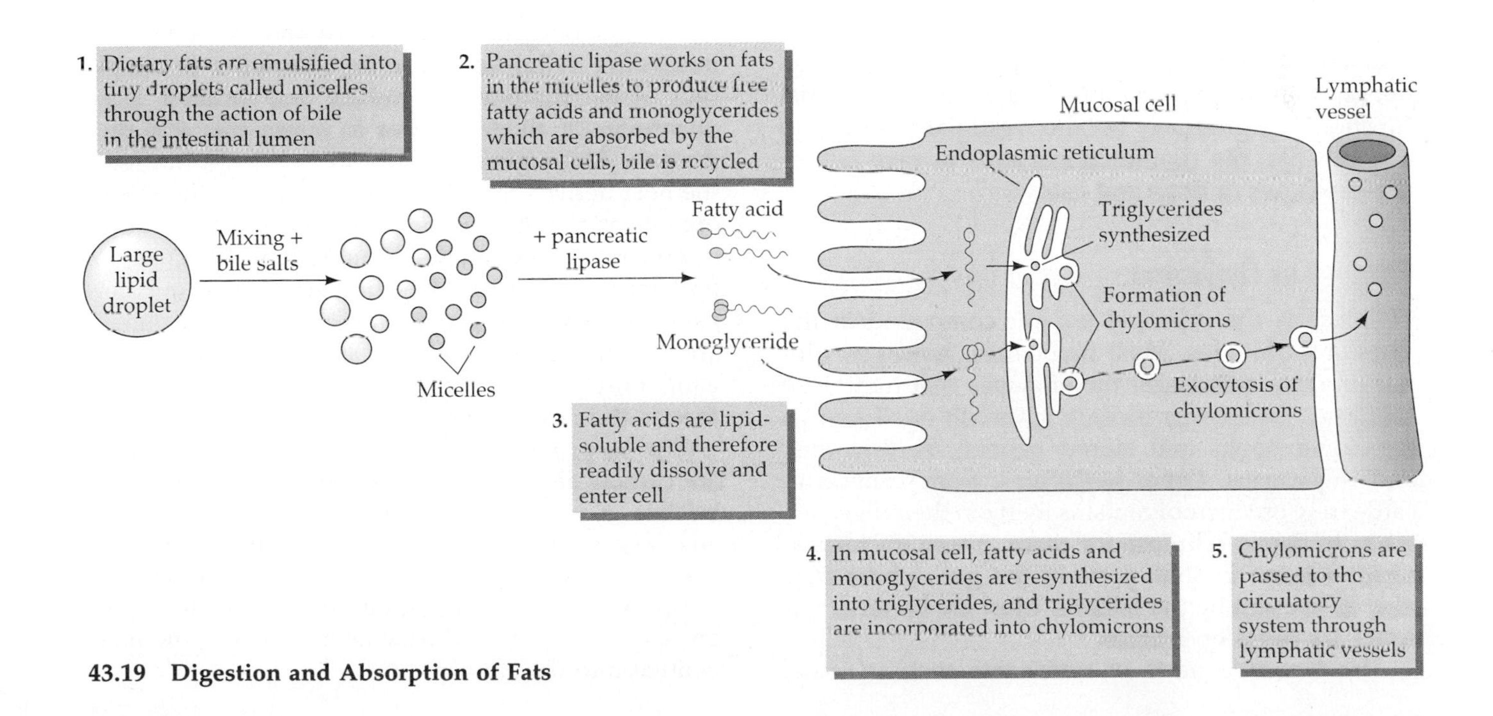

43.19 Digestion and Absorption of Fats

can result. The opposite condition, diarrhea, results if too little water is reabsorbed or if water is secreted into the colon. (Both constipation and diarrhea can be induced by toxins from microorganisms.) Feces are stored in the last segment of the colon and periodically excreted.

Immense populations of bacteria live within the colon. One of the resident species is *Escherichia coli*, the bacterium so popular with researchers in biochemistry, genetics, and molecular biology (see Box 4.B). This inhabitant of the colon lives on matter indigestible to humans and produces some products useful to the host. For example, vitamin K and biotin are synthesized by *E. coli* and absorbed across the wall of the colon. Many species of mammals maximize the nutritional benefits from such bacterial activity by reingesting their own feces, a behavior called **coprophagy**. Excessive or prolonged intake of antibiotics can lead to vitamin deficiency because the antibiotics kill the normal intestinal bacteria at the same time they are killing the disease-causing organisms for which they are intended. The intestinal bacteria produce gases such as methane and hydrogen sulfide as by-products of their largely anaerobic metabolism. Humans expel gas after eating beans because the beans are rich in carbohydrates that bacteria—but not humans—can break down.

The large intestine of humans has a small, fingerlike pouch called the **appendix**, which is best known for the trouble it causes when it becomes infected. The human appendix plays no essential role in digestion, but it does contribute to immune system function. It can be surgically removed without serious consequences. The part of the gut that forms the appendix in humans forms the much larger caecum in herbivores (see Figure 43.12), where it functions in cellulose digestion. As our primate ancestors evolved to exploit diets less rich in indigestible cellulose, the caecum no longer served an essential function and gradually became **vestigial** (reduced to a trace), like the nonfunctional eyes of cave fish or the dewclaws of dogs and cats.

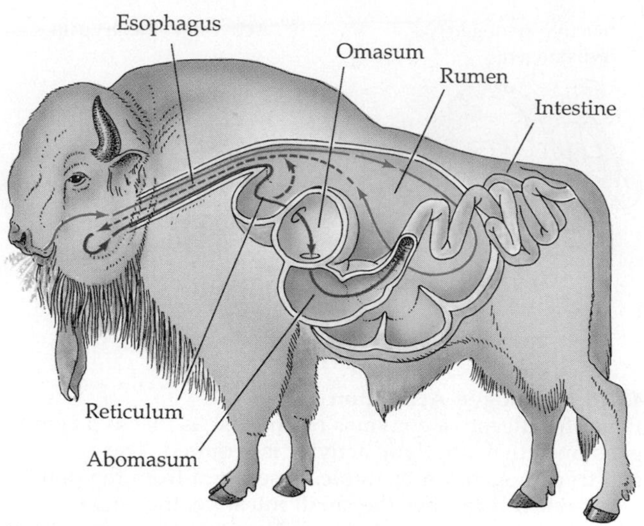

43.20 The Ruminant Stomach
Specialized stomach compartments—the rumen, reticulum, omasum, and abomasum—enable ruminants to digest and subsist on protein-poor plant material.

Digestion by Herbivores

Cellulose is the principal organic compound in the diets of herbivores. Most herbivores, however, cannot produce **cellulases**, the enzymes that hydrolyze cellulose. Exceptions include silverfish (well known for eating books and stored papers), earthworms, and shipworms. Other herbivores, from termites to cattle, rely on microorganisms living in their digestive tracts to digest cellulose for them. These microorganisms inhabit various parts of the gut, where they may be present by the billions. Most are bacteria, but some are fungi or protists.

The digestive tracts of **ruminants** such as cattle, goats, and sheep are specialized to maximize benefits from microorganisms. In place of the usual mammalian stomach, ruminants have a large, four-chambered organ (Figure 43.20). The first and largest of these chambers is the **rumen**, the second is the **reticulum**. Both are packed with anaerobic microorganisms that break down cellulose. These two chambers serve as fermentation vats for the digestion of cellulose. The ruminant periodically regurgitates the contents of the rumen (the cud) into the mouth for rechewing. When the more thoroughly ground-up vegetable fibers are swallowed again, they present more surface area to the microorganisms for their digestive actions.

The microorganisms in the rumen and reticulum metabolize cellulose and other nutrients to simple fatty acids. Ruminants produce and swallow large quantitites of alkaline saliva to buffer this acid production. Although fatty acids are the major nutrients the host derives from its microorganisms, the microorganisms themselves provide an important source of protein. A cow can derive more than 100 grams of protein per day from digestion of its own microorganisms. The plant materials ingested by a ruminant are a poor source of protein, but they contain inorganic nitrogen that the microorganisms use to synthesize their own amino acids.

The by-products of the fermentation of cellulose are carbon dioxide and methane, which the animal belches. A single cow can produce 400 liters of methane a day. Methane is the second most abundant "greenhouse gas," whose concentration in the atmosphere is increasing, and domesticated ruminants are second only to industry as a source of methane emitted into the atmosphere.

The food leaving the rumen carries with it enormous numbers of the cellulose-fermenting microorganisms. This mixture passes through the **omasum**, where it is concentrated by water reabsorption. It then enters the true stomach, the **abomasum**, which secretes hydrochloric acid and proteases. The microorganisms are killed by the acid, digested by the proteases, and passed on to the small intestine for further digestion and absorption. The rate of multiplication of microorganisms in the rumen is great enough to offset their loss, so a well-balanced, mutually beneficial relationship is maintained.

As we mentioned earlier, mammalian herbivores other than ruminants have microbial farms and cellulose fermentation vats in a branch off the large intestine called the caecum. Rabbits and hares are good examples (see Figure 43.12). Since the caecum empties into the large intestine, the absorption of the nutrients produced by the microorganisms is inefficient and incomplete. Therefore, these animals practice coprophagy. They frequently produce two kinds of feces, one of pure waste (which they discard), and one consisting mostly of caecal material, which they ingest directly from the anus. As this caecal material passes through the stomach and small intestine, the nutrients it contains are digested and absorbed.

CONTROL AND REGULATION OF DIGESTION

The vertebrate gut could be described as an assembly line in reverse. As with a standard assembly line, control and coordination of sequential processes is critical. Both neural and hormonal controls govern gut functions.

Neural Reflexes

Everyone has experienced salivation stimulated by the sight or smell of food. That response is a neural reflex, as is the act of swallowing following tactile stimulation at the back of the mouth. Many neural reflexes coordinate activities in different regions of the digestive tract so that it works in a properly timed, assembly line manner. For example, loading the stomach with food stimulates increased activity in the colon, which can lead to a bowel movement. This phenomenon is the **gastrocolic reflex**. The digestive tract is unusual in that it has an intrinsic (that is, its own) nervous system. So in addition to neural reflexes involving central nervous system, such as salivation and swallowing, neural messages can travel within the digestive tract without being processed by the central nervous system.

About one hundred years ago, the Russian physiologist Ivan Pavlov, in an effort to explain all regulation and control of the digestive tract in terms of neural reflexes, discovered a very basic form of learning, the **conditioned reflex**. When a dog is presented with food, it salivates—an unconditioned reflex. If a bell is rung whenever the dog is presented with food, after a number of trials the dog will salivate whenever the bell is rung, even if no food is present—a conditioned reflex. Pavlov tried to explain the control of the secretory activity of the pancreas in terms of a neural reflex. He showed that the presence in the duodenum of acidic chyme from the stomach stimulated the pancreas to secrete digestive juices. Even hydrochloric acid alone applied to the duodenum stimulated the pancreas to secrete. When Pavlov tried to discover the path of the neural reflex controlling this response, however, he ran into trouble. The response could not be eliminated by destroying neural connections between the gut, the central nervous system, and the pancreas. Other researchers even removed a section of duodenum from the digestive tract of an animal and sutured it into a closed loop with no neural connections. They placed the loop back in the abdominal cavity, and when acid was injected into it, the pancreas was stimulated to secrete.

You have probably guessed the explanation of Pavlov's dilemma, but the answer was first demonstrated by two British physiologists, William Bayliss and Ernest Starling, who were working at the same time as Pavlov. They removed a section of small intestine from an animal and scraped off its mucosal lining. They ground the mucosa with sand, filtered it, and injected the extract into the bloodstream of an animal. The recipient animal secreted pancreatic juices. This was the first demonstration of a chemical message traveling through the circulatory system and having an effect on a specific tissue. They named their newly discovered chemical message **secretin** and also coined the general term "hormone." The moral of the story is that control of the functions of the digestive tract is not by neural reflexes alone; it is also by hormonal mechanisms.

Hormonal Controls

Several hormones control the activities of the digestive tract and its accessory organs (Figure 43.21). Bayliss and Starling's mucosal cell extract probably contained multiple hormones, but the dominant one, secretin, is responsible primarily for stimulating the pancreas to secrete a solution rich in bicarbonate ions. In response to fats and proteins in the chyme, the mucosa of the small intestine also secretes **cholecystokinin**, a hormone that stimulates the gallbladder to release bile and the pancreas to release digestive enzymes. Cholecystokinin and secretin also slow down the movements of the stomach, which slows the delivery of chyme into the small intestine.

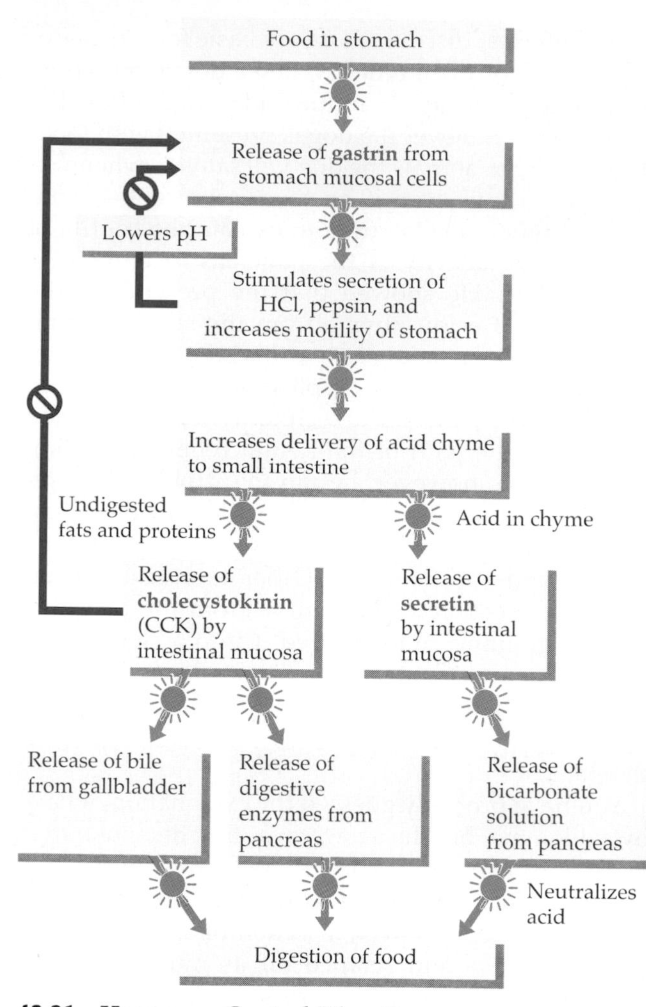

43.21 Hormones Control Digestion
Several hormones (blue type) are involved in feedback loops that control the sequential processing of food in the digestive tract.

The stomach secretes a hormone called **gastrin** into the blood. Cells in the lower region of the stomach release gastrin when they are stimulated by the presence of food. Gastrin circulates in the blood until it reaches cells in the upper areas of the stomach wall, where it stimulates the secretions and movements of the stomach. Gastrin release is inhibited when the stomach contents become too acidic—another example of negative feedback.

CONTROL AND REGULATION OF FUEL METABOLISM

Most animals do not eat continuously. There are times when food is in the gut and nutrients are being absorbed and are readily available to supply energy and molecular building blocks. When nutrients are not being absorbed, however, the continuous processes of energy metabolism and biosynthesis must run off of internal reserves. For this reason nutrient traffic must be controlled so that reserves accumulate during absorption and those reserves are used appropriately when the gut is empty.

The Role of the Liver

The liver directs the traffic of nutrient molecules used in energy metabolism (fuel molecules). When nutrients are abundant in the circulatory system, the liver can store them in the forms of glycogen (animal starch) and fat. The liver also synthesizes plasma proteins from circulating amino acids. When the availability of fuel molecules in the bloodstream declines, the liver delivers glucose and fats back to the blood. The liver has an enormous capacity to interconvert fuel molecules. Liver cells can convert monosaccharides into either glycogen or fat, and vice versa. Certain amino acids and some other molecules, such as pyruvate and lactate, can be converted into glucose—a process called **gluconeogenesis**. The liver is also the major controller of fat metabolism through its production of lipoproteins (Box 43.C).

Hormonal Control of Fuel Metabolism

The **absorptive period** refers to the time that food is in the gut and nutrients are being absorbed and circulated in the blood. During this time the liver converts glucose to glycogen and fat, the body fat tissues convert glucose and fatty acids to stored fat, and the cells of the body preferentially use glucose for metabolic fuel. When food is no longer in the gut—the **postabsorptive period**—these processes reverse. The liver breaks down glycogen to supply glucose to the blood; the liver and the fat tissues supply fatty acids to the blood; and most of the cells of the body preferentially use fatty acids for metabolic fuel. The major exception to this rule is the cells of the nervous system, which require a constant supply of glucose for their energetic needs. Although the nervous system can use other fuels to a limited extent, its overall dependence on glucose is the reason it is so important for other cells of the body to shift to fat metabolism during the postabsorptive period. This shift preserves the available glucose and glycogen stores for the nervous system for as long as possible.

What directs the traffic in fuel molecules? Two hormones produced and released by the pancreas, **insulin** and **glucagon**, are largely responsible for controlling the metabolic directions fuel molecules take (Figure 43.22). The most important of these hormones is insulin, which is produced in response to high blood glucose levels. The pancreas releases insulin into the circulatory system when blood glucose rises above the normal postabsorptive level of about 90 milligrams of glucose per 100 milliliters of plasma.

BOX 43.C

Lipoproteins: The Good, the Bad, and the Ugly

In the intestine, bile solves the problem of processing hydrophobic fats in an aqueous medium. The *transportation* of fats in the circulatory system presents the same problem, but the solution is lipoproteins. A **lipoprotein** is a particle made up of a core of fat and cholesterol and a covering of protein that makes it hydrophilic. The largest lipoprotein particles are the chylomicrons produced by the cells lining the intestine to transport dietary fat and cholesterol into the circulation (see Figure 43.19). As lipoproteins circulate through the liver and the fat tissues around the body, receptors on the capillary walls recognize the protein coat, and lipases begin to hydrolyze the fats, which are then absorbed into fat or liver cells. Thus the protein coat of the lipoprotein makes it water-soluble and serves as an "address" that targets the lipoprotein to a specific tissue.

Other lipoproteins originate in the liver and are classified according to their density. Because fat has a low density (it floats in water), the more fat a lipoprotein contains, the lower its density. Very-low-density lipoprotein (VLDL) produced by the liver contains mostly triglyceride fats that are being transported to fat cells in tissues around the body. Low-density lipoproteins (LDL) contain mostly cholesterol, which they transport to tissues around the body for use in biosynthesis and to be deposited. High-density lipoproteins (HDL) serve as acceptors of cholesterol and are believed to remove cholesterol from tissues and return it to the liver, where it can be used to synthesize bile. Because of their differing functions in cholesterol regulation, LDL is sometimes called "bad cholesterol" and HDL "good cholesterol," but those designations are somewhat controversial at present. We do know, however, that a high ratio of LDL to HDL in a person's blood is a risk factor for atherosclerotic heart disease. Cigarette smoking lowers HDL levels, and regular exercise increases HDL levels.

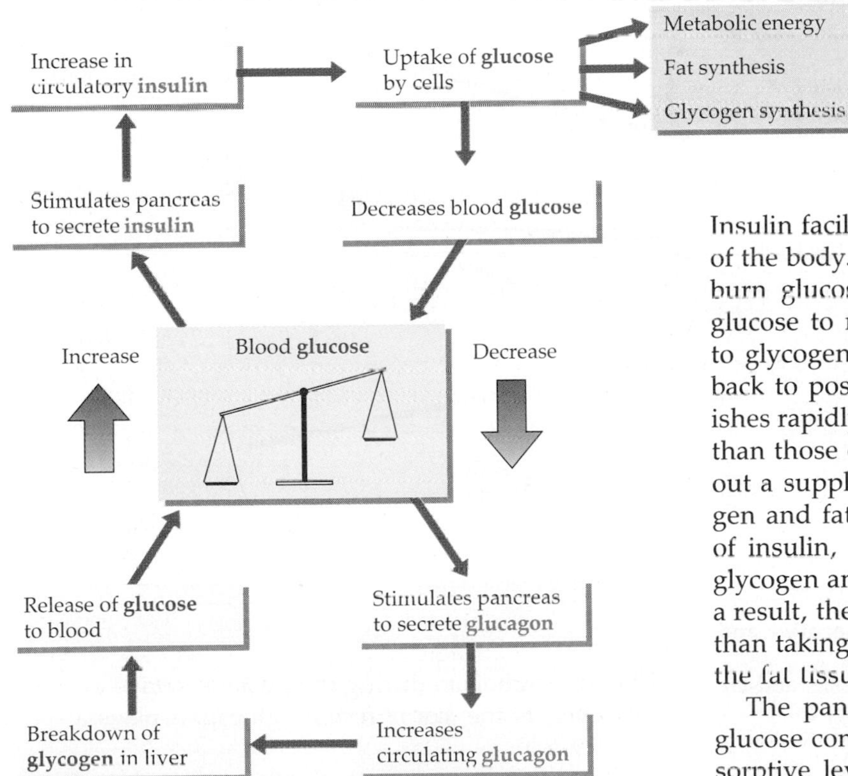

43.22 Regulation of Blood Glucose
Insulin and glucagon maintain the homeostasis of blood glucose.

Insulin facilitates the entry of glucose into most cells of the body. Thus when insulin is present, most cells burn glucose as their metabolic fuel, fat cells use glucose to make fat, and liver cells convert glucose to glycogen and fat. As soon as blood glucose falls back to postabsorptive levels, insulin release diminishes rapidly, and the entry of glucose into cells other than those of the nervous system is inhibited. Without a supply of glucose, cells switch to using glycogen and fat as their metabolic fuels. In the absence of insulin, the liver and fat cells stop synthesizing glycogen and fat and begin breaking them down. As a result, the liver supplies glucose to the blood rather than taking it from the blood, and both the liver and the fat tissues supply fatty acids to the blood.

The pancreas releases glucagon when the blood glucose concentration falls below the normal postabsorptive level. Glucagon has the opposite effect of insulin; it stimulates liver cells to break down glycogen and carry out gluconeogenesis, thus releasing glucose into the blood. The major hormonal control

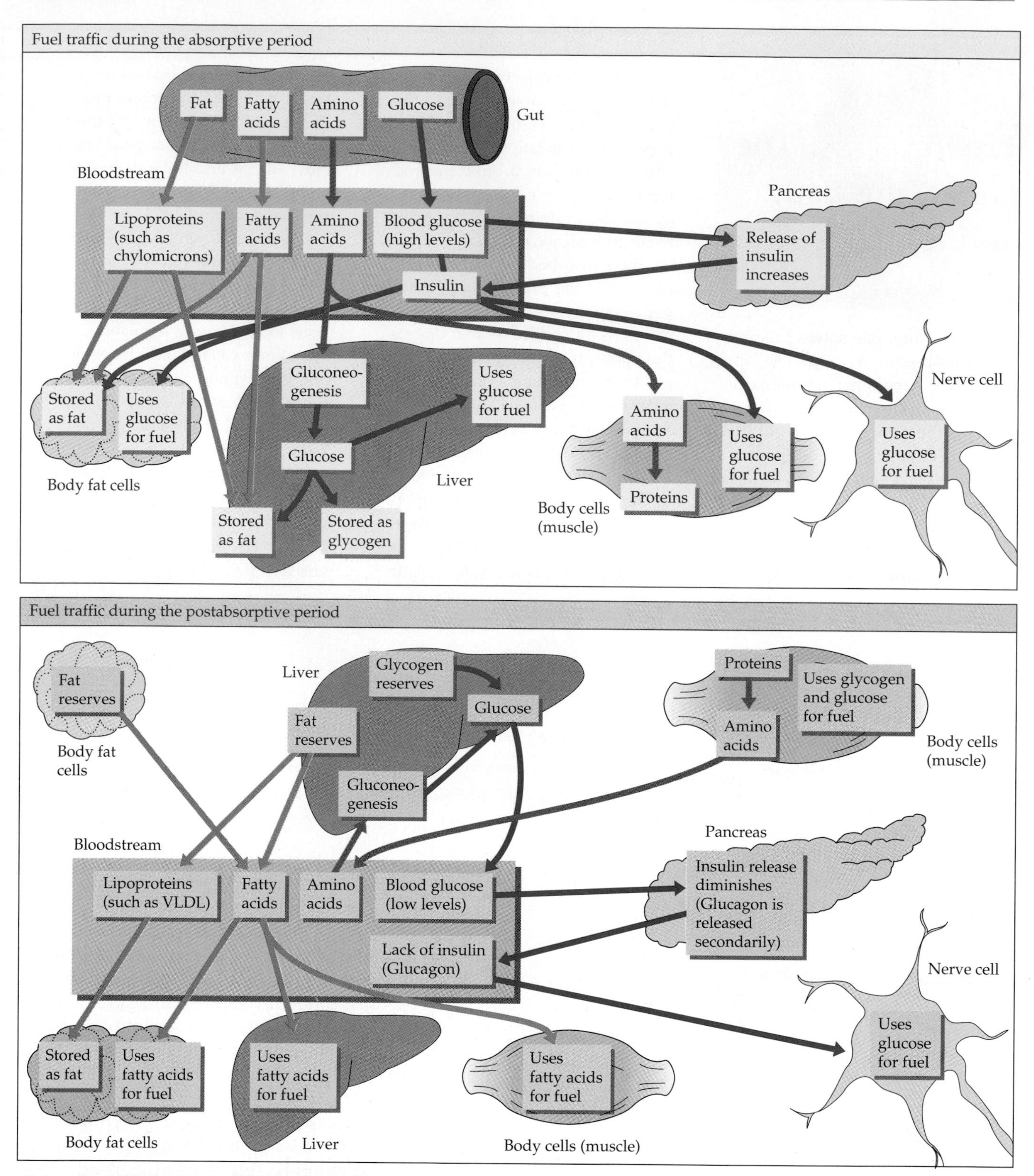

43.23 Fuel Molecule Traffic during the Absorptive and Postabsorptive Periods

Insulin promotes glucose uptake by liver, muscle, and fat cells during the absorption period. During the postabsorptive period the lack of insulin blocks glucose uptake by these same tissues and promotes fat and glycogen breakdown to supply metabolic fuel. Red arrows = carbohydrate traffic; blue arrows = protein/amino acid traffic; green arrows = fat/fatty acid traffic.

of fuel metabolism during the postabsorptive period, however, is the *lack of insulin*; glucagon plays a secondary role.

Two other hormones that help preserve blood glucose levels during the postabsorptive period are epinephrine and the glucocorticoid cortisol. Low blood glucose is a stress that triggers glucose-sensitive cells in the hypothalamus to signal the adrenal medulla, through the sympathetic nervous system, to secrete epinephrine. Epinephrine has an effect on liver cells lates the adrenal cortex to secrete cortisol. Cortisol inhibits glucose metabolism in many cells while promoting metabolism of fats and proteins.

The traffic of fuel molecules during the absorptive and postabsorptive periods is summarized in Figure similar to that of glucagon. Through the cAMP second-messenger cascade (see Chapter 36), it increases the breakdown of glycogen and increases gluconeogenesis. The stress of low blood glucose also stimu- 43.23. The steps controlled by insulin and glucagon are indicated. Note that during the absorptive period, the direction of traffic in all fuel molecules is toward storage, and glucose is the preferred energy source for all cells. During the postabsorptive period, most cells switch to metabolizing fat so that blood glucose reserves are saved for the nervous sytem. The level of circulating glucose is maintained through glycogen breakdown and gluconeogenesis.

SUMMARY of Main Ideas about Animal Nutrition

Carbohydrates, fats, and proteins in food supply animals with metabolic energy and carbon skeletons for biosynthesis.
Review Figures 43.2 and 43.4

An animal with insufficient caloric intake is undernourished and must metabolize its own carbohydrate, fat, and finally its own protein for energy.
Review Figure 43.3

Proper nutrition requires specific molecules such as essential amino acids, essential fatty acids, and vitamins.
Review Figure 43.5 and Table 43.2

Certain mineral elements are required nutrients.
Review Table 43.1

Inadequate supplies of required nutrients result in malnutrition and may lead to deficiency diseases.

Animals have evolved a great variety of adaptations for exploiting sources of food.

Carnivores eat other animals, herbivores eat plants, omnivores eat other animals and plants, and detritivores eat the decomposition products of dead organisms.

Some animals are deposit feeders, some are fluid feeders, and others are filter feeders.

Teeth are a general but highly variable feeding adaptation of vertebrates.
Review Figures 43.9 and 43.10

Guts are cavities or tubes where digestion and absorption occur.

Guts are usually divided into separate compartments with specialized functions but with the same basic tissue structure throughout.
Review Figures 43.12 and 43.14

Some compartments of the gut may have large populations of microorganisms that aid in digesting molecules which otherwise would be indigestible to the host.
Review Figure 43.20

In vertebrate guts, food passes from the mouth through the esophagus to the stomach, and then to the small intestine, followed by the large intestine.
Review Figure 43.16

The stomach stores food and initiates some aspects of digestion.

Digestive enzymes reduce carbohydrate, fat, and protein molecules to the simplest units that can be absorbed by cells of the small intestine and circulated to the body.
Review Figure 43.19 and Table 43.3

Bile from the liver, bicarbonate ions from the pancreas, and enzymes from the pancreas (in the form of inactive zymogens) are secreted into the small intestine through the common bile duct.
Review 43.17 and 43.18

Most nutrients are absorbed in the small intestine.

In the large intestine, or colon, water and ions are reabsorbed into the body, and feces are formed and stored.

The activities of the various compartments of the gut are controlled and coordinated by hormones and by neural reflexes involving the central nervous system and the intrinsic nervous system of the gut.
Review Figure 43.21

Fuel metabolism is regulated mainly by the pancreatic hormones insulin and glucagon.
Review Figure 43.22

The liver manages the traffic in fuel molecules.
Review Figure 43.23

During the absorptive period the liver absorbs the excess fuel molecules, interconverting the different types, and stores them as glycogen and fat.

During the postabsorptive period the liver breaks down its glycogen and fat to supply the blood with glucose and fatty acids.

SELF-QUIZ

1. Most of the metabolic energy a bird requires for a long-distance migratory flight is stored as
- *a.* glycogen.
- *b.* fat.
- *c.* protein.
- *d.* carbohydrate.
- *e.* ATP.

2. Which of the following statements about essential amino acids is true?
- *a.* They are not found in vegetarian diets.
- *b.* They are stored by the body for when they are needed.
- *c.* Without them one is undernourished.
- *d.* All animals require the same ones.
- *e.* Humans can acquire all of theirs by eating milk, eggs, and meat.

3. Which of the following statements about vitamins is true?
- *a.* They are essential inorganic nutrients.
- *b.* They are required in larger amounts than are essential amino acids.
- *c.* Many serve as coenzymes.
- *d.* Vitamin D can be acquired only by eating meat or dairy products.
- *e.* When vitamin C is eaten in large quantities, the excess is stored in fat for later use.

4. The digestive enzymes of the small intestine
- *a.* do not function best at a low pH.
- *b.* are produced and released in response to circulating secretin.
- *c.* are produced and released under neural control.
- *d.* are all secreted by the pancreas.
- *e.* are all activated by an acidic environment.

5. Which of the following statements about nutrient absorption across the gut epithelium is true?
- *a.* Carbohydrates are absorbed as disaccharides.
- *b.* Fats are absorbed as fatty acids and monoglycerides.
- *c.* Amino acids move across only by diffusion.
- *d.* Bile salts transport fats across.
- *e.* Most nutrients are absorbed in the duodenum.

6. Chylomicrons are like the tiny particles of dietary fat in the lumen of the small intestine in that
- *a.* both are coated with bile salts.
- *b.* both are lipid-soluble.
- *c.* both travel in lacteals.
- *d.* both contain triglyceride.
- *e.* both are coated with lipoproteins.

7. Microbial fermentation in the guts of cattle
- *a.* produces fatty acids as a major nutrient for the cattle.
- *b.* occurs in specialized regions of the small intestine.
- *c.* occurs in the caecum, from which food is regurgitated to be chewed again and swallowed into the true stomach.
- *d.* produces methane as a major nutrient.
- *e.* is possible because the stomach wall does not secrete hydrochloric acid.

8. Which of the following is stimulated by cholecystokinin?
- *a.* Stomach motility
- *b.* Release of bile
- *c.* Secretion of hydrochloric acid
- *d.* Secretion of bicarbonate ions
- *e.* Secretion of mucus

9. During the absorptive period
- *a.* breakdown of glycogen supplies glucose to blood.
- *b.* glucagon secretion is high.
- *c.* the number of circulating lipoproteins is low.
- *d.* glucose is the major metabolic fuel.
- *e.* synthesis of fats and glycogen in muscle is inhibited.

10. During the postabsorptive period
- *a.* glucose is the major metabolic fuel.
- *b.* glucagon stimulates the liver to produce glycogen.
- *c.* insulin facilitates the uptake of glucose by brain cells.
- *d.* the major metabolic fuel is fatty acids.
- *e.* liver functions slow down because of low insulin levels.

FOR STUDY

1. From what you have learned about nutrition in this chapter, discuss some of the problems with "crash" or "fad" diets. What should one take into account when considering or planning a diet aimed at weight reduction?

2. The digestive tract must move food slowly enough to enable digestion and absorption but fast enough to supply the animal's energetic needs. Describe controls that speed up and slow down the activities of the digestive tract.

3. Describe the role of the liver in the homeostasis of blood glucose. What are the controlling factors?

4. Why is obstruction of the common bile duct so serious? Consider in your answer the multiple functions of the pancreas and the way in which digestive enzymes are processed.

5. Trace the history of a fatty acid molecule from being on a piece of buttered toast to being in a plaque on a coronary artery. What possible forms and structures could it have passed through in the body? Describe a direct and an indirect route it could have taken.

READINGS

Atkinson, M. A. and N. K. MacLaren. 1990. "What Causes Diabetes?" *Scientific American*, July. The body's own immune system can cause this serious disease that destroys the ability to regulate fuel metabolism.

Brown, M. S. and J. L. Goldstein. 1984. "How LDL Receptors Influence Cholesterol and Atherosclerosis." *Scientific American*, November. A detailed discussion of what does and does not happen to the cholesterol and fatty acids we consume.

Davenport, H. 1972. "Why the Stomach Does Not Digest Itself." *Scientific American*, January. The stomach contains hydrochloric acid but usually does not digest its own lining.

Degabriele, R. 1980. "The Physiology of the Koala." *Scientific American*, July. An amazing adaptation to an unusual and limited diet.

Lienhard, G. E., J. W. Slot, D. E. James and M. M. Mueckler. 1992. "How Cells Absorb Glucose." *Scientific American*, January. Research on the glucose transporter and how it is regulated by insulin.

Moog, F. 1981. "The Lining of the Small Intestine." *Scientific American*, November. How nutrients are absorbed.

Orci, L., J.-D. Vassalli and A. Perrelet. 1988. "The Insulin Factory." *Scientific American*, September. The molecular biology of the synthesis and release of insulin by beta cells of the pancreas.

Sanderson, S. L. and R. Wassersug. 1990. "Suspension-Feeding Vertebrates." *Scientific American*, March. By filtering small particles out of the water, some animals exploit an abundant food resource.

Schmidt-Nielsen, K. 1990. *Animal Physiology: Adaptation and Environment*, 4th Edition. Cambridge University Press, New York. An outstanding textbook, emphasizing the comparative approach.

Scrimshaw, N. S. 1991. "Iron Deficiency." *Scientific American*, October. A discussion of one of the most common dietary deficiencies in humans and its consequences.

Uvnas-Moberg, K. 1989. "The Gastrointestinal Tract in Growth and Reproduction." *Scientific American*, July. A concise treatment of hormonal controls of digestive tract function and how they change to accommodate special needs of pregnancy and lactation.

Vander, A. J., J. H. Sherman and D. S. Luciano. 1994. *Human Physiology: The Mechanisms of Body Function*, 6th Edition. McGraw-Hill, New York. Chapters 17 and 18 provide a readily understandable treatment of digestion and absorption of food.

Water Is Essential for Life

During the dry season on the African savanna, animals must congregate around the scarce sources of water.

44

Salt and Water Balance and Nitrogen Excretion

Blood, sweat, and tears all taste salty, like seawater. Life evolved in the sea, and if a complex animal is to exist elsewhere, it must carry its own internal sea to bathe the cells of its body. This statement sometimes is taken to mean that the mineral ion, or salt, composition of our blood is the same as that of the ancient seas. That conclusion is wrong. Although all animals on Earth share the same distant origins, there is great diversity in the composition of their body fluids, and no animal has an internal environment just like the environment in which life evolved. However, all animals do require both water and salts.

The availability of salts and water can be the critical characteristic of an environment that determines which organisms can live there, in what numbers, and when. Some animals with extreme adaptations can live in the hottest, driest deserts on Earth, rarely, if ever, experiencing liquid water in their environment, but their populations are very sparse. Many species living in habitats that are seasonally dry migrate long distances to find water. During the dry season on the African plains, predators and prey alike congregate at scarce water holes. The availability of salts and water in the external environment is only part of the story, however. Most animals must regulate the salt and water composition of their internal environments within narrow limits. How they do so is the focus of this chapter.

INTERNAL ENVIRONMENT

The extracellular fluids (including interstitial fluid and the blood plasma) that we carry with us service all our cells (see Chapter 35). In addition to supplying cells with oxygen and nutrients and carrying away waste products, these extracellular fluids determine the water balance of the cells. To understand what is meant by water balance, recall that cell membranes are permeable to water and that the movement of water across membranes depends on differences in osmotic potential. (You may find it useful to review the discussion of osmosis in Chapter 5.) If the osmotic potential of the extracellular fluid is less negative (that is, the fluid contains fewer solutes) than that of the intracellular fluids, water moves into the cells, causing them to swell and possibly burst. If the osmotic potential of the extracellular fluid is more negative (the fluid contains more solutes) than that of the intracellular fluids, the cells lose water and shrink. The osmotic potential of the extracellular environment determines both the volume and osmotic potential of the intracellular environment.

Excretory systems consist of the organs that help regulate the osmotic potential and the volume of the extracellular fluids. In addition, excretory systems regulate the composition of the extracellular fluids by excreting molecules that are in excess (such as NaCl when we eat lots of salty popcorn) and conserving those that are valuable or in short supply (such as glucose and amino acids). In terrestrial organisms, excretory systems also eliminate the toxic waste products of nitrogen metabolism.

Exactly what the excretory system of a particular species must do depends on the environment in which that species lives. In this chapter we will examine excretory systems that maintain salt and water balance and eliminate nitrogen in marine, freshwater, and terrestrial habitats. In spite of the evolutionary diversity of the anatomical and physiological details, all these systems obey a common rule and employ common mechanisms. The common rule is that there is no active transport of water; water must be moved either by pressure or by a difference in osmotic potential. The common mechanisms derive from the fact that most excretory organs consist of systems of tubules that receive extracellular fluid and alter its composition to produce **urine**—the fluid waste product that is excreted. The extracellular fluid enters the excretory tubules by **filtration**, and its composition is changed by processes of active **secretion** and **reabsorption** of specific solute molecules by the cells of the tubules. These same three mechanisms—filtration, secretion, and reabsorption—are used both in systems that excrete water and conserve salts and in systems that do the opposite, conserving water and excreting salts. The intestine (see Chapter 43) is an important excretory organ. It absorbs water and controls the loss of water with the solid wastes of digestion. Intestinal cells influence the solute composition of the internal environment through the selective transport of ions.

WATER, SALTS, AND THE ENVIRONMENT

We think of marine and freshwater environments as being distinctly different, one salty and the other not. In reality, aqueous environments grade continuously from fresh to extremely salty. Consider a place where a river enters the sea through a bay or a marsh. Aqueous environments within that bay or marsh range in salinity (salt content) from the fresh water of the river to the open sea. Evaporating tide pools can become much saltier than the sea. Animals live in all these environments.

Most marine invertebrates can adjust to a wide range of environmental salinities by allowing their body fluids to have the same osmotic potential as the environment; they thus avoid the risk of being burst or shrunk by osmotic movement of water. Such animals are called **osmoconformers**. There are limits to osmoconformity, however. No animal could have the same osmotic potential as fresh water and survive; nor could animals survive with internal salt concentrations as high as those that may be reached in an evaporating tide pool. Such concentrations cause proteins to denature. Animals that maintain an osmotic potential of their internal fluids different from that of their environment are called **osmoregulators**. Even animals that osmoconform over a wide range of osmotic potentials must osmoregulate at the extremes of environmental salinity.

To osmoregulate in fresh water, animals must continuously excrete the water that invades their bodies by osmosis, but while doing so they must conserve solutes; hence they produce large amounts of dilute urine. To osmoregulate in salt water, animals must conserve water and excrete salts, thus tending to produce small amounts of urine.

The brine shrimp *Artemia* is adapted to live in environments of almost any salinity. *Artemia* are found in huge numbers in the most saline environments possible, such as the Great Salt Lake in Utah or coastal evaporation ponds where salt is obtained for commercial purposes (Figure 44.1). The salinity of such water reaches 300 grams per liter (normal seawater contains about (35 g/l). *Artemia* are harvested from these environments and sold for fish food. *Artemia* cannot survive for long in fresh water, but they can live in very dilute seawater, in which they maintain the osmotic potential of their body fluids above the osmotic potential of the environment. Under these conditions, *Artemia* is a **hypertonic osmoregulator**, meaning that it regulates the concentration (the osmolarity) of its body fluids so that they are more concentrated than (hypertonic to) the environment (Figure 44.2). At high environmental salinities, *Artemia* is an exceptionally effective hypotonic osmoregulator, keeping the osmotic potential of its body fluids well below that of the water in which it is living. Very few organisms can survive in the crystallizing brine in which *Artemia* thrives. The main mechanism this small crustacean uses for osmoregulation is the active transport of NaCl across its gill membranes.

Osmoconformers can be **ionic conformers**; they can control ionic composition, as well as the osmolarity, of their body fluids to match that of the environment. Most osmoconformers, however, are **ionic regulators** to some degree: They employ active transport mechanisms to maintain specific ions in their body fluids at concentrations different from those in the environment.

The terrestrial environment presents entirely different sets of problems for salt and water balance. Because the terrestrial environment is extremely desiccating (drying), most terrestrial animals must conserve water. Exceptions are animals such as muskrats and beavers that practically live in water. Terrestrial

44.1 Evaporating Salt Ponds Vary in Salinity
(a) Impoundments of water around San Francisco Bay yield salt when they evaporate. Because different species of bacteria and algae grow best at different salinities, the colors of the salt ponds change as they evaporate. *(b)* One animal that can live in these ponds at all salinities is a tiny crustacean, the brine shrimp (*Artemia*).

animals obtain their salts from their food. Plants generally have low concentrations of sodium, so most herbivores must be very effective in conserving sodium ions. As we mentioned in Chapter 43, some terrestrial herbivores travel long distances to naturally occurring salt licks to supplement their dietary intake of sodium. By contrast, birds that feed on marine animals must excrete the large excess of sodium they ingest with their food. Their **nasal salt glands** excrete a concentrated solution of sodium chloride via a duct that empties into the nasal cavity. Birds such as penguins and sea gulls that have nasal salt glands can be seen frequently sneezing or shaking their heads to get rid of the very salty droplets that form (Figure 44.3).

THE EXCRETION OF NITROGENOUS WASTES

The end products of the metabolism of carbohydrates and fats are water (H_2O) and carbon dioxide (CO_2); they present no problems for excretion. Proteins and nucleic acids, however, contain nitrogen in addition to carbon, hydrogen, and oxygen. The metabolism of proteins and nucleic acids produces nitrogenous waste in addition to H_2O and CO_2. Most of that waste is ammonia (NH_3). Ammonia is highly toxic and must be excreted continuously to prevent accumulation, or detoxified by conversion into other molecules for excretion. Those molecules are principally **urea** and **uric acid** (Figure 44.4).

Ammonia excretion is relatively simple for aquatic animals. Ammonia diffuses and is highly soluble in water. Animals that breathe water continuously lose ammonia from their blood to the environment by diffusion across their gill membranes. Animals that excrete nitrogen as ammonia are called **ammonotelic**;

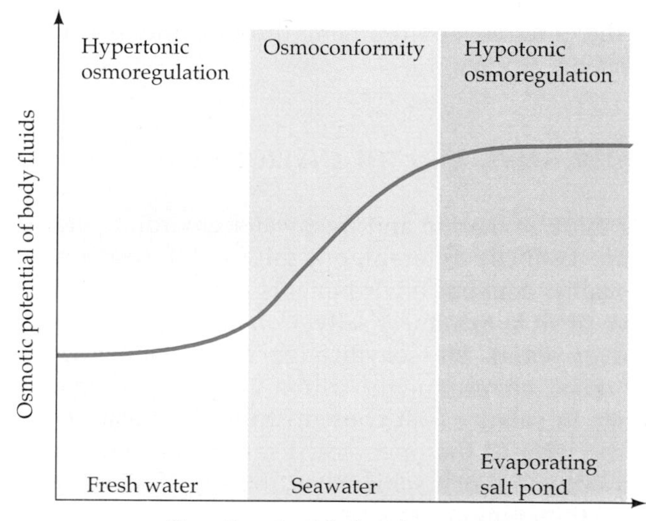

44.2 Osmoregulation and Osmoconformity
Over a broad range of environmental salinities, many marine invertebrates are osmoconformers. Animals that live at the extremes of environmental salinities, however, can display osmoregulatory abilities. They become hypertonic osmoregulators in very dilute water, or hypotonic osmoregulators in very saline water.

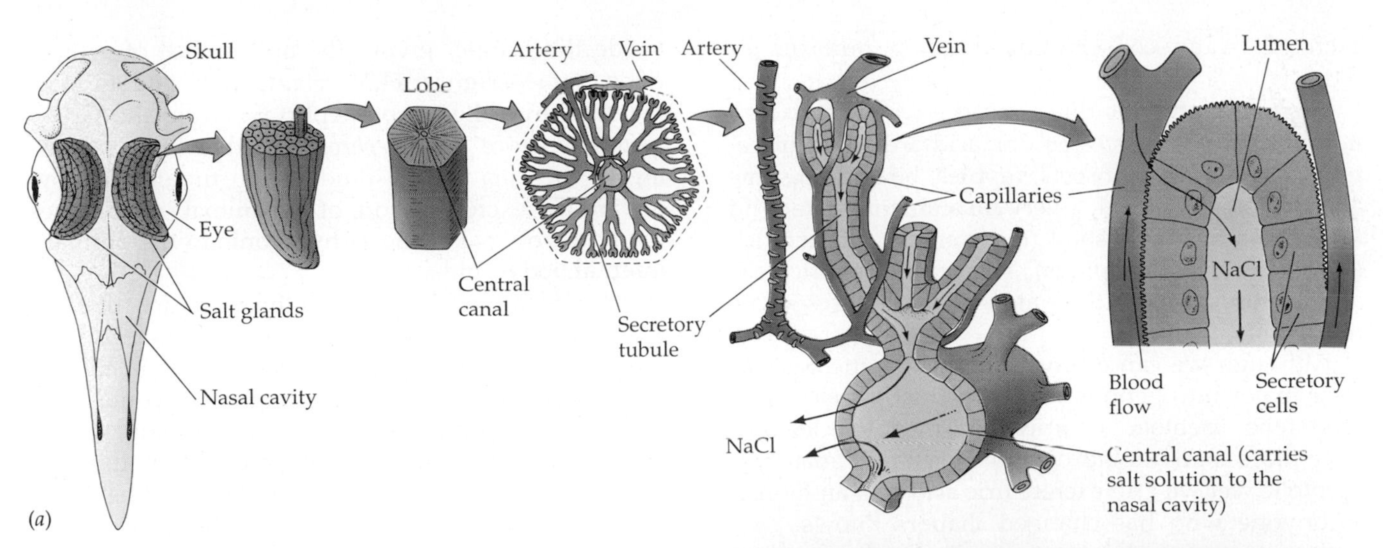

44.3 Nasal Glands Excrete Excess Salt
(a) Marine birds such as sea gulls and penguins have nasal salt glands adapted to excrete the excess salt from the seawater they consume with their food. Salt glands lie in grooves in the skull above the eyes. *(b)* This Adélie penguin has returned from a feeding trip at sea and is excreting salt through its nasal salt gland. A patch of evaporated salt lies on the rock below the bird's beak.

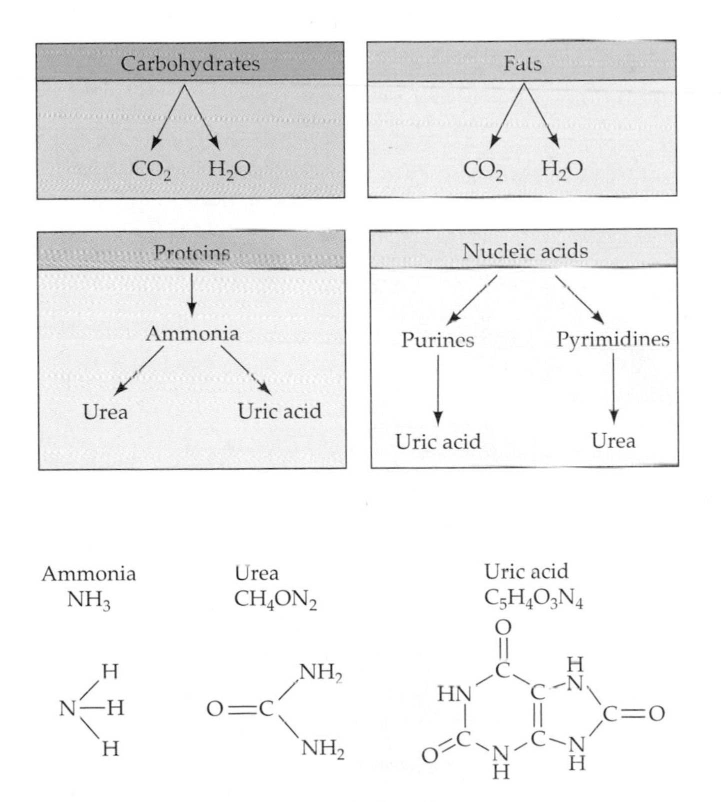

44.4 Waste Products of Metabolism
Whereas the metabolism of carbohydrates and fats yields only water and carbon dioxide, the metabolism of proteins and nucleic acids produces the nitrogenous wastes ammonia, uric acid, and urea. The metabolism of nucleic acids begins with their breakdown into their constituent bases—pyrimidines and purines (see Chapter 3).

they include aquatic invertebrates and bony fishes. Crocodiles and amphibian tadpoles are also ammonotelic.

Ammonia is a more dangerous metabolite for terrestrial animals that have limited access to water. In mammals, ammonia is lethal when it reaches only 5 milligrams per 100 milliliters of blood. Therefore, terrestrial (and some aquatic) animals convert ammonia into either urea or uric acid. **Ureotelic** animals, such as mammals, amphibians, and cartilaginous fishes (sharks and rays), excrete urea as their principal nitrogenous waste product. Urea is quite soluble in water, but excretion of urea solutions at low concentrations could result in a large loss of water that many terrestrial animals could ill afford. Later in the chapter we will see that mammals have evolved excretory systems that produce urine, which contains urea and is hypertonic to their body, thereby conserving water while excreting the urea. The cartilaginous fishes are another story. These marine species maintain their body fluids hypertonic to the marine environment by retaining high concentrations of urea. Because water moves into their bodies by osmosis and must be excreted, water conservation is not a problem for them.

Terrestrial animals that conserve water by excreting nitrogenous wastes as uric acid are called **uricotelic**. These include insects, reptiles, birds, and some amphibians. Uric acid is very insoluble in water and is excreted as a semisolid (for example, the whitish material in bird droppings). Therefore, the uricotelic animal loses very little water as it disposes of its nitrogenous waste.

Although we can classify animals on the basis of their major nitrogenous waste product as being ammonotelic, ureotelic, or uricotelic, most species produce more than one nitrogenous waste. Humans are ureotelic, yet we also excrete uric acid and ammonia, as anyone who has changed diapers knows. (Actually, most of the ammonia in diapers is produced by the bacterial breakdown of urea. The bacterium that performs this reaction was first isolated by a microbiologist from a diaper of his child.) The uric acid in human urine comes largely from the metabolism of nucleic acids and of caffeine. In the classical disease gout, uric acid levels in the body fluids increase, and uric acid precipitates in the joints and elsewhere, causing swelling and pain.

In some species, different developmental forms live in quite different habitats and have different forms of nitrogen excretion. Tadpoles of frogs and toads, for example, excrete ammonia across their gill membranes, but when they develop into adult frogs or toads they generally excrete urea. Some adult frogs and toads that live in arid habitats excrete uric acid. These examples show the considerable evolutionary flexibility in how nitrogenous wastes are excreted.

INVERTEBRATE EXCRETORY SYSTEMS

Most marine invertebrates are osmoconformers, so they have few adaptations for salt and water balance other than the active transport of specific ions. For nitrogen excretion they can passively lose ammonia by diffusion to the seawater. Freshwater and terrestrial invertebrates, however, display a variety of fascinating adaptations for maintaining salt and water balance and excreting nitrogen. Although diverse, all these adaptations are based on the same basic principles: filtration of body fluids and active secretion and reabsorption of specific ions.

Protonephridia

Many flatworms, such as **Planaria** (see Figure 43.11), live in fresh water and excrete water through an elaborate network of tubules running throughout their bodies. The tubules end in **flame cells**, so called because each flame cell has a tuft of cilia beating inside the tubule, giving the appearance of a flickering flame (Figure 44.5). Flame cells and tubules together are called **protonephridia** (from the Greek *proto* = "before" and *nephros* = "kidney"). The beating of the cilia causes fluid in the tubules to flow toward the excretory pore of the animal. As it leaves the flatworm, this fluid is hypotonic to the animal's internal body fluids.

How does the fluid get into the excretory tubules? Two possibilities exist. The first is that the beating of the cilia creates a slight negative pressure in the tubule. Water from the surrounding tissue could then filter into the tubule because of the pressure difference between the tissues and the excretory tubules. The fluid that filtered into the tubule would have the same osmolarity as the tissue fluids, but as it flowed down the tubules the cells of the tubules could actively reabsorb solutes from this tubular fluid or urine. The second possibility is that ions are actively transported into the excretory tubules at their upper ends. Water would follow these ions because of the increased osmotic potential of the tubular fluid. Then, as the fluid moved down the tubules, ions could be actively reabsorbed. In either case, the urine leaving the protonephridia would be hypotonic to the internal fluids of the flatworm.

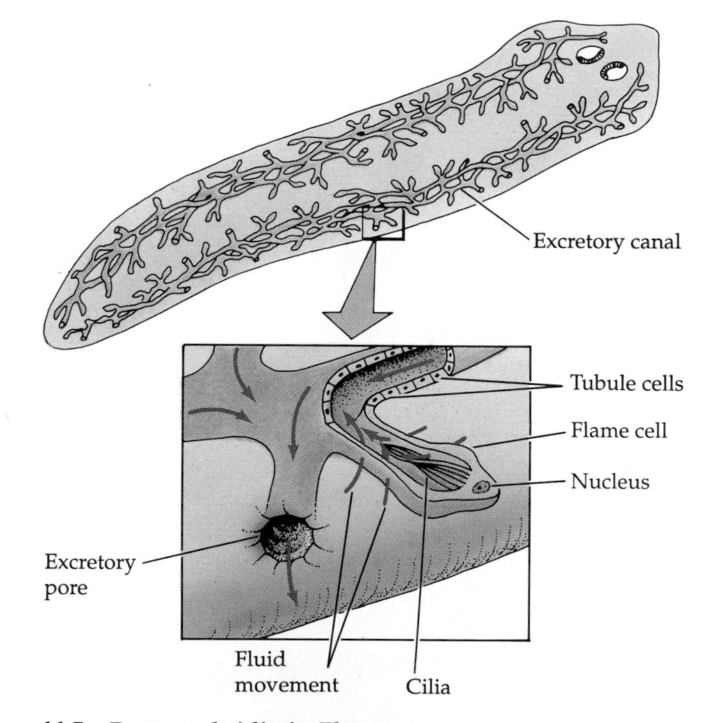

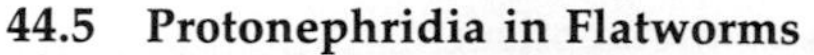

44.5 Protonephridia in Flatworms
The protonephridia of the flatworm *Planaria* consist of tubules ending in flame cells. Body fluids enter the space enclosed by the flame cell and are driven down the tubules toward the excretory pore by the beating of the cilia in the flame cells. The tubule cells modify the composition of this fluid.

Metanephridia

Filtration of body fluids and tubular processing of urine are highly developed in annelid worms, such as the earthworm. Recall that annelids have fluid-filled body cavities called coeloms (see Figure 26.24) and closed circulatory systems through which blood is pumped under pressure (see Figure 42.3). The pressure causes the blood to be filtered across the thin, permeable capillary walls into the coelom. This process is called filtration because the cells and large protein molecules of the blood stay behind in the capillaries while water and small molecules leave the capillaries and enter the coelom. Where does this coelomic fluid go?

Each segment of the earthworm contains a pair of **metanephridia** (*meta* is Greek for "akin to"). Each metanephridium begins in one segment as a ciliated, funnel-like opening in the coelom called a **nephrostome**, which leads into a tubule in the next segment. The tubule ends in a pore called the nephridiopore, which opens to the outside of the animal (Figure 44.6). Coelomic fluid enters the metanephridia through the nephrostomes. As the coelomic fluid passes through the tubules, the cells of the tubules actively reabsorb certain molecules from it and actively secrete other molecules into it. What leaves the animal through the nephridiopores is a hypotonic urine containing nitrogenous wastes, among other solutes.

Why should earthworms, which are terrestrial, excrete a hypotonic urine? Earthworms live in moist soil, an environment with 100 percent relative humidity, and they are in constant contact with a film of water covering the soil particles. Thus, like freshwater aquatic animals, an earthworm must excrete the excess water that continually enters its body by osmosis.

In the metanephridium we see all the basic processes used in the excretory systems of vertebrates that will be discussed later in this chapter: filtration of the body fluids, and tubular processing of the filtrate by active secretion and reabsorption of ions and molecules.

Malpighian Tubules

Insects have remarkable systems for excreting nitrogenous wastes with very little loss of water. These animals can live in the driest habitats on Earth. The insect excretory system consists of blind tubules (from 2 to more than 100) attached to the gut between the midgut and hindgut and projecting into the fluid-filled coelom (Figure 44.7). The cells of these **Malpighian tubules** actively transport uric acid, potassium ions, and sodium ions from the coelomic fluid into the tubules. As solutes are secreted into the tubules, water follows because of the difference in osmotic potential. The walls of the Malpighian tubules have muscle fibers whose contractions help to move the contents of the tubules toward the hindgut.

In the hindgut the tubular fluid continues to change in composition. The hindgut contents are more acidic than the tubular fluids, and as a result the uric acid becomes less soluble and precipitates out of solution as it approaches and enters the rectum. The cells of the walls of the hindgut and rectum actively transport sodium and potassium ions from the gut contents back into the coelom. Because the uric acid molecules have precipitated out of solution, water is free to follow the reabsorbed salts back into the coelom through osmosis. Remaining in the rectum are crystals of uric acid mixed with undigested food, and this dry matter is all that the insect eliminates. The Malpighian tubule system is a highly effective mechanism for excreting nitrogenous wastes and some salts without giving up any significant fraction of the animal's precious water supply.

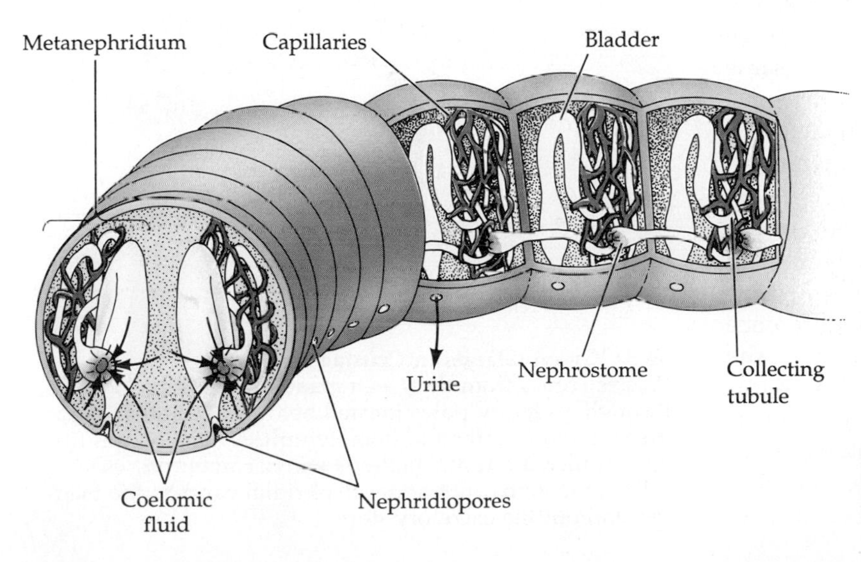

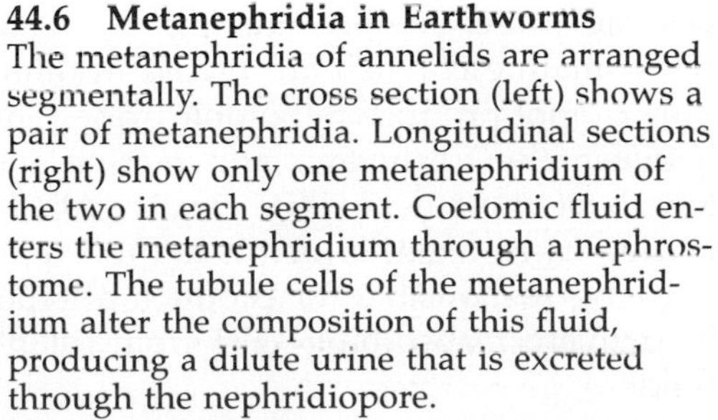
44.6 Metanephridia in Earthworms
The metanephridia of annelids are arranged segmentally. The cross section (left) shows a pair of metanephridia. Longitudinal sections (right) show only one metanephridium of the two in each segment. Coelomic fluid enters the metanephridium through a nephrostome. The tubule cells of the metanephridium alter the composition of this fluid, producing a dilute urine that is excreted through the nephridiopore.

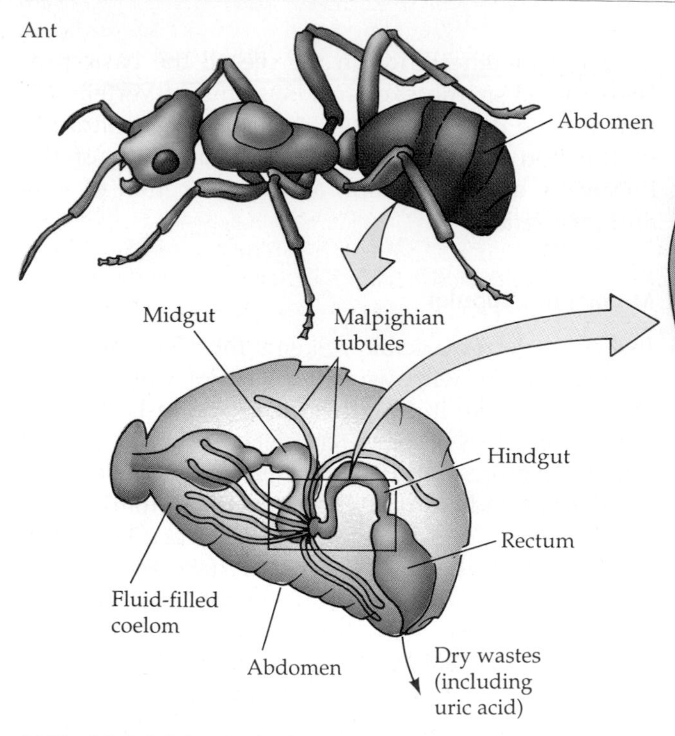

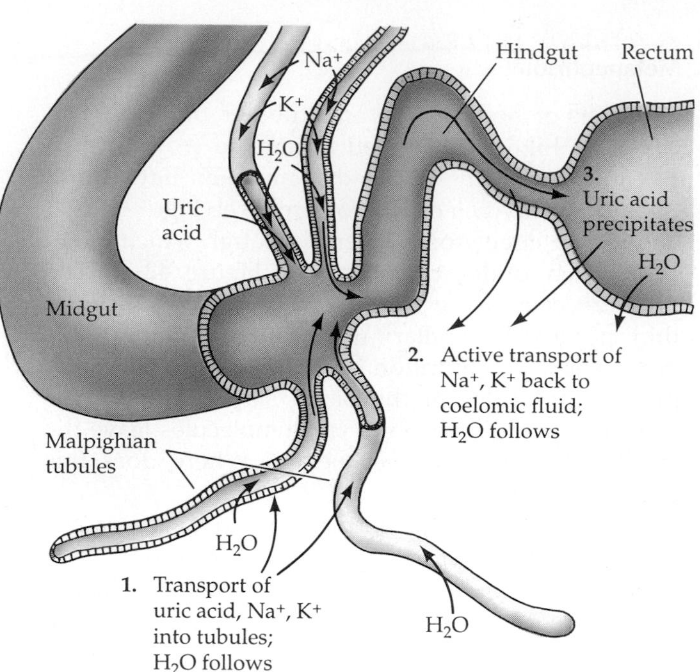

44.7 Malpighian Tubules in Insects
Blind, thin-walled Malpighian tubules are attached to the junction of the insect's mid- and hindgut.

Green Glands

Malpighian tubules are found in some arthropods, such as insects and spiders, but not in others, such as the crustaceans (crabs, lobsters, crayfish, and their relatives). The crustacean excretory system consists of an **end sac** and **labyrinth** connected by a **nephridial canal** to a bladder that empties to the outside by way of an excretory pore (Figure 44.8). This entire assembly in a crayfish or lobster is called a **green gland** and is regarded by many people as a gourmet treat. A key feature of this excretory system is that extracellular fluid (called hemolymph in these animals with open circulatory systems) is filtered into the end sac by the high pressure within the coelom. The volume and composition of the filtrate may be altered in the labyrinth, but it remains at the same osmolarity as the coelomic fluid. From the labyrinth, the fluid passes through a nephridial canal into a bladder, where it is stored until it is excreted through the excretory pore.

Freshwater crayfish produce large volumes of hypotonic urine; marine crayfish do not. Moreover, when marine crayfish are placed in dilute seawater, they cannot produce hypotonic urine. The difference between the freshwater and marine species is the length of their nephridial canals. All of this information taken together points to the nephridial canals as being responsible for producing hypotonic urine by actively reabsorbing salts from filtered coelomic fluids.

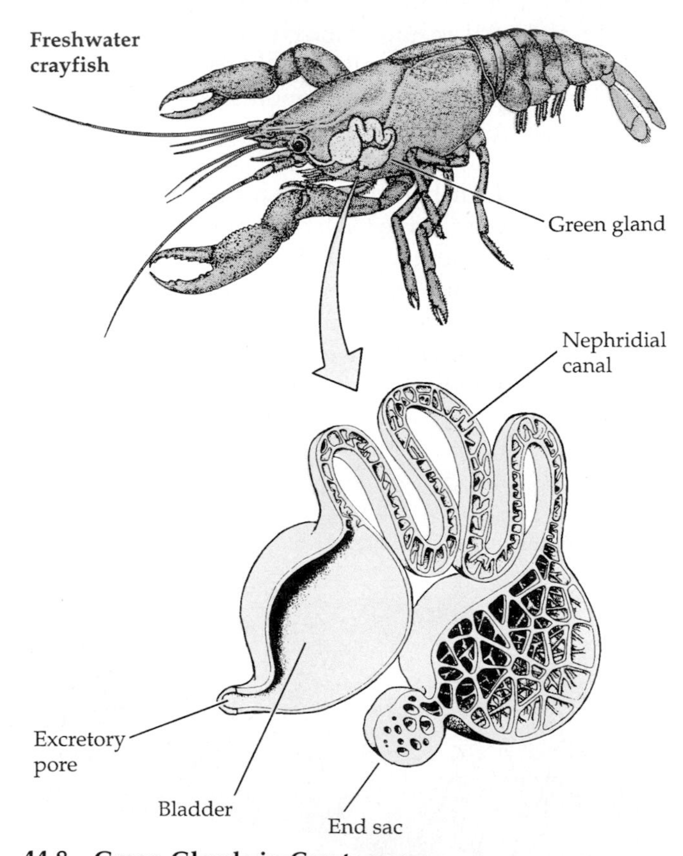

44.8 Green Glands in Crustaceans
Wastes empty from the green glands to the exterior through excretory pores located below each antenna. Each green gland is bathed in hemolymph. A hemolymph filtrate is forced into the bulbous end sac, then passes through the labyrinth and a nephridial canal to the bladder and out the excretory pore.

VERTEBRATE EXCRETORY SYSTEMS

The major vertebrate organ for salt and water balance and nitrogen excretion is the **kidney**. The functional unit of the kidney is the **nephron**. Each human kidney has about 1 million nephrons. To understand how the kidney works, you must understand the structure and function of the nephron. A remarkable fact about the kidneys of vertebrates is that in different species the same basic organ serves different needs. The kidneys of freshwater fish excrete water, whereas the kidneys of most mammals conserve water. To understand how the kidney can fulfill opposite functions in different animals, we need to look at the nephron.

The Structure and Functions of the Nephron

Each nephron has a vascular and a tubular component (Figure 44.9). The vascular component is unusual in consisting of two capillary beds between the arteriole that supplies it and the venule that drains it. The first capillary bed is a dense knot of very permeable vessels called a **glomerulus** (Figure 44.10*a*). Blood enters the glomerulus through what is called an **afferent arteriole** and exits through an **efferent arteriole**. The efferent arteriole gives rise to the second set of capillaries, the **peritubular capillaries**, which surround the tubular component of the nephron.

The tubular component of the nephron begins with **Bowman's capsule**, which encloses the glomerulus. The glomerulus appears to be pushed into Bowman's capsule much like a fist pushed into an inflated balloon. Together, the glomerulus and its surrounding Bowman's capsule are called the **renal corpuscle**. The cells of the capsule that come into direct contact with the glomerular capillaries are called **podocytes**. These highly specialized cells have numerous armlike extensions, each with hundreds of fine, fingerlike projections. The podocytes wrap around the capillaries so that their fingerlike projections cover the capillaries completely (Figure 44.10*b*,*c*).

The glomerulus filters the blood to produce a tubular fluid without cells and large molecules. The cells of the capillaries and the podocytes of Bowman's capsule participate in filtration. The walls of the capillaries have pores that allow water and small molecules to leave the capillary but that are too small to permit red blood cells and very large protein molecules to pass. Even smaller than the pores in the capillaries are the narrow slits between the fingerlike projections of the podocytes. The result is that water and small molecules pass from the capillary blood and enter the tubule of the nephron (Figure 44.10*d*), but red blood cells and proteins remain in the capillaries.

The force that drives filtration in the glomerulus is the pressure of the arterial blood. As in every

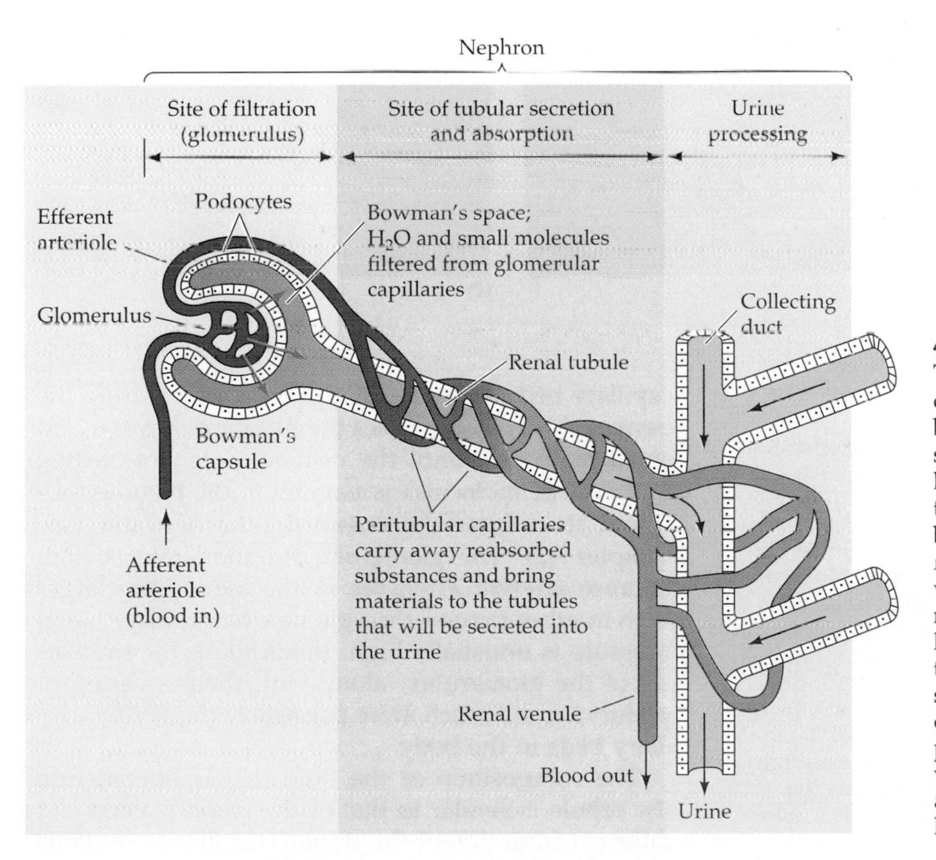

44.9 The Vertebrate Nephron
The nephron is a system of tubules closely associated with a system of blood vessels. An afferent arteriole supplies blood to the glomerulus, a knot of capillaries. The glomerulus is the site of blood filtration. It is drained by an efferent arteriole, which gives rise to the peritubular capillaries, which surround the tubules of the nephron. A venule drains the peritubular capillaries. The end of the renal tubule system envelops the glomerulus so that the filtrate from the glomerular capillaries enters the tubules. The processed filtrate (urine) of the individual nephrons enters collecting ducts and is delivered to a common duct leaving the kidney.

(a)

(b)

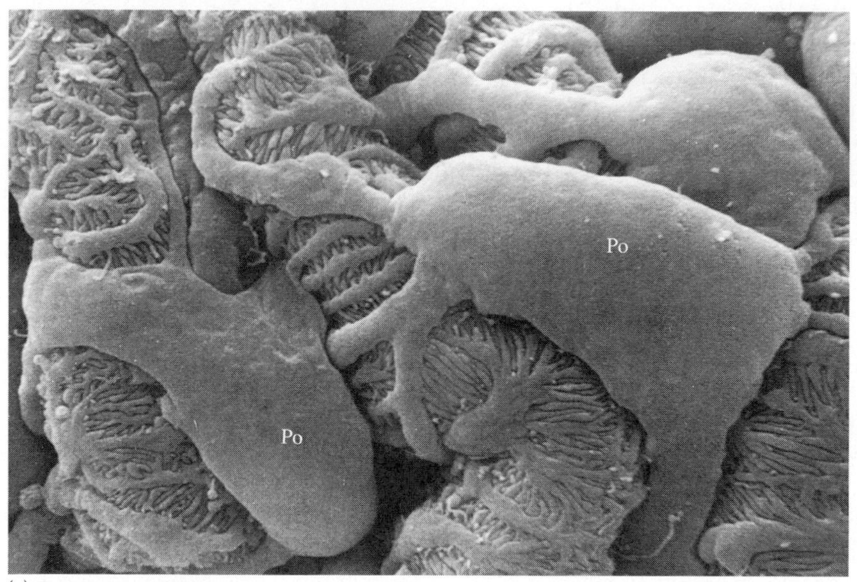

(c)

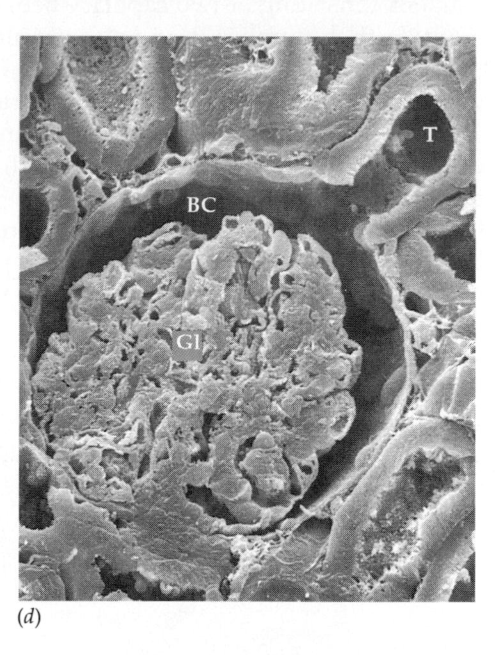

(d)

44.10 An SEM Tour of the Nephron
These scanning electron micrographs show the anatomical bases for kidney function. *(a)* When the blood vessels are filled with latex and all tissue etched away, we are left with a cast of the blood vessels in the kidney showing the knots of capillaries that are the glomeruli (Gl). Each glomerulus has an afferent and an efferent arteriole (Ar). Peritubular capillaries (Pt) are looser networks surrounding the tubules of the nephron. *(b)* In a live organism the capillaries of the glomeruli are tightly wrapped by specialized tubule cells called podocytes (Po) derived from the cells of the inner wall of Bowman's capsule. Here and in parts *(c)* and *(d)* we are looking at the glomerulus from inside Bowman's space. *(c)* Each podocyte has hundreds of tiny, fingerlike projections that create filtration slits between them. Anything passing from the glomerular capillaries into the tubule of the nephron must pass through these slits. (d) Bowman's capsule (BC) surrounds the glomerulus (Gl), collects the filtrate, and funnels it into the tubule (T) of the nephron.

capillary bed, the pressure of the blood entering the permeable capillary causes the filtration of water and small molecules until the osmotic potential created by the large molecules remaining in the blood is sufficient to counter the outward flow of water (see Chapter 42). The glomerular filtration rate is high because afferent arterioles in the kidney are larger than in other tissues; thus glomerular capillary blood pressure is unusually high. In addition, the capillaries of the glomerulus, along with their covering of podocytes, are much more permeable than other capillary beds in the body.

The composition of the fluid that is filtered into the tubule is similar to that of the blood plasma and different from that of the urine. This filtrate contains glucose, amino acids, ions, and nitrogenous wastes in the same concentrations as in the blood plasma, but it lacks the plasma proteins. As this fluid passes down the tubule, its composition changes as the cells of the tubule actively reabsorb certain molecules from the tubular fluid and secrete other molecules into it. When the tubular fluid leaves the kidney as urine, its composition is very different from that of the original filtrate. The function of the **renal tubules**—the tubules of the nephrons (see Figure 44.9)—is to control the composition of the urine through active secretion and reabsorption of specific molecules. The peritubular capillaries serve the needs of the renal tubules by bringing to them the molecules to be secreted into the tubules and carrying away the molecules that are reabsorbed from the tubules back into the blood.

The Evolution of the Nephron

It is believed that the earliest vertebrates lived in fresh water and that the nephron evolved as a structure to excrete the excess water constantly entering these animals by osmosis. Does a study of the evolution of the vertebrate nephron support this view? Since those earliest vertebrates are long extinct and the soft tissues of kidneys are not preserved in the fossil record, we must employ indirect means to reveal the evolution of the nephron. We can study the kidneys of present-day members of the oldest of the vertebrate groups with the expectation that they may retain some primitive features of structure and organization. Another approach is to study the embryonic stages of development in such species, because those developmental stages frequently reveal primitive structure and organization not evident in the adult organism. The embryonic development of the kidneys of fish and amphibians offers suggestions about the sequence of evolutionary stages of the vertebrate nephron.

The earliest nephrons to appear in embryonic fish and amphibians open directly into the coelomic cavity through a nephrostome (a ciliated opening; Figure 44.11*a*). This arrangement is very similar to the organization of the annelid metanephridium (see Figure 44.6). Near the nephrostome is a knot of capillaries that protrudes up under the membrane lining the coelomic cavity, where the blood filtrate can pass into the coelom. The resulting coelomic fluid flows through the nephrostome into tubules that convert the fluid into urine.

In a slightly more advanced stage of kidney development, the knot of capillaries (the glomerulus) does not protrude up under the coleomic lining; instead it is encapsulated by an elaboration of the renal tubule (Bowman's capsule) as it leaves the nephrostome (Figure 44.11*b*). In the most advanced stage of nephron development, the nephrostome is lost and all of the tubular fluid is derived directly from filtration in the glomerulus (Figure 44.11*c*).

The function of the earliest vertebrate nephron apparently was to eliminate coelomic fluid while conserving important molecules. Subsequent stages in the evolution of the nephron enhanced its ability to handle a large volume of filtrate derived directly from the blood. From this evidence we can conclude that the original function of the nephron was most likely to bail excess water out of animals while conserving valuable molecules. This conclusion supports the notion that the earliest vertebrates lived in fresh water

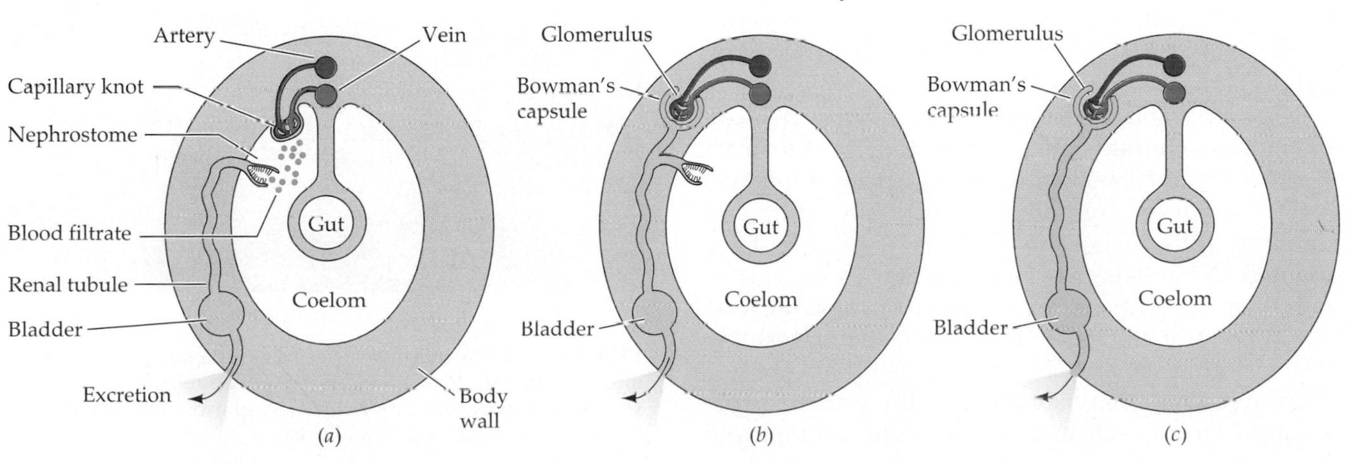

44.11 The Evolution of the Nephron
This model of the evolution of the vertebrate nephron is based on studies of the kidneys (especially embryological stages of those kidneys) of present-day descendants of the oldest vertebrate groups. *(a)* A schematic cross section of a vertebrate body shows that the most primitive nephron had a tubule with the nephrostome opening into the coelom; blood filtrate entered the coelom from knots of capillaries along its borders. *(b)* The next evolutionary stage probably involved a specialization of the tubule to enable filtration directly into the tubule, but the nephrostome remained. *(c)* The final stage eliminates the nephrostome entirely.

and that some of their descendents had to evolve secondary adaptations to enable them to live in habitats where it was necessary to conserve water.

Water Conservation in Vertebrates

If the vertebrate nephron evolved as a structure to excrete water while conserving salts and essential small molecules, how then have vertebrates adapted to environments where water must be conserved and salts excreted? The answer to this question differs for each vertebrate group. Even among marine fish, the bony fish have a different set of adaptations than do the cartilaginous fishes.

Marine bony fish cannot produce urine more concentrated than their body fluids, but they osmoregulate their body fluids to only one-fourth to one-third the osmotic potential of seawater. They prevent excessive loss of water by producing very little urine. Their urine production is low because their kidneys have fewer glomeruli than do the kidneys of freshwater fish. In some species of marine bony fish, the kidneys have no glomeruli at all! Even though the glomeruli are reduced or absent, renal tubules with closed ends are retained for active excretion of ions and certain molecules. Marine bony fish meet their water needs by drinking seawater, but this practice results in a large salt load. The fish handle salt loads by actively excreting ions from gill membranes and from the renal tubules. Nitrogenous wastes are lost as ammonia from the gill membranes.

Cartilaginous fish are osmoconformers but not ion conformers. Unlike marine bony fish, cartilaginous fish convert nitrogenous waste to urea and retain large amounts of that urea in their body fluids so that their body fluids have the same osmotic concentration as seawater. In some cases these fluids are even slightly hypertonic to seawater, causing the water to move into the body of the fish by osmosis. These species have adapted to a concentration of urea in the body fluids that would be fatal to other vertebrates. Sharks and rays have the problem of excreting the large amounts of salts that they take in with their food. They have several sites of active secretion of NaCl, but the major one is a salt-secreting **rectal gland**.

Most amphibians live in or near fresh water and are limited to humid habitats when they venture from the water. Like freshwater fish, typical amphibian species produce large amounts of dilute urine and conserve salts. Some amphibians, however, have adapted to habitats that require water conservation, and their adaptations are diverse. There is at least one species of saltwater amphibian, the crab-eating frog of Southeast Asia. For marine fish there are two different evolutionary solutions to the osmotic problems of living in salt water. Which one is employed by the crab-eating frog? Like the cartilaginous fish, the crab-eating frog retains urea in its body fluids to the extent that it is slightly hypertonic to the seawater in the mangrove swamps where it lives. Only adult frogs have this adaptation, however, so the species requires fresh water in order to reproduce.

Amphibians from very dry terrestrial environments have been studied by Vaughan Shoemaker at the University of California. An important adaptation in these species is a reduction in the water permeability of their skins. Some secrete a waxy substance that they spread over the skin to waterproof it (Figure 44.12). Several species of frogs that live in arid regions of Australia have remarkable adaptations. These animals burrow deep in the ground and estivate during long dry periods. **Estivation** is a state of very low metabolic activity. When it rains, these frogs come out of estivation, feed, and reproduce. Their most interesting adaptation, however, is that they have enormous urinary bladders. Prior to entering estivation, they fill their bladders with very dilute urine which may make up one-third of their body weight. This dilute urine serves as a water reservoir that they use gradually during the long periods of estivation. Australian aboriginal peoples use these estivating frogs as an emergency source of drinking water.

Reptiles occupy habitats ranging from aquatic to extremely hot and dry. Three major adaptations have freed the reptiles from maintaining the close association with water that is necessary for amphibians. First, reptiles do not need fresh water to reproduce because they employ internal fertilization and lay eggs with shells that retard evaporative water loss.

44.12 Waxy Frogs
Phyllomedusa is a tree frog that lives in a seasonally dry and hot habitat. It reduces its evaporative water loss by secreting waxes and fats from skin glands and spreading these secretions all over its body.

Second, they have scaly, dry skins that are much less permeable to water than is the skin of most amphibians. Third, they excrete nitrogenous wastes as uric acid solids and lose little water in the process.

Birds have the same adaptations for water conservation as reptiles have: internal fertilization, shelled eggs, skin that retards water loss, and uric acid as the nitrogenous waste product. In addition, some birds can produce a urine that is hypertonic to their body fluids. This last ability is much more highly developed in mammals.

STRUCTURE AND FUNCTION OF THE MAMMALIAN KIDNEY

The ability of mammals and birds to produce urine that is hypertonic to their body fluids represents a major step in kidney evolution. In these species we see for the first time the kidney playing the major role in water conservation. A structure that originally evolved to excrete water has been converted to a structure to do the opposite, conserve water. To understand how this evolutionary switch occurred, we must examine the structure and function of the whole kidney.

Anatomy

We will focus on humans as an example of the mammalian excretory system. Humans have two kidneys at the rear of the abdominal cavity at the level of the midback (Figure 44.13*a*). Each kidney releases the urine it produces into a tube (the **ureter**) that leads to the **urinary bladder**, where the urine is stored until it is excreted through the urethra. The **urethra** is a short tube opening to the outside at the end of the penis in males or just anterior to the vagina in females. Two sphincter muscles surrounding the base of the urethra control the timing of urination. One of these sphincters is a smooth muscle and is controlled by the parasympathetic division of the autonomic nervous system. When the bladder is full, it activates stretch sensors in its wall and a spinal reflex relaxes this sphincter. This reflex is the only control of urination in infants, but the reflex gradually comes under the influence of higher centers in the nervous system as a child grows older. The other sphincter is a skeletal muscle and is controlled by the voluntary, or conscious, nervous system. When the bladder is *very* full, only serious concentration prevents urination.

The kidney is shaped like a kidney bean; when cut down its long axis and split open as a bean splits open, its important anatomical features are revealed (Figure 44.13*b*). The ureter and the **renal artery** and **renal vein** enter the kidney on its concave (punched-in) side. The ureter divides into several branches, the ends of which envelop projections of kidney tissue called **renal pyramids**. The renal pyramids make up the internal core, or **medulla**, of the kidney. The medulla is capped by a distinctly different tissue called the **cortex**. The renal artery and vein give rise to many arterioles and venules in the region between the cortex and the medulla.

The secret of the ability of the mammalian kidney to produce concentrated urine is in the relationship between the structures of the nephron (the functional unit of the kidney) and the anatomy of the kidney. Each human kidney contains about 1 million nephrons, and their organization within the kidney is very regular (see Figure 44.13*b*). All of the glomeruli are located in the cortex. The initial segment of the tubule of a nephron is called the **proximal convoluted tubule**—"proximal" because it is closest to the glomerulus and "convoluted" because it is twisted (Figure 44.13*c*) All the proximal convoluted tubules are also located in the cortex. At a certain point the proximal tubule takes a dive directly down into the medulla, giving rise to the portion of the tubule called the **loop of Henle**, which runs straight down into the medulla, makes a hairpin turn, and comes straight back to the cortex. This ascending limb of the loop of Henle becomes the **distal convoluted tubule** in the cortex. The distal convoluted tubules of many nephrons join a common **collecting duct** in the cortex. The collecting ducts then run parallel with the loops of Henle down through the medulla, and empty into the ureter at the tips of the renal pyramids.

The organization of the blood vessels of the kidney closely parallels the organization of the nephrons (Figure 44.13*c*). Arterioles branch from the renal arteries and radiate into the cortex. An *afferent* arteriole carries blood to each glomerulus. Draining each glomerulus is an *efferent* arteriole that gives rise to the peritubular capillaries, which surround mostly the cortical portions of the tubules. A few peritubular capillaries run into the medulla in parallel with the loops of Henle and the collecting ducts. These capillaries are the **vasa recta**. All the peritubular capillaries from a nephron join back together into a venule that joins with venules from other nephrons and eventually leads to the renal vein, which takes blood from the kidney. Remember that anything coming into the kidney comes through the renal artery, and everything that comes into the kidney must leave either through the renal vein or the ureter (there is also some drainage of lymph from the kidney, but it is minor). Only a small, selective percentage of everything that is filtered leaves the kidney in the urine. To understand kidney function, you must understand how most of the substances and water filtered from the blood in the glomerulus return to the venous blood draining the kidney.

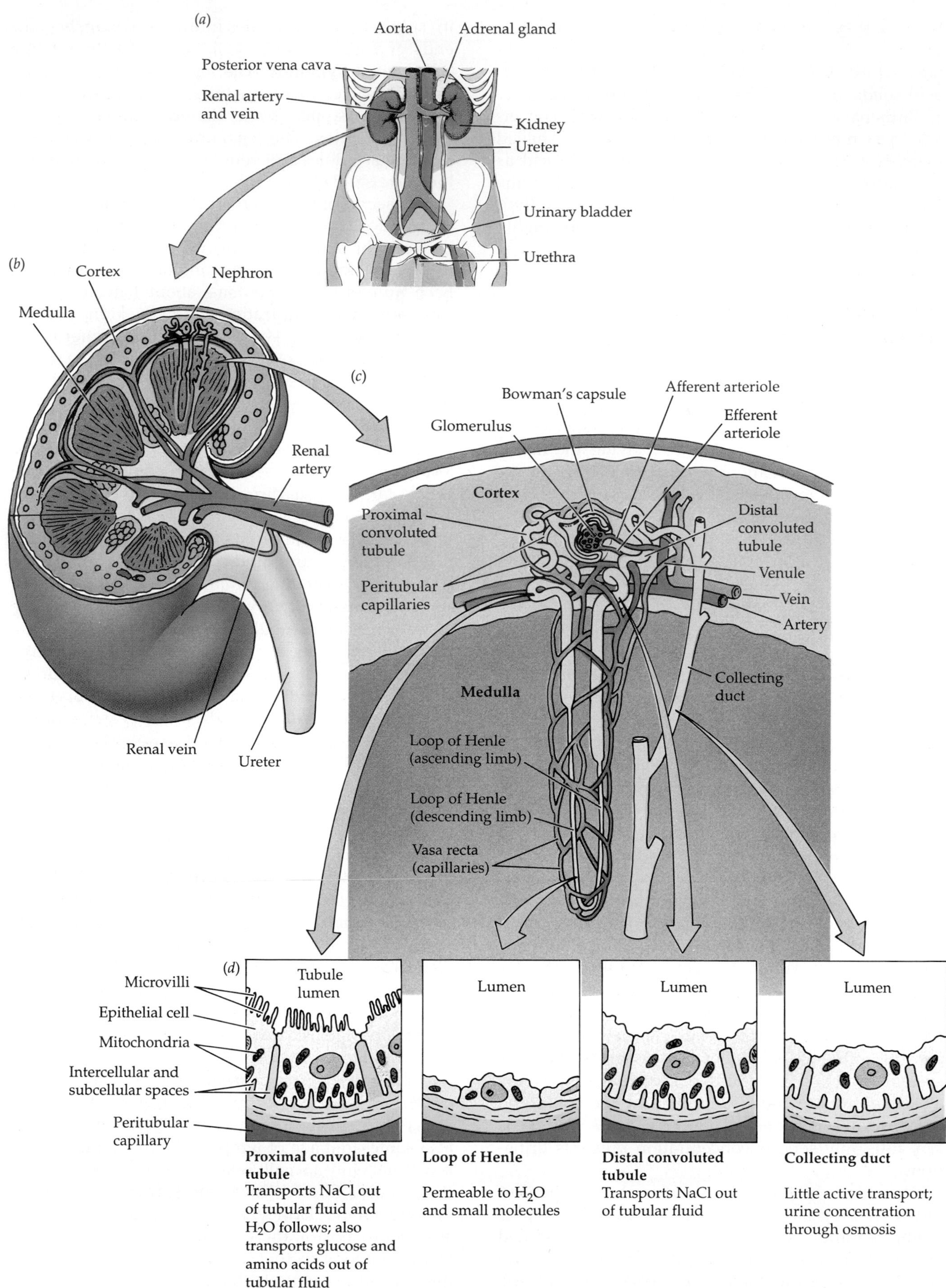
(a)
Aorta
Adrenal gland
Posterior vena cava
Renal artery and vein
Kidney
Ureter
Urinary bladder
Urethra
(b)
Cortex
Nephron
Medulla
Renal artery
Renal vein
Ureter
(c)
Bowman's capsule
Afferent arteriole
Efferent arteriole
Glomerulus
Cortex
Proximal convoluted tubule
Distal convoluted tubule
Peritubular capillaries
Venule
Vein
Artery
Collecting duct
Medulla
Loop of Henle (ascending limb)
Loop of Henle (descending limb)
Vasa recta (capillaries)
(d)
Microvilli
Epithelial cell
Mitochondria
Intercellular and subcellular spaces
Peritubular capillary
Tubule lumen
Lumen
Lumen
Lumen
Proximal convoluted tubule
Transports NaCl out of tubular fluid and H_2O follows; also transports glucose and amino acids out of tubular fluid
Loop of Henle
Permeable to H_2O and small molecules
Distal convoluted tubule
Transports NaCl out of tubular fluid
Collecting duct
Little active transport; urine concentration through osmosis

◀ **44.13 The Human Excretory System**
(a) The human excretory system consists of two kidneys positioned in the upper rear of the abdominal cavity. The urine they produce is conducted to the urinary bladder through the ureters. The urethra drains the bladder. *(b)* A longitudinal section of the kidney reveals an internal structure that includes a cortex and, beneath it, a medulla. Urine leaves the kidney from the inner surface of the medulla and is collected in branches of the ureter. *(c)* A closer look at the cortex and medulla reveals that the glomeruli, the proximal convoluted tubules, and the distal convoluted tubules are in the cortex. The loops of Henle and the vasa recta are in the medulla. Collecting ducts run from the cortex to the tips of the medulla. *(d)* The structures of the cells of the renal tubules reflect the functions of the different tubule segments. The cells of the proximal convoluted tubule have many mitochondria; a well-developed border of microvilli lines the inside of the tubule to increase the surface area available for the absorption of substances from the filtrate. Intercellular spaces and indentations at the basal end of the cells increase the area of cell contact with interstitial fluids. Distal tubule cells also have many mitochondria and extensively folded basal surfaces. Collecting-duct cells are less adapted for active secretion and reabsorption, and cells in the thin regions of the loop of Henle are flat, with few mitochondria or surface indentations.

Glomerular Filtration Rate

Most of the water and solutes filtered in the glomerulus are reabsorbed and do not appear in the urine. We reach this conclusion by comparing the huge daily filtration volume with the volume of urine produced each day. The kidneys receive about 20 percent of the blood pumped into arteries by the heart. The cardiac output of a human at rest is about 5 liters per minute, so the kidney receives over 1,400 liters of blood per day—an enormous volume. How much of this huge volume is filtered? The answer is about 12 percent. This is still a large number—180 liters per day! Since we normally urinate 2 to 3 liters per day, 98 to 99 percent of the fluid volume that is filtered in the glomerulus is reabsorbed into the blood. Where and how is this enormous fluid volume reabsorbed from the renal tubules back into the blood?

Tubular Reabsorption

Most of the water and solutes in the glomerular filtrate are reabsorbed in the proximal convoluted tubule. The cells of this section of the renal tubule are cuboidal, and their surfaces facing into the tubule have thousands of **microvilli**, which increase their surface area for reabsorption (Figure 44.13*d*). These cells have lots of mitochondria—an indication that they are biochemically active. They transport NaCl and other solutes, such as glucose and amino acids, out of the tubular fluid. Virtually all glucose molecules and amino acid molecules that are filtered from the blood are actively reabsorbed across the cells of the proximal convoluted tubules and into the interstitial fluids. This movement of solutes into the interstitial fluid makes it hypertonic to the tubular fluid, and water flows from the tubular fluid in response to this difference. The water and solutes that are moved across the tubular cells by this process are taken up by the peritubular capillaries and returned to the venous blood leaving the kidney.

In spite of the large volume of reabsorption of water and solutes by the proximal convoluted tubule, the overall concentration, or osmotic potential, of the fluid that enters the loop of Henle is not different from that of the blood plasma, even though their compositions are quite different. Next we have to consider how the kidney produces urine that is hypertonic to the blood plasma.

The Countercurrent Multiplier

Humans can produce urine that is four times more concentrated than their plasma. Some mammals that live in very dry deserts, such as kangaroo rats, are able to conserve water so well that their urine may be 12 to 15 times more concentrated than their blood plasma. How does the structure of the mammalian kidney enable it to produce urine that is more concentrated than the blood plasma? This remarkable ability is due to the loops of Henle, which function as a **countercurrent multiplier system**. "Countercurrent" refers to the direction of urine flow in the descending versus the ascending limbs of the loop (see Box 41.A) and "multiplier" refers to the ability of this system to create a concentration gradient in the renal medulla. The loops of Henle *do not* concentrate the tubular fluid, but they increase the osmotic concentration of the surrounding extracellular environment in the renal medulla. Let's see how they do it.

The cells of the descending limb of the loop of Henle, and the initial cells of the ascending limb, are unspecialized. These cells are flat, with no microvilli and few mitochondria. The part of the tubule made up of these cells is permeable to water and small molecules. Partway up the ascending limb, however, the cells become specialized for transport again. They are cuboidal, with lots of mitochondria, and have some microvilli on their surfaces facing into the tubule (see Figure 44.13*d*). The portion of the ascending limb made up of these cells is impermeable to water, but the cells actively transport NaCl out of the tubular fluid. As a result, the urine becomes more dilute as it flows toward the distal convoluted tubule (Figure 44.14). Where does the NaCl that is transported out of the ascending limb go? It enters the interstitial fluid in the renal medulla, from which it can diffuse back into the descending limb of the loop of Henle. The NaCl that enters the descending limb flows around the loop and back up the ascending limb, where it is

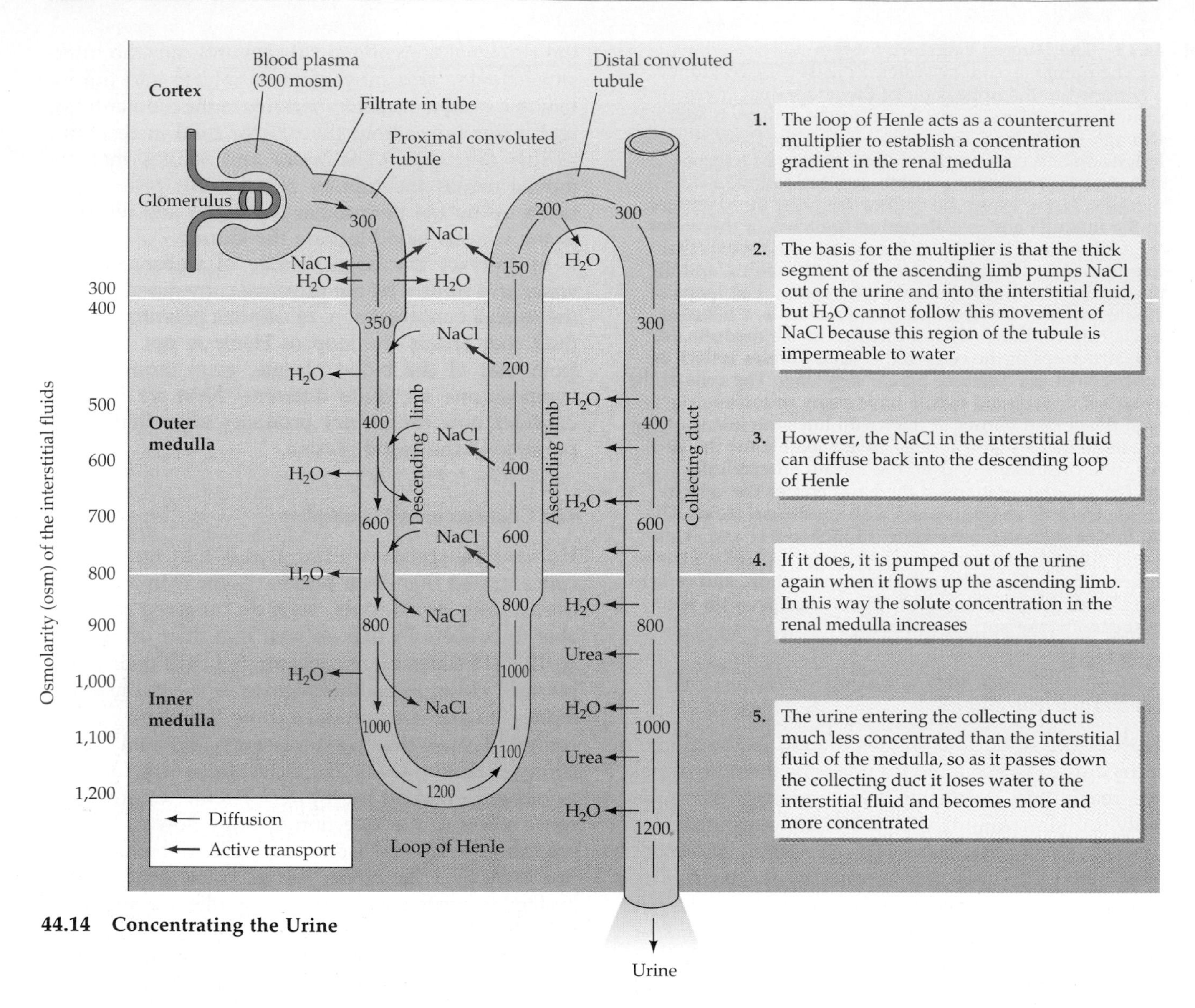

44.14 Concentrating the Urine

transported out of the urine once again. As a result, NaCl accumulates in the interstitial fluid of the renal medulla, with the highest concentration near the tips of the renal pyramids.

How does this action of the loop of Henle concentrate the urine? In Figure 44.13, we can see that the urine is less concentrated when it leaves the loop than when it entered. The active secretion of substances to be excreted and reabsorption of substances to be conserved continues in the distal convoluted tubules. The urine becomes concentrated in the collecting duct. The collecting duct runs from the cortex, where it receives filtrate from the distal tubules, down through the medulla, to the tip of the renal pyramids, where it discharges into the ureter. Over this distance it is surrounded by increasingly concentrated interstitial fluid. The collecting duct is permeable to water, but not to ions, so the osmotic potential of the interstitial environment draws water from the fluid in the collecting duct and leaves behind an increasingly concentrated solution. The urine that leaves the collecting duct at the tip of a renal pyramid can be almost as concentrated as the highest interstitial concentration established by the countercurrent multiplier system. The water reabsorbed from the collecting duct exits the renal medulla via the vasa recta, which are highly permeable to salts and water.

In summary, the mammalian kidney works in the following manner: The glomeruli filter large volumes of blood plasma. The proximal convoluted tubules reabsorb most of this volume along with valuable molecules such as glucose and amino acids, actively transporting NaCl and other solutes from the tubular fluid. Water follows because of the local difference in osmotic potential created by the transport of the solutes. The loops of Henle create a concentration gradient in the medulla of the kidney. As the urine flows in the collecting ducts through this concentration gradient, water is reabsorbed, thus creating a urine hypertonic to the blood plasma.

REGULATION OF KIDNEY FUNCTIONS

The function of kidneys is to regulate the volume, the osmolarity, and the chemical composition of extracellular fluids. If the fluids do not have the right composition to meet the needs of the body cells, the cells cannot survive. Thus, if the kidneys fail, death will ensue—unless the victim has access to an artificial kidney machine, which can cleanse the blood through **dialysis** (Box 44.A). Multiple systems control and regulate kidney functions. Although we will discuss them separately, they are always working together in an integrated fashion to match kidney function to the needs of the body.

BOX 44.A

Artificial Kidneys

Sudden and complete loss of kidney function is called acute renal failure. It results in the retention of salts and water (leading to high blood pressure), as well as in the retention of urea and metabolic acids. A patient with acute renal failure will die in one to two weeks if not treated. It is now possible to compensate for renal failure and even surgical removal of the kidneys by using artificial kidneys. In an artificial kidney, or dialysis unit, the blood of the patient and a dialyzing fluid come into very close contact, separated only by a semipermeable membrane. This membrane allows small molecules to diffuse from the patient's blood into the dialysis fluid. Because molecules and ions diffuse from an area of high concentration to an area of lower concentration, the composition of the dialysis fluid is crucial. The concentrations of the molecules or ions we want to conserve, such as glucose or sodium, must be the same in the dialysis fluid as in the plasma. The concentrations of molecules and ions we want to clear from the plasma, such as urea and sulfate, must be zero in the dialysis fluid. The total must equal the osmotic potential of the plasma.

The figure shows a schematic drawing of a dialysis machine. Arterial blood flows between semipermeable membranes, which are surrounded with dialysis fluid at body temperature. The "cleansed" blood is returned to the body through a vein and the used dialysis fluid is discarded. At any one time only about 500 milliliters of blood are in the dialysis unit, and it processes several hundred milliliters of blood per minute. A patient with no kidney function must be on the dialysis machine for 4 to 6 hours three times a week.

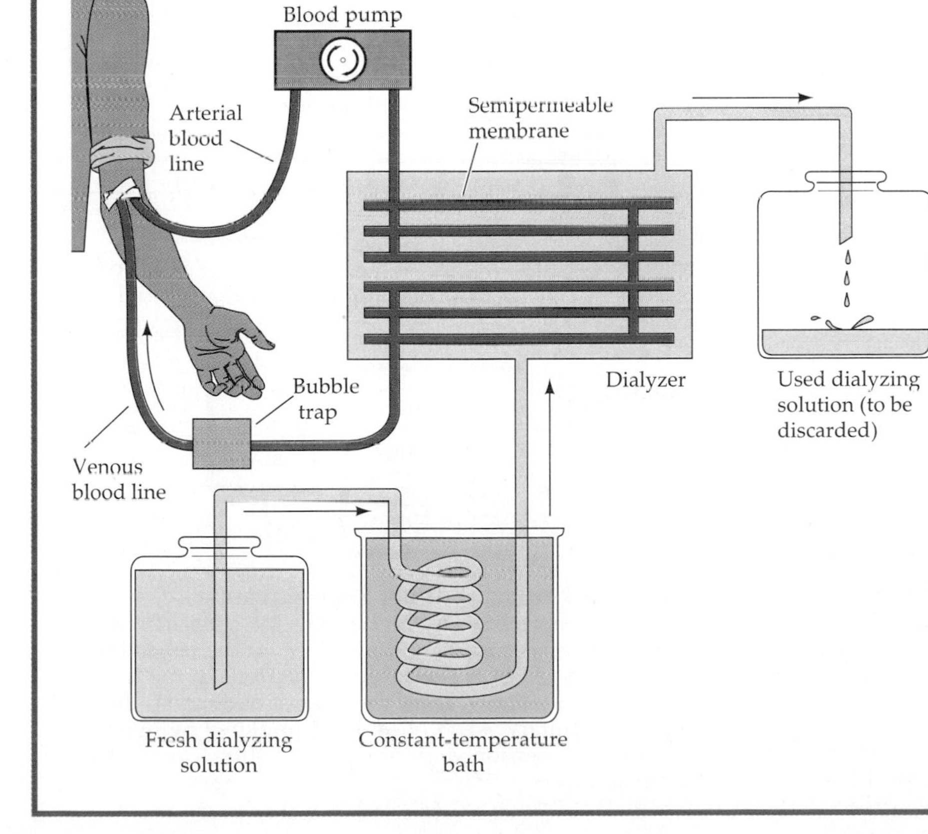

This man is monitoring the flow of his blood through a dialysis unit that eliminates metabolic waste products normally removed from the blood by the kidneys. The mechanisms of the dialysis unit are illustrated in the diagram.

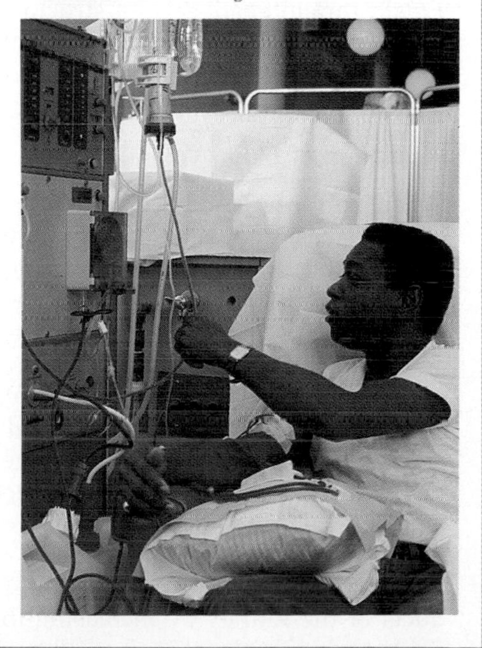

Autoregulation of the Glomerular Filtration Rate

If the kidneys stop filtering the blood, they cannot accomplish any of their functions. Therefore, there are mechanisms that keep the blood filtering through the glomeruli at a constant high rate, regardless of what is happening elsewhere in the body. Because these adaptations of the kidney are to maintain its own functions, these mechanisms are called autoregulatory. The glomerular filtration rate (GFR) depends on an adequate blood supply to the kidneys at an adequate blood pressure. The autoregulatory mechanisms compensate for decreases in cardiac output or decreases in blood pressure so that the GFR remains high (Figure 44.15).

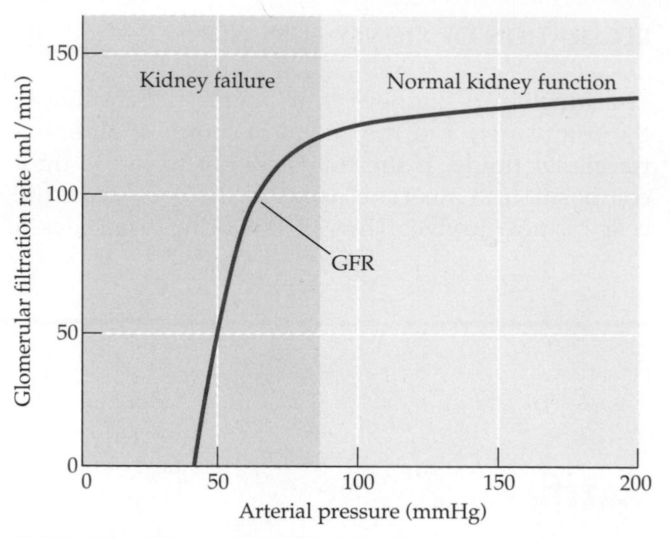

44.15 The Glomerular Filtration Rate Remains Constant
Glomerular filtration is driven by arterial pressure, yet because of autoregulatory mechanisms in the kidney the glomerular filtration rate (GFR) is independent of arterial pressure over a wide range. When arterial pressure falls too low, however, the kidney fails to produce urine.

One autoregulatory mechanism is the dilation (expansion) of the afferent renal arterioles when blood pressure falls. This dilation decreases the resistance in the arterioles and helps maintain blood pressure in the glomerular capillaries. If that response does not keep the GFR from falling, then the kidney releases an enzyme, **renin**, into the blood. Renin acts on a circulating protein to begin converting this protein into an active hormone called **angiotensin**. Angiotensin has several effects that help restore the GFR to normal. First, angiotensin causes the efferent renal arteriole to constrict, which elevates blood pressure in the glomerular capillaries. Second, angiotensin causes peripheral blood vessels all over the body to constrict—an action that elevates central blood pressure. Third, angiotensin stimulates the adrenal cortex to release the hormone **aldosterone**. Aldosterone stimulates sodium reabsorption by the kidney, thereby making the reabsorption of water more effective. Enhanced water reabsorption helps maintain blood volume and therefore central blood pressure. Finally, angiotensin acts on structures in the brain to stimulate thirst. Increased water intake in response to thirst increases blood volume and blood pressure.

Regulation of Blood Volume

When you lose blood, your blood pressure tends to fall. Besides activating the kidney autoregulatory mechanisms described in the previous section, a drop in blood pressure decreases the activity of stretch sensors in the walls of the large arteries such as the aorta and the carotids. These stretch sensors provide information to cells in the hypothalamus that produce **antidiuretic hormone** (also called vasopressin) and send it down their axons to the posterior pituitary gland (see Chapter 36). As stretch sensor activity falls, the production and release of this hormone increases (Figure 44.16).

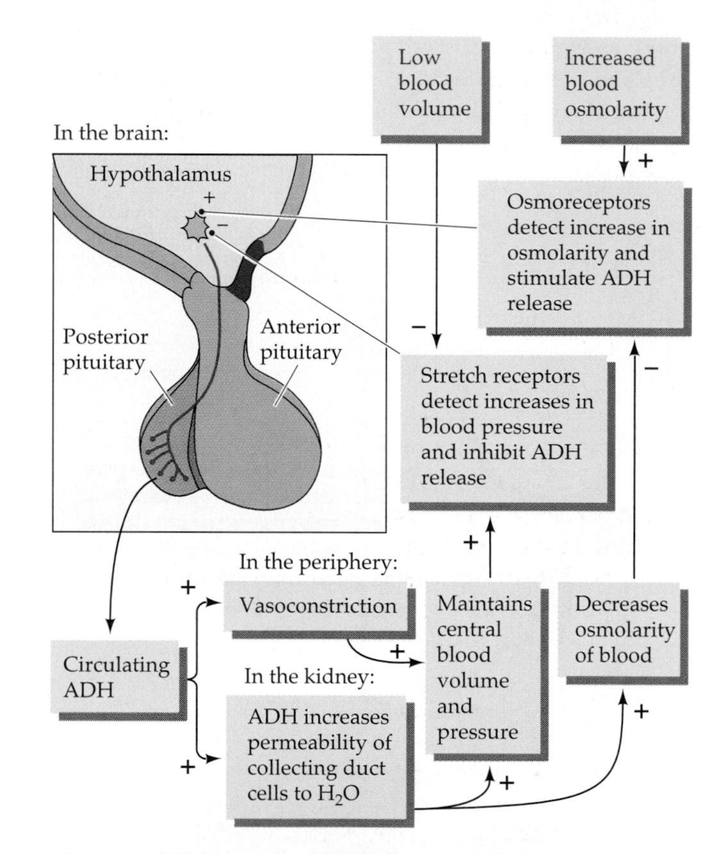

44.16 ADH Promotes Water Reabsorption
Antidiuretic hormone (ADH) controls the concentration of urine by increasing the permeability of the collecting duct to water. ADH is produced by neurons in the hypothalamus and released from their nerve endings in the posterior pituitary. ADH release is stimulated by hypothalamic osmosensors and inhibited by stretch sensors in the great arteries.

Antidiuretic hormone (ADH) acts on the collecting ducts of the kidney by increasing their permeability to water. When there is a high circulating level of ADH, the collecting ducts are very permeable to water, more water is reabsorbed from the urine, and only small quantities of concentrated urine are produced, thus conserving blood volume and blood pressure. Without ADH, water cannot be reabsorbed from the collecting ducts, and lots of very dilute urine is produced. Diabetes insipidus is a disease that results from a lack of ADH ("insipidus" derives from the dilute, or "tasteless" character of the urine). Diabetes mellitus, caused by an inability of cells to take up glucose from the blood, also causes copious urine production, but the urine tastes sweet ("mellitus" means honeylike). Caffeine and alcohol inhibit the actions of ADH and increase the production of urine. The resulting dehydrating effect of alcohol creates "hangover" headaches by decreasing the amount of cerebrospinal fluid that is providing a cushion (waterbed) for the brain.

Regulation of Blood Osmolarity

Sensors in the hypothalamus monitor the osmotic potential of the blood. If blood osmolarity increases,

BOX 44.B

Water Balance in the Vampire Bat

The Australian vampire bat (*Desmodus*) is a small mammal that feeds at night on the blood of sleeping large mammals, such as cattle. Blood is a liquid, high-protein diet. To process this diet, the renal system of the vampire bat must shift from drought conditions to flood conditions and back to drought conditions in minutes. At sunset, when the bat has not had a meal for many hours, it is producing a highly concentrated urine at a low rate to conserve its precious body water. If it is successful in finding prey, the bat must process as much blood in as short a time as possible, before the victim wakes up. To maximize its nutrient intake, the bat concentrates its blood meal by rapidly excreting its water content. Accordingly, within minutes the bat produces copious amounts of very dilute urine. The warm fluid running down the victim's neck and waking it is not blood!

As soon as the meal is ended—usually abruptly—the bat begins to digest the concentrated blood in its gut. Because the concentrated blood is mostly protein, a large amount of nitrogenous waste is produced and must be excreted as urea in solution. But now water is in short supply. The bat must limit its water loss because a long time may pass before the next meal. Consequently, the bat's kidneys produce small amounts of extremely concentrated urine. This urine can be more than 20 times the concentration of the bat's plasma. Humans, in comparison, can produce a urine only about 4 times as concentrated as their plasma. In this way the remarkable regulatory abilities of the vampire bat kidney enable the animal to process its unusual diet.

The graph shows the changes in a vampire bat's urine concentration and urine flow rate before and after its meal of blood. In the photograph, two vampire bats roost on the ceiling of their cave during the day.

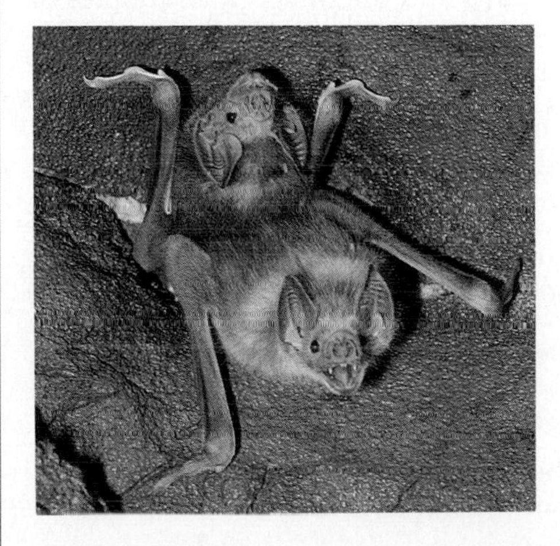

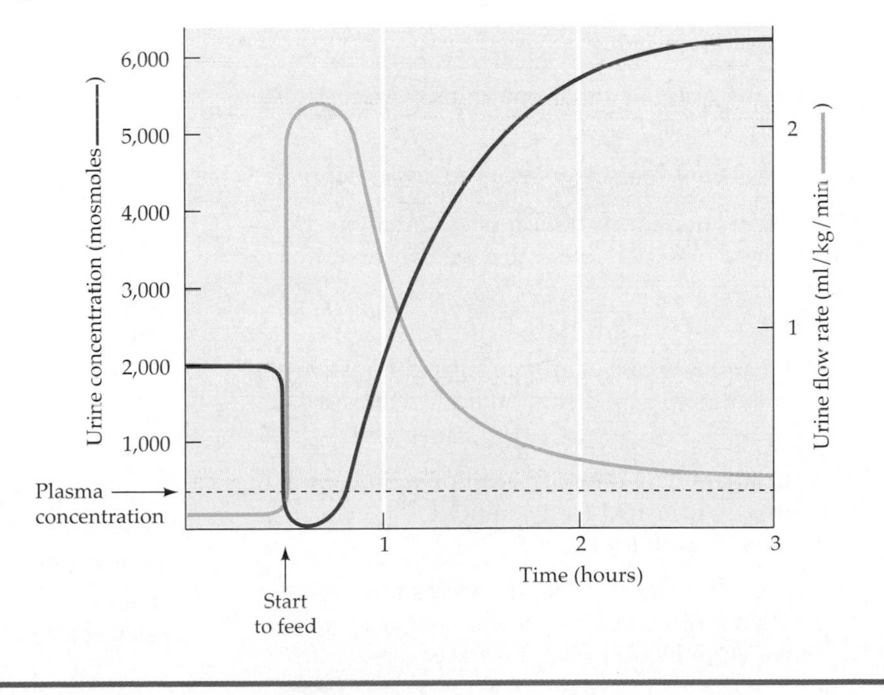

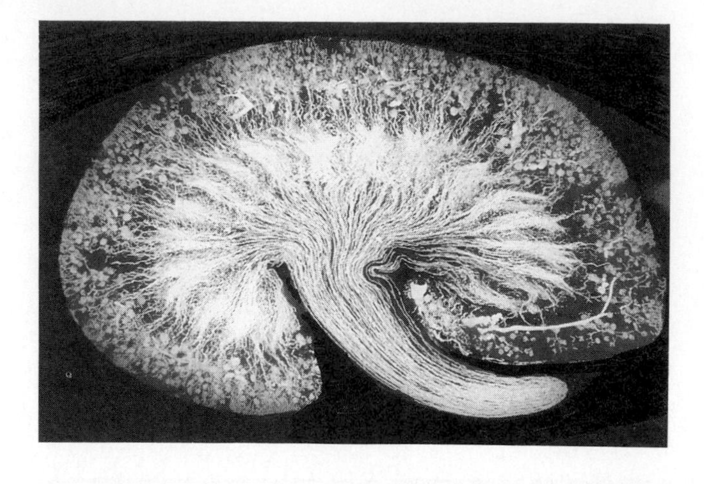

44.17 The Ability to Concentrate
The ability of the mammalian kidney to concentrate urine depends on the lengths of its loops of Henle relative to the overall size of the kidney. Some desert rodents have single renal pyramids that are so long they protrude out of the kidney and into the ureter, as shown here.

these **osmosensors** stimulate increased release of antidiuretic hormone to enhance water reabsorption from the kidney. The osmosensors also stimulate thirst. The resulting increased water intake dilutes the blood as it expands blood volume.

Regulatory Flexibility

The ability of the mammalian kidney to produce a concentrated urine has made it possible for mammals to inhabit some of the most arid habitats on Earth. Some of these animals, such as the desert gerbil, have such extremely long loops of Henle that their renal pyramid (each of their kidneys has only one in contrast to ours) extends far out of the concave surface of the kidney (Figure 44.17). These animals are so effective in conserving water that they can survive on the water released by the metabolism of their dry food; they do not need to drink! The concentrating ability of the mammalian kidney, coupled with the remarkable flexibility of its regulatory systems, enables it to adapt to rapidly changing conditions. This regulatory flexibility is quite pronounced in the vampire bat, which can display the full range of extremes of mammalian salt and water balance mechanisms in a matter of minutes (Box 44.B).

SUMMARY of Main Ideas about Salt and Water Balance and Nitrogen Excretion

The problems of salt and water balance and nitrogen excretion that animals face depend on their environments.

Marine animals can be osmoconformers or osmoregulators.
Review Figure 44.2

Freshwater animals must continually excrete water and conserve salts.

All animals are ionic regulators to some degree.

On land, water conservation is essential, and diet determines whether salts must be conserved or excreted.
Review Figure 44.3

Aquatic animals can eliminate nitrogenous wastes such as ammonia by diffusion across their gill membranes.

Terrestrial animals detoxify ammonia by converting it to urea or uric acid for excretion.
Review Figure 44.4

In excretory systems pressure drives filtration of body fluids into a system of tubules, whose cells alter the composition of the filtrate.

Protonephridia of flatworms consist of flame cells and excretory tubules.
Review Figure 44.5

Metanephridia of segmented worms take in coelomic fluid and alter its composition through active secretion and active reabsorption of solutes by tubule cells.
Review Figure 44.6

Arthropod excretory organs include the green glands of crustaceans and the Malpighian tubules of insects.
Review Figures 44.7 and 44.8

The vertebrate nephron originally evolved as an adaptation for excreting water and conserving solutes.
Review Figures 44.9 and 44.11

Mammals and birds produce urine that is hypertonic to their body fluids.

The organization of the renal tubules in the mammalian kidney is the basis for its ability to produce concentrated urine.

Glomeruli in the cortex of the kidney filter water and solutes from the blood.

Proximal convoluted tubules reabsorb most of the kidney-filtered water and many of the solutes.
Review Figure 44.13

The loop of Henle creates a concentration gradient in the tissues of the renal medulla, and the collecting duct concentrates the urine by allowing the osmotic loss of water to the surrounding interstitial fluid of the medulla.
Review Figure 44.14

Kidney function in mammals is regulated by autoregulatory mechanisms for maintaining a constant high glomerular filtration rate even if blood pressure varies.
Review Figure 44.15

An important autoregulatory mechanism is the release of renin by the kidney when blood pressure falls; renin activates angiotensin, which causes peripheral vasoconstriction, causes release of aldosterone, and stimulates thirst.

Changes in blood pressure and blood osmolarity influence the release of antidiuretic hormone, which controls the permeability of the collecting duct to water.
Review Figure 44.16

SELF-QUIZ

1. Which of the following statements about osmoregulators is true?
a. Most marine invertebrates are osmoregulators.
b. All freshwater invertebrates are hypertonic osmoregulators.
c. Cartilaginous fish are hypotonic osmoregulators.
d. Bony marine fish are hypertonic osmoregulators.
e. Mammals are hypotonic osmoregulators.

2. The excretion of nitrogenous wastes
a. by humans can be in the form of urea and uric acid.
b. by mammals is never in the form of uric acid.
c. by marine fish is in the form of urea.
d. does not contribute to the osmotic potential of the urine.
e. requires more water if the waste product is the rather insoluble uric acid.

3. How are earthworm metanephridia like mammalian nephrons?
a. Both process coelomic fluid.
b. Both take in fluid through a ciliated opening.
c. Both produce hypertonic urine.
d. Both employ tubular secretion and reabsorption to control urine composition.
e. Both deliver urine to a urinary bladder.

4. What is the role of renal podocytes?
a. They control the glomerular filtration rate by changing resistances of renal arterioles.
b. They reabsorb most of the glucose that is filtered from the plasma.
c. They prevent red blood cells and large molecules from entering the renal tubules.
d. They provide a large surface area for tubular secretion and reabsorption.
e. They release renin when the glomerular filtration rate falls.

5. Which of the following are *not* found in a renal pyramid?
a. Collecting ducts
b. Vasa recta
c. Peritubular capillaries
d. Convoluted tubules
e. Loops of Henle

6. Which part of the nephron is responsible for most of the difference in mammals between the glomerular filtration rate and the urine production rate?
a. The glomerulus
b. The proximal convoluted tubule
c. The loop of Henle
d. The distal convoluted tubule
e. The collecting duct

7. For mammals of the same size, what feature of their excretory systems would give them the greatest ability to produce a hypertonic urine?
a. Higher glomerular filtration rate
b. Longer convoluted tubules
c. Increased number of nephrons
d. More-permeable collecting ducts
e. Longer loops of Henle

8. Which of the following would *not* be a response stimulated by a large drop in blood pressure?
a. Constriction of afferent renal arterioles
b. Increased release of renin
c. Increased release of antidiuretic hormone
d. Increased thirst
e. Constriction of efferent renal arterioles

9. Which of the following statements about angiotensin is true?
a. It is secreted by the kidney when the glomerular filtration rate falls.
b. It is released by the posterior pituitary when blood pressure falls.
c. It stimulates thirst.
d. It increases permeability of the collecting ducts to water.
e. It decreases glomerular filtration rate when blood pressure rises.

10. Birds that feed on marine animals ingest a lot of salt, but excrete most of it by means of
a. Malpighian tubules.
b. rectal salt glands.
c. green glands.
d. hypertonic urine.
e. nasal salt glands.

FOR STUDY

1. What do marine fish, reptiles, mammals, and insects have in common with respect to water balance? Compare their physiological adaptations for dealing with their common problem.
2. What are the relative advantages and disadvantages of ammonia, urea, and uric acid as nitrogenous waste products of animals?
3. Explain how the kidney is able to maintain a constant glomerular filtration rate over a wide range of arterial blood pressures. Referring back to what you learned about regulation of circulatory function in Chapter 42, how can a decrease in glomerular filtration rate cause an increase in cardiac output?
4. Inulin is a molecule that is filtered in the glomerulus, but it is not secreted or reabsorbed by the renal tubules. If you injected inulin into a subject and after a brief time measured the concentration of inulin in the blood and in the urine of the subject, how could you determine the subject's glomerular filtration rate? Assume that the rate of urine production is 1 milliliter per minute.
5. Explain the roles of the loop of Henle and the collecting duct in producing a hypertonic urine in mammals. How is this mechanism controlled in response to changes in osmolarity of the blood and in blood pressure?

READINGS

Cantin, M. and J. Genest. 1986. "The Heart as an Endocrine Gland." *Scientific American*, February. Heart tissue secretes a hormone that helps control salt and water balance.

Eckert, R., D. Randall and G. Augustine. 1988. *Animal Physiology: Mechanisms and Adaptations*, 3rd Edition. W. H. Freeman, New York. An outstanding textbook of animal physiology. Chapter 12 covers water and salt balance and excretion.

Heatwole, H. 1978. "Adaptations of Marine Snakes." *American Scientist*, vol. 66, pages 594–604. A variety of adaptations, including means of maintaining salt and water balance, allows several groups of snakes to exploit the marine environment.

McClanahan, L. L., R. Ruibal and V. H. Shoemaker. 1994. "Frogs and Toads in Deserts." *Scientific American*, March. Recent research on amphibians adapted to arid environments.

Schmidt-Nielsen, K. 1990. *Animal Physiology: Adaptation and Environment*, 4th Edition. Cambridge University Press, New York. An excellent textbook, emphasizing the comparative approach.

Smith, H. W. 1961. *From Fish to Philosopher*. Doubleday, Garden City, NJ. A classic, using salt balance and excretory physiology as the organizing principle for a survey of vertebrate evolution.

Stricker, E. M. and J. G. Verbalis. 1988. "Hormones and Behavior: The Biology of Thirst and Sodium Appetite." *American Scientist*, vol. 76, page 261. The control of water and salt intake is an important part of osmoregulation.

Vander, A. J., J. H. Sherman and D. S. Luciano. 1994. *Human Physiology: The Mechanisms of Body Function*, 6th Edition. McGraw-Hill, New York. Chapter 16 deals with the regulation of water and salt balance.

Spider webs are objects of beauty and marvels of engineering. The construction of a classic web used to capture prey requires complex behavior. For example, a garden spider, as immortalized in the children's story *Charlotte's Web* by E. B. White, spins a new web every day in the early morning hours before dawn. From an initial attachment point, she strings a horizontal thread. From the middle of that thread she drops a vertical thread to a lower attachment point. Pulling it taut creates a Y, the center of which will be the hub of the finished web. The spider adds a few more radial supports and a few surrounding "framing" threads. Then she fills in all the radial spokes according to a set of rules. Finally, she lays down a spiral of sticky threads with regular spacing and attachment points to the radial spokes. This remarkable feat of construction takes only half an hour, but it requires thousands of specific movements performed in just the right sequence. Where is the blueprint for Charlotte's web? How does she acquire the construction skills needed to build it?

The blueprint is coded in the genes and built into the spider's nervous system as a motor "program," or score. Learning plays no role in the expression of that complex blueprint. Newly hatched spiders disperse to new locations and usually spin their first webs without ever having experienced a web built by an adult of their species. Nevertheless, they build perfect webs the first time; each of the thousands of movements happens in just the right sequence. It is remarkable that the genetic code and the simple nervous system of a spider can contain and express behavior as complex as spinning an orb web.

Web spinning by spiders is an animal behavior—an act or set of acts performed with respect to another animal or the environment. Behavior falls into three general classes: acts to acquire food, acts to avoid environmental threats, and acts to reproduce. In studying any behavior we can ask *what*, *how*, and *why*. *What* questions focus on the details of behavior, including the circumstances that influence when an animal acts in a certain way. Some *how* questions refer to the underlying neural, hormonal, and anatomical mechanisms that we have been studying in Part Six. Other *how* questions refer to the means by which an animal acquires a behavior. Some behaviors, such as web spinning, are genetically determined; others are learned. Many behaviors involve complex interactions of inheritance and learning.

Animal behavior is a large field of study. For this chapter we have selected some interesting examples of approaches to answering *what* and *how* questions. *Why* questions have to do with the evolution of behavior and will be the major focus of the next chapter. After a discussion of genetically determined behavior and behavior that results from a combination of genetics and learning, such as bird song, we will turn

An Orb-Weaving Garden Spider and Its Web

45 Animal Behavior

to animal communication. Studies of animal communication reveal the constraints that environment places on behavior. We will continue with a look at the timing of behaviors, or biological rhythms. Finally, we will discuss how animals find their way through unfamiliar territory. Throughout the chapter use what you read to raise your own questions about human behavior.

GENETICALLY DETERMINED BEHAVIOR

A behavior that is genetically determined rather than learned is also called a **fixed action pattern**. Such behavior is highly stereotypic, that is, performed the same way every time. It is also species-specific; there is very little variation in the way different individuals of the same species perform the behavior. The behavior is expressed differently, however, in even closely related species. For example, different species of spiders spin webs of different designs.

Fixed action patterns require no learning or prior experience for their expression, and they are generally not modified by learning. Another spider example illustrates this point. Spiders spin other structures in addition to webs for capturing prey. Most spiders lay their eggs in a cocoon that they form by spinning a base plate, building up the walls (inside of which they lay the eggs), and spinning a lid to close the cocoon. Although this behavior requires thousands of individual movements, it is performed exactly the same way every time and is not modified by experience. If the spider is moved to a new location after she finishes the base plate, she will continue to spin the sides of the cocoon, lay her eggs (which fall out the bottom), and spin the lid. If she is placed on her previously completed base plate the next time she is ready to begin a cocoon, she will spin a new base plate over the old one as if it were not there. If she is nutritionally deprived and runs out of silk in the middle of spinning a cocoon, she will complete all the thousands of movements in a pantomime of cocoon building. Once started, the cocoon-building motor score runs from beginning to end, and it can be started only at the beginning.

Fixed action patterns are good material for studying the mechanisms of animal behavior. We can study their genetics and the sequence of events whereby gene expression eventually results in a behavior; the influence of hormones on the development and expression of the behavior; and the detailed neurophysiology that underlies the behavior. First, however, we must demonstrate that a given behavior *is* genetically determined. One powerful way of proving genetic determination is to deprive the animal of any opportunity to learn the behavior in question, then see if that behavior is expressed.

Deprivation Experiments

In a **deprivation experiment**, an animal is reared so that it is deprived of all experience relevant to the behavior under study. For example, a tree squirrel was reared in isolation, on a liquid diet, and in a cage without soil or other particulate matter. When the young squirrel was given a nut, it put the nut in its mouth and ran around the cage. Eventually it oriented toward a corner of the cage and made stereotypic digging movements, placed the nut in the corner, went through the motions of refilling the imaginary hole, and ended by tamping the nonexistent soil with its nose. The squirrel had never handled a food object and had never experienced soil, yet the fixed action patterns involved in burying its nut were fully expressed.

Deprivation experiments occur naturally. Many species, especially insects living in seasonal environments, have life cycles of one year and the generations do not overlap: The adults lay eggs and die before the eggs hatch or the young mature into adults. Learning from adults of the parental generation is impossible in such species, so the complex behavior necessary for survival and reproductive success must be genetically programmed. Web spinning by spiders is an example of complex behavior in species that may have no opportunity to learn from other members of their species.

The courtship behavior of the triangular web spider is a similar example. The male spider must approach a female in her web. If he simply blundered into the web, he would give the same signals as a prey item caught in the web and the female would probably kill him and eat him. To avoid having his reproductive effort cut short, the male is genetically programmed to approach an anchor strand of the web and pluck it in just the right way to send a courtship message to the female. If the message is correct and the female is receptive, he can enter the web and mate with the female rather than serve merely as her dinner. In some species, the female eats the male anyway after they mate, but at least he has achieved reproductive success before becoming nutriment for his own offspring, and he is supplying energy for eggs fertilized only by him.

Triggering Fixed Action Patterns

A behavior that is not expressed during a deprivation experiment may nonetheless be genetically programmed. The right conditions may not have been available to stimulate the behavior during the experiment. Thus, the squirrel had to be given a nut to trigger its digging and burying behaviors. Specific stimuli are usually required to elicit the expression of fixed action patterns. These stimuli are called sign

45.1 Triggering Aggressive Behavior
A mounted immature male European robin (right) with no red feathers does not stimulate aggression from a territorial adult male. However, the formless clump of red feathers on the left elicits strong aggressive attacks from the territorial adult.

stimuli, or **releasers**. Konrad Lorenz and Niko Tinbergen are two European **ethologists** (scientists who study animal behavior) who did pioneering work on releasers and fixed action patterns. They received Nobel prizes for their studies, which are still considered as classics in the field of animal behavior.

Releasers are usually a simple subset of all the sensory information available to an animal. One example of the simplicity of a releaser comes from the work of the British ecologist David Lack on territorial aggression in the European robin. Male European robins have red feathers on their breasts; females and juveniles do not. During the breeding season, the sight of a male robin stimulates another male robin to sing, make aggressive displays, and attack the intruder if he does not heed the warnings. An immature male robin, or a female whose feathers are all brown, does not elicit aggressive behavior. A tuft of red feathers on a stick, however, will elicit an attack (Figure 45.1). A patch of red in certain locations is a sufficient releaser for male aggressive behavior in robins.

Just as the motor score of the fixed action pattern is genetically programmed, so is the information that enables the animal to recognize the releaser for that fixed action pattern. Evolving a genetic mechanism to respond to a simple stimulus is more feasible than evolving a mechanism to recognize a complex set of stimuli. The simplicity of most releasers has resulted in some curious discoveries.

Tinbergen and A. C. Perdeck carefully examined the releasers and fixed action patterns involved in the interactions between herring gulls and their chicks during feeding. The adult gull has a red dot at the end of its bill (Figure 45.2). When it returns to its nest, the chicks peck at the red dot, and the adult regurgitates food for the chicks to eat. Tinbergen and Perdeck asked what were the essential characteristics of the parent gull that released food solicitation behavior in the chicks. They made paper cutout models of gull heads and bills but varied the colors and the shapes. Then they rated each model according to how many pecks it received from chicks. The surprising results were that the shape or color of the head made no difference. In fact, a head was not even necessary; the chicks responded just as well to models of bills alone. The color of the bill and the dot also were not critical as long as there was a contrast between the two. Surprisingly, the most effective releaser for chick pecking was a long, thin object with a dark tip that had no resemblance to an adult herring gull (Figure 45.3).

The simplicity of the properties of releasers makes possible the existence of a **supernormal releaser**, one that is more effective than the natural condition in eliciting a fixed action pattern. In the case of a bird called the oystercatcher, the sight of its clutch of eggs releases incubation behavior. If an oystercatcher is given the choice between its own clutch of two eggs and a clutch of three artificial eggs, it will sit on the larger clutch of artificial eggs. When given the choice of its own clutch of two eggs or one very large artificial egg, it will try to incubate the large egg, even if it can hardly straddle it. The abnormally large clutch and egg are supernormal releasers. These choices would not occur in nature, so counterselection has not prevented the evolution of such maladaptive behavior.

Supernormal releasers, which are created by scientists, are a curiosity, but natural selection has produced some dramatic results by favoring the exaggeration of releasers. Many of the elaborate behavior

45.2 The Dot Marks the Spot
This gull is incubating eggs. When the young hatch they will peck at the parent's red bill spot, stimulating the parent to regurgitate food.

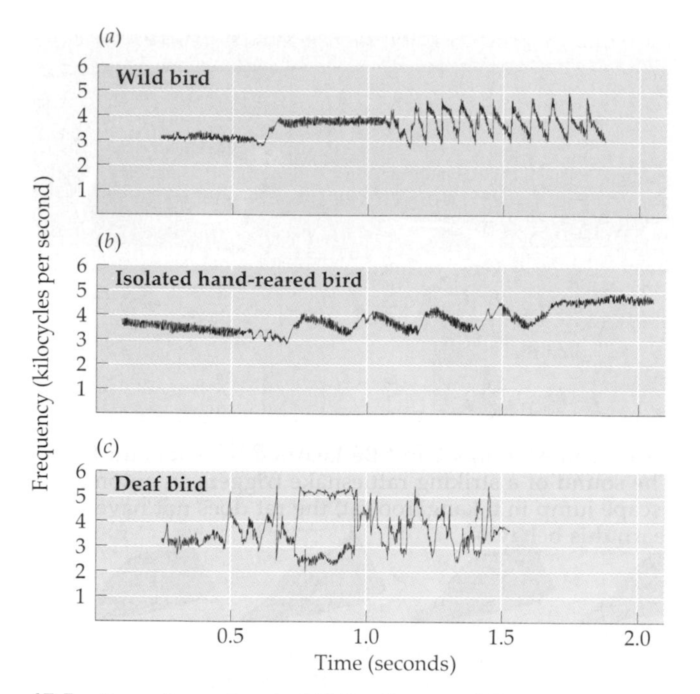

45.5 Song Learning in White-Crowned Sparrows These sonograms visually record sound frequencies and plot them over time. *(a)* The species-specific song of a male white-crowned sparrow in its natural state. *(b)* In a deprivation experiment, a bird hatched in an incubator and reared in isolation from other birds does not learn the species-specific song. *(c)* In a variation on the experiment, a young bird that heard the correct song but was deafened before it reached maturity could not reproduce the song.

learn, however, reveal strong genetic limits to the modifiability of their behavior through experience. In the case of a white-crowned sparrow, it must hear its species' song within a narrow **critical period** during its development. Once this critical period has passed, the bird cannot learn to sing its species-specific song, regardless of how many role models it experiences. What a bird can learn during its critical period is also severely limited, as revealed by experiments on hand-reared chaffinches that were played various tape recordings of bird song during their critical periods. If exposed to the songs of other species, the chaffinches did not learn them. They also did not learn a chaffinch song played backward or with the elements scrambled. Even if they heard a chaffinch song played in pure tones, they did not form a template. If they heard a normal chaffinch song along with all these other sounds, however, they developed templates and learned to sing the proper song the following spring. Thus the chaffinch is genetically programmed to recognize the appropriate song to learn and when to learn it.

What advantage are genetic limits on what a bird can learn and when? A bird's acoustic environment can be quite complex. Many species of songbirds may be singing in the same area. The critical period limits learning to the period of most intimate contact between the young bird and its parents to ensure that the father's song is the one experienced most intensively. Further limits on nestling song sensitivity help guarantee that the template it forms is not contaminated with other sounds it hears.

The learning of a song template by a nestling bird is an example of **imprinting**. We mentioned earlier that releasers generally are simple subsets of available information because there are limits to what can be programmed genetically. Imprinting makes it possible to encode complex information in the nervous system rather than in the genes. Offspring can imprint on their parents and parents on their offspring to ensure individual recognition, even in a crowded situation such as a colony or a herd. If a mother goat does not nuzzle and lick her newborn within 5 to 10 minutes after birth, she will not recognize it as her own later. In this case imprinting depends on olfactory cues, and the critical period is determined by the high levels of circulating oxytocin at the time of birth.

HORMONES AND BEHAVIOR

All behavior depends on the nervous system for initiation, coordination, and execution. In addition, as we have discussed, fixed action patterns are built into the nervous system as a motor score. Yet that motor score is expressed only under certain conditions. The endocrine system, through its controlling influences on development and physiological state of the animal, has a large role in determining when a particular motor pattern can be and is performed. In the previous section we saw that learning can play a role in the acquisiton of a species-specific behavior, but there may be narrow developmental windows or critical periods during which certain defined learning can occur. Hormones control the complex interaction between genetically determined behavior and learning during specific stages of development. We have already seen one example of hormones controlling behavior: high levels of oxytocin in the female goat's blood at the time she gives birth determine a window of time during which she can imprint on her infant. Next we will study two more-complex cases in which hormones control the development, learning, and expression of species-specific behavior.

Sexual Behavior in Rats

Differences in the behavior of males and females of a species are clear examples of genetically determined behavior. Such differences in sexual behavior have been shown to be due to the actions of the sex steroids on the developing brain and on the mature brain. Like most other animals, rats behave sexually in accord with fixed action patterns. When a female rat is in estrus (receptive to males), she responds to a tactile stimulus of her hindquarters with a stereotypic posture called **lordosis**. She lowers her front legs, extends her hind legs, arches her back, and deflects her tail to one side. When a male encounters a female in estrus, he copulates with her with the following sequence of behaviors: He mounts her from the rear, clasps her hindquarters, inserts his penis into her vagina, and thrusts. The roles of genotype and sex hormones in the development of the fixed action patterns of lordosis and copulation have been investigated through manipulating the exposure of the developing rat brain to sex steroids.

If a female rat has her ovaries removed (that is, is spayed), *either as a newborn or as an adult,* she will not show lordosis unless she is injected with female sex steroids. The hormones are necessary for the expression of the female sexual behavior. If this same adult female is injected with testosterone, she does not show male sexual behavior (first two panels of Figure 45.6). There is a surprising variation on this experiment. If a *newborn* female rat has her ovaries removed and is injected with testosterone, she will not show lordosis when treated with female sex hormones as

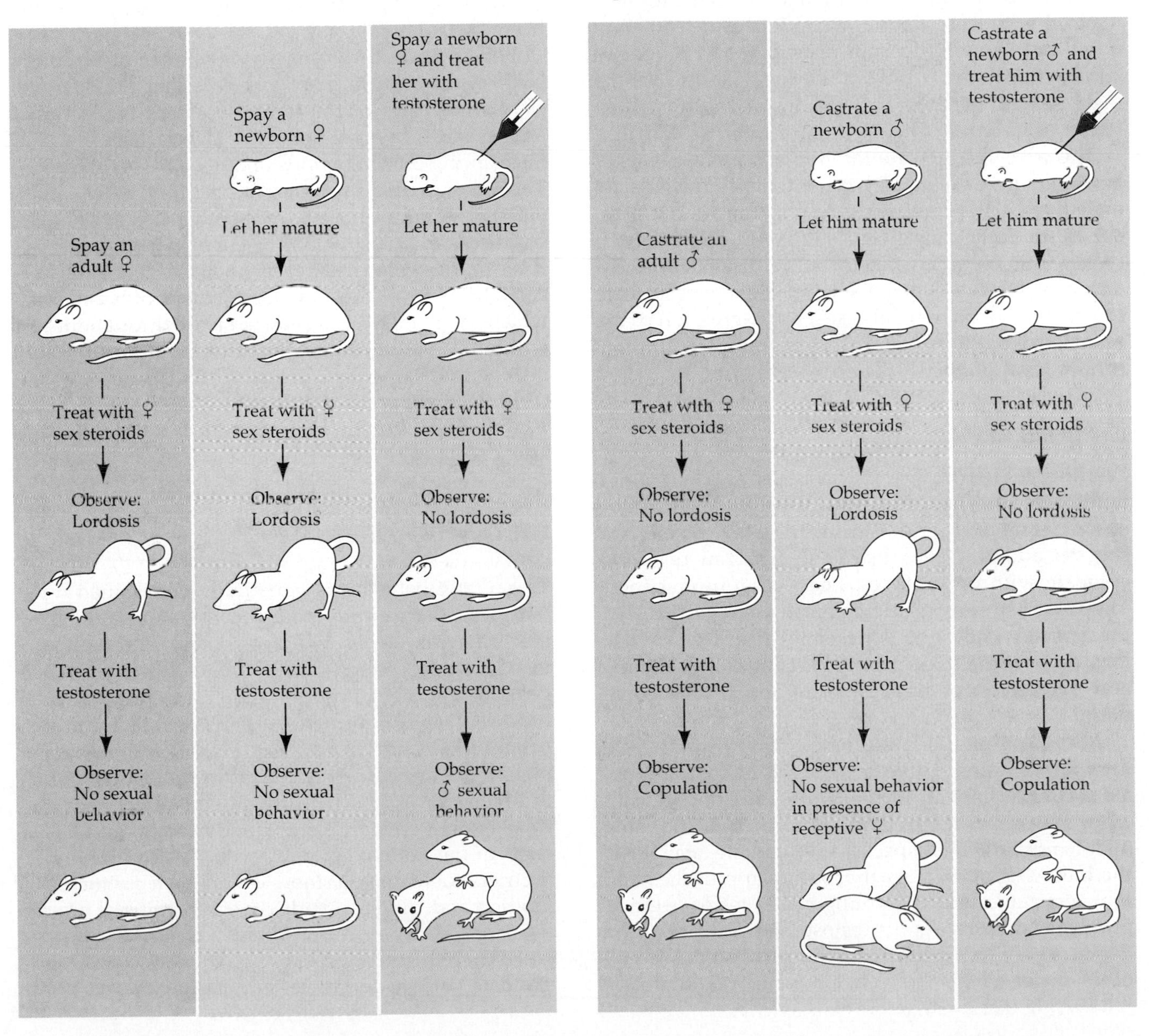

45.6 Hormonal Control of Sexual Behavior
Sex steroids control both the development and expression of sexual behavior in rats. The presence of testosterone in newborn rats of both sexes whose reproductive organs (ovaries or testes) have been removed establishes male behavior patterns, and its absence establishes female patterns. Injections of sex steroids in gonadectomized adult rats stimulate expression of the sexual behavior pattern that developed in response to genotype and early steroid exposure.

an adult. But if this *genetically female* adult rat is treated with testosterone, she will mount other females in estrus and show the male fixed action patterns associated with copulation. The presence of testosterone in the newborn ovariectomized female masculinizes her developing nervous system. When she reaches adulthood, her nervous system responds to male rather than to female steroids and generates male fixed action patterns (third panel of Figure 45.6).

Similar experiments on genetic males do not yield entirely reciprocal results. Castration (removal of the testes) of an adult male does not alter its response to treatment with sex steroids. Such a castrated male does not show lordosis when treated with female hormones, and it does show male sexual behavior when injected with testosterone. If a *newborn* male is castrated, it *will* show lordosis when injected with female hormones as an adult, but it will *not* show male sexual behavior when injected with testosterone as an adult. If the newborn male is castrated *and* injected with testosterone, when it becomes an adult it will not show lordosis in response to injections of female hormones. It will, however, show normal male sexual behavior in response to testosterone (three right-hand panels of Figure 45.6).

These results indicate that the nervous systems of both genetic males and genetic females develop female fixed action patterns if testosterone is not present at an early stage. Testosterone in the newborn causes the nervous system to develop male fixed action patterns whether the animal is a genetic male or a genetic female. In all cases, the expression of the sexual fixed action patterns in the adult requires a certain level of appropriate sex steroids.

Bird Brains and Bird Song

Learning is essential for the acquisition of bird song. Both male and female nestlings hear their species-specific song, but for most songbirds, only the males sing as adults. Males use song to claim territory, compete with other males, and declare dominance. They also use song to attract females, suggesting that the females know the song of their species even if they do not sing. Do sex steroids control the learning and expression of song in male and female songbirds?

After leaving the nest where they experienced their father's singing, young songbirds from temperate and arctic habitats may migrate and associate with other species in mixed flocks, but they do not sing and do not hear their species-specific song again until the following spring. As that spring approaches and the days get longer, the young male's testes begin to grow and mature. As his testosterone level rises, he begins to try to sing. Even if he is isolated from all other males of his species, his song will gradually improve until it is a proper rendition of his species-specific song. At that point the song is **crystallized**—the bird expresses it in similar form every spring thereafter. The juvenile bird's brain learns the pattern of the song by hearing the father. During the subsequent spring, under the influence of testosterone, the bird learns to express that song—a behavior that then becomes rigidly fixed in its nervous system.

Why don't the females sing? Can't they learn the patterns of their species-specific song? Don't they have the muscular or nervous system capabilities necessary to sing? Or do they simply lack the hormonal stimulus for developing the behavior? To answer these questions, investigators injected female songbirds with testosterone in the spring. In response to these injections, the females developed their species-specific song and sang just as the males do. Apparently females learn the song pattern of their species when they are nestlings and have the capability to express it, but they normally lack the hormonal stimulation.

What does testosterone do to the brain of the songbird? A remarkable discovery was that testosterone causes the parts of the brain necessary for learning and expressing song to grow larger (Figure 45.7). Each spring certain regions of the males' brains grow. The individual cells increase in size, they grow longer extensions and—most surprisingly—the *numbers* of brain cells in those regions of the bird brain increase. Prior to these discoveries, newborn vertebrates were thought to have their full complement of brain cells, which they would lose progressively throughout life without replacing them. Research on the neurobiology of bird song has revealed that hormones can control behavior by influencing brain structure as well as brain functioning on both a developmental and a seasonal basis.

THE GENETICS OF BEHAVIOR

To say that behavior is genetically determined does not mean there are specific genes that code for specific behavior. Genes code for protein, and there are many complex steps between the expression of a gene as a protein product and the expression of a behavior. Many intermediate steps exist between any gene and its phenotypic expression, but the complexity is especially great when the phenotypic trait is a behavior. A specific protein may affect behavior by playing a critical role in the development of patterns in the nervous system or in the functioning of the nervous or endocrine system; such influence is indirect and difficult to discover. In no case are all the steps between a gene and a behavior known. Nevertheless, the approaches of genetics clearly substantiate genetic components of behavior and bring

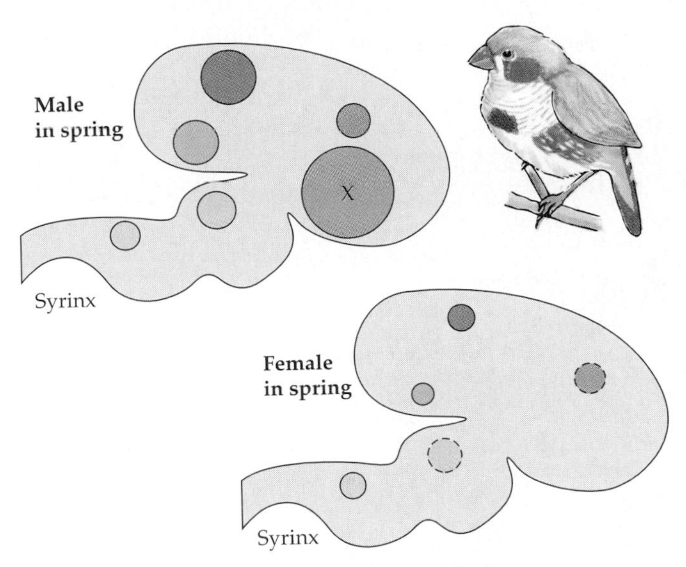

45.7 Effects of Testosterone on Bird Brains
Testosterone induces growth in the regions (colored circles) of a songbird's brain that are responsible for song. During the nonbreeding season, the brains of male and female zebra finches are similar, but in spring rising testosterone levels in the male cause the song regions of its brain to develop. The sizes of the circles are proportional to the volume of the brain occupied by that region; dashed circles indicate estimated volumes. The area labeled "X" is not found in the brains of female finches.

us closer to understanding the underlying mechanisms of inheritance of fixed action patterns. The genetic approaches we will discuss are hybridization, artificial selection, crossing of selected strains, and molecular analysis of genes and gene products.

Hybridization Experiments

The material for genetic analysis is variability, and variability is most pronounced between species. Closely related species frequently show large differences in fixed action patterns, and if such species can be hybridized, the offspring reveal interesting disruptions of their behavior. A classic case is nest building in lovebirds of the genus *Agapornis*. One species, *A. roseicollis*, carries nesting material tucked under its tail feathers. Another species, *A. fischeri*, carries nesting material in its beak (Figure 45.8). Are these simple behavior patterns learned or are they genetically programmed? When the two species are crossed, the hybrid offspring display a maladaptive combination of the two carrying methods. The hybrid picks up nesting material and tucks it into its tail feathers, but does not release the object immediately. As a result, the hybrid inevitably pulls the nesting material out of its tail feathers and drops it. With years of experience, hybrids learn to carry material in their beaks, but they always make an intention movement toward their tail feathers when they pick up nesting material. This hybridization study indicates that the ways in which birds of the two species carry nesting materials are genetically determined.

Konrad Lorenz conducted hybridization experiments on ducks to investigate the genetic determinants of their elaborate courtship. Dabbling ducks such as mallards, teals, pintails, and gadwalls are closely related and can interbreed, but they rarely interbreed in nature because of the specificity of their courtship displays. Each male duck performs a carefully choreographed water ballet (Figure 45.9), and the female probably will not accept his advances unless the entire display is successfully completed. The displays of dabbling duck species consist of about 20 components altogether. The display of each species includes a subset of these components put together in a certain sequence. When Lorenz crossbred the species, he found that the hybrids expressed some components of the display of each parent in new combinations. Most interesting, the hybrids sometimes showed display components that were not in the repertoire of either parent, but were characteristic of the displays of other species. These hybridization studies demonstrated that the motor patterns of the courtship displays were genetically programmed. Fe-

(a)

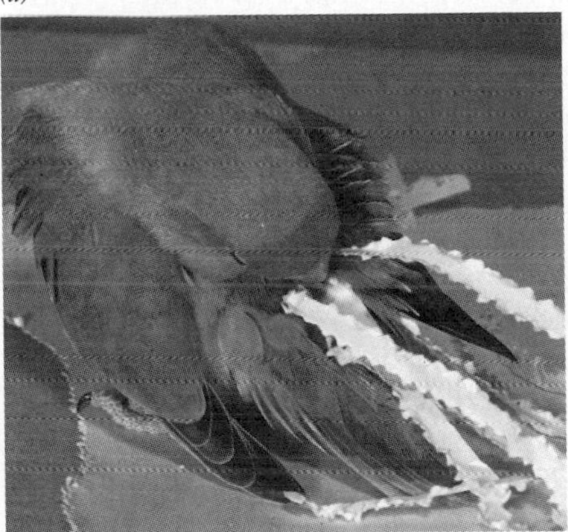

(b)

45.8 Nest Building Behavior Is in the Genes
(a) Peach-faced lovebirds (*Agapornis roseicollis*) carry nest-building materials tucked in their back feathers. *(b)* Fischer's lovebirds (*A. fischeri*) carry the objects in their bills. Hybrid offspring of the two species display a confused combination of the two behaviors, indicating that the behaviors are genetically programmed.

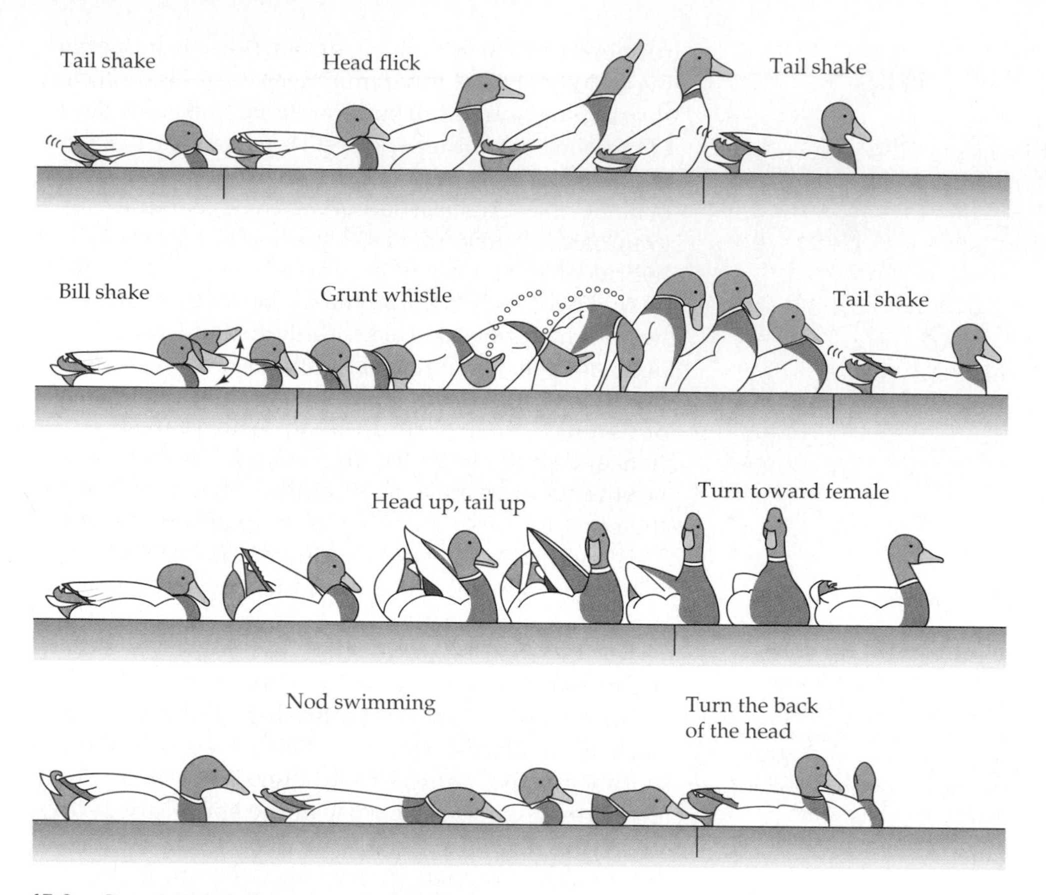

45.9 Courtship Ballet of the Mallard
The courtship display of the male mallard duck contains about ten elements. Closely related duck species may display some of the same ten elements, but they will have other elements not displayed by mallards. The elements of the courtship display and their sequence are species-specific and act to prevent hybridization.

males were not interested in males showing the hybrid displays, thus demonstrating the adaptive significance of the species-specific fixed action patterns.

Selection and Crossing of Selected Strains

Domesticated animals provide abundant evidence that artificial selection of mating pairs on the basis of their behavior can result in strains with distinct behavioral as well as anatomical characteristics. Among dogs, consider retrievers, pointers, and shepherds. Each has a particular behavioral tendency that can be honed to a fine degree by training, whereas other strains cannot be trained in this way. Dogs and other large animals, however, are not the best subjects for genetic studies. Most controlled selection experiments in behavioral genetics have been done on more convenient laboratory animals with short life cycle times and high numbers of offspring. A favorite subject for such studies has been the fruit fly, genus *Drosophila*. Artificial selection has been successful in shaping a variety of behavior patterns in fruit flies, especially aspects of their courtship and mating behavior. Crossing of selected strains reveals that behavioral differences produced by artificial selection are usually due to multiple genes that probably influence the behavior indirectly by altering general properties of the nervous system.

Few behavioral genetic studies reveal simple Mendelian segregation of behavioral traits. One case that does is nest cleaning behavior in honeybees. Nest cleaning counteracts a bacterium that infects and kills the larvae of honeybees. Hives or colonies of one strain of honeybee that is resistant to this disease practices hygienic behavior; when a larva dies, workers uncap its brood cell and remove the carcass from the hive. Another strain of honeybee does not show hygienic behavior and is, therefore, more susceptible to the spread of the disease. When these two strains were crossed, the results indicated that the hygienic behavior was controlled by two recessive genes. Colonies of the F_1 generation are all nonhygienic, indicating that the behavior is controlled by recessive genes. Backcrossing the F_1 with the pure hygienic strain produced the typical 3:1 ratio expected for a two-gene trait. The behavior of the nonhygienic colonies is very interesting. One-third of them show no hygienic behavior at all; one-third uncap the cells of dead larvae but do not remove them; and one-third do not uncap cells but will remove carcasses if the cells are open (Figure 45.10).

Even though these results appear to indicate a gene for uncapping and a gene for removal, these behavior patterns are complex. They involve sensory mechanisms, orientation movements, and motor patterns, each of which depends on multiple properties of many cells. The genetic deficits of nonhygienic bees could influence very small, specific, yet critical properties of some cells. Lacking one critical property, such as a critical synapse or a particular sensory receptor, the whole behavior would not be expressed. The responsible gene, then, is not a specific gene that codes for the entire behavior.

Molecular Genetics of Behavior

The powerful techniques of molecular genetics enable investigation of the functions of specific genes that influence behavior. For example, in the marine mollusk *Aplysia*, egg laying involves a sequence of fixed action patterns. The eggs are extruded from the animal in long strings by contractions of the muscles of the reproductive duct. The animal stops whatever it is doing (usually eating or crawling) and takes the egg string in its mouth. With a series of stereotypic head movements, it pulls the egg string from the duct and coils it into a mass glued together by secretions from its mouth. Finally, with a strong head movement, it affixes the entire mass of eggs to a solid substrate. *Aplysia* has a very simple nervous system, and it was discovered that specific cells in its nervous system produce a peptide that can elicit certain aspects of egg laying. This peptide is called egg-laying hormone. The amino acid sequence of egg-laying hormone was determined, and then molecular genetic techniques were used to find the gene that coded for it. The surprising discovery was that the gene codes for a precursor molecule that has almost 300 amino acids, whereas egg-laying hormone has only 36 amino acids. The precursor molecule also contains other peptides that function as neural signals controlling aspects of egg-laying behavior. One gene could thus code for a set of neural signals necessary to elicit the coordinated motor scores involved in egg-laying behavior. This example is about as close as we can get to making connections between a specific gene and a specific behavior, but the connections depend on the existence of a highly organized nervous system, which of course is a product of many genes.

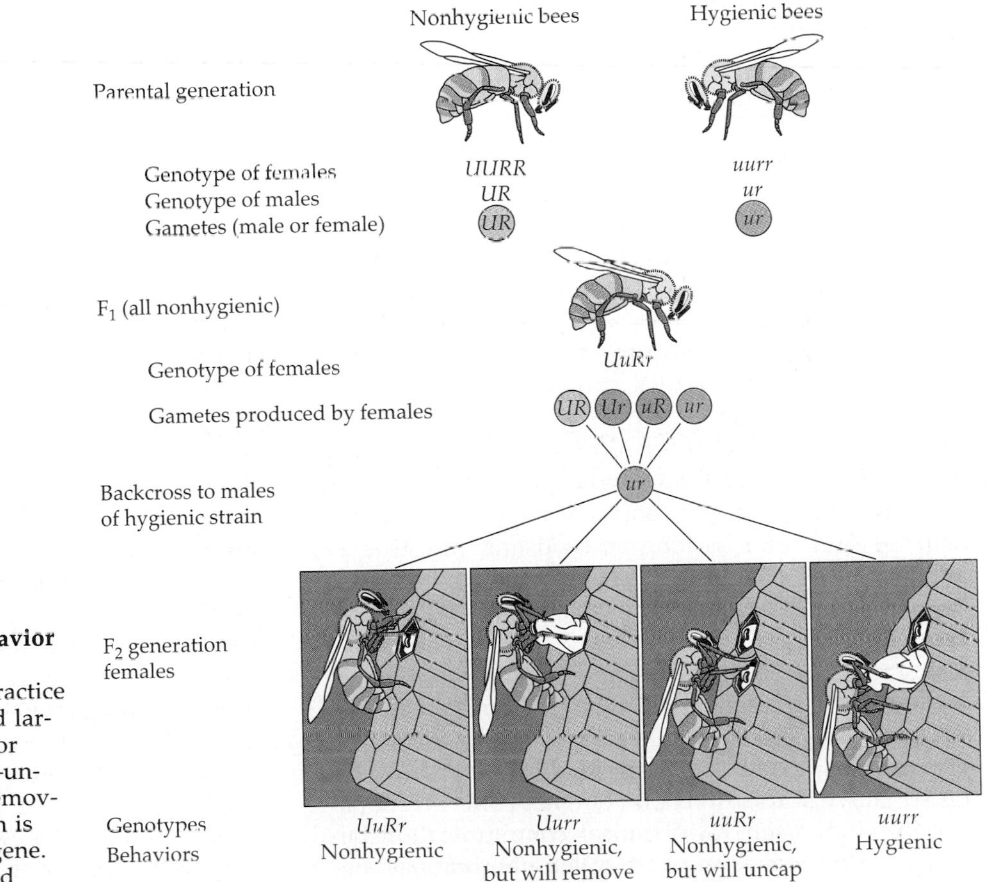

45.10 Genes and Hygienic Behavior in Honeybees
Some honeybee strains make a practice of removing the carcasses of dead larvae from their nests. This behavior seems to have two components—uncapping the larval cell (*u*), and removing the carcass (*r*)—each of which is under the control of a recessive gene. Honeybee females are diploid and males are haploid. The workers in a colony are all females.

COMMUNICATION

Having discussed methods of investigating animal behavior, let's focus now on specific types of behavior. As investigations into animal communication illustrate, a diversity of issues can arise in studies of even a specific type of behavior.

Communication is behavior that influences the actions of other individuals. It consists of **displays** or **signals** that can be perceived by other individuals. Displays and signals are behaviors, anatomical features, or physiological responses that convey information to another individual. The information they convey may be secondary or even incidental to their original function. If the transfer of information benefits the animal generating the signal or display, selection can shape it to enhance its information content. The displays or signals that an animal can generate depend of course on its physiology and anatomy. Its ability to perceive displays or signals depends on its sensory physiology and on the environment through which the display or signal must be transmitted. We learned in Chapter 39 that sensory physiology includes chemosensation, tactile sensation, audition, vision, and electrosensation. These are the forms of animal communication. Studies of communication can be complex because they must take into account the sender, the receiver, and the environment.

Chemical Communication

Chemosensation, which is based on receptor molecules that bind signaling molecules and initiate cellular responses, is probably the oldest form of animal communication. Intracellular signaling involves molecules and receptors, and hormonal integration preceded neural integration in animal evolution. It is a very short evolutionary jump from internal coordination by signal molecules and receptors to the release of molecules into the environment for detection by other individuals. Even protists use such means of communication. Slime molds spend most of their lives as individual amoeboid cells moving through soil and leaf litter as long as food and moisture is adequate (see Chapter 23). When the environment becomes less favorable, the individuals aggregate, form a fruiting body, and release spores. The chemical signal that coordinates aggregation is cAMP, a well-known intracellular signaling molecule (see Chapter 36). Stressed cells release cAMP into the environment, and other individuals sensing the cAMP move in the direction of higher concentrations.

Molecules used for chemical communication between individual animals are called **pheromones**. Because of the diversity of their molecular structures, pheromones can communicate very specific messages that contain a great deal of information. For example, when a female gypsy moth is ready to be inseminated, she releases a pheromone called gyplure. Male gypsy moths downwind by as much as thousands of meters are informed by these molecules that a female of their species is sexually receptive. By orienting to the wind direction and the concentration gradient, they can find her (Figure 45.11*a*). Territory marking is another example in which detailed information is conveyed by chemical communication (Figure 45.11*b*). The receiver of a message from a male cheetah detects information about the animal leaving the message: species, individual identity, reproductive status, height of message (indicating size of the animal that left it), and strength of scent (indicating time elapsed since the message was left).

An important feature of pheromones is that once they are released, they remain in the environment for a long time. A pheromonal message can act as a territorial marker long after the animal claiming the territory has left. By contrast, the signals of a vocal or visual territorial display of a song bird disappear as soon as the bird stops singing and displaying. The durability of pheromonal signals enables them to be used to mark trails, as ants do, or to indicate directionality, as in the case of the gypsy moth sex attractant. The chemical nature and the size of the pheromonal molecule determine its diffusion coefficient. The greater its diffusion coefficient the more rapidly a pheromone diffuses, the farther the message will reach, but the sooner it will disappear. Trail-marking and territory-marking pheromones tend to be relatively large molecules with low diffusion coefficients; sex attractants tend to be small molecules with high diffusion coefficients. A disadvantage of pheromonal communication is that the message cannot be changed rapidly. A discussion based on smells would surely be less effective than one using speech or sign language.

Visual Communication

Many species use visual communication. Visual signals are easy to produce, come in an endless variety, can be changed very rapidly, and clearly indicate the position of the signaler. The extreme directionality of visual signals means, however, that they are not the best for getting the attention of a receiver. The sensors of the receiver must be focused on the signaler or the message will be missed. Most animals are sensitive to light and can therefore receive visual signals, but sharpness of vision limits the detail of the information that can be transmitted. Birds are highly visual and have evolved a vast diversity of patterns of colored feathers and body appendages that can be incorporated into complex displays used in communication (Figure 45.12).

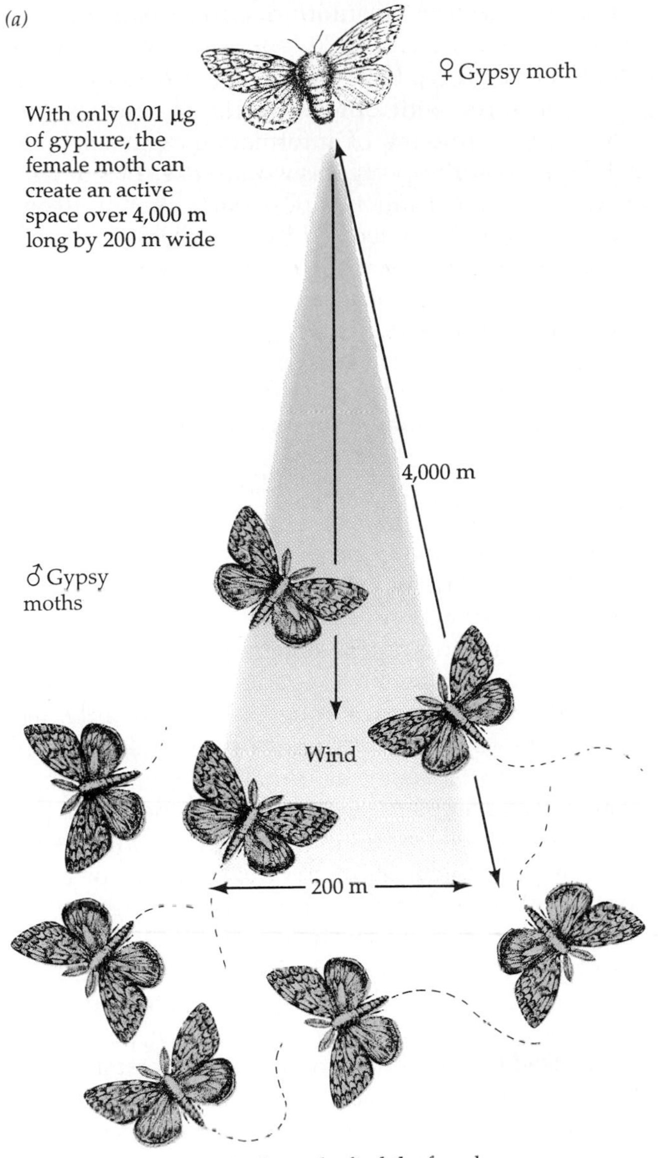

45.11 Many Animals Communicate with Pheromones
(a) A female gypsy moth secretes the pheromone gyplure which can attract males thousands of meters downwind when it binds to sensors on their antennas. *(b)* To mark his territory, this male cheetah is spraying pheromonal secretions from a scent gland in his hindquarters onto a tree. Other cheetahs passing the spot will know that the area is "claimed," and they will know something about who claimed it.

Because visual communication requires light, it is not useful for many species at night or in environments that lack light, such as caves and the ocean depths. Some species have surmounted this constraint on visual communication by evolving their own light-emitting mechanisms. Fireflies use a luciferin/luciferase mechanism to create flashes of light. By emitting flashes in species-specific patterns, fireflies can advertise for mates at night by sending visual signals.

Firefly communication raises another interesting issue. Any system of communication is vulnerable to

45.12 Selection for Effective Display Has Greatly Modified the Anatomy of Some Animals
Male peafowl—peacocks—have evolved brilliant tail plumage, which they display to the females (peahens) during courtship. These elaborate tail feathers are folded when the bird is not displaying.

exploitation by illicit senders or receivers. Predators are commonly illicit receivers. When an animal emits a message in any form, it tends to signal its own position, and the information can be used by predators. Predators of fireflies can locate them in the dark by the flashes of light they emit. Some species of fireflies are themselves predators of other species of fireflies and have evolved the "deceitful" behavior of emitting signal patterns that mimic the mating messages of other species. When a prospective suitor homes in on the signal, it is eaten.

Auditory Communication

Humans are very familiar with communicating by sound. In Chapter 39 we discussed the physical properties of sound and the sensory structures that transduce sound pressure waves into neural signals. Compared with visual communication, auditory communication has several obvious advantages and disadvantages. Sound can be used at night and in dark environments. Sound can go around objects that would interfere with visual signals. Sound is better than visual signals in getting the attention of a receiver because the sensors do not have to be focused on the signaler for the message to be received. Like visual signals, sound can provide directional information, as long as the receiver has at least two sensors spaced somewhat apart. Differences in the sound intensity, in the time of arrival, and in the phase of the pressure wave reaching the two sensors can provide information about the direction of the sound source. By maximizing or minimizing these features of the sounds they emit, animals can make their location easier or more difficult to determine. Pure tones with sudden onsets and offsets are much easier for the receiver to localize than complex sounds with gradual onsets and offsets.

Sound is good for communicating over long distances. Even though the intensity of sound decreases with distance from the source, loud sounds can be used to communicate over distances much greater than those possible with visual signals. An extreme example is the communication of whales. Some whales, such as the humpback, have very complex songs. When these sounds are produced at a certain depth (around 1000 meters), the sound waves are channeled between the thermocline (a sudden temperature change in the water column) and much deeper waters so they can be heard hundreds of kilometers away. In this way, humpback whales use sound communication to locate each other over vast areas of ocean.

Auditory signals cannot convey complex information as rapidly as visual signals can. A well-known expression states, "A picture is worth a thousand words." When individuals are in visual contact, an enormous amount of information is exchanged instantaneously (for example, species, sex, individual identity, maturity, level of motivation, dominance, vigor, alliances with other individuals, and so on). Coding that amount of information with all of its subtleties as auditory signals would take considerable time, thus increasing the possibility that the communicators could be located by predators.

Remarkably few species produce sound or communicate by sound. The animal world is relatively silent. Most invertebrates do not produce sound; cicadas and crickets are marvelous exceptions. Most fish, amphibians, and reptiles do not produce sound, and therefore do not communicate via this form.

Tactile Communication

Communication by touch is extremely common, although not always obvious. Animals in close contact use tactile interactions extensively, especially under conditions that do not favor visual communication. When social insects such as ants, termites, or bees meet, they contact each other with their antennas and front legs. Some of these contacts may involve the exchange of chemical signals as well as tactile ones. In Chapter 39 we discussed how the lateral line organs of fish are useful in sensing vibrations in the environment. Such vibrations can be used for intraspecific communication. At the beginning of this chapter we discussed how male spiders communicate with females by creating vibrations in their webs.

One of the most remarkable and best-studied uses of tactile communication is the dance of honeybees that is used to convey information about distance and direction to a food source. Dancing bees make sounds and carry odors on their bodies, but they convey a great deal of information by dance movement. The dances are monitored by other bees, who follow and touch the dancer. When a foraging bee finds food, she returns to the hive and communicates her discovery by dancing in the dark on the vertical surface of the honeycomb. If the food is less than 80 to 100 meters from the hive, she performs a **round dance**, running rapidly in a circle and reversing her direction after each circumference. The odor on her body indicates the flower to be looked for, but the dance contains no information about the direction in which to go. If the food source is farther than 80 meters, she performs a waggle dance, which conveys information about both the direction and the distance of the food source. In the waggle dance, a bee repeatedly traces out a figure-eight pattern as she runs on the vertical surface. She alternates half circles to the left and right, with vigorous wagging of her abdomen in the short, straight crossover between turns. The angle of the straight line indicates the direction of the food sources relative to the direction of the sun (Fig-

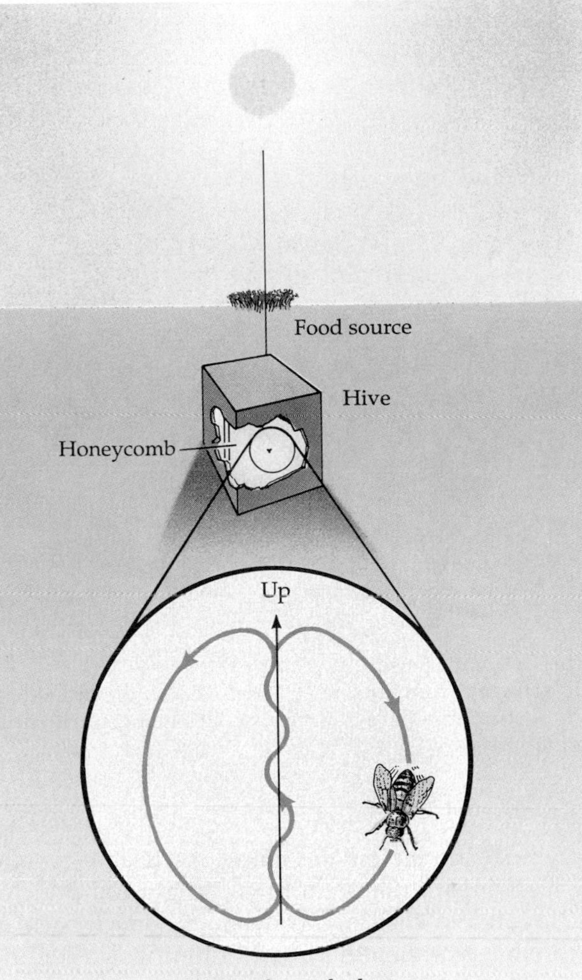

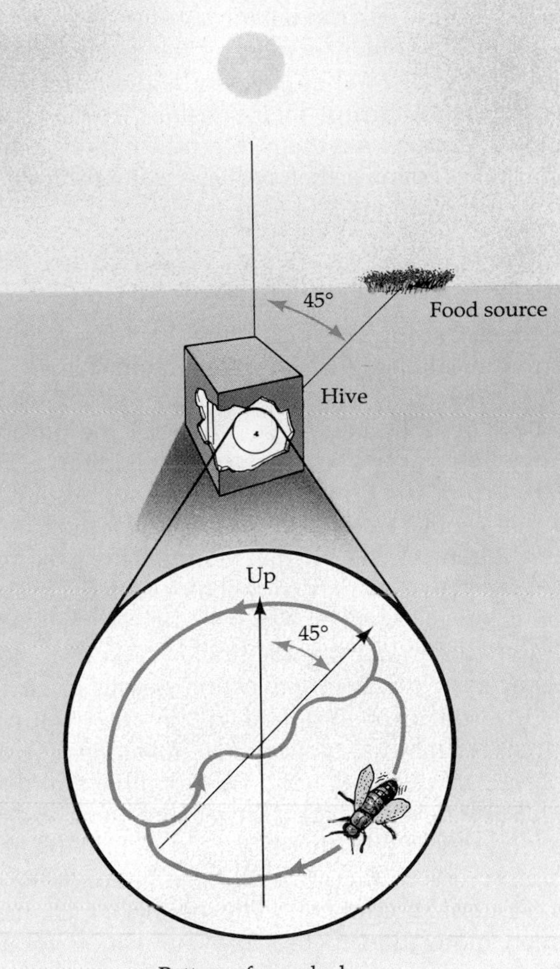

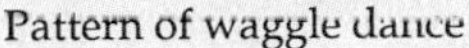

45.13 The Waggle Dance of the Honeybee
By running straight up on the surface of the honeycomb in a dark hive, a honeybee tells her hive mates that there is a food source in the direction of the sun and at least 80 meters from the hive. The rate of circling indicates distance to the food source, while the intensity of the waggling indicates the richness of the food source. If the food source were in the opposite direction from the sun, she would orient her waggle runs straight down. When her waggle dance is run at an angle from the vertical, the other bees know that the same angle separates the direction of the food source from the direction of the sun.

ure 45.13). The speed of the dancing indicates the distance to the food source: the farther away it is, the slower the waggle run. The dances of honeybees are unusual because they are based on an arbitrary convention: straight down could just as well indicate the direction of the sun as straight up. Arbitrary, symbolic conventions like this have been developed to an extreme degree in human language.

Electrocommunication

In Chapter 39 we learned about electrosensors of fish species living in murky water and about electric fish that generate electric fields by emitting series of electric pulses. We discussed how these trains of electrical pulses can be used for sensing objects in the immediate surroundings of the fish, but they can also be used for communication. An electrode connected to an amplifier and a speaker can be used to "listen" to the signal generated by each fish in a tank holding numerous electric fish. The amplifier reveals that each fish emits a pulse at a different frequency, and the frequency each fish uses relates to its status in the population. Glass knife fish (*Eigenmannia*) males emit lower frequencies than females. The most dominant male has the lowest frequency, and the most dominant female has the highest frequency. When a new individual is introduced into the population, the other individuals adjust their frequencies so that they do not overlap, and the signal of the new individual indicates its position in the population. In their natural environment—the murky waters of tropical rainforests—these fish can tell the identity, sex, and social position of another member of the population by its electric signals. When two individuals interact directly, they interrupt their constant frequency emissions and modulate them to produce more-complex, "chirplike" signals that perhaps communicate even more information.

Origins of Communication Signals

Included among the constraints on the evolution of communication signals are the anatomical and physiological characteristics of a species that are available to be shaped by natural selection for the purpose of conveying information to other individuals. Charles Darwin was the first person to give serious attention to this problem of the origin and evolution of communication signals. In 1872 he published the results of his detailed studies in a book entitled *The Expression of the Emotions in Man and Animals*. In this perceptive book Darwin identified several important means by which communication signals originate.

One source of raw material for the evolution of signals is **intention movements**: movements that precede a particular behavior. For example, a bird ready to take off flexes its legs, sleeks its feathers, and raises its wings slightly at the shoulders. A bird in a threatening situation, such as facing a neighbor challenging its territorial boundary, will experience a conflict between motivation to flee and motivation to attack. Thus its intention movements are exaggerated and mixed. Raising its wings at the shoulders might make it appear bigger to its adversary, thus augmenting its attack intention movements. The combination of behavior patterns will convey information to the adversary about the degree of motivation of the defender. Selection can work to enhance this threat display that originated in intention movements. In red-winged blackbirds, for example, the red shoulder patches are most effectively displayed when the wings are slightly raised in the threat posture (see Figure 46.15*b*). Darwin pointed out the threat posture of the domestic cat, which appears to result from conflicting motivations. The forelegs seem to push back while the hind legs push forward, resulting in the hunched back that makes the cat look bigger and more threatening (Figure 45.14).

45.14 Threat Display of a Cat
This drawing from Charles Darwin shows the arched back posture of a cat when it is experiencing conflicting motivations to attack or to flee.

Autonomic responses (see Chapter 38) are another possible source of material upon which selection can operate to produce displays. Urination and defecation are autonomic responses that have been extensively used as signals. The erection of fur or feathers under sympathetic nervous system control is another example. A particularly picturesque example is the mating display of the male frigate bird. Frigate birds are large and black and nest on oceanic islands near the equator. The male builds a nest and then sits on it, trying to attract females flying overhead by spreading his wings, inflating a huge, bright red pouch in his throat, and shaking his head and pouch back and forth while vocalizing. Throat fluttering is a common thermoregulatory response of birds involving rapid movements of air in and out of pouches of skin on the front of the neck. The bizarre and dramatic courtship display of the male frigate bird probably originated through exaggeration of autonomic thermoregulatory behaviors (Figure 45.15).

Displacement behavior is a third class of behavior that can be shaped by natural selection into communication displays. In a tense situation that involves highly conflicting motivations such as attack and escape, an animal sometimes does something completely irrelevant. It might groom itself, feed, or attack some object in the vicinity. If such behavior enhances the display of the animal, it can be favored by selection and incorporated into the display. For example, male three-spined stickleback fish defend small territories when they build a nest to attract potential mates. At the boundary of his territory, the male stickleback can experience equally strong motivations to attack and to flee from a neighboring male. During boundary disputes, the males engage in head-down threat displays that resemble nest-building postures, but no nest building takes place. This threat display probably evolved from nest-building displacement activities resulting from the approach–avoidance conflicts of boundary disputes.

This discussion of animal communication has fo-

45.15 Frigate Bird Display
A male frigate bird displays his red throat pouch for the female next to him. This courtship display may have evolved from the throat-fluttering behavior birds use in thermoregulation.

cused on examples of how natural selection shapes the behavior of a species. In Chapter 46 we will study in more detail the evolution and functions of displays in the context of social behavior.

THE TIMING OF BEHAVIOR: BIOLOGICAL RHYTHMS

An important aspect of behavior is its temporal organization. The neurophysiology of most behavior is poorly understood, but the study of biological rhythms has led to major discoveries about brain mechanisms.

Circadian Rhythms

Our planet turns on its axis once every 24 hours, creating a cycle of environmental conditions that has existed throughout the evolution of life. Many organisms thus evolved rhythmicity. In Chapter 34 we encountered rhythmicity in plants. Indeed, daily cycles are a characteristic of almost all organisms. What is surprising, however, is that daily rhythmicity does not depend on the 24-hour cycle of light and dark. If animals are kept under absolutely constant environmental conditions, such as constant dark and constant temperature, with food and water available all the time, they still demonstrate daily cycles of activity: sleeping, eating, drinking, and just about anything else that can be measured. This persistence of the cycle in the absence of light/dark cycles suggests that the animal has an internal (endogenous) clock. Without time cues from the environment, however, these daily cycles are not exactly 24 hours. They are therefore called **circadian rhythms** (*circa* = "about," *dies* = "day").

To discuss biological rhythms, we must introduce some terminology. A rhythm can be thought of as a series of cycles, and the length of one of those cycles is the **period** of the rhythm. Any point on the cycle is a **phase** of that cycle. Hence when two rhythms completely match, they are in phase, and if a rhythm is shifted (as in resetting a clock), it is phase-advanced or phase-delayed. Since the period of a circadian rhythm is not exactly 24 hours, it must be phase-advanced or phase-delayed every day to remain in phase with the daily cycle of the environment. This process of the resetting of the rhythm by environmental cues is called **entrainment**. An animal kept in constant conditions will not be entrained to the 24-hour cycle of the environment, and its circadian clock will free-run with its natural periodicity. If its period is less than 24 hours, the animal will begin its activity a little earlier each day (Figure 45.16).

Animals that have free-running circadian rhythms can be used in experiments to investigate stimuli that phase-shift or entrain the clock. Under natural conditions, environmental cues such as the onset of light or dark entrain the free-running rhythm to the 24-hour cycle of the real world. In the laboratory it is possible to entrain circadian rhythms in animals held under constant conditions with short pulses of light or dark administered every 24 hours. Researchers can also entrain animals to light or dark pulses given at intervals not equal to 24 hours, as long as those intervals are not too short or too long. The range of entrainment of the endogenous clock is limited.

When you fly across several time zones, your circadian clock is out of phase with the real world at your destination and jet lag results. Gradually your endogenous rhythm synchronizes itself with the real world as it is reentrained every day by environmental cues. Since your internal rhythm cannot be shifted by more than 30 to 60 minutes each day, it takes a number of days to reentrain your endogenous clock to real time in your new location. This period of reentrainment is when you have the symptoms of jet lag, because your endogenous rhythm is waking you up, making you sleepy, initiating activities in your digestive tract, and stimulating many other physiological functions at inappropriate times of the day.

Where is the circadian clock? In mammals the master circadian clock is located in two tiny groups of cells just above the optic chiasm, the place where the two optic nerves cross (see Figure 39.29). Hence these structures are called the **suprachiasmatic nuclei**. If these two little groups of cells are destroyed, the animal loses circadian rhythmicity. Under constant conditions the animal is equally likely to be active or

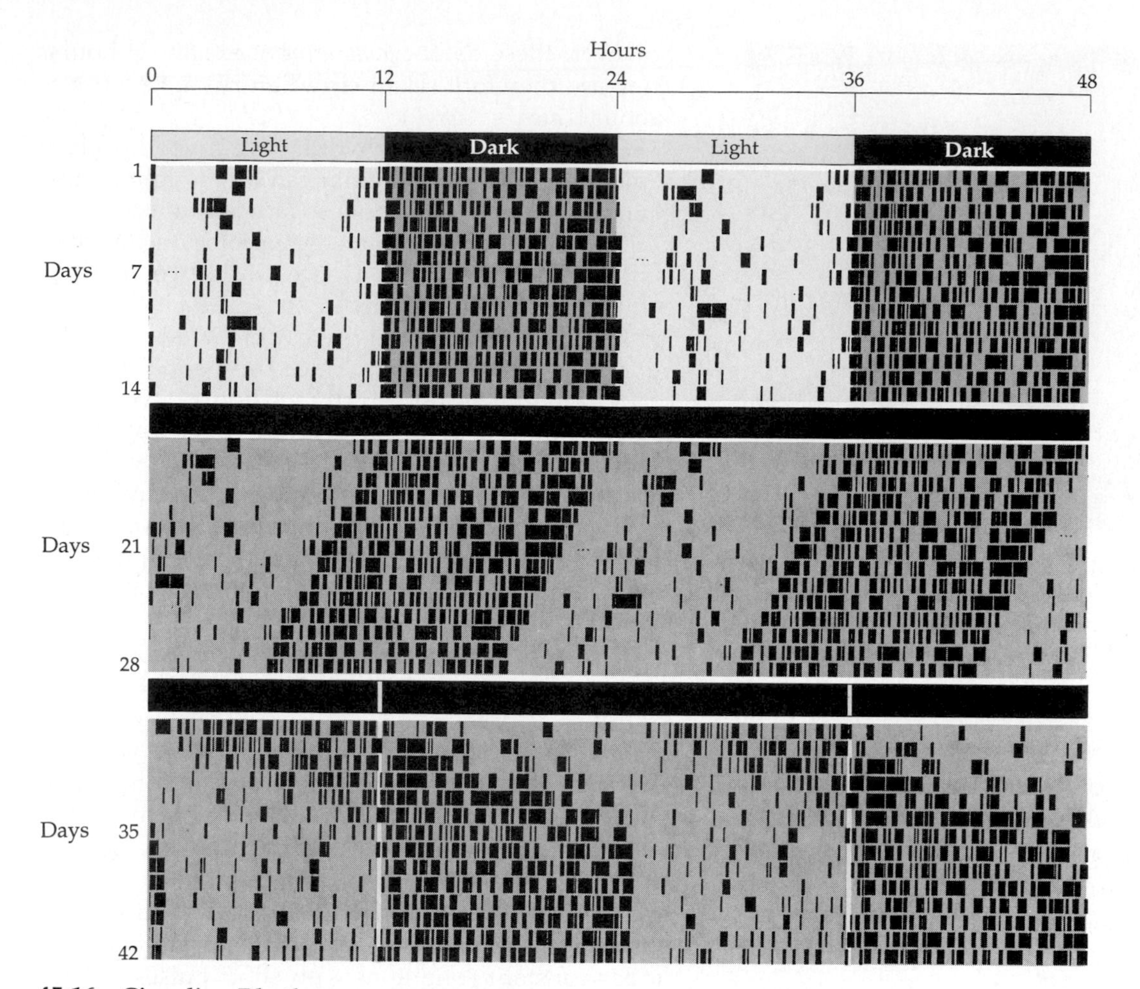

45.16 Circadian Rhythms
The black marks indicate that a mouse is running on an activity wheel. Two days are recorded on each line, so the data for each day are plotted twice, once to the right side of the 24-hour mark and again at the start of the next line down. This double plotting makes patterns easier to see. Changes in the schedule of light and dark are indicated by shading. First the mouse sees 12 hours of light and 12 hours of dark every day, then it is placed in total darkness, and finally it is given a 10-minute exposure to light each day. In constant darkness the circadian rhythm is free-running, but a 10-minute flash of light can entrain it. This figure is idealized but represents results from real experiments.

asleep, eat or drink, at any time of day. Its activities are randomly distributed (Figure 45.17). Experiments first done by Patricia deCoursey at the University of South Carolina show that circadian rhythmicity can be restored in an animal whose suprachiasmatic nuclei have been destroyed by transplanting those nuclei from another animal. In no other known case can a brain tissue transplant restore a complex behavior! Since the restored rhythm has the period of the animal donating the tissue, the transplant clearly controls the recipient's behavior and does not just provide a permissive factor.

Because circadian rhythms are found in virtually every animal group, as well as in protists, plants, and fungi, the molecular mechanisms for generating circadian rhythms must be very general properties of cells. A diversity of master clocks using these mechanisms have been produced by natural selection. Invertebrates, for example, do not have suprachiasmatic nuclei, but they certainly have circadian rhythmicity. In the mollusk *Aplysia,* cells in the eyes are the circadian clock. Circadian control systems are diverse even among vertebrates. The circadian clock of birds resides in the pineal gland, a mass of tissue between the cerebral hemispheres that produces the hormone melatonin. If the pineal gland of a bird is removed, the bird loses circadian rhythmicity. In mammals, light entrains the circadian clock via photosensors in the eyes, but in birds, the pineal gland itself is sensitive to light and is sometimes called the third eye. If a small amount of black ink is injected under the skin on the top of a bird's head to blacken

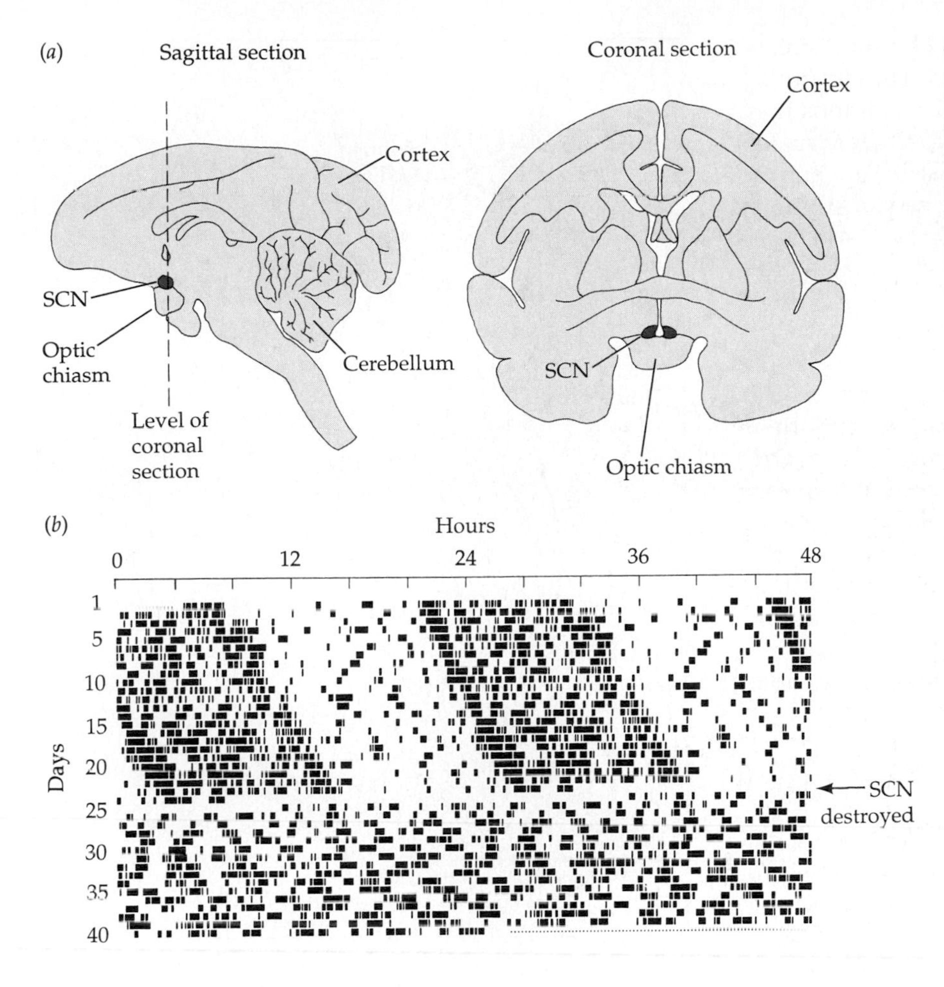

45.17 Where the Clock Is *(a)* The circadian clock of mammals is in the suprachiasmatic nuclei (SCN) of the brain. These nuclei are located just above the site on the bottom of the brain where the optic nerves join. *(b)* If its suprachiasmatic nuclei are destroyed, a mammal loses its circadian rhythm.

the skull over the pineal gland, the bird will not entrain to the cycle of light and dark, but will have a free-running circadian rhythm.

Circannual Rhythms

In addition to turning on its axis every 24 hours, our planet also revolves around the sun once every 365 days. Earth is tilted on its axis, so its revolution around the sun results in seasonal changes in day length at all locations except at the equator. These changes in day length secondarily create seasonal changes in temperature, humidity, weather, and other variables. Because the behavior of animals must adapt to these seasonal changes, animals must be able to anticipate seasons and adjust their behavior accordingly. Most animals should not come into reproductive condition and mate just prior to winter, because their offspring would be born during a time of little food and harsh weather conditions. For many species, the change in day length is an excellent and absolutely reliable indicator of seasonal changes to come. If photoperiod has a direct effect on the physiology and behavior of a species, that species is said to be photoperiodic. For example, if male deer are held in captivity and subjected to two cycles of day-length change in one year, they will grow and drop their antlers twice during that year. Many species of birds are also photoperiodic.

For some animals, changing day length is not a reliable cue. Hibernators, for example, spend long months in dark burrows underground but have to be physiologically prepared to breed almost as soon as they emerge in the spring. The timing of their breeding is important because their young must have time to grow and fatten before the next winter. Other examples of animals that receive little or ambiguous information from changes in day length are resident tropical species, migratory species that overwinter near the equator, or migratory species (mostly birds) that migrate across the equator. For a bird overwintering in the tropics, there is no change in photoperiod that it can use to time its migration north to the breeding grounds. A bird that crosses the equator must fly south as day length decreases at one time of year but fly north when day length decreases at another time of year. If change in day length is not a reliable cue for seasonal behavior, what else could be used as a calendar?

Hibernators and equatorial migrants have endogenous annual rhythms, called **circannual rhythms**. Their nervous systems have built-in calendars. Just as circadian rhythms are not exactly 24 hours, circannual rhythms are not exactly 365 days. The circannual rhythm of an animal under constant conditions may be 360 days, or 345 days. Rarely is it longer than 365 days, however, because being late for an annual event such as breeding would be a very costly mistake.

HOW DO THEY FIND THEIR WAY?

Within a local environment, finding your way is not a problem. You remember landmarks and orient yourself with respect to those reference points. **Orientation** is a very common animal behavior. It simply means that the animal organizes its activity spatially with respect to reference points such as objects in the environment, a predator or prey, a mate or offspring, a nest or food source, or even a signal such as the call or display of another individual. But what if a destination is at a considerable distance? How does the animal orient to it and find its way?

Piloting

In most cases the answer is quite simple: The animal knows and remembers the structure of its environment. It uses landmarks to find its nest, a safe hiding place, or a food source. Orienting by means of landmarks is called **piloting**. Even long-distance migrations of animals can be achieved by piloting that does not depend on specific landmarks. For example, the gray whales that spend the summer feeding in the Gulf of Alaska and the Bering Sea and migrate south in the winter to breed in lagoons on the Pacific coast of Baja California can find their way by following two simple rules (Figure 45.18): Keep the land to the left in the fall and to the right in the spring. By following the west coast of North America, they can travel from summer to winter areas and back again by piloting. Coastlines, mountain chains, rivers, water currents, and wind patterns serve as piloting cues for many species. Yet there are remarkable cases of long-distance orientation and movement that cannot be explained on the basis of piloting by landmarks.

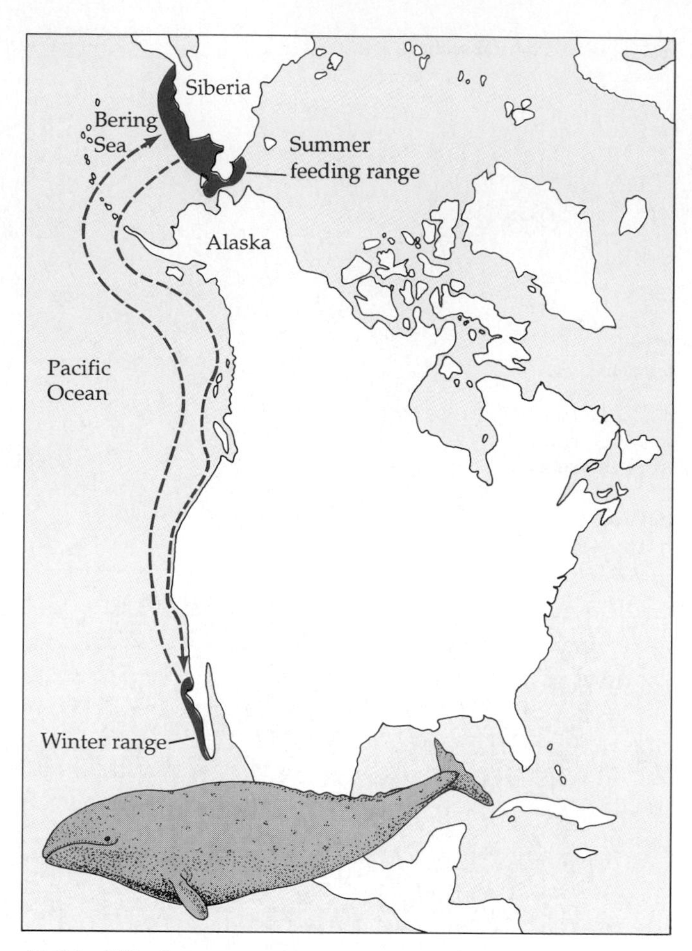

45.18 Piloting
Gray whales migrate south in winter, from the Bering Sea to the coast of Baja California. They follow a landmark: the west coast of North America. Such navigation is called piloting.

Homing

The ability of an animal to return to a nest site, burrow, or any other specific location is **homing**. In most cases homing is merely piloting in a known environment, but some animals are capable of much more sophisticated feats of navigation. People who breed and race homing pigeons take the pigeons from their home loft and release them at a remote site where they have never been before. The first pigeon home wins. Data on departure directions, known flying speed, and distance traveled show that the pigeons fly fairly directly from the point of release to home. They do not randomly search until they encounter familiar territory. Scientists have used homing pigeons to investigate the mechanisms of animal navigation. In one series of experiments the pigeons were fitted with frosted contact lenses. They could see no details other than degree of light and dark. These pigeons still homed and fluttered down to the ground in the vicinity of their loft. They were able to navigate without visual images of the landscape.

Marine birds provide many dramatic cases of homing over great distances in an environment where landmarks are rare. In daily feeding trips, many marine birds fly over hundreds of miles of featureless ocean and then return directly to a nest site on a tiny island. Remarkable feats of homing are demonstrated by albatrosses. When a young albatross leaves its nest on an oceanic island, it flies widely over the southern oceans for eight or nine years before it reaches reproductive maturity. At that time it flies back to the

45.19 Coming Home
A pair of black-browed albatrosses engage in courtship display over their partially completed mud nest. Many albatrosses return to the site of their own birth to find a mate, and will return to that site year after year.

island where it was raised to select a mate and build a nest (Figure 45.19). After the first mating season the pair separates, and each bird resumes its solitary wanderings over the oceans. The next year they return to the same nest site at the same time, reestablish their pair bond, and breed. Thereafter they return to the nest to breed every other year, spending many months in between at sea. These long-distance, synchronous homing trips are amazing feats of navigation and timing.

Migration

Ever since humans inhabited temperate and subpolar latitudes, they must have been aware of the fact that whole populations of animals, especially birds, disappear and reappear seasonally. It was not until the early nineteenth century, however, that patterns of migration were established by marking individual birds with identification bands around their legs. Being able to identify individual birds in a population made it possible to demonstrate that the same birds and their offspring returned to the same breeding grounds year after year, and that these same birds were found during the nonbreeding season at distant locations hundreds or even thousands of kilometers from the breeding grounds.

How do migrants find their way over such great distances? A reasonable hypothesis is that young birds on their first migration follow experienced birds and learn the landmarks by which they pilot in subsequent years. However, adult birds of many species leave the breeding grounds before the young have finished fattening and are ready to begin their first migration. These naive birds must be able to navigate accurately on their own and with little room for mistakes. Some species of small songbirds breed in the high latitudes of North America, fly to the coast for fattening, and then fly over the North Atlantic Ocean on a direct route to South America (Figure 45.20). They cannot land on water and their fuel reserves are limited by their small size. Considering distance, flight speed, and metabolic rate, they must be extremely efficient and accurate in navigating to their landfall on the coast of South America.

Navigation

Homing and migrating animals find their way by several mechanisms of navigation. Although piloting is a type of navigation, it cannot explain the abilities of many species to take direct routes to their destinations through areas they have never experienced.

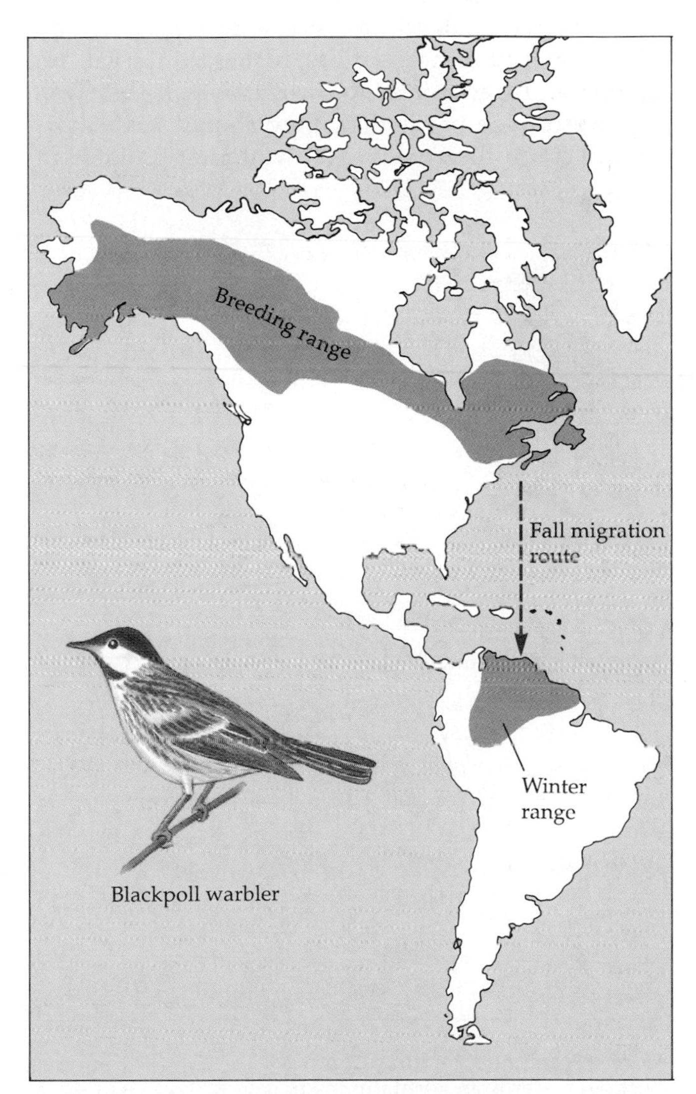

45.20 Songbirds Migrate over the Atlantic
The blackpoll warbler is one of many species that breeds over the northern United States and Canada and winters in South America. Its fall migration is first to the northeast coast of North America, where it feeds in preparation for the nonstop overwater flight to South America.

Humans use two systems of navigation that differ in complexity: distance and direction navigation and bicoordinate navigation. Distance and direction navigation involves knowing the direction to reach the destination and knowing how far away that destination is. With a compass to determine direction and a means of measuring distance, humans can navigate. Bicoordinate navigation, also known as true navigation, involves knowing the latitude and longitude (the map coordinates) of position and destination. From that information a route can be plotted to the destination. Do animals have these sophisticated abilities to navigate?

Researchers conducted an experiment with European starlings to determine their method of navigation. These short-distance migrants travel between breeding grounds in the Netherlands and northern Germany and wintering grounds to the southwest, in southern England and western France (Figure 45.21). The birds were captured before the fall migration and transported to Switzerland. If they were capable of true navigation, they should have flown northwest to their traditional wintering ground. Instead they flew in their normal southwest direction and landed in Spain. The researchers concluded that the starlings used distance and direction for navigation.

45.22 Raring to Go
A captive bird ready to migrate shows migratory restlessness in a circular cage. The cage is lined with a paper funnel, and on the floor is an ink pad. The bird's feet mark the orientation of its activity.

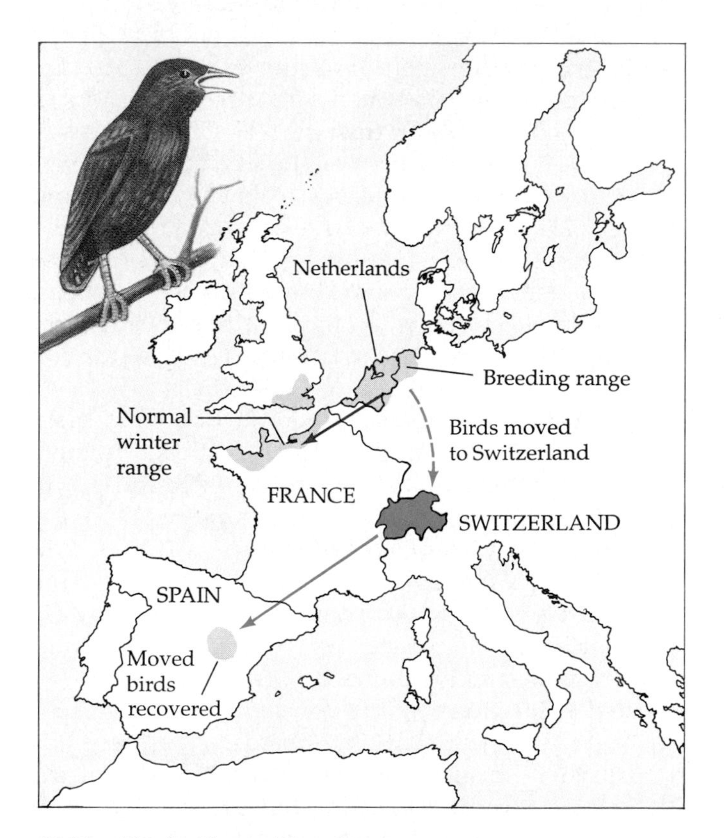

45.21 Navigation with a Compass
European starlings normally make a short winter migration in a southwesterly direction, from the Netherlands to coastal France and southern England (red arrow). Experimental populations of starlings moved to a site in Switzerland did not fly northwest to their traditional grounds, but followed the same southwesterly route (blue arrow), which took them to Spain.

How do animals determine distance and direction? In many instances, distance is not a problem as long as the animal recognizes its destination. Homing animals recognize landmarks and can pilot once they reach familiar areas. Some evidence suggests that biological rhythms play a role in determining migration distances for some species. Birds kept in captivity display increased and oriented activity at the time of year when they would normally migrate (Figure 45.22). Such **migratory restlessness** has a definite duration, which corresponds to the usual duration of migration of the species. Since distance is determined by how long an animal moves in a given direction, the programming of the duration of migratory restlessness can set the distance for its migration.

Two obvious candidates for determining direction are the sun and the stars. During the day the sun is an excellent compass, as long as time is known. In the northern hemisphere the sun rises in the east, sets in the west, and points south at noon. Animals can tell the time of day by means of their circadian clocks. Furthermore, clock-shifting experiments demonstrate that animals use circadian clocks to determine direction from the position of the sun. Researchers placed birds in a circular cage that enabled them to see the sun and sky but no other visual cues. Food bins were arranged around the sides of the cage, and the birds were trained to expect food in the bin in one particular direction, for example south.

After training, no matter when they were fed, and even with the cage rotated between feedings, they always went to the bin at the southern end of the cage for food (Figure 45.23*a*). Next the birds were placed in a room with a controlled light cycle and their circadian rhythms were phase-shifted. For example, in the controlled light room the lights were turned on at midnight. After a couple of weeks the circadian clocks of the birds were phase-advanced by six hours. The birds were returned to the circular cage under natural light conditions with sunrise at 6 A.M. Because of the shift in their circadian rhythms, their endogenous clocks were indicating noon at the time the sun came up. If food was always in the south, and it was sun-up, they should have oriented 90° to the right of the direction of the sun. But since their circadian clocks were telling them it was noon, they looked for food in the direction of the sun—the east bin (Figure 45.23*b*). The six-hour phase shift in their circadian clocks resulted in a 90° error in their orientation. These types of experiments on many species have shown that animals can orient by means of a time-compensated solar compass.

Many animals are normally active at night; in addition, many day-active species of birds migrate at night and cannot use the sun to determine direction. Two sources of information about direction are available from the stars. The positions of constellations change because the Earth is rotating. With a star map and a clock, direction can be determined from any constellation. One point in the sky, however, does not change position during the night: the point directly over the axis on which the Earth turns. In the northern hemisphere, a star called Polaris or the North Star lies in that position and always indicates north. Stephen Emlen at Cornell University investigated whether birds use these sources of directional information from the stars. Emlen raised young birds in a planetarium, where star patterns are projected on the ceiling of a large, domed room. The star patterns in the planetarium could be slowly rotated to simulate the rotation of Earth. When the star patterns were not rotated, birds caught in the wild could still orient perfectly well in the planetarium, but birds raised in the planetarium under a nonmoving sky could not. If the star patterns in the planetarium were rotated each night as the young birds matured, they were able to orient in the planetarium, showing that birds can learn to use star patterns for orientation if the sky rotates. No evidence was found, however, that the birds used their circadian clocks to derive directional information from the star patterns. Experienced birds were not confused by a still sky or a sky that rotated faster than normal. The birds were orienting to the fixed point in the sky, the North Star. Young birds raised under a sky that rotated around a different star imprinted on that star and oriented to it as if it were the North Star (Figure 45.24). These studies showed that birds raised in the northern hemisphere learn a star map that they can use for orientation at night by imprinting on the fixed point in the sky.

Animals cannot use sun and star compasses when the sky is overcast, yet they still home and migrate under such conditions. Are there other sources of information they can use for orientation? There appears to be considerable redundancy in animals' abilities to sense direction. Pigeons home perfectly well on overcast days, but this ability is severely impaired if small magnets are attached to their heads. These experiments and subsequent ones with more-sophisticated ways of disrupting the magnetic field around the bird have demonstrated a magnetic sense. Cells have been found that contain small particles of the magnetic mineral magnetite, but the neurophysiology of the magnetic sense is largely unknown. Another cue is the plane of polarization of light, which can give directional information even under heavy cloud cover. Very low frequencies of sound can give information about coastlines and mountain chains. Weather patterns can also provide considerable directional information.

(*a*) Pigeon placed in a circular cage from which it can see the sky (but not the horizon) can be trained to seek food in one direction, even when cage is rotated between trials

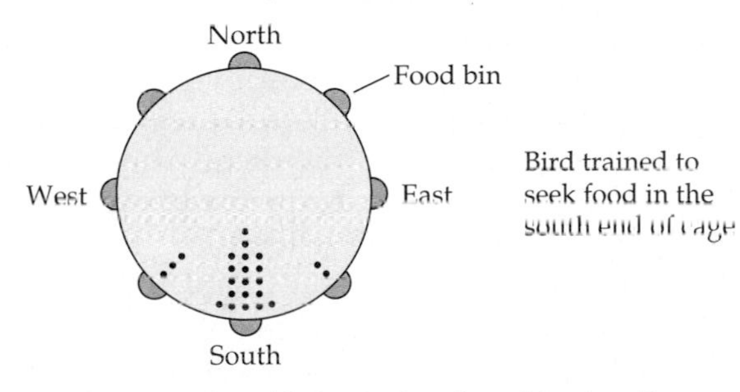

(*b*) Pigeon placed on altered light–dark cycle and its circadian rhythm phase-advanced by 6 hours. Bird is then returned to training cage under natural sky

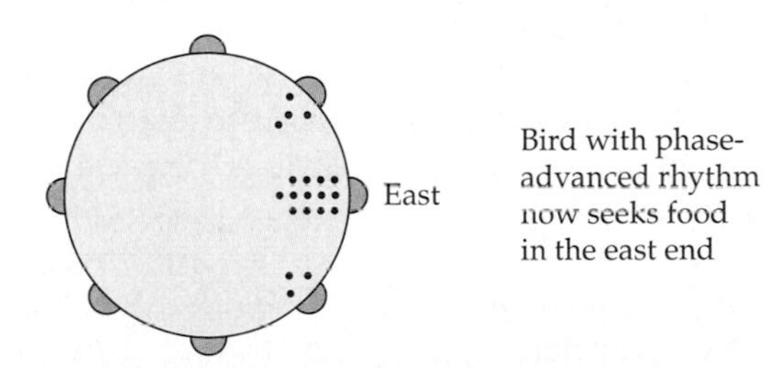

45.23 The Time-Compensated Solar Compass
(a) In the circular cage experiments, pigeons were trained to search for food in the south by filling only the southernmost food bin. Each dot represents a peck in search of food. *(b)* Birds whose circadian rhythms were phase-shifted forward by 6 hours oriented as though the dawn sun was at its noon position, searching for the south food bin in the east.

45.24 Star Patterns Can Be Altered in a Planetarium This scientist has placed birds in orientation cages (see Figure 45.22) in a planetarium. By changing the positions or movements of the stars projected on the planetarium ceiling, he can investigate what information the birds use to orient their migratory restlessness.

Much less is known about the mechanisms of bicoordinate navigation than about distance and direction navigation, but some animals have the ability both to sense their geographical positions and to know where they should go. In other words, they have a map sense. Distance and direction capabilities do little good without a map. Information about longitude and latitude is available from natural cues, but the evidence that animals can or do use those sources of information is meager. Longitude can be determined by position of the sun and time of day: If the sun comes up earlier than expected, the animal must be east of home, and if the sun comes up later than expected, then it is west of home. Time and sun position also give information about latitude. At a given time of day in the northern hemisphere, a sun position higher in the sky than expected indicates that an animal is south of home, and if it is lower in the sky than expected the animal is north of home. Other information about longitude and latitude can come from sensing Earth's magnetic lines of force and from the positions of the stars. To pinpoint home using these sources of information, an animal would have to have extremely precise and accurate sensory capabilities that have yet to be demonstrated. Perhaps, however, pinpointing home is not required of an animal's bicoordinate navigational abilities. If an animal gets anywhere near home, piloting can take over, and piloting can use a variety of long-distance cues—such as coastlines, smells, and low-frequency sounds—that have greater ranges than specific visual landmarks have.

HUMAN BEHAVIOR

The behavior of an animal is a mixture of components that are genetically programmed and components that can be molded by learning. Even some aspects of learned behavior patterns, however, may have genetic determinants in terms of what can be learned and when it can be learned. Thus natural selection shapes not only the physiology and morphology of a species, but also its behavior. In some situations natural selection favors fixed action patterns; in others learned behavior is favored. In many cases a mixture of fixed and learned behavioral components is the optimal adaptation. Given these considerations, how would we characterize human behavior?

An important characteristic of human behavior is the extent to which it can be modified by experience. Transmission of learned behavior from generation to generation is culture, the hallmark of humans. Nevertheless, the structure and many functions of our brain are coded in our genome, including drives, limits to and propensities for learning, and even some motor patterns. Biological drives such as hunger, thirst, sexual desire, and sleepiness are inherent to our nervous systems. Is it reasonable, therefore, to expect that emotions such as anger, aggression, fear, love, hate, and jealousy are solely the consequences of learning? Our sensory systems enable us to use certain subsets of information from the environment; similarly, the structure of our nervous system makes it more or less possible to process certain types of information. Consider, for example, how basic and simple it is for an infant to learn spoken language, yet how many years that same child must struggle to master reading and writing. Verbal communication is deeply rooted in our evolutionary past, whereas reading and writing are relatively recent products of human culture.

Finally, the evidence indicates that some motor patterns are programmed into our nervous systems. Studies of diverse human cultures from around the world reveal basic similarities of facial expressions and body language in human populations that have had little or no contact with one another. Infants born blind smile, frown, and show other facial expressions at appropriate times, even though they have never observed such expressions in others. Acknowledging that our behavior has been shaped through evolution in no way detracts from the value we place on the learning abilities of humans. The genetic determination of human behavior is in terms of its broad outline rather than its fine detail.

SUMMARY of Main Ideas about Animal Behavior

Numerous behavior patterns of many species are genetically determined and expressed without prior experience; they are called fixed action patterns.

Deprivation experiments demonstrate whether a given behavior is a fixed action pattern.

An animal must be in the appropriate stage of development or motivation, and required stimuli or releasers must be present for a fixed action pattern to be expressed.

A releaser is a stimulus that elicits a fixed action pattern; since the effective features of a releaser are simple, supernormal releasers are possible.
Review Figure 45.3

Fixed action patterns are adaptive in situations where there are no opportunities to learn, where it is possible to learn the wrong behavior, and where mistakes are costly and dangerous.

Learned behavior, such as bird song, may be shaped by natural selection in terms of what can be learned and when it can be learned.
Review Figure 45.5

Hormones play a part in controlling behavior.

Exposure to sex steroids during development can determine whether male or female fixed action patterns develop for sexual behavior, and sex steroid levels in the adult control the expression of those fixed action patterns.
Review Figure 45.6

Genetic experiments demonstrate the heritability of behaviors, the fact that artificial selection can change behaviors, and possible molecular mechanisms connecting genes to behavior.
Review Figures 45.9 and 45.10

Communicative behaviors convey information to other individuals, who then alter their behavior.

All sensory systems have been exploited by selection for purposes of communication.
Review Figures 45.11 and 45.13

Endogenous circadian rhythms drive daily patterns of behavior.

Although the free-running period of a circadian rhythm is not precisely 24 hours, the rhythm can be entrained to a 24-hour cycle.
Review Figure 45.16

Circadian clock mechanisms have been localized to specific brain structures.
Review Figure 45.17

Long-distance movements of animals require abilities to navigate. Mechanisms of navigation used by animals include piloting, distance and direction navigation, and bicoordinate navigation.
Review Figure 45.21

Time-compensated solar compasses and star maps are used for directional information.
Review Figures 45.23 and 45.24

Most human behavior is due to or influenced strongly by learning, but behavioral drives, emotions, some motor patterns, and propensities and abilities to learn certain types of information may be genetically determined.

SELF-QUIZ

1. The building of a web by a spider is an example of
a. a fixed action pattern.
b. a releaser.
c. displacement behavior.
d. imprinting.
e. a learned behavior.

2. If you do not see courtship behavior in a deprivation experiment, you can conclude that
a. the animal is not sexually mature.
b. the animal has low sexual drive.
c. it is the wrong time of year.
d. the appropriate releaser is not present.
e. None of the above

3. Which of the following statements about releasers is true?
a. The appropriate releaser always triggers a fixed action pattern.
b. A releaser is a simple subset of sensory cues available to the animal.
c. Releasers are learned through imprinting.
d. A releaser triggers a learned behavior pattern.
e. An animal responds to a releaser only when it is sexually mature.

4. Which of the following statements about the genetics of behavior is true?
a. About 20 genes control the courtship displays of male dabbling ducks.
b. One gene can code for several chemical signals involved in controlling a behavior.
c. Genes for retrieving, pointing, and herding have been described in dogs.
d. A single gene causes lovebirds to carry nesting material tucked in their tail feathers.
e. Hygienic behavior in bees has been shown to be due to two dominant genes.

5. A display or signal is a behavior that
   a. has evolved to influence the behavior of other individuals.
   b. stimulates one or more types of sensors.
   c. stimulates the endocrine or reproductive systems of other individuals.
   d. began as an intention movement.
   e. began as a displacement behavior.
6. If the sun were to come up earlier than expected on the basis of a circadian rhythm
   a. it could cause symptoms of jet lag.
   b. it could phase-advance the circadian rhythm.
   c. the animal could be east of home.
   d. it could entrain the circadian rhythm.
   e. All of the above
7. To have the ability to pilot, an animal must
   a. have a time-compensated solar compass.
   b. orient to a fixed point in the night sky.
   c. be able to know the distance between two points.
   d. know landmarks.
   e. know its longitude and latitude.
8. Birds that migrate at night
   a. inherit a star map.
   b. determine direction by knowing the time and the position in the sky of a star constellation.
   c. orient to the fixed point in the sky.
   d. imprint on one or more key constellations.
   e. determine distance, but not direction, from the stars.
9. The most likely explanation for the observation that humans from entirely different societies smile when they greet a friend is that
   a. they share a common culture.
   b. they have imprinted on smiling faces when they were infants.
   c. they have learned that smiling does not stimulate aggression.
   d. smiling is a fixed action pattern.
   e. smiling is a behavior that has spread around the world.
10. If (1) a bird is trained to seek food on the western side of a cage open to the sky, (2) the bird's circadian rhythm is then phase-delayed by 6 hours, and (3) after phase-shifting the bird is returned to the open cage at noon real time, it seek food in the
    a. north.
    b. south.
    c. east.
    d. west.

FOR STUDY

1. Critique this statement: Hygienic behavior of bees is controlled by two genes as demonstrated by hybridization and backcrossing experiments.
2. Photoperiod (day length) can provide information about season (time of year), so why do some birds have circannual rhythms?
3. If you raised a songbird in a deprivation experiment and it did not sing the song of its species the following fall, what possible hypotheses could you formulate about this result and how could you test them?
4. Male dogs lift a hind leg when they urinate; female dogs squat. If a male puppy receives an injection of estrogen when it is a newborn, it will never lift its leg to urinate for the rest of its life; it will squat. How might this result be explained?
5. Pick an animal (other than a human) that you think would have mostly learned behavior and another animal that you think would have mostly genetically determined behavior. What differences in their biological characteristics could account for the differences in their behavioral repertoires?

READINGS

Alcock, J. 1993. *Animal Behavior*, 5th Edition. Sinauer Associates, Sunderland, MA. A balanced textbook, recommended to readers searching for a good next step into the subject.

Emlen, S. 1975. "The Stellar-Orientation System of a Migratory Bird." *Scientific American*, August. Experiments on stellar-orientation mechanisms of birds done in a planetarium.

Gould, J. L. and P. Marler. 1987. "Learning by Instinct." *Scientific American*, January. The interactions of learning and instinct, focusing on bees and on bird song.

Gwinner, P. 1986. "Internal Rhythms in Bird Migration." *Scientific American*, April. Circannual rhythms play critical roles in long-distance migration.

Kirchner, W. H. and W. F. Towne. 1994. "The Sensory Basis of the Honeybee's Dance Language." *Scientific American*, June. New experiments test what components of the honeybee's dance communicate information.

Lorenz, K. 1958. "The Evolution of Behavior." *Scientific American*, December. An essay on the evolution of releasers.

Scheller, R. H. and R. Axel. 1984. "How Genes Control an Innate Behavior." *Scientific American*, March. One gene codes for a number of neural signals.

Tinbergen, N. 1960. *The Herring Gull's World*. Doubleday, Garden City, NJ. A delightful account of the behavior of one species from the pen of one of the founders of modern ethology.

Tinbergen, N. 1952. "The Curious Behavior of the Stickleback." *Scientific American*, December. A classic study of releasers and fixed action patterns.

GLOSSARY

Abdomen (ab´ duh mun) [L.: belly] In arthropods, the posterior portion of the body; in mammals, the part of the body containing the intestines and most other internal organs, posterior to the thorax.

Abomasum (ab´ oh may´ sum) The true stomach of ruminants (animals such as cattle, sheep, and goats).

Abscisic acid (ab sighs´ ik) [L. *abscissio*: breaking off] A plant growth substance having growth-inhibiting action. Causes stomata to close.

Abscission (ab sizh´ un) [L. *abscissio*: breaking off] The process by which leaves, petals, and fruits separate from a plant.

Absolute temperature scale A temperature scale in which the degree is the same size as in the Celsius (centigrade) scale, and zero is the state of no molecular motion. Absolute zero is –273° on the Celsius scale.

Absorption (1) Of light: complete retention, without reflection or transmission. (2) Of liquids: soaking up (taking in through pores or cracks).

Absorption spectrum A graph of light absorption versus wavelength of light; shows how much light is absorbed at each wavelength.

Abyssal zone (uh biss´ ul) [Gr. *abyssos*: bottomless] That portion of the deep ocean where no light penetrates.

Abzyme An immunoglobulin (antibody) with catalytic activity.

Accessory fruit A fruit derived from parts in addition to the ovary and seeds. (Contrast with simple fruit, aggregate fruit, multiple fruit.)

Accessory pigments Pigments that absorb light and transfer energy to chlorophylls for photosynthesis.

Acclimatization Changes in an organism that improve its ability to tolerate seasonal changes in its environment.

Acellular Not composed of cells.

Acetylcholine A neurotransmitter substance that carries information across vertebrate neuromuscular junctions and some other synapses. **Acetylcholinesterase** is an enzyme that breaks down acetylcholine.

Acetyl CoA (acetyl coenzyme A) Compound that reacts with oxaloacetate to produce citrate at the beginning of the citric acid cycle; a key metabolic intermediate in the formation of many compounds.

Acid [L. *acidus*: sharp, sour] A substance that can release a proton. (Contrast with base.)

Acid precipitation Precipitation that has a lower pH than normal as a result of acid-forming precursors introduced into the atmosphere by human activities.

Acidic Having a pH of less than 7.0 (a hydrogen ion concentration greater than 10^{-7} molar).

Acoelomate Lacking a coelom.

Acquired Immune Deficiency Syndrome See AIDS.

Acrosome (a´ krow soam) [Gr. *akros*: highest or outermost + *soma*: body] The structure at the forward tip of an animal sperm which is the first to fuse with the egg membrane and enter the egg cell.

ACTH (adrenocorticotropin) A pituitary hormone that stimulates the adrenal cortex.

Actin [Gr. *aktis*: a ray] One of the two major proteins of muscle; it makes up the thin filaments. Forms the microfilaments found in most eukaryotic cells.

Action potential An impulse in a neuron taking the form of a wave of depolarization or hyperpolarization imposed on a polarized cell surface.

Action spectrum A graph of biological activity versus wavelength of light. It compares the effectiveness of light of different wavelengths.

Activation energy The energy barrier that blocks the tendency for a set of chemical substances to react. A reaction is speeded up if this energy barrier is surmounted by adding heat energy, or if the barrier is lowered by providing a different reaction pathway with the aid of a catalyst. Designated by the symbol E_a.

Active site The region on the surface of an enzyme where the substrate binds, and where catalysis occurs.

Active transport The transport of a substance across a biological membrane against a concentration gradient—that is, from a region of low concentration (of that substance) to a region of high concentration. Active transport requires the expenditure of energy and is a saturable process. (Contrast with facilitated diffusion, free diffusion; see primary active transport, secondary active transport.)

Adaptation (a dap tay´ shun) In evolutionary biology, a particular structure, physiological process, or behavior that makes an organism better able to survive and reproduce. Also, the evolutionary process that leads to the development or persistance of such a trait.

Adenosine triphosphate See ATP.

Adenylate cyclase Enzyme catalyzing the formation of cyclic AMP from ATP.

Adhesion molecules See cell adhesion molecules.

Adrenal (a dree´ nal) [L. *ad-*: toward + *renes*: kidneys] An endocrine gland located near the kidneys of vertebrates, consisting of two glandular parts, the cortex and medulla.

Adrenaline See epinephrine.

Adrenocorticotropin See ACTH.

Adsorption Binding of a gas or a solute to the surface of a solid.

Aerenchyma (air eng´ kyma) [Gr. *aer*: air + *enchyma*: infusion] Modified parenchyma tissue, with many air spaces, found in shoots of some aquatic plants. (See parenchyma.)

Aerobic (air oh´ bic) [Gr. *aer*: air + *bios*: life] In the presence of oxygen, or requiring oxygen.

Afferent (af´ ur unt) [L. *ad*: to + *ferre*: to bear] To or toward, as in a neuron that carries impulses to the central nervous system, or a blood vessel that carries blood to a structure. (Contrast with efferents.)

Age distribution The proportion of individuals in a population belonging to each of the age categories into which the population has been divided. The number of divisions is arbitrary.

Aggregate fruit A fruit developing from several carpels of a single flower. (Contrast with simple fruit, accessory fruit, multiple fruit.)

AIDS (Acquired immune deficiency syndrome) Condition in which the body's helper T lymphocytes are destroyed, leaving the victim subject to opportunistic diseases. Caused by the HIV-I virus.

Air sacs Structures in the avian respiratory system that facilitate unidirectional flow of air through the lungs.

Alcohol An organic compound with one or more hydroxyl (–OH) groups.

Aldehyde (al´ duh hide) A compound with a –CHO functional group. Many sugars are aldehydes. (Contrast with ketone.)

Aldosterone (al dahs´ ter own) A steroid hormone produced in the adrenal cortex of mammals. Promotes secretion of potassium and reabsorption of sodium in the kidney.

Aleurone layer (al´ yur own) [Gr. *aleuron*: wheat flour] In grass seeds, a specialized cell layer just between the seed coat and the endosperm, synthesizing hydrolytic enzymes under the influence of gibberellin, and thus helping mobilize reserves for the developing embryo.

Alga (al´ gah) (plural: algae) [L.: seaweed] Any one of a wide diversity of protists belonging to the phyla Pyrrophyta, Chrysophyta, Phaeophyta, Rhodophyta, and Chlorophyta (and, formerly, Cyanophyta—"blue-green algae"). Most live in the water, where they are the dominant autotrophs; most are unicellular, but a minority are multicellular ("seaweeds" and similar protists).

Allele (a leel´) [Gr. *allos*: other] The alternate forms of a genetic character found at a given locus on a chromosome.

Allele frequency The relative proportion of a particular allele in a specific population.

Allergy [Ger. *allergie*: altered reaction] An overreaction to an antigen in amounts that do not affect most people; often involves IgE antibodies.

Allometric growth A pattern of growth in which some parts of the body of an organism grow faster than others, resulting in a change in body proportions as the organism grows.

Allopatric (al´ lo pat´ rick) [Gr. *allos*: other + *patria*: fatherland] Pertaining to populations that occur in different places.

Allopatric speciation See geographical speciation.

Allostery (al´ lo steer´ y) [Gr. *allos*: other + *stereos*: structure] Regulation of the activity of an enzyme by binding, at a site other than the catalytic active site, of an effector molecule that does not have the same structure as any of the enzyme's substrates.

Alpha helix Type of protein secondary structure; a right-handed spiral.

Alternation of generations The succession of haploid and diploid phases in a sexually reproducing organism. In most animals (male wasps and honey bees are notable exceptions), the haploid phase consists only of the gametes. In fungi, algae, and plants, however, the haploid phase may be the more prominent phase (as in fungi and mosses) or may be as prominent as the diploid phase (see the life cycle of *Ulva*, for example). In vascular plants, the diploid phase is more prominent.

Altruistic act A behavior whose performance harms the actor but benefits other individuals.

Alveolus (al ve´ o lus) (plural: alveoli) [L. *alveus*: cavity] A small, baglike cavity, especially the blind sacs of the lung.

Amensalism (a men´ sul ism) Interaction in which one animal is harmed and the other is unaffected. (Contrast with commensalism, mutualism.)

Amine An organic compound with an amino group (see Amino acid).

Amino acid An organic compound of the general formula H_2N–CH*R*–COOH, where *R* can be one of 20 or more different side groups. An amino acid is so named because it has both a basic amine group, $-NH_2$, and an acidic carboxyl group, –COOH. Proteins are polymers of amino acids.

Ammonotelic (am moan´ o teel´ ic) [Gr. *telos*: end] Describes an organism in which the final product of breakdown of nitrogen-containing compounds (primarily proteins) is ammonia. (Contrast with ureotelic, uricotelic.)

Amniocentesis A medical procedure in which cells from the fetus are obtained from the amniotic fluid. The genetic material of the cells is then examined. (Contrast with chorionic villus sampling.)

Amniotic egg The eggs of birds and reptiles, which can be incubated in air because the embryo is enclosed by a fluid-filled sac.

Amoeba (a mee´ bah) [Gr. *amoibe*: change] Any one of a large number of different kinds of unicellular protists belonging to the phylum Rhizopoda, characterized among other features by its ability to change shape frequently through the protrusion and retraction of cytoplasmic extensions called pseudopods.

Amoeboid (a mee´ boid) Like an amoeba; constantly changing shape by the protrusion and retraction of pseudopodia.

Amphi- [Gr.: both] Prefix used to denote a character or kind of organism that occupies two or more states. For example, amphibian (an animal that lives both on the land and in the water).

Amphibian (am fib´ ee an) A member of the vertebrate class Amphibia, such as a frog, toad, or salamander.

Amphipathic (am´ fi path´ ic) [Gr. *amphi*: both + *pathos*: emotion] Of a molecule, having both hydrophilic and hydrophobic regions.

amu (atomic mass unit, or **dalton)** The basic unit of mass on an atomic scale, defined as one-twelfth the mass of a carbon-12 atom. There are 6.023×10^{23} amu in one gram. This number is known as Avogadro's number.

Amylase (am´ ill ase) Any of a group of enzymes that digest starch.

Anabolism (an ab´ uh liz´ em) [Gr. *ana*: up, throughout + *ballein*: to throw] Synthetic reactions of metabolism, in which complex molecules are formed from simpler ones. (Contrast with catabolism.)

Anaerobic (an ur row´ bic) [Gr. *an*: not + *aer*: air + *bios*: life] Occurring without the use of molecular oxygen, O_2.

Anagenesis See vertical evolution.

Analogy (a nal´ o jee) [Gr. *analogia*: resembling] A resemblance in function, and often appearance as well, between two structures which is due to convergence in evolution rather than to common ancestry. (Contrast with homology.)

Anaphase (an´ a phase) [Gr. *ana*: indicating upward progress] The stage in nuclear division at which the first separation of sister chromatids (or, in the first meiotic division, of paired homologues) occurs. Anaphase lasts from the moment of first separation to the time at which the moving chromosomes converge at the poles of the spindle.

Anaphylactic shock A precipitous drop in blood pressure caused by loss of fluid from capillaries because of an increase in their permeability stimulated by an allergic reaction.

Ancestral trait Trait shared by a group of organisms as a result of descent from a common ancestor.

Androgens (an´ dro jens) The male sex steroids.

Aneuploid (an´ you ploy dee) A condition in which one or more chromosomes or pieces of chromosomes are either lacking or present in excess.

Angiosperm (an´ jee oh spurm) [Gr. *angion*: vessel + *sperma*: seed] One of the flowering plants; literally, one whose seed is carried in a "vessel," which is the fruit. (See fruit.)

Angiotensin (an´ jee oh ten´ sin) A peptide hormone that raises blood pressure by causing peripheral vessels to constrict; maintains glomerular filtration by constricting efferent glomerular vessels; stimulates thirst; and stimulates the release of aldosterone.

Animal [L. *animus*: breath, soul] A member of the kingdom Animalia. In general, a multicellular eukaryote that obtains its food by ingestion.

Animal pole In some eggs, zygotes, and embryos, the pole away from the bulk of the yolk (contrast with vegetal pole).

Anion (an´ eye one) An ion with one or more negative charges. (Contrast with cation.)

Anisogamy (an´ eye sog´ a mee) [Gr. *aniso*: unequal + *gamos*: marriage] The existence of two dissimilar gametes (egg and sperm).

Annelid (an´ el id) A member of the phylum Annelida; one of the segmented worms, such as an earthworm or leech.

Annual Referring to a plant whose life cycle is completed in one growing season. (Contrast with biennial, perennial.)

Anorexia nervosa (an or ex´ ee ah) [Gr. *an*: not + *orexis*: appetite] Severe malnutrition and body wasting brought on by a psychological aversion to food.

Anterior Toward the front.

Anterior pituitary The portion of the vertebrate pituitary gland that derives from gut epithelium and produces tropic hormones.

Anther (an´ thur) [Gr. *anthos*: flower] A pollen-bearing portion of the stamen of a flower.

Antheridium (an´ thur id´ ee um) (plural: antheridia) [Gr. *antheros*: blooming] The multicellular structure that produces the sperm in bryophytes and ferns.

Antibody One of millions of blood proteins, produced by the immune system, that specifically recognizes a foreign substance and initiates its removal from the body.

Anticodon A "triplet" of three nucleotides in transfer RNA that is able to pair with a complementary triplet (a codon) in messenger RNA, thus aligning the transfer RNA on the proper place on the messenger. The codon (and, reciprocally, the anticodon) codes for a specific amino acid.

Antidiuretic hormone A hormone that controls water reabsorption in the mammalian kidney. Also called vasopressin.

Antigen (an´ ti jun) Any substance that stimulates the production of an antibody or antibodies upon introduction into the body of a vertebrate.

Antigenic determinant A specific region of an antigen, which is recognized by and binds to a specific antibody.

Antiparallel Parallel but running in opposite directions. The two strands of DNA are antiparallel.

Antipodals (an tip´ o dulls) [Gr. *anti*: against + *podus*: foot] Cells (usually three) of the mature embryo sac of a flowering plant, located at the end opposite the egg (and micropyle).

Antiport A membrane transport protein that carries one substance in one direction and another in the opposite direction. (Contrast with symport.)

Antisense nucleic acid A single-stranded RNA or DNA complementary to and thus targeted against the mRNA transcribed from a harmful gene such as an oncogene.

Anus (a´ nus) Opening through which digestive wastes are expelled, located at the posterior end of the gut.

Aorta (a or´ tuh) [Gr. *aorte*: aorta] The main trunk of the arteries leading to the systemic (as opposed to the pulmonary) circulation.

Apex (a´ pecks) The tip or highest point of a structure, as the apex of a growing stem or root.

Apical (a´ pi kul) Pertaining to the apex, as the apical meristem, which is the actively growing tissue at the tip of a stem or root.

Apomixis (ap oh mix´ is) [Gr. *apo*: away from + *mixis*: sexual intercourse] The asexual production of seeds.

Apoplast (ap´ oh plast) in plants, the continuous meshwork of cell walls and extracellular spaces through which material can pass without crossing a plasma membrane. (Contrast with symplast.)

Appendix A vestigial portion of the human gut at the junction of the ileum with the colon.

Apterous Lacking wings. (Contrast with alate: having wings.)

Aquatic [L. *aqua*: water] Living in or on water, or taking place in or on water.

Aqueous [L. *aqua*: water] Containing water, or dissolved in water.

Archaebacteria (ark´ ee bacteria) [Gr. *archaios*: ancient] One of the two kingdoms of prokaryotes; the archaebacteria possess distinctive lipids and lack peptidoglycan. Most live in extreme environments. (Contrast with eubacteria.)

Archegonium (ar´ ke go´ nee um) [Gr. *archegonos*: first of a kind] The multicellular structure that produces eggs in bryophytes, ferns, and gymnosperms.

Archenteron (ark en´ ter on) [Gr. *archos*: beginning + *enteron*: bowel] The earliest primordial animal digestive tract.

Arteriole One of the branches of an artery.

Arteriosclerosis See atherosclerosis.

Artery A muscular blood vessel carrying oxygenated blood away from the heart to other parts of the body. (Contrast with vein.)

Artifact [L. *ars, artis*: art + *facere*: to make] Something made by human effort or intervention. In biology, something that was not present in the living cell or organism, but was unintentionally produced by an experimental procedure.

Ascospore (ass´ ko spor) A fungus spore produced within an ascus.

Ascus (ass´ cuss) [Gr. *askos*: bladder] In fungi belonging to the class Ascomycetes (sac fungi), the club-shaped sporangium within which spores are produced by meiosis.

Asexual Without sex.

Associative learning "Pavlovian" learning, in which an animal comes to associate a previously neutral stimulus (such as the ringing of a bell) with a particular reward or punishment.

Assortative mating A breeding system under which mates are selected on the basis of a particular trait or group of traits. Results in more pairs of individuals sharing traits than would be the case if mating were random.

Assortment (genetic) The random separation during meiosis of nonhomologous chromosomes and of genes carried on nonhomologous chromosomes. For example, if genes *A* and *B* are borne on nonhomologous chromosomes, meiosis of diploid cells of genotype *AaBb* will produce haploid cells of the following types in equal numbers: *AB, Ab, aB,* and *ab*.

Asymmetric The state of lacking any plane of symmetry.

Asymmetric carbon atom In a molecule, a carbon atom to which four different atoms or groups are bound.

Atherosclerosis (ath´ er oh sklair oh´ sis) A disease of the lining of the arteries characterized by fatty, cholesterol-rich deposits in the walls of the arteries. When fibroblasts infiltrate these deposits and calcium precipitates in them, the disease become arteriosclerosis, or "hardening of the arteries."

Atmosphere The gaseous mass surrounding our planet. Also: a unit of pressure, equal to the normal pressure of air at sea level.

Atom [Gr. *atomos*: indivisible] The smallest unit of a chemical element. Consists of a nucleus and one or more electrons.

Atomic mass unit See amu.

Atomic number The number of protons in the nucleus of an atom, also equal to the number of electrons around the neutral atom. Determines the chemical properties of the atom.

Atomic weight The average weight of an atom of an element on the amu scale. (The average depends upon the relative amounts of different isotopes of an element on Earth.)

ATP (adenosine triphosphate) A compound containing adenine, ribose, and three phosphate groups. When it is formed, useful energy is stored; when it is broken down (to ADP or AMP), energy is released to drive endergonic reactions. ATP is a universal energy storage compound.

Atrium (a´ tree um) A body cavity, as in the hearts of vertebrates. The thin-walled chamber(s) entered by blood on its way to the ventricle(s). Also, the outer ear.

Autocatalysis An enzymatic reaction in which the inactive form of an enzyme is converted into its active form by the enzyme itself.

Autoimmune disease A disorder in which the immune system attackes the animal's own body.

Autonomic nervous system The system (which in vertebrates comprises sympathetic and parasympathetic subsystems) that controls such involuntary functions as those of guts and glands.

Autoradiography The detection of a radioactive substance in a cell or organism by putting it in contact with a photographic emulsion and allowing the material to "take its own picture." The emulsion is developed, and the location of the radioactivity in the cell is seen by the presence of silver grains in the emulsion.

Autoregulatory mechanism A feedback mechanism that enables a structure to regulate its own function.

Autosome Any chromosome (in a eukaryote) other than a sex chromosome.

Autotroph (au´ tow trow´ fik) [Gr. *autos*: self + *trophe*: food] An organism that is capable of living exclusively on inorganic materials, water, and some energy source such as sunlight or chemically reduced matter. (Contrast with heterotroph.)

Auxin (awk´ sin) [Gr. *auxein*: increase] In plants, a substance (indoleacetic acid) that regulates growth and various aspects of development.

Auxotroph (awks´ o trofe) [Gr. *auxanein*: to grow + *trophe*: food] A mutant form of an organism that requires a nutrient or nutrients not required by the wild-type, or reference, form of the organism. (Contrast with prototroph.)

Avogadro's number The conversion factor between atomic mass units and grams. More usefully, the number of atoms in that quantity of an element which, expressed in grams, is numerically equal to the atomic weight in amu; 6.023×10^{23} atoms. (See mole.)

Axon [Gr.: axle] Fiber of a neuron which can carry action potentials. Carries impulses away from the cell body of the neuron; releases a neurotransmitter substance.

Axon hillock The junction between an axon and its cell body; where action potentials are generated.

Axon terminals The endings of an axon; they form synapses and release neurotransmitter.

Axoneme (ax´ oh neem) The complex of microtubules and their crossbridges that forms the motile apparatus of a cilium.

Bacillus (buh sil´ us) [L.: little rod] Any of various rod-shaped bacteria.

Bacteriophage (bak teer´ ee o fayj) [Gr. *bakterion*: little rod + *phagein*: to eat] One of a group of viruses that infect bacteria and ultimately cause their disintegration.

Bacterium (bak teer´ ee um) (plural: bacteria) [Gr. *bakterion*: little rod] A prokaryote. An organism with chromosomes not contained in nuclear envelopes.

Balanced polymorphism [Gr. *polymorphos*: having many forms] The maintenance of more than one form, or the maintenance at a given locus of more than one allele, at frequencies of greater than one percent in a population. Often results when heterozygotes are superior to both homozygotes.

Baroreceptor [Gr. *baros*: weight] A pressure-sensing cell or organ.

Barr body In mammals, an inactivated X chromosome.

Basal body Centriole found at the base of a eukaryotic flagellum or cilium.

Basal metabolic rate The minimum rate of energy turnover in an awake (but resting) bird or mammal that is not expending energy for thermoregulation.

Base A substance which can accept a proton (H^+). (Contrast with acid.) In nucleic acids, a nitrogen-containing base (purine or pyrimidine) is attached to each sugar in the backbone.

Base pairing See complementary base pairing.

Basic having a pH greater than 7.0 (having a hydrogen ion concentration lower than 10^{-7} molar).

Basidium (bass id´ ee yum) In fungi of the class Basidiomycetes, the characteristic sporangium in which four spores are formed by meiosis and then borne externally before being shed.

Batesian mimicry Mimicry by a relatively harmless kind of organism of a more dangerous one, by which the mimic enjoys protection from predators that mistake it for the dangerous model. (Contrast with Müllerian mimicry.)

B cell A type of lymphocyte involved in the humoral immune response of vertebrates. Upon recognizing an antigenic determinant, a B cell develops into a plasma cell, which secretes an antibody. (Contrast with a T cell.)

Benefit An improvement in survival and reproductive success resulting from a behavior. (Contrast with cost.)

Benthic zone [Gr. *benthos*: bottom of the sea] The bottom of the ocean. (Contrast with pelagic zone.)

Beta-pleated sheet Type of protein secondary structure; results from hydrogen bonding between polypeptide regions running antiparallel to each other.

Biennial Referring to a plant whose life cycle includes vegetative growth in the first year and flowering and senescence in the second year. (Contrast with annual, perennial.)

Bilateral symmetry The condition in which only the right and left sides of an organism, divided exactly down the back, are mirror images of each other. (Contrast with radial symmetry.)

Bile A secretion of the liver delivered to the small intestine via the common bile duct. In the intestine, bile emulsifies fats.

Binocular cells Neurons in the visual cortex that respond to input from both retinas; involved in depth perception.

Binomial (bye nome´ ee al) Consisting of two names; for example, the binomial nomenclature of biology which gives the name of the genus followed by the name of the species.

Biodiversity crisis The current high rate of loss of species, caused primarily by human activities.

Biogenesis [Gr. *bios*: life + *genesis*: source] The origin of living things from other living things.

Biogeochemical cycles Movement of elements through living organisms and the physical environment.

Biogeography The scientific study of the geographic distribution of organisms. Ecological biogeography is concerned with the habitats in which organisms live, historical biogeography with the complete geographic ranges of organisms and the historical circumstances that determine the ranges.

Biological species concept The view that a species is most usefully defined as a population or series of populations within which there is a significant amount of gene flow under natural conditions, but which is genetically isolated from other populations.

Biology [Gr. *bios*: life + *logos*: discourse] The scientific study of life in all its forms.

Bioluminescence The production of light by biochemical processes in an organism.

Biomass The total weight of all the living organisms, or some designated group of living organisms, in a given area.

Biome (bye´ ome) A major division of the ecological communities of Earth; characterized by distinctive vegetation.

Biota (bye oh´ tah) All of the organisms, including animals, plants, fungi, and microorganisms, found in a given area.

Biotic (bye ah´ tik) Pertaining to any aspect of life, especially to characteristics of entire populations or ecosystems.

Bipedal locomotion (by ped´ ul) [L. *bipes*: two-footed] Walking on two feet.

Biradial symmetry Radial symmetry modified so that only two planes can divide the animal into similar halves.

Blastocoel (blass´ toe seal) [Br. *blastos*: sprout + *koilos*: hollow] The central, hollow cavity of a blastula.

Blastodisc (blass´ toe disk) A disk of cells forming on the surface of a large yolk mass, comparable to a blastula, but occurring in forms in which the massive yolk restricts cleavage to one side of the egg only.

Blastomere A cell produced by the division of a fertilized egg.

Blastopore The opening from the archenteron to the exterior of a gastrula.

Blastula (blass´ chu luh) [Gr. *blastos*: sprout] An early stage in animal embryology; in many species, a hollow sphere of cells surrounding a central cavity.

Blood–brain barrier A property of the blood vessels of the brain that prevents most chemicals from diffusing from the blood into the brain.

Bloom A sudden increase in the density of phytoplankton, especially in a freshwater lake.

Body plan An entire animal, its organ systems, and the integrated functioning of its parts.

Bohr effect (boar) The reduction in affinity of hemoglobin for oxygen caused by acidic conditions, usually as a result of increased CO_2.

Bolting In rosetted angiosperms, a dramatic elongation of the stem, usually followed by flowering.

Bottleneck A combination of environmental conditions that causes a serious reduction in the size of the population.

Bowman's capsule An elaboration of kidney tubule cells that surrounds a know of capillaries (the glomerulus). Blood is filtered across the walls of these capillaries and the filtrate is collected into Bowman's capsule.

Brain A structure of nervous systems that provides the highest level of integration, control, and regulation.

Brain stem The portion of the vertebrate brain between the spinal cord and the forebrain.

Bronchus (plural: bronchi) The major airway(s) branching off the trachea into the vertebrate lung.

Browser An animal that feeds on the tissues of woody plants.

Bryophyte (bri´ uh fite´) [Gr. *bruon*: moss + *phyton*: plant] Any nonvascular plant, including mosses, liverworts, and hornworts.

Bud primordium [L. *primordium*: the beginning] In plants, a small mass of potentially meristematic tissue found in the angle between the leaf stalk and the shoot apex. Will give rise to a lateral branch under appropriate conditions.

Budding Asexual reproduction in which a more or less complete new organism simply grows from the body of the parent organism and eventually detaches itself.

Buffering A process by which a system resists change—particularly in pH, in which case added acid or base is partially converted to another form.

Bulb In plants, an underground storage organ composed principally of enlarged and fleshy leaf bases.

Bundle sheath In C_4 plants, a layer of photosynthetic cells between the mesophyll and a vascular bundle of a leaf.

C_3 photosynthesis The form of photosynthesis in which 3-phosphoglycerate is the first stable product, and ribulose bisphosphate is the CO_2 receptor.

C_4 photosynthesis The form of photosynthesis in which oxaloacetate is the first stable product, and phosphoenolpyruvate is the CO_2 acceptor. C_4 plants also perform the reactions of C_3 photosynthesis.

Caecum (see´ cum) [L. *caecus*: blind] A blind branch off the large intestine. In many nonruminant mammals, the caecum contains a colony of microorganisms that contribute to the digestion of food.

Calcitonin A hormone produced by the thyroid gland; it lowers blood calcium and promotes bone formation. (Contrast with parathormone.)

Callus [L. *calleo*: thick-skinned] In plants, wound tissue, of relatively undifferentiated proliferating cell mass, frequently maintained in cell culture.

Calmodulin (cal mod´ joo lin) A calcium-binding protein found in all animal and plant cells; mediates many calcium-regulated processes.

Calorie [L. *calor*: heat] The amount of heat required to raise the temperature of one gram of water by one degree Celsius (1°C) from 14.5°C to 15.5°C. In nutrition studies, "Calorie" (spelled with a capital C) refers to the kilocalorie (1 kcal = 1,000 cal), the amount of heat required to raise the temperature of one kilogram of water by 1°C.

Calvin–Benson cycle The stage of photosynthesis in which CO_2 reacts with RuBP to form 3PG, 3PG is reduced to a sugar, and RuBP is regenerated, while other products are released to the rest of the plant.

Calyptra (kuh lip´ tra) [Gr. *kalyptra*: covering for the head] A hood or cap found partially covering the apex of the sporophyte capsule in many moss species, formed from the expanded wall and neck of the archegonium.

Calyx (kay´ licks) [Gr. *kalyx*: cup] All of the sepals of a flower, collectively.

CAM See crassulacean acid metabolism.

Cambium (kam´ bee um) [L. *cambiare*: to exchange] A meristem that gives rise to radial rows of cells in stem and root, increasing them in girth; commonly applied to the vascular cambium which produces wood and phloem, and the cork cambium, which produces bark.

cAMP (cyclic AMP) A compound, formed from ATP, that mediates the effects of numerous animal hormones. Also needed for the transcription of catabolite-repressible operons in bacteria. Used for communication by cellular slime molds.

Canopy The leaf-bearing part of a tree. Collectively the aggregate of the leaves and branches of the larger woody plants of an ecological community.

Capacitance vessels Refers to veins because of their variable capacity to hold blood.

Capillaries [L. *capillaris*: hair] Very small tubes, especially the smallest blood-carrying vessels of animals between the termination of the arteries and the beginnings of the veins.

Capping In eukaryote RNA processing, the addition of a modified G at the 5´ end of the molecule.

Capsid The protein coat of a virus.

Capsule In bryophytes, the spore case. In some bacteria, a gelatinous layer exterior to the cell wall.

Carbohydrates Organic compounds with the general formula $C_nH_{2m}O_m$. Common examples are sugars, starch, and cellulose.

Carbon budget The amount of atmospheric carbon (from carbon dioxide) incorporated into organic molecules by a plant.

Carboxylic acid (kar box sill´ ik) An organic acid containing the carboxyl group, –COOH, which dissociates to the carboxylate ion, $-COO^-$.

Carcinogen (car sin´ oh jen) A substance that causes cancer.

Cardiac (kar´ dee ak) [Gr. *kardia*: heart] Pertaining to the heart and its functions.

Carnivore [L. *carn*: flesh + *vovare*: to devour] An organism that feeds on animal tissue. (Contrast with detritivore, herbivore, omnivore.)

Carotenoid (ka rah´ tuh noid) [L. *carota*: carrot] A yellow, orange, or red lipid pigment commonly found as an accessory pigment in photosynthesis; also found in fungi.

Carpel (kar´ pel) [Gr. *karpos*: fruit] The organ of the flower that contains one or more ovules.

Carrier In facilitated diffusion, a membrane protein that binds a specific molecule and transports it through the membrane. In genetics, a person heterozygous for a recessive trait. In respiratory and photosynthetic electron transport, a participating substance such as NAD that exists in both oxidized and reduced forms.

Carrying capacity In ecology, the largest number of organisms of a particular species that can be maintained indefinitely in a given part of the environment.

Cartilage In vertebrates, a tough connective tissue found in joints, the outer ear, and elsewhere. Forms the entire skeleton in some animal groups.

Casparian strip A band of cell wall containing suberin and lignin, found in the endodermis. Restricts the movement of water across the endodermis.

Catabolism [Ge. *kata*: down + *ballein*: to throw] Degradational reactions of metabolism, in which complex molecules are broken down. (Contrast with anabolism.)

Catabolite repression The decreased synthesis of many enzymes that tend to provide glucose for a cell; caused by the presence of excellent carbon sources, particularly glucose.

Catalyst (cat´ a list) [Gr. *kata-*, implying the breaking down of a compound] A chemical substance that accelerates a reaction without itself being consumed in the overall course of the reaction. Catalysts lower the activation energy of a reaction. Enzymes are biological catalysts.

Cation (cat´ eye on) An ion with one or more positive charges. (Contrast with anion.)

Caudal [L. *cauda*: tail] Pertaining to the tail, or to the posterior part of the body.

cDNA See complementary DNA.

Cell adhesion molecules Molecules on animal cell surfaces that affect the selective association of cells during development of the embryo.

Cell cycle The stages through which a cell passes between one division and the next. Includes all stages of interphase and mitosis.

Cell theory The theory, well established, that organisms consist of cells, and that all cells come from preexisting cells.

Cell wall A relatively rigid structure that encloses cells of plants, fungi, many protists, and most bacteria. The cell wall gives these cells their shape and limits their expansion in hypotonic media.

Cellular immune system That part of the immune system that is based on the activities of T cells. Directed against parasites, fungi, intracellular viruses, and foreign tissues (grafts). (Contrast with humoral immune system.)

Cellular respiration See respiration.

Cellulose (sell´ you lowss) A straight-chain polymer of glucose molecules, used by plants as a structural supporting material. **Cellulase** is an enzyme that hydrolyzes cellulose.

Central dogma of molecular biology The statement that information flows from DNA to RNA to polypeptide (in retroviruses, there is also information flow from RNA to cDNA).

Central nervous system That part of the nervous system which is condensed and centrally located, e.g., the brain and spinal cord of vertebrates; the chain of cerebral, thoracic and abdominal ganglia of arthropods.

Centrifuge [L. *fugere*: to flee] A device in which a sample can be spun around a central axis at high speed, creating a centrifugal force that mimics a very strong gravitational force. Used to separate mixtures of suspended materials.

Centriole (sen´ tree ole) A paired organelle that helps organize the microtubules in animal and protist cells during nuclear division.

Centromere (sen´ tro meer) [Gr. *centron*: center + *meros*: part] The region where sister chromatids join.

Cephalization (sef´ uh luh zay´ shun) [Gr. *kephale*: head] The evolutionary trend toward increasing concentration of brain and sensory organs at the anterior end of the animal.

Cephalopod (sef´ a low pod) A member of the mollusk class Cephalopoda, such as a squid or an octopus.

Cerebellum (sair´ uh bell´ um) [L.: diminutive of *cerebrum*: brain] The brain region that controls muscular coordination; located at the anterior end of the hindbrain.

Cerebral cortex The thin layer of gray matter (neuronal cell bodies) that overlays the cerebrum.

Cerebrum (su ree´ brum) [L.: brain] The dorsal anterior portion of the forebrain, making up the largest part of the brain of mammals. In mammals, the chief coordination center of the nervous system; consists of two **cerebral hemispheres**.

Cervix (sir´ vix) [L.: neck] The opening of the uterus into the vagina.

cGMP (cyclic guanosine monophosphate) An intracellular messenger that is part of signal transmission pathways involving G-proteins. (See G-protein.)

Channel A membrane protein that forms an aqueous passageway though which specific solutes may pass by simple diffusion; some channels are gated: they open and close in response to binding of specific molecules.

Character In taxonomy, any trait of an organism used in creating a classification system.

Chemical bond An attractive force stably linking two atoms.

Chemiosmotic mechanism According to this model, ATP formation in mitochondria and chloroplasts results from a pumping of protons across a membrane (against a gradient of electrical charge and of pH), followed by the return of the protons through a protein channel with ATPase activity.

Chemoautotroph An organism that uses carbon dioxide as a carbon source and obtains energy by oxidizing inorganic substances from its environment. (Contrast with chemoheterotroph, photoautotroph, photoheterotroph.)

Chemoheterotroph An organism that must obtain both carbon and energy from organic substances. (Contrast with chemoautotroph, photoautotroph, photoheterotroph.)

Chemosensor A cell or tissue that senses specific substances in its environment.

Chemosynthesis Synthesis of food substances, using the oxidation of reduced materials from the environment as a source of energy.

Chiasma (kie az´ muh) (plural: chiasmata) [Gr.: cross] An "x"-shaped connection between paired homologous chromosomes in prophase I of meiosis. A chiasma is the visible manifestation of crossing-over between homologous chromosomes.

Chitin (kye´ tin) [Gr. *chiton*: tunic] The characteristic tough but flexible organic component of the exoskeleton of arthropods, consisting of a complex, nitrogen-containing polysaccharide. Also found in cell walls of fungi.

Chlorophyll (klor´ o fill) [Gr. *chloros*: green + *phyllon*: leaf] Any of a few green pigments associated with chloroplasts or with certain bacterial membranes; responsible for trapping light energy for photosynthesis.

Chloroplast [Gr. *chloros*: green + *plast*: a particle] An organelle bounded by a double membrane containing the enzymes and pigments that perform photosynthesis. Chloroplasts occur only in eukaryotes.

Choanocyte (cho´ an oh cite) The collared, flagellated feeding cells of sponges.

Cholecystokinin (ko´ lee sis to kai nin) A hormone produced and released by the lining of the duodenum when it is stimulated by undigested fats and proteins. It stimulates the gallbladder to release bile and slows stomach activity.

Chorion (kor´ ee on) [Gr. *khorion*: afterbirth] The outermost of the membranes protecting mammal, bird, and reptile embryos; in mammals it forms part of the placenta.

Chorionic villus sampling A medical procedure that extracts a portion of the chorion from a pregnant woman to enable genetic and biochemical analysis of the embryo. (Contrast with amniocentesis.)

Chromatid (kro´ ma tid) Each of a pair of new sister chromosomes from the time at which the molecular duplication occurs until the time at which the centromeres separate at the anaphase of nuclear division.

Chromatin The nucleic acid–protein complex found in eukaryotic chromosomes.

Chromatography Any one of several techniques for the separation of chemical substances, based on differing relative tendencies of the substances to associate with a mobile phase or a stationary phase.

Chromatophore (krow mat´ o for) [Gr. *chroma*: color + *phoreus*: carrier] A pigment-bearing cell that expands or contracts to change the color of the organism.

Chromosomal aberration Any large change in the structure of a chromosome, including duplication or loss of chromosomes or parts thereof, usually gross enough to be detected with the light microscope.

Chromosome (krome´ o sowm) [Gr. *chroma*: color = *soma*: body] In bacteria and viruses, the DNA molecule that contains most or all of the genetic information of the cell or virus. In eukaryotes, a structure composed of DNA and proteins that bears part of the genetic information of the cell.

Chromosome walking A technique based on recognition of overlapping fragments; used as a step in DNA sequencing.

Chylomicron (ky low my´ cron) Particles of lipid coated with protein, produced in the gut from dietary fats and secreted into the extracellular fluids.

Chyme (kime) [Gr. *chymus*, juice] Created in the stomach; a mixture of ingested food with the digestive juices secreted by the salivary glands and the stomach lining.

Ciliate (sil´ ee ate) A member of the protist phylum Ciliophora, unicellular organisms that propel themselves by means of cilia.

Cilium (sil´ ee um) (plural: cilia) [L. *cilium*: eyelash] Hairlike organelle used for locomotion by many unicellular organisms and for moving water and mucus by many multicellular organisms. Generally shorter than a flagellum.

Circadian rhythm (sir kade´ ee an) [L. *circa*: approximately + *dies*: day] A rhythm in behavior, growth, or some other activity that recurs about every 24 hours under constant conditions.

Circannual rhythm (sir can´ you al) [L. *circa*: approximately + *annus*: year) A rhythm of behavior, growth, or some other activity that recurs on a yearly basis.

Citric acid cycle A set of chemical reactions in cellular respiration, in which acetyl CoA reacts with oxaloacetate to form citric acid, and oxaloacetate is regenerated. Acetyl CoA is oxidized to carbon dioxide, and hydrogen atoms are stored as NADH and $FADH_2$.

Clade (clayd) [Gr. *klados*: branch] All of the organisms, both living and fossil, descended from a particular common ancestor.

Cladistic classification A classification based entirely on the phylogenetic relationships among organisms.

Cladogenesis (clay doh jen´ e sis) [Gr. *klados*: branch + *genesis*: source] The formation of a new species by the splitting of an evolutionary lineage.

Cladogram Graphic representation of a cladistic relationship.

Class In taxonomy, the category below the phylum and above the order; a group of related, similar orders.

Clathrin A fibrous protein on the inner surfaces of animal cell membranes that strengthens coated vesicles and thus participates in receptor-mediated endocytosis.

Clay A soil constituent comprising particles smaller than 2 micrometers in diameter.

Cleavages First divisions of the fertilized egg of an animal.

Climax In ecology, a community that terminates a succession and which tends to replace itself unless it is further disturbed or the physical environment changes.

Climograph (clime´ o graf) Graph relating temperature and precipitation with time of year.

Cline A gradual change in the traits of a species over a geographical gradient.

Clitoris (klit´ er us, kilte´ er us) A structure in the human female reproductive system that is homologous with the male penis and is involved in sexual stimulation.

Cloaca (klo ay´ kuh) [L. *cloaca*: sewer] In some invertebrates, the posterior part of the gut; in many vertebrates, a cavity receiving material from the digestive, reproductive, and excretory systems.

Clonal deletion In immunology, the inactivation or destruction of lymphocyte clones that would produce immune reactions against the animal's own body.

Clonal selection The mechanism by which exposure to antigen results in the activation of selected T-cell or B-cell clones, resulting in an immune response.

Clone [Gr. *klon*: twig, shoot] Genetically identical cells or organisms produced from a common ancestor by asexual means.

Clutch The number of offspring produced in a given batch.

Coacervate (ko as´ er vate) [L. *coacervare*: to heap up] An aggregate of colloidal particles in suspension.

Coacervate drop Drops formed when a mixture of large proteins and polysaccharides is shaken in water. The interiors of these drops, which are often very stable, contain most of the proteins and polysaccharides.

Coated vesicle Vesicle, sometimes formed from a coated pit, with characteristic "bristly" surface; its membrane contains distinctive proteins, including clathrin.

Coccus (kock´ us) [Gr. *kokkos*: berry, pit] Any of various spherical or spheroidal bacteria.

Cochlea (kock´ lee uh) [Gr. *kokhlos*: a land snail] A spiral tube in the inner ear of vertebrates; it contains the sensory cells involved in hearing.

Codominance A condition in which two alleles at a locus produce different phenotypic effects and both effects appear in heterozygotes.

Codon A "triplet" of three nucleotides in messenger RNA that directs the placement of a particular amino acid into a polypeptide chain. (Contrast with anticodon.)

Coefficient of relatedness The probability that an allele in one individual is an identical copy, by descent, of an allele in another individual.

Coelom (see´ lum) [Gr. *koiloma*: cavity] The body cavity of certain animals, which is lined with cells of mesodermal origin.

Coelomate Having a coelom.

Coenocyte (seen´ a sight) [Gr.: common cell] A "cell" bounded by a single plasma membrane, but containing many nuclei.

Coenzyme A nonprotein molecule that plays a role in catalysis by an enzyme. The coenzyme may be part of the enzyme molecule or free in solution. Some coenzymes are oxidizing or reducing agents, others play different roles.

Coevolution Concurrent evolution of two or more species that are mutually affecting each other's evolution.

Cohort (co´ hort) [L. *cohors*: company of soldiers] A group of similar-age organisms, considered as it passes through time.

Coitus (koe´ i tus) [L. *coitus*: a coming together] The act of sexual intercourse.

Coleoptile (koe´ lee op´ til) [Gr. *koleos*: sheath + *ptilon*: feather] A pointed sheath covering the shoot of grass seedlings.

Collagen [Gr. *kolla*: glue] A fibrous protein found extensively in bone and connective tissue.

Collecting duct In vertebrates, a tubule that receives urine produced in the nephrons of the kidney and delivers that fluid to the ureter for excretion.

Collenchyma (cull eng´ kyma) [Gr. *kolla*: glue + *enchyma*: infusion] A type of plant cell, living at functional maturity, which lends flexible support by virtue of primary cell walls thickened at the corners. (Contrast with parenchyma, sclerenchyma.)

Colon [Gr. *kolon*: large intestine] The large intestine.

Colostrum (koh los´ trum) Substance secreted by the mammary glands around the time of an infant's birth. It contains protein and lactose but little fat, and its rate of production is less than the rate of milk production two or three days after birth.

Commensalism The form of symbiosis in which one species benefits from the association, while the other is neither harmed nor benefited.

Common bile duct A single duct that delivers bile from the gallbladder and secretions from the pancreas into the small intestine.

Communication Action on the part of one organism (or cell) that alters the pattern of behavior in another organism (or cell) in an adaptive fashion.

Community Any ecologically integrated group of species of microorganisms, plants, and animals inhabiting a given area.

Companion cell Specialized cell found adjacent to a sieve tube element in some flowering plants.

Comparative analysis An approach to studying evolution in which hypotheses are tested by measuring the distribution of states among a large number of species.

Compensation point The light intensity at which the rates of photosynthesis and of cellular respiration are equal.

Competitive inhibitor A substance, similar in structure to an enzyme's substrate, that binds the active site and thus inhibits a reaction.

Competition In ecology, use of the same resource by two or more species, when the resource is present in insufficient supply for the combined needs of the species.

Competitive exclusion A result of competition between species for a limiting resource in which one species completely eliminates the other.

Competitive inhibitor A substance, similar in structure to an enzyme's substrate, that binds the active site and inhibits a reaction.

Complement system A group of eleven proteins that play a role in some reactions of the immune system. The complement proteins are not immunoglobulins.

Complementary base pairing The A–T (or A–U), T–A (or U–A), C–G and G–C pairing of bases in double-stranded DNA, in transcription, and between tRNA and mRNA.

Complementary DNA (cDNA) DNA formed by reverse transcriptase acting with an RNA template; essential intermediate in the reproduction of retroviruses; used as a tool in recombinant DNA technology; lacks introns.

Complete metamorphosis A change of state during the life cycle of an organism in which the body is almost completely rebuilt to produce an individual with a completely different body form. Characteristic of insects such as butterflies, moths, beetles, ants, wasps, and flies.

Compound (1) A substance made up of atoms of more than one element. (2) Made up of many units, as the compound eyes of arthropods (as opposed to the simple eyes of the same group of organisms).

Compression wood See reaction wood.

Condensation reaction A reaction in which two molecules become connected by a covalent bond, and a molecule of water is released. ($AH + BOH \rightarrow AB + H_2O$.)

Cones (1) In the vertebrate retina: photoreceptors responsible for color vision. (2) In gymnosperms: reproductive structures consisting of many sporophylls packed relatively tightly.

Conidium (ko nid´ ee um) [Gr. *konis*: dust] An asexual fungus spore borne singly or in chains either apically or laterally on a hypha.

Conifer (kahn´ e fer) [Gr. *konos*: cone + *phero*: carry] One of the cone-bearing gymnosperms, mostly trees, such as pines and firs.

Conjugation (kahn´ jew gay´ shun) [L. *conjugare*: yoke together] The close approximation of two cells during which they exchange genetic material, as in *Paramecium* and other ciliates, or during which DNA passes from one to the other through a tube, as in bacteria.

Connective tissue An animal tissue that connects or surrounds other tissues; its cells are embedded in a collagen-containing matrix.

Connexon In a gap junction, a protein channel linking adjacent animal cells.

Consensus sequences Short stretches of DNA that appear, with little variation, in many different genes.

Constitutive enzyme An enzyme that is present in approximately constant amounts in a system, whether its substrates are present or absent. (Contrast with inducible enzyme.)

Consumer An organism that eats the tissues of some other organism.

Continental climate A pattern, typical of the interiors of large continents at high latitudes, in which bitterly cold winters alternate with hot summers. (Contrast with maritime climate.)

Continental drift The gradual drifting apart of the world's continents that has occurred over a period of billions of years.

Contractile vacuole An organelle, often found in protists, which pumps excess water out of the cell and keeps it from being "flooded" in hypotonic environments.

Cooperative act Behavior in which two or more individuals interact to their mutual benefit. No conscious awareness by the actors of the effects of their behavior is implied.

Cooption The act of capturing something for a particular use. In ecology refers to the diversion of ecological production for human use. Such production is said to be coopted.

Copulation Reproductive behavior that results in a male depositing sperm in the reproductive tract of a female.

Corepressor A low molecular weight compound that unites with a protein (the repressor) to prevent transcription in a repressible operon.

Cork A waterproofing tissue in plants, with suberin-containing cell walls. Produced by a cork cambium.

Corm A conical, underground stem that gives rise to a new plant. (Contrast with bulb.)

Corolla (ko role´ lah) [L.: diminutive of *corona*: wreath, crown] All of the petals of a flower, collectively.

Coronary (kor´ oh nair ee) Referring to the blood vessels of the heart.

Corpus luteum (kor´ pus loo´ tee um) [L. *corpus*: body + *luteum*: yellow] A structure formed from a follicle after ovulation; it produces hormones important to the maintenance of pregnancy.

Cortex [L.: bark or rind] (1) In plants: the tissue between the epidermis and the vascular tissue of a stem or root. (2) In animals: the outer tissue of certain organs, such as the adrenal cortex and cerebral cortex.

Corticosteroids Steroid hormones produced and released by the cortex of the adrenal gland.

Cost See energetic cost, opportunity cost, risk cost.

Cotyledon (kot´ ul lee´ dun) [Gr. *kotyledon*: a hollow space] A "seed leaf." An embryonic organ which stores and digests reserve materials; may expand when seed germinates.

Covalent bond A chemical bond that arises from the sharing of electrons between two atoms. Usually a strong bond.

Crassulacean acid metabolism (CAM) A metabolic pathway enabling the plants that possess it to store carbon dioxide at night and then perform photosynthesis during the day with stomata closed.

Crista (plural: cristae) A small, shelflike projection of the inner membrane of a mitochondrion; the site of oxidative phosphorylation.

Critical night length In the photoperiodic flowering response of short-day plants, the length of night above which flowering occurs and below which the plant remains vegetative. (The reverse applies in the case of long-day plants.)

Critical period The age during which some particular type of learning must take place or during which it occurs much more easily than at other times. Typical of song learning among birds.

Cross-pollination The pollination of one plant by pollen from another plant. (Contrast with self-pollination.)

Cross (transverse) section A section taken perpendicular to the longest axis of a structure.

Crossing over The mechanism by which linked markers undergo recombination. In general, the term refers to the reciprocal exchange of corresponding segments between two homologous chromatids. However, the reciprocity of crossing-over is problematical in prokaryotes and viruses; and even in eukaryotes, very closely linked markers often recombine by a nonreciprocal mechanism.

CRP The cAMP receptor protein that interacts with the promoter to enhance transcription; a lowered cAMP concentration results in catabolite repression.

Crustacean (crus tay´ see an) A member of the phylum Crustacea, such as a crab, shrimp, or sowbug.

Cryptic appearance The resemblance of an animal to some part of its environment, which helps it to escape detection by predators.

Culture A laboratory association of organisms under controlled conditions. Also the collection of knowledge, tools, values, and rules that characterize a human society.

Cuticle A waxy layer on the outer surface of a plant or an insect, tending to retard water loss.

Cutin (cue´ tin) [L. *cutis*: skin] A mixture of long, straight-chain hydrocarbons and waxes secreted by the plant epidermis, providing a water-impermeable coating on aerial plant parts.

Cyanobacteria (sigh an´ o bacteria) [Gr. *kuanos*: the color blue] A division of photosynthetic bacteria, formerly referred to as blue-green algae; they lack sexual reproduction, and they use chlorophyll *a* in their photosynthesis.

Cyclic AMP See cAMP.

Cyclins Proteins that activate maturation-promoting factor, bringing about transitions in the cell cycle.

Cyst (sist) [Gr. *kystis*: pouch] (1) A resistant, thick-walled cell formed by some protists and other organisms. (2) An abnormal sac, containing a liquid or semisolid substance, produced in response to injury or illness.

Cytochromes (sy´ toe chromes) [Gr. *kytos*: container + *chroma*: color] Iron-containing red proteins, components of the electron-transfer chains in photophosphorylation and respiration.

Cytokinesis (sy´ toe kine ee´ sis) [Gr. *kytos*: container + *kinein*: to move] The division of the cytoplasm of a dividing cell. (Contrast with mitosis.)

Cytokinin (sy´ toe kine´ in) [Gr. *kytos*: container + *kinein*: to move] A member of a class of plant growth substances playing roles in senescence, cell division, and other phenomena.

Cytoplasm The contents of the cell, excluding the nucleus.

Cytoplasmic determinants In animal development, gene products whose spatial distribution may determine such things as embryonic axes.

Cytoskeleton The network of microtubules and microfilaments that gives a eukaryotic cell its shape and its capacity to arrange its organelles and to move.

Cytosol The fluid portion of the cytoplasm, excluding organelles and other solids.

Cytotoxic T cells Cells of the cellular immune system that recognize and directly eliminate virus-infected cells. (Contrast with helper T cells, suppressor T cells.)

Dalton See amu.

Deciduous (de sid´ you us) [L. *decidere*: fall off] Referring to a plant that sheds its leaves at certain seasons. (Contrast with evergreen.)

Degeneracy The situation in which a single amino acid may be represented by any of two or more different codons in messenger RNA. Most of the amino acids can be represented by more than one codon.

Degradative succession Ecological succession occuring on the dead remains of the bodies of plants and animals, as when leaves or animal bodies rot.

Dehydration See condensation reaction.

Deletion (genetic) A mutation resulting from the loss of a continuous segment of a gene or chromosome. Such mutations never revert to wild-type. (Contrast with duplication, point mutation.)

Deme (deem) [Gr. *demos*: common people] Any local population of individuals belonging to the same species and among which mating is random.

Demographic processes The events—such as births, deaths, immigration, and emigration—that determine the number of individuals in a population.

Demographic stochasticity Random variations in the factors influencing the size, density, and distribution of a population.

Demography The study of dynamical changes in the sizes, densities, and distributions of populations.

Denaturation Loss of activity of an enzyme or nucleic acid molecule as a result of structural changes induced by heat or other means.

Dendrite [Gr. *dendron*: a tree] A fiber of a neuron which often cannot carry action potentials. Usually much branched and relatively short compared with the axon, and commonly carries information to the cell body of the neuron.

Denitrification Metabolic activity by which inorganic nitrogen-containing ions are reduced to form nitrogen gas and other products; carried on by certain soil bacteria.

Density dependence Change in the severity of action of agents affecting birth and death rates within populations. Such changes may be directly or inversely related to population density.

Density independence The state where the severity of action of agents affecting birth and death rates within a population does not change with the density of the population.

Deoxyribonucleic acid See DNA.

Depolarization A change in the electric potential across a membrane from a condition in which the inside of the cell is more negative than the outside to a condition in which the inside is less negative, or even positive, with reference to the outside of the cell. (Contrast with hyperpolarization.)

Desmosome (dez´ mo sowm) [Gr. *desmos*: bond + *soma*: body] An adhering junction between animal cells.

Derived trait A trait found among members of a lineage that was not present in the ancestors of that lineage.

Dermal tissue system The outer covering of a plant, consisting of epidermis in the young plant and periderm in a plant with extensive secondary growth. (Contrast with ground tissue system and vascular tissue system.)

Determinate cleavage A pattern of early embryological development in which the potential of cells is determined very early such that separated cells develop only into partial embryos. (Contrast with indeterminate cleavage.

Determination Process whereby an embryonic cell or group of cells becomes fixed into a predictable developmental pathway.

Detritivore (di try´ ti vore) [L. *detritus*: worn away + *vorare*: to devour] An organism that eats the dead remains of other organisms.

Deuterium An isotope of hydrogen possessing one neutron in its nucleus. Deuterium oxide is called "heavy water."

Deuterostome One of two major lines of evolution in animals, characterized by radial cleavage, enterocoelous development, and other traits.

Deuterium An isotope of hydrogen, possessing one neutron in its nucleus; deuterium oxide is called "heavy water."

Development Progressive change, as in structure or metabolism; in most kinds of organisms, development continues throughout the life of the organism.

Dialysis (dye ahl´ uh sis) [Gr. *dialyein*: separation] The removal of ions or small molecules from a solution by their diffusion across a semipermeable membrane to a solvent where their concentration is lower.

Diaphragm (dye´ uh fram) [Gr. *diaphrassein*, to barricade] (1) A sheet of muscle that separates the thoracic and abdominal cavities in mammals; responsible for the action of breathing. (2) A method of birth control in which a sheet of rubber is fitted over the woman's cervix, blocking the entry of sperm.

Diastole (dye ahs´ toll ee) [Gr.: dilation] The portion of the cardiac cycle when the heart muscle relaxes. (Contrast with systole.)

Dicot (short for dicotyledon) [Gr. *dis*: two + *kotyledon*: a cup-shaped hollow] Any member of the angiosperm class Dicotyledones, flowering plants in which the embryo produces two cotyledons prior to germination. Leaves of most dicots have major veins arranged in a branched or reticulate pattern.

Differentiation Process whereby originally similar cells follow different developmental pathways. The actual expression of determination.

Diffuse coevolution The situation in which the evolution of a lineage is influenced by its interactions with a number of species, most of which exert only a small influence on the evolution of the focal lineage.

Diffusion Random movement of molecules or other particles, resulting in even distribution of the particles when no barriers are present.

Digestion Enzyme-catalyzed process by which large, usually insoluble, molecules (foods) are hydrolyzed to form smaller molecules of soluble substances.

Dihybrid cross A mating in which the parents differ with respect to the alleles of two loci of interest.

Dikaryon (di care´ ee ahn) [Gr. *dis*: two + *karyon*: kernel] A cell or organism carrying two genetically distinguishable nuclei. Common in fungi.

Dioecious (die eesh´ us) [Gr.: two houses] Organisms in which the two sexes are "housed" in two different individuals, so that eggs and sperm are not produced in the same individuals. Examples: humans, fruit flies, oak trees, date palms. (Contrast with monoecious.)

Diploblastic Having two cell layers. (Contrast with triploblastic.)

Diploid (dip´ loid) [Gr. *diploos*: double] Having a chromosome complement consisting of two copies (homologues) of each chromosome. A diploid individual (or cell) usually arises as a result of the fusion of two gametes, each with just one copy of each chromosome. Thus, the two homologues in each chromosome pair in a diploid cell are of separate origin, one derived from the female parent and one from the male parent.

Diplontic life cycle A life cycle in which every cell except the gametes is diploid.

Directional selection Selection in which phenotypes at one extreme of the population distribution are favored. (Contrast with disruptive selection; stabilizing selection.)

Disaccharide A carbohydrate made up of two monosaccharides (simple sugars).

Dispersal stage Stage in its life history at which an organism moves from its birthplace to where it will live as an adult.

Displacement activity Apparently irrelevant behavior performed by an animal under conflict situations, especially when tendencies to attack and escape are closely balanced.

Display A behavior that has evolved to influence the actions of other individuals.

Disruptive selection Selection in which phenotypes at both extremes of the population distribution are favored. (Contrast with directional selection; stabilizing selection.)

Distal Away from the point of attachment or other reference point. (Contrast with proximal.)

Disturbance A short-term event that disrupts populations, communities, or ecosystems by changing the environment.

Diverticulum (di ver tic´ u lum) [L. *divertere*: turn away] A small cavity or tube that connects to a major cavity or tube.

Division A term used by some microbiologists and formerly by botanists, corresponding to the term phylum.

DNA (deoxyribonucleic acid) The fundamental hereditary material of all living organisms. In eukaryotes, stored primarily in the cell nucleus. A nucleic acid using deoxyribose rather than ribose.

DNA hybridization A process by which DNAs from two species are mixed and heated so that interspecific double helixes are formed.

DNA ligase Enzyme that unites Okazaki fragments of the lagging strand during DNA replication; also mends breaks in DNA strands. It connects pieces of a DNA strand and is used in recombinant DNA technology.

DNA methylation Addition of methyl groups to DNA; plays role in regulation of gene expression; protects a bacterium's DNA against its restriction endonucleases.

DNA polymerase Any of a group of enzymes that catalyze the formation of DNA strands from a DNA template.

Dominance In genetic terminology, the ability of one allelic form of a gene to determine the phenotype of a heterozygous individual, in which the homologous chromosome carries both it and a different allele. For example, if *A* and *a* are two allelic forms of a gene, *A* is said to be dominant to *a* if *AA* diploids and *Aa* diploids are phenotypically identical and are distinguishable from *aa* diploids. The *a* allele is said to be recessive.

Dominance hierarchy The set of relationships within a group of animals, usually established and maintained by aggression, in which one individual has precedence over all others in eating, mating, and other activities; a second individual has precedence over all but the highest-ranking individual, and so on down the line.

Dormancy A condition in which normal activity is suspended, as in some seeds and buds.

Dorsal [L. *dorsum*: back] Pertaining to the back or upper surface. (Contrast with ventral.)

Double fertilization Process virtually unique to angiosperms in which one sperm nucleus combines with the egg to produce a zygote, and the other sperm nucleus combines with the two polar nuclei to produce the first cell of the triploid endosperm.

Double helix Of DNA: molecular structure in which two complementary polynucleotide strands, antiparallel to each other, form a right-handed spiral.

Duodenum (doo´ uh dee´ num) The beginning portion of the vertebrate small intestine. (Contrast with ileum, jejunum.)

Duplication (genetic) A mutation resulting from the introduction into the genome of an extra copy of a segment of a gene or chromosome. (Contrast with deletion, point mutation.)

Dynein [Gr. *dunamis*: power] A protein that undergoes conformational changes and thus plays a part in the movement of eukaryotic flagella and cilia.

Ear pinnae (pin´ ee) [L. wings] External ear structures that surround the auditory canals.

Ecdysone (eck die´ sone) [Gr. *ek*: out of + *dyo*: to clothe] In insects, a hormone that induces molting.

Echinoderm (e kine´ oh durm) A member of the phylum Echinodermata, such as a seastar or sea urchin.

Ecological biogeography The study of the distributions of organisms from an ecological perspective, usually concentrating on migration, dispersal, and species interactions.

Ecological community The species living together at a particular site.

Ecological niche (nitch) [L. *nidus*: nest] The functioning of a species in relation to other species and its physical environment.

Ecology [Gr. *oikos*: house + *logos*: discourse, study] The scientific study of the interaction of organisms with their environment, including both the physical environment and the other organisms that live in it.

Ecosystem (eek´ oh sis tum) The organisms of a particular habitat, such as a pond or forest, together with the physical environment in which they live.

Ecto- (eck´ toh) [Gr.: outer, outside] A prefix used to designate a structure on the outer surface of the body. For example, ectoderm. (Contrast with endo- and meso-.)

Ectoderm [Gr. *ektos*: outside + *derma*: skin] The outermost of the three embryonic tissue layers first delineated during gastrulation. Gives rise to the skin, sense organs, nervous system, etc.

Ectotherm [Gr. *ektos*: outside + *thermos*: heat] An animal unable to control its body temperature. (Contrast with endotherm.)

Edema (i dee´ mah) [Gr. *oidema*: swelling] Tissue swelling caused by the accumulation of fluid.

Edge effect The changes in ecological processes in a community caused by physical and biological factors originating in an adjacent community.

Effector Any organ, cell, or organelle that moves the organism through the environment or else alters the environment to the organism's advantage. Examples include muscle, bone, and a wide variety of exocrine glands.

Efferent [L. *ex*: out + *ferre*: to bear] Away from, as in neurons that conduct action potentials out from the central nervous system, or arterioles that conduct blood away from a structure. (Contrast with afferent.)

Egg In all sexually reproducing organisms, the female gamete; in birds, reptiles, and some other vertebrates, a structure witin which early embryonic development occurs.

Elasticity The property of returning quickly to a former state after a disturbance.

Electrocardiogram (EKG) A graphic recording of electrical potentials from the heart.

Electroencephalogram (EEG) A graphic recording of electrical potentials from the brain.

Electromyogram (EMG) A graphic recording of electrical potentials from muscle.

Electron (e lek´ tron) [L. *electrum*: amber (associated with static electricity), from Gr. *slektor*: bright sun (color of amber)] One of the three most important fundamental particles of matter, with mass approximately 0.00055 amu and charge –1.

Electron microscope An instrument that uses an electron beam to form images of minute structures; the transmission electron microscope is useful for thinly-sliced material, and the scanning electron microscope gives surface views of cells and organisms.

Electrophoresis (e lek´ tro fo ree´ sis) [L. *electrum*: amber + Gr. *phorein*: to bear] A separation technique in which substances are separated from one another on the basis of their electric charges and molecular weights.

Electrotonic potential In neurons, a hyperpolarization or small depolarization of the membrane potential induced by the application of a small electric current. (Contrast with action potential, resting potential.)

Elemental substance A substance composed of only one type of atom.

Embolus (em´ buh lus) [Gr. *embolos*: inserted object; stopper] A circulating blood clot. Blockage of a blood vessel by an embolus or by a bubble of gas is referred to as an **embolism**. (Contrast with thrombus.)

Embryo [Gr. *en-*: in + *bryein*: to grow] A young animal, or young plant sporophyte, while it is still contained within a protective structure such as a seed, egg, or uterus.

Embryo sac In angiosperms, the female gametophyte. Found within the ovule, it consists of eight or fewer cells, membrane bounded, but without cellulose walls between them.

Emergent property A property of a complex system that is not exhibited by its individual component parts.

Emigration The deliberate and usually oriented departure of an organism from the habitat in which it has been living.

Endemic (en dem´ ik) [Gr. *endemos*: dwelling in a place] Confined to a particular region, thus often having a comparatively restricted distribution.

Endergonic reaction One for which energy must be supplied. (Contrast with exergonic reaction.)

Endo- [Gr.: within, inside] A prefix used to designate an innermost structure. For example, endoderm, endocrine. (Contrast with ecto-, meso-.)

Endocrine gland (en´ doh krin) [Gr. *endon*: inside + *krinein*: to separate] Any gland, such as the adrenal or pituitary gland of vertebrates, that secretes certain substances, especially hormones, into the body through the blood.

Endocrinology The study of hormones and their actions.

Endocytosis A process by which liquids or solid particles are taken up by a cell through invagination of the plasma membrane. (Contrast with exocytosis.)

Endoderm [Gr. *endon*: within + *derma*: skin] The innermost of the three embryonic tissue layers first delineated during gastrulation. Gives rise to the digestive and respiratory tracts and structures associated with them.

Endodermis [Gr. *endon*: within + *derma*: skin] In plants, a specialized cell layer marking the inside of the cortex in roots and some stems. Frequently a barrier to free diffusion of solutes.

Endomembrane system Endoplasmic reticulum plus Golgi apparatus plus, when present, lysosomes; thus, a system of membranes that exchange material with one another.

Endometrium (en do mee´ tree um) [Gr. *endon*: within + *metrios*: womb] The epithelial cells lining the uterus of mammals.

Endoplasmic reticulum [Gr. *endon*: within + L. *plasma*: form; L. *reticulum*: little net] A system of membrane-bounded tubes and flattened sacs, often continuous with the nuclear envelope, found in the cytoplasm of eukaryotes. Exists as rough ER, studded with ribosomes, and smooth ER, lacking ribosomes.

Endorphins Naturally occurring, opiatelike substances in the mammalian brain.

Endoskeleton A skeleton covered by other, soft body tissues. (Contrast with exoskeleton.)

Endosperm [Gr. *endon*: within + *sperma*: seed] A specialized triploid seed tissue found only in angiosperms; contains stored food for the developing embryo.

Endosymbiosis [Gr. *endon*: within + *syn*: together + *bios*: life] The living together of two species, with one living inside the body (or even the cells) of the other.

Endosymbiotic theory Theory that the eukaryotic cell evolved from a prokaryote that contained other, endosymbiotic prokaryotes.

Endotherm [Gr. *endon*: within + *thermos*: hot] An animal that can control its body temperature by the expenditure of its own metabolic energy. (Contrast with ectotherm.)

Energetic cost The difference between the energy an animal would have expended had it rested, and that expended in performing a behavior.

Energy The capacity to do work.

Enhancer In eukaryotes, a DNA sequence, lying on either side of the gene it regulates, that stimulates a specific promoter.

Enkephalins [Gr. *en-*: in + *kephale*: head] Two of the endorphins. (See endorphin.)

Enterocoelous development A pattern of development in which the coelum is formed by an outpocketing of the embryonic gut (enteron).

Enterokinase (ent uh row kine´ ase) An enzyme secreted by the mucosa of the duodenum. It activates the zymogen trypsinogen to create the active digestive enzyme trypsin.

Entrainment With respect to circadian rhythms, the process whereby the period is adjusted to match the 24-hour environmental cycle.

Entropy (en´ tro pee) [Gr. *en*: in + *tropein*: to change] A measure of the degree of disorder in any system. A perfectly ordered system has zero entropy; increasing disorder is measured by positive entropy. Spontaneous reactions in a closed system are always accompanied by an increase in disorder and entropy.

Environment An organism's surroundings, both living and nonliving; includes temperature, light intensity, and all other species that influence the focal organism.

Enzyme (en´ zime) [Gr. *en*: in + *zyme*: yeast] A protein, on the surface of which are chemical groups so arranged as to make the enzyme a catalyst for a chemical reaction.

Eon The largest division of geological time.

Epi- [Gr.: upon, over] A prefix used to designate a structure located on top of another; for example: epidermis, epiphyte.

Epicotyl (epp´ i kot´ il) [Gr. *epi*: upon + *kotyle*: something hollow] That part of a plant embryo or seedling that is above the cotyledons.

Epidermis [Gr. *epi*: upon + *derma*: skin] In plants and animals, the outermost cell layers. (Only one cell layer thick in plants.)

Epididymis (epuh did´ uh mus) [Gr. *epi*: upon + *didymos*: testicle] Coiled tubules in the testes that store sperm and conduct sperm from the seiminiferous tubules to the vas deferens.

Epinephrine (ep i nef´ rin) [Gr. *epi*: upon + *nephros*: a kidney] The "fight or flight" hormone. Produced by the medulla of the adrenal gland, it also functions as a neurotransmitter. Also known as adrenaline.

Epiphyte (ep´ e fyte) [Gr. *epi*: upon + *phyton*: plant] A specialized plant that grows on the surface of other plants but does not parasitize them.

Episome A plasmid that may exist either free or integrated into a chromosome. (See plasmid.)

Epistasis An interaction between genes, in which the presence of a particular allele of one gene determines whether another gene will be expressed.

Epithelium In animals, a layer of cells covering or lining an external surface or a cavity.

Equilibrium (1) In biochemistry, a state in which forward and reverse reactions are proceeding at counterbalancing rates, so there is no observable change in the concentrations of reactants and products. (2) In evolutionary genetics, a condition in which allele and genotype frequencies in a population are constant from generation to generation.

Era The second largest division of geological time.

Error signal In physiology, the difference between a set-point and a feedback signal that results in a corrective response.

Erythrocyte (ur rith´ row sight) [Gr. *erythros*: red + *kytos*: hollow vessel] A red blood cell.

Esophagus (i soff´ i gus) [Gr. *oisophagos*: gullet] That part of the gut between the pharynx and the stomach.

Essential amino acid An amino acid an animal cannot synthesize for itself and must obtain from its diet.

Essential element An irreplaceable mineral element without which normal growth and reproduction cannot proceed.

Estivation (ess tuh vay´ shun) [L. *aestivalis*: summer] A state of dormancy and hypometabolism that occurs during the summer; usually a means of surviving drought and/or intense heat. Contrast with hibernation.

Estrogen Any of several steroid sex hormones, produced chiefly by the ovaries in mammals.

Estrous cycle The cyclical changes in reproductive physiology and behavior in female mammals (other than some primates), culminating in estrus.

Estrus (es´ truss) [L. *oestrus*: frenzy] The period of heat, or maximum sexual receptivity, in some female mammals. Ordinarily, the estrus is also the time of release of eggs in the female.

Ethology (ee thol´ o jee) [Gr. *ethos*: habit, custom + *logos*: discourse] The study of whole patterns of animal behavior in natural environments, stressing the analysis of adaptation and evolution of the patterns.

Ethylene One of the plant hormones, the gas $H_2C{=}CH_2$.

Etiolation Plant growth in the absence of light.

Eubacteria (yew bacteria) Kingdom including the great majority of bacteria, such as the gram negative bacteria, gram positive bacteria, mycoplasmas, etc. (Contrast with Archaebacteria.)

Euchromatin Chromatin that is diffuse and non-staining during interphase; may be transcribed. (Contrast with heterochromatin.)

Eukaryotes (yew car´ ry otes) [Gr. *eu*: true + *karyon*: kernel or nucleus] Organisms whose cells contain their genetic material inside a nucleus. Includes all life other than the viruses, Archaebacteria, and Eubacteria.

Eusocial Term applied to insects, such as termites, ants, and many bees and wasps, in which individuals cooperate in the care of offspring, there are sterile castes, and generations overlap.

Eutrophication (yoo trofe´ ik ay´ shun) [Gr. *eu-*: well + *trephein*: to flourish] The addition of nutrient materials to water. Especially in lakes, the subsequent flourishing of algae and microorganisms can result in oxygen depletion and the eventual stifling of life in the water.

Evergreen A plant that retains its leaves through all seasons. (Contrast with deciduous.)

Evolution Any gradual change. Organic evolution, often referred to as evolution, is any genetic and resulting phenotypic change in organisms from generation to generation.

Evolutionary agent Any factor that influences the direction and rate of evolutionary changes.

Evolutionary biology The collective branches of biology that study evolutionary process and their products—the diversity and history of living things.

Evolutionary conservative Traits of organisms that evolve very slowly.

Evolutionary radiation The proliferation of species within a single evolutionary lineage.

Excitatory postsynaptic potential (EPSP) A change in the resting potential of a postsynaptic membrane in a positive (depolarizing) direction. (Contrast with inhibitory postsynaptic potential.)

Excretion Release of metabolic wastes by an organism.

Exergonic reaction A reaction in which free energy is released. (Contrast with endergonic reaction.)

Exo- (eks´ oh) Same as ecto-.

Exocrine gland (eks´ oh krin) [Gr. *exo*: outside + *krinein*: to separate] Any gland, such as a salivary gland, that secretes to the outside of the body or into the gut.

Exocytosis A process by which a vesicle within a cell fuses with the plasma membrane and releases its contents to the outside. (Contrast with endocytosis.)

Exon A portion of a DNA molecule, in eukaryotes, that codes for part of a polypeptide. (Contrast with intron.)

Exoskeleton (eks´ oh skel´ e ton) A hard covering on the outside of the body; the exoskeleton of insects and other arthropods has many of the same functions as the bony internal skeleton of vertebrates. (Contrast with endoskeleton.)

Experiment A scientific method in which particular factors are manipulated while other factors are held constant so that the potential influences of the manipulated factors can be determined.

Exploitation competition Competition that occurs because resources are depleted. (Contrast with interference competition.)

Exponential growth Growth, especially in the number of organisms in a population, which is a simple function of the size of the growing entity: the larger the entity, the faster it grows. (Contrast with logistic growth.)

Expressivity The degree to which a genotype is expressed in the phenotype— may be affected by the environment.

Extensor A muscle the extends an appendage.

Extinction The termination of a lineage of organisms.

Extrinsic protein A membrane protein found only on the surface of the membrane. (Contrast with intrinsic protein.)

F_1 generation The immediate progeny of a mating; the first filial generation.

F_2 generation The immediate progeny of a mating between members of the F_1 generation.

F-duction Transfer of genes from one bacterium to another, using the F-factor as a vehicle.

F-factor In some bacteria, the fertility factor; a plasmid conferring "maleness" on the cell that contains it.

Facilitated diffusion Passive movement through a membrane involving a specific carrier protein; does not proceed against a concentration gradient. (Contrast with active transport, free diffusion.)

Facultative Capable of occurring or not occurring, as in facultative aerobes. (Contrast with obligate.)

Family In taxonomy, the category below the order and above the genus; a group of related, similar genera.

Fat A triglyceride that is solid at room temperature. (Contrast with oil.)

Fatty acid A molecule with a long hydrocarbon tail and a carboxyl group at the other end. Found in many lipids.

Fauna (faw´ nah) All of the animals found in a given area. (Contrast with flora.)

Feces [L. *faeces*: dregs] Waste excreted from the digestive system.

Feedback control Control of a particular step of a multistep process, induced by the presence or absence of a product of one of the later steps. A thermostat regulating the flow of heating oil to a furnace in a home is a negative feedback control device.

Fermentation (fur men tay´ shun) [L. *fermentum*: yeast] The degradation of a substance such as glucose to smaller molecules with the extraction of energy, without the use of oxygen (i.e., anaerobically). Involves the glycolytic pathway.

Fertilization Union of gametes. Also known as syngamy.

Fertilization membrane A membrane surrounding an animal egg which becomes rapidly raised above the egg surface within seconds after fertilization, serving to prevent entry of a second sperm.

Fetus The latter stages of an embryo that is still contained in an egg or uterus; in humans, the unborn young from the eighth week of pregnancy to the moment of birth.

Fiber An elongated and tapering cell of vascular plants, usually with a thick cell wall. Serves a support function.

Fibrin A protein that polymerizes to form long threads that provide structure to a blood clot.

Filter feeder An organism that feeds upon much smaller organisms, that are suspended in water or air, by means of a straining device.

Filtration In the excretory physiology of some animals, the process by which the initial urine is formed; water and most solutes are transferred into the excretory tract, while proteins are retained in the blood or hemolymph.

First law of thermodynamics Energy can be neither created nor destroyed.

Fission Reproduction of a prokaryote by division of a cell into two comparable progeny cells.

Fitness The contribution of a genotype or phenotype to the composition of subsequent generations, relative to the contribution of other genotypes or phenotypes. (See inclusive fitness.)

Fixed action pattern A behavior that is genetically programmed.

Flagellate (flaj´ el late) A member of the phylum Mastigophora, unicellular eukaryotes that propel themselves by flagella.

Flagellin (fla jell´ in) The protein from which prokaryotic (but not eukaryotic) flagella are constructed.

Flagellum (fla jell´ um) (plural: flagella) [L. *flagellum*: whip] Long, whiplike appendage that propels cells. Prokaryotic flagella differ sharply from those found in eukaryotes.

Flexor A muscle that flexes an appendage.

Flora (flore´ ah) All of the plants found in a given area. (Contrast with fauna.)

Florigen A plant hormone (not yet isolated) involved in the conversion of a vegetative shoot apex to a flower.

Flower The total reproductive structure of an angiosperm; its basic parts include the calyx, corolla, stamens, and carpels.

Fluorescence The emission of a photon of visible light by an excited atom or molecule.

Follicle [L. *folliculus*: little bag] In female mammals, an immature egg surrounded by nutritive cells.

Follicle-stimulating hormone A gonadotropic hormone produced by the anterior pituitary.

Food chain A portion of a food web, most commonly a simple sequence of prey species and the predators that consume them.

Food web The complete set of food links between species in a community; a diagram indicating which ones are the eaters and which are consumed.

Forb Any broad-leaved (dicotyledonous), herbaceous plant. Especially applied to such plants growing in grasslands.

Fossil Any recognizable structure originating from an organism, or any impression from such a structure, that has been preserved over geological time.

Founder effect Random changes in allele frequencies resulting from establishment of a population by a very small number of individuals.

Fovea [L. *fovea*; a small pit] The area, in the vertebrate retina, of most distinct vision.

Frame-shift mutation A mutation resulting from the addition or deletion of a single base pair in the DNA sequence of a gene. As a result of this, mRNA transcribed from such a gene is translated normally until the ribosome reaches the point at which the mutation has occurred. From that point on, codons are read out of proper register and the amino acid sequence bears no resemblance to the normal sequence. (Contrast with missense mutation, nonsense mutation.)

Free diffusion Diffusion directly across a membrane without the involvement of carrier molecules. Free diffusion is not saturable and cannot cause the net transport from a region of low concentration to a region of higher concentration. (Contrast with facilitated diffusion and active transport.)

Free energy That energy which is available for doing useful work, after allowance has been made for the increase or decrease of disorder. Designated by the symbol G (for Gibbs free energy), and defined by: $G = H - TS$, where H = heat, S = entropy, and T = absolute (Kelvin) temperature.

Frequency-dependent selection Selection that changes in intensity when the proportion of individuals under selection increases or decreases.

Fruit In angiosperms, a ripened and mature ovary (or group of ovaries) containing the seeds. Sometimes applied to reproductive structures of other groups of plants, and includes any adjacent parts which may be fused with the reproductive structures.

Fruiting body A structure that bears spores.

Fundamental niche The range of condition under which an organism could survive if it were the only one in the environment. (Contrast with realized niche.)

Fungus (fung´ gus) A member of the kingdom Fungi, a (usually) multicellular eukaryote with absorptive nutrition.

G_1 phase In the cell cycle, the gap between the end of mitosis and the onset of the S phase.

G_2 phase In the cell cycle, the gap between the S (synthesis) phase and the onset of mitosis.

G-protein A membrane protein involved in signal transduction; characterized by binding guanyl nucleotides. The activation of certain receptors activates the G-protein, which in turn activates adenylate cyclase. G-protein activation involves binding a GTP molecule in place of a GDP molecule.

Gametangium (gam i tan´ gee um) [Gr. *gamos*: marriage + *angeion*: vessel or reservoir] Any plant or fungal structure within which a gamete is formed.

Gamete (gam´ eet) [Gr. *gamete*: wife, *gametes*: husband] The mature sexual reproductive cell: the egg or the sperm.

Gametocyte (ga meet´ oh site) [Gr. *gamete*: wife, *gametes*: husband + *kytos*: cell] The cell that gives rise to sex cells, either the eggs or the sperm. (See oocyte and spermatocyte.)

Gametogenesis (ga meet´ oh jen´ e sis) [Gr. *gamete*: wife, *gametes*: husband + *genesis*: source] The specialized series of cellular divisions that leads to the production of sex cells (gametes). (Contrast with oogenesis and spermatogenesis.)

Gametophyte (ga meet´ oh fyte) In plants with alternation of generations, the haploid phase that produces the gametes. (Contrast with sporophyte.)

Ganglion (gang´ glee un) [Gr.: tumor] A group or concentration of neuron cell bodies.

Gap junction A 2.7-nanometer gap between plasma membranes of two animal cells, spanned by protein channels. Gap junctions allow chemical substances or electrical signals to pass from cell to cell.

Gas exchange In animals, the process of taking up oxygen from the environment and releasing carbon dioxide to the environment.

Gastrovascular cavity Serving for both digestion (gastro) and circulation (vascular); in particular, the central cavity of the body of jellyfish and other cnidarians.

Gastrula (gas´ true luh) [Gr. *gaster*: stomach] An embryo forming the characteristic three cell layers (ectoderm, endoderm, and mesoderm) which will give rise to all of the major tissue systems of the adult animal.

Gastrulation Development of a blastula into a gastrula.

Gated channel A channel (membrane protein) that opens and closes in response to binding of specific molecules or to changes in membrane potential.

Gene [Gr. *gen*: to produce] A unit of heredity. Used here as the unit of genetic function which carries the information for a single polypeptide.

Gene amplification Creation of multiple copies of a particular gene, allowing the production of large amounts of the RNA transcript (as in rRNA synthesis in oocytes).

Gene cloning Formation of a clone of bacteria or yeast cells containing a particular foreign gene.

Gene family A set of identical, or once-identical, genes, derived from a single parent gene; need not be on the same chromosomes; classic example is the globin family in vertebrates.

Gene flow The exchange of genes between different species (an extreme case referred to as hybridization) or between different populations of the same species caused by migration following breeding.

Gene pool All of the genes in a population.

Gene therapy Treatment of a genetic disease by providing patients with cells containing wild-type alleles for the genes that are nonfunctional in their bodies.

Generative nucleus In a pollen tube, a haploid nucleus that undergoes mitosis to produce the two sperm nuclei that participate in double fertilization. (Contrast with tube nucleus.)

Generator potential A stimulus-induced change in membrane resting potential in the direction of threshold for generating action potentials.

Genet The genetic individual of a plant that is composed of a number of nearly identical but repeated units.

Genetic drift Changes in gene frequencies from generation to generation in a small population as a result of random processes.

Genetic stochasticity Variation in the frequencies of alleles and genotypes in a population over time.

Genetics The study of heredity.

Genetic structure The frequencies of alleles and genotypes in a population.

Genome (jee´ nome) The genes in a complete haploid set of chromosomes.

Genome project An effort to map and sequence the entire genome of a species.

Genotype (jean´ oh type) [Gr. *gen*: to produce + *typos*: impression] An exact description of the genetic constitution of an individual, either with respect to a single trait or with respect to a larger set of traits. (Contrast with phenotype.)

Genus (jean´ us) (plural: genera) [Gr. *genos*: stock, kind] A group of related, similar species.

Geographical (allopatric) speciation Formation of two species from one by the interposition of (or crossing of) a physical barrier. (Contrast with parapatric, sympatric speciation.)

Geotropism See gravitropism.

Germ cell A reproductive cell or gamete of a multicellular organism.

Germination The sprouting of a seed or spore.

Gestation (jes tay´ shun) [L. *gestare*: to bear] The period during which the embryo of a mammal develops within the uterus. Also known as **pregnancy**.

Gibberellin (jib er el´ lin) [L. *gibberella*: hunchback (refers to shape of a reproductive structure of a fungus that produces gibberellins)] One of a class of plant growth substances playing roles in stem elongation, seed germination, flowering of certain plants, etc. Named for the fungus *Gibberella*.

Gill An organ for gas exchange in aquatic organisms.

Gill arch A skeletal structure that supports gill filaments and the blood vessels that supply them.

Gizzard (giz´ erd) [L. *gigeria*: cooked chicken parts] A very muscular port of the stomach of birds that grinds up food, sometimes with the aid of fragments of stone.

Gland An organ or group of cells that produces and secretes one or more substances.

Glans penis Sexually sensitive tissue at the tip of the penis.

Glia (glee´ uh) [Gr.: glue] Cells, found only in the nervous system, which do not conduct action potentials.

Glomerulus (glo mare´ yew lus) [L. *glomus*: ball] Sites in the kidney where blood filtration takes place. Each glomerulus consists of a knot of capillaries served by afferent and efferent arterioles.

Glucocorticoids Steroid hormones produced by the adrenal medulla. Secreted in response to ACTH, they inhibit glucose uptake by many tissues in addition to mediating other stress responses.

Glucagon A hormone produced and released by cells in the islets of Langerhans of the pancreas. It stimulates the breakdown of glycogen in liver cells.

Gluconeogenesis The biochemical synthesis of glucose from other substances, such as amino acids, lactate, and glycerol.

Glucose (glue´ kose) [Gr. *gleukos*: sweet wine mash for fermentation] The most common sugar, one of several monosaccharides with the formula $C_6H_{12}O_6$.

Glycerol (gliss' er ole) A three-carbon alcohol with three hydroxyl groups, the linking component of phospholipids and triglycerides.

Glycogen (gly´ ko jen) A branched-chain polymer of glucose, similar to starch (which is less branched and may be of lower molecular weight). Exists mostly in liver and muscle; the principal storage carbohydrate of most animals and fungi.

Glycolysis (gly kol´ li sis) [from glucose + Gr. *lysis*: loosening] The enzymatic breakdown of glucose to pyruvic acid. One of the oldest energy-yielding machanisms in living organisms.

Glycosidic linkage The connection in an oligosaccharide or polysaccharide chain, formed by removal of water during the linking of monosaccharides.by root pressure.

Glyoxysome (gly ox´ ee soam) A type of microbody, found in plants, in which stored lipids are converted to carbohydrates.

Golgi apparatus (goal´ jee) A system of concentrically folded membranes found in the cytoplasm of eukaryotic cells. Plays a role in the production and release of secretory materials such as the digestive enzymes manufactured in the pancreas. First described by Camillo Golgi (1844–1926).

Gonad (go´ nad) [Gr. *gone*: seed, that which produces seed] An organ that produces sex cells in animals: either an ovary (female gonad) or testis (male gonad).

Gonadotropin A hormone that stimulates the gonads.

Grade The level of complexity found in an animal's body plan.

Gram stain A differential stain useful in characterizing bacteria.

Granum Within a chloroplast, a stack of thylakoids.

Gravitropism A directed plant growth response to gravity.

Grazer An animal that eats the vegetative tissues of herbaceous plants.

Green gland An excretory organ of crustaceans.

Gross morphology The sizes and shapes of the major body parts of a plant or animal.

Gross primary production The total energy captured by plants growing in a particular area.

Ground meristem That part of an apical meristem that gives rise to the ground tissue system of the primary plant body.

Ground tissue system Those parts of the plant body not included in the dermal or vascular tissue systems. Ground tissues function in storage, photosynthesis, and support.

Groundwater Water present deep in soils and rocks; may be stationary or flow slowly eventually to discharge into lakes, rivers, or oceans.

Group transfer The exchange of atoms between molecules.

Growth Irreversible increase in volume (probably the most accurate definition, but at best a dangerous oversimplification).

Growth factors A group of proteins that circulate in the blood and trigger the normal growth of cells. Each growth factor acts only on certain target cells.

Growth stage That stage in the life history of an organism in which it grows to its adult size.

Guard cells In plants, paired epidermal cells which surround and control the opening of a stoma (pore).

Gut An animal's digestive tract.

Guttation The extrusion of liquid water through openings in leaves, caused by root pressure.

Gymnosperm (jim´ no sperm) [Gr. *gymnos*: naked + *sperma*: seed] A plant, such as a pine or other conifer, whose seeds do not develop within an ovary (hence, the seeds are "naked").

Habit The form or pattern of growth characteristic of an organism.

Habitat The environment in which an organism lives.

Habituation (ha bich´ oo ay shun) The simplest form of learning, in which an animal presented with a stimulus without reward or punishment eventually ceases to respond.

Hair cell A type of mechanosensor in animals.

Half-life The time required for half of a sample of a radioactive isotope to decay to its stable, nonradioactive form.

Halophyte (hal´ oh fyte) [Gr. *halos*: salt + *phyton*: plant] A plant that grows in a saline (salty) environment.

Haploid (hap´ loid) [Gr. *haploeides*: single] Having a chromosome complement consisting of just one copy of each chromosome. This is the normal "ploidy" of gametes or of asexual spores produced by meiosis or of organisms (such as the gametophyte generation of plants) that grow from such spores without fertilization.

Haplontic life cycle A life cycle in which the zygote is the only diploid cell.

Hardy–Weinberg rule The rule that the basic processes of Mendelian heredity (meiosis and recombination) do not alter either the frequencies of genes or their diploid combinations. The Law also states how the percentages of diploid combinations can be predicted from a knowledge of the proportions of alleles in the population.

Haustorium (haw stor´ ee um) [L. *haustus*: draw up] A specialized hypha or other structure by which fungi and some parasitic plants draw food from a host plant.

Haversian systems Units of organization in compact bone that reflect the action of intercommunicating osteoblasts.

Helper T cells T cells that participate in the activation of B cells and of other T cells; targets of the HIV-I virus, the agent of AIDS. (Contrast with cytotoxic T cells, suppressor T cells.)

Hematocrit (heme at o krit) [Gr. *haima*: blood + *krites*: judge] The proportion of 100 cc of blood that consists of red blood cells.

Hemizygous(hem´ ee zie´ gus) [Gr. *hemi*: half + *zygotos*: joined] In a diploid organism, having only one allele for a given trait, typically the case for X-linked genes in male mammals and Z-linked genes in female birds. (Contrast with homozygous, heterozygous.)

Hemoglobin (hee´ mo glow´ bin) [Gr. *haima*: blood + L. *globus*: globe] The colored protein of vertebrate blood (and blood of some invertebrates) which transports oxygen.

Hepatic (heh pat´ ik) [Gr. *hepar*: liver] Pertaining to the liver.

Hepatic duct The duct that conveys bile from the liver to the gallbladder.

Herbicide (ur´ bis ide) A chemical substance that kills plants.

Herbivore [L. *herba*: plant + *vorare*: to devour] An animal which eats the tissues of plants. (Contrast with carnivore, detritivore, omnivore.)

Heritable Able to be inherited; in biology usually refers to genetically determined traits.

Hermaphroditism (her maf´ row dite´ ism) [Gr. *hermaphroditos*: a person with both male and female traits] The coexistence of both female and male sex organs in the same organism.

Hertz (abbreviated as Hz) Cycles per second.

Hetero- [Gr.: other, different] A prefix used in biology to mean that two or more different conditions are involved; for example, heterotroph, heterozygous.

Heterochromatin Chromatin that retains its coiling during interphase; generally not transcribed. (Contrast with euchromatin.)

Heterocyst A large, thick-walled cell in the filaments of certain cyanobacteria; performs nitrogen fixation.

Heterogeneous nuclear RNA (hnRNA) The product of transcription of a eukaryotic gene, including transcripts of introns.

Heterokaryon (het´ er oh care´ ee ahn) [Gr. *heteros*: different + *karyon*: kernel] A cell or organism carrying a mixture of genetically distinguishable nuclei. A heterokaryon is usually the result of the fusion of two cells without fusion of their nuclei.

Heteromorphic (het´ er oh more´ fik) [Gr. *heteros*: different + *morphe*: form] having a different form or appearance, as two heteromorphic life stages of a plant. (Contrast with isomorphic.)

Heterosporous (het´ er os´ por us) Producing two types of spores, one of which gives rise to a female megaspore and the other to a male microspore. Heterosporous plants produce distinct female and male gametophytes. (Contrast with homosporous.)

Heterotherm An animal that regulates its body temperature at a constant level at some times but not others, such as a hibernator.

Heterotroph (het´ er oh trof) [Gr. *heteros*: different + *trophe*: food] An organism that requires preformed organic molecules as food. (Contrast with autotroph.)

Heterozygous (het´ er oh zie´ gus) [Gr. *heteros*: different + *zygotos*: joined] Of a diploid organism having different alleles of a given gene on the pair of homologues carrying that gene. (Contrast with homozygous.)

Hexose A six-carbon sugar, such as glucose or fructose.

Hfr (for "high frequency of recombination") Donor bacterium in which the F-factor has been integrated into the chromosome. This produces a bacterium that transfers its chromosomal markers at a very high frequency to recipient (F^-) cells.

Hibernation [L. *hibernus*: winter] The state of inactivity of some animals during winter; marked by a drop in body temperature and metabolic rate.

Hippocampus A part of the forebrain that takes part in long-term memory formation.

Histamine (hiss; tah meen) A substance released within a damaged tissue by a type of white blood cell. Histamines are responsible for aspects of allergice reactions, including the increased vascular permeability that leads to edema (swelling).

Histology The study of tissues.

Histone Any one of a group of basic proteins forming the core of a nucleosome, the structural unit of a eukaryotic chromosome. (See nucleosome.)

Historical biogeography The study of the distributions of organisms from a long-term, historical perspective.

hnRNA See heterogeneous nuclear RNA.

Holdfast In many large attached algae, specialized tissue attaching the plant to its substratum.

Homeobox A segment of DNA, found in a few genes, perhaps regulating the expression of other genes and thus controlling large-scale developmental processes.

Homeostasis (home´ ee o sta´ sis) [Gr. *homos*: same + *stasis*: position] The maintenance of a steady state, such as a constant temperature or a stable social structure, by means of physiological or behavioral feedback responses.

Homeotherm (home´ ee o therm) [Gr. *homos*: same + *therme*: heat] An animal that maintains a constant body temperature by virtue of its own heating and cooling mechanisms. (Contrast with heterotherm, poikilotherm.)

Homeotic genes (home´ ee ott´ ic) Genes that determine what entire segments of an animal become.

Homeotic mutation A drastic mutation causing the transformation of body parts in *Drosophila* metamorphosis. Examples include the *Antennapedia* and *ophthalmoptera* mutants.

Homolog (home´ o log´) [Gr. *homos*: same + *logos*: word] One of a pair, or larger set, of chromosomes having the same overall genetic composition and sequence. In diploid organisms, each chromosome inherited from one parent is matched by an identical (except for mutational changes) chromosome—its homolog—from the other parent.

Homology (ho mol´ o jee) [Gr. *homologi(a)*: agreement] A similarity between two structures that is due to inheritance from a common ancestor. The structures are said to be homologous. (Contrast with analogy.)

Homoplasy (home´ uh play zee) [Gr. *homos*: same + *plastikos*: to mold] The presence in several species of a trait not present in their most common ancestor. Can result from convergent evolution, reverse evolution, or parallel evolution.

Homosporous Producing a single type of spore that gives rise to a single type of gametophyte, bearing both female and male reproductive organs. (Contrast with heterosporous.)

Homozygous (home´ o zie´ gus) [Gr. *homos*: same + *zygotos*: joined] Of a diploid organism having identical alleles of a given gene on both homologous chromosomes. An organism may be a "homozygote" with respect to one gene and, at the same time, a "heterozygote" with respect to another. (Contrast with heterozygous.)

Hormone (hore´ mone) [Gr. *hormon*: excite, stimulate] A substance produced in one part of a multicellular organism and transported to another part where it exerts its specific effect on the physiology or biochemistry of the target cells.

Host An organism that harbors a parasite and provides it with nourishment.

Host–parasite interaction The dynamic interaction between populations of a host and the parasites that attack it.

Humoral immune system The part of the immune system mediated by B cells; it is mediated by circulating antibodies and is active against extracellular bacterial and viral infections.

Humus (hew´ muss) The partly decomposed remains of plants and animals on the surface of a soil. Its characteristics depend primarily upon climate and the species of plants growing on the site.

Hyaluronidase (hill yew ron´ uh dase) An enzyme that digests proteoglycans. Found in sperm cells, it helps digest the coatings surrounding an egg so the sperm can penetrate the egg cell membrane.

Hybrid (high´ brid) [L. *hybrida*: mongrel] The offspring of genetically dissimilar parents.

Hybridoma A cell produced by the fusion of an antibody-producing cell with a myeloma cell; it produces monoclonal antibodies.

Hydrocarbon A compound containing only carbon and hydrogen atoms.

Hydrogen bond A chemical bond which arises from the attraction between the slight positive charge on a hydrogen atom and a slight negative charge on a nearby fluorine, oxygen, or nitrogen atom. Weak bonds, but found in great quantities in proteins, nucleic acids, and other biological macromolecules.

Hydrological cycle The sum total of movement of water from the oceans to the atmosphere, to the soil, and back to the oceans. Some water is cycled many times within compartments of the system before completing one full circuit.

Hydrolyze (hi´ dro lize) [Gr. *hydro*: water + *lysis*: cleavage] To break a chemical bond, as in a peptide linkage, with the insertion of the components of water, –H and –OH, at the cleaved ends of a chain. The digestion of proteins is a hydrolysis.

Hydrophilic [Gr. *hydro*: water + *philia*: love] Having an affinity for water. (Contrast with hydrophobic.)

Hydrophobic [Gr. *hydro*: water + *phobia*: fear] Molecules and amino acid side chains, which are mainly hydrocarbons (compounds of C and H with no charged groups or polar groups), have a lower energy when they are clustered together than when they are distributed through an aqueous solution. Because of their attraction for one another and their reluctance to mix with water they are called "hydrophobic." Oil is a hydrophobic substance; phenylalanine is a hydrophobic animo acid in a protein. (Contrast with hydrophilic.)

Hydrophobic interaction A weak attraction between highly nonpolar molecules or parts of molecules suspended in water.

Hydrostatic skeleton The incompressible internal liquids of some animals that transfer forces from one part of the body to another when acted upon by the surrounding muscles.

Hydroxyl group The –OH group, characteristic of alcohols.

Hymenopteran (high´ man op´ ter an) A member of the insect order Hymenoptera, such as a wasp, bee, or ant.

Hyperpolarization A change in the resting potential of a membrane so the inside of a cell becomes more electronegative. (Contrast with depolarization.)

Hypertension High blood pressure.

Hypertonic [Gr.: higher tension] Having a more negative osmotic potential, as a result of having a higher concentration of osmotically active particles. Said of one solution as compared with another. (Contrast with hypotonic, isotonic.)

Hypha (high´ fuh) (plural: hyphae) [Gr. *hyphe*: web] In the fungi, any single filament. May be multinucleate (zygomycetes, ascomycetes) or multicellular (basidiomycetes).

Hypocotyl That part of the embryonic or seedling plant shoot that is below the cotyledons.

Hypothalamus The part of the brain lying below the thalamus; it coordinates water balance, reproduction, temperature regulation, and metabolism.

Hypothetico-deductive method A method of science in which hypotheses are erected, predictions are made from them, and experiments and observations are performed to test the predictions. The process may be repeated many times in the course of answering a question.

Hypotonic [Gr.: lower tension] Having a less negative osmotic potential, as a result of having a lower concentration of osmotically active particles. Said of one solution as compared with another. (Contrast with hypertonic, isotonic.)

Imaginal disc In insect larvae, groups of cells that develop into specific adult organs.

Imbibition [L. *imbibo*: to drink] The binding of a solvent to another molecule. Dry starch and protein will imbibe water.

Immunoglobulins A class of proteins, with a characteristic structure, active as receptors and effectors in the immune system.

Immunological tolerance A mechanism by which an animal does not mount an immune response to the antigenic determinants of its own macromolecules.

Imprinting A rapid form of learning, in which an animal comes to make a particular response, which is maintained for life, to some object or other organism.

Inclusive fitness The sum of an individual's own fitness (the effect of producing its own offspring: the individual selection component) plus its influence on fitness in relatives other than direct descendants (the kin selection component).

Incomplete dominance Condition in which the heterozygous phenotype is intermediate between the two homozygous phenotypes.

Incomplete metamorphosis Insect development in which changes between instars are gradual.

Incus (in´ kus) [L. *incus*: anvil] The middle of the three bones that conduct movements of the eardrum to the oval window of the inner ear. (See malleus, stapes.)

Indeterminate cleavage A pattern of development in which individual cells retain the potential to develop into complete organisms if separated from one another well into development.

Individual fitness That component of inclusive fitness that results from an organism producing its own offspring. (Contrast with kin selection component.)

Indoleacetic acid See auxin.

Induced fit A change in the tertiary structures of some enzymes, caused by binding of substrate to the active site.

Inducer In enzyme systems, a small molecule which, when added to a growth medium, causes a large increase in the level of some enzyme. Generally it acts by binding to repressor and changing its conformation so that the repressor does not bind to the operator. In embryology, a substance that causes a group of target cells to differentiate in a particular way.

Inducible enzyme An enzyme that is present in much larger amounts when a particular compound (the inducer) has been added to the system. (Contrast with constitutive enzyme.)

Inflammation A nonspecific defense against pathogens; characterized by redness, swelling, pain, and increased temperature.

Inflorescence A structure composed of several flowers.

Ingestion Taking in of food by swallowing.

Inhibitor A substance which binds to the surface of an enzyme and interferes with its action on its substrates.

Inhibitory postsynaptic potential A change in the resting potential of a postsynaptic membrane in the hyperpolarizing (negative) direction.

Initiation complex Combination of a ribosomal light subunit, an mRNA molecule, and the tRNA charged with the first amino acid coded for by the mRNA; formed at the onset of translation.

Inositol triphosphate (IP3) An intracellular second messenger derived from membrane phospholipids.

Insertion sequence A large piece of DNA that can give rise to copies at other loci; a type of transposable genetic element.

Instar (in´ star) [L.: image, form] An immature stage of an insect between molts.

Instinct Behavior that is relatively highly sterotyped and self-differentiating, that develops in individuals unable to observe other individuals performing the behavior or to practice the behavior in the presence of the objects toward which it is usually directed.

Insulin (in´ su lin) [L. *insula*: island] A hormone, synthesized in islet cells of the pancreas, that promotes the conversion of glucose to the storage material, glycogen.

Integrase An enzyme that integrates retroviral cDNA into the genome of the host cell.

Integrated pest management A method of control of pests in which natural predators and parasites are used in conjunction with sparing use of chemical methods to achieve control of a pest without causing serious adverse environmental side effects.

Integument [L. *integumentum*: covering] A protective surface structure. In gymnosperms and angiosperms, a layer of tissue around the ovule which will become the seed coat. Gymnosperm ovules have one integument, angiosperm ovules two.

Intention movement The preparatory motions that animals go through prior to a complete behavior response; for example, the crouch before flying, the snarl before biting, etc.

Intercalary meristem A meristematic region in plants which occurs not apically, but between two regions of mature tissue. Intercalary meristems occur in the nodes of grass stems, for example.

Intercostal muscles Muscles between the ribs that can augment breathing movements by elevating and suppressing the rib cage.

Interference competition Competition resulting from direct behavioral interactions between organisms. (Contrast with exploitation competition.)

Interferon A glycoprotein produced by virus-infected animal cells; increases the resistance of neighboring cells to the virus.

Interkinesis The phase between the first and second meiotic divisions.

Interleukins Regulatory proteins, produced by macrophages and lymphocytes, that act upon other lymphocytes and direct their development.

Intermediate filaments Fibrous proteins that stabilize cell structure and resist tension.

Internode Section between two nodes of a plant stem.

Interphase The period between successive nuclear divisions during which the chromosomes are diffuse and the nuclear envelope is intact. It is during this period that the cell is most active in transcribing and translating genetic information.

Interspecific competition Competition between members of two or more species.

Interstitial fluid In vertebrates, the fluid filling the spaces between cells.

Intertropical convergence zone The tropical region where the air rises most strongly; moves north and south with the passage of the sun overhead.

Intraspecific competition Competition among members of a single species.

Intrinsic protein A membrane protein that is embedded in the phospholipid bilayer of the membrane. (Contrast with extrinsic protein.)

Intrinsic rate of increase The rate at which a population can grow when its density is low and environmental conditions are highly favorable.

Intron A portion of a DNA molecule that, because of RNA splicing, is not involved in coding for part of a polypeptide molecule. (Contrast with exon.)

Invagination An infolding.

Invasiveness Ability of a bacterium to multiply within the body of a host.

Inversion (genetic) A rare mutational event that leads to the reversal of the order of genes within a segment of a chromosome, as if that segment had been removed from the chromosome, turned 180°, and then reattached.

Invertebrate Any animal that is not a vertebrate, that is, whose nerve cord is not enclosed in a backbone of bony segments.

In vitro [L.: in glass] In a test tube, rather than in a living organism. (Contrast with in vivo.)

In vivo [L.: in the living state] In a living organism. Many processes that occur in vivo can be reproduced in vitro with the right selection of cellular components. (Contrast with in vitro.)

Ion (eye´ on) [Gr.: wanderer] An atom or group of atoms with electrons added or removed, giving it a negative or positive electrical charge.

Ionic channel A membrane protein that can let ions pass across the membrane. The channel can be ion-selective, and it can be voltage-gated or ligand-gated.

Ionic bond A chemical bond which arises from the electrostatic attraction between positively and negatively charged ions. Usually a strong bond.

Iris (eye´ ris) [Gr. *iris*: rainbow] The round, pigmented membrane that surrounds the pupil of the eye and adjusts its aperture to regulate the amount of light entering the eye.

Irruption A rapid increase in the density of a population. Often followed by massive emigration.

Islets of Langerhans Clusters of hormone-producing cells in the pancreas.

Isogamy (eye sog´ ah mee) [Gr. *isos*: equal + *gamos*: marriage] A kind of sexual reproduction in which the gametes (or gametangia) are not distinguishable on the basis of size or morphology.

Isolating mechanism Geographical, physiological, ecological, or behavioral mechanisms that lead to a reduction in the frequency of hybrid matings.

Isomers Molecules consisting of the same numbers and kinds of atoms, but differing in the way in which the atoms are combined.

Isomorphic (eye´ so more´ fik) [Gr. *isos*: equal + *morphe*: form] having the same form or appearance, as two isomorphic life stages. (Contrast with heteromorphic.)

Isotonic [Gr.: same tension] Having the same osmotic potential. Said of two solutions. (Contrast with hypertonic, hypotonic.)

Isotope (eye´ so tope) [Gr. *isos*: equal + *topos*: place] Two isotopes of the same chemical element have the same number of protons in their nuclei, but differ in the number of neutrons.

Isozymes Chemically different enzymes that catalyze the same reaction.

Jejunum (jih jew´ num) The middle division of the small intestine, where most absorption of nutrients occurs. (See duodenum, ileum.)

Joule (jool, or jowl) A unit of energy, equal to 0.24 calories.

Juvenile hormone In insects, a hormone maintaining larval growth and preventing maturation or pupation.

Karyogamy (care´ ee og´ uh me) [Gr. *karyon*: kernel, nut + *gamos*: marriage] Fusion of gamete nuclei.

Karyotype The number, forms, and types of chromosomes in a cell.

Kelvin temperature scale See absolute temperature scale.

Keratin (ker´ a tin) [Gr. *keras*: horn] A protein which contains sulfur and is part of such hard tissues as horn, nail, and the outermost cells of the skin.

Ketone (key' tone) A compound with a C=O group attached to two other groups, neither of which is an H atom. Many sugars are ketones. (Contrast with aldehyde.)

Keystone species A species that exerts a major influence on the composition and dynamics of the community in which it lives.

Kidneys A pair of excretory organs in vertebrates.

Kin selection component The component of inclusive fitness resulting from helping the survival of relatives containing the same alleles by descent from a common ancestor.

Kinase (kye´ nase) An enzyme that transfers a phosphate group from ATP to another molecule. Protein kinases transfer phosphate from ATP to specific proteins, playing important roles in cell regulation.

Kinesis (ki nee´ sis) [Gr.: movement] Orientation behavior in which the organism does not move in a particular direction with reference to a stimulus but instead simply moves at an increasing or decreasing rate until it ends up farther from the object or closer to it. (Contrast with taxis.)

Kinetochore (kin net´ oh core) [Gr. *kinetos*: moving + *khorein*: to move] Specialized structure on a centromere to which microtubules attach.

Kingdom The highest taxonomic category in the Linnaean system.

Knockout mouse A genetically engineered mouse in which one or more functioning alleles have been replaced by defective alleles.

Lactic acid The end product of fermentation in vertebrate muscle and some microorganisms.

Lagging strand In DNA replication, the daughter strand that is synthesized discontinuously.

Lamella Layer.

Larynx (lar´ inks) A structure between the pharynx and the trachea that includes the vocal cords.

Larva (plural: larvae) [L.: ghost, early stage] An immature stage of any invertebrate animal that differs dramatically in appearance from the adult.

Lateral Pertaining to the side.

Lateral inhibition In visual information processing in the arthropod eye, the mutual inhibition of optic nerve cells; results in enhanced detection of edges.

Laterization (lat´ ur iz ay shun) The formation of a nutrient-poor soil that is rich in insoluble iron and aluminum compounds.

Law of independent assortment Alleles of different, unlinked genes assort independently of one another during gamete formation, Mendel's second law.

Law of segregation Alleles segregate from one another during gamete formation, Mendel's first law.

Leader sequence A sequence of amino acids at the N-terminal end of a newly synthesized protein, determining where the protein will be placed in the cell.

Leading strand In DNA replication, the daughter strand that is synthesized continuously.

Leaf axil The upper angle between a leaf and the stem, site of lateral buds which under appropriate circumstances become activated to form lateral branches.

Leaf primordium [L.: the beginning] A small mound of cells on the flank of a shoot apical meristem that will give rise to a leaf.

Lek A traditional courtship display ground, where males display to females.

Lenticel Spongy region in a plant's periderm, allowing gas exchange.

Leucoplast A colorless plastid that stores starch or fat.

Leukocyte (loo´ ko sight) [Gr. *leukos*: clear + *kutos*: hollow vessel] A white blood cell.

Leuteinizing hormone A peptide hormone produced by pituitary cells that stimulates follicle maturation in females.

Lichen (lie´ kun) [Gr. *leikhen*: licker] An organism resulting from the symbiotic association of a true fungus and either a cyanobacterium or a unicellular alga.

Life cycle The entire span of the life of an organism from the moment of fertilization (or asexual generation) to the time it reproduces in turn.

Life history The stages an individual goes through during its life.

Life table A table showing, for a group of equal-aged individuals, the proportion still alive at different times in the future and the number of offspring they produce during each time interval.

Ligament A band of connective tissue linking two bones in a joint.

Ligand (lig´ and) A molecule that binds to a receptor site of another molecule.

Light compass reaction A reaction of many invertebrates in which the angle between the direction of movement and the direction of the sun is kept constant.

Lignin The principal noncarbohydrate component of wood, a polymer that binds together cellulose fibrils in some plant cell walls.

Limiting resource The required resource whose supply most strongly influences the size of a population.

Linkage In genetics, association between markers on the same chromosome such that they do not show random assortment. Linked markers recombine with one another at frequencies less than 0.5; the closer the markers on the chromosome, the lower the frequency of recombination.

Lipase (lip´ ase; lye´ pase) An enzyme that digests fats.

Lipids (lip´ ids) [Gr. *lipos*: fat] Substances in a cell which are easily extracted by organic solvents; fats, oils, waxes, steroids, and other large organic molecules, including those which, with proteins, make up the cell membranes. (See phospholipids.)

Litter The partly decomposed remains of plants on the surface and in the upper layers of the soil.

Littoral zone The coastal zone from the upper limits of tidal action down to the depths where the water is thoroughly stirred by wave action.

Liver A large digestive gland. In vertebrates, it secretes bile and is involved in the formation of blood.

Lobes Regions of the human cerebral hemispheres; includes the temporal, frontal, parietal, and occipital lobes.

Locus In genetics, a specific location on a chromosome. May be considered to be synonymous with "gene."

Logistic growth Growth, especially in the size of an organism or in the number of organisms that constitute a population, which slows steadily as the entity approaches its maximum size. (Contrast with exponential growth.)

Loop of Henle (hen´ lee) Long, hairpin loop of the mammalian renal tubule that runs from the cortex down into the medulla, and back to the cortex. Creates a concentration gradient in the interstitial fluids in the medulla.

Lophophore A U-shaped fold of the body wall with hollow, ciliated tentacles that encircles the mouth of animals in several different phyla. Used for filtering prey from the surrounding water.

Lordosis (lor doe´ sis) [Gk. *lordosis*: curving forward] A posture assumed by females of some mammalian species (especially rodents) to signal sexual receptivity.

Lumen (loo´ men) [L.: light] The cavity inside any tubular part of an organ, such as a piece of gut or a kidney tubule.

Lungs A pair of saclike chambers within the bodies of some animals, functioning in gas exchange.

Luteinizing hormone A gonadotropin produced by the anterior pituitary. It stimulates the gonads to produce sex hormones.

Lymph [L. *lympha*: water] A clear, watery fluid that is formed as a filtrate of blood; it contains white blood cells; it collects in a series of special vessels and is returned to the bloodstream.

Lymphocyte A major class of white blood cells. Includes T cells, B cells, and other cell types important in the immune response.

Lysis (lie´ sis) [Gr.: a loosening] Bursting of a cell.

Lysogenic The condition of a bacterium that carries the genome of a virus in a relatively stable form. (Contrast with lytic.)

Lysosome (lie´ so soam) [Gr. *lysis*: a loosening + *soma*: body] A membrane-bounded inclusion found in eukaryotic cells (other than plants). Lysosomes contain a mixture of enzymes that can digest most of the macromolecules found in the rest of the cell.

Lysozyme (lie´ so zyme) An enzyme in saliva, tears, and nasal secretions that attacks bacterial cell walls, as one of the body's nonspecific defense mechanisms.

Lytic Condition in which a bacterium lyses shortly after infection by a virus; the viral genome does not become stabilized within the bacterial cell. (Contrast with lysogenic.)

Macro- (mack´ roh) [Gr. *makros*: large, long] A prefix commonly used to denote something large. (Contrast with micro-.)

Macroevolution Evolutionary changes occurring over long time spans and usually involving changes in many traits. (Contrast with microevolution.)

Macroevolutionary time The time required for macroveolutionary changes in a lineage.

Macromolecule A giant polymeric molecule. The macromolecules are proteins, polysaccharides, and nucleic acids.

Macronutrient A mineral element required by plant tissues in concentrations of at least 1 milligram per gram of their dry matter.

Major histocompatibility complex (MHC) A complex of linked genes, with multiple alleles, that control a number of immunological phenomena; it is important in graft rejection.

Malleus (mal´ ee us) [L. *malleus*: hammer] The first of the three bones that conduct movements of the eardrum to the oval window of the inner ear. (See incus, stapes.)

Malpighian tubule (mal pee´ gy un) A type of protonephridium found in insects.

Mammal [L. *mamma*: breast, teat] Any animal of the class Mammalia, characterized by the production of milk by the female mammary glands and the possession of hair for body covering.

Mantle A sheet of specialized tissues that covers most of the viscera of mollusks; provides protection to internal organs and secretes the shell.

Mapping In genetics, determining the order of genes on a chromosome and the distances between them.

Marine [L. *mare*: sea, ocean] Pertaining to or living in the ocean. (Contrast with aquatic, terrestrial.)

Maritime climate Weather pattern typical of coasts of continents, particularly those on the western sides at mid latitudes, in which the difference between summer and winter is relatively small. (Contrast with continental climate.)

Marsupial (mar soo´ pee al) A mammal belonging to the subclass Metatheria, such as opossums and kangaroos. Most have a pouch (marsupium) that contains the milk glands and serves as a receptacle for the young.

Mass extinctions Geological periods during which rates of extinction were much higher than during intervening times.

Mass number The sum of the number of protons and neutrons in an atom's nucleus.

Maternal inheritance (cytoplasmic inheritance) Inheritance in which the phenotype of the offspring depends on factors, such as mitochondria or chloroplasts, that are inherited from the female parent through the cytoplasm of the female gamete.

Mating type In some bacteria, fungi, and protists, sexual reproduction can occur only between partners of different mating type. "Mating type" is not the same as "sex," since some species have as many as 8 mating types; mating may also be between hermaphroditic partners of opposite mating type, with both partners acting as both "male" and "female" in terms of donating and receiving genetic information.

Maturation The automatic development of a pattern of behavior, which becomes increasingly complex or precise as the animal matures. Unlike learning, the development does not require experience to occur.

Mechanosensor A cell that is sensitive to physical movement and generates action potentials in response.

Medulla (meh dull´ luh) [L.: narrow] (1) The inner, core region of an organ, as in the adrenal medulla (adrenal gland) or the renal medulla (kidneys). (2) The portion of the brain stem that connects to the spinal cord.

Medusa (meh doo´ suh) The tentacle-bearing, jellyfish-like, free-swimming sexual stage in the life cycle of a cnidarian.

Mega- [Gr. *megas*: large, great] A prefix often used to denote something large. (Contrast with micro-.)

Megareserve A large park or reserve; usually has associated buffer areas in which human use of the environment is restricted to activities that do not destroy the functioning of the ecosystem.

Megasporangium The special structure (sporangium) that produces the megaspores.

Megaspore [Gr. *megas*: large + *spora*:seed] In plants, a haploid spore that produces a female gametophyte. In many cases the megaspore is larger than the male-producing microspore.

Meiosis (my oh´ sis) [Gr.: diminution] Division of a diploid nucleus to produce four haploid daughter cells. The process consists of two successive nuclear divisions with only one cycle of chromosome replication.

Membrane potential The difference in electrical charge between the inside and the outside of a cell, caused by a difference in the distribution of ions.

Mendelian population A local population of individuals belonging to the same species and exchanging genes with one another.

Menopause The time in a human female's life when the ovarian and menstrual cycles cease.

Menstrual cycle The monthly sloughing off of the uterine lining if fertilization does not occur in the female. Occurs between puberty and menopause.

Meristem [Gr. *meristos*: divided] Plant tissue made up of actively dividing cells.

Mesenchyme (mez´ en kyme) [Gr. *mesos*: middle + *enchyma*: infusion] Embryonic or unspecialized cells derived from the mesoderm.

Meso- (mez´ oh) [Gr.: middle] A prefix often used to designate a structure located in the middle, or a stage that appears at some intermediate time. For example, mesoderm, Mesozoic.

Mesoderm [Gr. *mesos*: middle + *derma*: skin] The middle of the three embryonic tissue layers first delineated during gastrulation. Gives rise to skeleton, circulatory system, muscles, excretory system, and most of the reproductive system.

Mesoglea The jelly-like middle layer that constitutes the bulk of the bodies of the medusae of many cnidarians; not a true cell layer.

Mesophyll (mez´ a fill) [Gr. *mesos*: middle + *phyllon*: leaf] Chloroplast-containing, photosynthetic cells in the interior of leaves.

Mesosome (mez´ o soam´) [Gr. *mesos*: middle + *soma*: body] A localized infolding of the plasma membrane of a bacterium.

Messenger RNA (mRNA) A transcript of one of the strands of DNA, it carries information (as a sequence of codons) for the synthesis of one or more proteins.

Meta- [Gr.: between, along with, beyond] A prefix used in biology to denote a change or a shift to a new form or level; for example, as used in metamorphosis.

Metabolic compensation Changes in biochemical properties of an organism that render it less sensitive to temperature changes.

Metabolic pathway A series of enzyme-catalyzed reactions so arranged that the product of one reaction is the substrate of the next.

Metabolism (meh tab´ a lizm) [Gr. *metabole*: to change] The sum total of the chemical reactions that occur in an organism, or some subset of that total (as in "respiratory metabolism").

Metamorphosis (met´ a mor´ fo sis) [Gr. *meta*: between + *morphe*: form, shape] A radical change occurring between one developmental stage and another, as for example from a tadpole to a frog or an insect larva to the adult.

Metaphase (met´ a phase) [Gr. *meta*: between] The stage in nuclear division at which the centromeres of the highly supercoiled chromosomes are all lying on a plane (the metaphase plane or plate) perpendicular to a line connecting the division poles.

Metastasis (meh tass´ tuh sis) The spread of cancer cells from their original site to other parts of the body.

Methanogen Any member of a group of Archaebacteria that release methane as a metabolic product. This group is considered to be an extremely ancient one.

MHC See major histocompatibility complex.

Micelles (my sells´) [L. *mica*: grain, crumb] The small particles of fat in the small intestine, resulting from the emulsification of dietary fat by bile.

Micro- (mike´ roh) [Gr. *mikros*: small] A prefix often used to denote something small. (Contrast with macro-, mega-.)

Microbiology [Gr. *mikros*: small + *bios*: life + *logos*: discourse] The scientific study of microscopic organisms, particularly bacteria, unicellular algae, protists, and viruses.

Microbody A small organelle, bounded by a single membrane and possessing a granular interior. Peroxisomes and glyoxysomes are types of microbodies.

Microevolution The small evolutionary changes typically occurring over short time spans; generally involving a small number of traits and minor genetic changes. (Contrast with macroevolution.)

Microevolutionary time The time required for microevolutionary changes within a lineage of organisms.

Microfilament Minute fibrous structure generally composed of actin found in the cytoplasm of eukaryotic cells. They play a role in the motion of cells.

Micromorphology The structure of the macromolecules of an organism.

Micronutrient A mineral element required by plant tissues in concentrations of less than 100 micrograms per gram of their dry matter.

Microorganism Any microscopic organism, such as a bacterium or one-celled alga.

Micropyle (mike´ roh pile) [Gr. *mikros*: small + *pyle*: gate] Opening in the integument(s) of a seed plant ovule through which pollen grows to reach the female gametophyte within.

Microsporangium The special structure (sporangium) that produces the microspores.

Microspores [Gr. *mikros*: small + *spora*: seed] In plants, a haploid spore that produces a male gametophyte. In many cases the microspore is smaller than the female-producing megaspore.

Microtubules Minute tubular structures found in centrioles, spindle apparatus, cilia, flagella, and other places in the cytoplasm of eukaryotic cells. These tubules play roles in the motion and maintenance of shape of eukaryotic cells.

Microvilli (singular: microvillus) The projections of epithelial cells, such as the cells lining the small intestine, that increase their surface area.

Middle lamella A layer of derivative polysaccharides that separates plant cells; a common middle lamella lies outside the primary walls of the two cells.

Migration The regular, seasonal movements of animals between breeding and nonbreeding ranges.

Mimicry (mim´ ik ree) The resemblance of one kind of organism to another, or to some inanimate object; serves the function of making the organism difficult to find, of discouraging potential enemies or of attracting potential prey. (See Batesian mimicry and Müllerian mimicry.)

Mineral An inorganic substance other than water.

Mineralocorticoid A hormone produced by the adrenal cortex that influences mineral ion balance; aldosterone.

Minimal medium A medium for the growth of bacteria, fungi, or tissue cultures, containing only those nutrients absolutely required for the growth of wild-type cells.

Minimum viable population. The smallest number of individuals required for a population to persist in a region.

Missense mutation A mutation that changes a codon for one amino acid to a codon for a different amino acid. (Contrast with frame-shift mutation, nonsense mutation.)

Mitochondrial matrix The fluid interior of the mitochondrion, enclosed by the inner mitochondrial membrane.

Mitochondrion (my´ toe kon´ dree un) (plural: mitochondria) [Gr. *mitos*: thread + *chondros*: cartilage, or grain] An organelle that occurs in eukaryotic cells and contains the enzymes of the ctric acid cycle, the respiratory chain, and oxidative phosphorylation. A mitochondrion is bounded by a double membrane.

Mitosis (my toe´ sis) [Gr. *mitos*: thread] Nuclear division in eukaryotes leading to the formation of two daughter nuclei each with a chromosome complement identical to that of the original nucleus.

Mitotic center Cellular region that organizes the microtubules for mitosis. In animals a centrosome serves as the mitotic center.

Mobbing Gathering of calling animals around a predator; their calls and the confusion they create reduce the probability that the predator can hunt successfully in the area.

Modular organism An organism which grows by producing additional units of body construction that are very similar to the units of which it is already composed.

Mole A quantity of a compound whose weight in grams is numerically equal to its molecular weight expressed in atomic mass units. Avogadro's number of molecules: 6.023×10^{23} molecules.

Molecular clock See radiometric clock.

Molecular formula A representation that shows how many atoms of each element are present in a molecule.

Molecular weight The sum of the atomic weights of the atoms in a molecule.

Molecule A particle made up of two or more atoms joined by covalent bonds or ionic attractions.

Mollusk (mol´ lusk) A member of the phylum Mollusca, such as a snail, clam, or octopus.

Molting The process of shedding part or all of an outer covering, as the shedding of feathers by birds or of the entire exoskeleton by arthropods.

Monecious (mo nee´ shus) [Gr.: one house] Organisms in which both sexes are "housed" in a single individual, which produces both eggs and sperm. (In some plants, these are found in different flowers within the same plant.) Examples: corn, peas, earthworms, hydras. (Contrast with dioecious, perfect flower.)

Moneran (moh neer´ un) A bacterium. This term was coined when both archaebacteria and eubacteria were considered to be members of a single kingdom, Monera.

Mono- [Gr. *monos*: one] Prefix denoting a single entity. (Contrast with poly.)

Monoclonal antibody Antibody produced in the laboratory from a clone of hybridoma cells, each of which produces the same specific antibody.

Monocot (short for monocotyledon) [Gr. *monos*: one + *kotyledon*: a cup-shaped hollow] Any member of the angiosperm class Monocotyledones, plants in which the embryo produces but a single cotyledon (seed leaf). Leaves of most monocots have their major veins arranged parallel to each other.

Monohybrid cross A mating in which the parents differ with respect to the alleles of only one locus of interest.

Monomer A small molecule, two or more of which can be combined to form oligomers (consisting of a few monomers) or polymers (consisting of many monomers).

Monophyletic (mon´ oh fih leht´ ik) [Gk. *monos*: single + *phylon*: tribe] Being descended from a single ancestral stock.

Monosaccharide A simple sugar. Oligosaccharides and polysaccharides are made up of monosaccharides.

Monosynaptic reflex A neural reflex that begins in a sensory neuron and makes a single synapse before activating a motor neuron.

Morphogens Diffusible substances whose concentration gradients determine patterns of development in animals and plants.

Morphogenesis (more´ fo jen´ e sis) [Gr. *morphe*: form + *genesis*: origin] The development of form. Morphogenesis is the overall consequence of determination, differentiation, and growth.

Morphology (more fol´ o jee) [Gr. *morphe*: form + *logos*: discourse] The scientific study of organic form, including both its development and function.

Mosaic development Pattern of animal embryonic development in which each blastomere contributes a specific part of the adult body. (Contrast with regulative development.)

Motor end plate The modified area on a muscle cell membrane where a synapse is formed with a motor neuron.

Motor neuron A neuron carrying information from the central nervous system to an effector such as a muscle fiber.

Motor unit A motor neuron and the set of muscle fibers it controls.

mRNA (See messenger RNA.)

Mucosa (mew koh´ sah) An epithelial membrane containing cells that secrete mucus. The inner cell layers of the digestive and respiratory tracts.

Müllerian mimicry The resemblance of two or more unpleasant or dangerous kinds of organisms to each other; the mimicry gives each added protection because potential enemies that learn to avoid members of one group tend to avoid members of the others even though they lack prior experience with them.

Multicellular [L. *multus*: much + *cella*: chamber] Consisting of more than one cell, as for example a multicellular organism. (Contrast with unicellular.)

Multiple fruit A fruit formed from an inflorescence. (Contrast with accessory fruit, aggregate fruit, simple fruit.)

Muscle fiber A single muscle cell. In the case of striated muscle, a syncitial, multinucleate cell.

Muscle spindle Modified muscle fibers encased in a connective sheat and functioning as stretch sensors.

Muscle tissue Contractile tissue containing actin and myosin organized into polymeric chains called microfilaments. In vertebrates, the tissues are either cardiac muscle, smooth muscle, or striated (skeletal) muscle.

Mutagen (mute´ ah jen) [L. *mutare*: change + Gr. *genesis*: source] An agent, especially a chemical, that increases the mutation rate.

Mutation In the broad sense, any discontinuous change in the genetic constitution of an organism. In the narrow sense, the word usually refers to a "point mutation," a change along a very narrow portion of the nucleic acid sequence.

Mutation pressure Evolution (change in gene proportions) by different mutation rates alone.

Mutualism The type of symbiosis, such as that exhibited by fungi and algae or cyanobacteria in forming lichens, in which both species profit from the association.

Mycelium (my seel´ ee yum) [Gr. *mykes*: fungus] In the fungi, a mass of hyphae.

Mycorrhiza (my´ ka rye´ za) [Gr. *mykes*: fungus + *rhiza*: root] An association of the root of a plant with the mycelium of a fungus.

Myelin (my´ a lin) A material forming a sheath around some axons. It is formed by Schwann cells that wrap themselves about the axon. It serves to insulate the axon electrically and to increase the rate of transmission of a nervous impulse.

Myofibril (my´ oh fy´ bril) [Gr. *mys*: muscle + L. *fibrilla*: small fiber] A polymeric unit of actin or myosin in a muscle.

Myogenic (my oh jen´ ik) [Gr. *mys*: muscle + *genesis*: source] Originating in muscle.

Myoglobin (my´ oh globe´ in) [Gr. *mys*: muscle + L. *globus*: sphere] An oxygen-binding molecule found in muscle. Consists of a heme unit and a single globiin chain, and carrys less oxygen than hemoglobin.

Myosin [Gr. *mys*: muscle] One of the two major proteins of muscle, it makes up the thick filaments. (See actin.)

NAD (nicotinamide adenine dinucleotide) A compound found in all living cells, existing in two interconvertible forms: the oxidizing agent NAD^+ and the reducing agent NADH.

NADP (nicotinamide adenine dinucleotide phosphate) Like NAD, but possessing another phosphate group; plays similar roles but is used by different enzymes.

Natal group The group into which an individual was born.

Natural killer cell A small leukocyte that nonspecifically kills certain tumor cells and virus-infected cells in tissue cultures.

Natural selection The differential contribution of offspring to the next generation by various genetic types belonging to the same population. The mechanism of evolution proposed by Charles Darwin.

Nauplius (no´ plee us) [Gk. *nauplios*: shellfish] The typical larva of crustaceans. Has three pairs of appendages and a median compound eye.

Negative control The situation in which a regulatory macromolecule (generally a repressor) functions to turn off transcription. In the absence of a regulatory macromolecule, the structural genes are turned on.

Negative feedback A pattern of regulation in which a change in a sensed variable results in a correction that opposes the change.

Nekton [Gr. *nekhein*: to swim] Animals, such as fish, that can swim against currents of water. (Contrast with plankton.)

Nematocyst (ne mat´ o sist) [Gr. *nema*: thread + *kystis*: cell] An elaborate, threadlike structure produced by cells of jellyfish and other cnidarians, used chiefly to paralyze and capture prey.

Nephridium (nef rid´ ee um) [Gr. *nephros*: kidney] An organ which is involved in excretion, and often in water balance, involving a tube that opens to the exterior at one end.

Nephron (nef´ ron) [Gr. *nephros*: kidney] The basic component of the kidney, which is made up of numerous nephrons. Its form varies in detail, but it always has at one end a device for receiving a filtrate of blood, and then a tubule that absorbs selected parts of the filtrate back into the bloodstream.

Nephrostome (nef´ ro stome) [Gr. *nephros*: kidney + *stoma*: opening] An opening in a nephridium through which body fluids can enter.

Nerve A structure consisting of many neuronal axons and connective tissue.

Net primary production Total photosynthesis minus respiration by plants.

Neural plate A thickened strip of ectoderm along the dorsal side of the early vertebrate embryo; gives rise to the central nervous system.

Neurohormone A hormone produced and secreted by neurons.

Neuron (noor´ on) [Gr. *neuron*: nerve, sinew] A cell derived from embryonic ectoderm and characterized by a membrane potential that can change in response to stimuli, generating action potentials. Action potentials are generated along an extension of the cell (the axon), which makes junctions (synapses) with other neurons, muscle cells, or gland cells.

Neurotransmitter A substance, produced in and released by one neuron, that diffuses across a synapse and excites or inhibits the postsynaptic neuron.

Neurula (nure´ you la) [Gr. *neuron*: nerve] Embryonic stage during formation of the dorsal nerve cord by two ectodermal ridges.

Neutral alleles Alleles that differ so slightly that the proteins for which they code function identically.

Neutron (new´ tron) [E.: neutral] One of the three most fundamental particles of matter, with mass approximately 1 amu and no electrical charge.

Nicotinamide adenine dinucleotide (See NAD.)

Nicotinamide adenine dinucleotide phosphate (See NADP.)

Nitrification The oxidation of ammonia to nitrite and nitrate ions, performed by certain soil bacteria.

Nitrogenase In nitrogen-fixing organisms, an enzyme complex that mediates the stepwise reduction of atmospheric N_2 to ammonia.

Nitrogen fixation Conversion of nitrogen gas to ammonia, which makes nitrogen available to living things. Carried out by certain prokaryotes, some of them free-living and others living within plant roots.

Node [L. *nodus*: knob, knot] In plants, a (sometimes enlarged) point on a stem where a leaf or bud is or was attached.

Node of Ranvier A gap in the myelin sheath covering an axons, where the axonal membrane can fire action potentials.

Noncompetitive inhibitor An inhibitor that binds the enzyme at a site other than the active site. (Contrast with competitive inhibitor.)

Nondisjunction Failure of sister chromatids to separate in meiosis II or mitosis, or failure of homologous chromosomes to separate in meiosis I. Results in aneuploidy.

Nonpolar molecule A molecule whose electric charge is evenly balanced from one end of the molecule to the other.

Nonsense (chain-terminating) mutation Mutations that change a codon for an amino acid to one of the codons (UAG, UAA, or UGA) that signal termination of translation. The resulting gene product is a shortened polypeptide that begins normally at the amino terminal end and ends at the position of the altered codon. (Contrast with frame-shift mutation, missense mutation.)

Nonvascular plants Those plants lacking well-developed vascular tissue; the liverworts, hornworts, and mosses. (Contrast with vascular plants.)

Normal flora The bacteria and fungi that live on animal body surfaces without causing disease.

Norepinephrine A neurotransmitter found in the central nervous system and also at the postganglionic nerve endings of the sympathetic nervous system. Also called noradrenaline.

Notochord (no´ tow kord) [Gr. *notos*: back + *chorde*: string] A flexible rod of gelatinous material serving as a support in the embryos of all chordates and in the adults of tunicates and lancelets.

Nuclear envelope The surface, consisting of two layers of membrane, that encloses the nucleus of eukaryotic cells.

Nucleic acid (new klay´ ik) [F.: nucleus of a cell] A long-chain alternating polymer of deoxyribose or ribose and phosphate groups, with nitrogenous bases—adenine, thymine, uracil, guanine, or cytosine (A, T, U, G, or C)—as side chains. DNA and RNA are nucleic acids.

Nucleoid (new´ klee oid) The region that harbors the chromosomes of a prokaryotic cell. Unlike the eukaryotic nucleus, it is not bounded by a membrane.

Nucleolar organizer (new klee´ o lar) A region on a chromosome that is associated with the formation of a new nucleolus following nuclear division. The site of the genes that code for ribosomal RNA.

Nucleolus (new klee´ oh lus) [from L. diminutive of *nux*: little kernel or little nut] A small, generally spherical body found within the nucleus of eukaryotic cells. The site of synthesis of ribosomal RNA.

Nucleoplasm (new´ klee o plazm) The fluid material within the nuclear envelope of a cell, as opposed to the chromosomes, nucleoli, and other particulate constituents.

Nucleosome A portion of a eukaryotic chromosome, consisting of part of the DNA molecule wrapped around a group of histone molecules, and held together by another type of histone molecule. The chromosome is made up of many nucleosomes.

Nucleotide The basic chemical unit (monomer) in a nucleic acid. A nucleotide in RNA consists of one of four nitrogenous bases linked to ribose, which in turn is linked to phosphate. In DNA, deoxyribose is present instead of ribose.

Nucleus (new´ klee us) [from L. diminutive of *nux*: kernel or nut] (1) In chemistry, the dense central portion of an atom, made up of protons and neutrons, with a positive charge. Surrounded by a cloud of negatively charged electrons. (2) In cells, the centrally located chamber of eukaryotic cells that is bounded by a double membrane and contains the chromosomes. The information center of the cell.

Nutrient A food substance; or, in the case of mineral nutrients, an inorganic element required for completion of the life cycle of an organism.

Obligate (ob´ li gut) Necessary, as in obligate anaerobe. (Contrast with facultative.)

Obligate anaerobe An animal that can live only in oxygenated environments.

Observational analysis A scientific method in which data are gathered in unmanipulated situations to test hypotheses. Often employed in the field where experimental manipulations are difficult or impossible.

Oceanic zone The deeper ocean basins.

Oil A triglyceride that is liquid at room temperature. (Contrast with fat.)

Okazaki fragments Newly formed DNA strands making up the lagging strand in DNA replication. DNA ligase links the Okazaki fragments to give a continuous strand.

Olfactory Having to do with the sense of smell.

Oligomer A compound molecule of intermediate size, made up of two to a few monomers. (Contrast with monomer, polymer.)

Omasum (oh may´ sum) The third division of the ruminant stomach. Its function is mostly the absorption of wastes. (Contrast with abomasum, rumen.)

Ommatidium [Gr. *omma*: an eye] One of the units which, collected into groups of up to 20,000, make up the compound eye of arthropods.

Omnivore [L. *omnis*: all, everything + *vorare*: to devour] An organism that eats both animal and plant material. (Contrast with carnivore, detritivore, herbivore.)

Oncogenic (ong´ co jen´ ik) [Gr. *onkos*: mass, tumor + *genes*: born] Causing cancer.

Ontogeny (on toj´ e nee) [Gr. *onto*: from "to be" + *genesis*: source] The development of a single organism in the course of its life history.

Oocyte (oh´ eh site) [Gr. *oon*: egg + *kytos*: cell] The cell that gives rise to eggs in animals.

Oogenesis (oh´ eh jen e sis) [Gr. *oon*: egg + *genesis*: source] Female gametogenesis, leading to production of the egg.

Oogonium (oh´ eh go´ nee um) In some algae and fungi, a cell in which an egg is produced.

Operator The region of an operon that acts as the binding site for the repressor.

Operon A genetic unit of transcription, typically consisting of several structural genes that are transcribed together; the operon contains at least two control regions: the promoter and the operator.

Opportunity cost The sum of the benefits an animal forfeits by not being able to perform some other behavior during the time when it is performing a given behavior.

Opsin (op´ sin) [Gr. *opsis*: sight] The protein protion of the visual pigment rhodopsin. (See rhodopsin.)

Optic chiasm Stucture on the lower surface of the vertebrate brain where the two optic nerves come together.

Optical isomers Isomers that differ in the configuration of the four different groups attached to a single carbon atom; so named because solutions of the two isomers rotate the plane of polarized light in opposite directions. The two isomers are mirror images of one another.

Order In taxonomy, the category below the class and above the family; a group of related, similar families.

Organ A body part, such as the heart, liver, brain, root, or leaf, composed of different tissues integrated to perform a distinct function for the body as a whole.

Organ of Corti Structure in the inner ear that transforms mechanical forces produced from pressure waves ("sound waves") into action potentials that are sensed as sound.

Organelles (or´ gan els´) [L.: little organ] Organized structures that are found in or on cells. Examples: ribosomes, nuclei, mitochrondria, chloroplasts, cilia, and contractile vacuoles.

Organic Pertaining to any aspect of living matter, e.g., to its evolution, structure, or chemistry. The term is also applied to any chemical compound that contains carbon.

Organism Any living creature.

Organizer, embryonic A region of an embryo which directs the development of nearby regions. In amphibian early gastrulas, the dorsal lip of the blastopore.

Osmoregulation Regulation of the chemical composition of the body fluids of an organism.

Osmosensor A neuron that converts changes in the osmotic potential of interstial fluids into action potentials.

Osmosis (oz mo´ sis) [Gr. *osmos*: to push] The movement of water through a differentially permeable membrane from one region to another where the water potential is more negative. This is often a region in which the concentration of dissolved molecules or ions is higher, although the effect of dissolved substances may be offset by hydrostatic pressure in cells with semi-rigid walls.

Osmotic potential A property of any solution, resulting from its solute content; it may be zero or have a negative value. A negative osmotic potential tends to cause water to move into the solution; it may be offset by a positive pressure potential in the solution or by a more negative water potential in a neighboring solution. (Contrast with pressure potential.)

Ossicle (ah´ sick ul) [L. *os*: bone] The calcified construction unit of echinoderm skeletons.

Osteoblasts Cells that lay down the protein matrix of bone. (Contrast with osteoclasts.)

Osteoclasts Cells that dissolve bone. (Contrast with osteoblasts.)

Otolith (oh´ tuh lith) [Gk.*otikos*: ear + *lithos*: stone[Structures in the vertebrate vestibular apparatus that mechanically stimulate hair cells when the head moves or changes position.

Outgroup A taxon that separated from another taxon, whose lineage is to be inferred, before the latter underwent evolutionary radiation.

Oval window The flexible membrane which, when moved by the bones of the middle ear, produces pressure waves in the inner ear

Ovary (oh´ var ee) Any female organ, in plants or animals, that produces an egg.

Oviduct [L. *ovum*: egg + *ducere*: to lead] In mammals, the tube serving to transport eggs to the uterus or to outside of the body.

Oviparous (oh vip´ uh rus) Reproduction in which eggs are released by the female and development is external to the mother's body. (Contrast with viviparous.)

Ovulation The release of an egg from an ovary.

Ovule (oh´ vule) [L. *ovulum*: little egg] In plants, an organ that contains a gametophyte and, within the gametophyte, an egg; when it matures, an ovule becomes a seed.

Ovum (oh´ vum) [L.: egg] The egg, the female sex cell.

Oxidation (ox i day´ shun) Relative loss of electrons in a chemical reaction; either outright removal to form an ion, or the sharing of electrons with substances having a greater affinity for them, such as oxygen. Most oxidation, including biological ones, are associated with the liberation of energy. (Contrast with reduction.)

Oxidative phosphorylation ATP formation in the mitochondrion, associated with flow of electrons through the respiratory chain. (Contrast with substrate-level phosphorylation.)

Oxidizing agent A substance that can accept electrons from another. The oxidizing agent becomes reduced; its partner becomes oxidized.

P generation The individuals that mate in a genetic cross. Their immediate offspring are the F_1 generation.

Pacemaker That part of the heart which undergoes most rapid spontaneous contraction, thus setting the pace for the beat of the entire heart. In mammals, the sinoatrial (SA) node. Also, an artificial device, implanted in the heart, that initiates rhythmic contraction of the organ.

Pacinian corpuscle A sensory neuron surrounded by sheaths of connective tissue. Found in the deep layers of the skin, where it senses touch and vibration.

Paleobiology The study of fossil evidence, the comparative biochemistry of living organisms, and conditions on the early Earth to determine the stages in the evolution of life.

Paleobotany The scientific study of fossil plants and all aspects of extinct plant life.

Paleontology (pale´ ee on tol´ oh jee) [Gr. *palaios*: ancient, old + *logos*: discourse] The scientific study of fossils and all aspects of extinct life.

Palisade parenchyma In leaves, one or several layers of tightly packed, columnar photosynthetic cells, frequently found just below the upper epidermis.

Pancreas (pan´ cree us) A gland, located near the stomach of vertebrates, that secretes digestive enzymes into the small intestine and releases insulin into the bloodstream.

Pangaea (pan jee´ uh) [Gk. *pan*: all, every] The single land mass formed when all the continents came together in the Permian period.

Parabronchi Passages in the lungs of birds through which air flows.

Paradigm A general framework within which some scientific discipline (or even the whole Earth) is viewed and within which questions are asked and hypotheses are developed. Scientific revolutions usually involve major paradigm changes.

Parapatric speciation Development of reproductive isolation among members of a continuous population in the absence of a geographical barrier. (Contrast with geographic, sympatric speciation.)

Paraphyletic taxon A taxon that includes some, but not all, of the descendants of a single ancestor.

Parasite An organism that attacks and consumes parts of an organism much larger than itself. Parasites sometimes, but not always, kill the host.

Parasitoid A parasite that is so large relative to its host that only one individual or at most a few individuals can live within a single host.

Parasympathetic nervous system A portion of the autonomic (involuntary) nervous system. Activity in the parasympathetic nervous system produces effects such as decreased blood pressure and decelerated heart beat. The neurotransmitter for this system is acetylcholine. (Contrast with sympathetic nervous system.)

Parathormone Hormone secreted by the parathyroid glands. Stimulates osteoclast activity and raises blood calcium levels.

Parathyroids Four glands on the posterior surface of the thyroid that produce and release parathormone.

Parenchyma (pair eng´ kyma) [Gr. *para*: beside + *enchyma*: infusion] A plant tissue composed of relatively unspecialized cells without secondary walls.

Parental investment Investment in one offspring or group of offspring that reduces the ability of the parent to assist other offspring.

Parsimony The principle of preferring the simplest among a set of plausible explanations of a phenomenon. Commonly employed in evolutionary and biogeographic studies.

Parthenocarpy Formation of fruit from a flower without fertilization.

Parthenogenesis (par´ then oh jen´ e sis) [Gr. *parthenos*: virgin + *genesis*: source] The production of an organism from an unfertilized egg.

Partial pressure The portion of the barometric pressure of a mixture of gases that is due to one component of that mixture. For example, the partial pressure of oxygen at sea level is 20.9% of barometric pressure.

Parturition (part uh rish un) [L. *parturire*, to give birth] Childbirth.

Pasteur effect The sharp decrease in rate of glucose utilization when

Pastoralism A nomadic form of human culture based on the tending of herds of domestic animals.conditions become aerobic.

Patch clamping A technique for isolating a tiny patch of membrane to allow the study of ion movement through a particular channel.

Pathogen (path´ o jen) [Gr. *pathos*: suffering + *gignomai*: causing] An organism that causes disease.

Pattern formation In animal embryonic development, the organization of differentiated tissues into specific structures such as wings.

Pedigree The pattern of transmission of a genetic trait in a family.

Pelagic zone (puh ladj´ ik) [Gr. *pelagos*: the sea] The open waters of the ocean.

Pellicle (pell´ ik el) [L. *pellis*: skin] A thin, filmy covering.

Penetrance Of a genotype, the proportion of individuals with that genotype who show the expected phenotype.

Penis (pee´ nis) [L.: tail, penis] The male organ inserted into the female during coitus (sexual intercourse).

Pentose (pen´ tose) [Gk. *penta*: five] A five-carbon sugar, such as ribose or deoxyribose.

PEP carboxylase The enzyme that combines carbon dioxide with PEP to form a 4-carbon dicarboxylic acid at the start of C_4 photosynthesis or of Crassulacean acid metabolism (CAM).

Pepsin [Gr. *pepsis*: digestion] An enzyme, in gastric juice, that digests protein.

Peptide linkage The connecting group in a protein chain, –CO–NH–, formed by removal of water during the linking of amino acids, –COOH to $-NH_2$. Also called an amide linkage.

Peptidoglycan The cell wall material of many prokaryotes, consisting of a single enormous molecule that surrounds the entire cell.

Perennial (per ren´ ee al) [L. *per*: through + *annus*: a year] Referring to a plant that lives from year to year. (Contrast with annual, biennial.)

Perfect flower A flower with both stamens and carpels, therefore hermaphroditic.

Pericycle [Gr. *peri*: around + *kyklos*: ring or circle] In plant roots, tissue just within the endodermis, but outside of the root vascular tissue. Meristematic activity of pericycle cells produces lateral root primordia.

Periderm The outer tissue of the secondary plant body, consisting primarily of cork.

Period (1) A minor category in the geological time scale. (2) The duration of a cyclical event, such as a circadian rhythm.

Peripheral nervous system Neurons that transmit information to and from the central nervous system and whose cell bodies reside outside the brain or spinal cord.

Peristalsis (pair´ i stall´ sis) [Gr. *peri*: around + *stellein*: place] Wavelike muscular contractions proceeding along a tubular organ, propelling the contents along the tube.

Peritoneum The mesodermal lining of the coelom among coelomate animals.

Permafrost Soil that remains frozen for many years.

Permease A protein in membranes that specifically transports a compound or family of compounds across the membrane.

Peroxisome A microbody that houses reactions in which toxic peroxides are formed. The peroxisome isolates these peroxides from the rest of the cell.

Petal In an angiosperm flower, a sterile modified leaf, nonphotosynthetic, frequently brightly colored, and often serving to attract pollinating insects.

Petiole (pet´ ee ole) [L. *petiolus*: small foot] The stalk of a leaf.

pH The negative logarithm of the hydrogen ion concentration; a measure of the acidity of a solution. A solution with pH = 7 is said to be neutral; pH values higher than 7 characterize basic solutions, while acidic solutions have pH values less than 7.

Phage (fayj) Short for bacteriophage.

Phagocyte A white blood cell that ingests microorganisms by endocytosis.

Phagocytosis [Gr.: *phagein* to eat; cell-eating] A form of endocytosis, the uptake of a solid particle by forming a pocket of plasma membrane around the particle and pinching off the pocket to form an intracellular particle bounded by membrane. (Contrast with pinocytosis.)

Pharyngeal slits Slits in the pharynx that originally functioned in gas exchange but became modified for other purposes among vertebrates.

Pharynx [Gr.: throat] The part of the gut between the mouth and the esophagus.

Phenogram Graphic representation of phenetic similarities.

Phenotype (fee´ no type) [Gr. *phanein*: to show + *typos*: impression] The observable properties of an individual as they have developed under the combined influences of the genetic constitution of the individual and the effects of environmental factors. (Contrast with genotype.)

Pheromone (feer´ o mone) [Gr. *phero*: carry + *hormon*: excite, arouse] A chemical substance used in communication between organisms of the same species.

Phloem (flo´ um) [Gr. *phloos*: bark] In vascular plants, the food-conducting tissue. It consists of sieve cells or sieve tubes, fibers, and other specialized cells.

Phosphate group The functional group $-OPO_3H_2$; the transfer of energy from one compound to another is often accomplished by the transfer of a phosphate group.

Phosphodiester linkage The connection in a nucleic acid strand, formed by linking two nucleotides.

3-Phosphoglycerate The first product of photosynthesis, produced by the reaction of ribulose bisphosphate with carbon dioxide.

Phospholipids Cellular materials that contain phosphorus and are soluble in organic solvents. An example is lecithin (phosphatidyl choline). Phospholipids are important constituents of cellular membranes. (See lipids.)

Phosphorylation The addition of a phosphate group.

Photoautotroph An organism that obtains energy from light and carbon from carbon dioxide. (Contrast with chemoautotroph, chemoheterotroph, photoheterotroph.)

Photoheterotroph An organism that obtains energy from light but must obtain its carbon from organic compounds. (Contrast with chemoautotroph, chemoheterotroph, photoautotroph.)

Photon (foe´ tohn) [Gr. *photos*: light] A quantum of visible radiation; a "packet" of light energy.

Photoperiod (foe´ tow peer´ ee ud) The duration of a period of light, such as the length of time in a 24-hour cycle in which daylight is present. The regulation of pro cesses such as flowering by the changing length of day (or of night) is known as **photoperiodism**.

Photophosphorylation Photosynthetic reactions in which light energy trapped by chlorophyll is used to produce ATP and, in noncyclic photophosphorylation, is used to reduce $NADP^+$ to NADPH.

Photorespiration Light-driven uptake of oxygen and release of carbon dioxide, the carbon being derived from the early reactions of photosynthesis.

Photosensor A cell that senses and responds to light energy.

Photosynthesis (foe tow sin´ the sis) [literally, "synthesis out of light"] Metabolic processes, carried out by green plants, by which visible light is trapped and the energy used to synthesize compounds such as ATP and glucose.

Phototropism [Gr. *photos*: light + *trope*: a turning] A directed plant growth response to light.

Phylogenetic tree Graphic representation of lines of descent among organisms.

Phylogeny (fy loj´ e nee) [Gr. *phylon*: tribe, race + *genesis*: source] The evolutionary history of a particular group of organisms; also, the diagram of the "family tree" that shows genetic linkages between ancestors and descendants.

Phylum [Gr. *phylon*: tribe, stock] In taxonomy, a high-level category just beneath kingdom and above the class; a group of related, similar classes.

Physiological time The time required for significant changes in the physiological processes or states within an organism.

Physiology (fiz´ ee ol´ o jee) [Gr. *physis*: natural form + *logos*: discourse, study] The scientific study of the functions of living organisms and the individual organs, tissues, and cells of which they are composed.

Phytoalexins Substances toxic to fungi, produced by plants in response to fungal infection.

Phytochrome (fy´ tow krome) [Gr. *phyton*: plant + *chroma*: color] A plant pigment regulating a large number of developmental and other phenomena in plants; can exist in two different forms, one of which is active and the other is not. Different wavelengths of light can drive it from one form to the other.

Phytoplankton (fy´ tow plangk´ ton) [Gr. *phyton*: plant + *planktos*: wandering] The autotrophic portion of the plankton, consisting mostly of algae.

Pigment A substance that absorbs visible light.

Piloting Finding one's way by means of landmarks.

Pilus (pill´ us) [Lat. *pilus*: hair] A surface appendage by which some bacteria adhere to one another during conjugation.

Pinocytosis [Gr.: drinking cell] A form of endocytosis; the uptake of liquids by engulfing a sample of the external medium into a pocket of the plasma membrane followed by pinching off the pocket to form an intracellular vesicle. (Contrast with phagocytosis and endocytosis.)

Pistil [L. *pistillum*: pestle] The female structure of an angiosperm flower, within which the ovules are borne. May consist of a single carpel, or of several carpels fused into a single structure. Usually differentiated into ovary, style, and stigma.

Pith In plants, relatively unspecialized tissue found within a cylinder of vascular tissue.

Pituitary A small gland attached to the base of the brain in vertebrates. Its hormones control the activities of other glands. Also known as the hypophysis.

Placenta (pla sen´ ta) [Gr. *plax*: flat surface] The organ, found in most mammals, that provides for the nourishment of the fetus and elimination of the fetal waste products. It is formed by the union of membranes of the mother's uterine lining with the membranes from the fetus.

Placental (pla sen´ tal) Pertaining to mammals of the subclass Eutheria, a group that is characterized by the presence of a placenta and that contains the majority of living species of mammals.

Plankton [Gr. *planktos*: wandering] The free-floating organisms of the sea and fresh water that for the most part move passively with the water currents. Consisting mostly of microorganisms and small plants and animals. (Contrast with nekton.)

Plant A member of the kingdom Plantae. Multicellular, gaining its nutrition by photosynthesis.

Planula (plan´ yew la) [L. *planum*: something flat] The free-swimming, ciliated larva of the cnidarians.

Plaque (plack) [Fr.: a metal plate or coin] (1) A circular clearing in a turbid layer (lawn) of bacteria growing on the surface of a nutrient agar gel. Produced by successive rounds of infection initiated by a single bacteriophage. (2) An accumulation of prokaryotic organisms on tooth enamel. Acids produced by the metabolism of these microorganisms can cause tooth decay.

Plasma (plaz´ muh) [Gr. *plassein*: to mold] The liquid portion of blood, in which blood cells and other particulates are suspended.

Plasma cell An antibody-secreting cell that developed from a B cell. The effector cell of the humoral immune system.

Plasma membrane The membrane that surrounds the cell, regulating the entry and exit of molecules and ions. Every cell has a plasma membrane.

Plasmid A DNA molecule distinct from the chromosome(s); that is, an extrachromosomal element. May replicate independently of the chromosome.

Plasmodesma (plural: plasmodesmata) [Gr. *plasma*: formed or molded + *desmos*: band] A cytoplasmic strand connecting two adjacent plant cells.

Plasmodium In the noncellular slime molds, a multinucleate mass of protoplasm surrounded by a membrane; characteristic of the vegetative feeding stage.

Plasmolysis (plaz mol´ i sis) Shrinking of the cytoplasm and plasma membrane away from the cell wall, resulting from the osmotic outflow of water. Occurs only in cells with rigid cell walls.

Plastid Organelle in plants that serves for food manufacture (by photosynthesis) or food storage; bounded by a double membrane.

Platelet A membrane-bounded body without a nucleus, arising as a fragment of a cell in the bone marrow of mammals. Important to blood-clotting action.

Pleiotropy (plee´ a tro pee) [Gr. *pleion*: more] The determination of more than one character by a single gene.

Pleural membrane [Gk. *pleuras*: rib, side] The membrane lining the outside of the lungs and the walls of the thoracic cavity. Inflammation of these membranes is a condition known as **pleurisy.**

Podocytes Cells of Bowman's capsule of the nephron that cover the capillaries of the glomerulus, forming filtration slits.

Poikilotherm (poy´ kill o therm) [Gr. *poikilos*: varied + *therme*: heat] An animal whose body temperature tends to vary with the surrounding environment. (Contrast with homeotherm, heterotherm.)

Point mutation A mutation that results from a small, localized alteration in the chemical structure of a gene. Such mutations can give rise to wild-type revertants as a result of reverse mutation. In genetic crosses, a point mutation behaves as if it resided at a single point on the genetic map. (Contrast with deletion.)

Polar body A nonfunctional nucleus produced by meiosis, accompanied by very little cytoplasm. The meiosis which produces the mammalian egg produces in addition three polar bodies.

Polar molecule A molecule in which the electric charge is not distributed evenly in the covalent bonds.

Polar nucleus One of two nuclei derived from each end of the angiosperm embryo sac, both of which become centrally located. They fuse with a male nucleus to form the primary triploid nucleus that will prduce the endosperm tissue of the angiosperm seed.

Pollen [L.: fine powder, dust] The fertilizing element of seed plants, containing the male gametophyte and the gamete, at the stage in which it is shed.

Pollination Process of transferring pollen from the anther to the receptive surface (stigma) of the ovary in plants.

Poly- [Gr. *poly*: many] A prefix denoting multiple entities.

Polygamy [Gr. *poly*: many + *gamos*: marriage] A breeding system in which an individual acquires more than one mate. In polyandry, a female mates with more than one male, in polygyny, a male mates with more than one female.

Polygenes Multiple loci whose alleles increase or decrease a continuously variable phenotypic trait.

Polymer A large molecule made up of similar or identical subunits called monomers. (Contrast with monomer, oligomer.)

Polymerase chain reaction (PCR) A technique for the rapid production of millions of copies of a particular stretch of DNA.

Polymerization reactions Chemical reactions that generate polymers by means of condensation reactions.

Polymorphism (pol´ lee mor´ fiz um) [Gr. *poly*: many + *morphe*: form, shape] (1) In genetics, the coexistence in the same population of two distinct hereditary types based on different alleles. (2) In social organisms such as colonial cnidarians and social insects, the coexistence of two or more functionally different castes within the same colony.

Polyp The sessile, asexual stage in the life cycle of most cnidarians.

Polypeptide A large molecule made up of many amino acids joined by peptide linkages. Large polypeptides are called proteins.

Polyploid (pol´ lee ploid) A cell or an organism in which the number of complete sets of chromosomes is greater than two.

Polysaccharide A macromolecule composed of many monosaccharides (simple sugars). Common examples are cellulose and starch.

Polysome A complex consisting of a threadlike molecule of messenger RNA and several (or many) ribosomes. The ribosomes move along the mRNA, synthesizing polypeptide chains as they proceed.

Polytene (pol´ lee teen) [Gr. *poly*: many + *taenia*: ribbon] An adjective describing giant interphase chromosomes, such as those found in the salivary glands of fly larvae. The characteristic, reproducible pattern of bands and bulges seen on these chromosomes has provided a method for preparing detailed chromosome maps of several organisms.

Pons [L. *pons*: bridge] Region of the brain stem anterior to the medulla.

Population Any group of organisms coexisting at the same time and in the same place and capable of interbreeding with one another.

Population density The number of individuals (or modules) of a population in a unit of area or volume.

Population dynamics The sum of the activities of the members of a population.

Population structure The proportions of individuals in a population belonging to different age classes (age structure). Also, the distribution of the population in space and the amount of migration between subpopulations.

Population vulnerability analysis A determination of the risk of extinction of a population given its current size and distribution.

Portal vein A vein connecting two capillary beds, as in the hepatic portal system.

Positive control The situation in which a regulatory macromolecule is needed to turn transcription of structural genes on. In its absence, transcription will not occur.

Positive cooperativity Occurs when a molecule can bind several ligands and each one that binds alters the conformation of the molecule so that it can bind the next ligand more easily. The binding of four molecules of O_2 by hemoglobin is an example of positive cooperativity.

Positive feedback A regulatory system in which an error signal stimulates responses that increase the error.

Postabsorptive period When there is no food in the gut and no nutrients are being absorbed.

Posterior Toward or pertaining to the rear.

Postsynaptic cell The cell whose membranes receive the neurotransmitter released at a synapse.

Postzygotic isolating mechanism Any factor that reduces the viability of zygotes resulting from matings between individuals of different species.

Predator An organism that kills and eats other organisms. Predation is usually thought of as involving the consumption of animals by animals, but in the broad usage of ecology it can also mean the eating of plants.

Pressure potential The actual physical (hydrostatic) pressure within a cell. (Contrast with osmotic potential, water potential.)

Presynaptic excitation/inhibition Occurs when a neuron modifies activity at a synapse by releasing a neurotransmitter onto the presynaptic nerve terminal.

Prey [L. *praeda*: booty] An organism hunted or caught as an energy source.

Prezygotic isolating mechanism A mechanism that reduces the probability that individuals of different species will mate.

Primary active transport Form of active transport in which ATP is hydrolyzed, yielding the energy required to transport ions against their concentration gradients. (Contrast with secondary active transport.)

Primary growth In plants, growth produced by the apical meristems. (Contrast with secondary growth.)

Primary producer A photosynthetic or chemosynthetic organism that synthesizes complex organic molecules from simple inorganic ones.

Primary succession Succession that begins in an areas initially devoid of life, such as on recently exposed glacial till or lava flows.

Primary structure The specific sequence of amino acids in a protein.

Primary wall Cellulose-rich cell wall layers laid down by a growing plant cell.

Primate (pry´ mate) A member of the order Primates, such as a lemur, monkey, ape, or human.

Primer A short, single-stranded segment of DNA serving as the necessary starting material for the synthesis of a new DNA strand, which is synthesized from the 3´ end of the primer.

Primitive streak A line running axially along the blastodisc, the site of inward cell migration during formation of the three-layered embryo. Formed in the embryos of birds and fish.

Primordium [L. *primordium*: origin] The most rudimentary stage of an organ or other part.

Principle of superposition The generalization that younger rocks lie on top of older rocks unless Earth movements have altered their positions.

Pro- [L.: first, before, favoring] A prefix often used in biology to denote a developmental stage that comes first or an evolutionary form that appeared earlier than another. For example, prokaryote, prophase.

Probe A segment of single stranded nucleic acid used to identify DNA molecules containing the complementary sequence.

Procambium Primary meristem that produces the vascular tissue.

Progesterone [L. *pro*: favoring + *gestare*: to bear] A vertebrate female sex hormone that maintains pregnancy.

Prokaryotes (pro kar´ ry otes) [L. *pro*: before + Gk. *karyon*: kernel, nucleus] Organisms whose genetic material is not contained within a nucleus. The bacteria. Considered an earlier stage in the evolution of life than the eukaryotes.

Prometaphase The phase of nuclear division that begins with the disintegration of the nuclear envelope.

Promoter The region of an operon that acts as the initial binding site for RNA polymerase.

Prophage (pro´ fayj) The noninfectious units that are linked with the chromosomes of the host bacteria and multiply with them but do not cause dissolution of the cell. Prophage can later enter into the lytic phase to complete the virus life cycle.

Prophase (pro´ phase) The first stage of nuclear division, during which chromosomes condense from diffuse, threadlike material to discrete, compact bodies.

Proplastid [Gr. *pro*: before + *plastos*: molded] A plant cell organelle which under appropriate conditions will develop into a plastid, usually the photosynthetic chloroplast. If plants are kept in the dark, proplastids may become quite large and complex.

Prostaglandin Any one of a group of specialized lipids with hormone-like functions. It is not clear that they act at any considerable distance from the site of their production.

Prosthetic group Any nonprotein portion of an enzyme.

Protease (pro´ tee ase) See proteolytic enzyme.

Protein (pro´ teen) [Gr. *protos*: first] One of the most fundamental building substances of living organisms. A long-chain polymer of amino acids with twenty different common side chains. Occurs with its polymer chain extended in fibrous proteins, or coiled into a compact macromolecule in enzymes and other globular proteins.

Proteolytic enzyme An enzyme whose main catalytic function is the digestion of a protein or polypeptide chain. The digestive enzymes trypsin, pepsin, and carboxypeptidase are all proteolytic enzymes (proteases).

Protist A member of the kingdom Protista, which consists of those eukaryotes not included in the kingdoms Animalia, Fungi, or Plantae. Many protists are unicellular. The kingdom Protista includes protozoa, algae, and fungus-like protists.

Protoderm Primary meristem that gives rise to epidermis.

Proton (pro´ ton) [Gr. *protos*: first] One of the three most fundamental particles of matter, with mass approximately 1 amu and an electrical charge of +1.

Proton motive force The proton gradient and electric charge difference produced by chemiosmotic proton pumping. It drives protons back across the membrane, with the concomitant formation of ATP.

Protonema (pro´ tow nee´ mah) [Gr. *protos*: first + *nema*: thread] The hairlike growth form that constitutes an early stage in the development of a moss gametophyte.

Proto-oncogenes The normal alleles of genes possessing oncogenes (cancer-causing genes) as mutant alleles. Proto-oncogenes encode growth factors and receptor proteins.

Protoplast A cell that would normally have a cell wall, from which the wall has been removed by enzymatic digestion or by special growth conditions.

Protostome One of two major lines of animal evolution, characterized by spiral, determinate cleavage of the egg, and by schizocoelous development. (Contrast with deuterostome.)

Prototroph (pro´ tow trofe´) [Gr. *protos*: first + *trophein*: to nourish] The nutritional wild-type, or reference form, of an organism. Any deviant form that requires growth nutrients not required by the prototrophic form is said to be a nutritional mutant, or auxotroph.

Protozoa A group of single-celled organisms classified by some biologists as a single phylum; includes the flagellates, amoebas, and ciliates. This textbook follows most modern classifications in elevating the protozoans to a distinct kingdom (Protista) and each of their major subgroups to the rank of phylum.

Provincialized A biogeographic term referring to the separation, by environmental barriers, of the biota into units with distinct species compositions.

Provirus See prophage.

Proximal Near the point of attachment or other reference point. (Contrast with distal.)

Pseudocoelom A body cavity not surrounded by a peritoneum. Characteristic of nematodes and rotifers.

Pseudogene A DNA segment that is homologous to a functional gene but contains a nucleotide change that prevents its expression.

Pseudoplasmodium [Gr. *pseudes*: false + *plasma*: mold or form] In the cellular slime molds such as *Dictyostelium*, an aggregation of single amoeboid cells. Occurs prior to formation of a fruiting structure.

Pseudopod (soo´ do pod) [Gr. *pseudes*: false + *podos*: foot] A temporary, soft extension of the cell body that is used in location, attachment to surfaces, or engulfing particles.

Pulmonary Pertaining to the lungs.

Pupa (pew´ pa) [L.: doll, puppet] In certain insects (the Holometabola), the encased developmental stage that intervenes between the larva and the adult.

Pupil The opening in teh vertebrate eye through which light passes.

Purine (pure´ een) A type of nitrogenous base. The purines adenine and guanine are found in nucleic acids.

Purkinje fibers Specialized heart muscle cells that conduct excitation throughout the ventricular muscle.

Pyramid of biomass Graphical representation of the total masses at different trophic levels in an ecosystem.

Pyramid of energy Graphical representation of the total energy contents at different trophic levels in an ecosystem.

Pyrimidine (peer im´ a deen) A type of nitrogenous base. The pyrimidines cytosine, thymine, and uracil are found in nucleic acids.

Pyrogen A substance that causes fever.

Pyruvate A three-carbon acid; the end product of glycolysis and the raw material for the citric acid cycle.

Q_{10} A value that compares the rate of a biochemical process or reaction over a 10°C range of temperature. A process that is not temperature-sensitive has a Q_{10} of 1. Values of 2 or 3 mean the reaction speeds up as temperature increases.

Quantum (kwon´ tum) [L. *quantus*: how great] An indivisible unit of energy.

Quaternary structure Of aggregating proteins, the arrangement of polypeptide subunits.

R factor (resistance factor) A plasmid that contains one or more genes that encode resistance to antibiotics.

Radial symmetry The condition in which two halves of a body are mirror images of each other regardless of the angle of the cut, providing the cut is made along the center line. Thus, a cylinder cut lengthwise down its center displays this form of symmetry. (Contrast with bilateral symmetry.)

Radioisotope A radioactive isotope of an element. Examples are carbon-14 (^{14}C) and hydrogen-3, or tritium (^{3}H).

Radiometric clock The use of the regular, known rates of decay of radioisotopes of elements to determine dates of events in the distant past.

Radiotherapy Treatment, as of cancer, with X- or gamma rays.

Radula The toothed feeding organ of many mollusks. Used to scrape prey from hard substrates.

Rain shadow A region of low precipitation on the leeward side of a mountain range.

Ramet The repeated morphological units of sessile, modular organisms. (Contrast with genet.)

Random drift Evolution (change in gene proportions) by chance processes alone.

Rate constant Of a particular chemical reaction, a constant which, when multiplied by the concentration(s) of reactant(s), gives the rate of the reaction.

Reactant A chemical substance that enters into a chemical reaction with another substance.

Reaction, chemical A process in which atoms combine or change bonding partners.

Reaction wood Modified wood produced in branches in response to gravitational stimulation. Gymnosperms produce compression wood that tends to push the branch up; angiosperms produce tension wood that tends to pull the branch up.

Realized niche The actual niche occupied by an organism; it differs from the fundamental niche because of the presence of other species.

Receptacle [L. *receptaculum*: reservoir] In an angiosperm flower, the end of the stem to which all of the various flower parts are attached.

Receptive field Of a neuron, the area on the retina from which the activity of that neuron can be influenced.

Receptor-mediated endocytosis A form of endocytosis in which macromolecules in the environment bind specific receptor proteins in the plasma membrane and are brought into the cell interior in coated vesicles.

Receptor potential The change in the resting potential of a sensory cell when it is stimulated.

Recessive See dominance.

Reciprocal altruism The exchange of altruistic acts between two or more individuals. The acts may be separated considerably in time.

Reciprocal crosses A pair of crosses, in one of which a female of genotype A mates with a male of genotype B and in the other of which a female of genotype B mates with a male of genotype A.

Recombinant An individual, meiotic product, or single chromosome in which genetic materials originally present in two individuals end up in the same haploid complement of genes. The reshuffling of genes can be either by independent segragation, or by crossing over between homologous chromosomes. For example, a human may pass on genes from both parents in a single haploid gamete.

Recombinant DNA technology The application of genetic tools (restriction endonucleases, plasmids, and transformation) to the production of specific proteins by biological "factories" such as bacteria.

Rectum The terminal portion of the gut, ending at the anus.

Redirected activity The direction of some behavior, such as aggression, away from the primary target and toward another, less appropriate object.

Redox reaction A chemical reaction in which one reactant becomes oxidized and the other becomes reduced.

Reducing agent A substance that can donate electrons to another substance. The reducing agent becomes oxidized, and its partner becomes reduced.

Reduction (re duk´ shun) Gain of electrons; the reverse of oxidation. Most reductions lead to the storage of chemical energy, which can be released later by an oxidation reaction. Energy storage compounds such as sugars and fats are highly reduced compounds. (Contrast with oxidation.)

Reflex An automatic action, involving only a few neurons (in vertebrates, often in the spinal cord), in which a motor response swiftly follows a sensory stimulus.

Refractory period Of a neuron, the time interval after an action potential, during which another action potential cannot be elicited.

Region In biogeography, a major division of the world distinguished by its peculiar animals or plants. For example, Africa south of the Sahara is recognized as constituting the Ethiopian region.

Regulative development A pattern of animal embryonic development in which the fates of the first blastomeres are not absolutely fixed. (Contrast with mosaic development.)

Regulatory gene A gene that contains the information for making a regulatory macromolecule, often a repressor protein.

Releaser A sensory stimulus that triggers a fixed action pattern.

Releasing hormone One of several hypothalamic hormones that stimulates the secretion of anterior pituitary hormone.

REM sleep A sleep state characterized by dreaming, skeletal muscle relaxation, and rapid eye movements.

Renal [L. *renes*: kidneys] Relating to the kidneys.

Replica plating A technique used in the selection of colonies of cells with a desired genotype.

Replication fork A point at which a DNA molecule is replicating. The fork forms by the unwinding of the parent molecule.

Repressible enzyme An enzyme whose synthesis can be decreased or prevented by the presence of a particular compound.

Repressor A protein coded by the regulatory gene. The repressor can bind to a specific operator and prevent transcription of the operon.

Reproductive isolating mechanism Any trait that prevents individuals from two different populations from producing fertile hybrids.

Reproductive isolation The condition in which a population is not exchanging genes with other populations of the same species.

Reproductive value The expected contribution of an individual of a particular age to the future growth of the population to which it belongs.

Resolving power Of an optical device such as a microscope, the smallest distance between two lines that allows the lines to be seen as separate from one another.

Resource Something in the environment required by an organism for its maintenance and growth that is consumed in the process of being used.

Resource defense polygamy A breeding system in which individuals of one sex (usually males) defend resources that are attractive to individuals of the other sex (usually females); individuals holding better resources attract more mates.

Respiration (res pi ra´ shun) [L. *spirare*: to breathe] (1) Cellular respiration; the oxidation of the end products of glycolysis with the storage of much energy in ATP. The oxidant in the respiration of eukaryotes is oxygen gas. Some bacteria can use nitrate or sulfate instead of O_2. (2) Breathing.

Respiratory chain The terminal reactions of cellular respiration, in which electrons are passed from NAD or FAD, through a series of intermediate carriers, to molecular oxygen, with the concomitant production of ATP.

Respiratory uncoupler A substance that allows protons to cross the inner mitochondrial membrane without the concomitant formation of ATP, thus uncoupling respiration from phosphorylation.

Resting potential The membrane potential of a living cell at rest. In cells at rest, the interior is negative to the exterior. (Contrast with action potential, electrotonic potential.)

Restoration ecology The science and practice of restoring damaged or degraded ecosystems.

Restriction endonuclease Any one of several enzymes, produced by bacteria, that break foreign DNA molecules at very specific sites. Some produce "sticky ends." Extensively used in recombinant DNA technology.

Restriction map A partial genetic map of a DNA molecule, showing the points at which particular restriction endonuclease recognition sites reside.

Retina (rett´ in uh) [L. *rete*: net] The light-sensitive layer of cells in the vertebrate or cephalopod eye.

Retinal The light-absorbing portion of visual pigment molecules. Derived from β-carotene.

Retrovirus An RNA virus that contains reverse transcriptase. Its RNA serves as a template for cDNA production, and the cDNA is integrated into a chromosome of the mammalian host cell.

Reverse transcriptase An enzyme that catalyzes the production of DNA (cDNA), using RNA as a template; essential to the reproduction of retroviruses.

Reversion (genetic) A mutational event that restores wild-type phenotype to a mutant.

RFLP (Restriction fragment length polymorphism) Coexistence of two or more patterns of restriction fragments (patterns produced by restriction enzymes), as revealed by a probe. The polymorphism reflects a difference in DNA sequence on homologous chromosomes.

Rhizoids (rye´ zoids) [Gr. *rhiza*: root] Hairlike extensions of cells in mosses, liverworts, and a few vascular plants that serve the same function as roots and root hairs in vascular plants. The term is also applied to branched, rootlike extensions of some fungi and algae.

Rhizome (rye´ zome) [Gr. *rhizoma*: mass of roots] A special underground stem (as opposed to root) that runs horizontally beneath the ground.

Rhodopsin A photopigment used in the visual process of transducing photons of light into changes in the membrane potential of photosensory cells.

Ribonucleic acid See RNA.

Ribose (rye´ bose) A sugar of chemical formula $C_5H_{10}O_5$, one of the building blocks of ribonucleic acids.

Ribosomal RNA (rRNA) Several species of RNA that are incorporated into the ribosome.

Ribosome A small organelle that is the site of protein synthesis.

Ribozyme An RNA molecule with catalytic activity.

Ribulose 1,5-bisphosphate (RuBP) The compound in chloroplasts which reacts with carbon dioxide in the first reaction of the Calvin-Benson cycle.

Risk cost The increased chance of being injured or killed as a result of performing a behavior, compared to resting.

RNA (ribonucleic acid) A nucleic acid using ribose. Various classes of RNA are involved in the transcription and translation of genetic information. RNA serves as the genetic storage material in some viruses.

RNA polymerase An enzyme that catalyzes the formation of RNA from a DNA template.

RNA splicing The last stage of RNA processing in eukaryotes, in which the transcripts of introns are excised through the action of small nuclear ribonucleoprotein particles (snRNP).

Rods Light-sensitive cells (photosensors) in the retina. (Contrast with cones.)

Root cap A thimble-shaped mass of cells, produced by the root apical meristem, that protects the meristem and that is the organ that perceives the gravitational stimulus in root gravitropism.

Root hair A specialized epidermal cell with a long, thin process that absorbs water and minerals from the soil solution.

Round dance The dance performed on the vertical surface of a honeycomb by a returning honeybee forager when she has discovered a food source less than 100 meters from the hive.

Round window A flexible membrane between the middle and inner ear that distributes pressure waves in the fluid of the inner ear.

rRNA See ribosomal RNA.

Rubisco (RuBP carboxylase) Enzyme that combines carbon dioxide with ribulose bisphosphate to produce 3-phosphoglycerate, the first product of C_3 photosynthesis. The most abundant protein on Earth.

Rumen (rew´ mun) The first division of the ruminant stomach. It stores and initiates bacterial fermentation of food. Food is regurgitated from the rumen for further chewing. (Contrast with abomasum, omasum.)

Ruminant An herbivorous, cud-chewing mammal such as a cow, sheep, or deer, having a stomach consisting of four compartments.

S phase In the cell cycle, the stage of interphase during which DNA is replicated. (Contrast with G_1 phase, G_2 phase.)

Sap An aqueous solution of nutrients, minerals, and other substances that passes through the xylem of plants.

Saprobe [Gr. *sapros*:rotten + *bios*: life] An organism (usually a bacterium or fungus) that obtains its carbon and energy directly from dead organic matter.

Sarcomere (sark´ o meer) [Gr. *sark*: flesh + *meros*: a part] The contractile unit of a skeletal muscle.

Saturated hydrocarbon A compound consisting only of carbon and hydrogen, with the hydrogen atoms connected by single bonds.

Schizocoelous development Formation of a coelom during embryological development by a splitting of mesodermal masses.

Schwann cell A glial cell that wraps around part of the axon of a peripheral neuron, creating a myelin sheath.

Sclereid A type of sclerenchyma cell, commonly found in nutshells, that is not elongated.

Sclerenchyma (skler eng´ kyma) A plant tissue composed of cells with heavily thickened cell walls, dead at functional maturity. The principal types of sclerenchyma cells are fibers and sclereids.

Scrapie-associated fibril A type of protein fibril found in nervous tissues of mammals infected with certain diseases, notably scrapie, kuru, and Creutzfeld-Jacob disease. Little is known about these fibrils, including whether they are the causal agents of the diseases.

Scrotum (skrote´ um) A sac of skin that contains the testicles in most species of mammals.

Secondary active transport Form of active transport in which ions or molecules are transported against their concentration gradient using energy obtained by relaxation of a gradient of sodium ion concentration rather than directly from ATP. (Contrast with primary active transport.)

Secondary compound A compound synthesized by a plant that is not needed for basic cellular metabolism. Typically has an antiherbivore or antiparasite function.

Secondary growth In plants, growth produced by vascular and cork cambia, contributing to an increase in girth. (Contrast with primary growth.)

Secondary structure Of a protein, localized regularities of structure, such as the α-helix and the β-pleated sheet.

Secondary wall Wall layers laid down by a plant cell that has ceased growing; often impregnated with lignin or suberin.

Second law of thermodynamics States that in any real (irreversible) process, there is a decrease in free energy and an increase in entropy.

Second messenger A compound, such as cyclic AMP, that is released within a target cell after a hormone or other "first messenger" has bound to a surface receptor on a cell; the second messenger triggers further reactions within the cell.

Secretin (si kreet´ in) A peptide hormone secreted by the upper region of the small intestine when acidic chyme is present. Stimulates the pancreatic duct to secrete bicarbonate ions.

Section A thin slice, usually for microscopy, as a tangential section or a transverse section.

Seed A fertilized, ripened ovule of a gymnosperm or angiosperm. Consists of the embryo, nutritive tissue, and a seed coat.

Seed crop The number of seeds produced by a plant during a particular bout of reproduction.

Seedling A young plant that has grown from a seed (rather than by grafting or by other means.)

Segmentation genes In insect larvae, genes that determine the number and polarity of larval segments.

Segregation (genetic) The separation of alleles, or of homologous chromosomes, from one another during meiosis so that each of the haploid daughter nuclei produced by meiosis contains one or the other member of the pair found in the diploid mother cell, but never both.

Selective permeability A characteristic of a membrane, allowing certain substances to pass through while other substances are excluded.

Self-differentiating Behavior that develops without experience with the normal objects toward which it is usually directed and without any practice. (See also instinct.)

Selfish act A behavioral act that benefits its performer but harms the recipients.

Self-pollination The fertilization of a plant by its own pollen. (Contrast with cross-pollination.)

Semelparous organism An organism that reproduces only once in its lifetime. (Contrast with iteroparous.)

Semen (see´ men) [L.: seed] The thick, whitish liquid produced by the male reproductive organ in mammals, containing the sperm.

Semicircular canals Part of the vestibular system of mammals.

Semiconservative replication The common way in which DNA is synthesized. Each of the two partner strands in a double helix acts as a template for a new partner strand. Hence, after replication, each double helix consists of one old and one new strand.

Seminiferous tubules The tubules within the testes within which sperm production occurs.

Senescence [L. *senescere*: to grow old] Aging; deteriorative changes with aging.

Sensor A sensory cell; a cell transduces a physical or chemical stimulus into a membrane potential change.

Sensory neuron A neuron leading from a sensory cell to the central nervous system. (Contrast with motor neuron.)

Sepal (see´ pul) One of the outermost structures of the flower, usually protective in function and enclosing the rest of the flower in the bud stage.

Septum [L.: partition] A membrane or wall between two cavities.

Sertoli cells Cells in the seminiferous tubules that nuture the developing sperm.

Serum That part of the blood plasma that remains after clots have formed and been removed.

Sessile (sess´ ul) [L. *sedere*: to sit] Permanently attached; not moving.

Sertoli cells Cells in the seminiferous tubules that nuture the developing sperm.

Set point In a regulatory system, the threshold sensitivity to the feedback stimulus.

Sex chromosome In organisms with a chromosomal mechanism of sex determination, one of the chromosomes involved in sex determination. One sex chromosome, the X chromosome, is present in two copies in one sex and only one copy in the other sex. The autosomes, as opposed to the sex chromosomes, are present in two copies in both sexes. In many organisms, there is a second sex chromosome, the Y chromosome, that is found in only one sex—the sex having only one copy of the X.

Sexduction See F-duction.

Sex linkage The pattern of inheritance characteristic of genes located on the sex chromosomes of organisms having a chromosomal mechanism for sex determination. The sex that is diploid with respect to sex chromosomes can assume three genotypes: homozygous wild-type, homozygous mutant, or heterozygous carrier. The other sex, haploid for sex chromosomes, is either hemizygous wild-type or hemizygous mutant.

Sexuality The ability, by any of a multitude of mechanisms, to bring together in one individual genes that were originally carried by two different individuals. The capacity for genetic recombination.

Sexual selection Selection by one sex of characteristics in individuals of the opposite sex. Also, the favoring of characteristics in one sex as a result of competition among individuals of that sex for mates.

Shoot The aerial part of a vascular plant, consisting of the leaves, stem(s), and flowers.

Sibling A brother or sister.

Sieve plate In sieve tubes, the highly specialized end walls in which are concentrated the clusters of pores through which the protoplasts of adjacent sieve tube elements are interconnected.

Sieve tube A column of specialized cells found in the phloem, specialized to conduct organic matter from sources (such as photosynthesizing leaves) to sinks (such as roots). Found principally in flowering plants.

Sieve tube element A single cell of a sieve tube, containing cytoplasm but relatively few organelles, with highly specialized perforated end walls leading to elements above and below.

Sign stimulus The single stimulus, or one out of a very few stimuli, by which an animal distinguishes key objects, such as an enemy, or a mate, or a place to nest, etc.

Signal A component of a behavior that transmits information to another individual that influences the future behavior of the receiver.

Signal sequence N-terminal sequence of a protein that directs the protein through a particular cellular membrane.

Simple development In insects and other arthorpods, development in which eggs hatch into juveniles that are similar in form to adults.

Simple fruit A fruit that develops from a single ovary. (Contrast with accessory fruit, aggregate fruit, multiple fruit.)

Sinoatrial node (sigh´ no ay´ tree al) The pacemaker of the mammalian heart.

Sinus (sigh´ nus) [L. *sinus*: a bend, hollow] A cavity in a bone, a tissue space, or an enlargement in a blood vessel.

Skeletal muscle See striated muscle.

Sliding filament theory A proposed mechanism of muscle contraction based on formation and breaking of crossbridges between actin and myosin filaments, causing them to slide together.

Small intestine The portion of the gut between the stomach and the colon, consisting of the duodenum, the jejunum, and the ileum.

Small nuclear ribonucleoprotein particle (snRNP) A complex of an enzyme and a small nuclear RNA molecule, functioning in RNA splicing.

Smooth muscle One of three types of muscle tissue. Usually consists of sheets of mononucleated cells innervated by the autonomic nervous system.

Social insect One of the kinds of insect that form colonies with reproductive castes and worker castes; in particular, the termites, ants, social bees, and social wasps.

Society A group of individuals belonging to the same species and organized in a cooperative manner; in the broadest sense, includes parents and their offspring.

Sodium cotransport Carrier-mediated transport of molecules across membranes driven by sodium ions binding to the same carrier and moving down their concentration gradient.

Sodium–potassium pump The complex protein in plasma membranes that is responsible for primary active transport; it pumps sodium ions out of the cell and potassium ions into the cell, both against their concentration gradients.

Solute A substance that is dissolved in a liquid (solvent).

Solution A liquid (solvent) and its dissolved solutes.

Solvent A liquid that has dissolved or can dissolve one or more solutes.

Somatic [Gr. *soma*: body] Pertaining to the body, or body cells (rather than to germ cells).

Somite (so´ might) One of the segments into which an embryo becomes divided longitudinally, leading to the eventual segmentation of the animal as illustrated by the spinal column, ribs, and associated muscles.

Southern blotting Transfer of DNA fragments from an electrophoretic gel to a sheet of paper or other absorbent material for analysis with a probe.

Spatial summation In the production or inhibition of action potentials in a postsynaptic neuron, the interaction of depolarizations and hyperpolarizations produced by several terminal boutons.

Spawning The direct release of sex cells into the water.

Speciation (spee´ shee ay´ shun) The process of splitting one population into two populations that are reproductively isolated from one another.

Species (spee´ shees) [L.: kind] The basic lower unit of classification, consisting of a population or series of populations of closely related and similar organisms. The more narrowly defined "biological species" consists of individuals capable of interbreeding freely with each other but not with members of other species.

Species diversity A weighted representation of the species of organisms living in a region; large and common species are given greater weight than are small and rare ones. (Contrast with species richness.)

Species pool All the species potentially available to colonize a particular habitat.

Species richness The number of species of organisms living in a region. (Contrast with species diversity.)

Specific heat The amount of energy that must be absorbed by a gram of a substance to raise its temperature by one degree centigrade. By convention, water is assigned a specific heat of one.

Sperm [Gr. *sperma*: seed] A male reproductive cell.

Spermatocyte (spur mat´ oh site) [Gr. *sperma*: seed + *kytos*: cell] The cell that gives rise to the sperm in animals.

Spermatogenesis (spur mat´ oh jen´ e sis) [Gr. *sperma*: seed + *genesis*: source] Male gametogenesis, leading to the production of sperm.

Spermatogonia Undifferentiated germ cells that give rise to primary spermatocytes and hence to sperm.

Spermatophore A package of sperm deposited in the environment by an invertebrate male, and then either inserted by him into the reproductive tract of the female or taken up by the female herself.

Sphincter (sfingk´ ter) [Gr. *sphinkter*: that which binds tight] A ring of muscle that can close an orifice, for example at the anus.

Spindle apparatus An array of microtubules stretching from pole to pole of a dividing nucleus and playing a role in the movement of chromosomes at nuclear division. Named for its shape.

Spiracle (spy´ rih kel) [L. *spirare*: to breathe] An opening of the treacheal respiratory system of terrestrial arthorpods.

Spiteful act A behavioral act that harms both the actor and the recipient of the act.

Spongy parenchyma In leaves, a layer of loosely packed photosynthetic cells with extensive intercellular spaces for gas diffusion. Frequently found between the palisade parenchyma and the lower epidermis.

Spontaneous generation The idea that life is generated continually from nonliving matter. Usually distinguished from the current idea that life evolved from nonliving matter under primordial conditions at an early stage in the history of earth.

Spontaneous reaction A chemical reaction which will proceed on its own, without any outside influence. A spontaneous reaction need not be rapid.

Sporangiophore [Gr. *phore*: to bear] Any branch bearing one or more sporangia.

Sporangium (spor an´ gee um) [Gr. *spora*: seed + *angeion*: vessel or reservoir] In plants and fungi, any specialized stucture within which one or more spores are formed.

Spore [Gr. *spora*: seed] Any asexual reproductive cell capable of developing into an adult plant without gametic fusion. Haploid spores develop into gametophytes, diploid spores into sporophytes. In prokaryotes, a resistant cell capable of surviving unfavorable periods.

Sporophyll (spor´ o fill) [Gr. *spora*: seed + *phyllon*: leaf] Any leaf or leaflike structure that bears sporangia; refers to carpels and stamens of angiosperms and to sporangium-bearing leaves on ferns, for example.

Sporophyte (spor´ o fyte) [Gr. *spora*: seed + *phyton*: plant] In plants with alternation of generations, the diploid phase that produces the spores. (Contrast with gametophyte.)

Stabilizing selection Selection against the extreme phenotypes in a population, so that the intermediate types are favored. (Contrast with disruptive selection.)

Stamen (stay´ men) [L.: thread] A male (pollen-producing) unit of a flower, usually composed of an anther, which bears the pollen, and a filament, which is a stalk supporting the anther.

Starch [O.E. *stearc*: stiff] An α-linked polymer of glucose; used by plants as a means of storing energy and carbon atoms.

Stasis Period during which little or no evolutionary change takes place within a lineage or groups of lineages.

Statocyst (stat´ oh sist) [Gk. *statos*: stationary + *kystos*: pouch] An organ of equilibrium in some invertebrates.

Statolith(stat´ oh lith) [Gk. *statos*: stationary + *lithos*: stone] A solid object that responds to gravity or movement and stimulates the mechanosensors of a statocyst.

Stele (steel) [Gr. *stele*: pillar] The central cylinder of vascular tissue in a plant stem.

Stem cell A cell capable of extensive proliferation, generating more stem cells and a large clone of differentiated progeny cells, as in the formation of red blood cells.

Step cline A sudden change in one or more traits of a species along a geographical gradient.

Steroid Any of numerous lipids based on a 17-carbon atom ring system.

Sticky ends On a piece of two-stranded DNA, short, complementary, one-stranded regions produced by the action of a restriction endonuclease. Sticky ends allow the joining of segments of DNA from different sources.

Stigma [L.: mark, brand] The part of the pistil at the apex of the style, which is receptive to pollen, and on which pollen germinates.

Stimulus Something causing a response; something in the environment detected by a receptor.

Stolon A horizontal stem that forms roots at intervals.

Stoma (plural: stomata) [Gr. *stoma*: mouth, opening] Small opening in the plant epidermis that permits gas exchange; bounded by a pair of guard cells whose osmotic status regulates the size of the opening.

Stratosphere The part of the atmosphere above the troposphere; extends upward to approximately 50 kilometers above the surface of the earth; contains very little water.

Stratum (plural strata) A layer or sedimentary rock laid down at a particular time in a past.

Striated muscle Contractile tissue characterized by multinucleated cells containing highly ordered arrangements of actin and myosin microfilaments. Also known as **skeletal muscle**.

Strobilus (strobe´ a lus) [Gr. *strobilos*: a cone] The cone, or characteristic multiple fruit, of the pine and other gymnosperms. Also, a cone-shaped mass of sprophylls found in club mosses.

Stroma The fluid contents of an organelle, such as a chloroplast.

Stromatolite A composite, flat-to-domed structure composed of successive mineral layers. Some are known to be produced by the action of bacteria in salt or fresh water, and some ancient ones are considered to be evidence for early life on the earth.

Structural formula A representation of the positions of atoms and bonds in a molecule.

Structural gene A gene that encodes the primary structure of a protein.

Style [Gr. *stylos*: pillar or column] In flowering plants, a column of tissue extending from the tip of the ovary, and bearing the stigma or receptive surface for pollen at its apex.

Sub- [L.: under] A prefix often used to designate a structure that lies beneath another or is less than another. For example, subcutaneous, subspecies.

Suberin A waxy material serving as a waterproofing agent in cork and in the Casparian strips of the endodermis in plants.

Submucosa (sub mew koe´ sah) The tissue layer just under the epithelial lining of the lumen of the digestive tract. (Contrast with mucosa.)

Substrate (sub´ strayte) The molecule or molecules on which an enzyme exerts catalytic action.

Substrate level phosphorylation ATP formation resulting from direct transfer of a phosphate group to ADP from an intermediate in glycolysis. (Contrast with oxidative phosphorylation.)

Succession In ecology, the gradual, sequential series of changes in species composition of a community following a disturbance.

Sulfhydryl group The –SH group.

Summation The ability of a neuron to fire action potentials in response to numerous subthreshold postsynaptic potentials arriving simultaneously at differentiated places on the cell, or arriving at the same site in rapid succession.

Supercoiling Coiling on coiling, as in DNA during prophase.

Supernormal stimulus Any stimulus, or any intensity of a variable stimulus, that is preferred by animals over the natural sign stimulus.

Suppressor T cells T cells that inhibit the res-ponses of B cells and other T cells to antigens. (Contrast with cytotoxic T cells, helper T cells.)

Surface tension A measure of the cohesiveness of the surface of a liquid. As a result of hydrogen bonding, water has a very high surface tension, allowing some insects to walk on the water surface.

Surface-to-volume ratio For any cell, organism, or geometrical solid, the ratio of surface area to volume; this is an important factor in setting an upper limit on the size a cell or organism can attain.

Surfactant A substance that decreases the surface tension of a liquid. Lung surfactant, secreted by cells of the alveoli, is mostly phospholipid and decreases the amount of work necessary to inflate the lungs.

Survivorship curve A plot of the logarithm of the fraction of individuals still alive, as a function of time.

Suspensor In plants, a cell or group of cells derived from the zygote, but not actually part of the embryo proper, which in some seed plants pushes the young embryo deeper into nutritive gametophyte tissue or endosperm by its growth.

Swim bladder An internal gas-filled organ that helps fishes maintain their position in the water column; later evolved into an organ for gas exchange in some lineages.

Symbiosis (sim´ bee oh´ sis) [Gr.: to live together] The living together of two or more species in a prolonged and intimate ecological relationship. (See parasitism, commensalism, mutualism.)

Symmetry In biology, the property that two halves of an object are mirror images of each other. (See bilateral symmetry and radial symmetry.)

Sympathetic nervous system A division of the autonomic (involuntary) nervous system. Its activities include increasing blood pressure and acceleration of the heartbeat. The neurotransmitter at the sympathetic terminals is epinephrine or norepinephrine. (Contrast with parasympathetic nervous system.)

Sympatric (sim pat´ rik) [Gr. *syn*: together + *patria*: homeland] Referring to populations whose geographic regions overlap at least in part.

Sympatric speciation Formation of new species even though members of the daughter species overlap in their distribution during the speciation process. (Contrast with geographic, parapatric speciation.)

Symplast The continuous meshwork of the interiors of living cells in the plant body, resulting from the presence of plasmodesmata. (Contrast with apoplast.)

Symport A membrane transport protein that carries two substances in the same direction across the membrane. (Contrast with antiport.)

Synapse (sin´ aps) [Gr. *syn*: together + *haptein*: to fasten] The narrow gap between the terminal bouton of one neutron and the dendrite or cell body of another.

Synapsis (sin ap´ sis) The highly specific parallel alignment (pairing) of homologous chromosomes during the first division of meiosis.

Synaptic vesicle A membrane-bounded vesicle, containing neurotransmitter, which is produced in and discharged by the presynaptic neuron.

Synergids (sin nur´ jids) Two cells found close to the egg cell in the angiosperm embryo sac; they disappear shortly after fertilization.

Syngamy (sing´ guh mee) [Gr. *syn-*: together + *gamos*: marriage] Union of gametes. Also known as fertilization.

Syrinx (sear´ inks) [Gr.: pipe, cavity] A specialized structure at the junction of the trachea and the primary bronchi leading to the lungs. The vocal organ of birds.

Systematics The scientific study of the diversity of organisms.

Systemic circulation The part of the circulatory system serving those parts of the body other than the lungs or gills.

Systole (sis´ tuh lee) [Gr.: contraction] Contraction of a chamber of the heart, driving blood forward in the circulatory system.

T cell A type of lymphocyte, involved in the cellular immune response. The final stages of its development occur in the thymus gland. (Contrast with B cell; see also cytotoxic T cell, helper T cell, suppressor T cell.)

T cell receptor A protein on the surface of a T cell that recognizes the antigenic determinant for which the cell is specific.

Target cell A cell which has the appropriate receptors to bind and respond to a particular hormone or other chemical mediator.

Taste bud A structure in the epithelium of the tongue that includes a cluster of chemosensors innervated by sensory neurons.

TATA box An eight-base-pair sequence, found about 25 base pairs before the starting point for transcription in many eukaryotic promoters, that binds a transcription factor and thus helps initiate transcription.

Taxis (tak´ sis) [Gr. *taxis*: arrange, put in order] The movement of an organism in a particular direction with reference to a stimulus. A taxis usually involves the employment of one sense and a movement directly toward or away from the stimulus, or else the maintenance of a constant angle to it. Thus a positive phototaxis is movement toward a light source, negative geotaxis is movement upward (away from gravity), and so on.

Taxon A unit in a taxonomic system.

Taxonomy (taks on´ oh me) [Gr. *taxis*: arrange, classify] The science of classification of organisms.

Telophase (tee´ lo phase) [Gr. *telos*: end] The final phase of mitosis or meiosis during which chromosomes became diffuse, nuclear envelopes reform, and nucleoli begin to reappear in the daughter nuclei.

Template In biochemistry, a molecule or surface upon which another molecule is synthesized in complementary fashion, as in the replication of DNA. In the brain, a pattern that responds to a normal input but not to incorrect inputs.

Template strand In a stretch of double-stranded DNA, the strand that is transcribed.

Temporal summation In the production or inhibition of action potentials in a postsynaptic neuron, the interaction of depolarizations or hyperpolarizations produced by rapidly repeated stimulation of a single point.

Tendon A collagen-containing band of tissue that connects a muscle with a bone.

Tension wood See reaction wood.

Tepal In an angiosperm flower, a sterile modified leaf. This term is used to refer to such flower parts when one is unable to distinguish between petals and sepals.

Terrestrial (ter res´ tree al) [L. *terra*: earth] Pertaining to the land. (Contrast with aquatic, marine.)

Territory A fixed area from which an animal or group of animals excludes other members of the same species by aggressive behavior or display.

Tertiary structure In reference to a protein, the relative locations in three-dimensional space of all the atoms in the molecule. The overall shape of a protein. (Contrast with primary, secondary, and quaternary structures.)

Test cross A cross of a dominant-phenotype individual (which may be either heterozygous or homozygous) with a homozygous-recessive individual.

Testis (tes´ tis) (plural: testes) [L.: witness] The male gonad; that is, the organ that produces the male sex cells.

Testosterone (tes toss´ tuhr own) A male sex steroid hormone.

Tetanus [Gr. *tetanos*: stretched] (1) In physiology, a state of sustained, maximal muscular contraction caused by rapidly repeated stimulation. (2) In medicine, an often-fatal disease ("lockjaw") caused by the bacterium *Clostridium tetani.*

Thalamus A region of the vertebrate forebrain; involved in integration of sensory input.

Thallus (thal´ us) [Gr.: sprout] Any algal body which is not differentiated into root, stem, and leaf.

Thermocline In a body of water, the zone where the temperatures change abruptly to about 4°C.

Thermoneutral zone The range of temperatures over that an endotherm does not have to expend extra energy to thermoregulate.

Thermosensor A cell or structure that responds to changes in temperature.

Thoracic cavity The portion of the mammalian body cavity bounded by the ribs, shoulders, and diaphragm. Contains the heart and the lungs.

Thorax In an insect, the middle region of the body, between the head and abdomen. In mammals, the part of the body between the neck and the diaphragm.

Thrombin An enzyme that converts fibrinogen to fibrin, thus triggering the formation of blood clots.

Thrombus (throm´ bus) [Gk. *thrombos*: clot] A blood clot that forms within a blood vessel and remains attached to the wall of the vessel. (Contrast with embolus.)

Thylakoid A flattened sac within a chloroplast. The membranes of the numerous thylakoids contain all of the chlorophyll in a plant, in addition to the electron carriers of photophosphorylation. Thylakoids stack to form grana.

Thymus A ductless, glandular portion of the lymphoid system, involved in development of the immune system of vertebrates.

Thyroid [Gr. *thyreos*: door-shaped] A two-lobed gland in vertebrates. Produces the hormone thyroxin.

Thyrotropic hormone A hormone that is produced in the pituitary gland of amphibia such as frogs and transported in the bloodstream to the thyroid gland, inducing the thyroid gland to produce the thyroid hormone that regulates metamorphosis from tadpole to adult frog.

Tight junction A junction between epithelial cells, in which there is no gap whatever between the adjacent cells. Materials may get through a tight junction only by entering the epithelial cells themselves.

Tissue A group of similar cells organized into a functional unit and usually integrated with other tissues to form part of an organ such as a heart or leaf.

Tonus A low level of muscular tension that is maintained even when the bodyis at rest.

Tornaria (tor nare´ e ah) [L. *tornus*: lathe] The free-swimming ciliated larva of certain echinoderms and hemichordates; its existence indicates the evolutionary relationship of these two groups.

Totipotency In a cell, the condition of possessing all the genetic information and other capacities necessary to form an entire individual.

Toxigenicity The ability of a bacterium to produce chemical substances injurious to the tissues of the host organism.

Trachea (tray´ kee ah) [Gr. *trakhoia*: a small rough artery] A tube that carries air to the bronchi of the lungs of vertebrates, or to the cells of arthropods.

Tracheid (tray´ kee id) A distinctive conducting and supporting cell found in the xylem of nearly all vascular plants, characterized by tapering ends and walls that are pitted but not perforated.

Trade winds The winds that blow toward the intertropical convergence zone from the northeast and southeast.

Transcription The synthesis of RNA, using one strand of DNA as the template.

Transcription factors Proteins that assemble on a eukaryotic chromosome, allowing RNA polymerase II to perform transcription.

Transduction (1) Transfer of genes from one bacterium to another, with a bacterial virus acting as the carrier of the genes. (2) In sensory cells, the transformation of a stimulus (e.g., light energy, sound pressure waves, chemical or electrical stimulants) into action potentials.

Transfection Uptake, incorporation, and expression of recombinant DNA.

Transfer cells A modified parenchyma cell that transports mineral ions from its cytoplasm into its cell wall, thus moving the ions from the symplast into the apoplast.

Transfer RNA (tRNA) A category of relatively small RNA molecules (about 75 nucleotides). Each kind of transfer RNA is able to accept a particular activated amino acid from its specific activating enzyme, after which the amino acid is added to a growing polypeptide chain.

Transformation Mechanism for transfer of genetic information in bacteria in which pure DNA extracted from bacteria of one genotype is taken in through the cell surface of bacteria of a different genotype and incorporated into the chromosome of the recipient cell. By extension, the term has come to be applied to phenomena in other organisms in which specific genetic alterations have been produced by treatment with purified DNA from genetically marked donors.

Transgenic Containing recombinant DNA incorporated into its genetic material.

Translation The synthesis of a protein (polypeptide). This occurs on ribosomes, using the information encoded in messenger RNA.

Translocation (1) In genetics, a rare mutational event that moves a portion of a chromosome to a new location, generally on a nonhomologous chromosome. (2) In vascular plants, movement of solutes in the phloem.

Transpiration [L. *spirare*: to breathe] The evaporation of water from plant leaves and stem, driven by heat from the sun, and providing the motive force to raise water (plus ions) from the roots.

Transposable element A segment of DNA that can move to, or give rise to copies at, another locus on the same or a different chromosome. May be a single insertion sequence or a more complex structure (transposon) consisting of two insertion sequences and one or more intervening genes.

Trichocyst (trick´ o sist) [Gr. *trichos*: hair + *kystis*: cell] A threadlike organelle ejected from the surface of ciliates, used both as a weapon and as an anchoring device.

Triglyceride A simple lipid in which three fatty acids are combined with one molecule of glycerol.

Triplet See codon.

Triplet repeat Occurrence of repeated triplet of bases in a gene, often leading to genetic disease, as does excessive repetition of CGG in the gene responsible for fragile-X syndrome.

Triploblastic Having three cell layers. (Contrast with diploblastic.)

Trisomic Containing three, rather than two members of a chromosome pair.

tRNA See transfer RNA.

Trochophore (troke´ o fore) [Gr. *trochos*: wheel + *phoreus*: bearer] The free-swimming larva of some annelids and mollusks, distinguished by a wheel-like band of cilia around the middle, and indicating an evolutionary relationship between these two groups.

Trophic level A group of organisms united by obtaining their energy from the same part of the food web of a biological community.

Tropic hormones Hormones of the anterior pituitary that control the secretion of hormones by other endocrine glands.

Tropism [Gr. *tropos*: to turn] In plants, growth toward or away from a stimulus such as light (phototropism) or gravity (gravitropism).

Tropomyosin (troe poe my´ oh sin) A protein that, along with actin, constitutes the thin filaments of myofibrils. It controls the interactions of actin and myosin necessary for muscle contraction.

Troposphere The atmospheric zone reaching upward approximately 17 km in the tropics and subtropics but only to about 10 km at higher latitudes. The zone in which virtually all the water vapor in the atmosphere is located.

Trypsin A protein-digesting enzyme. Secreted by the pancreas in its inactive form (trypsinogen), it becomes active in the duodenum of the small intestine.

T-tubules A set of transverse tubes that penetrates skeletal muscle fibers and terminates in the sarcoplasmic reticulum. The T-system transmits impulses to the sacs, which then release CA^{2+} to initiate muscle contraction.

Tube foot In echinoderms, a part of the water vascular system. It grasps the substratum, prey, or other solid objects.

Tube nucleus In a pollen tube, the haploid nucleus that does not participate in double fertilization. (Contrast with generative nucleus.)

Tuber [L.: swelling] A short, fleshy underground stem, usually much enlarged, and serving a storage function, as in the case of the potato.

Tubulin A protein that polymerizes to form microtubules.

Tumor A disorganized mass of cells, often growing out of control. Malignant tumors spread to other parts of the body.

Turgor See pressure potential.

Twitch A single unit of muscle contraction.

Tympanic membrane [Gr. *tympanum*: drum] The eardrum.

Umbilical cord Tissue made up of embryonic membranes and blood vessels that connects the embryo to the placenta in eutherian mammals.

Uncoupler See respiratory uncoupler.

Understory The aggregate of smaller plants growing beneath the canopy of dominant plants in a forest.

Unicellular (yoon´ e sell´ yer ler) [L. *unus*: one + *cella*: chamber] Consisting of a single cell; as for example a unicellular organism. (Contrast with multicellular.)

Uniport A membrane transport protein that carries a single substance. (Contrast with antiport, symport.)

Unitary organism An organism that consists of only one module.

Unsaturated hydrocarbon A compound containing only carbon and hydrogen atoms. One or more pairs of carbon atoms are connected by double bonds.

Upwelling The upward movement of nutrient-rich, cooler water from deeper layers of the ocean.

Urea A compound serving as the main excreted form of nitrogen by many animals, including mammals.

Ureotelic Describes an organism in which the final product of the breakdown of nitrogen-containing compounds (primarily proteins) is urea. (Contrast with ammonotelic, uricotelic.)

Ureter (your´ uh tur) [Gr. *ouron*: urine] A long duct leading from the vertebrate kidney to the urinary bladder or the cloaca.

Urethra (you ree´ thra) [Gr. *ouron*: urine] In most mammals, the canal through which urine is discharged from the bladder and which serves as the genital duct in males.

Uric acid A compound that serves as the main excreted form of nitrogen in some animals, particularly those which must conserve water, such as birds, insects, and reptiles.

Uricotelic Describes an organism in which the final product of the breakdown of nitrogen-containing compounds (primarily proteins) is uric acid. (Contrast with ammonotelic, ureotelic.)

Urinary bladder A structure structure that receives urine from the kidneys via the ureter, stores it, and expels it periodically through the urethra.

Urine (you´ rin) [Gk. *ouron*: urine] In vertebrates, the fluid waste product containing the toxic nitrogenous by-products of protein and amino acid metabolism.

Uterus (yoo´ ter us) [L.: womb] The uterus or womb is a specialized portion of the female reproductive tract in certain mammals. It receives the fertilized egg and nurtures the embryo in its early development.

Vaccination Injection of virus or bacteria or their proteins into the body, to induce immunization. The injected material is usually attenuated (weakened) before injection.

Vacuole (vac´ yew ole) [Fr.: small vacuum] A liquid-filled cavity in a cell, enclosed within a single membrane. Vacuoles play a wide variety of roles in cellular metabolism, some being digestive chambers, some storage chambers, some waste bins, and so forth.

Vagina (vuh jine´ uh) [L.: sheath] In female mammals, the passage leading from the external genital orifice to the uterus; receives the copulatory organ of the male in mating.

Van der Waals interaction A weak attraction between atoms resulting from the interaction of the electrons of one atom with the nucleus of the other atom. This attraction is about one-fourth as strong as a hydrogen bond.

Vascular (vas´ kew lar) Pertaining to organs and tissues that conduct fluid, such as blood vessels in animals and phloem and xylem in plants.

Vascular bundle In vascular plants, a strand of vascular tissue, including conducting cells of xylem and phloem as well as thick-walled fibers.

Vascular plants Those plants with xylem and phloem, including psilophytes, club mosses, horsetails, ferns, gymnosperms, and angiosperms. (Contrast with nonvascular plants.)

Vascular ray In vascular plants, radially oriented sheets of cells produced by the vascular cambium, carrying materials laterally between the wood and the phloem.

Vascular tissue system The conductive system of the plant, consisting primarily of xylem and phloem. (Contrast with dermal tissue system, ground tissue system.)

Vasopressin See antidiuretic hormone.

Vector (1) An agent, such as an insect, that carries a pathogen affecting another species. (2) A plasmid or virus that carries an inserted piece of DNA into a bacterium for cloning purposes in recombinant DNA technology.

Vegetal pole In some eggs, zygotes, and embryos, the pole near the bulk of the yolk. (Contrast with animal pole.)

Vegetative Nonreproductive, or nonflowering, or asexual.

Vein [L. *vena*: channel] A blood vessel that returns blood to the heart. (Contrast with artery.)

Vena cava [L.: hollow vein] One of a pair of large veins that carry blood from the systemic circulatory system into the heart.

Ventral [L. *venter*: belly, womb] Toward or pertaining to the belly or lower side. (Contrast with dorsal.)

Ventricle A muscular heart chamber that pumps blood through the body.

Vernalization [L. *vernalis*: belonging to spring] Events occurring during a required chilling period, leading eventually to flowering. Vernalization may require many weeks of below-freezing temperatures.

Vertebral column The jointed, dorsal column that is the primary support structure of vertebrates.

Vertebrate An animal whose nerve cord is enclosed in a backbone of bony segments, called vertebrae. The principal groups of vertebrate animals are the fishes, amphibians, reptiles, birds, and mammals.

Vertical evolution Evolutionary change in a single lineage over time. Also called anagenesis.

Vessel [L. *vasculum*: a small vessel] In botany, a tube-shaped portion of the xylem consisting of hollow cells (vessel elements) placed end to end and connected by perforations. Together with tracheids, vessel elements conduct water and minerals in the plant.

Texas. 4.10: R. Rodewald, Univ. of Virginia / BPS. 4.11: Jim Solliday / BPS. 4.14: Hilton Mollenhauer, U.S.D.A. Research Unit, College Station, TX. 4.15: Runk / Schoenberger from Grant Heilman Photography, Inc. 4.16: E. H. Newcomb & W. P. Wergin, Univ. of Wisconsin / BPS. 4.17: D. J. Wrobel, Monterey Bay Aquarium / BPS. 4.19: B. F. King, Univ. of California, Davis, School of Medicine / BPS. 4.20*a*: G. T. Cole, Univ. of Texas, Austin / BPS. 4.21*c*: H. S. Pankratz, Michigan State Univ. / BPS. 4.22*a*: R. Rodewald, Univ. of Virginia / BPS. 4.23: E. H. Newcomb & S. E. Frederick, Univ. of Wisconsin / BPS. 4.24: M. C. Ledbetter, Brookhaven National Laboratory. 4.25 *left*: Gopal Murti, Science Photo Library / Photo Researchers, Inc. 4.25 *center*: R. Alexley / Peter Arnold, Inc. 4.25 *right*: Gopal Murti, Science Photo Library / Photo Researchers, Inc. 4.26*a,b*: W. L. Dentler, Univ. of Kansas / BPS. 4.27*a*: B. F. King, Univ. of California, Davis, School of Medicine / BPS. 4.28*a*: J. R. Waaland, Univ. of Washington / BPS. 4.28*b*: E. H. Newcomb, Univ. of Wisconsin / BPS. Box 4.A: After N. Campbell, 1990, Biology, 2nd Ed., Benjamin Cummings Publishing Co.

Chapter 5 *Opener*: J. David Robertson, Duke Univ. Medical Center. 5.2: After L. Stryer, 1981, Biochemistry, 2nd Ed., W. H. Freeman. 5.4*a*: L. A. Staehelin, Univ. of Colorado / BPS. 5.4*b*: J. D. Robertson, Duke Univ. 5.7*c*: G. T. Cole, Univ. of Texas, Austin / BPS. 5.8 *top*: D. S. Friend, Univ. of California, San Francisco. 5.8 *center*: Darcy E. Kelly, Univ. of Washington. 5.8 *bottom*: Courtesy of C. Peracchia. 5.19*a–d*: M. M. Perry, *J. Cell Sci.* 39, p. 266, 1979.

Chapter 6 *Opener*: Jim Merli. 6.1: Nuridsany et Perennou / Photo Researchers, Inc. 6.9, 6.16*b*, 6.17*b*: Richard Alexander, Univ. of Pennsylvania

Chapter 7 *Opener*: Runk / Schoenberger from Grant Heilman Photography, Inc. 7.10*a*: Hilton Mollenhauer, U.S.D.A. Research Unit, College Station, Texas. 7.10*b*: E. Racker, Cornell Univ. Box 7.A*a*: Ken Lucas / BPS. Box 7.A*b*: Michael Fogden DRK PHOTO. Box 7.A*c*: G. M. Thomas & G. Poinar, Univ. of California, Berkeley. Box 7.A*d*: K. V. Wood, courtesy of M. DeLuca, Univ. of California, San Diego.

Chapter 8 *Opener*: Art Wolfe. 8.1, *bottom*: J. H. Troughton and L. A. Donaldson. 8.1, *top*: Runk / Schoenberger from Grant Heilman Photography, Inc. 8.19: J. R. Waaland, Univ. of Washington / BPS. 8.20*a*: J. A. Bassham, Lawrence Berkeley Lab., Univ. of California. 8.20*b*, 8.27*b*: E. H. Newcomb & S. E. Frederick, Univ. of Wisconsin / BPS.

Chapter 9 *Opener*: Scott Spiker / Adventure Photo. 9.1*a*: Phil Gates, Univ. of Durham / BPS. 9.1*b*: R. Rodewald, Univ. of Virginia / BPS. 9.2*a*: G.F. Bahr, Armed Forces Inst. of Pathology. 9.3*b*: A. L. Olins, Univ. of Tennessee, Oak Ridge Grad. School of Biomedical Science. 9.5 *insert*: David Ward, Yale Univ. School of Medicine. 9.8: Andrew S. Bajer, Univ. of Oregon. 9.9: J. B. Rattner & S. G. Phillips, *J. Cell Biol.* 57, p. 359, 1973. 9.10*b*, *c*: C. L. Rieder, New York State Dept. of Health / BPS. 9.11*a*: T. E. Schroeder, Univ. of Washington / BPS. 9.11*b*: B. A. Palevitz & E. H. Newcomb, Univ. of Wisconsin / BPS. 9.12: G. T. Cole, Univ. of Texas, Austin / BPS. 9.14*a,b*: David Ward, Yale Univ. School of Medicine. 9.15, 9.17: C. A. Hasenkampf, Univ. of Toronto / BPS. 9.19: B. Schuh, Monmouth Medical Center. 9.20: © Ruth Kavenoff, Designergenes Ltd., P.O. Box 100, Del Mar, CA 90214. 9.21: J.J. Cardamone, Jr., Univ. of Pittsburgh / BPS.

Chapter 10 *Opener*: Runk / Schoenberger from Grant Heilman Photography, Inc. 10.12: Carl W. May / BPS. 10.17: Namboori B. Raju, Stanford Univ., *Eur. J. Cell Biol.* 23, p. 208, 1980. 10.20: Peter J.Bryant / BPS. 10.23: After N. Campbell, 1990, Biology, 2nd Ed., Benjamin Cummings Publishing Co. 10.24: © Walter Chandoha, 1991.

Chapter 11 *Opener*: Dan Richardson. 11.3 *right*, 11.5: Dan Richardson. 11.15: G. W. Willis, Ochsner Medical Instution / BPS. 11.19*a*: Dan Richardson. 11.22*b*: Courtesy of J. E. Edstrom and *EMBO J.*

Chapter 12 *Opener*: A. B. Dowsett, Science Photo Library / Photo Researchers, Inc. 12.1: Richard Humbert / BPS. 12.4: L. Caro & R. Curtiss. 12.17: Brian Matthews, Univ. of Oregon.

Chapter 13 *Opener*: Ken Edward, Science Source / Photo Researchers Inc. 13.7: Karen Dyer, Vivigen. 13.8: Joseph Gall, Carnegie Institution of Washington. 13.9: O. L. Miller, Jr., & B. R. Beatty. 13.12: After W. T. Keeton and J. L. Gould, *Biological Science*, 5th Edition, W. W. Norton & Co.

Chapter 14 *Opener*: Hank Morgan / Photo Researchers, Inc. 14.5*a*: N.Y. State Agricultural Experiment Station, Cornell Univ. 14.5*b*: J. S. Yun & T.E. Wagner, Ohio Univ. 14.11*b*: Mike Tincher, courtesy of Agracetus, Inc. (a subsidiary of W.R. Grace & Co.) 14.13: M. L. Pardue & J. G. Gall, *Chromosomes Today* 3, p. 47, 1972. 14.17*a*: Phil Gates, Univ. of Durham / BPS. 14.18: Paul F. Umbeck, courtesy of Agracetus, Inc. (a subsidiary of W.R. Grace & Co.) 14.19: N.Y. State Agricultural Experiment Station, Cornell Univ. 14.20: Advanced Genetic Sciences. 14.21: Larry Lefever from Grant Heilman Photography, Inc.

Chapter 15 *Opener*: Chip Mitchell. 15.1: From C. Harrison et al., *J. Med. Genet.* 20, p. 280, 1983. 15.9: Elaine Rebman / Photo Researchers, Inc. 15.10: P. P. H. DeBruyn, Univ. of Chicago.

Chapter 16 *Opener*: Painting by Keith Haring. 16.1: Z. Skobe, Forsyth Dental Center / BPS. 16.2: Courtesy of Lennart Nilsson. © Boehringer Ingelheim GmbH. 16.4: G. W. Willis, Ochsner Medical Institution / BPS. 16.9: R. Rodewald, Univ. of Virginia / BPS. 16.10*b*: Arthur J. Olson, Scripps Research Institute. 16.14: L. Winograd, Stanford Univ. 16.20: A. Liepins, Sloan-Kettering Research Inst. 16.25: A. Calin, Stanford Univ. School of Medicine. 16.26: After R. C. Gallo, The AIDS Virus, © 1987: by Scientific American, Inc.

Chapter 17 *Opener*: Norbert Wu. 17.1 *sea urchin embryo*: George Watchmaker. 17.1 *sea urchin*: D. J. Wrobel, Monterey Bay Aquarium / BPS. 17.1 *tadpole, frog, chick embryo*: © E. R. Degginger. 17.1 *rooster*: John Colwell from Grant Heilman Photography, Inc. 17.3*a*: After J. E. Sulston and H. R. Horvitz, *Dev. Biol.* 56, p. 110, 1977. 17.10: George M. Malacinski and A. W. Neff. 17.12: Peter J. Bryant / BPS. 17.19: Susan Strome. 17.28: C. Rushlow and M. Levine. 17.23*a*: From B. Alberts et al., 1983, *Molecular Biology of the Cell*, Garland Publishing Co. 17.24: F. R. Turner, Indiana Univ. 17.26: E. B. Lewis.

Chapter 18 *Opener*: Larry Ulrich / DRK PHOTO. 18.3: Stanley M. Awramik, U. of California / BPS.

Chapter 19 *Opener*: Frans Lanting / Minden Pictures. 19.6: Frank S. Balthis, Nature's Design. 19.7: Harold W. Pratt / BPS. 19.11*a,b*: Gary J.James / BPS. 19.12: After D. Futuyma, *Evolutionary Biology*, 2nd Ed., Sinauer Associates, Inc., 1987. 19.16*a,b*: Richard Alexander, Univ. of Pennsylvania.

Chapter 20 *Opener*: Raymond A. Mendez. 20.1: Des and Jen Bartlett, Bruce Coleman, Inc. 20.2*a*: Edward Ely / BPS. 20.2*b*: © John Shaw / NHPA. 20.4: Anthony D. Bradshaw, Univ. of Liverpool. 20.6: © E. R. Degginger. 20.7*a–c*: R.W. VanDevender. 20.8: Paul A. Johnsgard, Univ. of Nebraska. 20.9*a*: Peter J. Bryant / BPS. 20.9*b*: Kenneth Y. Kaneshiro, Univ. of Hawaii at Manoa. 20.9*c*: Peter J. Bryant / BPS. 20.11*b*: Heather Angel, BIOFOTOS. 20.11*a*: Virginia P. Weinland / Photo Researchers, Inc. 20.12*a*: Gary J. James / BPS. 20.12*b,c*: © Jim Denny.

Chapter 21 *Opener*: Joe McDonald. 21.1*a*: Helen E. Carr / BPS. 21.1*b*: Barbara J. Miller / BPS. 21.1*c*: © L. Campbell / NHPA. 21.5*a*: Jon Stewart / BPS. 21.5*b*: Barbara J. Miller / BPS. 21.9*a,b*: Peter J. Bryant / BPS. 21.10: Illustration by Marianne Collins.

21.12*a,b*: Paul A. Johnsgard, Univ. of Nebraska. 21.14*a,b*: Art Wolfe.

Chapter 22 *Opener*: Tony Brain, Science Photo Library/Photo Researchers, Inc. 22.1: Alfred Pasieka/Science Photo Library Photo Researchers, Inc. 22.2: H. W. Jannasch, Woods Hole Oceanographic Institution. 22.3, 22.4, 22.5: T. J. Beveridge, Univ. of Guelph/ BPS. 22.6*a*: Leonard Lessin , Peter Arnold Inc. 22.7*b*: C. Forsberg & T. J. Beveridge, Univ. of Guelph/BPS. 22.7*a*: Paul W. Johnson/BPS. 22.7*b*: K. Stephens, Stanford Univ./BPS. 22.8*a*: J.A. Breznak & H. S. Pankratz, Michigan State Univ./BPS. 22.8*b*: G. W. Willis, Ochsner Medical Institution/ BPS. 22.9: T. J. Beveridge, Univ. of Guelph/ BPS. 22.10: G. W. Willis, Ochsner Medical Institution/BPS. 22.11: D. A. Glawe, Univ. of Illinois/BPS. 22.12: G. W. Willis, Ochsner Medical Institution/BPS. 22.13*a*: W. Burgdorfer, Rocky Mountain Lab. 22.13*b*: Nat. Animal Disease Center, Ames, IA. 22.14: S. C. Holt, Univ. of Texas Health Science Center, San Antonio/BPS. 22.15*a*: Paul W. Johnson/BPS. 22.15*b*: H. S. Pankratz, Michigan State Univ./BPS. 22.15*c*: © E. R. Degginger. 22.16*a,b*: Leon J. Le Beau/BPS. 22.17*a*: G. W. Willis, Ochsner Medical Institution/BPS. 22.17*b*: G. W. Willis, Ochsner Medical Institution/BPS. 22.18: Centers for Disease Control, Atlanta. 22.19: M. G. Gabridge, cytoGraphics, Inc./BPS. 22.20: Arthur J. Olson, Scripps Research Institute. 22.21: D. T. Brown et al., *J. Virol.* 10, p. 524, 1972. 22.22*a*: D. L. D. Caspar, Brandeis Univ. 22.22*b*: D. S. Goodsell & A. J. Olson, Scripps Research Institute. 22.22*c*: S. C. Holt, Univ. of Texas Health Science Center, San Antonio/BPS. 22.22*d*: F. A. Murphy, Centers for Disease Control, Atlanta. Box 22.A1 *upper left*: Centers for Disease Control, Atlanta. Box 22.A *upper center*: S. C. Holt, Univ. of Texas Health Science Center, San Antonio/BPS. Box 22.A *lower left*: Leon J. Le Beau/BPS. Box 22.A *lower center*: A J. J. Cardamone, Jr., Univ. of Pittsburgh/BPS.

Chapter 23 *Opener*: Jan Hinsch, Science Photo Library/Photo Researchers, Inc. 23.3*a*: Animals Animals/Oxford Scientific Films. 23.3*b*: © E. R. Degginger. 23.3*c*: James Solliday/BPS. 23.4: Eric V. Gravé, Science Source/Photo Researchers, Inc. 23.6: G. W. Willis, M.D./BPS. 23.26*a*: Dennis D. Kunkel/ BPS. 23.7: Paul W.Johnson/BPS. 23.9*a*: Jim Solliday/BPS.23.9*b*: Eric V. Gravé/Photo Researchers, Inc. 23.11*a*: Jim Solliday/BPS. 23.11*b–d*: Paul W. Johnson/BPS. 23.12*b*: M. A. Jakus, NIH.23.14: Eric V. Gravé. 23.16*a*: Barbara J. Miller/BPS. 23.16*b*: Henry Aldrich, Institute of Food and Agricultural Sciences, Univ. of Florida. 23.17*a*: D. W. Francis, Univ. of Delaware. 23.17*b*: © David Scharf. 23.18: J. R. Waaland, Univ. of Washington/BPS. 23.19: Dwight R. Kuhn. 23.20*a*: Paul W. Johnson/ BPS. 23.20*b*: Gary J.James/BPS. 23.21*a*: Charles Gellis/Photo Researchers, Inc. 23.21*b*: V. Cassie. 23.24*a*: J.N. A. Lott, McMaster Univ./BPS. 23.24*b*: J. R. Waaland, Univ. of Washington/BPS. 23.25*a*: Maria Schefter/BPS. 23.25*b*: J. N. A. Lott, McMaster Univ./BPS. 23.26*a*: Dennis D. Kunkel/BPS. 23.26*b*: Harold W. Pratt/BPS. 23.26*c*: J. R. Waaland, Univ. of Washington/BPS. Box 23.A*a*: Gerald Corsi, Tom Stack and Associates. Box 23.A*b*: J. N. A. Lott, McMaster Univ./BPS.

Chapter 24 *Opener*: Stephen J. Kraseman/ DRK PHOTO. 24.1*a*: D. A. Glawe, Univ. of Illinois/BPS. 24.1*b*: L. E. Gilbert, Univ. of Texas, Austin/BPS. 24.1*c*: G. L. Barron, Univ. of Guelph/BPS. 24.2*b*: G. T.Cole, Univ. of Texas, Austin/BPS. 24.3: D. A. Glawe, Univ. of Illinois/BPS. 24.4: G. L. Barron, Univ. of Guelph/BPS. 24.5: Barbara J. Miller/BPS. 24.7 *upper & lower left*: W. F.Schadel, Small World Enterprises/BPS. 24.8: J. R. Waaland, Univ. of Washington/BPS. 24.9*b*: D. A. Glawe, Univ. of Illinois/BPS. 24.10: W. F. Schadel, Small World Enterprises/BPS. 24.11*a*: Jim Solliday/BPS. 24.11*b*: D. A. Glawe, Univ. of Illinois/BPS. 24.11*c*: Richard Humbert/BPS. 24.12: Centers for Disease Control, Atlanta. 24.13*a*: Michael Fogden DRK PHOTO. 24.13*b*: Photography by Rannels, Grant Heilman Photography, Inc. 24.13*c*: M. Graybill and J. Hodder/BPS. 24.14 *inset*: © Biophoto Associates. 24.15*a*: E. I. Friedmann, Florida State Univ. 24.15*b*: Barbara J. O'Donnell/BPS. 24.16*a*: Grant Heilman, Grant Heilman Photography, Inc. 24.16*b*: J. N. A. Lott, McMaster Univ./BPS. 24.16*c*: Barbara J. Miller/BPS. 24.17*a*: J. N. A. Lott, McMaster Univ./BPS. Box 24.A: R. L. Peterson, Univ. of Guelph/BPS.

Chapter 25 *Opener*: Art Wolfe. 25.1: J. Robert Stottlemeyer/BPS. 25.3: Gary J. James/BPS. 25.4*a,b*: J. R. Waaland/BPS. 25.5*a*: © E. R. Degginger. 25.5*b*: Runk/ Schoenberger from Grant Heilman Photography, Inc. 25.5*c*: J. N. A Lott, McMaster Univ./BPS. 25.7: J. H. Troughton. 25.8: © E. R. Degginger. 25.9: Fig. information provided by Prof. Hermann Pfefferkorn, Dept. of Geology, Univ. of Pennsylvania. Original oil painting by John Woolsey. 25.14*a*: Runk/ Schoenberger from Grant Heilman Photography, Inc. 25.14*b*: J. N. A. Lott, McMaster Univ./BPS. 25.15*a*: Carl W. May/BPS. 25.15*b*, 25.16: J. N. A. Lott, McMaster Univ./BPS. 25.17*a*: Barbara J. Miller/BPS. 25.17*b*: J. N.A. Lott, McMaster Univ./BPS. 25.17*c*: Art Wolfe. 25.18: J. N. A. Lott, McMaster Univ./BPS. 25.21: Phil Gates, Univ. of Durham/BPS. 25.22*a*: John Cancalosi/DRK PHOTO. 25.22*b*: Ken Lucas/BPS. 25.22*c*: Gary J. James/BPS. 25.22*d*: Joel Simon. 25.25*a*: B. Miller/BPS. 25.25*b*: Roger de la Harpe/BPS. 25.25*c*: Grant Heilman Photography 25.26*a*: Barbara J. Miller/BPS. 25.26*b*: Barbara J. Miller/BPS. 25.28*a*: J. N.A. Lott, McMaster Univ./BPS. 25.28*b*: J. N.A. Lott, McMaster Univ./BPS. 25.28*c*: Catherine M. Pringle/BPS. 25.28*d*: C.S. Lobban/BPS. 25.30*a*: Jon Mark Stewart/BPS. 25.30*b*: Lara Hartley, TERRAPHOTOGRAPHICS/BPS. 25.30*c*: © E. R. Degginger. 25.31*a*: Jon Mark Stewart/BPS. 25.31*b*: Lefever/Grushow from Grant Heilman Photography, Inc. 25.31*c*: Jon Mark Stewart/BPS. 25.32: Barbara J. O'Donnell/BPS.

Chapter 26 *Opener*: Norbert Wu. 26.8*a*: Ken Lucas/BPS. 26.8*b*: Robert Brons/BPS. 26.8*c*: Joel Simon. 26.10, 26.11, 26.12, 26.13: Adapted from F. M. Bayerand, H. B. Owre, 1968, *The Free-Living Lower Invertebrates*, Macmillan Pubishing Co. 26.14: After G. and R. Brusca, 1990, *Invertebrates*, Sinauer Associates, Inc. 25.14*a*: Andrew J. Martinez/Photo Researchers,Inc. 25.14*b*: Douglas Faulkner/Photo Researchers, Inc. 26.16*a*: Robert Brons/BPS. 26.16*b,c*: After G. and R. Brusca, 1990, *Invertebrates*, Sinauer Associates, Inc. 26.17*b*, 26.18*a*: D. J. Wrobel, Monterey Bay Aquarium/BPS. 26.18*c*, 26.21*b*: Robert Brons/BPS. 26.21*c*: Jim Solliday/BPS. 26.22*a*: After G. and R. Brusca, 1990, *Invertebrates*, Sinauer Associates, Inc. 26.22*b*: Jim Solliday/BPS. 26.23: R. R. Hessler, Scripps Institute of Oceanography. 26.25*a*: Andrew J. Martinez/Photo Researchers, Inc. 26.25*c*: Roger K. Burnard/BPS. 26.25*d*: © Robert & Linda Mitchell. 26.27*a*: M. P. L. Fogden/Bruce Coleman, Inc. 26.27*b*: From Kristensen and Hallas, 1980. 26.29*a*: Joel Simon. 26.28: J. N. A. Lott, McMaster Univ./ BPS. 26.29*b*: Barbara J. Miller/BPS. 26.30*a*: Ken Lucas/BPS. 26.30*b*: Peter J. Bryant/BPS. 26.30*c*: L. E. Gilbert, Univ. of Texas, Austin/ BPS. 26.30*d*: Robert Brons/BPS. 26.32*a*: Gregory Ochocki/Photo Researchers, Inc. 26.32*b*: Peter J.Bryant/BPS. 26.32*c*: D. J. Wrobel, Monterey Bay Aquarium/BPS. 26.32*d*: C. R. Wyttenbach, Univ. of Kansas/ BPS. 26.33*a*: Ken Lucas/BPS. 26.33*b*: Roger K. Burnard/BPS. 26.34*a*: Richard Humbert/ BPS. 26.34*b*: Peter J. Bryant/BPS. 26.34*c*- *g*: Peter J. Bryant/BPS. 26.34*h*: © E. R. Degginger. 26.36*a*: Ken Lucas/BPS. 26.36*b*: Harold W. Pratt/BPS. 26.36*c*: D. J. Wrobel, Monterey Bay Aquarium/BPS. 26.36*d,e,g*: Ken Lucas/BPS. 26.36*f*: J. W. Porter, Univ. of Georgia/BPS. 26.37: D. J. Wrobel, Monterey Bay Aquarium/BPS.

Chapter 27 *Opener*: Heather Angel, BIOFOTOS. 27.2*a*: D. J. Wrobel, Monterey Bay Aquarium/BPS. 27.2*b*: Robert Brons/BPS. 27.4: D. J. Wrobel, Monterey Bay Aquarium/BPS. 27.5*b*: C.R. Wyttenbach, Univ. of Kansas/BPS. 27.8*a*: Doug Perrine/ DRK PHOTO. 27.8*b*: Joel Simon 27.8*c–e*: D. J. Wrobel, Monterey Bay Aquarium/BPS. 27.9: Robert Brons/BPS. 27.10*b*: © Heather Angel, BIOFOTOS. 27.12: Tom Stack/Tom Stack and Associates. 27.14*a*: Tom McHugh/Photo Researchers, Inc. 27.14*b*: D. J. Wrobel, Monterey Bay Aquarium/BPS. 27.15*a*: Ken Lucas/BPS. 27.16*a*: Peter Scoones, Planet Earth Pictures. 27.15*b–d*: Ken Lucas/BPS.

27.15*e*: Animals Animals/Breck P. Kent. 27.16*a*: Peter Scoones/Planet Earth Pictures. 27.17*a*: Ken Lucas/BPS. 27.17*b*: Art Wolfe. 27.17*c*: E.D. Brodie, Jr., Univ. of Texas, Arlington/BPS. 27.20*a*: Doug Perrine DRK PHOTO. 27.20*b*: Carl Gans, Univ. of Michigan/BPS. 27.20*c*: Joe McDonald, Bruce Coleman Inc. 27.20*d*: Michael P. Fogden, Bruce Coleman Inc. 27.21*a*: © E. R. Degginger. 27.21*b*: Wayne Lankinen/DRK PHOTO. 27.22: Courtesy of Carnegie Museum of Natural History, Pittsburgh. 27.24*a*: Johnny Johnson/DRK PHOTO. 27.24*b*: D. Cavagnaro/DRK PHOTO. 27.24*c,d*: Stephen J. Kraseman/DRK PHOTO. 27.26*a*: John Cancalosi/Tom Stack and Associates. 27.26*b*: Joel Simon. 27.26*c*: M. P. L. Fogden, Bruce Coleman, Inc. 27.27*a*: J. N. A. Lott, McMaster Univ./BPS. 27.27*b*: Merlin D. Tuttle, Bat Conservation International. 27.27*c*: Stephen J. Kraseman/DRK PHOTO. 27.27*d*: Robert Stottlemyer/BPS. 27.29*a*: Art Wolfe. 27.29*b*: Stanley Breeden/DRK PHOTO. 27.29*c*: Frans Lanting/Minden Pictures. 27.30*a,b*: Steve Kaufman/DRK PHOTO. 27.31*a*: © E. R. Degginger. 27.31*b*: © Peter Drowne, E. R. Degginger. 27.31*c*: Kennan Ward/DRK PHOTO. 27.31*d*: © Peter Drowne, E. R. Degginger. 27.34*a*: Edward S. Ross. 27.34*b,c*: Robert E. Ford, TERRAPHOTOGRAPHICS/BPS. Box 27A: Anne Marie Weber/Adventure Photo.

Chapter 28 *Opener*: Frans Lanting/Minden Pictures. 28.3: Peter Ward, Univ. of Washington 28.4: Mark R. Meyer. 28.7*a*: Ken Lucas/BPS. 28.7*b*: S. M. Awramik, Univ. of California/BPS. 28.8*a*: S. Conway Morris. 23.8*b*: From S. J.Gould, 1989, Wonderful Life, © W. W. Norton. 28.9: Courtesy of the Natural History Museum of London. 28.10: Courtesy of the Smithsonian Institution. 28.11: Ken Lucas/BPS. 28.12: Painting by Rudolph Zallinger; courtesy of the Peabody Museum of Natural History, Yale Univ. 28.14: Painting by Chip Clark; courtesy of the Smithsonian Institution.

Chapter 29 *Opener*: Ed Reschke, Peter Arnold, Inc. 29.1: Joel Simon. 29.2: Michael P. Gadomski/Bruce Coleman, Inc. 29.3: Gary J. James/BPS. 29.4: Thomas Hovland from Grant Heilman Photography, Inc. 29.7*a*: Runk/Schoenberger from Grant Heilman Photography, Inc. 29.7*b*: © E. R. Degginger. 29.8: Grant Heilman Photography. 29.9*a*: J. R. Waaland, Univ. of Washington/BPS. 29.9*b*: © E. R. Degginger. 29.9*c*: Carl W. May/BPS. 29.13*a,b*: Phil Gates, Univ. of Durham/BPS. 29.13*c*: Runk/Schhoenberger from Grant Heilman Photography, Inc. 29.13*d*: Phil Gates, Univ. of Durham/BPS. 29.13*e*: © E. R. Degginger. 29.13*f*, 29.15*b*, 29.17 *top & bottom*: J. R. Waaland, Univ. of Washington/BPS. 29.18*a*: L. Elkin, Hayward State Univ./BPS. 29.18*b,c*: Jim Solliday/BPS. 29.18*d*: Phil Gates, Univ. of Durham/BPS. 29.20 *top*: Dwight R. Kuhn. 29.20 *bottom*: © E. R. Degginger. 29.20: J. R. Waaland, Univ. of Washington/BPS. 29.21*a,b*: Phil Gates, Univ. of Durham/BPS. 29.23: J. N.A. Lott, McMaster Univ./BPS. 29.24: Jim Solliday/BPS. 29.25: J. N. A. Lott, McMaster Univ., BPS. 29.26*a,b*: Phil Gates, Univ. of Durham/BPS. 29.27*b*: W. F. Schadel, Small World Enterprises/BPS. 29.27*c*: E. J. Cable/Tom Stack and Associates.

Chapter 30 *Opener*: Gary Gray/DRK PHOTO. 30.2*a,b*: Runk/Schoenberger from Grant Heilman Photography, Inc. 30.7: Phil Gates, Univ. of Durham/BPS. 30.10: J. H. Troughton & L. A. Donaldson. 30.12: Jon Stewart/BPS. 30.13: Thomas Eisner, Cornell Univ. 30.17: Larry Lefever from Grant Heilman Photography, Inc.

Chapter 31 *Opener*: Thomas Eisner, Cornell Univ. 31.1: Jon Mark Stewart/BPS. 31.2: J. N. A. Lott, McMaster Univ./BPS. 31.3*a*: J. N. A. Lott, McMaster Univ./BPS. 31.3*b*: J. N. A. Lott, McMaster Univ./BPS. 31.4: Carl W. May/BPS. 31.5: Gary J. James/BPS. 31.6: Joel Simon. 31.7: J. N. A. Lott, McMaster Univ./BPS. 31.8: © Robert & Linda Mitchell. 31.9: J. Antonovics, Duke Univ. 31.10: Jane Grushow from Grant Heilman Photography, Inc. 31.13: Richard Alexander, Univ. of Pennsylvania.

Chapter 32 *Opener*: Runk/Schoenberger from Grant Heilman Photography, Inc. 32.1: Lou Jacobs, Jr. from Grant Heilman Photography, Inc. 32.4: © William E. Ferguson. 32.7: Runk/Schoenberger from Grant Heilman Photography, Inc. 32.8: Barbara J. O'Donnell/BPS. 32.10: E. H. Newcomb & S. R. Tandon, Univ. of Wisconsin/BPS. 32.12, 32.13: Runk/Schoenberger from Grant Heilman Photography, Inc. Box 32.A: Aladar A. Szalay, Univ. of Alberta.

Chapter 33 *Opener*: Andrew Taylor Photography. 33.4: Art Wolfe. 33.5: Barbara J. Miller/BPS. 33.6: Barry L. Runk from Grant Heilman Photography, Inc. 33.10: Runk/Schoenberger from Grant Heilman Photography, Inc. 33.18 *inset*: Biophoto Associates/Photo Researchers, Inc. 33.20: Grant Heilman Photography. 33.22: J. N. A. Lott, McMaster Univ./BPS. 33.21: Grant Heilman from Grant Heilman Photography, Inc. 33.23: J. N. A. Lott, McMaster Univ./BPS. 33.28: R. Last, Cornell Univ. Courtesy of the Society for Plant Physiology.

Chapter 34 *Opener*: Art Wolfe. 34.1 *bottom*: J. R. Waaland, Univ. of Washington/BPS. 34.1 *top*: Jim Solliday/BPS. 34.2: Dennis D. Kunkel/BPS. 34.3: J. N. A. Lott, McMaster Univ./BPS. 34.7*a*: Gary J. James/BPS. 34.7*b*: John Colwell from Grant Heilman Photography, Inc. 34.16*a*: J. N. A. Lott, McMaster Univ./BPS. 34.16*b*: Phil Gates, Univ. of Durham/BPS. 34.18: Plant Genetics, Inc.

Chapter 35 *Opener*: Animals Animals/Gerald L. Kooyman. 35.14*a*: © Cherry Alexander/NHPA. 35.14*b*: Belinda Wright/DRK PHOTO. 35.19*b,c*: G. W. Willis, Ochsner Medical Institution/BPS. 35.19d Fran Thomas, Stanford Univ. 35.20*a*: Stephen J. Kraseman/DRK PHOTO. 35.20*b*: Art Wolfe.

Chapter 36 *Opener*: R. D. Fernald, Stanford Univ. 36.7*a*: AP Wide World Photos. 36.7*b*: The Bettmann Archive Inc. 36.9: S. H. Ingbar, Harvard Medical School. Box 36.B: James Sugar, Black Star.

Chapter 37 *Opener*: Belinda Wright/DRK Photo. 37.1*a*: © M. Walker/NHPA. 37.1*b*: J. Greenfield, Planet Earth Pictures. 37.1*c*: Geoff du Feu, Planet Earth Pictures. 37.2: David M. Phillips/Photo Researchers, Inc. 37.6, *center insert*: P. Motta, Univ. La Sapienza, Rome, Science Photo Library/Photo Researchers, Inc. 37.6, *bottom insert*: David M. Phillips/Photo Researchers, Inc. 37.7: P. Bagavandoss. 37.12*a*: Animals Animals/Fritz Prenzel. 37.12*b (embryo)*: John Cancalosi/DRK PHOTO. 37.12*b (adult)*: Animals Animals/Mickey Gibson. 37.16*a,b*: From *A Child Is Born*. Photos © Lennart Nilsson, Bonnier Fakta.

Chapter 38 *Opener*: Scott Camazine/Photo Researchers, Inc. 38.3: From R.G. Kessel and R. H. Kardon, *Tissues and Organs*. © 1979, W. H. Freeman and Co. 38.21: Dan McCoy, Rainbow. Boxes 38.A and 38.B: William F. Gilly, Hopkins Marine Station.

Chapter 39 *Opener*: Shelly Katz/Time Magazine. 39.4*a*: R. A. Steinbrecht. 39.4*b*: Animals Animals/G. I. Bernard, Oxford Scientific Films. 39.5 *center*: Peter J. Bryant/BPS. 39.8 *top*: P. Motta, Univ. La Sapienza, Rome, Science Photo Library/Photo Researchers, Inc. 39.17 *right*: S. Fisher, Univ. of California, Santa Barbara. 39.21*a*: Animals Animals/G. I. Bernard, Oxford Scientific Films. 39.24: © E. R. Lewis, Y. Y. Zeevi & F. S. Werblin, Univ. of California, Berkeley/BPS. 39.30: Michael Fogden/DRK PHOTO. 39.31: © Stephen Dalton/NHPA.

Chapter 40 *Opener*: Mik Dakin/Bruce Coleman, Inc. 40.1: D. J. Wrobel, Monterey Bay Aquarium/BPS. 40.2: CNRI Science Photo Library Photo Researchers. 40.5*a*: Secchi-Lecaque-Roussel, UCLAF/CNRI, Science Photo Library/Photo Researchers, Inc. 40.5*b*: P. Motta, Univ. La Sapienza, Rome, Science Photo Library/Photo Researchers, Inc. 40.7*a*: M. I. Walker, Science Source/Photo Researchers, Inc. 40.7*b*: Michael Abbey, Science Source/Photo Researchers, Inc. 40.7*c*: CNRI, Science Photo Library/Photo Researchers, Inc. 40.8*b*: F. A. Pepe, Univ. of Pennsylvania School of Medicine/BPS. 40.11: G.W. Willis, Ochsner Medical Institution/BPS. 40.14: John Dudak/Phototake. 40.19: G. Mili. 40.20: Robert Brons/BPS. 40.24*a*: David J. Wrobel/BPS.

Chapter 41 *Opener*: Courtesy of NASA. 41.1*a*: © Sea Studios, Inc. 41.1*b*: Robert Brons/BPS. 41.1*c*: © Robert & Linda Mitchell. 41.3: © 1991, Eric Reynolds/Adventure Photo. 41.5*b*: Peter J. Bryant/BPS. 41.5*c*: Thomas Eisner, Cornell Univ. 41.9: Walt Tyler, Univ. of California, Davis. 41.13 *top insert*: Science Photo Library, Science Source/Photo Researchers, Inc. 41.13 *bottom insert*: P. Motta, Univ. La Sapienza, Rome, Science Photo Library/Photo Researchers, Inc. 41.16: George Holton/Photo Researchers, Inc.

Chapter 42 *Opener*: Larry Mulvehill, Science Source/Photo Researchers, Inc. 42.11: © Ed Reschke. 42.13: UNICEF, Maggie Murray-Lee. 42.15: After N. Campbell, 1990, Biology, 2nd Ed., Benjamin Cummings Publishing Co. 42.16*a*: From R. G. Kessel & R. H. Kardon, *Tissues and Organs*, © 1979, W. H. Freeman and Co. 42.17*b*: Secchi-Lecaque-Roussel, UCLAF/CNRI, Science Photo Library/Photo Researchers, Inc. Box 42.A: Jon Feingersh/Tom Stack and Associates.

Chapter 43 *Opener*: Animals Animals/Carl Roessler. 43.1*a*: Timothy O'Keefe/Tom Stack and Associates. 43.1*b*: Animals Animals/A. G.(Bert) Wells, Oxford Scientific Films. 43.1*c*: Animals Animals/Bruce Watkins. 43.1*d*: Jack Stein Grove/Tom Stack and Associates. 43.6: D. J. Wrobel, Monterey Bay Aquarium/BPS. 43.7: © Robert & Linda Mitchell. 43.8: © James D. Watt. 43.11: © Ed Reschke. 43.14 *insert*: E. S. Strauss.

Chapter 44 *Opener*: Art Wolfe. 44.1*a*: Helen E. Carr/BPS. 44.1*b*: © E. R. Degginger. 44.3*b*: Marc Chappell, Univ. of California, Riverside. 44.10 *a–d*: From R. G. Kessel & R. H. Kardon, *Tissues and Organs*, © 1979, W.H. Freeman and Co. 44.12: Gregory G. Dimijian, M.D./Photo Researchers, Inc. 44.17: Lise Bankir, I.N.S.E.R.M. Unit, Hopital Necker, Paris. Box 44.A: © Custom Medical Stock Photos. Box 44.B: © Robert & Linda Mitchell.

Chapter 45 *Opener*: John Gerlach/DRK PHOTO. 45.2: Marc Chappell, Univ. of California, Riverside. 45.8: W. C. Dilger. 45.11*b*: Anup & Manos Shah/Planet Earth Pictures. 45.12: Anthony Mercieca/Photo Researchers, Inc. 45.15: H. Craig Heller. 46.17: Animals Animals/Michael Dick. 45.19: H. Craig Heller. 45.22, 45.24: © Jonathan Blair, Woodfin Camp & Associates.

Chapter 46 *Opener*: © E. R. Degginger. 46.1: Kennan Ward/DRK PHOTO. 46.4: Aileen N. C. Morse, Marine Science Institute, Univ of California at Santa Barbara. 46.5: T. G. Whitham, Northern Arizona Univ. 46.7*a*: Art Wolfe. 46.7*b*: Robert and Jean Pollock/BPS. 46.7*c*: Drawing by Sally Landry. 46.8: Erwin and Peggy Bauer. 46.10: John Alcock, Arizona State Univ. 46.12: John Alcock, Arizona State Univ. 46.11: Natalie J. Demong. 46.13: Michael Fogden/DRK PHOTO. 46.15*a*: Elizabeth N. Orians. 46.16: Clem Haagner/Bruce Coleman, Inc. 46.19: S. G. Hoffman. 46.20: J. Erckmann. 46.21: Georgette Douwma/Planet Earth Pictures. 46.23: Patricia Moehlman. 46.24*a*: Jeremy Burgess, Science Photo Library/Photo Researchers, Inc. 46.24*b*: D. Houston/Bruce Coleman, Inc. 46.25: Stanley Breeden/DRK PHOTO. 46.26: Art Wolfe. 46.27: Jonathan Scott, Planet Earth Pictures. 46.28: K. and K. Ammann/Bruce Coleman, Inc.

Chapter 47 *Opener*: © E. R. Degginger. 47.2: Frans Lanting/Minden Pictures. 47.5: © Brian Hawkes/NHPA. 47.9: Dennis Johns. 47.14: Richard Root/Cornell Univ. 47.16: Elizabeth N. Orians. 47.17: Douglas Sprugel, College of Forest Resources, Univ. of Washington. 47.18: Michio Hoshino/Minden Pictures. 47.19: G.Tortoli, Food and Agriculture Organization of the United Nations. 47.20: Animals Animals Earth Scenes/George H. H. Huey.

Chapter 48 *Opener*: Michael Fogden/DRK PHOTO. 48.2*a*: S. K. Webster, Monterey Bay Aquarium/BPS. 48.2*b*: E. S. Ross. 48.3: Mark Mattock/Planet Earth Pictures. 48.7: John R. Hosking, NSW Agriculture, Australia. 48.8*a*: Thomas Eisner, Cornell Univ. 48.8*b*: Ken Lucas/BPS. 48.9: © James Carmichael/NHPA. 48.10: Thomas Eisner, Cornell Univ. 48.11: M.Freeman/Bruce Coleman, Inc. 48.12: Kim Heacox Photography/DRK PHOTO. 48.15: Charlie Ott/Photo Researchers, Inc. 48.16: G. T. Bernard, Oxford Scientific Films/Animals Animals Earth Scenes 48.17: Jonathan Scott/Planet Earth Pictures. 48.18: A. Kerstich/Planet Earth Pictures. 48.19*a*: Larry E. Gilbert, Univ. of Texas, Austin. 48.19*b,c*: Daniel Janzen, Univ. of Pennsylvania. 48.20*a*: S. Nielsen/DRK PHOTO. 48.20*b*: © Raymond A. Mendez. 48.21: John Alcock, Arizona State Univ. 48.22*a*: © Larry Ulrich. 48.22*b*: E. S. Ross. 48.23*a*: Thomas Eisner, Cornell Univ. 48.24: Joel Simon. 48.26: After M. Begon, J. Harper, and C. Townsend, 1986, *Ecology*, Blackwell Scientific Publications. 48.28*a–c*: Robert and Jean Pollock/BPS.

Chapter 49 *Opener*: Courtesy of NASA. 49.6: Animals Animals/Len Rue, Jr. 49.17*a*: Brian A. Whitton. 49.17*b*: J. N. A. Lott, McMaster Univ./BPS.

Chapter 50 *Opener*: Norman Owen Tomalin/Bruce Coleman, Inc. 50.10: E. O. Wilson, Harvard Univ. 50.14*a,b*: Art Wolfe. 50.15*a*: J. N. A. Lott, McMaster Univ./BPS. 50.15*b*: M. Sutton/Tom Stack and Associates. 50.16*a,b*: Paul W. Johnson/BPS. 50.17: D. W. Kaufman, Kansas State Univ. 50.18: Edward Ely/BPS. 50.19: Kim Heacox Photography/DRK PHOTO. 50.21*a*: Elizabeth N. Orians. 50.21*b*: Frans Lanting/Minden Pictures. 50.22: Elizabeth N. Orians. 50.23: Art Wolfe. 50.24*a*: © E. R. Degginger. 50.24*b*: Jeff Britnell/DRK PHOTO. 50.26*a*: Norbert Wu. 50.26*b*: M. C. Chamberlain/DRK PHOTO. 50.28: Gary Gray/DRK PHOTO. 50.29: Art Wolfe.

Chapter 51 *Opener*: Kevin Schafer, Tom Stack and Associates. 51.1: Paintings by H. Douglas Pratt. Courtesy of the Bernice P. Bishop Museum, Honolulu, Hawaii. 51.2: David S. Boynton. 51.4: Joe McDonald/Tom Stack and Associates. 51.5: Danny R. Billings, Missouri Dept. of Conservation. 51.6: Animals Animals/Fred Whitehead. 51.8: Kenneth W. Fink/Bruce Coleman, Inc. 51.9: Edward S. Ross. 51.10*a*: John Gerlach/DRK PHOTO. 51.11: Richard P. Smith, Tom Stack and Associates. 51.12*a*: © Walt Anderson. 51.12*b*: Frans Lanting/Minden Pictures. 51.12*c*: Animals Animals/Paul Freed. 51.14*a,b*: Art Wolfe. 51.15, 51. 16: Courtesy of the Smithsonian Institution, Office of Environmental Awareness/Richard Bierregaard, photographer. 51.18: Animals Animals/Peter Weimann. 51.19*a,b*: © Bill Gabriel, BIOGRAPHICS. Box 51.A *top*: Steven L. Hilty/Bruce Coleman, Inc. Box 51.A *bottom*: N. H. Cheatham/DRK PHOTO.

INDEX

Numbers in *italic* refer to information in an illustration, caption, table, or box.